Study Guide

Anatomy & Physiology

Fourth Edition

Elaine N. Marieb, R.N., Ph.D.
Holyoke Community College

Benjamin Cummings
Boston Columbus Indianapolis New York San Francisco Upper Saddle River
Amsterdam Cape Town Dubai London Madrid Milan Munich Paris Montréal Toronto
Delhi Mexico City São Paulo Sydney Hong Kong Seoul Singapore Taipei Tokyo

Editor-in-Chief: Serina Beauparlant
Project Editors: Sabrina Larson and Katy German
Assistant Editor: Nicole Graziano
Managing Editor: Deborah Cogan
Production Manager: Michele Mangelli
Production Supervisor: Leslie Austin
Compositor: Cecelia G. Morales
Interior Designer: Cecelia G. Morales
Cover Design: Riezebos Holzbaur Design Group
Senior Manufacturing Buyer: Stacey Weinberger
Marketing Manager: Derek Perrigo

Cover Credit: Electronic Publishing Services, Inc.

Benjamin Cummings
is an imprint of

www.pearsonhighered.com

ISBN 10: 0-321-66174-5; ISBN 13: 978-0-321-66174-6

1 2 3 4 5 6 7 8 9 10–BRR–13 12 11 10 09

Preface

This *Study Guide* is intended to help health professional and life science students master the basic concepts of human anatomy and physiology through review and reinforcement exercises. The order of topics reflects the organization of *Anatomy & Physiology*, Fourth Edition, by Elaine N. Marieb and Katja Hoehn, and each chapter begins with a listing of *Student Objectives* taken from that textbook. However, this *Study Guide* may be used with other human anatomy and physiology textbooks as well.

Scope

Exercises in this *Study Guide* review human anatomy from microscopic to macroscopic levels. Topics range from simple chemistry to entire body systems. Pathophysiology is briefly introduced with each system so that students can apply their learning. Where relevant, clinical aspects of study are covered—for example, muscles used as injection sites, the role of ciliated cells in protecting the respiratory tract, and reasons for skin ulcer formation. Developmental stages of youth, adulthood, and old age are emphasized to encourage conceptualization of the human body as a dynamic and continually changing organism.

Learning Aids

Each chapter is divided into three major sections. The first section, called **Building the Framework**, walks the student through the chapter material more or less in the order in which it is presented in the corresponding chapter of the textbook. This section is intentionally inviting and uses a variety of techniques to involve the student in the learning process. The use of coloring exercises helps promote visualization of key structures and processes. Coloring exercises have proven to be a unique motivating approach for learning and reinforcement. Each illustration has been carefully prepared to show sufficient detail for learning key concepts without overwhelming the student with complexity or tedious repetition of coloring. There are over 120 coloring exercises in this *Study Guide*. When completed, the colored diagrams provide an ideal reference and review tool.

Other question formats in this section include selecting from key choices, matching terms with appropriate descriptions, defining important terms, and labeling diagrams. Elimination questions require the student to discover similarities and dissimilarities among a number of structures or processes. Correcting true/false questions adds a new dimension to this traditional exercise format. In addition, students are asked to construct graphs and complete tables, exercises that not only reinforce learning but also provide a handy study aid. If applicable, each **Building the Framework** section ends with a visualization exercise, another unique feature of this *Study Guide*.

These *Incredible Journey* exercises ask students to imagine themselves in miniature traveling through various organs and systems within the body. Each organ system (as well as the topic of chemistry) is surveyed by a visualization exercise that summarizes the content of the chapter. These exercises allow the student to consider what has been learned from another point of view in an unusual and dynamic way.

The second major section of each chapter, called **Challenging Yourself**, typically consists of two groups of questions: Questions in *At the Clinic* focus on applying knowledge to clinical situations, whereas those in *Stop and Think* stress comprehension of principles pertaining to nonclinical situations. For the most part, the clinical questions are written to approximate real situations and require short answers. The *Stop and Think* questions require critical thinking. They cross the lines between topics and prod the student to put two and two together, to synthesize old and new information, and to think logically.

The final section of each chapter is called **Covering All Your Bases**. This section reintroduces concepts already covered but tests them via different formats. It includes multiple-choice questions and a section called *Word Dissection* that asks the student to define word roots encountered in the chapter and to present examples. Although this section is perhaps less interesting than the first two, knowing word roots is an advantage in studying A&P, and the multiple-choice questions are much more challenging than they might at first appear. Most have more than one correct answer, and the student is asked to choose *all* the correct responses. When this third and final section has been completed, the student will have been exposed to all of the basic and necessary information of the related textbook chapter.

Answers for all of the exercises are provided at the back of the book in the Appendix.

In addition to using this *Study Guide*, students will benefit from the additional study tools made available at no additional cost by Benjamin Cummings. First is the *Interactive Physiology® 10-System Suite* CD-ROM, which students will find in the back of each new text. This CD-ROM uses detailed animations and engaging quizzes to help students advance beyond memorization to a genuine understanding of complex A&P topics. Covering ten body systems, this tutorial series encourages active learning through quizzes, activities, and review exercises that are oriented toward making the difficult task of learning A&P more interesting and fun.

Also available to students is access to *myA&P™*, which organizes Benjamin Cummings media into one online location (www.myaandp.com). Content is organized into learning units, which follow the topic coverage of the chapters in *Anatomy & Physiology,* Fourth Edition. Available media includes chapter quizzes and tests, *A&P Flix* animations with quizzes, *Interactive Physiology® 10-System Suite* tutorials with quizzes, interactive learning activities, *MP3 Tutor Sessions, Practice Anatomy Lab™ 2.0* clinical case studies, an eText version of the text, *PhysioEx™ 8.0* laboratory simulations, and more. Students must have an instructor course ID number to access this material.

Instructions for the Student: How to Use This Book

This study guide to accompany *Human Anatomy & Physiology* is the outcome of years of personal attempts to find and create exercises most helpful to my own students when they study and review for a lecture test or laboratory quiz.

Although I never cease to be amazed at how remarkable the human body is, I would never try to convince you that studying it is easy. The study of human anatomy and physiology has its own special terminology. It requires that you become familiar with the basic concepts of chemistry to understand physiology, and often it requires rote memorization of facts. It is my hope that this *Study Guide* will help simplify your task. To make the most of the exercises, carefully read these descriptions of what to expect before starting work.

This *Study Guide* has three sections—each with a special focus and purpose—as explained here.

1. Building the Framework

This first section of each chapter is the most varied. It begins with a list of *Student Objectives* (keyed to the text) and covers virtually all of the major points you are expected to know about the information in the corresponding chapter. The types of exercises you can expect to see are described below.

• **Labeling and Coloring.** Some of these questions ask you only to label a diagram, but most also ask that you do some coloring of the figure. You can usually choose whichever colors you prefer. Soft, colored pencils are recommended so that the underlying diagram shows through. Because most figures have multiple parts to color, you will need a variety of colors—18 should suffice. In the coloring exercises, you are asked to choose a particular color for each structure to be colored. That color is then used for both a color coding circle found next to the name of the structure or organ and the structure or organ within the figure. This allows you to quickly identify the colored structure by name in cases where the diagram is not labeled. In a few cases, you are given specific coloring instructions to follow.

• **Matching.** Here you are asked to match a term denoting a structure or physiological process with a descriptive phrase or sentence. In many exercises, some terms are used more than once and others are not used at all.

• **Completion.** You are asked to select the correct term or phrase to answer a specific question, or to fill in blanks to complete a sentence or a table or chart.

• **Definitions.** You are asked to provide a brief definition of a particular structure or process.

• **True/False.** One word or one phrase is underlined in a sentence. You decide if the sentence is true as it is written. If if is not, you are asked to correct the underlined word or phrase.

• **Elimination.** Here you are asked to find the term that does not belong in a particular grouping of related terms. In this type of exercise, you must analyze how each term is like or different from the others.

• **Graphs.** You are given data and asked to plot them on a graph. Graphs are important because they enable you to pinpoint important variations in some physiological factor.

• **Visualization.** *The Incredible Journey* is a special type of completion exercise. For these exercises, you are asked to imagine that you have been miniaturized and injected into the body of a human being (your host). Anatomical landmarks and physiological events are described from your miniaturized viewpoint, and you are then asked to identify your observations. My students have found these exercises fun to complete, and I hope you will, too.

Each exercise has complete instructions. Read through them before beginning. When there are multiple instructions, complete them in the order given.

2. Challenging Yourself

This second section of each chapter consists of short-answer essay questions. Don't worry; as long as you've studied, you do have the information you need in your memory bank to cope with this section. Its two major parts, *At the Clinic* and *Stop and Think*, ask you to apply your new learning to both clinical and nonclinical situations. At first, you may have to stretch a little to answer some of these questions, but stick with it—success builds. In cases where you have been unable to come up with the correct answer, pay particular attention to just what the right answer is, and try to analyze how that answer was arrived at (that is, the problem-solving route). This approach will help you develop those needed problem-solving skills.

3. Covering All Your Bases

This third section of each chapter is for the student "darned and determined" to do well in the course. Using multiple-choice questions, this section retests information already tested in the first two sections. But watch out, because the multiple-choice questions are special. You are not just looking for that one right answer. There may be only one right answer, but it is more likely that there will be two, three, or even four right answers to a given question. If you can "catch" them all, you really have this chapter in your pocket! The other part of this section is the *Word Dissection* exercise, which tests your understanding of word roots used in the chapter. Do not ignore this section. Once you learn word roots, your (A&P) life becomes a lot easier.

At times, it may appear that information is being duplicated in the different types of exercises. Although there *is* some duplication, the concepts being tested reflect different vantage points or approaches in the different exercises. Remember, when you understand a concept from several different perspectives, you have mastered that concept.

I sincerely hope that this *Study Guide* challenges you to increase your knowledge, comprehension, retention, and appreciation of the structure and function of the human body. Please write and let me know what you like or dislike about it (and why). This will help me tailor the next edition even more to the needs of "my" A&P students.

Good luck!

Elaine Marieb

Applied Sciences
Benjamin Cummings Science
1301 Sansome Street
San Francisco, California 94111

Contents

20

The Immune System: Innate and Adaptive Body Defenses 437

21

The Respiratory System 461

22

The Digestive System 490

23

Nutrition, Metabolism, and Body Temperature Regulation 516

24

25

26

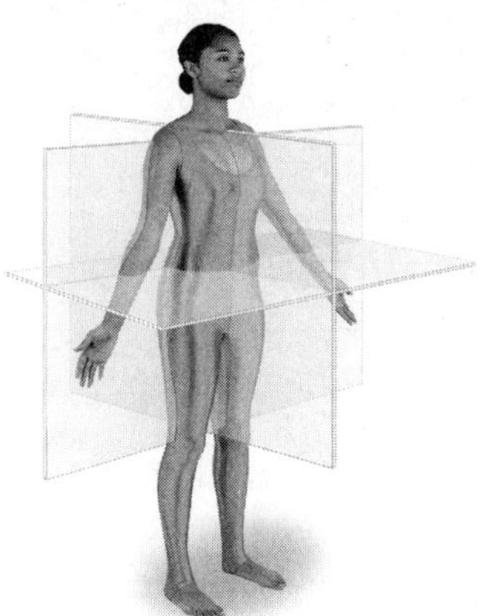

1 The Human Body: An Orientation

Student Objectives

When you have completed the exercises in this chapter, you will have accomplished the following objectives:

An Overview of Anatomy and Physiology

1. Define anatomy and physiology and describe their subdivisions.
2. Explain the principle of complementarity.

Levels of Structural Organization

3. Name the different levels of structural organization that make up the human body, and explain their relationships.
4. List the 11 organ systems of the body, identify their components, and briefly explain the major function(s) of each system.

Maintaining Life

5. List the functional characteristics necessary to maintain life in humans.
6. List the survival needs of the body.

Homeostasis

7. Define homeostasis and explain its importance.
8. Describe how negative and positive feedback maintain body homeostasis.
9. Describe the relationship between homeostatic imbalance and disease.

The Language of Anatomy

10. Describe the anatomical position.
11. Use correct anatomical terminology to describe body directions, body regions, and body planes or sections.
12. Locate and name the major body cavities, and their subdivisions and associated membranes, and list the major organs contained within them.
13. Name the four quadrants or nine regions of the abdominopelvic cavity and list the organs they contain.

Most of us have a natural curiosity about our bodies, and the study of anatomy and physiology elaborates on this interest. Anatomists have developed a universally acceptable set of reference terms that allows body structures to be located and identified with a high degree of clarity. Initially, students might have difficulties with the language used to describe anatomy and physiology, but without such a special vocabulary, confusion is inevitable.

The topics in Chapter 1 enable students to test their mastery of terminology commonly used to describe the body and the relationships among its various parts. Concepts concerning the functions vital for life and homeostasis are also reviewed. Additional topics include body organization from simple to complex levels and an introduction to organ systems.

BUILDING THE FRAMEWORK

An Overview of Anatomy and Physiology

1. Contrast briefly the following pairs of anatomical terms.

1. Regional anatomy versus systemic anatomy ______________________

____

____

____

2. Gross anatomy versus microscopic anatomy ______________________

____

____

____

3. Developmental anatomy versus embryology ______________________

____

____

____

4. Histology versus cytology ______________________

____

____

____

2. Circle all the terms or phrases that correctly relate to the study of physiology; use a highlighter to identify those terms or phrases that pertain to the study of anatomy.

A. Measuring an organ's size, shape, and weight
B. Can be studied in dead specimens
C. Often studied in living subjects
D. Chemistry principles
E. Measuring the acid content of the stomach
F. Principles of physics
G. Observing a heart in action
H. Dynamic
I. Dissection
J. Experimentation
K. Observation
L. Directional terms
M. Static

3. "Function always follows structure." This statement is referred to as the principle of ______________________.

Levels of Structural Organization

1. The structures of the body are organized in successively larger and more complex structures. Fill in the answer blanks with the correct terms for these increasingly larger structures.

Chemicals → ______________ → ______________ → ______________ → ______________ → Organism

2. Circle the term that does not belong in each of the following groupings.

1. Atom Cell Tissue Alive Organ
2. Brain Stomach Heart Liver Epithelium
3. Epithelium Heart Muscle tissue Nervous tissue Connective tissue
4. Human Digestive system Horse Pine tree Amoeba

3. Figures 1.1–1.6 on pp. 4–6 represent some of the various body organ systems. First identify and name each organ system illustrated by filling in the blank directly under the illustration. Then select a different color for each organ and use it to color the coding circles and corresponding structures in the illustrations.

◯ Blood vessels

◯ Heart

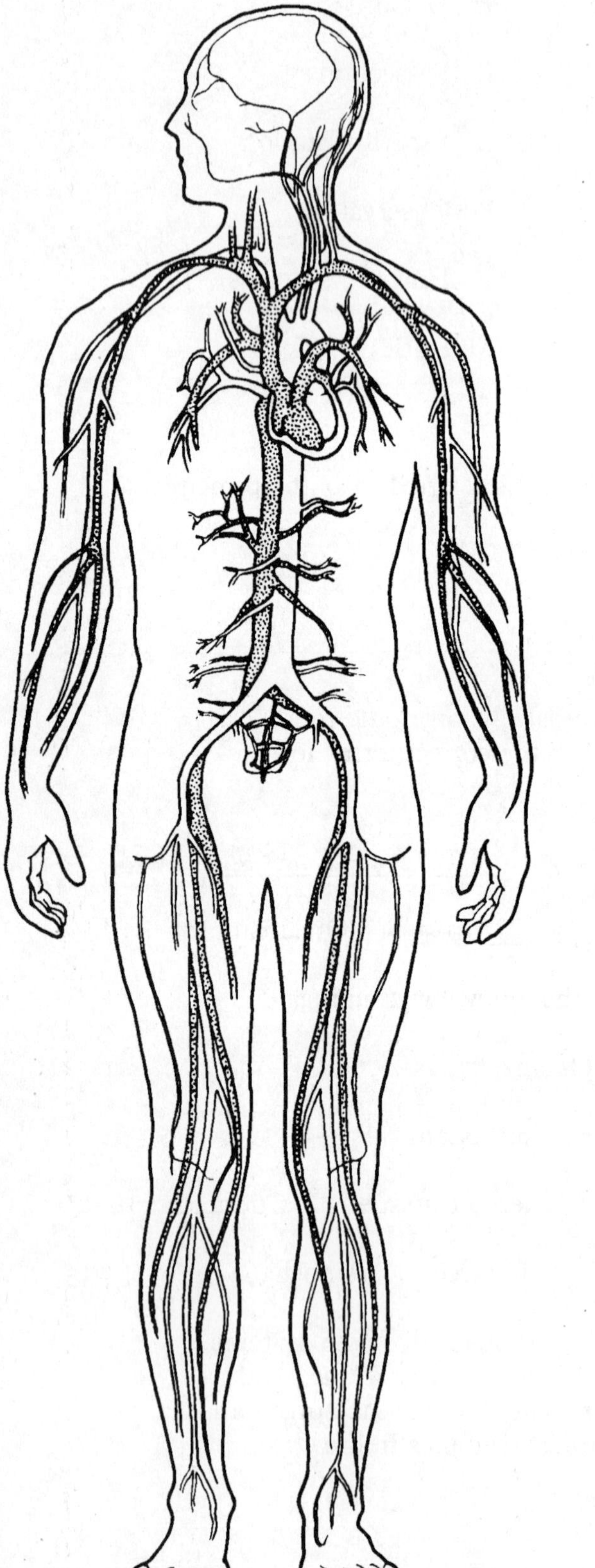

Figure 1.1

◯ Nasal cavity

◯ Lungs

◯ Trachea

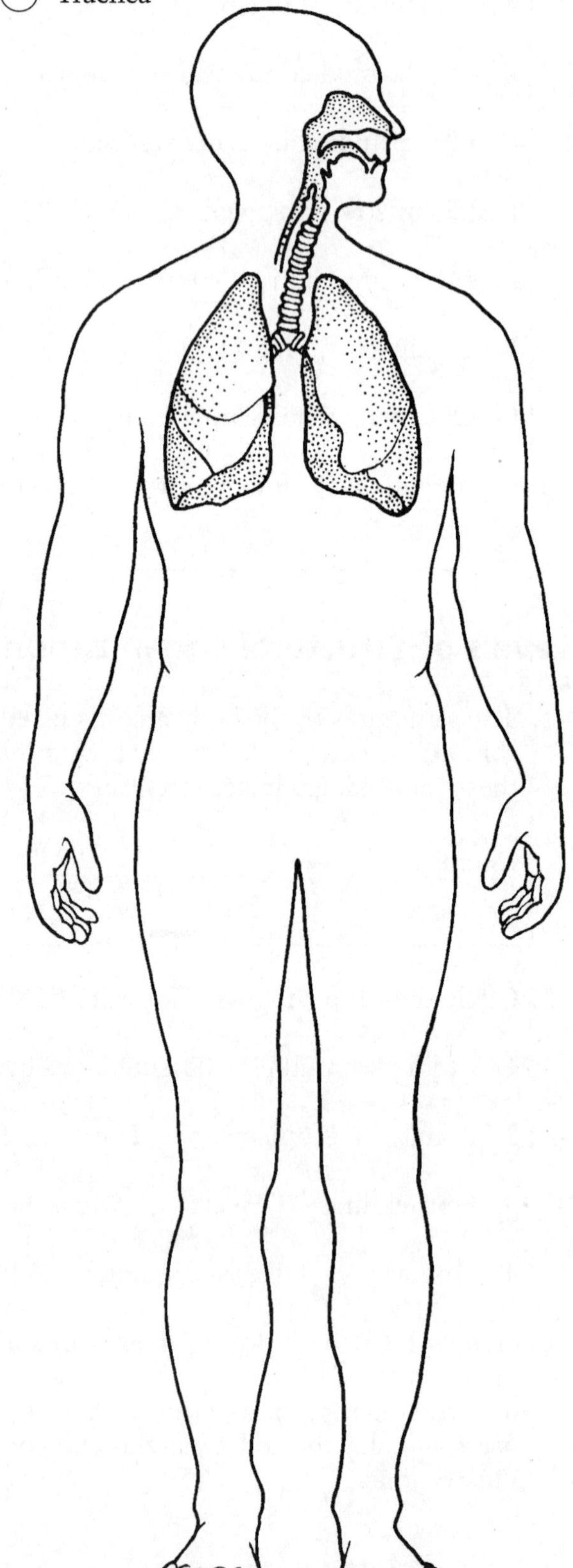

Figure 1.2

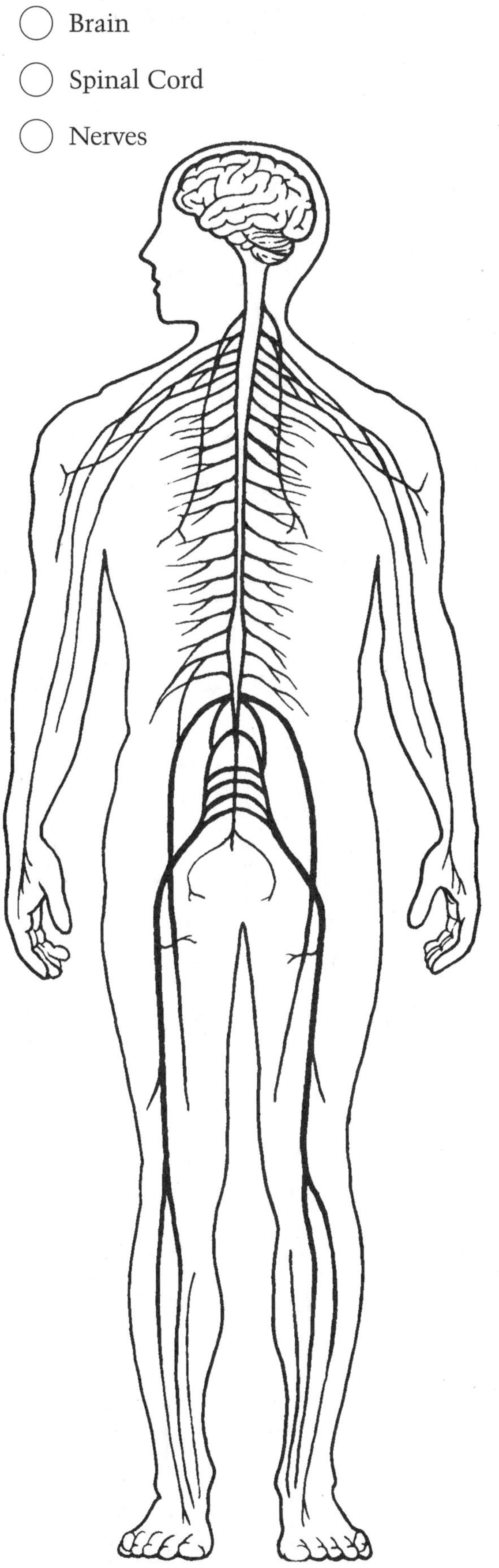

◯ Brain

◯ Spinal Cord

◯ Nerves

◯ Kidneys

◯ Ureters

◯ Urethra

◯ Bladder

Figure 1.3

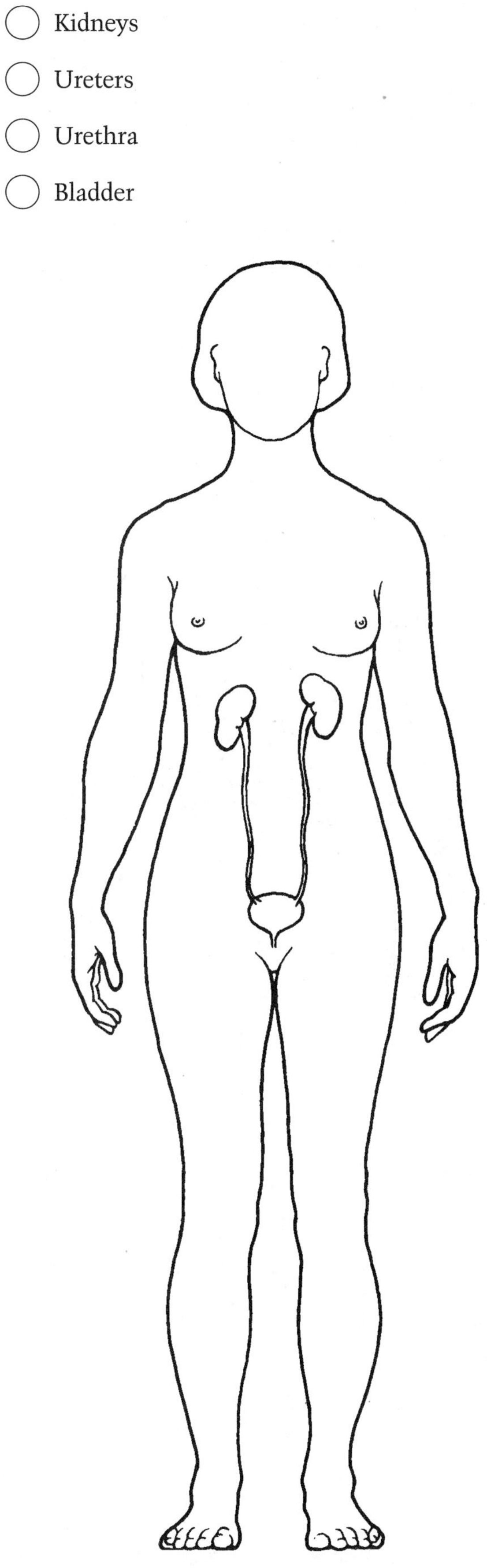

Figure 1.4

- ◯ Oral cavity
- ◯ Stomach
- ◯ Intestines
- ◯ Esophagus
- ◯ Rectum
- ◯ Ovaries
- ◯ Uterus

Figure 1.5

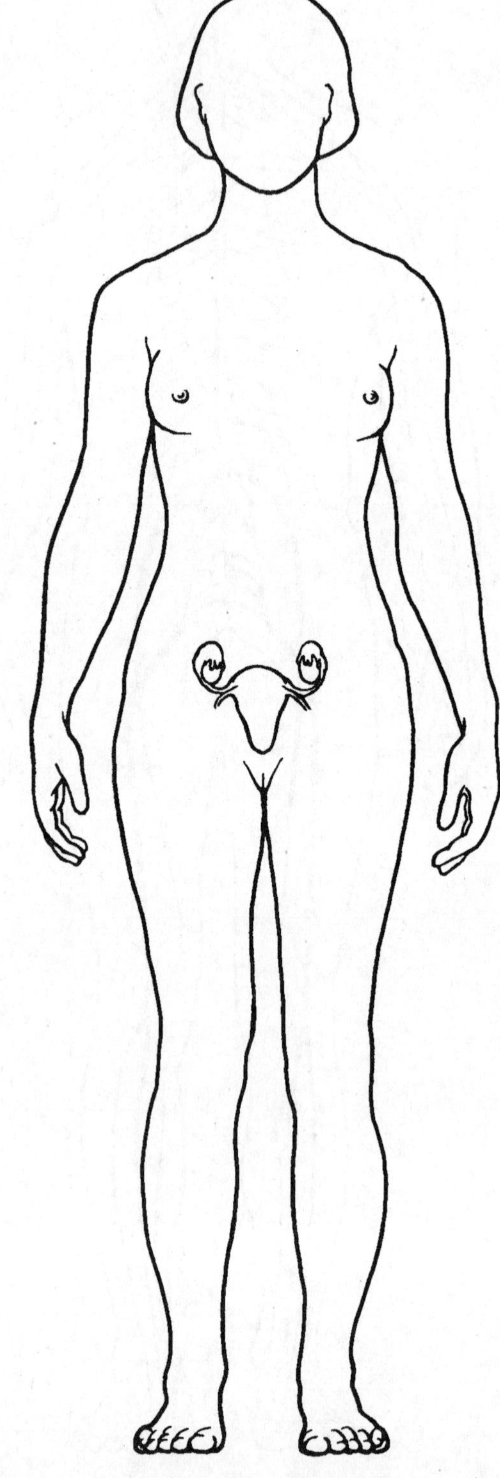

Figure 1.6

4. Using the key choices, identify the body systems to which the following organs or functions belong. Insert the correct answers in the answer blanks.

Key Choices

A. Cardiovascular	D. Integumentary	F. Muscular	I. Respiratory
B. Digestive	E. Lymphatic/ Immune	G. Nervous	J. Skeletal
C. Endocrine		H. Reproductive	K. Urinary

______________ 1. Rids the body of nitrogen-containing wastes

______________ 2. Is affected by the removal of the thyroid gland

______________ 3. Provides support and levers on which the muscular system can act

______________ 4. Includes the heart

______________ 5. Protects underlying organs from drying out and mechanical damage

______________ 6. Protects the body; destroys bacteria and tumor cells

______________ 7. Breaks down foodstuffs into small particles that can be absorbed

______________ 8. Removes carbon dioxide from the blood

______________ 9. Delivers oxygen and nutrients to the body tissues

______________ 10. Moves the limbs; allows facial expression

______________ 11. Conserves body water or eliminates excesses

______________ 12. Allows conception and childbearing

______________ 13. Controls the body with chemicals called hormones

______________ 14. Is damaged when you cut your finger or get a severe sunburn

5. Using the key choices from Exercise 4, choose the organ system to which each of the following sets of organs belongs. Enter the correct letters in the answer blanks.

_____ 1. Blood vessels, heart

_____ 2. Pancreas, pituitary, adrenal glands

_____ 3. Kidneys, bladder, ureters

_____ 4. Testis, vas deferens, urethra

_____ 5. Esophagus, large intestine, stomach

_____ 6. Breastbone, vertebral column, skull

_____ 7. Brain, nerves, sensory receptors

Maintaining Life

1. Match the terms pertaining to functional characteristics of organisms in Column B with the appropriate descriptions in Column A. Fill in the blanks with the appropriate answers.

	Column A	Column B
________________	1. Keeps the body's internal environment distinct from the external environment	A. Digestion
		B. Excretion
________________	2. Provides new cells for growth and repair	C. Growth
________________	3. Occurs when constructive activities occur at a faster rate than destructive activities	D. Maintenance of boundaries
________________	4. The tuna sandwich you have just eaten is broken down to its chemical building blocks	E. Metabolism
		F. Movement
________________	5. Elimination of carbon dioxide by the lungs and of nitrogenous wastes by the kidneys	G. Responsiveness
________________	6. Ability to react to stimuli; a major role of the nervous system	H. Reproduction
________________	7. Walking, throwing a ball, riding a bicycle	
________________	8. All chemical reactions occurring in the body	
________________	9. At the cellular level, membranes; for the whole organism, the skin	

2. Using the key choices, correctly identify the survival needs that correspond to the following descriptions. Insert the correct answers in the answer blanks.

Key Choices

A. Appropriate body temperature
C. Nutrients
E. Water
B. Atmospheric pressure
D. Oxygen

________________ 1. Includes carbohydrates, proteins, fats, and minerals

________________ 2. Essential for normal operation of the respiratory system and breathing

________________ 3. Single substance accounting for over 60% of body weight

________________ 4. Required for the release of energy from foodstuffs

________________ 5. Provides the basis for body fluids of all types

________________ 6. When too high or too low, physiological activities cease, primarily because molecules are destroyed or become nonfunctional

Homeostasis

1. Define *homeostasis*. ______________________________

____

2. For each of the following sets of data, graph the provided data as shown in the example just below. Then, determine whether the data show negative or positive feedback.

EXAMPLE:

Body temperature (T_B), °C:	36.5	37.0	37.5	38.0	37.5
Time, hours, AM:	6	7	8	9	10

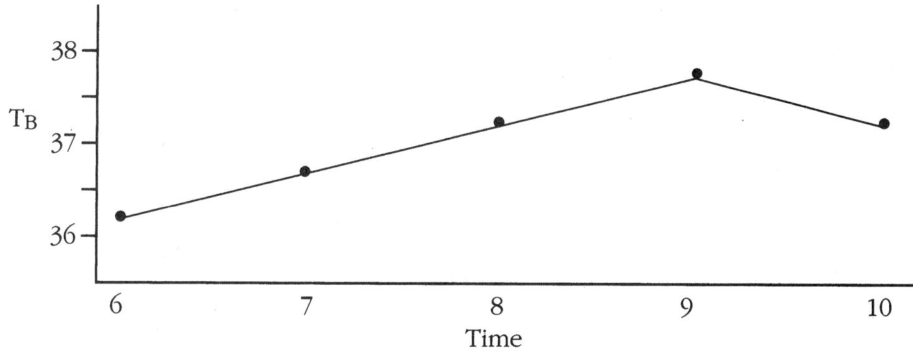

Type of feedback: negative

1. Blood glucose (G_{BL}), mg/100 ml:

Blood glucose (G_{BL}), mg/100 ml:	70	73	78	87	90	88	76
Time, hours, AM:	5	6	7	8	9	10	11

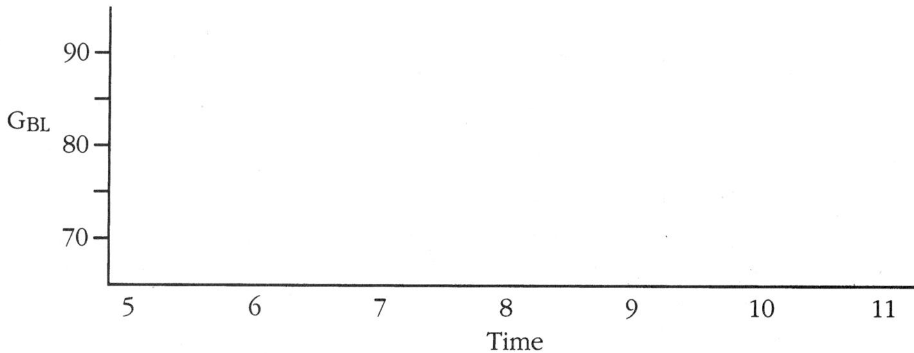

Type of feedback: ______________________________

2. Blood pH:

Blood pH:	7.36	7.38	7.42	7.43	7.41	7.39	7.37	7.40
Time:	1 P.M.	2 P.M.	3 P.M.	4 P.M.	5 P.M.	6 P.M.	7 P.M.	8 P.M.

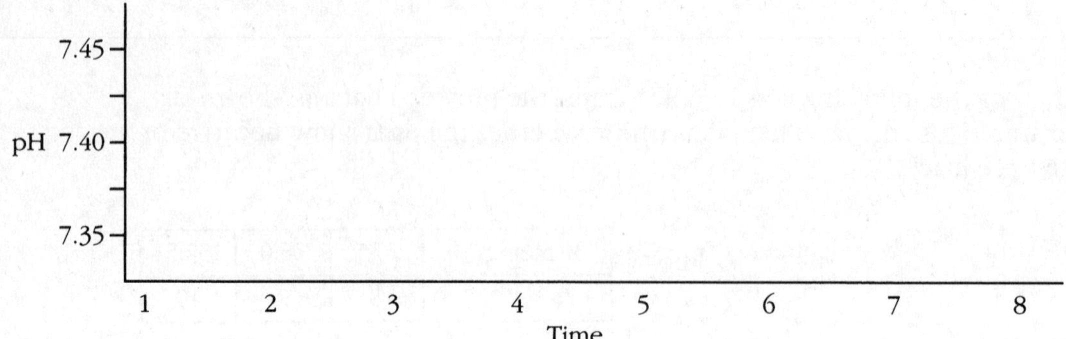

Type of feedback: ______________________

3. Blood pressure (BP), mm Hg:

Blood pressure (BP), mm Hg:	95	89	86	80	73	66	58
Time, minutes:	0	10	15	20	25	30	35

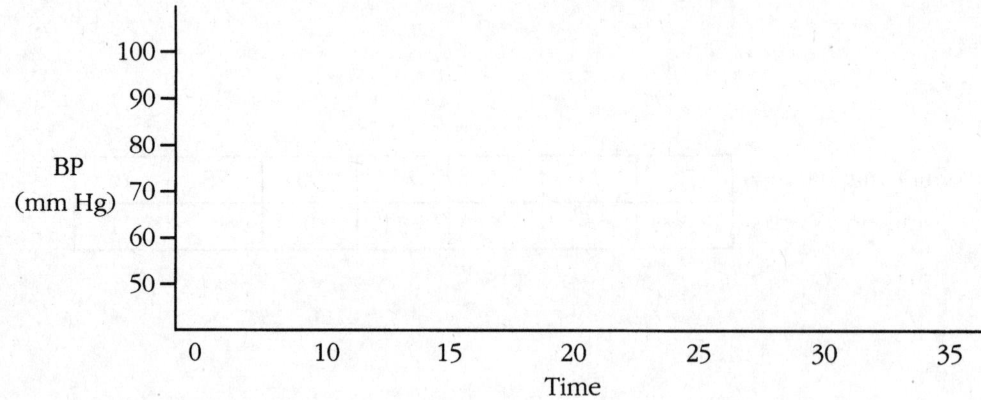

Type of feedback: ______________________

The Language of Anatomy

1. Complete the following statements by choosing an anatomical term from the key choices. Enter the appropriate answers in the answer blanks.

Key Choices

A. Anterior	D. Inferior	G. Posterior	J. Superior
B. Distal	E. Lateral	H. Proximal	K. Transverse
C. Frontal	F. Medial	I. Sagittal	

__________ 1.
__________ 2.
__________ 3.
__________ 4.
__________ 5.

In the anatomical position, the nose and belly button are on the __(1)__ body surface, the calves and shoulder blades are on the __(2)__ body surface, and the soles of the feet are the most __(3)__ part of the body. The nipples are __(4)__ to the shoulders and __(5)__ to the armpits. The heart is __(6)__ to the spine and __(7)__ to the lungs. The knee is __(8)__ to the toes but __(9)__ to the thigh. In humans, the ventral surface can also be called the __(10)__ surface; however, in four-legged animals, the ventral surface is the __(11)__ surface.

__________ 6.
__________ 7.
__________ 8.
__________ 9.

If an incision cuts the brain into right and left parts, the section is a __(12)__ section, but if the brain is cut so that superior and inferior parts result, the section is a __(13)__ section. You are told to cut an animal along two planes so that the paired lungs are observable in both sections. The two sections that meet this requirement are the __(14)__ and __(15)__ sections.

__________ 10.
__________ 11.

The thoracic cavity is __(16)__ to the abdominopelvic cavity and __(17)__ to the spinal cavity.

__________ 12. __________ 15.
__________ 13. __________ 16.
__________ 14. __________ 17.

2. In 1–6 below, a directional term (e.g., distal in 1) is followed by terms indicating different body structures or locations (e.g., the elbow/the shoulder). In each case, choose the structure or location that best matches the directional term given.

1. Distal—the elbow/the shoulder
2. Lateral—the shoulder/the breastbone
3. Superior—the forehead/the chin
4. Superficial—the skeleton/the muscles
5. Proximal—the knee/the foot
6. Inferior—the liver/the small intestine

3. Circle the term or phrase that does not belong in each of the following groupings.

1. Transverse Distal Frontal Sagittal
2. Lumbar Flank Antecubital Abdominal Scapular
3. Calf Brachial Femoral Popliteal
4. Epigastric Hypogastric Lumbar Left upper quadrant
5. Orbital cavity Nasal cavity Ventral cavity Oral cavity

4. Select the key choices that identify the following body parts or areas. Enter the appropriate answers in the blanks.

Key Choices

A. Abdominal	E. Calf	I. Gluteal	M. Popliteal
B. Axillary	F. Cervical	J. Groin	N. Pubic
C. Brachial	G. Digital	K. Lumbar	O. Scapular
D. Buccal	H. Femoral	L. Occipital	P. Umbilical

__________ 1. Armpit

__________ 2. Thigh region

__________ 3. Buttock area

__________ 4. Neck region

__________ 5. "Belly button" area

__________ 6. Pertaining to the toes

__________ 7. Posterior aspect of head

__________ 8. Area where trunk meets thigh

__________ 9. Back area from ribs to hips

__________ 10. Pertaining to the cheek

5. Correctly label all body areas indicated with leader lines on Figure 1.7. Some of the terms you will need are in the key of Exercise 4; others will have to be provided from memory. In addition, identify the sections labeled A and B in the figure.

Section A: ______________________ Section B: ______________________

The body position indicated by the diagrams in Figure 1.7 are fairly close to the anatomical position. What body part(s) are incorrectly depicted? ______________________________

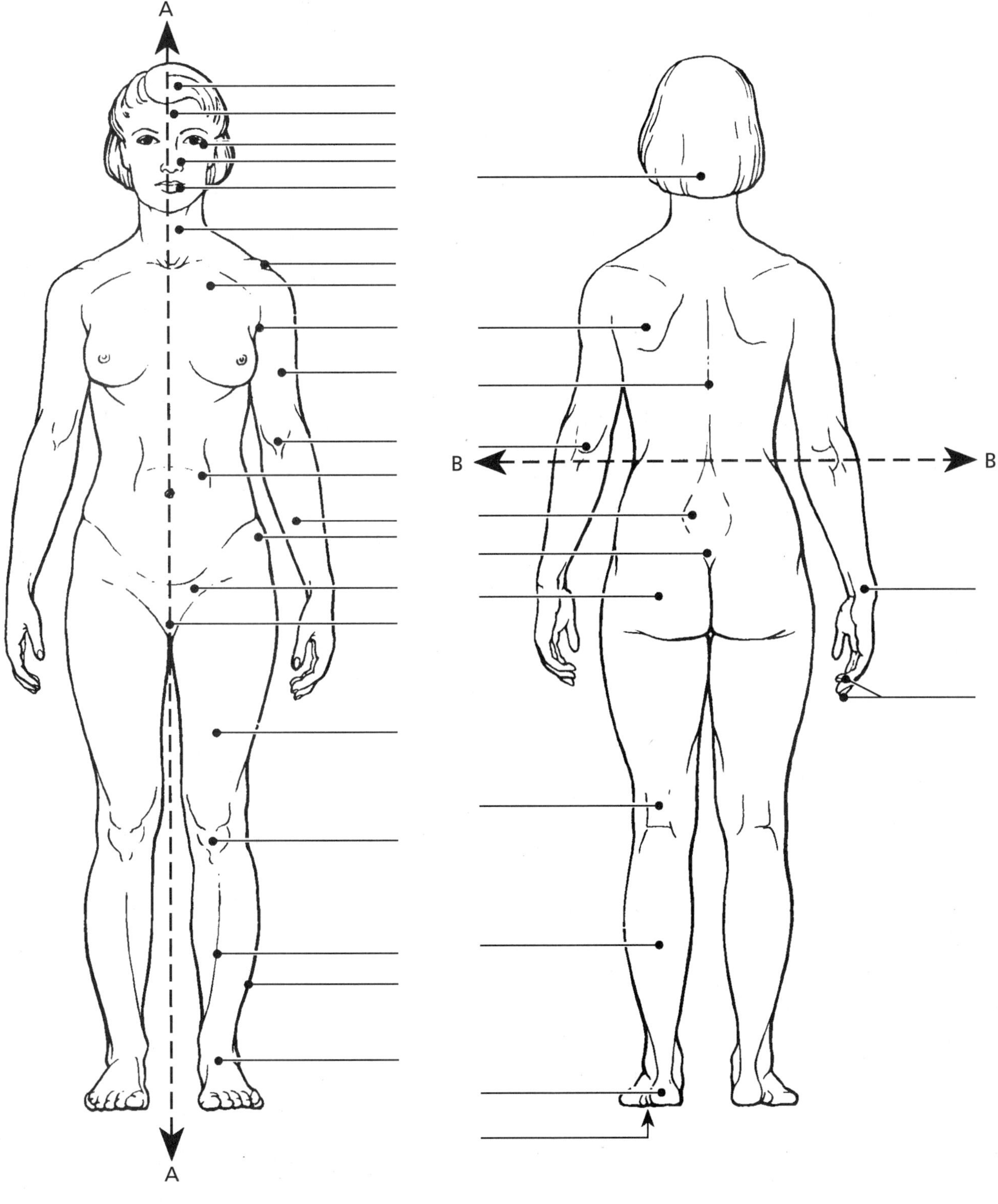

Figure 1.7

6. Complete the following statements by filling in the answer blanks with the correct terms.

______ 1. The abdominopelvic and thoracic cavities are subdivisions of the __(1)__ body cavity; the cranial and spinal cavities are
______ 2. parts of the __(2)__ body cavity. The __(3)__ body cavity is totally surrounded by bone and provides very good
______ 3. protection to the structures it contains.

7. Identify the following serous membranes. Insert your answers in the answer blanks.

1. Covers the lungs and cavity of the thorax ______
2. Forms a slippery sac around the heart ______
3. Surrounds the abdominopelvic cavity organs ______

8. Using the key choices, identify the body cavities where the following body organs are located. Enter the appropriate answers in the answer blanks.

Key Choices

A. Abdominopelvic B. Cranial C. Spinal D. Thoracic

______ 1. Stomach ______ 7. Bladder

______ 2. Small intestine ______ 8. Heart

______ 3. Large intestine ______ 9. Lungs

______ 4. Spleen ______ 10. Brain

______ 5. Liver ______ 11. Rectum

______ 6. Spinal cord ______ 12. Ovaries

9. Refer to the organs listed in Exercise 8. In the answer blanks below, record the numbers of those that would be found in each abdominal region. Some organs may be found in more than one abdominal region.

______ 1. Hypogastric region

______ 2. Right lumbar region

______ 3. Umbilical region

______ 4. Epigastric region

______ 5. Left iliac region

10. Select different colors for the dorsal and ventral body cavities. Color the coding circles below and the corresponding cavities in part A of Figure 1.8. Complete the figure by labeling those body cavity subdivisions that have a leader line. Complete part B by labeling each of the abdominal regions indicated by a leader line.

◯ Dorsal body cavity ◯ Ventral body cavity

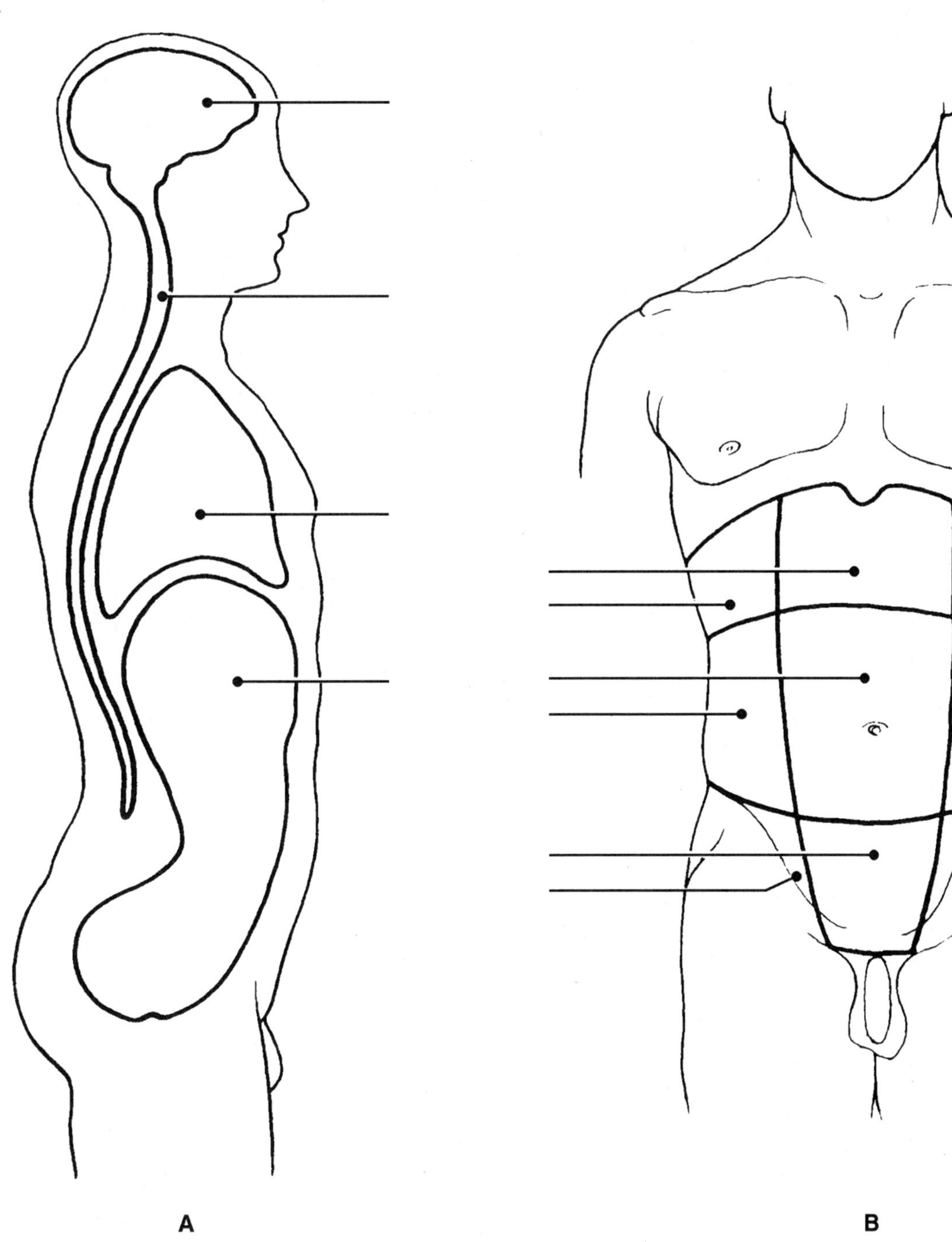

Figure 1.8

11. Choose different colors for the cavities listed with color coding circles. Color the coding circles and the corresponding cavities in Figure 1.9. Complete this exercise by identifying the structure provided with a leader line.

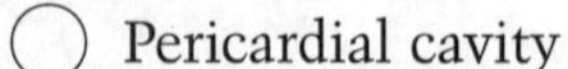

- ◯ Pleural cavities
- ◯ Pericardial cavity
- ◯ Abdominal cavity
- ◯ Pelvic cavity

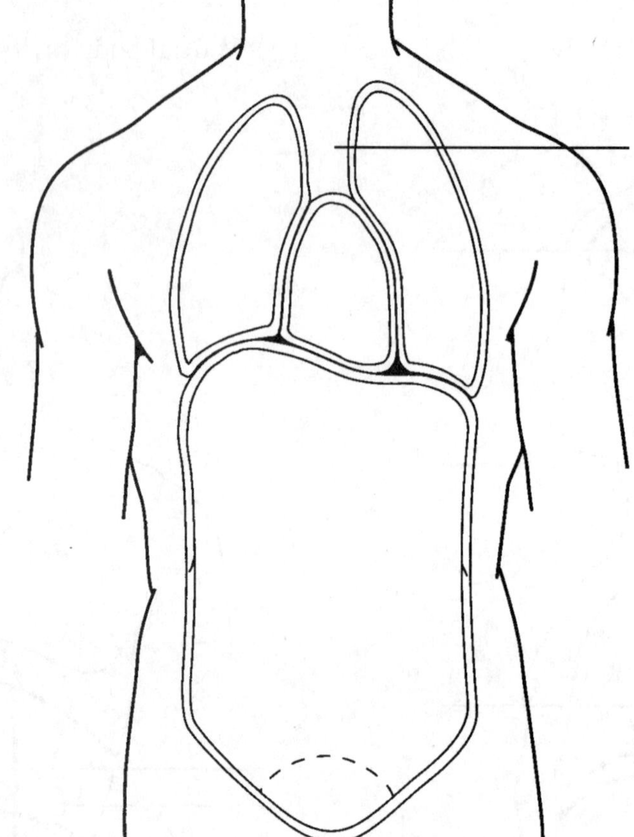

Figure 1.9

12. From the key choices, select the body cavities where the following surgical procedures would occur. Insert the correct answers in the answer blanks. Be precise; select the name of the cavity subdivision if appropriate.

Key Choices

A. Abdominal	D. Pelvic
B. Cranial	E. Thoracic
C. Dorsal	F. Ventral

________________ 1. Surgery to remove a cancerous prostate gland

________________ 2. Coronary bypass surgery (heart surgery)

________________ 3. Removal of a brain tumor

________________ 4. Surgery to remove the distal part of the colon and establish a colostomy

________________ 5. Gallbladder removal

CHALLENGING YOURSELF

At the Clinic

1. A jogger has stepped in a pothole and sprained his ankle. What organ systems have suffered damage?
2. A newborn baby is unable to hold down any milk. Examination reveals a developmental disorder in which the esophagus fails to connect to the stomach. Which survival needs are most immediately threatened?
3. In congestive heart failure, the weakened heart is unable to pump with sufficient strength to empty its own chambers. As a result, blood backs up in the veins, blood pressure rises, and circulation is impaired. Describe what will happen as this situation worsens owing to positive feedback. Then, predict how a heart-strengthening medication will reverse the positive feedback.
4. The control center for regulating body temperature is in a region of the brain called the hypothalamus. When the body is fighting an infection, the hypothalamus allows body temperature to climb higher than the normal value of about 38°C. Explain how this can be considered a negative feedback mechanism.
5. A patient reports stabbing pains in the right hypochondriac region. The medical staff suspects gallstones. What region of the body will be examined?
6. Number the following structures, from darkest (black) to lightest (white), as they would appear on an X ray. Number the darkest one 1, the next darkest 2, etc.

   ______________________ (a) Soft tissue

   ______________________ (b) Femur (bone of the thigh)

   ______________________ (c) Air in lungs

   ______________________ (d) Gold (metal) filling in a tooth

7. Mrs. Bigda is three months pregnant, and her obstetrician wants to measure the head size of the fetus in her womb to make sure development is progressing normally. Which medical imaging technique will the physician probably use and why?
8. Mr. Petros is behaving abnormally, and doctors strongly suspect he has a brain tumor. Which medical imaging device—conventional X ray, DSA, PET, ultrasound, or MRI—would be best for precisely locating a tumor within the brain? Explain your choice.
9. The Chan family was traveling in their van and had a minor accident. The children in the backseat were wearing seat belts but they still sustained bruises around the abdomen and had some internal organ injuries. Why is this area more vulnerable to damage than others?
10. The hormone testosterone is released by the interstitial cells of the testes in response to interstitial cell–stimulating hormone (ICSH) released by the anterior pituitary gland. The hypothalamus regulates the pituitary's release of ICSH by secreting a releasing factor. As testosterone levels rise in the blood, it exerts negative feedback on the hypothalamus. What is the result of this negative feedback on the amount of releasing factor secreted? On the amount of ICSH released, and subsequently on the amount of testosterone released?

Stop and Think

1. Circle the *most likely* structure or function of the items listed below:
   1. Thick outer layer of skin: *protection* or *sensation*
   2. Pigment of eyes: *absorb light* or *attract a mate*
   3. Thin lining of intestine: *protection* or *absorption*
   4. Pumping blood: *thick, hollow, muscular organ* or *thin, narrow vessels*
   5. Breathing: *thin, elastic tissue* or *dense, tough tissue*
2. Name the organ systems that come in contact with the external environment and then list some common infections of these systems.
3. List the functional characteristics that are *essential* to maintaining life.
4. Which of the survival needs must be renewed routinely? Of these, which can be stored for later use?
5. Why is the type of feedback that maintains homeostasis referred to as *negative* feedback?
6. Explain how the two-layered condition of serous membranes comes about.
7. Why do anatomists prefer abdominopelvic regions, while medical personnel tend to favor quadrants?
8. Make a simple line drawing of a chair. Then, add three different (common) planes to divide the chair structure. Identify each plane by name. Which of these planes could yield two parts, one consisting of the back of the chair and the other consisting mostly of the seat? Which would give you two mirror images of chair parts? Which could provide sections passing only through the legs, seat, or back of the chair?
9. Dominic has a hernia in his inguinal region, pain from an infected kidney in his lumbar region, and hemorrhoids in his perineal area. Explain where each of these regions is located.

COVERING ALL YOUR BASES

Multiple Choice

Select the best answer or answers from the choices given.

1. Which of the following is not a discipline of anatomy?
   A. Developmental
   B. Regional
   C. Microscopic
   D. Homeostatic
   E. Systemic
2. Histology is the same as:
   A. light microscopy
   B. ultrastructure
   C. functional morphology
   D. surface anatomy
   E. microscopic anatomy
3. Which of the following activities would not represent an anatomical study?
   A. Making a section through the heart to observe its interior
   B. Drawing blood from recently fed laboratory animals at timed intervals to determine their blood sugar levels
   C. Examining the surface of a bone
   D. Viewing muscle tissue through a microscope
4. Which subdivision of anatomy involves the study of organs that function together?
   A. Regional
   B. Developmental
   C. Systemic
   D. Histology
   E. Radiographic

5. Consider the following listed levels: (1) chemical; (2) tissue; (3) organ; (4) cellular; (5) organismal; (6) systemic. Which of the following choices has the levels listed in order of increasing complexity?

A. 1, 2, 3, 4, 5, 6
B. 1, 4, 2, 5, 3, 6
C. 3, 1, 2, 4, 6, 5
D. 1, 4, 2, 3, 6, 5
E. 4, 1, 3, 2, 6, 5

6. Which of the following is (are) characteristic of a living organism?

A. Maintenance of boundaries
B. Metabolism
C. Waste disposal
D. Reproduction
E. Response to environment

7. Which is not essential to survival?

A. Water
B. Oxygen
C. Gravity
D. Atmospheric pressure
E. Nutrients

8. Two of the fingers on Sally's left hand are fused together by a web of skin. Twenty-year-old Joey has a small streak of white hair on the top of his head, although the rest of his hair is black. Both Sally and Joey exhibit:

A. rare but minor diseases
B. serious birth defects
C. examples of normal anatomical variation

9. Which of the following is (are) involved in maintaining homeostasis?

A. Effector
B. Control center
C. Receptor
D. Feedback
E. Lack of change

10. In a negative feedback mechanism:

A. the response of the effector is to depress or end the original stimulus
B. the response of the effector is to enhance the original stimulus
C. the effect is usually damaging to the body
D. the physiological function is maintained within a narrow range

11. When a capillary is damaged, a platelet plug is formed. The process involves platelets sticking to each other. The more platelets that stick together, the more the plug attracts additional platelets. This is an example of:

A. negative feedback
B. positive feedback

12. In the preceding question, the overall effect is to maintain total blood volume within the normal range of 4.5 to 5.5 liters. Thus, the occurrence of a platelet plug preserves homeostasis and is in fact:

A. negative feedback
B. positive feedback

13. A coronal plane through the head:

A. could pass through both the nose and the occiput
B. could pass through both ears
C. must pass through the mouth
D. could lie in a horizontal plane

14. Which of these organs would not be cut by a section through the midsagittal plane of the body?

A. Urinary bladder
B. Gallbladder
C. Small intestine
D. Heart

15. The dorsal aspect of the human body is also its:

A. anterior surface
B. posterior surface
C. lateral aspect
D. superior aspect

16. A neurosurgeon orders a spinal tap for a patient. Into what body cavity will the needle be inserted?

A. Ventral
B. Thoracic
C. Dorsal
D. Cranial
E. Pelvic

17. An accident victim has a collapsed lung. Which cavity has been entered?

A. Mediastinal
B. Pericardial
C. Pleural
D. Vertebral
E. Ventral

18. A patient has a ruptured appendix. What condition does the medical staff watch for?

A. Peritonitis
B. Pleurisy

19. The membrane that forms the outermost layer of the heart is the:

A. visceral pleura
B. visceral peritoneum
C. parietal pericardium
D. visceral pericardium
E. parietal pleura

20. A patient complains of pain in the lower right quadrant. Which system is most likely to be involved?
A. Respiratory
B. Digestive
C. Urinary
D. Skeletal
E. Muscular

21. Which of the following groupings of the abdominopelvic regions is medial?
A. Hypochondriac, hypogastric, umbilical
B. Hypochondriac, lumbar, inguinal
C. Hypogastric, umbilical, epigastric
D. Lumbar, umbilical, iliac
E. Iliac, umbilical, hypochondriac

22. Which body system would be affected by degenerative cartilage?
A. Muscular
B. Nervous
C. Cardiovascular
D. Skeletal
E. Lymphatic

23. The position of the heart relative to the structures around it would be described accurately as:
A. deep to the sternum (breastbone)
B. lateral to the lungs
C. superior to the diaphragm
D. inferior to the ribs
E. anterior to the vertebral column

24. What term(s) could be used to describe the position of the nose?
A. Intermediate to the eyes
B. Inferior to the brain
C. Superior to the mouth
D. Medial to the ears
E. Anterior to the ears

25. Which of the following statements is correct?
A. The brachium is proximal to the antebrachium.
B. The femoral region is superior to the tarsal region.
C. The orbital region is inferior to the buccal region.
D. The axillary region is lateral to the sternal region.
E. The crural region is posterior to the sural region.

26. Which of the following body regions is (are) found on the torso?
A. Gluteal
B. Inguinal
C. Popliteal
D. Acromial
E. Olecranal

27. Which of the following body regions would be moved by a muscle called the abductor pollicis longus?
A. Arm
B. Forearm
C. Wrist
D. Thumb
E. Fingers

28. Harry was sweating profusely as he ran the 10k race. The sweat glands producing the sweat would be considered which part of the feedback system?
A. Stimulus
B. Effectors
C. Control center
D. Receptors

Word Dissection

For each of the following word roots, fill in the literal meaning and give an example, using a word found in this chapter.

Word root	Translation	Example
1. ana		
2. chondro		
3. corona		
4. cyto		
5. epi		
6. gastr		
7. histo		
8. homeo		
9. hypo		
10. lumbus		
11. meta		
12. ology		
13. org		
14. para		
15. parie		
16. pathy		
17. peri		
18. stasis		
19. tomy		
20. venter		
21. viscus		

2

Chemistry Comes Alive

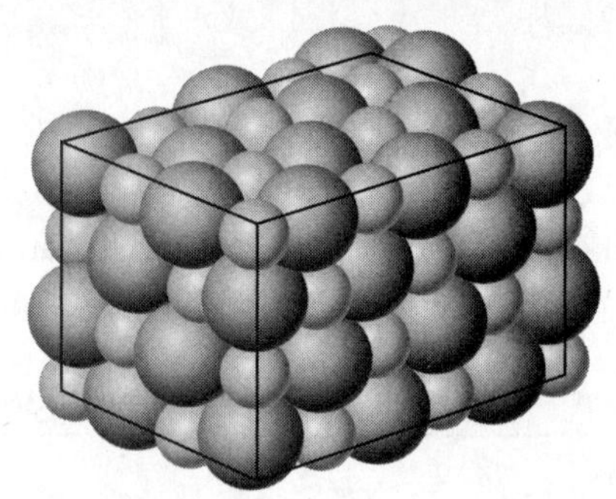

Student Objectives

When you have completed the exercises in this chapter, you will have accomplished the following objectives:

PART 1: BASIC CHEMISTRY

Definition of Concepts: Matter and Energy

1. Differentiate between matter and energy and between potential energy and kinetic energy.
2. Describe the major energy forms.

Composition of Matter: Atoms and Elements

3. Define chemical element and list the four elements that form the bulk of body matter.
4. Define atom. List the subatomic particles; describe their relative masses, charges, and positions in the atom.
5. Define atomic number, atomic mass, atomic weight, isotope, and radioisotope.

How Matter Is Combined: Molecules and Mixtures

6. Define molecule, and distinguish between a compound and a mixture.
7. Compare solutions, colloids, and suspensions.

Chemical Bonds

8. Explain the role of electrons in chemical bonding and in relation to the octet rule.
9. Differentiate among ionic, covalent, and hydrogen bonds.
10. Compare and contrast polar and nonpolar compounds.

Chemical Reactions

11. Define the three major types of chemical reactions: synthesis, decomposition, and exchange. Comment on the nature of oxidation-reduction reactions and their importance.
12. Explain why chemical reactions in the body are often irreversible.
13. Describe factors that affect chemical reaction rates.

PART 2: BIOCHEMISTRY

Inorganic Compounds

14. Explain the importance of water and salts to body homeostasis.
15. Define acid and base, and explain the concept of pH.

Organic Compounds

16. Describe and compare the building blocks, general structures, and biological functions of carbohydrates and lipids.
17. Explain the role of dehydration synthesis and hydrolysis in the formation and breakdown of organic molecules.
18. Describe the four levels of protein structure.
19. Indicate the function of molecular chaperones.
20. Describe enzyme action.
21. Compare and contrast DNA and RNA.
22. Explain the role of ATP in cell metabolism.

Everything in the universe is composed of one or more elements, the unique building blocks of all matter. Although over 100 elemental substances exist, only four of these (carbon, hydrogen, oxygen, and nitrogen) make up over 96% of all living material.

Student activities in Chapter 2 test mastery of the basic concepts of both inorganic and organic chemistry. Chemistry is the science of the composition, properties, and reactions of matter. Inorganic chemistry is the science of nonliving substances such as salts and minerals, which generally do not contain carbon. Organic chemistry is carbon-based chemistry. Biochemistry is the science of living organisms, whether they are plants, animals, or microorganisms.

Student understanding of atomic structure, the bonding behavior of elements, and the structure and activities of the most abundant biochemical molecules (protein, fats, carbohydrates, and nucleic acids) is tested in various ways. Mastering these concepts is necessary to understanding the functions of the body.

BUILDING THE FRAMEWORK

PART 1: BASIC CHEMISTRY

Definition of Concepts: Matter and Energy

1. Select *all* terms that apply to each of the following statements and insert the letters in the answer blanks.

_____ 1. The energy located in the bonds of food molecules:

A. is called thermal energy
B. is a form of potential energy
C. causes molecular movement
D. can be transformed to the bonds of ATP

_____ 2. Heat is:

A. thermal energy
B. infrared radiation
C. directly related to kinetic energy
D. a result of all energy conversions

_____ 3. Whenever energy is transformed:

A. the amount of useful energy decreases
B. some energy is lost as heat
C. some energy is created
D. some energy is destroyed

2. Use the key choices to identify the energy *form* in use in each of the following examples.

Key Choices

A. Chemical B. Electrical C. Mechanical D. Radiant

__________ 1. Chewing food

__________ 2. Vision (two types, please—think!)

__________ 3. Bending your fingers to make a fist

__________ 4. Breaking the bonds of ATP molecules to energize your muscle cells to make that fist

__________ 5. Lying under a sunlamp

Composition of Matter: Atoms and Elements

1. Complete the following table by inserting the missing words.

Particle	Location	Electrical charge	Mass
		+1	
Neutron			
	Orbitals		

2. Insert the *chemical symbol* (the chemist's shorthand) in the answer blanks for each of the following elements.

_____ 1. Oxygen	_____ 4. Iodine	_____ 7. Calcium	_____ 10. Magnesium
_____ 2. Carbon	_____ 5. Hydrogen	_____ 8. Sodium	_____ 11. Chloride
_____ 3. Potassium	_____ 6. Nitrogen	_____ 9. Phosphorus	_____ 12. Iron

3. Using the key choices, select the correct responses to the following descriptive statements. Insert the appropriate answers in the answer blanks.

Key Choices

A. Atom C. Element E. Ion G. Molecule I. Protons

B. Electrons D. Energy F. Matter H. Neutrons J. Valence

__________ 1. An electrically charged atom or group of atoms

__________ 2. Anything that takes up space and has mass (weight)

__________ 3. A unique substance composed of atoms having the same atomic number

________________ 4. Negatively charged particles, forming part of an atom

________________ 5. Subatomic particles that determine an atom's chemical behavior, or bonding ability

________________ 6. The ability to do work

________________ 7. The smallest particle of an element that retains the properties of the element

________________ 8. The smallest particle of a compound, formed when atoms combine chemically

________________ 9. Positively charged particles forming part of an atom

________________ 10. The combining power of an atom

________________ 11. ________________ 12. Subatomic particles responsible for most of an atom's mass

4. From the list below, select the element or elements that match the descriptions. Insert their chemical symbols in the answer blanks.

Oxygen	Iodine	Calcium	Magnesium
Carbon	Hydrogen	Sodium	Chloride
Potassium	Nitrogen	Phosphorus	Iron

________________ 1. Found as a salt in bones and teeth

________________ 2. Make up more than 96% of the mass of a living cell

________________ 3. Essential for transport of oxygen in red blood cells

________________ 4. Essential cations in muscle contraction

________________ 5. Essential for production of thyroid hormones

________________ 6. Present in nucleic acids (in addition to C, H, O, and N)

________________ 7. The most abundant negative ion in extracellular fluids

How Matter Is Combined: Molecules and Mixtures

1. Match the following terms to the appropriate descriptions below:

Mixture	Solution	Molarity	Colloid	Suspension

________________ 1. Concentration expressed in moles

________________ 2. Components are physically, not chemically, combined

__________ 3. Homogeneous combination of solvent and solute(s)

__________ 4. Large particles can settle out unless constantly mixed

__________ 5. Large particles will not settle out

__________ 6. Exhibits the sol-gel phenomenon

__________ 7. An example is sand and water

2. Briefly explain how the following pairs of chemical species differ. Identify each substance as a molecule of an element, a molecule of a compound, an atom, or an ion.

1. H_2O_2 and $2OH^-$ ______________________________

____

2. $2O^{2-}$ and O_2 ______________________________

____

3. $2H^+$ and H_2 ______________________________

____

3. Referencing the periodic table of the elements, determine the molecular weight of each of the following:

_____ 1. Water (H_2O) _____ 2. Ammonia (NH_3) _____ 3. Carbonic acid (H_2CO_3)

4. Circle the term that does not belong in each of the following groupings.

1. Cholesterol $C_6H_{12}O_6$ Perspiration H_2O NH_3 Compound

2. Urine $Ca_3(PO_4)_2$ Inhaled air Exhaled air Plasma

3. Scatters light Transparent True solution Saline

4. Blood cells Water Solute Salt (NaCl) Glucose molecules

5. Weight/volume method 5 g NaCl and 100 ml H_2O 5% NaCl One-molar NaCl

6. Heterogeneous mixture Scattering of light Salt water Milky Colloid

Chemical Bonds

1. Figure 2.1 is a diagram of an atom. Select two different colors and use them to color the coding circles and the subatomic structures on the diagram. Identify each valence electron by inserting an X in the correct location on the diagram. Then respond to the questions that follow, referring to this diagram. Insert your answers in the answer blanks.

○ Nucleus ○ Electrons

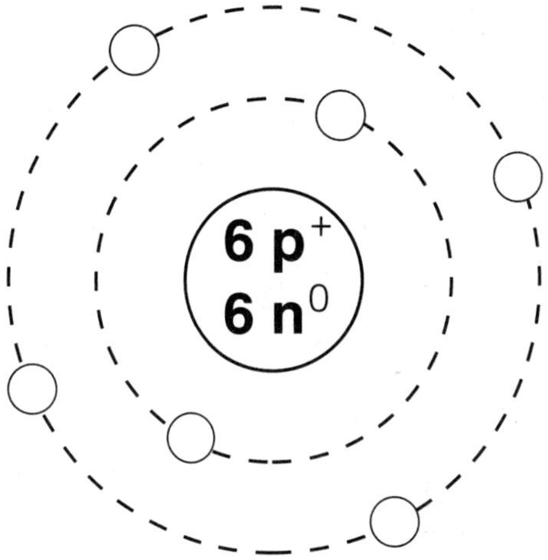

______________ 1. What is the atomic number?

______________ 2. Give the atomic mass of this atom.

______________ 3. Name the atom represented here.

______________ 4. How many electrons are needed to fill its outer shell?

______________ 5. What is the valence of this atom?

______________ 6. Is the atom chemically active *or* inert?

______________ 7. Imagine this atom with one additional neutron. What is the name given to this slightly different form of the same element?

______________ 8. Imagine this atom with two additional neutrons and a nucleus that can eject beta particles. What is the name given to this type of atom?

______________ 9. The electrons in the second electron shell are more energetic. True or false?

$6\ p^+$
$6\ n^0$

Figure 2.1

2. Two types of chemical bonding are shown in Figure 2.2. Select three different colors. Use two colors to color the electrons named at each coding circle. Use the third color to add an arrow to show the direction of transfer of any electron. Then insert the name of the type of bond and the name of the compound in the blanks provided below the figure.

○ Electrons shared ○ Electron(s) transferred

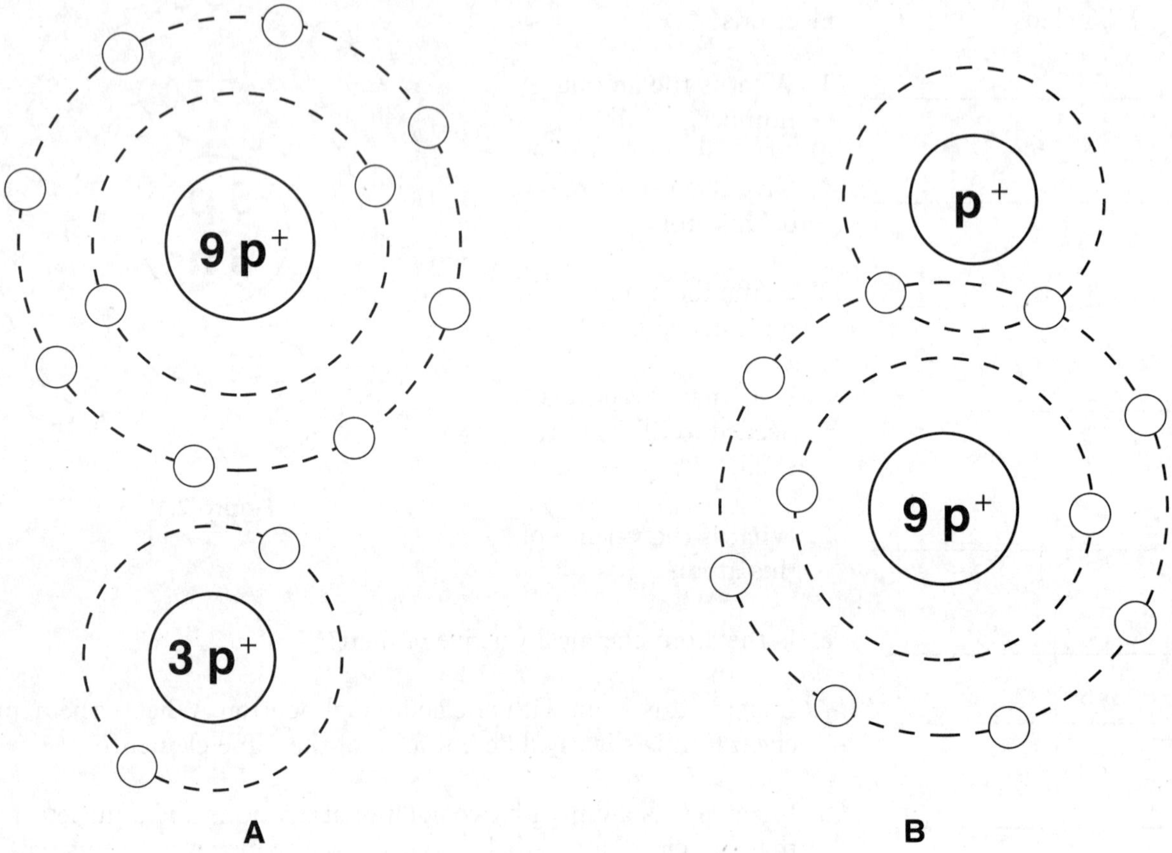

9 p$^+$

3 p$^+$

A

p$^+$

9 p$^+$

B

Figure 2.2

A. Type of bond ____________________

Name of compound ____________________

B. Type of bond ____________________

Name of compound ____________________

3. Complete the following table by filling in the missing parts.

Chemical symbol	Atomic number	Atomic mass	Electron distribution
	1		
		12	
N			
			2, 6, 0
		23	
	12		
			2, 8, 5
S			
	17		
		39	
Ca			

4. Using the information in the table you constructed in Exercise 3, complete the following table, which concerns atoms that form ionic bonds.

Chemical symbol	Loss/gain of electrons	Electrical charge
H		
Na		
Mg		
Cl		
K		
Ca		

5. Complete the following table relating to atoms that form covalent bonds.

Chemical symbol	Number of electrons shared	Number of bonds made
H		
C		
N		
O		

6. Figure 2.3 illustrates five water molecules held together by hydrogen bonds. First, correctly identify the oxygen and hydrogen atoms both by color and by inserting their atomic symbols on the appropriate circles (atoms). Then label the following structures in the figure:

○ Oxygen

○ Hydrogen

○ Positive pole

○ Negative pole

○ Hydrogen bonds

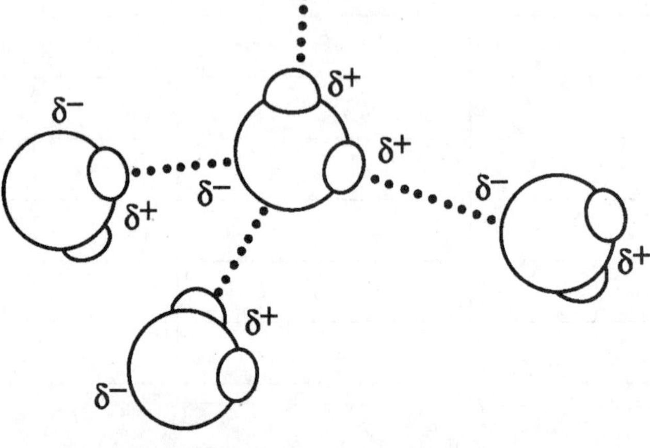

Figure 2.3

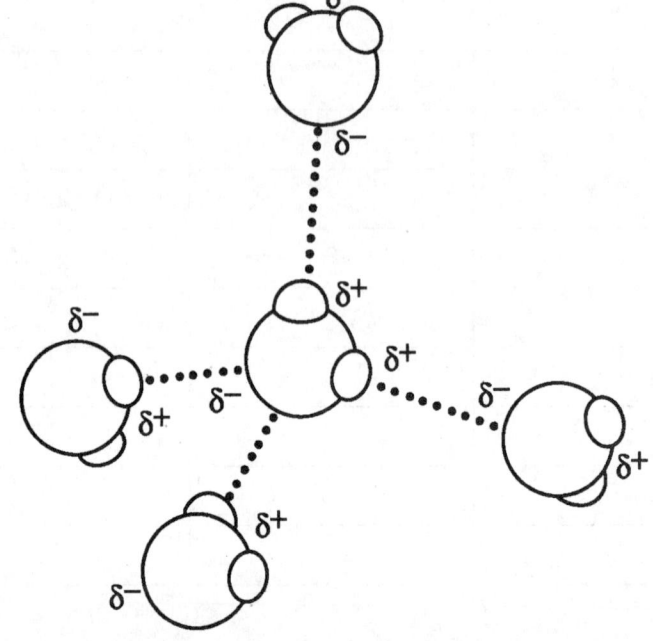

7. Circle each structural formula that is likely to be a polar covalent compound.

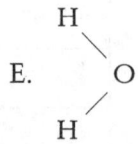

A. Cl–C(Cl)(Cl)–Cl   B. H – Cl   C. H–N(H)–H   D. Cl – Cl   E. H–O–H

Chemical Reactions

1. Match the terms in Column B with the equations listed in Column A. Enter the letters of all correct answers in the answer blanks.

	Column A	Column B
________	1. A + B → AB	A. Synthesis (combination)
________	2. AB + CD → AD + CB	B. Decomposition (dissociation)
________	3. XY → X + Y	C. Displacement (exchange)
________	4. AB + CD ⇄ AD + CB	D. Designates chemical equilibrium

2. Circle the term or phrase that does not belong with the grouping in each of the following cases.

1. High reaction rate   Low product concentration   High reactant concentration   Low reactant concentration
2. High reaction rate   Large particles   Small particles   Many forceful collisions
3. Catalyst   Enzyme   Reactant   Increase reaction rate
4. Oxidized substrate   Electron donor   Electron acceptor   Loss of hydrogen
5. Catabolic   Simple → complex   Complex → simple   Degradation

PART 2: BIOCHEMISTRY

1. Use an X to designate which of the following are organic compounds.

_____ 1. Carbon dioxide   _____ 3. Fats   _____ 5. Proteins   _____ 7. H_2O

_____ 2. Oxygen   _____ 4. KCl   _____ 6. Glucose   _____ 8. DNA

Inorganic Compounds

1. Complete the following statements concerning the properties and biological importance of water.

________________ 1.
________________ 2.
________________ 3.
________________ 4.
________________ 5.
________________ 6.
________________ 7.

The ability of water to maintain a relatively constant temperature and thus prevent sudden changes is because of its high __(1)__. Biochemical reactions in the body must occur in __(2)__. About __(3)__% of the volume of a living cell is water. Water molecules are bonded to other water molecules because of the presence of __(4)__ bonds. The cooling effect of perspiration as it evaporates from the skin is the result of the high __(5)__ of water. Water, as H^+ and OH^- ions, is essential in biochemical reactions such as __(6)__ and __(7)__ reactions.

2. Circle the term that does not belong in each of the following groupings.

1. HCl H_2SO_4 Vinegar   Milk of magnesia
2. NaOH   Blood (pH 7.4) $Al(OH)_2$ Urine (pH 6.2)
3. KCl   NaCl $NaHCO_3$ $MgSO_4$

4. Ionic Organic Salt Inorganic

5. pH 7 pH 4 Neutrality $OH^- = H^+$

6. Basi pH 8 Alkaline pH 2

3. Use the key choices to identify the substances described in the following statements. Insert the appropriate answers in the answer blanks.

Key Choices

A. Acids B. Bases C. Buffers D. Salts

________ 1. ________ 2. ________ 3. Substances that ionize in water; good electrolytes

________ 4. Proton (H^+) acceptors

________ 5. Substances that dissociate in water to release hydrogen ions and a negative ion other than hydroxide (OH^-)

________ 6. Substances that dissociate in water to release ions other than H^+ and OH^-

________ 7. Substances formed when an acid and a base are combined

________ 8. Substances such as lemon juice and vinegar

________ 9. Substances that prevent rapid or large swings in pH

________ 10. Substances such as ammonia and milk of magnesia

4. Define *pH*. ____

5. Using the key choices, fully characterize weak and strong acids.

Key Choices

A. Ionize completely in water

B. Ionize incompletely in water

C. Act as part of a buffer system

D. When placed in water, always act to change the pH

E. Ionize at high pH

F. Ionize at low pH

G. Ionize at pH 7

Weak acid: ______________________ Strong acid: ______________________

6. Complete the following statements concerning a particular solution—cola.

_______________ 1.
_______________ 2.
_______________ 3.

A can of cola consists mostly of sugar dissolved in water. It also contains carbon dioxide gas, which makes the cola "fizzy" and makes the solution's pH lower than 7. In chemical terminology, you could say that cola is an aqueous solution in which water is the __(1)__, sugar and carbon dioxide are __(2)__, and the dissolved carbon dioxide makes the solution __(3)__.

Organic Compounds

1. Match the terms in Column B with the descriptions in Column A. Enter the correct letters in the answer blanks.

	Column A	Column B
_______________	1. Building blocks of carbohydrates	A. Amino acids
_______________	2. Building blocks of fat	B. Carbohydrates
_______________	3. Building blocks of protein	C. Lipids (fats)
_______________	4. Building blocks of nucleic acids	D. Fatty acids
_______________	5. Cellular cytoplasm is primarily composed of this substance	E. Glycerol
		F. Nucleotides
_______________	6. The single most important fuel source for body cells	G. Monosaccharides
_______________	7. Not soluble in water	H. Proteins
_______________	8. Contains C, H, and O in the ratio CH_2O	
_______________	9. Contain C, H, and O, but have relatively small amounts of oxygen	
_______________	10. _______________ 11. These building blocks contain N in addition to C, H, and O	
_______________	12. Contain P in addition to C, H, O, and N	
_______________	13. Used to insulate the body and found in all cell membranes	
_______________	14. Primary components of meat and cheese	
_______________	15. Primary components of bread and lollipops	
_______________	16. Primary components of egg yolk and peanut oil	
_______________	17. Includes collagen and hemoglobin	
_______________	18. Class that usually includes cholesterol	

2. For each of the following statements that is true, insert T in the answer blank. For each false statement, correct the underlined word(s) and insert your correction in the answer blank.

______________ 1. Phospholipids are <u>polarized</u> molecules.

______________ 2. <u>Steroids</u> are the major form in which body fat is stored.

______________ 3. <u>Water</u> is the most abundant compound in the body.

______________ 4. <u>Nonpolar</u> molecules are generally soluble in water.

______________ 5. The bases of RNA are A, G, C, and <u>U</u>.

______________ 6. The universal energy currency of living cells is <u>RNA</u>.

______________ 7. RNA is <u>single stranded</u>.

______________ 8. The bond type linking the subunits of proteins together is commonly called the <u>hydrogen</u> bond.

______________ 9. The external fuel of choice used by cells as a ready source of energy is <u>starch</u>.

______________ 10. The nucleotide base complementary to G is <u>C</u>.

______________ 11. The backbone of a nucleic acid molecule consists of sugar and <u>base</u> units.

3. Five simplified diagrams of the structures of five biochemical molecules are shown in Figure 2.4.

First, identify the molecules and insert their correct names in the answer blanks on the figure (A through E).

Second, select a different color for each type of molecule listed below and use it to color the coding circle and the corresponding molecular structure in the illustration.

○ Fat ○ Nucleotide ○ Monosaccharide

○ Functional protein ○ Polysaccharide

Third, answer the questions relating to these diagrams by inserting your answers in the blanks below the figure.

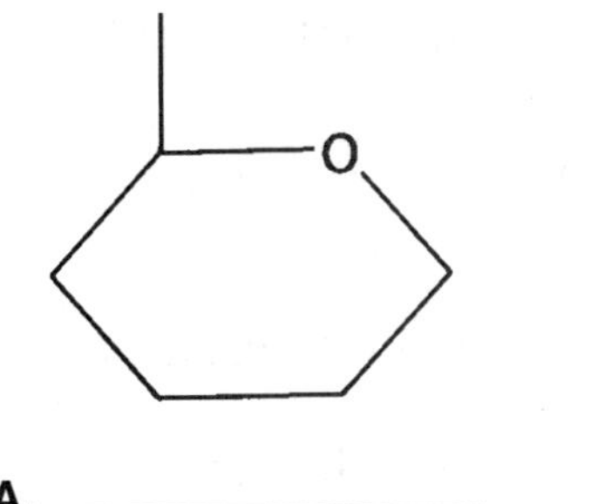

A. ____________

B. ____________

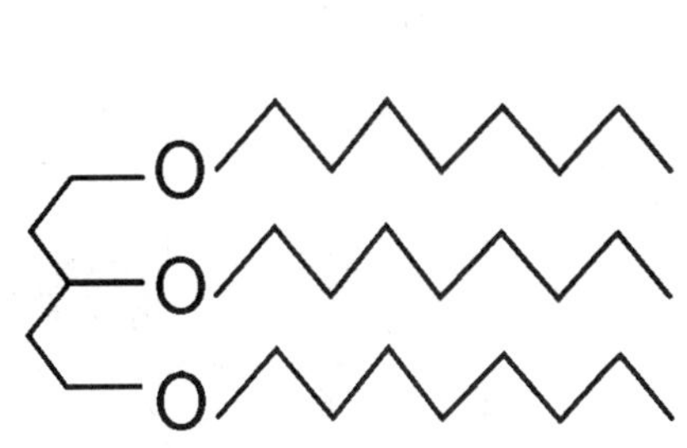

C. ____________

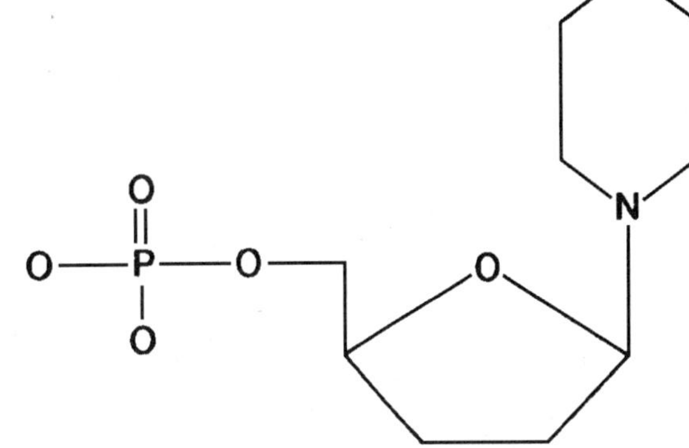

D. ____________

E. ____________

Figure 2.4

____________ 1. Give an example of a biochemical having a structure like diagram C.

____________ 2. Which two diagrams illustrate structures of monomers (building blocks)?

____________ 3. Name the level of structure illustrated in diagram B.

____________ 4. Which two diagrams illustrate molecules used for energy storage?

____________ 5. Which diagram shows a structure that most looks like a molecule of ATP?

4. If any of the following statements about enzymes is true, write T in the answer blank. If a statement is false, correct the underlined word and write the correction in the answer blank.

__________ 1. All enzymes are <u>proteins</u>.

__________ 2. The substances on which enzymes act are called <u>cofactors</u>.

__________ 3. The name of an enzyme usually ends in the suffix -<u>ide</u>.

__________ 4. <u>Coenzymes</u> are vitamins that assist the chemical action of enzymes.

__________ 5. Enzymes <u>increase</u> the activation energy of biochemical reactions.

__________ 6. The active site of an enzyme is the location of <u>substrate</u> attachment.

__________ 7. Changes in pH or temperature decrease enzyme activity because bonds break and the enzyme returns to its <u>tertiary</u> structure.

5. The biochemical reaction shown in Figure 2.5 represents the complete digestion of a polymer (as consumed in food) down to its constituent monomers (building blocks). Select two colors and color the coding circles and the structures. Then, select the one correct answer for each statement below and insert your answer in the answer blank.

○ Monomer ○ Polymer

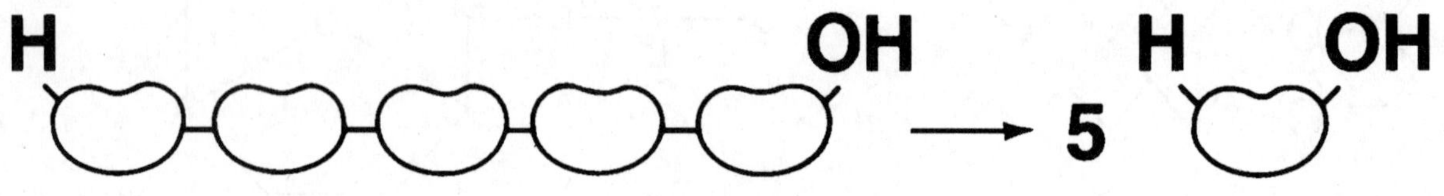

H OH → 5 H OH

Figure 2.5

__________ 1. If starch is the polymer, the monomer is:

A. glycogen B. amino acid C. glucose D. maltose

__________ 2. During polymer digestion, water as H^+ and OH^- ions would:

A. be a product of the reaction

B. act as a catalyst

C. enter between monomers, bond to them, and keep them separated

D. not be involved in this reaction

__________ 3. Another name for the chemical digestion of polymers is:

A. dehydration B. hydrolysis C. synthesis D. displacement

__________ 4. If the monomers are amino acids, they may differ from each other by their:

A. R group B. amino group C. acid group D. peptide bond

6. Place an X in the blank before each phrase that accurately describes molecular chaperones.

_____ 1. Crucial to correct/normal protein structure

_____ 2. Numbers increase during fever

_____ 3. Numbers decline when proteins are denatured

_____ 4. Originally called heat-shock proteins

_____ 5. Defensive molecules

7. This exercise concerns the way polar molecules—phospholipids, in this example—interact with water. As you can see in Figure 2.6, the phospholipid molecules have arranged themselves to form a hollow sphere within the aqueous environment of the beaker.

First, color all the water-containing areas blue, all areas that exclude water yellow, and the polar heads of the phospholipids gray-black.

Second, draw in a row of phospholipids on the surface of the water in the beaker, showing how they would orient themselves (heads vs. tails) to the water in the beaker.

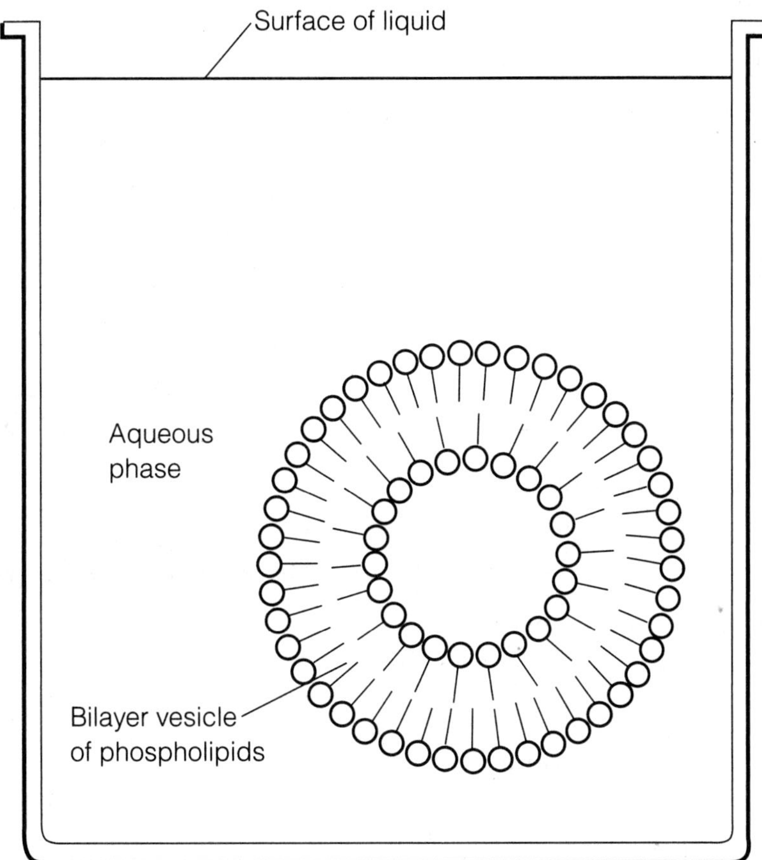

Figure 2.6

8. Figure 2.7 shows the molecular structure of DNA, a nucleic acid.

○ Deoxyribose sugar (d-R)

○ Phosphate unit (P)

○ Adenine (A)

○ Cytosine (C)

○ ______________ ( )

○ ______________ ( )

A. Identify the two unnamed nitrogen bases and insert their correct names and symbols in the two blanks beside the color coding.

B. Complete the identification of the bases on the diagram by inserting the correct symbols in the appropriate spaces on the right side of the diagram.

C. Select different colors and color the coding circles and the corresponding parts of the diagram.

D. Label one d-R sugar unit and one P unit of the "backbones" of the DNA structure by inserting leader lines and labels on the diagram.

E. Circle the associated nucleotide.

F. Answer the following questions by writing your answers in the answer blanks.

1. Name the bonds that help hold the two DNA strands together. ______________

2. Name the three-dimensional shape of the DNA molecule. ______________

3. How many base pairs are present in this segment of a DNA model? ______________

4. What is the term that means "base pairing"?

____

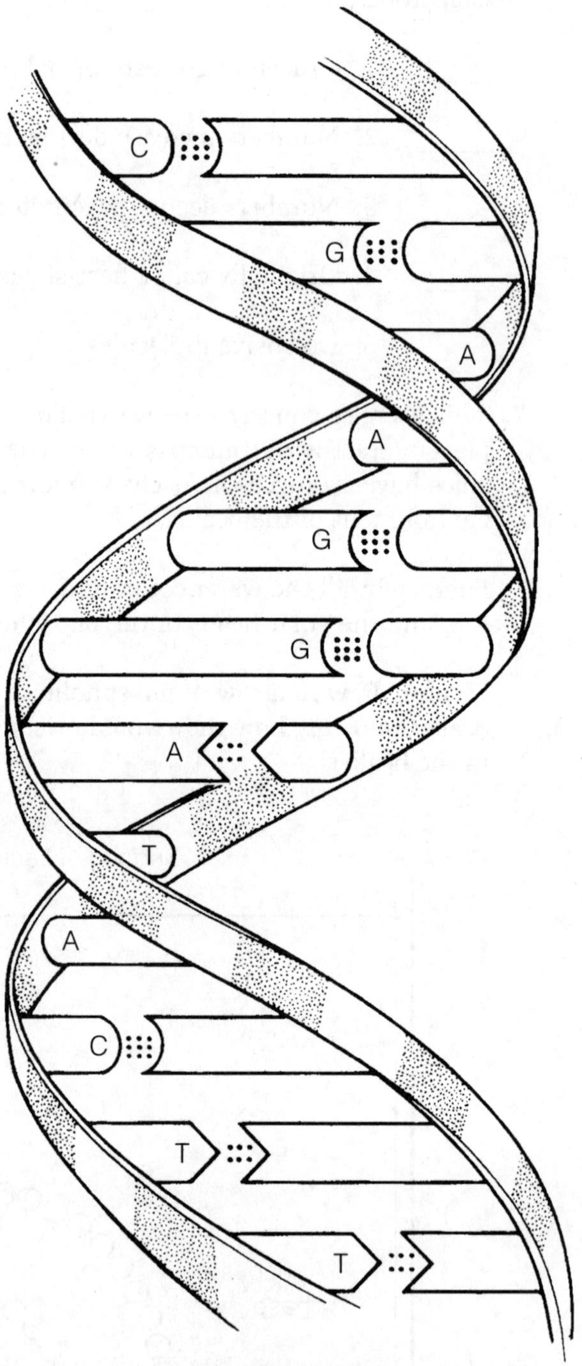

Figure 2.7
Part of a DNA molecule (coiled)

9. Circle the term that does not belong in each of the following groupings.

1. Adenine Guanine Glucose Thymine
2. DNA Ribose Phosphate Deoxyribose
3. Galactose Glycogen Fructose Glucose
4. Amino acid Polypeptide Glycerol Protein
5. Glucose Sucrose Lactose Maltose
6. Synthesis Dehydration Hydrolysis Water removal
7. Amino acid Compound Fatty acid Carbon
8. Enzyme activity Body temperature Denaturation Favorable pH

10. Two diagrams, A and B, are shown in Figure 2.8. These represent the stylized structures of ATP and cyclic AMP. Select a different color for each term with a color-coding circle and color the structures. Identify each biochemical by inserting its name in blanks A and B below. Label all high-energy bonds with corresponding colored arrows on the diagrams.

○ Phosphate ○ Ribose ○ Adenine ○ High-energy bond

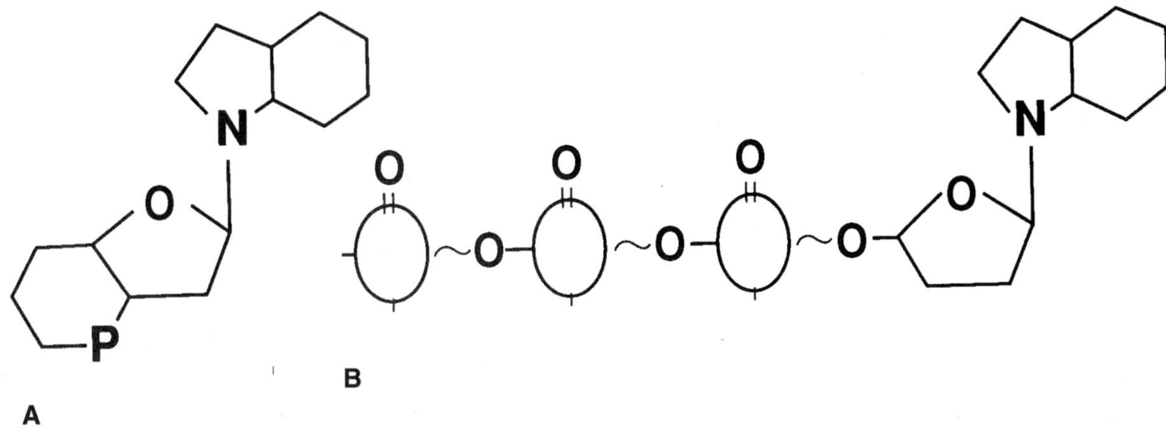

A. ______________________ B. ______________________

Figure 2.8

11. Complete the following chart on proteins by responding to the questions that are printed in the headings of the chart. Note the clues that are given! Insert your answers in the spaces provided.

Name of protein?	Secondary or tertiary structure?	Molecular shape: fibrous or globular?	Water-soluble or insoluble?	Function of protein?	Location in body?
Collagen					Cartilage and bone
Proteinase					Stomach
	Quaternary			Carries O_2	Red blood cells
Keratin	Secondary		Insoluble		

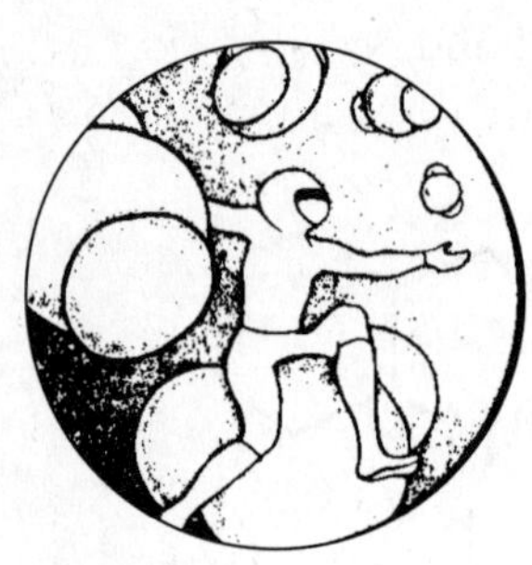

THE INCREDIBLE JOURNEY:

A Visualization Exercise for Biochemistry

. . . you are suddenly up-ended and are carried along in a sea of water molecules at almost unbelievable speed.

1. Complete the narrative by inserting the missing words in the answer blanks.

______________ 1.
______________ 2.
______________ 3.
______________ 4.

For this journey, you are miniaturized to the size of a very small molecule by colleagues who will remain in contact with you by radio. Your instructions are to play the role of a water molecule and to record any reactions that involve water molecules. Since water molecules are dipoles, you are outfitted with an insulated rubber wetsuit with one __(1)__ charge at your helmet and two __(2)__ charges, one at the end of each leg.

As soon as you are injected into your host's bloodstream, you feel as though you are being pulled apart. Some large, attractive forces are pulling at your legs from different directions! You look about but can see only water molecules. After a moment's thought, you remember the dipole construction of your wetsuit. You record that these forces must be the __(3)__ in water that are easily formed and easily broken.

After this initial surprise, you are suddenly up-ended and carried along in a sea of water molecules at almost unbelievable speed. You measure and record this speed as 1000 miles/sec. You are about to watch some huge, red, disk-shaped structures (probably __(4)__) taking up O_2 molecules, when you

5. ____________________

6. ____________________

7. ____________________

8. ____________________

9. ____________________

10. ____________________

11. ____________________

12. ____________________

13. ____________________

14. ____________________

15. ____________________

16. ____________________

17. ____________________

18. ____________________

19. ____________________

20. ____________________

are swept into a very turbulent environment. Your colleagues radio that you are in the small intestine. With difficulty, because of numerous collisions with other molecules, you begin to record the various types of molecules you see.

In particular, you notice a very long helical molecule made of units with distinctive R groups. You identify and record this type of molecule as a __(5)__, made of units called __(6)__ that are joined together by __(7)__ bonds. As you move too close to the helix during your observations, you are nearly pulled apart to form two ions, __(8)__, but you breathe a sigh of relief as two ions of another water molecule take your place. You watch as these two ions move between two units of the long helical molecule. Then, in a fraction of a second, the bond between the two units is broken. As you record the occurrence of this chemical reaction, called __(9)__, you are jolted into another direction by an enormous globular protein, the very same __(10)__ that controls and speeds up this chemical reaction.

Once again you find yourself in the bloodstream, heading into an organ identified by your colleagues as the liver. Inside a liver cell, you observe many small monomers, made up only of C, H, and O atoms. You identify these units as __(11)__ molecules because they are being bonded to form very long, branched polymers called __(12)__ by the liver cells. You record that this type of chemical reaction is called __(13)__, and you happily note that this reaction also produces __(14)__ molecules like you!

Once more in the bloodstream, you are ferried, according to your colleagues, into the cartilage of the knee. What an environment you have found! Long, glistening, parallel white fibrous structures indicate that this protein must be __(15)__. Before you fully appreciate and record your observations, you are immobilized along with thousands of other water molecules by this long twisting protein. Like them, you are trapped, all dipoles in the same orientation, in layers of __(16)__ around this protein. Your colleagues are quick to advise that movement of your host's knee will press water out of the cartilage and you will be released. And so you are with a great "woosh."

Via another speedy journey through the bloodstream, you next reach the skin. You move deep into the skin and finally gain access to a sweat gland. In the sweat gland, you collide with millions of water molecules and some ionized salt molecules that are continually attracted to your positive and negative charges. Suddenly, the internal temperature rises, and molecular collisions __(17)__ at an alarming rate, propelling you through the pore of the sweat gland onto the surface of the skin. As you watch, other water molecules on the surface begin to __(18)__, using __(19)__ energy from the body. You record that the cooling effect just observed is the result of a property of water called high heat of __(20)__. Saved from the fate of evaporating into thin air, you are speedily rescued by your colleagues.

CHALLENGING YOURSELF

At the Clinic

1. It is determined that a patient is in acidosis. What does this mean, and would you treat the condition with a chemical that would *raise* or *lower* the pH?
2. Carbon dioxide concentration influences blood pH in the following manner: High levels of CO_2 increase the rate of formation of carbonic acid. If a patient has difficulty ventilating the lungs (particularly in exhaling, as in emphysema), would you expect the patient to be in acidosis or alkalosis?
3. Hugo, a patient with kidney disease, is unable to excrete sufficient amounts of hydrogen ions. Would you expect Hugo to hyperventilate or hypoventilate?
4. Vitamin D is essential for absorption of calcium, which contributes to the hardness of bones. What would most likely correlate with a softening of the bones: an inability to absorb fat, or excess fat absorption?
5. A newborn is diagnosed with sickle-cell anemia, a genetic disease in which an amino acid substitution results in abnormal hemoglobin. Explain to the parents how the substitution can have such a drastic effect on the structure of the protein.
6. Chris has just dissolved pepsinogen in a test tube containing distilled water and albumin protein. When he checks for protein digestion later, he finds that none has occurred. What substance (inorganic) should he have added to the test tube to mimic conditions of protein digestion in the body? What exactly does this needed substance do to promote protein digestion?
7. Johnny's body temperature is spiking upward. When it reaches 104°F, his mother puts in a call to the pediatrician. She is advised to give Johnny children's aspirin and sponge his body with cool to tepid water to prevent a further rise in temperature. How might a fever (excessively high body temperature) be detrimental to Johnny's welfare?
8. Stanley has acute indigestion and is doubled over with pain. How could an antacid reduce his stomach discomfort?
9. Explain the effect of mineral and vitamin deficiency on enzyme activity.

Stop and Think

1. Explain why the formation of ATP from ADP and P_i requires more energy than the amount released for cellular use when ATP is broken down.
2. Can all molecular structures be explained by the concepts of chemical bonding presented in this chapter? What about ozone (O_3) and carbon monoxide (CO)?
3. Using the octet rule, predict the atomic number of the next two inert gases after helium (atomic number 2).
4. Proteins within a cell commonly have negative surface charges. Indicate whether aqueous mixtures containing proteins form colloids or suspensions, and explain why.
5. Explain why a one-molar solution of sodium chloride has twice as many solute particles as a one-molar solution of glucose. How many particles would a one-molar solution of *calcium chloride* have compared to a one-molar glucose solution?
6. Combining carbonic acid, H_2CO_3, and calcium hydroxide, $Ca(OH)_2$, results in the formation of two water molecules and what salt?
7. Draw the following molecules, showing the bonds as lines between the chemical symbols: N_2H_4, H_2CO_3, CH_3COOH.

8. Hydrogen bonding determines the spacing between water molecules in ice crystals. What would happen to lakes in the winter if water molecules were packed more tightly in the solid form than in the liquid form?

9. Explain, in terms of chemical bonding, how ammonia functions as a base.

10. What happens to the pH of water when its temperature is raised to 38°C? What does this imply about our definition of neutral pH?

11. What is "special" about valence shell electrons?

12. Triglycerides float on water. How does this explain the use of triglycerides as the main form of reserve body fuel?

13. Dissect the word "atom."

14. Can chlorine make covalent bonds? If so, how many will each chlorine atom make?

15. What can a cell do to actively control the direction of a reversible reaction?

16. Justify classifying water as both an acid and a base.

17. Given a hydrogen ion concentration $[H^+]$ of 0.000001 mol/liter, what is the pH? How does the relative acidity of this solution compare to that of a solution of pH 7?

18. Graph the following data relating to an enzyme-catalyzed reaction. Then answer the associated questions.

Substrate concentration (mol)	0.1	0.2	0.3	0.4	0.5
Reaction rate (mol product/min)	0.05	0.06	0.068	0.069	0.069

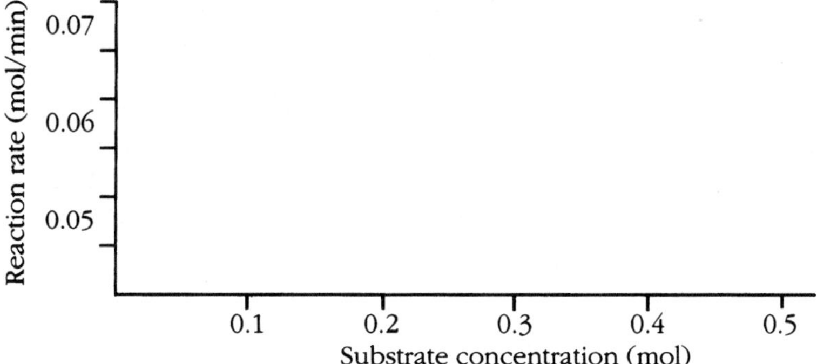

1. Why is there a plateau in the graph?
2. How would an alteration of the enzyme's active site, such as might occur in competitive or noncompetitive inhibition, affect the reaction rate?

19. Matter *occupies space* and *has mass*. Describe how energy *must* be described in terms of these two criteria. Then list the criteria usually used to define or describe energy.

20. How do saturated and unsaturated fats differ in molecular structure and gross appearance?

21. Relative to the chemical equation $X + YC \rightarrow XC + Y$, the reactants are

____________________; the products are ____________________. If interpreting this

reaction in terms of mole quantities, how many moles of each (reactants and products)

are indicated by the equation?____________________

22. It is said that all chemical reactions reach equilibrium if certain conditions are met. What is the effect of continually adding more reactants to the system?

23. Describe several uses of radioisotopes in biological research or medicine.

24. Evelyn is quite proud of her slender, model-like figure and boasts that she doesn't have an "ounce of excess body fat." Barbara, on the other hand, is grossly overweight. She complains of being "hot" most of the time, and on a hot day she is miserable. Evelyn generally feels chilled except on very hot days. Explain the relative sensitivity to environmental temperature of these two women on the basis of information you have been given.

COVERING ALL YOUR BASES

Multiple Choice

Select the best answer or answers from the choices given.

1. Energy:
A. has mass
B. does work
C. has weight
D. takes up space
E. causes motion

2. Energy contained in molecular bonds is:
A. electrical
B. mechanical
C. chemical
D. radiant
E. kinetic

3. Which of the following are among the four major elements in the body?
A. Iron
B. Carbon
C. Oxygen
D. Copper
E. Sulfur

4. Which of the following is (are) true concerning the atomic nucleus?
A. Contains the mass of the atom
B. The negatively charged particles are here
C. Particles can be ejected
D. Contains particles that determine atomic number
E. Contains particles that interact with other atoms

5. Which of the following would indicate an isotope of oxygen?
A. Atomic number 7
B. Atomic number 8
C. Atomic mass 16
D. Atomic mass 18
E. Gain of 2 electrons

6. An example of a suspension is:
A. sugar in tea
B. cornstarch in water
C. proteins in plasma
D. carbon dioxide in soda water
E. oil in water

7. Pick out the correct match(es) of element and valence number.
A. Oxygen-6
B. Chlorine-8
C. Calcium-2
D. Nitrogen-3
E. Carbon-4

8. In which mixtures would you expect to find hydrogen bonds?
A. Sugar in water
B. Oil in water
C. Oil in gasoline
D. Salt in water

9. Which of the following is (are) a synthesis reaction?
A. Glucose to glycogen
B. Glucose + fructose to sucrose
C. Starch to glucose
D. Polypeptide to dipeptides
E. Amino acids to dipeptide

10. Which of the following will have the greatest reaction rate, assuming the same temperature, concentration, and catalyst activity?
A. Carbon dioxide and water combining to form carbonic acid
B. Glucose and fructose forming sucrose
C. Amino acids combining to dipeptides
D. Fatty acids combining with glycerol
E. All of these will have the same reaction rate with the listed factors being equal.

11. Organic compounds include:
    A. water
    B. carbon dioxide
    C. oxygen
    D. carbonic acid
    E. glycerol

12. Important functions of water include:
    A. cushioning
    B. transport medium
    C. participation in chemical reactions
    D. solvent for sugars, salts, and other solutes
    E. reduces temperature fluctuations

13. Which of the following molecules does not have at least one double covalent bond?
    A. Carbon dioxide
    B. Formaldehyde (H_2CO)
    C. Oxygen gas
    D. Hydrogen cyanide (HCN)
    E. Ethanol (CH_3CH_2OH)

14. Which of the elements listed is the most abundant extracellular anion?
    A. Phosphorus
    B. Sulfur
    C. Potassium
    D. Chloride
    E. Calcium

15. The element essential for normal thyroid function is:
    A. Iodine
    B. Iron
    C. Copper
    D. Selenium
    E. Zinc

16. To prepare a 15% salt solution such as potassium chloride (KCl) by the weight-volume method, you would:
    A. weigh out 15 g of KCl and add water to obtain 100 g of solution
    B. weigh out 15 g of KCl and add water to obtain 100 ml of solution
    C. weigh out the number of grams equal to the molecular weight of KCl and add water to obtain 1 liter of solution
    D. weigh out 15 g of KCl and add 15 ml of water
    E. measure out 15 ml of KCl and add water to obtain 100 ml of solution

17. Alkaline substances include:
    A. gastric juice
    B. water
    C. blood
    D. orange juice
    E. ammonia

18. Which of the following is (are) not a monosaccharide?
    A. Glucose
    B. Fructose
    C. Sucrose
    D. Glycogen
    E. Deoxyribose

19. Which is a building block of neutral fats?
    A. Ribose
    B. Guanine
    C. Glycerol
    D. Glycine
    E. Glucose

20. What lipid type is the foundation of the cell membrane?
    A. Triglyceride
    B. Steroid
    C. Vitamin D
    D. Phospholipid
    E. Prostaglandin

21. Which of the following is (are) not derived from cholesterol?
    A. Vitamin A
    B. Vitamin D
    C. Steroid hormones
    D. Bile salts
    E. Prostaglandins

22. Which of the following is primarily responsible for the helical structure of a polypeptide chain?
    A. Hydrogen bonding
    B. Tertiary folding
    C. Peptide bonding
    D. Quaternary associations
    E. Complementary base pairing

23. Absence of which of the following nitrogen bases would prevent protein synthesis?
    A. Adenine
    B. Cytosine
    C. Guanine
    D. Thymine
    E. Uracil

24. Which of the following is (are) not true of RNA?
    A. Double stranded
    B. Contains cytosine
    C. Directs protein synthesis
    D. Found primarily in the nucleus
    E. Can act as an enzyme

25. DNA:
    A. contains uracil
    B. is a helix
    C. is the "genes"
    D. contains ribose

26. ATP is not associated with:
    A. a basic nucleotide structure
    B. high-energy phosphate bonds
    C. deoxyribose
    D. inorganic phosphate
    E. reversible reaction

27. Compared to a solution with a pH of 6, a solution with a pH of 8 has:
   A. 4 times less H^+
   B. 2 times more H^+
   C. 100 times more H^+
   D. 100 times less H^+
   E. 2 times less H^+

28. Glucose is to starch as:
   A. a steroid is to a lipid
   B. a nucleotide is to nucleic acid
   C. an amino acid is to a protein
   D. a polypeptide is to an amino acid

Word Dissection

For each of the following word roots, fill in the literal meaning and give an example, using a word found in this chapter.

Word root	Translation	Example
1. ana		
2. cata		
3. di		
4. en		
5. ex		
6. glyco		
7. hydr		
8. iso		
9. kin		
10. lysis		
11. mono		
12. poly		
13. syn		
14. tri		

3

Cells: The Living Units

Student Objectives

When you have completed the exercises in this chapter, you will have accomplished the following objectives:

Overview of the Cellular Basis of Life

1. Define cell.
2. List the three major regions of a generalized cell and indicate the function of each.

The Plasma Membrane: Structure

3. Describe the chemical composition of the plasma membrane and relate it to membrane functions.
4. Compare the structure and function of tight junctions, desmosomes, and gap junctions.

The Plasma Membrane: Membrane Transport

5. Relate plasma membrane structure to active and passive transport mechanisms.
6. Compare and contrast simple diffusion, facilitated diffusion, and osmosis relative to substances transported, direction, and mechanism.
7. Differentiate between primary and secondary active transport.
8. Compare and contrast endocytosis and exocytosis in terms of function and direction.
9. Compare and contrast pinocytosis, phagocytosis, and receptor-mediated endocytosis.

The Plasma Membrane: Generation of a Resting Membrane Potential

10. Describe the role of the plasma membrane when cells interact with their environment.

The Plasma Membrane: Cell-Environment Interactions

11. Describe the role of the glycocalyx when cells interact with their environment.
12. List several roles of membrane receptors and that of voltage-sensitive membrane channel proteins.

The Cytoplasm

13. Describe the composition of the cytosol. Define inclusions and list several types.
14. Discuss the structure and function of mitochondria.
15. Discuss the structure and function of ribosomes, the endoplasmic reticulum, and the Golgi apparatus, including functional interrelationships among these organelles.
16. Compare the functions of lysosomes and peroxisomes.
17. Name and describe the structure and function of cytoskeletal elements.
18. Describe the roles of centrioles in mitosis and in formation of cilia and flagella.
19. Describe how the two main types of cell extensions, cilia and microvilli, differ in structure and function.

The Nucleus

20. Outline the structure and function of the nuclear envelope, nucleolus, and chromatin.

Cell Growth and Reproduction

21. List the phases of the cell life cycle and describe the events of each phase.
22. Describe the process of DNA replication.
23. Define gene and genetic code and explain the function of genes.

24. Name the two phases of protein synthesis and describe the roles of DNA, mRNA, tRNA, and rRNA in each phase.

25. Contrast triplets, codons, and anticodons.

26. Describe the importance of ubiquitin-dependent degradation of soluble proteins.

Extracellular Materials

27. Name and describe the composition of extracellular materials.

The basic unit of structure and function in the human body is the cell. Each of a cell's parts, or organelles, as well as the entire cell, is organized to perform a specific function. Cells have the ability to metabolize, grow, reproduce, move, and respond to stimuli. The cells of the body differ in shape, size, and specific roles in the body. Cells that are similar in structure and function form tissues, which, in turn, form the various body organs.

Student activities in this chapter include questions relating to the structure and functional abilities of the generalized animal cell.

BUILDING THE FRAMEWORK

Overview of the Cellular Basis of Life

1. Answer the following questions by inserting your responses in the answer blanks.

1. List the four concepts of the cell theory. ______________________________

____

____

____

2. Describe three different cell shapes. ______________________________

____

3. Name the three major parts of any cell. ______________________________

____

4. Define *generalized* or *composite cell*. ______________________________

____

____

____

The Plasma Membrane: Structure and Functions

1. Figure 3.1 is a diagram of a portion of a plasma membrane. Select four different colors and color the coding circles and the corresponding structures in the diagram. Then respond to the questions that follow, referring to Figure 3.1 and inserting your answers in the answer blanks.

○ Phospholipid molecules ○ Carbohydrate molecules

○ Protein molecules ○ Cholesterol molecules

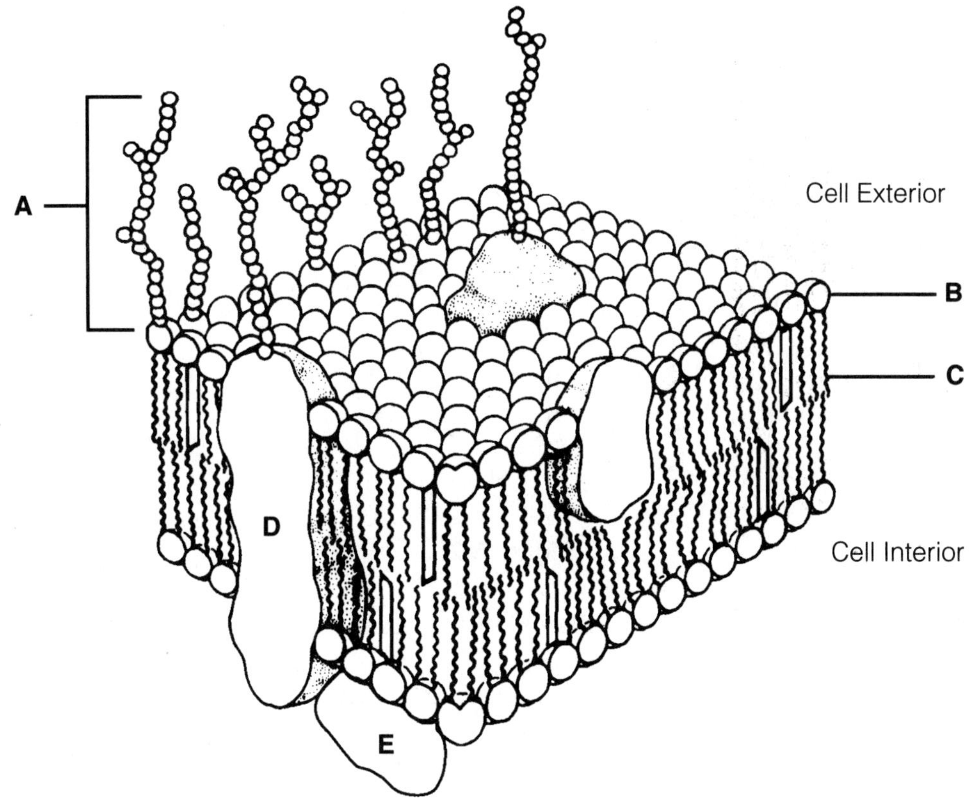

Figure 3.1

1. What name is given to this model of membrane structure? ____________________

2. What is the function of cholesterol molecules in the plasma membrane? ____________

____

3. Name the carbohydrate-rich area at the cell surface (indicated by bracket A). __________

4. Which label, B or C, indicates the nonpolar region of a phospholipid molecule? ________

5. Does nonpolar mean hydrophobic or hydrophilic? ____________________

6. Which label, D or E, indicates an integral protein and which a peripheral protein? ______

____

2. Label the specializations of the plasma membrane, shown in Figure 3.2, and color the diagram as you wish. Then, answer the questions provided that refer to this figure.

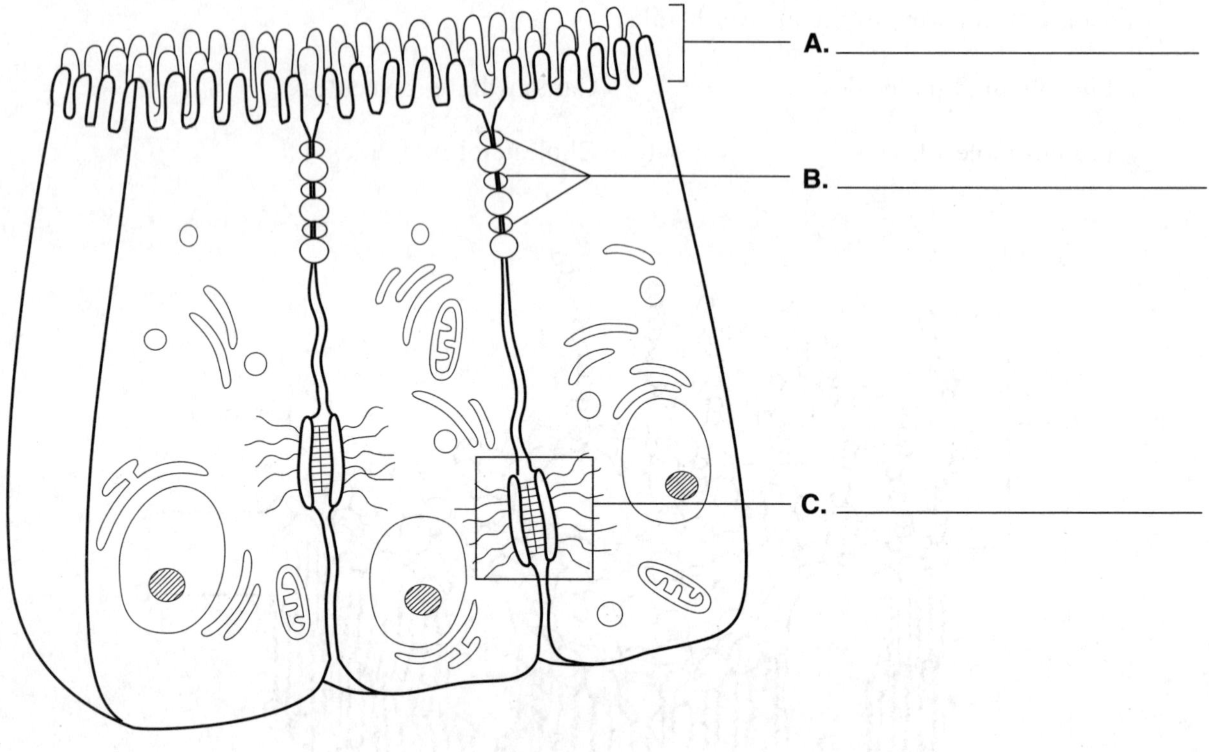

Figure 3.2

1. What is the structural significance of microvilli? ______

______

2. What type of cell function(s) does the presence of microvilli typically indicate? ______

3. What protein acts as a microvilli "stiffener"? ______

4. Name two factors in addition to special membrane junctions that help hold cells together. ______

5. Which cell junction forms an impermeable barrier? ______

6. Which cell junction is a buttonlike adhesion? ______

7. Which junction has linker proteins spanning the intercellular space? ______

8. Which cell junction (not shown) allows direct passage from one cell's cytoplasm to the next? ____________________

9. What name is given to the transmembrane proteins that allow this direct passage? ____________________

3. Figure 3.3 is a simplified diagram of the plasma membrane. Structure A represents channel proteins constructing a pore, structure B represents an ATP-energized solute pump, and structure C is a transport protein that does not depend on energy from ATP. Identify these structures and the membrane phospholipids by color before continuing.

○ Pore ○ Solute pump ○ Passive transport pump ○ Phospholipids

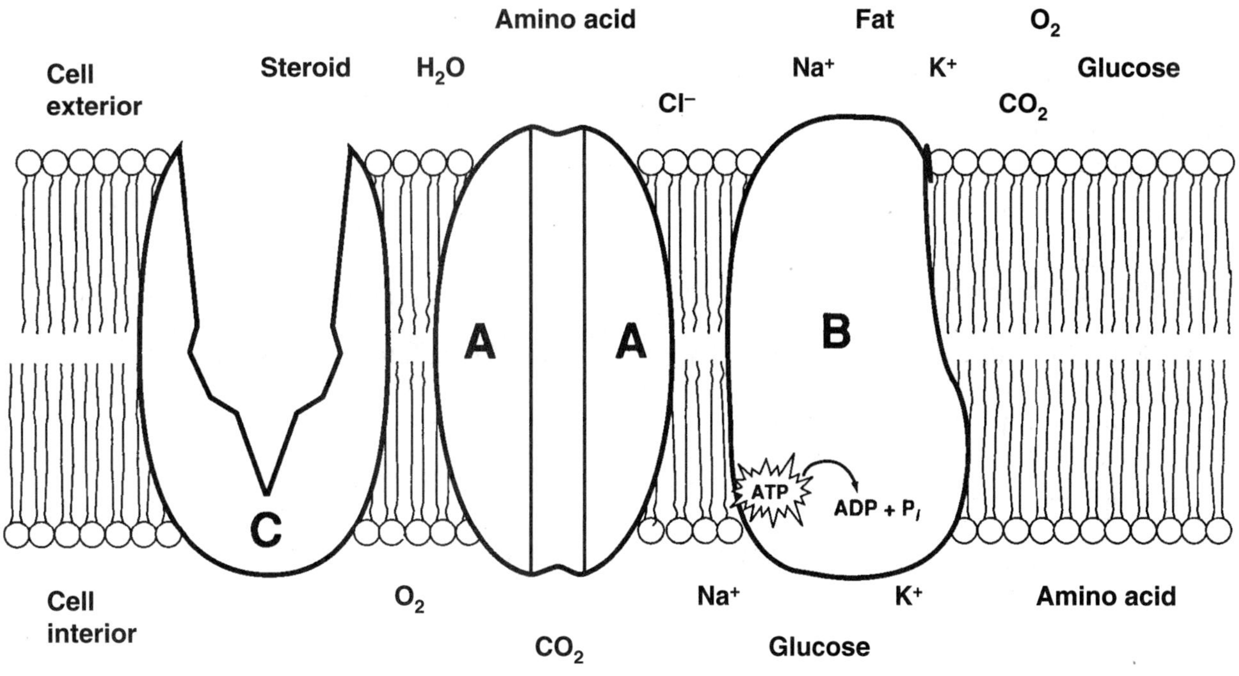

Figure 3.3

Now add arrows to Figure 3.3 as instructed next: For each substance that moves through the plasma membrane, draw an arrow indicating its (most likely) direction of movement (into or out of the cell). If it is moved actively, use a red arrow; if it is moved passively, use a blue arrow.

Finally, answer the following questions referring to Figure 3.3:

1. Which of the substances shown move passively *through the lipid* part of the membrane? ______________________________

2. Which of the substances shown enter the cell by attachment to a passive-transport protein carrier? ______________________________

3. Which of the substances shown moves passively through the membrane by moving through its pores? ______________________________

4. Which of the substances shown would have to use a solute pump to be transported through the membrane? ______________________________

4. Select the key choices that characterize each of the following statements. Insert the appropriate answers in the answer blanks.

Key Choices

A. Protein-coated pit	D. Exocytosis	G. Receptor-mediated endocytosis
B. Diffusion, simple	E. Phagocytosis	H. Solute pumping
C. Diffusion, osmosis	F. Pinocytosis	

______________ 1. Engulfment processes that require ATP

______________ 2. Driven by molecular energy

______________ 3. May provide signaling platforms

______________ 4. Moves down (with) a concentration gradient

______________ 5. Moves up (against) a concentration gradient; requires a carrier

______________ 6. Uses a clathrin-coated vesicle ("pit")

______________ 7. Typically involves *coupled systems*; that is, symports or antiports

______________ 8. Examples of vesicular transport

______________ 9. A means of bringing fairly large particles into the cell

______________ 10. Used to eject wastes and to secrete cell products

5. Figure 3.4 shows three microscope fields containing red blood cells. Arrows indicate the direction of net osmosis. Select three different colors and use them to color the coding circles and the corresponding cells in the diagrams. Then, respond to the questions below, referring to Figure 3.4 and inserting your answers in the spaces provided.

○ Water moves into the cells

○ Water enters and exits the cells at the same rate

○ Water moves out of the cells

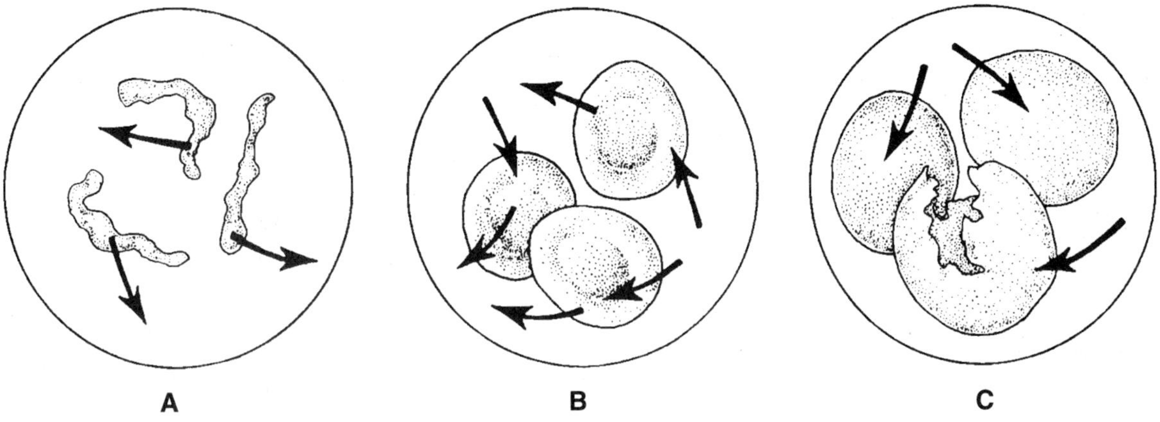

A B C

Figure 3.4

1. Name the type of tonicity illustrated in diagrams A, B, and C.

   A. ____________ B. ____________ C. ____________

2. Name the terms that describe the cellular shapes in diagrams A, B, and C.

   A. ____________ B. ____________ C. ____________

3. What does *isotonic* mean? ____________

4. Why are the cells in diagram C bursting? ____________

   ____

5. What is the difference between tonicity and osmolarity? ____________

   ____

6. The differential permeability of the plasma membrane to sodium (Na^+) and potassium (K^+) ions results in the development of a voltage (resting membrane potential) of about –70 mV across the membrane as indicated in the simple diagram in Figure 3.5.

   First, draw in some Na^+ and K^+ ions in the cytoplasm and extracellular fluid, taking care to indicate their *relative* abundance in the two sites.

   Second, add positive and negative signs to the inner and outer surfaces of the "see-through" cell's plasma membrane to indicate its electrical polarity.

   Third, draw in arrows and color them to match each of the coding circles associated with the conditions noted just below.

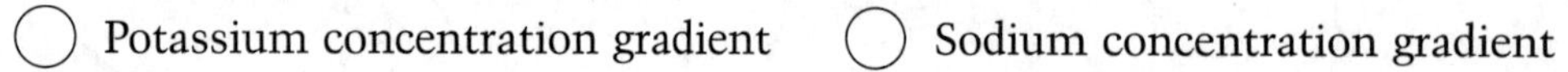

○ Potassium electrical gradient ○ Sodium electrical gradient

○ Potassium concentration gradient ○ Sodium concentration gradient

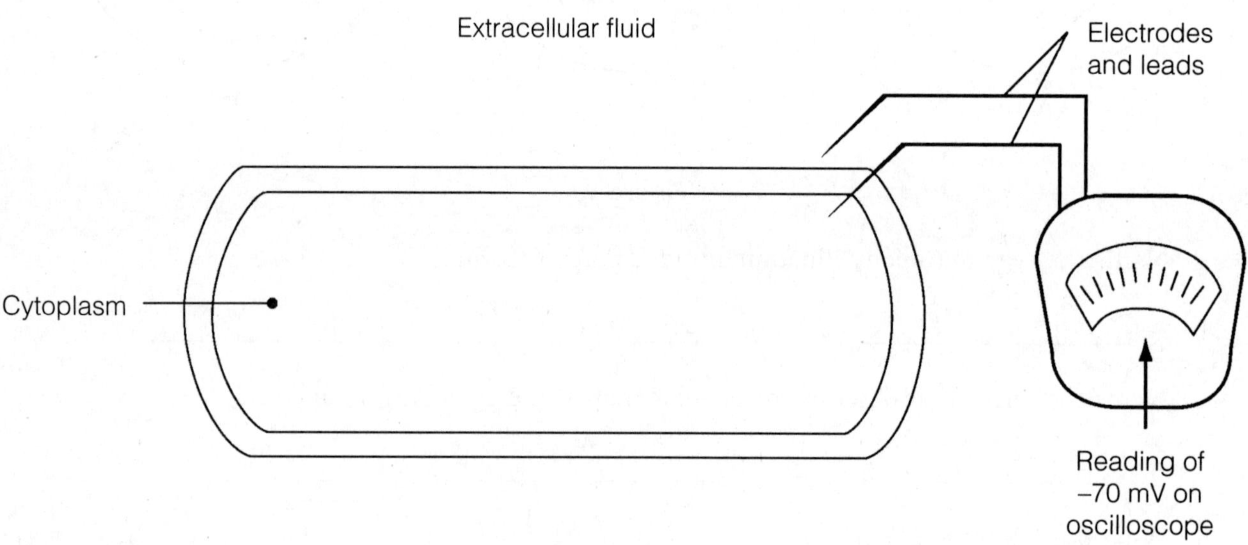

Figure 3.5

Which ion—Na^+ or K^+—is more important in determining the resting membrane potential? ______________________

7. Referring to plasma membranes, circle the term or phrase that does not belong in each of the following groupings.

   1. Fused protein molecules of adjacent cells   Tight junction   Lining of digestive tract   Communication between adjacent cells   No intercellular space

   2. Lipoprotein filaments   Binding of tissue layers   Heart muscle   Impermeable junction   Desmosomes

3. Impermeable intercellular space Molecular communication Embryonic cells

   Gap junction Protein channel

4. Resting membrane potential High extracellular potassium ion (K^+) concentration

   High extracellular sodium ion (Na^+) concentration Nondiffusible protein anions

5. –50 to –100 millivolts Electrochemical gradient Inside membrane negatively charged

   K^+ diffuses across membrane more rapidly than Na^+ Protein anions move out of cell

6. Active transport Sodium-potassium pump Polarized membrane

   More K^+ pumped out than Na^+ carried in ATP required

7. Carbohydrate chains on cytoplasmic side of membrane Cell adhesion

   Glycocalyx Recognition sites Antigen receptors

8. Facilitated diffusion Nonselective Glucose saturation Carrier molecule

9. Clathrin-coated pit Exocytosis Receptor-mediated High specificity

10. CAMs Membrane receptors G proteins Channel-linked proteins

11. Second messenger NO Ca^{2+} Cyclic AMP

12. Cadherins Glycoproteins Phospholipids Integrins

The Cytoplasm

1. Define *cytosol.* ____

____

2. Differentiate clearly between *organelles* and *inclusions.* ______________________

____

____

____

3. Using the following terms, correctly label all cell parts indicated by leader lines in Figure 3.6. Then select different colors for each structure and use them to color the coding circles and the corresponding structures in the illustration.

○ Plasma membrane ○ Mitochondrion ○ Nuclear membrane ○ Centrioles
○ Chromatin threads ○ Nucleolus ○ Golgi apparatus ○ Microvilli
○ Rough endoplasmic reticulum (rough ER) ○ Smooth endoplasmic reticulum (smooth ER)

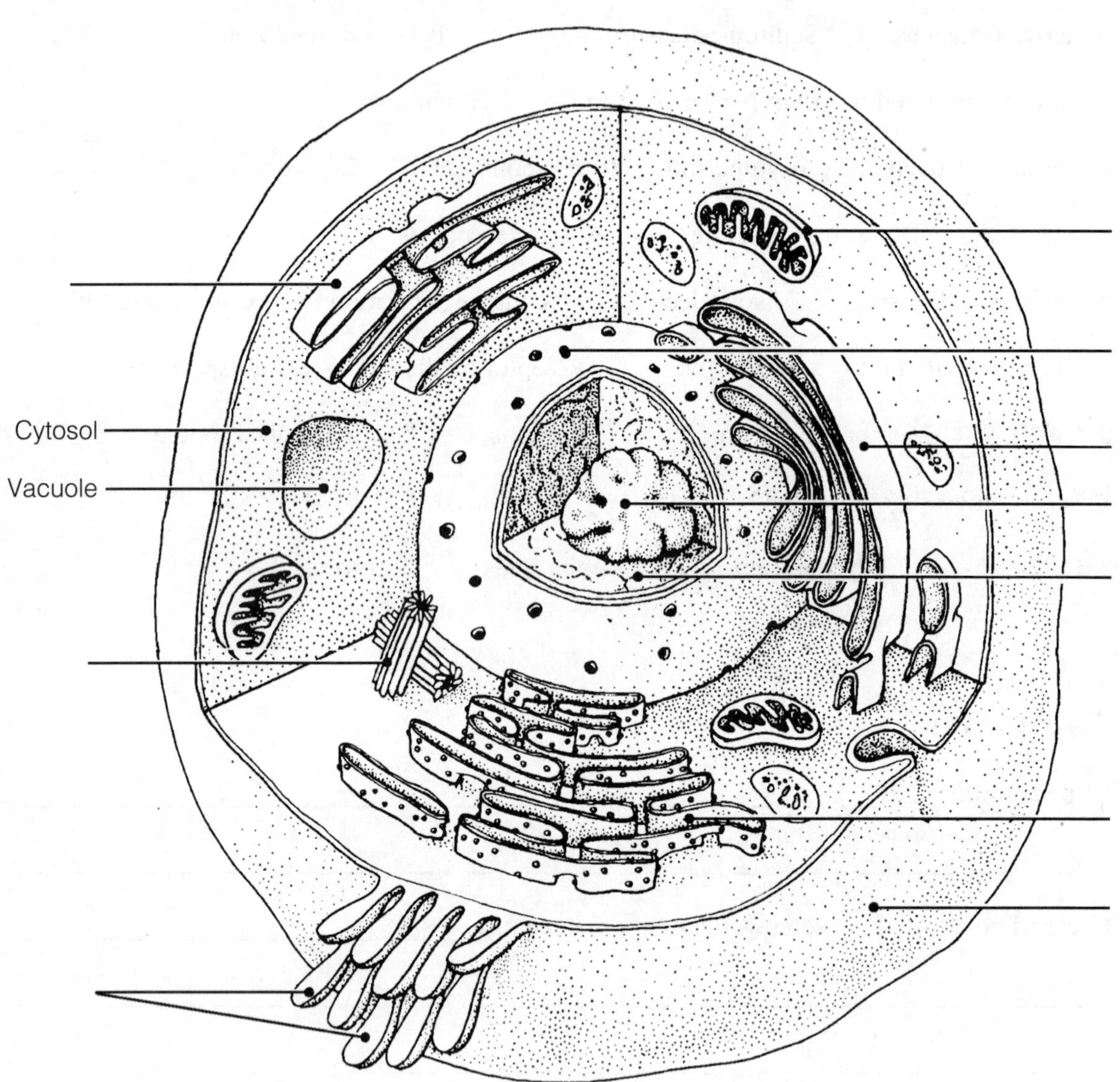

Figure 3.6

4. Complete the following table to fully describe the various cell parts. Insert your responses in the spaces provided under each heading.

Cell structure	Location	Function
	External boundary of the cell	Confines cell contents; regulates entry and exit of materials
Lysosome		
	Scattered throughout the cell	Controls release of energy from foods; forms ATP
	Projections of the plasma membrane	Increase the membrane surface area
Golgi apparatus		
	Two rod-shaped bodies near the nucleus	"Spin" the mitotic spindle
Smooth ER		
Rough ER		
	Attached to membranes or scattered in the cytoplasm	Synthesize proteins
		Act collectively to move substances across cell surface in one direction
	Internal structure of centrioles; part of the cytoskeleton	
Peroxisomes		
		Contractile protein (actin); moves cell or cell parts; core of microvilli
Intermediate filaments	Part of cytoskeleton	
Inclusions		

5. Relative to cellular organelles, circle the term or phrase that does not belong in each of the following groupings.

1. Peroxisomes Enzymatic breakdown Centrioles Lysosomes
2. Microtubules Intermediate filaments Cytoskeleton Cilia
3. Ribosomes Smooth ER Rough ER Protein synthesis
4. Mitochondrion Cristae Self-replicating Vitamin A storage
5. Centrioles Basal bodies Mitochondria Cilia Flagella
6. ER Endomembrane system Ribosomes Secretory vesicles
7. Nucleus DNA Lysosomes Mitochondria

6. Name the cytoskeletal element (microtubules, microfilaments, or intermediate filaments) described by each of the following phrases.

____________________ 1. give the cell its shape

____________________ 2. resist tension placed on a cell

____________________ 3. radiate from the cell center

____________________ 4. interact with myosin to produce contractile force

____________________ 5. are the most stable

____________________ 6. have the thickest diameter

7. Different organelles are abundant in different cell types. Match the cell types with their abundant organelles by selecting a letter from the key choices.

Key Choices

A. mitochondria C. rough ER E. microfilaments G. intermediate filaments

B. smooth ER D. peroxisomes F. lysosomes H. Golgi apparatus

____ 1. cell lining the small intestine (assembles fats)

____ 2. white blood cell; a phagocyte

____ 3. liver cell that detoxifies carcinogens

____ 4. muscle cell (contractile cell)

____ 5. mucus-secreting cell (secretes a protein product)

____ 6. cell at external skin surface (withstands friction and tension)

____ 7. kidney tubule cells (make and use large amounts of ATP)

8. Describe the components and importance of the endomembrane system.

____

____

____

____

____

The Nucleus

1. Complete the following brief table to describe the nucleus and its parts. Insert your responses in the spaces provided.

Nuclear structure	General location/appearance	Function
Nucleus		
Nucleolus		
Chromatin		
Nuclear membrane		

2. Figure 3.7 shows a portion of the proposed model of a chromatin fiber. Select two different colors for the coding circles and the corresponding structures on the figure. Then respond to the two questions that follow.

○ DNA helix ○ Nucleosome

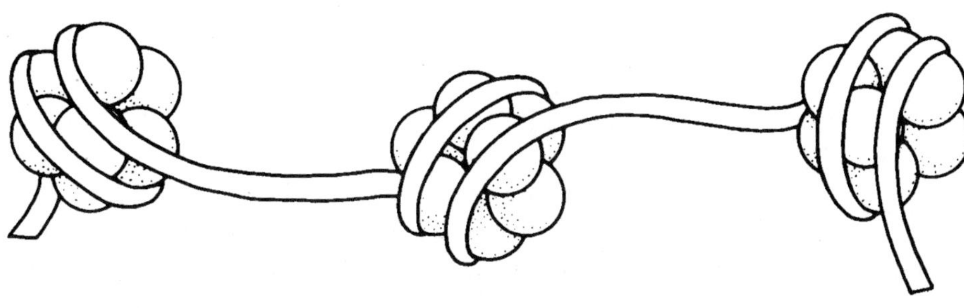

Figure 3.7

1. What is the chemical composition of a nucleosome? (Be specific.) ____________________

____

2. What is the function of the nucleosome components named above? ____________________

____

Cell Growth and Reproduction

1. The cell life cycle consists of interphase and ____________________, during which the cell ____________________. Name the three phases of interphase in order, and indicate the important events of each phase.

Phase of interphase	Important events

2. Diagram in Figure 3.8 (by making a pie-shaped graph) the relative lengths of time of the G_1, S, G_2, and M phases for the cell types listed:

A. Rapidly dividing cell type **B.** Slowly dividing cell type

Figure 3.8

3. Complete the following statements concerning control of cell division.

____________________ 1. A complex of two proteins called (1) gives the "OK" signal for a cell to begin mitosis. One of these proteins, called (2), is always present. The other, a regulatory protein called (3), is regenerated anew with each cycle.

____________________ 2.

____________________ 3.

4. The following statements describe events that occur during the different phases of mitosis. Identify the phase by choosing the correct responses from the key choices and inserting the answers in the answer blanks.

Key Choices

A. Anaphase B. Metaphase C. Prophase D. Telophase E. None of these

__________ 1. Chromatin coils and condenses to form deeply staining bodies.

__________ 2. Centromeres break, and chromosomes begin migration toward opposite poles of the cell.

__________ 3. The nuclear membrane and nucleoli reappear.

__________ 4. Chromosomes cease their poleward movement.

__________ 5. Chromosomes align on the equator of the spindle.

__________ 6. The nucleoli and nuclear membrane disappear.

__________ 7. The spindle forms through the migration of the centrioles.

__________ 8. DNA replication occurs.

__________ 9. Chromosomes obviously are duplex structures.

__________ 10. Chromosomes attach to the spindle fibers.

__________ 11. Cytokinesis occurs.

__________ 12. The nuclear membrane is absent during the entire phase.

__________ 13. This is the period during which a cell is not in the M phase.

__________ 14. Chromosomes (chromatids) are V shaped.

5. Identify the phases of mitosis depicted in Figure 3.9 by inserting the correct terms in the blanks under each diagram. Then select different colors to represent the structures below, and use them to color the coding circles and the corresponding structures in the illustration. When you have completed the work on Figure 3.9, identify all of the mitotic stages provided with leader lines in the photomicrograph of an onion root tip in Figure 3.10.

○ Nuclear membranes, if present ○ Centrioles ○ Chromosomes

○ Nucleoli, if present ○ Spindle fibers, if present

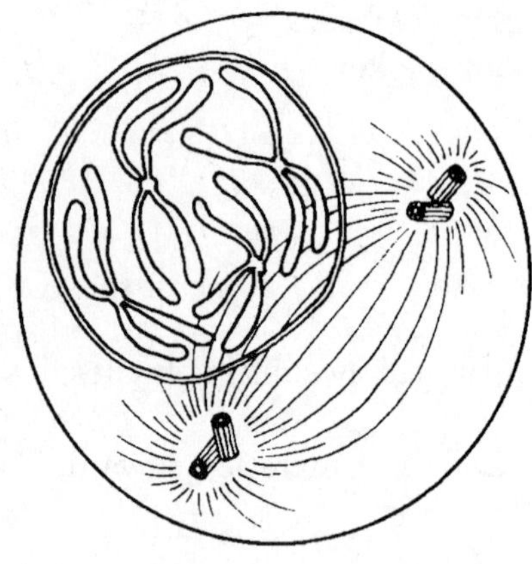

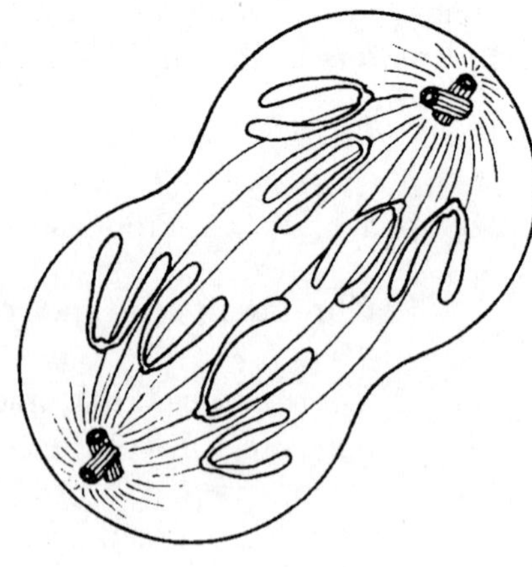

Figure 3.9

A. ______________________

B. ______________________

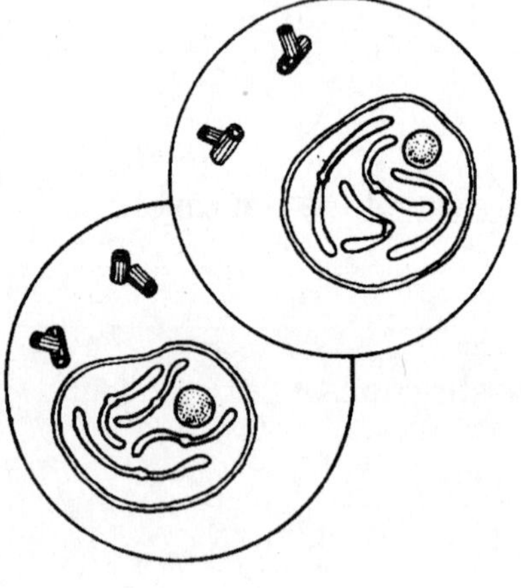

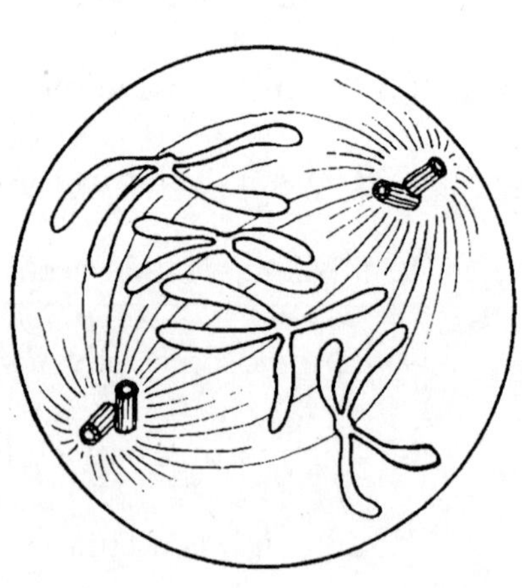

C. ______________________

D. ______________________

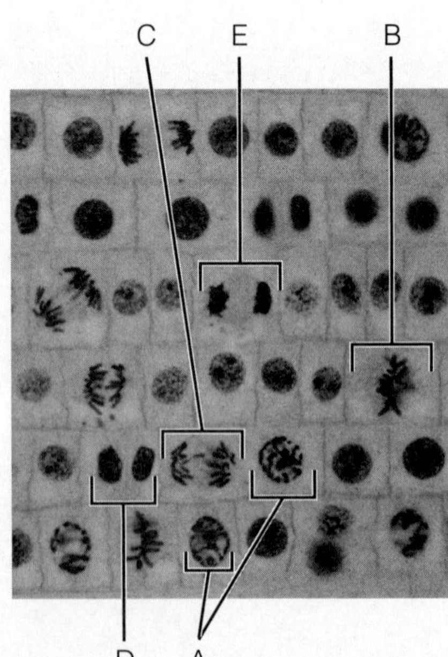

Figure 3.10

A. ________________

B. ________________

C. ________________

D. ________________

E. ________________

6. The following statements provide an overview of the structure of DNA (the genetic material) and its role in the body. Choose responses from the key choices that complete the statements. Insert the appropriate answers in the answer blanks.

Key Choices

A. Adenine	G. Enzymes	M. Nucleotides	S. Ribosome
B. Amino acids	H. Genes	N. Old	T. Sugar (deoxyribose)
C. Bases	I. Growth	O. Phosphate	U. Template, or model
D. Codons	J. Guanine	P. Proteins	V. Thymine
E. Complementary	K. Helix	Q. Replication	W. Transcription
F. Cytosine	L. New	R. Repair	X. Uracil

1. ____________________
2. ____________________
3. ____________________
4. ____________________
5. ____________________
6. ____________________
7. ____________________
8. ____________________
9. ____________________
10. ____________________
11. ____________________
12. ____________________
13. ____________________
14. ____________________
15. ____________________
16. ____________________
17. ____________________
18. ____________________

DNA molecules contain information for building specific __(1)__. In a three-dimensional view, a DNA molecule looks like a spiral staircase; this is correctly called a __(2)__. The constant parts of DNA molecules are the __(3)__ and __(4)__ molecules, forming the DNA-ladder uprights, or backbones. The information of DNA is actually coded in the sequence of nitrogen-containing __(5)__, which are bound together to form the "rungs" of the DNA ladder. When the four DNA bases are combined in different three-base sequences, called triplets, different __(6)__ of the protein are called for. It is said that the N-containing bases of DNA are __(7)__, which means that only certain bases can fit or interact together. Specifically this means that __(8)__ can bind with guanine, and adenine binds with __(9)__.

The production of proteins involves the cooperation of DNA and RNA. RNA is another type of nucleic acid that serves as a "molecular slave" to DNA. That is, it leaves the nucleus and carries out the instructions of the DNA for the building of a protein on a cytoplasmic structure called a __(10)__. When a cell is preparing to divide, in order for its daughter cells to have all its information, it must oversee the __(11)__ of its DNA so that a "double dose" of genes is present for a brief period. For DNA synthesis to occur, the DNA must uncoil, and the bonds between the N-bases must be broken. Then the two single strands of __(12)__ each act as a __(13)__ for the building of a whole DNA molecule. When completed, each DNA molecule formed is half __(14)__ and half __(15)__. The fact that DNA replicates before a cell divides ensures that each daughter cell has a complete set of __(16)__. Cell division, which then follows, provides new cells so that __(17)__ and __(18)__ can occur.

7. Figure 3.11 is a diagram illustrating protein synthesis.

First, select four different colors and use them to color the coding circles and the corresponding structures in the diagram.

Second, using the letters of the genetic code, label the nitrogen bases on the encoding strand (strand 2) of the DNA double helix, on the mRNA strands, and on the tRNA molecules.

Third, referring to Figure 3.11, answer the questions that follow, and insert your answers in the answer blanks.

○ Backbones of the DNA double helix ○ tRNA molecules

○ Backbone of the mRNA strands ○ Amino acid molecules

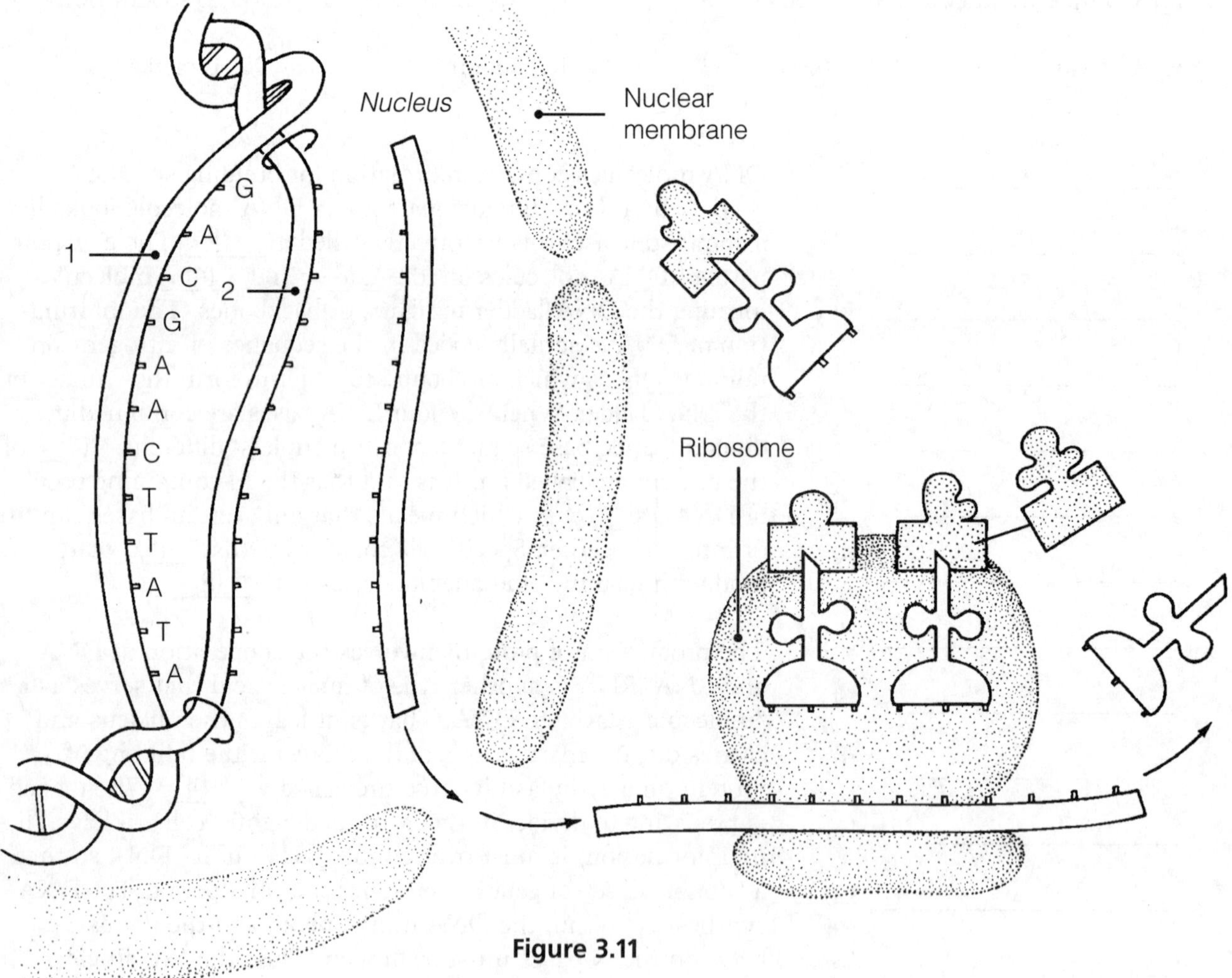

Figure 3.11

1. Transfer of the genetic message from DNA to mRNA is called ____________________.

2. Assembly of amino acids according to the genetic information carried by mRNA is called ____________________.

3. All types of RNA are made on the ____________________.

4. The set of three nitrogen bases on tRNA that is complementary to an mRNA codon is called a ______________________. The complementary three-base sequence on DNA is called a ______________________.

5. Define *gene*. ____

____

8. Complete the following statements. Insert your answers in the answer blanks.

____________________ 1.
____________________ 2.
____________________ 3.
____________________ 4.
____________________ 5.
____________________ 6.
____________________ 7.

Division of the (1) is referred to as mitosis. Cytokinesis is division of the (2). The major structural difference between chromatin and chromosomes is that the latter are (3). Chromosomes attach to the spindle fibers by undivided structures called (4). If a cell undergoes nuclear division but not cytoplasmic division, the product is a (5). The structure that acts as a scaffolding for chromosomal attachment and movement is called the (6). (7) is the period of cell life when the cell is not involved in division.

Extracellular Materials

1. Name the three major categories of extracellular materials in the body, provide examples of each class, and cite some of the important roles of these substances.

1. ____

2. ____

3. ____

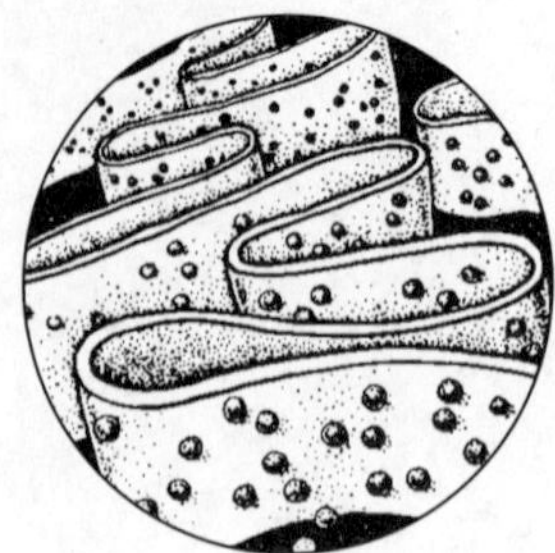

THE INCREDIBLE JOURNEY:

A Visualization Exercise for the Cell

A long, meandering membrane with dark globules clinging to its outer surface now comes into sight.

1. Complete the narrative by inserting the missing words in the answer blanks.

1. ____________________
2. ____________________
3. ____________________
4. ____________________
5. ____________________
6. ____________________
7. ____________________
8. ____________________
9. ____________________
10. ____________________
11. ____________________

For this journey, you will be miniaturized to the size of a small protein molecule and will travel in a microsubmarine, specially designed to enable you to pass easily through living membranes. You are injected into the intercellular space between two epithelial cells, and you are instructed to observe one cell firsthand and to identify as many of its structures as possible. You struggle briefly with the controls and then maneuver your microsub into one of these cells. Once inside the cell, you find yourself in a kind of "sea." This salty fluid that surrounds you is the (1) of the cell.

Far below looms a large, dark oval structure, much larger than anything else. You conclude that it is the (2) . As you move downward, you pass a cigar-shaped structure with strange-looking folds on its inner surface. Although you have a pretty good idea that it must be a (3) , you decide to investigate more thoroughly. After passing through the membrane of the structure, you are confronted with yet another membrane. Once past this membrane, you are inside the strange-looking structure. You activate the analyzer switch in your microsub for a readout on which molecules are in your immediate vicinity.

As suspected, there is an abundance of energy-rich (4) molecules. Having satisfied your curiosity, you leave this structure to continue the investigation.

A long, meandering membrane with dark globules clinging to its outer surface now comes into sight. You maneuver closer and sit back to watch the activity. As you watch, amino acids are joined together and a long, threadlike protein molecule is built. The globules must be (5) , and the membrane, therefore, is the (6) . Once again you head toward the large dark structure seen and tentatively identified earlier. On approach, you observe that this huge structure has very large openings in its outer wall; these openings must be the (7) . Passing through one of these openings, you discover that from the inside the color of this structure is a result of dark, coiled, intertwined masses of (8) , which your analyzer confirms contain genetic materials, or (9) molecules. Making your way through this tangled mass, you pass two round, dense structures that appear to be full of the same type of globules you saw outside. These two round structures are (10) . All this information confirms your earlier identification of this cellular structure, so now you move to its exterior to continue your observations.

Just ahead, you see what appears to be a mountain of flattened sacs with hundreds of small vesicles at its edges. The vesicles appear to be migrating away from this area and heading toward the outer edges of the cell. The mountain of sacs must be the (11) . Eventually you come upon a rather simple-looking membrane-bounded sac. Although it doesn't look too exciting and has few distinguishing marks, it does not resemble anything else you have seen so far. Deciding to obtain a chemical

____________________ 12. analysis before entering this sac, you activate the analyzer, and on the screen you see "Enzymes—Enzymes—Danger—Danger." There is little doubt that this apparently innocent structure is actually a (12).

Completing your journey, you count the number of organelles identified so far. Satisfied that you have observed most of them, you request retrieval from the intercellular space.

CHALLENGING YOURSELF

At the Clinic

1. An infant is brought in with chronic diarrhea, which her mother says occurs whenever the baby drinks milk. The doctor diagnoses lactose intolerance. She explains to the parents that their baby is unable to digest milk sugar and suggests adding lactase to the baby's milk. How would lactose intolerance lead to diarrhea? How does adding lactase prevent diarrhea?

2. Anaphylaxis is a systemic (bodywide) allergic reaction in which capillaries become excessively permeable. This results in increased filtration and fluid accumulation in the tissues, leading to edema. Why is this condition life-threatening even if no frank bleeding occurs?

3. Some people have too few receptors for the cholesterol-carrying low-density lipoprotein (LDL). As a result, cholesterol builds up in blood vessel walls, restricting blood flow and leading to high blood pressure. By what cellular transport process is cholesterol taken up from the blood in a person with normal numbers of LDL receptors?

4. Sugar (glucose) can appear in the urine in nondiabetics if sugar intake is exceptionally high (when you pig out on sweets!). What functional aspect of carrier-mediated transport does this phenomenon demonstrate?

5. Plasma proteins such as albumin have an osmotic effect. In normally circulating blood, the proteins cannot leave the bloodstream easily and, thus, tend to remain in the blood. But, if stasis (blood flow stoppage) occurs, the proteins will begin to leak out into the interstitial fluid (IF). Explain why this leads to edema.

6. Hydrocortisone is an anti-inflammatory drug that acts to stabilize lysosomal membranes. Explain how this effect reduces cell damage and inflammation. Why is this steroid hormone marketed in a cream (oil) base and used topically (applied to the skin)?

7. Streptomycin (an antibiotic) binds to the small ribosomal subunit of bacteria (but not to the ribosomes of the host cells infected by bacteria). The result is the misreading of bacterial mRNA and the breakup of polysomes. What process is being affected, and how does this kill the bacterial cells?

8. Phagocytes gather in the air sacs of the lungs, especially in the lungs of smokers. What is the connection?

Stop and Think

1. Think *carefully* about the chemistry of the plasma membrane, then answer this question: Why is minor damage to the membrane usually not a problem?

2. Knowing that diffusion rate is inversely proportional to molecular weight, predict the results of the following experiment: Cotton balls are simultaneously inserted in opposite ends of a 1-meter-long glass tube. One cotton ball is saturated with ammonium hydroxide (NH_4OH), the other with sulfuric acid (H_2SO_4). The two gases diffuse until they meet, at which point a white precipitate of ammonium chloride is formed. At what relative point along the tube does the precipitate form?

3. The upper layers of the skin constantly slough off. Predict the changes in the integrity of desmosomes as skin cells age (and move closer to the skin's surface).
4. Some cells produce lipid-soluble products. Can you deduce how such products are stored, that is, prevented from exiting the cell?
5. List three examples of folding of cellular membranes to increase membrane surface area.
6. Should the existence of mitochondrial ribosomes come as a complete surprise? Explain your response.
7. If a structure (such as the lens and cornea of the eye) contains no blood vessels (that is, is *avascular*), is it likely to be very thick? Why or why not?
8. Examine the organelles of the cells depicted in Figure 3.12 A and B. Predict the product of each cell and state your reasons.

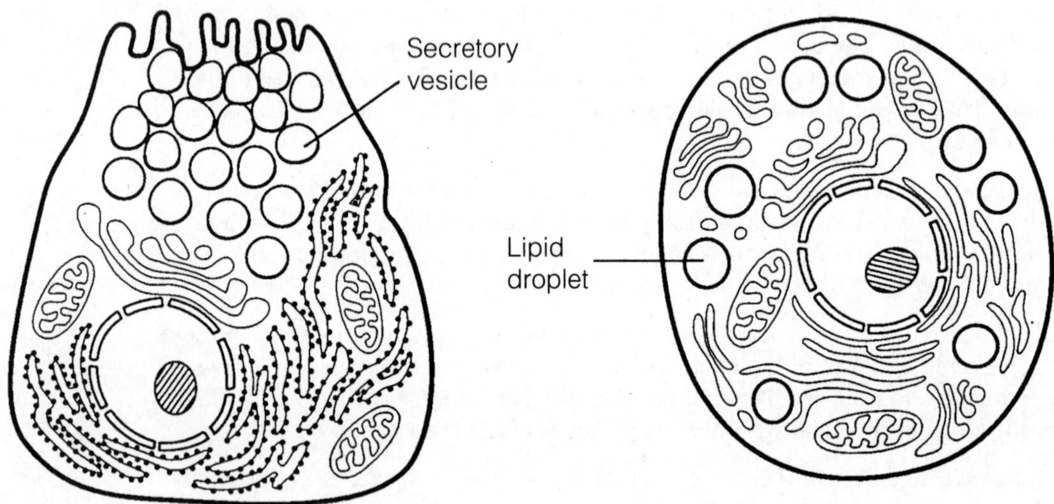

Figure 3.12

9. In Figure 3.13, an artificial cell with an aqueous solution enclosed in a selectively permeable membrane has just been immersed in a beaker containing a different solution. The membrane is permeable to water and to the simple sugars glucose and fructose, but is completely impermeable to the disaccharide sucrose. Answer the following questions pertaining to this situation.
   1. Which solute(s) will exhibit net diffusion into the cell?
   2. Which solute(s) will exhibit net diffusion out of the cell?
   3. In which direction will there be net osmotic movement of water?
   4. Will the cell crenate or swell?

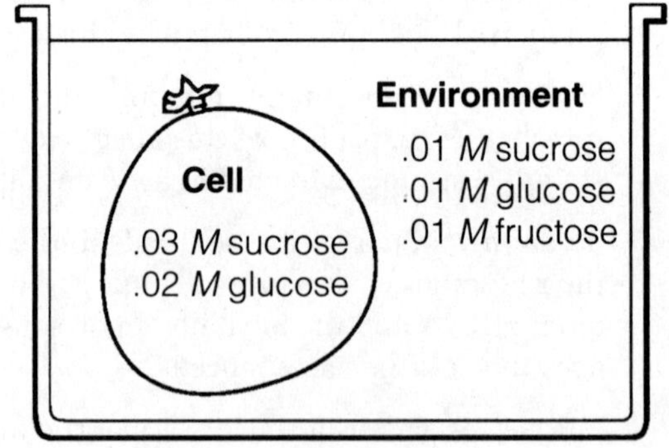

Figure 3.13

10. Using the information provided:

mRNA codon:	AUG	AAC	CGU	GAA	AGU	UAG
amino acid:	Met	Asn	Arg	Glu	Ser	Stop

and given the DNA sequence TA<u>CGCATCACTTTT</u>GATC:
   1. What is the amino acid sequence encoded?
   2. If the nucleotides that are underlined were deleted by mutation, what would the resulting amino acid sequence be?

COVERING ALL YOUR BASES

Multiple Choice

Select the best answer or answers from the choices given.

1. Which of the following is not a basic concept of the cell theory?
   A. All cells come from preexisting cells.
   B. Cellular properties define the properties of life.
   C. Organismic activity is based on individual and collective cellular activity.
   D. Cell structure determines and is determined by biochemical function.
   E. Organisms can exhibit properties that cannot be explained at the cellular level.

2. The dimension outside the observed range for human cells is:
   A. 10 micrometers
   B. 30 centimeters long
   C. 2 nanometers
   D. 1 meter long

3. A cell's plasma membrane would not contain:
   A. phospholipid
   B. nucleic acid
   C. protein
   D. cholesterol
   E. glycolipid

4. Which of the following would you expect to find in or on cells whose main function is absorption?
   A. Microvilli
   B. Cilia
   C. Desmosomes
   D. Gap junctions
   E. Secretory vesicles

5. Adult cell types expected to have gap junctions include:
   A. skeletal muscle
   B. bone
   C. heart muscle
   D. smooth muscle

6. For diffusion to occur, there must be:
   A. a selectively permeable membrane
   B. equal amounts of solute
   C. a concentration difference
   D. some sort of carrier system
   E. all of these

7. Fluid moves out of capillaries by filtration. If the plasma glucose concentration is 100 mg/dl, what will be the concentration of glucose in the filtered fluid?
   A. Less than 100 mg/dl
   B. 100 mg/dl
   C. More than 100 mg/dl
   D. The concentration cannot be determined.

8. Lack of cholesterol in the plasma membrane would result in the membrane's:
   A. excessive fluidity
   B. instability
   C. increased protein content
   D. excessive fragility
   E. reduced protein content

9. Which of the following membrane components is involved in glucose transport?
   A. Phospholipid bilayer
   B. Transmembrane protein
   C. Cholesterol
   D. Peripheral protein
   E. Glycocalyx

10. If a 10% sucrose solution within a semipermeable sac causes the fluid volume in the sac to increase a given amount when the sac is immersed in water, what would be the effect of replacing the sac solution with a 20% sucrose solution?
    A. The sac would lose fluid.
    B. The sac would gain the same amount of fluid.
    C. The sac would gain more fluid.
    D. There would be no effect.

11. Cyanide binds to at least one molecule that is involved with ATP synthesis. In cells exposed to cyanide, most of the cyanide would be found in:
    A. mitochondria
    B. lysosomes
    C. ribosomes
    D. ER

12. Intestinal cells absorb glucose, galactose, and fructose by carrier-mediated transport. Poisoning the cells' mitochondria inhibits the absorption of glucose and galactose but not that of fructose. By what processes are these sugars absorbed?
    A. All are absorbed by facilitated diffusion.
    B. All are absorbed by active transport.
    C. Glucose and galactose are actively transported, but fructose is moved by facilitated diffusion.
    D. Fructose is actively transported, but glucose and galactose are moved by facilitated diffusion.

13. In a polarized cell:
    A. sodium is being pumped out of the cell
    B. potassium is being pumped out of the cell
    C. sodium is being pumped into the cell
    D. potassium is being pumped into the cell

14. Which of the following are possible functions of the glycocalyx?
    A. Determination of blood groups
    B. Binding sites for toxins
    C. Aiding the binding of sperm to egg
    D. Guiding embryonic development
    E. Increasing the efficiency of absorption

15. A cell stimulated to increase steroid production will have abundant:
    A. ribosomes
    B. rough ER
    C. smooth ER
    D. Golgi apparatus
    E. secretory vesicles

16. A cell's ability to replenish its ATP stores has been diminished by a metabolic poison. What organelle is most likely to be affected?
    A. Nucleus
    B. Plasma membrane
    C. Centriole
    D. Microtubule
    E. Mitochondrion

17. Steroid hormones increase protein synthesis in their target cells. How would this stimulus be signified in a bone-forming cell, which secretes the protein collagen?
    A. Increase in heterochromatin
    B. Increase in endocytosis
    C. Increase in lysosome formation
    D. Increase in formation of secretory vesicles
    E. Increase in amount of rough ER

18. In certain nerve cells that sustain damage, the rough ER disbands and most ribosomes are free. What does this indicate?
    A. Decrease in protein synthesis
    B. Increase in protein synthesis
    C. Increase in synthesis of intracellular proteins
    D. Increase in synthesis of secreted proteins

19. What cellular inclusions increase in number in a light-skinned human after increased exposure to sunlight?
    A. Melanin granules
    B. Lipid droplets
    C. Glycogen granules
    D. Mucus
    E. Zymogen granules

20. A cell with abundant peroxisomes would most likely be involved in:
    A. secretion
    B. storage of glycogen
    C. ATP manufacture
    D. movement
    E. detoxification activities

21. Biochemical tests show a cell with replicated DNA, but incomplete synthesis of proteins needed for cell division. In what stage of the cell cycle is this cell most likely to be?
    A. M
    B. G_2
    C. S
    D. G_1

22.–24. Consider the following information for Questions 22–24:
A DNA segment has this nucleotide sequence:
A A G C T C T T A C G A A T A T T C

22. Which mRNA matches or is complementary?
    A. A A G C T C T T A C G A A T A T T C
    B. T T C G A G A A T G C T T A T A A G
    C. A A G C U C U U A C G A A U A U U C
    D. U U C G A G A A U G C U U A U A A G

23. How many amino acids are coded in this segment?
    A. 18
    B. 9
    C. 6
    D. 3

24. What is the tRNA anticodon sequence for the fourth codon from the left?
    A. G
    B. GC
    C. GCU
    D. CGA

25. The organelle that consists of a stack of 3–10 membranous discs associated with vesicles is:
    A. Mitochondrion
    B. Smooth ER
    C. Golgi apparatus
    D. Lysosome

26. Which statement concerning lysosomes is false?
    A. They have the same structure and function as peroxisomes.
    B. They form by budding off the Golgi apparatus.
    C. They are abundant in phagocytes.
    D. They contain their digestive enzymes to prevent general cytoplasmic damage.

27. The fundamental structure of the plasma membrane is determined almost exclusively by:
    A. phospholipid molecules
    B. peripheral proteins
    C. cholesterol molecules
    D. integral proteins

28. Centrioles:
    A. start to duplicate in G_1
    B. reside in the centrosome
    C. are made of microtubules
    D. lie parallel to each other

29. The *trans* face of the Golgi apparatus:
   A. is where products are dispatched in vesicles
   B. is its convex face
   C. receives transport vesicles from the rough ER
   D. is in the center of the Golgi stack

30. The protein that tags cytoplasmic proteins for destruction is:
   A. Ubiquitin
   B. Cyclin
   C. Proteosome
   D. Histone

31. It is impossible to see chromosomes in an interphase cell because:
   A. they must be moved to the center of the spindle before they become visible
   B. they are an extended threadlike form called chromatin
   C. they have left the nucleus
   D. DNA synthesis has not occurred yet

Word Dissection

For each of the following word roots, fill in the literal meaning and give an example, using a word found in this chapter.

Word root	Translation	Example
1. chondri		
2. chrom		
3. crist		
4. cyto		
5. desm		
6. dia		
7. dys		
8. flagell		
9. meta		
10. mito		
11. nucle		
12. onco		
13. osmo		
14. permea		
15. phag		
16. philo		
17. phobo		
18. pin		
19. plasm		
20. telo		
21. tono		
22. troph		
23. villus		

4

Tissue: The Living Fabric

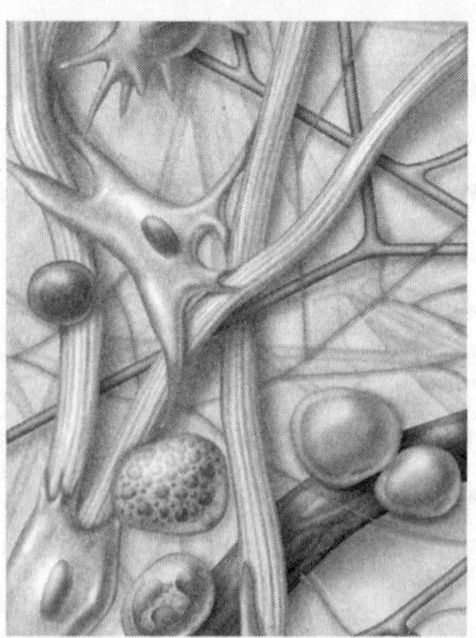

Student Objectives

When you have completed the exercises in this chapter you will have accomplished the following objectives:

Preparing Human Tissue for Microsocopy

1. List the steps involved in preparing animal tissue for microscopic viewing.

Epithelial Tissue

2. List several structural and functional characteristics of epithelial tissue.
3. Name, classify, and describe the various types of epithelia, and indicate their chief function(s) and location(s).
4. Define gland.
5. Differentiate between exocrine and endocrine glands, and between unicellular and multicellular glands.
6. Describe how multicellular exocrine glands are classified structurally and functionally.

Connective Tissue

7. Indicate common characteristics of connective tissue, and list and describe its structural elements.
8. Describe the types of connective tissue found in the body, and indicate their characteristic functions.

Nervous Tissue

9. Indicate the general characteristics of nervous tissue.

Muscle Tissue

10. Compare and contrast the structures and body locations of the three types of muscle tissue.

Covering and Lining Membranes

11. Describe the structure and function of cutaneous, mucous, and serous membranes.

Tissue Repair

12. Outline the process of tissue repair involved in normal healing of a superficial wound.

In the human body, as in any multicellular body, every cell has the ability to perform all activities necessary to remain healthy and alive. In the multicellular body, however, the individual cells are no longer independent. Instead, they aggregate, forming cell communities made of cells similar to one another in form and function. Once specialized, each community is committed to performing a specific activity that helps to maintain homeostasis and serves the body as a whole.

Fully differentiated cell communities are called tissue (from *tissu*, meaning "woven"). Tissues, in turn, are organized into functional units called organs (such as the heart and brain). Because individual tissues are unique in cellular shape and structure, they are readily recognized and are often named for the organ of origin—for example, muscle tissue and nervous tissue.

Student activities in Chapter 4 include questions relating to the structure and function of tissues, membranes, glands and glandular tissue, tissue repair, and the developmental aspects of tissues.

BUILDING THE FRAMEWORK

Overview of Body Tissues

1. Circle the term that does not belong in each of the following groupings.

1. Columnar Areolar Cuboidal Squamous
2. Collagen Cell Matrix Cell product
3. Cilia Flagellum Microvilli Elastic fibers
4. Glands Bones Epidermis Mucosae
5. Adipose Hyaline Osseous Nervous
6. Blood Smooth Cardiac Skeletal
7. Polarity Cell-to-cell junctions Regeneration possible Vascular
8. Matrix Connective tissue Collagen Keratin
9. Cartilage GAGs Vascular Water retention
10. Bone Fluid matrix Collagen Calcium salt

2. Twelve tissue types are diagrammed in Figure 4.1. Identify each tissue type by inserting the correct name in the blank below each diagram. Select different colors for the following structures and use them to color the coding circles and corresponding structures in the diagrams.

○ Epithelial cells

○ Nerve cells

○ Muscle cells

○ Matrix (Where found, matrix should be colored differently from the living cells of that tissue type. Be careful. This may not be as easy as it seems!)

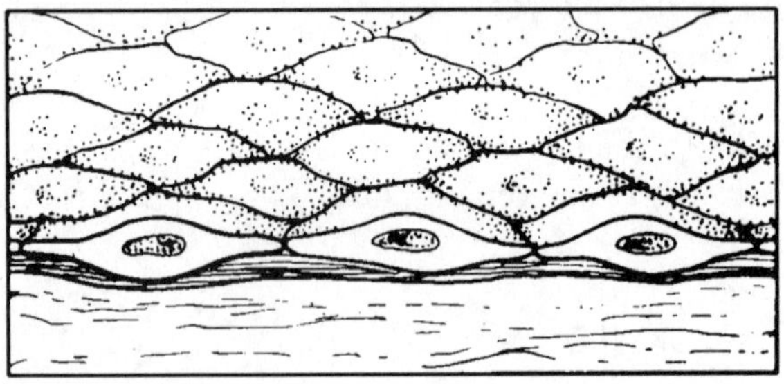

A ______________________

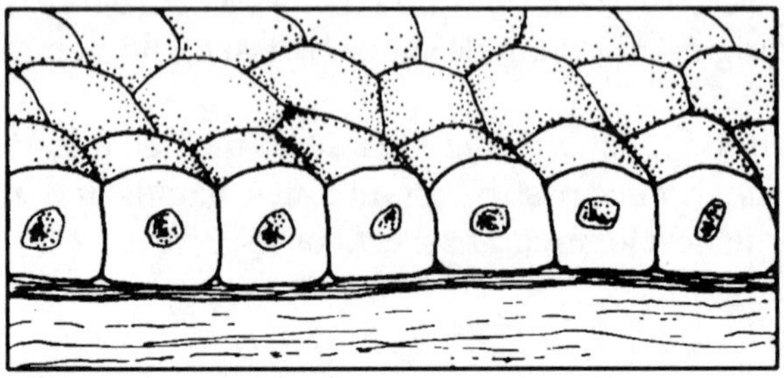

B ______________________

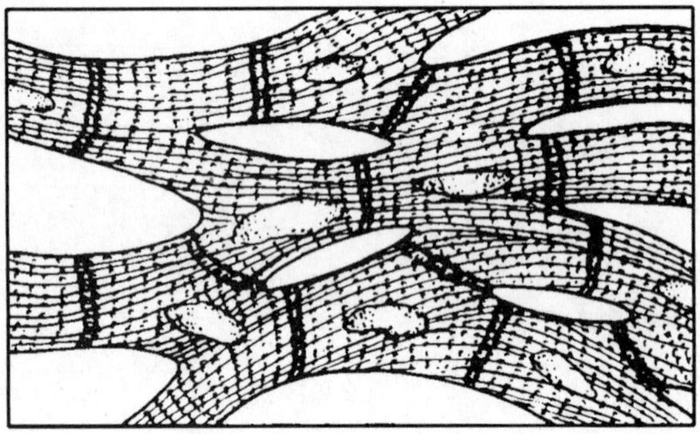

C ______________________

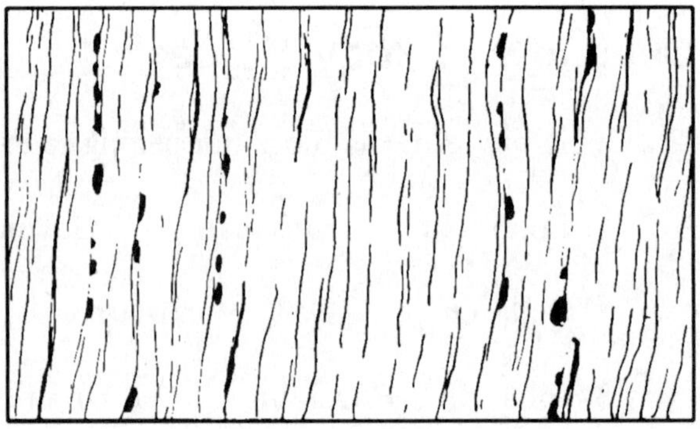

D ______________________

E ______________________

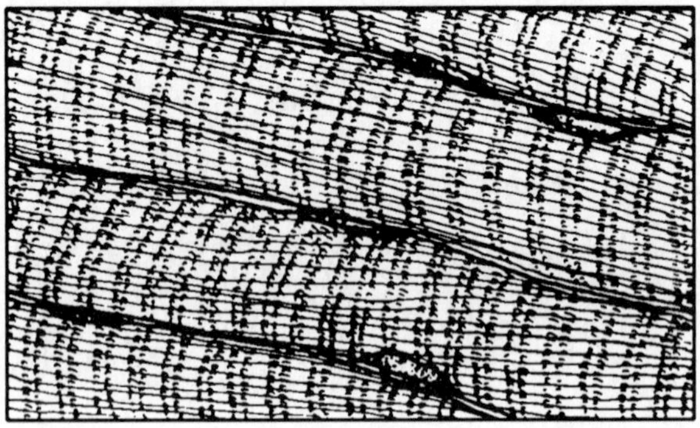

F ______________________

Figure 4.1

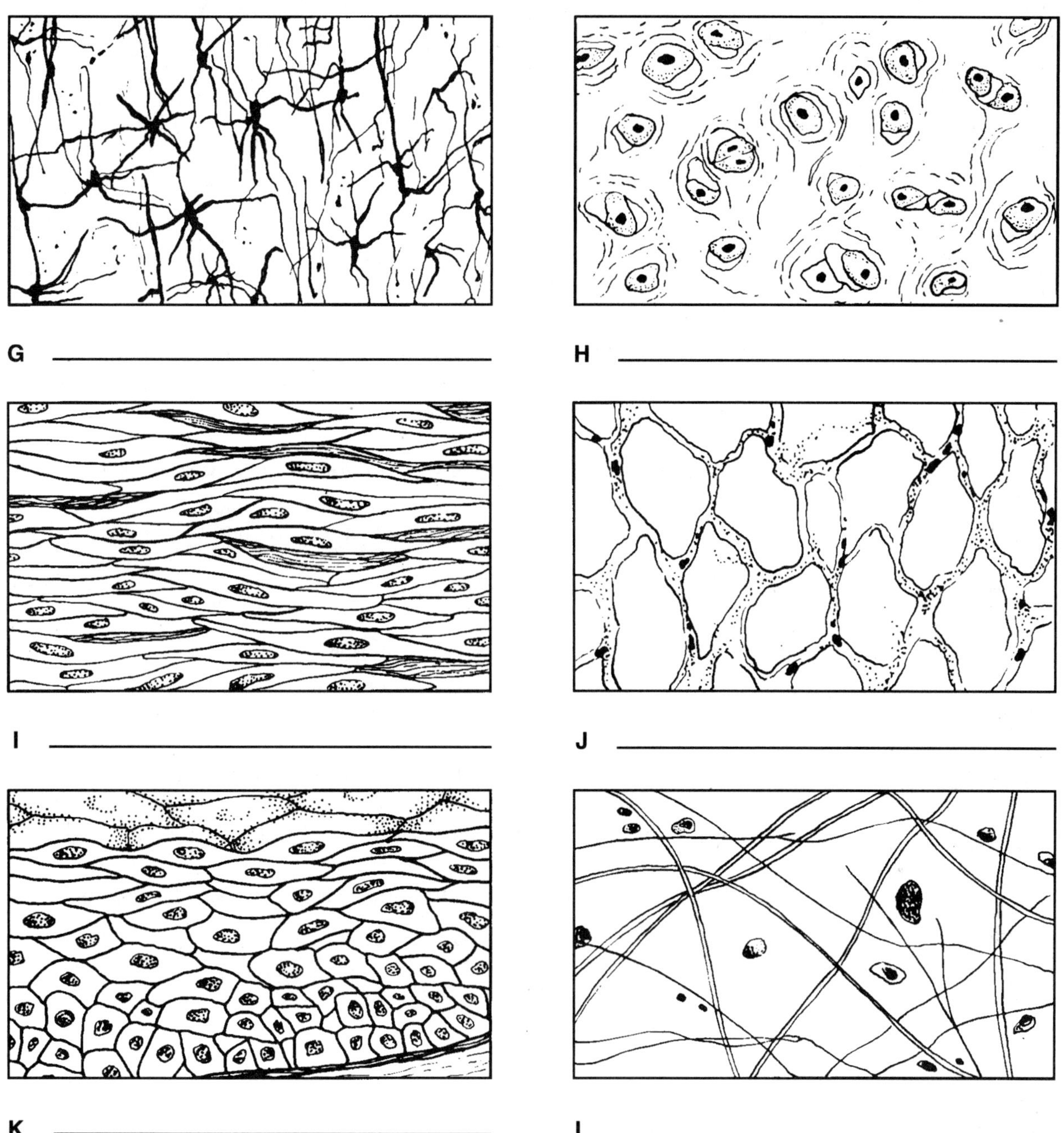

Figure 4.1 (continued)

3. Using the key choices, correctly identify the following *major* tissue types. Enter the appropriate answer in the answer blanks.

Key Choices

A. Connective B. Epithelium C. Muscle D. Nervous

__________ 1. Forms membranes

__________ 2. Allows for movement of limbs and for organ movements within the body

__________ 3. Uses electrochemical signals to carry out its functions

__________ 4. Supports and reinforces body organs

__________ 5. Cells of this tissue may absorb and/or secrete substances

__________ 6. Basis of the major controlling system of the body

__________ 7. Its cells shorten to exert force

__________ 8. Forms endocrine and exocrine glands

__________ 9. Surrounds and cushions body organs

__________ 10. Characterized by having large amounts of extracellular material

__________ 11. Allows you to smile, grasp, swim, ski, and throw a ball

__________ 12. Widely distributed; found in bones, cartilages, and fat depots

__________ 13. Forms the brain and spinal cord

Epithelial Tissue

1. List the six major functions of epithelium. __________

____

2. List six special characteristics of epithelium. __________

____

____

____

____

3. For 1–5, match the epithelial type named in Column B with the appropriate *location* in Column A.

	Column A	Column B
________	1. Lines the stomach and most of the intestines	A. Pseudostratified ciliated columnar
________	2. Lines the inside of the mouth	B. Simple columnar
________	3. Lines much of the respiratory tract	C. Simple cuboidal
________	4. Endothelium and mesothelium	D. Simple squamous
________	5. Lines the inside of the urinary bladder	E. Stratified columnar
		F. Stratified squamous
		G. Transitional

For 6–10, match the epithelium named in Column B with the appropriate *function* in Column A.

	Column A	Column B
________	6. Protection	H. Endothelium
________	7. Small molecules pass through rapidly	I. Simple columnar
________	8. Propel sheets of mucus	J. A ciliated epithelium
________	9. Absorption, secretion, or ion transport	K. Stratified squamous
________	10. Stretches	L. Transitional

4. Arrange the following types of epithelium from 1 to 5 in order of increasing *protectiveness*.

_____ A. Simple squamous
_____ B. Stratified squamous
_____ C. Simple cuboidal
_____ D. Pseudostratified
_____ E. Simple columnar

5. Arrange the following types of epithelium from 1 to 4 in order of increasing *absorptive* ability.

_____ A. Simple squamous
_____ B. Stratified squamous
_____ C. Simple cuboidal
_____ D. Simple columnar

6. Epithelium exhibits many of the plasma membrane modifications described in conjunction with our discussion of the composite cell in Chapter 3. Figure 4.2 depicts some of these modifications.

First: Choose a color for the coding circles and the corresponding structures in the figure.

- ◯ Epithelial cell cytoplasm
- ◯ Connective tissue
- ◯ Epithelial cell nucleus
- ◯ Blood vessel
- ◯ Nerve fibers

Second: Correctly identify the following structures or regions by labeling appropriate leader lines using terms from the list below:

A. Epithelium	E. Connective tissue	I. Cilia
B. Basal region	F. Basement membrane	J. Reticular lamina
C. Apical region	G. Basal lamina	K. Tight junctions
D. Capillary	H. Microvilli	L. Desmosome

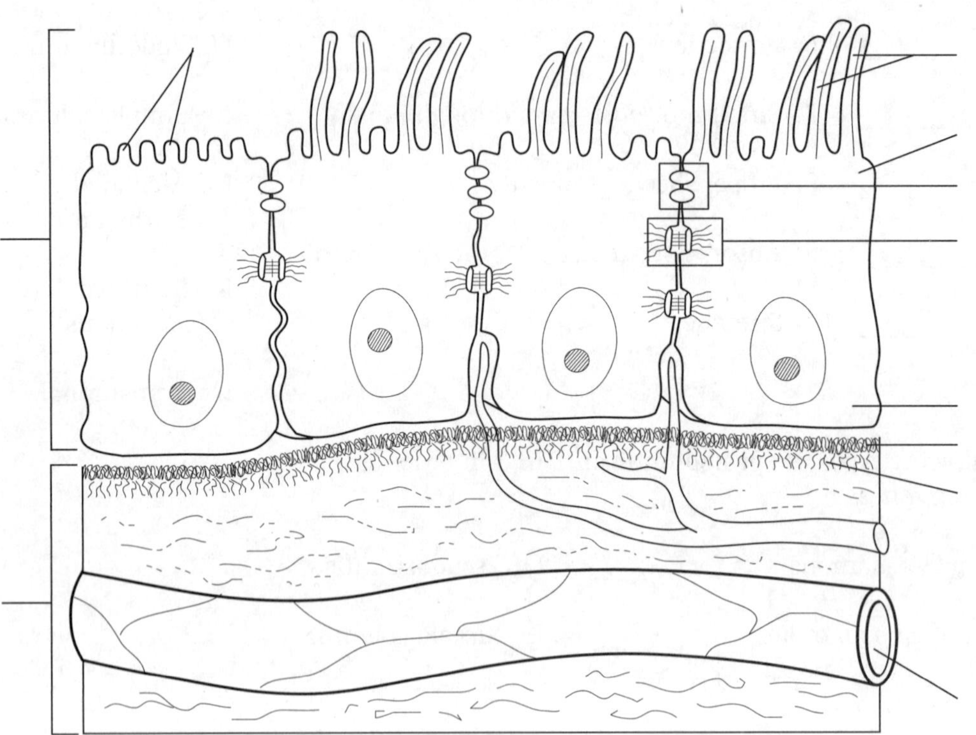

Figure 4.2

7. Identify the structural class of each gland described here:

________________ 1. Flask-shaped gland, unbranched ducts

________________ 2. Slender, straight gland, unbranched ducts

________________ 3. Combination of gland shapes

________________ 4. Branched ducts

8. Write T in the answer blank if a statement is true. If a statement is false, correct the underlined word(s) by writing the correct word(s) in the answer blank.

________________ 1. Exocrine glands are classified <u>functionally</u> as merocrine, holocrine, or apocrine.

________________ 2. The above classification refers to the way <u>ducts branch</u>.

________________ 3. Most exocrine glands are <u>apocrine</u>.

________________ 4. In <u>apocrine</u> glands, secretions are produced and released immediately by exocytosis.

________________ 5. <u>Holocrine</u> glands store secretions until the cells rupture. Ruptured cells are replaced through mitosis.

________________ 6. In apocrine glands, the secretory cells <u>die</u> when they pinch off at the apex to release secretions.

________________ 7. A sweat gland is an example of a <u>holocrine</u> gland.

________________ 8. The mammary gland is the most likely example of an <u>apocrine</u> gland.

9. Using the key choices, correctly identify the types of glands described below. Enter the appropriate answers in the answer blanks.

Key Choices

A. Endocrine B. Exocrine C. Neither of these

________________ 1. Duct from this gland type carries secretions to target organ or location

________________ 2. Examples are the thyroid and adrenal glands

________________ 3. Glands secrete regulatory hormones directly into blood or lymph

________________ 4. The more numerous of the two types of glands

________________ 5. Duct from ovary that carries ovum (egg) to uterus

________________ 6. Examples are the liver, which produces bile, and the pancreas, which produces digestive enzymes

10. Complete the following statements by filling in the appropriate answers.

1. Endocrine and exocrine glands are formed from __(1)__ tissue. Unicellular exocrine glands called __(2)__ are found in the intestinal mucosae, where they secrete __(3)__, a lubricating, water-soluble glycoprotein. Multicellular glands are composed of three structures: __(4)__, __(5)__, and __(6)__. Exocrine glands classified as compound tubular are glands with __(7)__ ducts and with secretory cells located in __(8)__ secretory units.

_______________ 1.

_______________ 2.

_______________ 3.

_______________ 4.

_______________ 5.

_______________ 6.

_______________ 7.

_______________ 8.

Connective Tissue

1. Using the key choices, identify the following connective tissue types. Insert the appropriate answers in the answer blanks.

Key Choices

A. Adipose connective tissue	G. Fibrocartilage
B. Areolar connective tissue	H. Hyaline cartilage
C. Dense regular connective	I. Mucous connective
D. Dense irregular connective	J. Osseous tissue
E. Elastic cartilage	K. Reticular connective tissue
F. Elastic connective tissue	L. Vascular tissue

_______________ 1. Parallel bundles of collagenic fibers provide strength; found in tendons

_______________ 2. Stores fat

_______________ 3. The skin dermis

_______________ 4. Hardest tissue of our "skull cap"

_______________ 5. Composes the basement membrane; surrounds and cushions blood vessels and nerves; its gel-like matrix contains all categories of fibers and many cell types

_______________ 6. Forms the embryonic skeleton; covers surfaces of bones at joints; reinforces the trachea

__________________ 7. Insulates the body

__________________ 8. Firm, slightly "rubbery" matrix; milky white and "glassy" in appearance

__________________ 9. Cells are arranged in concentric circles around a nutrient canal; matrix is hard due to calcium salts

__________________ 10. Contains collagenous fibers; found in intervertebral discs

__________________ 11. Makes supporting framework of lymphoid organs

__________________ 12. Found in umbilical cord

__________________ 13. Found in external ear and auditory tube

__________________ 14. Provides the medium for nutrient transport throughout the body

__________________ 15. Forms the "stretchy" ligaments of the vertebral column

2. Areolar connective tissue is often considered the prototype of connective tissue proper because of its variety of cell types and fibers. Figure 4.3 shows most of these elements. Identify all structures or cell types provided with leader lines. Color the diagram as your fancy strikes you.

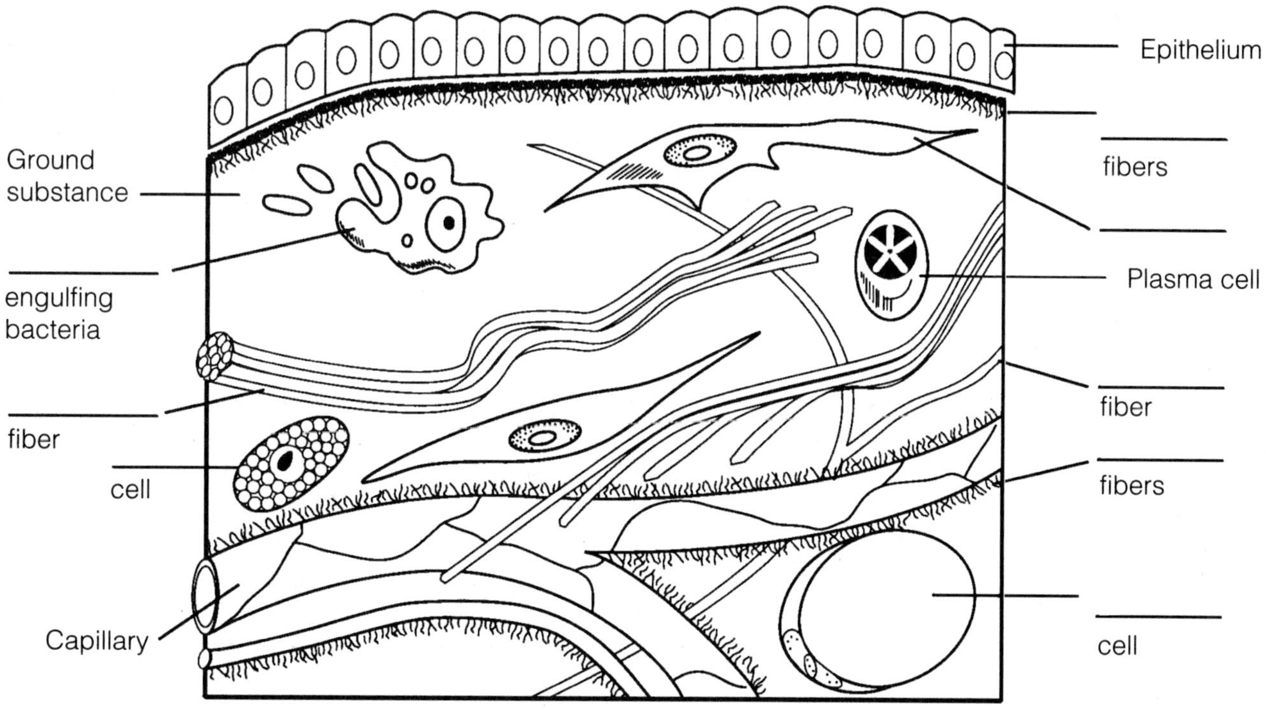

Figure 4.3

3. Arrange the following tissue types from 1 to 3 in order of *decreasing* vascularity.

_____ A. Cartilage

_____ B. Areolar connective

_____ C. Dense connective

4. Using the key choices, select the structural or related elements of connective tissue (CT) types that permit specialized functions. Insert the appropriate answers in the answer blanks.

Key Choices

A. Adipocytes	D. Elastic fibers	G. Macrophages	J. Osteocytes
B. Chondrocytes	E. Ground substance	H. Matrix	K. Osteoblasts
C. Collagen fibers	F. Hemocytoblast	I. Mesenchyme	L. Reticular fibers

_______________ 1. Composed of ground substance and structural protein fibers

_______________ 2. Composed of glycoproteins and water-binding glycosaminoglycans

_______________ 3. Tough protein fibers that resist stretching or longitudinal tearing

_______________ 4. Primary bone marrow cell type that remains actively mitotic

_______________ 5. Fine, branching protein fibers that construct a supportive network

_______________ 6. Large, irregularly shaped cells, widely distributed, often found in CT; they engulf cellular debris and foreign matter and are active in immunity

_______________ 7. The medium through which nutrients and other substances diffuse

_______________ 8. Living elements that maintain the firm, flexible matrix in cartilage

_______________ 9. Randomly coiled protein fibers that recoil after being stretched

_______________ 10. The structural element of areolar tissue that is fluid and provides a reservoir of water and salts for neighboring tissues

_______________ 11. In a loose CT, the nondividing cells that store nutrients

_______________ 12. The embryonic tissue that gives rise to all types of CT

_______________ 13. Cellular elements that produce the collagen fibers of bone matrix

Nervous Tissue

1. Describe briefly how the particular structure of a neuron relates to its function in the body. ____

____

2. Circle the word that does *not* apply to neuroglia: Support Insulate Conduct Protect

Muscle Tissue

1. The three types of muscle tissue exhibit certain similarities and differences. Insert Sk (skeletal), C (cardiac), or Sm (smooth) into the appropriate blanks to indicate which muscle type exhibits each characteristic.

Characteristic

_____ 1. Voluntarily controlled

_____ 2. Involuntarily controlled

_____ 3. Banded appearance

_____ 4. Uninucleate

_____ 5. Multinucleate

_____ 6. Found attached to bones

_____ 7. Enables you to swallow

_____ 8. Found in the walls of the small intestine, uterus, bladder, and veins

_____ 9. Contains spindle-shaped cells

_____ 10. Contains cylindrical cells with branching ends

_____ 11. Contains long, nonbranching cylindrical cells

_____ 12. Displays intercalated discs

_____ 13. Concerned with locomotion of the body as a whole

_____ 14. Changes the internal volume of an organ as it contracts

_____ 15. Tissue of the circulatory pump

2. Develop two criteria that would identify the three different muscle itssues in two steps.

Step 1: ____

Step 2: ____

Covering and Lining Membranes

1. Five simplified diagrams are shown in Figure 4.4. Select different colors for the membranes listed below and use them to color the coding circles and the corresponding structures.

◯ Mucosae

◯ Visceral pleura (serosa)

◯ Parietal pleura (serosa)

◯ Visceral pericardium (serosa)

◯ Parietal peritoneum (serosa)

◯ Visceral peritoneum (serosa)

◯ Mesentery

◯ Endothelium

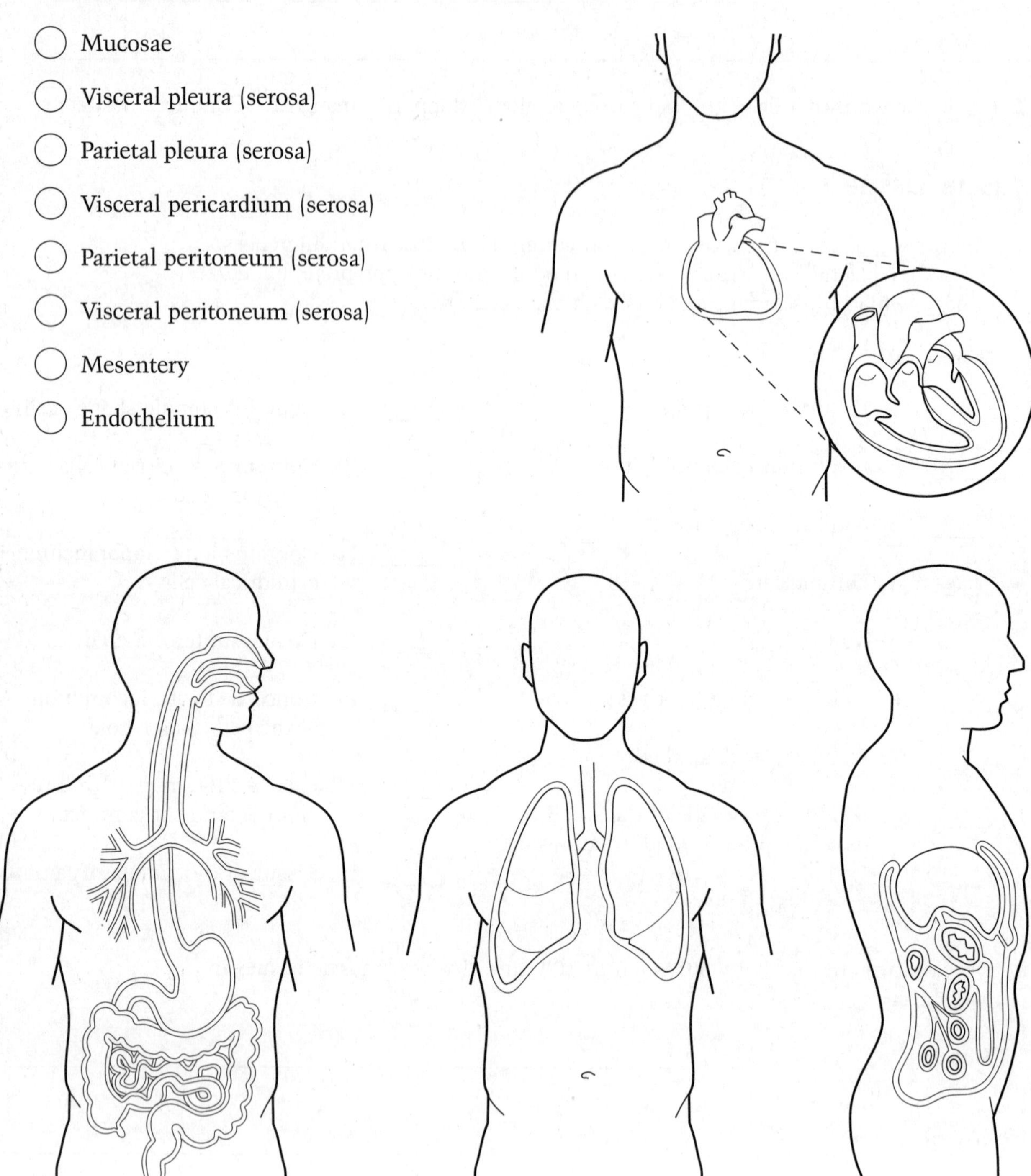

Figure 4.4

2. Complete the following table relating to epithelial membranes. Enter your responses in the areas left blank.

Membrane	Tissue type (epithelial/connective)	Common locations	Functions
Mucous	Epithelial sheet with underlying connective tissue (lamina propria)		Protection, lubrication, secretion, absorption
Serous		Lines internal ventral cavities and covers their organs	
Cutaneous			Protection from external insults; protection from water loss

Tissue Repair

1. For each of the following statements about tissue repair that is true, enter T in the answer blank. For each false statement, correct the underlined word(s) by writing the correct word(s) in the answer blank.

__________ 1. The nonspecific response of the body to injury is called regeneration.

__________ 2. Intact capillaries near an injury dilate, leaking plasma, blood cells, and antibodies, which cause the blood to clot. The clot at the surface dries to form a scab.

__________ 3. During organization, the first phase of tissue repair, capillary buds invade the clot, forming a delicate pink tissue called endodermal tissue.

__________ 4. Fibroblasts synthesize fibers that bridge the gap.

__________ 5. When damage is not too severe, the surface epithelium migrates beneath the dry scab and across the surface of the granulation tissue. This repair process is called proliferation.

__________ 6. If tissue damage is very severe, tissue repair is more likely to occur by fibrosis, or scarring.

__________ 7. During fibrosis, fibroblasts in the granulation tissue lay down keratin fibers, which form a strong, compact, but inflexible mass.

__________ 8. The repair of cardiac muscle and nervous tissue occurs only by fibrosis.

__________ 9. Organization is replacement of a blood clot by granulation tissue.

__________ 10. Granulation tissue resists infection by secreting virus-inhibiting substances.

__________ 11. Problems associated with regeneration include shrinking, loss of elasticity, and formation of adhesions.

CHALLENGING YOURSELF

At the Clinic

1. Since adipocytes are incapable of cell division, how is weight gain accomplished?
2. A woman sees her gynecologist because she is unable to become pregnant. The doctor discovers granulation tissue in her vaginal canal and explains that sperm are susceptible to some of the same chemicals as bacteria. What is inhibiting the sperm?
3. Bradley tripped and tore one of the tendons surrounding his ankle. In anguish, he asked his doctor how soon he could expect it to heal. What do you think the doctor's response was and why?
4. Why do cartilage and tendons take so long to heal?
5. In mesothelial cancer of the pleura, serous fluid is hypersecreted. How does this contribute to respiratory problems?
6. In cases of ruptured appendix, what serous membrane is likely to become infected? Why can this be life-threatening?

Stop and Think

1. Try to come up with some *advantages* of the avascularity of epithelium and cartilage. Also, try to deduce why tendons are poorly vascularized.
2. Sarah, the trainee of the electron microscopist at the local hospital, is reviewing some micrographs of muscle cells and macrophages (phagocytic cells). She notices that the muscle cells are loaded with mitochondria while the macrophages have abundant lysosomes. Why is this so?
3. What types of tissue can have microvilli? Cilia?
4. Why do endocrine glands start out having ducts?
5. On his anatomy test, Bruno answered two questions incorrectly. He confused a basal lamina with a basement membrane, and a mucous membrane with a sheet of mucus. What are the differences between each of these sound-alike pairs of structures or substances?
6. Time for an educated guess. Do you think the elastic connective tissue layer in the large arteries is regularly or irregularly arranged? Explain your reasoning.
7. What kind of tissue surrounds a bone shaft?
8. Why are skeletal muscle cells multinucleate?
9. Explain how the nervous system (obviously an internal organ system) can be derived from the same primary germ layer as the outer covering of the body.

COVERING ALL YOUR BASES

Multiple Choice

Select the best answer or answers from the choices given.

1. Scar tissue is a type of:
   A. epithelium
   B. connective tissue
   C. muscle
   D. nervous tissue

2. Which of the following terms/phrases could *not* be applied to epithelium?
   A. Basement membrane
   B. Free surface
   C. Desmosomes present
   D. Strong matrix
   E. Ciliated

3. Which is not a type of epithelium?
   A. Reticular
   B. Simple squamous
   C. Pseudostratified
   D. Transitional
   E. Stratified columnar

4. In which of the following tissue types might you expect to find goblet cells?
   A. Simple cuboidal
   B. Simple columnar
   C. Simple squamous
   D. Stratified squamous
   E. Transitional

5. Mesothelium is found in:
   A. kidney tubules
   B. mucous membranes
   C. serosae
   D. the liver
   E. the lining of cardiovascular system organs

6. An epithelium "built" to withstand friction is:
   A. simple squamous
   B. stratified squamous
   C. simple cuboidal
   D. simple columnar
   E. pseudostratified

7. Functions of keratin include:
   A. absorbing sunlight
   B. energy storage
   C. directing protein synthesis
   D. waterproofing
   E. providing toughness

8. What cellular specialization causes fluid to flow over the epithelial surface?
   A. Centrioles
   B. Flagella
   C. Cilia
   D. Microvilli
   E. Myofilaments

9. The gland type that secretes its product continuously, by exocytosis, into a duct is:
   A. merocrine
   B. holocrine
   C. endocrine
   D. apocrine

10. Which of the following is not a function of some kind of connective tissue?
    A. Binding
    B. Support
    C. Protection
    D. Sensation
    E. Repair

11. The original embryonic connective tissue is:
    A. mucous connective
    B. mesenchyme
    C. vascular
    D. areolar connective
    E. connective tissue proper

12. Components of connective tissue matrix include:
    A. hyaluronic acid
    B. basal lamina
    C. proteoglycans
    D. glycosaminoglycans

13. Which of the following fibrous elements gives a connective tissue high tensile strength?
    A. Reticular fibers
    B. Elastic fibers
    C. Collagen fibers
    D. Myofilaments

14. The cell that forms bone is the:
    A. fibroblast
    B. chondroblast
    C. hematopoietic stem cell
    D. osteoblast
    E. reticular cell

15. Which of the following cell types secretes histamine and perhaps heparin?
    A. Macrophage
    B. Mast cell
    C. Reticular cell
    D. Fibroblast
    E. Histiocyte

16. Resistance to stress applied in a longitudinal direction is provided best by:
    A. fibrocartilage
    B. elastic connective
    C. reticular connective
    D. dense regular connective
    E. areolar connective

17. What kind of connective tissue acts as a sponge, soaking up fluid when edema occurs?
    A. Areolar connective
    B. Adipose connective
    C. Dense irregular connective
    D. Reticular connective
    E. Vascular tissue

18. Viewed through the microscope, most cells in this type of tissue have only a rim of cytoplasm.
    A. Reticular connective
    B. Adipose connective
    C. Areolar connective
    D. Osseous tissue
    E. Hyaline cartilage

19. The major function of reticular tissue is:
    A. nourishment
    B. insulation
    C. stroma formation
    D. protection
    E. movement

20. What type of connective tissue prevents muscles from pulling away from bones during contraction?
    A. Dense irregular connective
    B. Dense regular connective
    C. Areolar
    D. Elastic connective
    E. Hyaline cartilage

21. The type of connective tissue that provides flexibility to the vertebral column is:
    A. dense irregular connective
    B. dense regular connective
    C. reticular connective
    D. areolar
    E. fibrocartilage

22. Phrases that describe cartilage include:
    A. highly vascularized
    B. holds large volumes of water
    C. has no nerve endings
    D. grows both appositionally and interstitially
    E. can get quite thick

23. Which type of cartilage is most abundant throughout life?
    A. Elastic cartilage
    B. Fibrocartilage
    C. Hyaline cartilage

24. Select the one false statement about mucous and serous membranes.
    A. The epithelial type is the same in all serous membranes, but there are different epithelial types in different mucous membranes.
    B. Serous membranes line closed body cavities, while mucous membranes line body cavities open to the outside.
    C. Serous membranes always produce serous fluid, and mucous membranes always secrete mucus.
    D. Both membranes contain an epithelium plus a layer of loose connective tissue.

25. Serous membranes:
    A. line the mouth
    B. have parietal and visceral layers
    C. consist of epidermis and dermis
    D. have a connective tissue layer called the lamina propria
    E. secrete a lubricating fluid

26. Which of the following terms describe cardiac muscle?
    A. Striated
    B. Intercalated discs
    C. Multinucleated
    D. Involuntary
    E. Branching

27. Events of tissue repair include:
    A. regeneration
    B. organization
    C. granulation
    D. fibrosis
    E. inflammation

28. The prototype connective tissue is:
    A. areolar connective tissue
    B. mesenchyme
    C. dense fibrous
    D. reticular

29. Examples of GAGs include:
    A. chondroitin sulfate
    B. hyaluronic acid
    C. laminin
    D. histamine

Word Dissection

For each of the following word roots, fill in the literal meaning and give an example, using a word found in this chapter.

Word root	Translation	Example
1. ap		
2. areola		
3. basal		
4. blast		
5. chyme		
6. crine		
7. endo		
8. epi		
9. glia		
10. holo		
11. hormon		
12. hyal		
13. lamina		
14. mero		
15. meso		
16. retic		
17. sero		
18. squam		
19. strat		

5

The Integumentary System

Student Objectives

When you have completed the exercises in this chapter, you will have accomplished the following objectives:

The Skin

1. Name the tissue types composing the epidermis and dermis. List the major layers of each and describe the functions of each layer.
2. Describe the factors that normally contribute to skin color. Briefly describe how changes in skin color may be used as clinical signs of certain disease states.

Appendages of the Skin

3. Compare the structure and locations of sweat and oil glands. Also compare the composition and functions of their secretions.
4. Compare and contrast eccrine and apocrine glands.
5. List the parts of a hair follicle and explain the function of each part. Also describe the functional relationship of arrector pili muscles to the hair follicles.
6. Name the regions of a hair and explain the basis of hair color. Describe the distribution, growth, replacement, and changing nature of hair during the life span.
7. Describe the structure of nails.

Functions of the Integumentary System

8. Describe how the skin accomplishes at least five different functions.

Homeostatic Imbalances of Skin

9. Explain why serious burns are life threatening. Describe how to determine the extent of a burn and differentiate first-, second-, and third-degree burns.
10. Summarize the characteristics of the three major types of skin cancers.

The integumentary system consists of the skin and its derivatives—glands, hairs, and nails. Although the skin is very thin, it provides a remarkably effective external shield that acts to protect our internal organs from what is outside the body.

This chapter reviews the anatomical characteristics of the skin (composed of the dermis and the epidermis) and its derivatives. It also reviews the manner in which the skin responds to both internal and external stimuli to protect the body.

BUILDING THE FRAMEWORK

The Skin

1. 1. Name the tissue type composing the epidermis.

2. Name the tissue type composing the dermis.

2. The more superficial cells of the epidermis become less viable and ultimately die. What two factors account for this natural demise of the epidermal cells?

1. ______________________________

2. ______________________________

____

3. Several types of skin markings may reveal structural characteristics of the dermis. Complete the following statements by inserting your responses in the answer blanks.

______________ 1. Skin cuts that run parallel to __(1)__ gape less than cuts running across these skin markings.

______________ 2. A more scientific term for "stretch marks" is __(2)__.

______________ 3. Skin markings that occur where the dermis is secured to deeper structures are called __(3)__.

______________ 4. __(4)__ appear when dermis elasticity declines from age or excessive sun exposure.

4. Figure 5.1 depicts a longitudinal section of the skin. Label the skin structures and areas indicated by leader lines and brackets on the figure. Select different colors for the structures below and color the coding circles and the corresponding structures on the figure.

- ◯ Arrector pili muscle
- ◯ Adipose tissue
- ◯ Hair follicle
- ◯ Nerve fibers
- ◯ Sweat (sudoriferous) gland
- ◯ Sebaceous gland

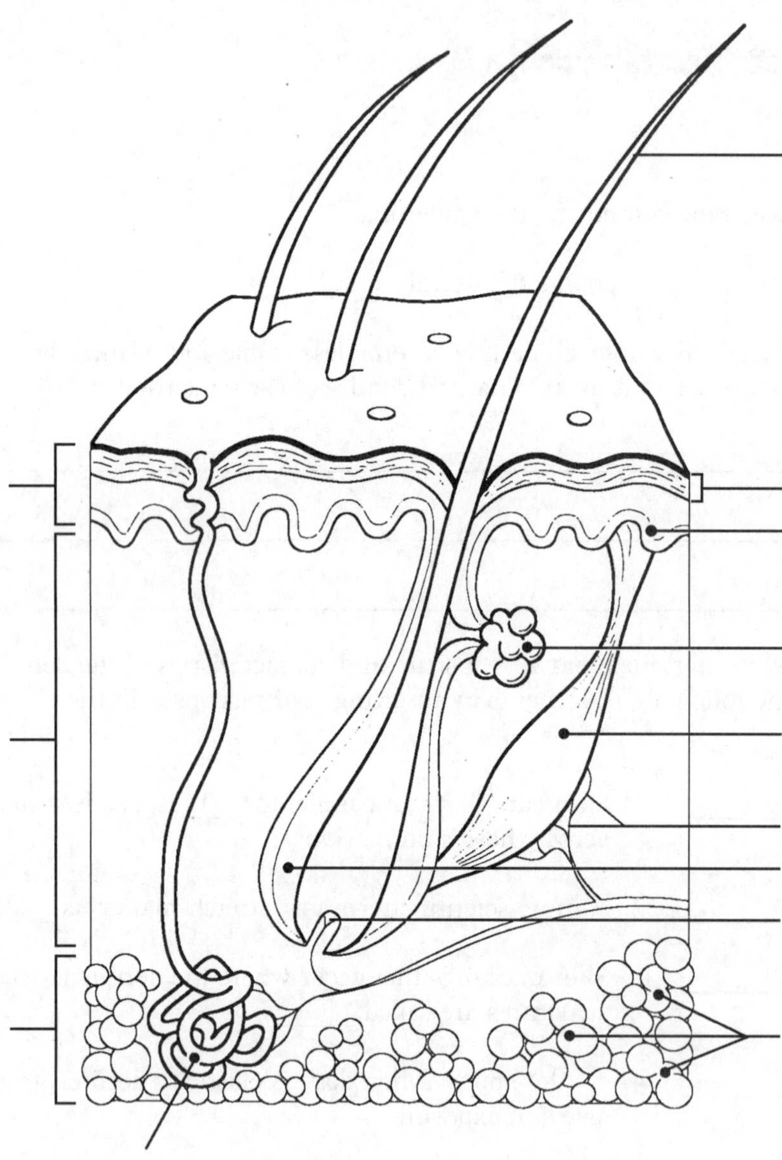

Figure 5.1

5. Using the key choices, choose all responses that apply to the following descriptions. Enter the appropriate letters and/or terms in the answer blanks. (Note: S. = stratum)

Key Choices

A. S. basale	D. S. lucidum	G. Reticular layer	J. Hypodermis
B. S. corneum	E. S. spinosum	H. Epidermis (as a whole)	
C. S. granulosum	F. Papillary layer	I. Dermis (as a whole)	

____________________ 1. Layer of translucent cells, absent in thin skin

____________________ 2. Strata containing all (or mostly) dead cells

____________________ 3. Dermal layer responsible for fingerprints

____________________ 4. Vascular region

____________________ 5. Actively mitotic epidermal region, the deepest epidermal layer

____________________ 6. Cells are flat, dead "bags" of keratin

____________________ 7. Site of elastic and collagen fibers

____________________ 8. General site of melanin formation

____________________ 9. Major skin area where derivatives (hair, nails) *reside*

____________________ 10. Largely adipose tissue; anchors the skin to underlying tissues

____________________ 11. The stratum germinativum

____________________ 12. Epidermal layer where most melanocytes are found

____________________ 13. Cells of this layer contain keratohyalin and lamellated granules

____________________ 14. Accounts for the bulk of epidermal thickness

____________________ 15. When tanned, becomes leather; provides mechanical strength to the skin

____________________ 16. Epidermal layer containing the "oldest" cells

6. Circle the term that does not belong in each of the following groupings.

1. Reticular layer   Keratin   Dermal papillae   Meissner's corpuscles
2. Melanin   Freckle   Wart   Malignant melanoma
3. Prickle cells   Stratum basale   Stratum spinosum   Cell shrinkage
4. Meissner's corpuscles   Pacinian corpuscles   Tactile cells   Arrector pili

5. Waterproof substance Elastin Lamellated granules Produced by keratinocytes

6. Mast cells Macrophages Fibroblasts Melanocytes

7. Intermediate filaments Keratin fibrils Keratohyaline Lamellated granules

8. Keratinocyte Fibroblast Tactile cell Epidermal dendritic cell

7. This exercise examines the relative importance of three pigments in determining skin color. Indicate which pigment is identified by the following descriptions by inserting the appropriate answer from the key choices in the answer blanks.

Key Choices

A. Carotene B. Hemoglobin C. Melanin

__________________ 1. Most responsible for the skin color of dark-skinned people

__________________ 2. Provides an orange cast to the skin

__________________ 3. Provides a natural sunscreen

__________________ 4. Most responsible for the skin color of Caucasians

__________________ 5. Phagocytized by keratinocytes

__________________ 6. Found predominantly in the stratum corneum

__________________ 7. Found within red blood cells in the blood vessels

8. Abnormalities of skin color can be helpful in alerting a physician to certain pathologies. Match the clinical terms in Column B with the possible-cause descriptions in Column A. Place the correct letter in each answer blank.

	Column A	Column B
__________________	1. A bluish cast of the skin resulting from inadequate oxygenation of the blood	A. Cyanosis
		B. Erythema
__________________	2. Observation of this condition might lead to tests for anemia or low blood pressure	C. Hematoma
__________________	3. Accumulation of bile pigments in the blood; may indicate liver disease	D. Jaundice
		E. Pallor
__________________	4. Clotted mass of blood that may signify bleeder's disease	
__________________	5. A common result of inflammation, allergy, and fever	

Appendages of the Skin

1. Figure 5.2 shows longitudinal and cross-sectional views of a hair follicle.

Part A

1. Identify and label all structures provided with leader lines.

2. Select different colors to identify the structures described below and color both the coding circles and the corresponding structures on the diagram.

   ◯ Contains blood vessels that nourish the growth zone of the hair

   ◯ Secretes sebum into the hair follicle

   ◯ Pulls the hair follicle into an upright position during fright or exposure to cold

   ◯ The follicle sheath that consists of dermal tissue

   ◯ The follicle sheath that consists of epidermal tissue

   ◯ The actively growing region of the hair

3. Draw in the nerve fibers and blood vessels that supply the follicle, the hair, the hair root, and the arrector pili.

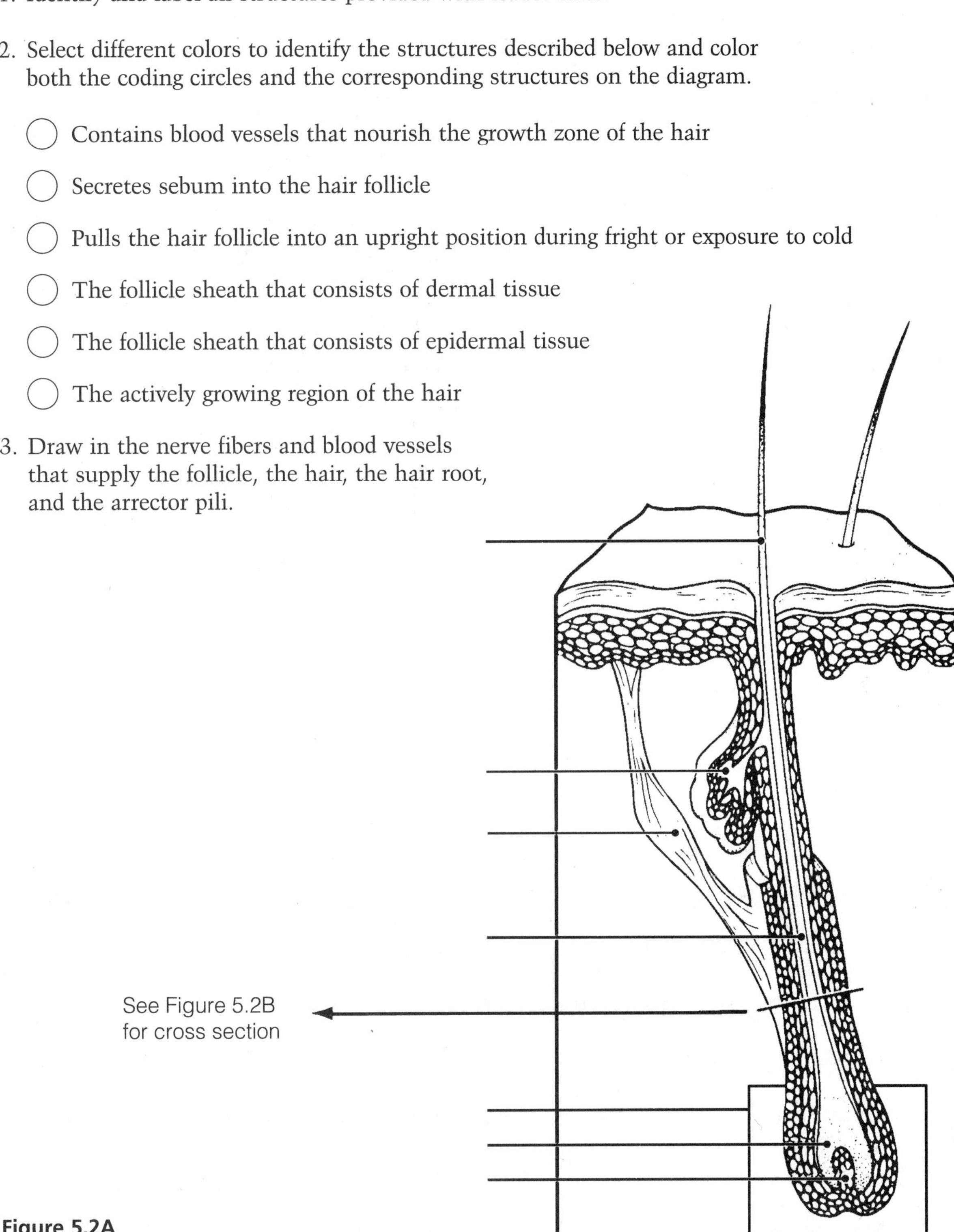

Figure 5.2A

Part B

4. Identify the two portions of the follicle wall by placing the correct name of the sheath at the end of the appropriate leader line.

5. Color these regions using the same colors used for the identical structures in Part A.

6. Label, color code, and color these three regions of the hair:

◯ Cortex ◯ Cuticle ◯ Medulla

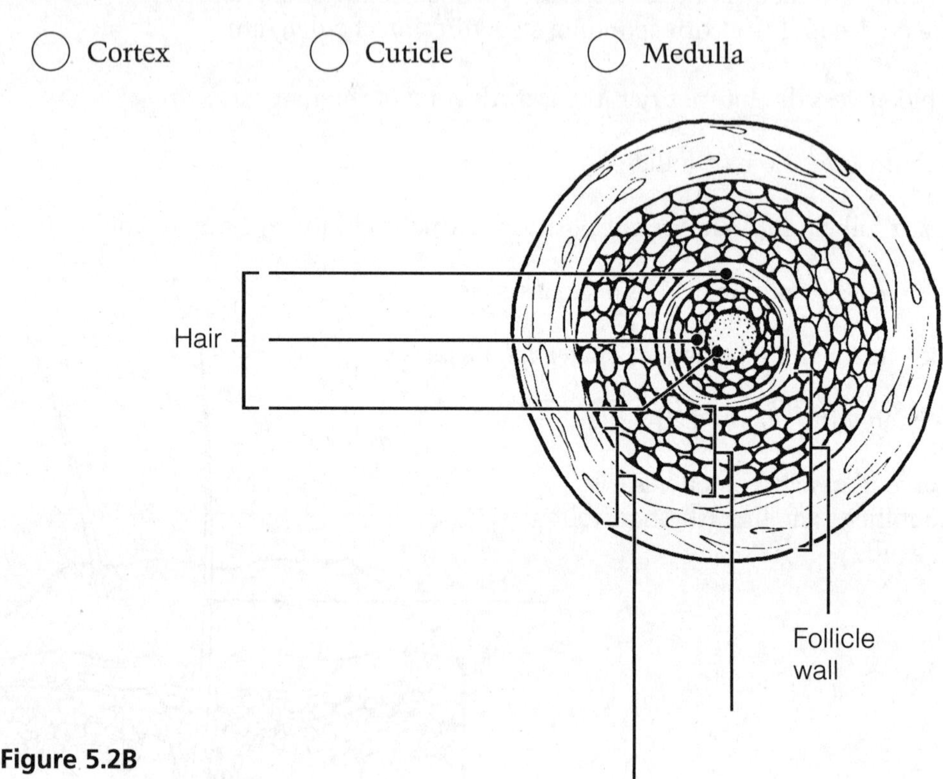

Figure 5.2B

2. Circle the term that does not belong in each of the following groupings.

1. Luxuriant hair growth Testosterone Poor nutrition Good blood supply
2. Vitamin D Cholesterol UV radiation Keratin
3. Stratum corneum Nail matrix Hair bulb Stratum basale
4. Scent glands Eccrine glands Apocrine glands Axilla
5. Terminal hair Vellus hair Dark, coarse hair Eyebrow hair
6. Hard keratin Hair shaft Desquamation Durable
7. Growth phase Resting phase Atrophy Inactive

3. What is the scientific term for baldness? ______________________________

4. Name four factors that can cause hair loss and hair thinning *other than* nutritional or circulatory factors. ______________________________

____

5. Draw a simple diagram of a fingertip *bearing a fingernail* in the space at the right. Identify and label the following nail regions on your sketch: free edge, body, lunule, lateral nail folds, proximal nail fold (eponychium).

1. What is the common name for the eponychium? ______________________________

2. Why does the lunule appear whiter than the rest of the nail? ______________________________

____

____

6. Using the key choices, complete the following statements. Insert the appropriate letters in the answer blanks.

Key Choices

A. Sebaceous glands B. Sweat glands (apocrine) C. Sweat glands (eccrine)

______________ 1. Their products are an oily mixture of lipids, cholesterol, and cell fragments.

______________ 2. Functionally, these are merocrine glands.

______________ 3. The less numerous variety of perspiration gland, their secretion (often milky in appearance) contains proteins and other substances that favor bacterial growth.

______________ 4. Their ducts open to the external environment via a pore.

______________ 5. These glands are found everywhere on the body except the palms of the hands and soles of the feet.

______________ 6. Their secretions contain bactericidal substances.

______________ 7. They become more active at puberty under the influence of androgens.

______________ 8. Their secretions, when oxidized, are seen on the skin surface as a blackhead.

______________ 9. The ceruminous glands that produce earwax are a modification of this gland variety.

______________ 10. These glands are involved in thermoregulation.

Functions of the Integumentary System

1. The skin protects the body by providing three types of barriers. Classify each of the protective factors listed below as an example of a chemical barrier (C), a biological barrier (B), or a mechanical (physical) barrier (M).

_____ 1. Epidermal dendritic cells and macrophages

_____ 2. Intact epidermis

_____ 3. Bactericidal secretions

_____ 4. Keratin

_____ 5. Melanin

_____ 6. Acid mantle

2. Substances that can penetrate the skin in limited amounts include (circle all that apply):

Fat-soluble vitamins Steroid hormones Water-soluble substances

Organic solvents Oxygen Mercury, lead, and nickel

3. In what way does a sunburn impair the body's ability to defend itself?

(Assume the sunburn is mild.) ____

____

4. Explain the role of sweat glands in maintaining body temperature homeostasis.

In your explanation, indicate how their activity is regulated. ______________________

____

____

____

5. Complete the following statements. Insert your responses in the answer blanks.

______________ 1.

______________ 2.

______________ 3.

______________ 4.

______________ 5.

The cutaneous sensory receptors that reside in the skin are actually part of the __(1)__ system. Four types of stimuli that can be detected by certain of the cutaneous receptors are __(2)__, __(3)__, __(4)__, and __(5)__.

Vitamin D is synthesized when modified __(6)__ molecules in the __(7)__ of the skin are irradiated by __(8)__ light. Vitamin D is important in the absorption and metabolism of __(9)__ ions.

______________ 6. ______________ 8.

______________ 7. ______________ 9.

Homeostatic Imbalances of Skin

1. Overwhelming infection is one of the most important causes of death in burn patients. What is the other major problem they face, and what are its possible consequences?

____

____

2. This section reviews the severity of burns. Using the key choices, select the correct burn type for each of the following descriptions. Enter the correct answers in the answer blanks.

Key Choices

A. First-degree burn B. Second-degree burn C. Third-degree burn

__________ 1. Full-thickness burn; epidermal and dermal layers destroyed; skin is blanched

__________ 2. Blisters form

__________ 3. Epidermal damage, redness, and some pain (usually brief)

__________ 4. Epidermal and some dermal damage; pain; regeneration is possible

__________ 5. Regeneration impossible; requires grafting

__________ 6. Pain is absent because nerve endings in the area are destroyed

3. What is the importance of the "rule of nines" in the treatment of burn patients?

____

4. Fill in the type of skin cancer that matches each of the following descriptions:

__________ 1. Cells of the stratum spinosum develop lesions; metastasizes to lymph nodes.

__________ 2. Cells of the lowest level of the epidermis invade the dermis and hypodermis; exposed areas develop an ulcer; slow to metastasize.

__________ 3. Rare but deadly cancer of pigment-producing cells.

5. What does ABCD mean in reference to examination of pigmented areas? __________

____

____

THE INCREDIBLE JOURNEY:

A Visualization Exercise for the Skin

Your immediate surroundings resemble huge, grotesquely twisted vines . . . you begin to climb upward.

1. Complete the narrative by inserting the missing words in the answer blanks.

1. ____________________
2. ____________________
3. ____________________
4. ____________________
5. ____________________
6. ____________________
7. ____________________
8. ____________________
9. ____________________

For this trip, you are miniaturized for injection into your host's skin. Your journey begins when you are injected into a soft gel-like substance. Your immediate surroundings resemble huge, grotesquely twisted vines. But when you peer carefully at the closest "vine," you realize you are actually seeing connective tissue fibers. Most of the fibers are fairly straight, although tangled together, and look like strong cables. You identify these as the __(1)__ fibers. Here and there are fibers that resemble coiled springs. These must be the __(2)__ fibers that help give skin its springiness. At this point, there is little question that you are in the __(3)__ region of the skin, particularly since you can also see blood vessels and nerve fibers around you.

Carefully, using the fibers as steps, you begin to climb upward. After climbing for some time and finding that you still haven't reached the upper regions of the skin, you stop for a rest. As you sit, a strange-looking cell approaches, moving slowly with parts alternately flowing forward and then receding. Suddenly you realize that this must be a __(4)__ that is about to dispose of an intruder (you) unless you move in a hurry! You scramble to your feet and resume your upward climb. On your right is a large fibrous structure that looks like a tree trunk anchored in place by muscle fibers. By scurrying up this __(5)__ sheath, you are able to escape from the cell and again scan your surroundings. Directly overhead are tall cubelike cells, forming a continuous sheetlike membrane. In your rush to escape you reached the __(6)__ region of the skin. As you watch the activity of the cells in this layer, you notice that many of the cells are pinching in two and that the daughter cells are being forced upward. Obviously, this is the specific layer that continually replaces cells that rub off the skin surface, and these cells are the __(7)__ cells.

Looking through the transparent cell membrane of one of the basal cells, you see a dark mass hanging over the nucleus. You wonder if this cell could have a tumor; but then, looking through the membranes of the neighboring cells, you find that they also have dark umbrella-like masses hanging over their nuclei. As you consider this matter, a black cell with long tentacles begins to pick its way carefully between the other cells. As you watch, one of the transparent cells engulfs the end of a tentacle of the black cell, and within seconds contains some of its black substance. Suddenly, you remember that one of the skin's protective functions is to protect the deeper layers from sun damage; the black substance must be the protective pigment __(8)__.

Once again you begin your upward climb and notice that the cells are becoming shorter and harder and are full of a tough, waxy substance. This substance has to be __(9)__, which would account for the increasing hardness of the cells. Climbing still higher, the cells become flattened like huge shingles.

____________________ 10. The only material apparent in the cells is the waxy substance; there is no nucleus, and there appears to be no activity in these cells. Considering the clues—shinglelike cells, no nuclei, full of the waxy substance, no activity—these cells are obviously (10) and therefore are very close to the skin surface.

Suddenly, you feel a strong agitation in your immediate area. The pressure is tremendous. Looking upward through the transparent cell layers, you see your host's fingertips vigorously scratching the area directly overhead. You wonder if you are causing his skin to sting or tickle. Then, within seconds, the cells around you begin to separate and fall apart, and you are catapulted out into the sunlight. Because the scratching fingers might descend once again, you quickly advise your host of your whereabouts.

CHALLENGING YOURSELF

At the Clinic

1. Xeroderma pigmentosum is a severe, genetically linked skin cancer in which DNA repair mechanisms are impaired. Why would sufferers of this condition need to stay out of the sun?
2. A new mother brings her infant to the clinic, worried about a yellowish, scummy deposit that has built up on the baby's scalp. What is this condition called, and is it serious?
3. During a diaper change, an alert day-care worker notices a dark, bruised-looking area at the base of a baby's spine. Worried about possible child abuse, she reports the spot to her supervisor, who tells her not to worry because it is a Mongolian spot. What is a Mongolian spot?
4. Hives are welts, or reddened "bumps," that indicate sites of local inflammation. They are often a sign of an allergic reaction. Recall from Chapter 3 the role of capillary permeability and plasma loss in causing edema. Would systemic hives be cause for worry?
5. After a worker in a furniture refinishing establishment fell into a vat of paint stripper, he quickly removed his clothes and rinsed off in the safety shower. Were his safety measures adequate? What vital organs might suffer early damage from poisoning through skin by organic solvents?
6. What two factors in the treatment of critical third-degree burn patients are absolutely essential?
7. Mr. Bellazono, a fisherman in his late 60s, comes to the clinic to complain of small ulcers on both forearms as well as on his face and ears. Although he has had them for several years, he has not had any other problems. What is the likely diagnosis, and what is the likely cause?
8. Both newborn and aged individuals have very little subcutaneous tissue. How does this affect their sensitivity to cold?
9. The hypodermis of the face is quite loose and has few connections to the deep fascia of the muscles. Explain how this relates to the greater need to suture cuts on the face compared to other body regions.
10. Martha, the mother of a 13-month-old infant, brings her child to the clinic because his skin has turned orange. Why does the pediatrician inquire about the child's diet?

11. Mrs. Ibañez volunteered to help at a hospital for children with cancer. When she first entered the cancer ward, she was upset by the fact that most of the children had no hair. What is the explanation for their baldness?

12. Carmen slipped on some ice and split open the skin of her chin on the sidewalk. As the physician in the emergency room was giving her six stitches, he remarked that the split was straight along a cleavage line. How cleanly is her wound going to heal, and is major scarring likely to occur?

Stop and Think

1. How can the skin be both a membrane and an organ?
2. Why does the border between the epidermis and the dermis undulate?
3. The skin covering your shins is *not* freely movable. Palpate (feel) your shins and compare that region to the other regions of the body. Then try to deduce why there is little free movement of the skin of the shins.
4. In terms of both function and benefit, why are surface keratinocytes dead?
5. What nerve endings in the skin respond to the lightest touch?
6. Why does sunburned skin peel in sheets?
7. Would increasing protein intake (such as by taking gelatin supplements) increase hair and nail strength in an otherwise healthy individual?
8. Kareem had a nervous habit of chewing on the inner lining of his lip with his front teeth. The lip grew thicker and thicker from years of continual irritation. Kareem's dentist noticed his greatly thickened lip and suggested he have it checked. A biopsy revealed hyperplasia and scattered areas of dysplasia, but no evidence of neoplasia. What do these terms mean? Did Kareem have cancer of the mouth?
9. Studies have shown that women who live or work together tend to develop synchronized monthly cycles. What aspect of the integumentary system might explain this sexual signaling?
10. If our cells and body fluids are hyperosmotic to the water of a swimming pool (and they *are*), then why do we not swell and pop when we go for a swim?
11. Mrs. Jones, a histologist, always makes her gelatin salad for bridge parties. How does that quivering lemon-flavored gelatin—laden with shredded cabbage, sliced almonds, and raisins—remind her of connective tissue?

COVERING ALL YOUR BASES

Multiple Choice

Select the best answer or answers from the choices given.

1. Which is *not* part of the skin?
   A. Epidermis
   B. Hypodermis
   C. Dermis
   D. Superficial fascia

2. Which of the following is *not* a tissue type found in the skin?
   A. Stratified squamous epithelium
   B. Loose connective tissue
   C. Dense irregular connective tissue
   D. Ciliated columnar epithelium
   E. Vascular tissue

3. Epidermal cells that aid in the immune response include:
   A. tactile cells
   B. epidermal dendritic cells
   C. melanocytes
   D. spinosum cells

4. Which organelle is most prominent in cells manufacturing the protein keratin?
A. Ribosomes
B. Golgi apparatus
C. Smooth endoplasmic reticulum
D. Lysosomes

5. Which epidermal layer has the highest concentration of epidermal dendritic cells, and has numerous desmosomes and thick bundles of keratin filaments?
A. Stratum corneum
B. Stratum lucidum
C. Stratum granulosum
D. Stratum spinosum

6. Fingerprints are caused by:
A. the genetically determined arrangement of dermal papillae
B. the conspicuous epidermal ridges
C. the sweat pores
D. all of these

7. Find the *false* statement concerning vitamin D.
A. Dark-skinned people make no vitamin D.
B. If vitamin D production is inadequate, one may develop weak bones.
C. If the skin is not exposed to sunlight, one may develop weak bones.
D. Vitamin D is needed for the uptake of calcium from food in the intestine.

8. Use logic to deduce the answer to this question. Given what you now know about skin color and skin cancer, which of the following groups would have the highest rate of skin cancer?
A. Blacks in tropical Africa
B. Scientists in research stations in Antarctica
C. Whites in northern Australia
D. Norwegians in the southern part of the U.S.
E. Blacks in the U.S.

9. Which structure is not associated with a hair?
A. Shaft
B. Cortex
C. Eponychium
D. Matrix
E. Cuticle

10. Which of the following hair colors is not produced by melanin?
A. Blonde
B. Brown
C. Black
D. Gray
E. White

11. Which is the origin site of cells that are directly responsible for growth of a hair?
A. Hair bulb
B. Hair follicle
C. Papilla
D. Hair bulge
E. Matrix

12. Concerning movement of hairs:
A. Movement is the function of the arrector pili.
B. Movement is sensed by the root hair plexus.
C. Muscle contraction flattens the hair against the skin.
D. Muscle contraction is prompted by cold or fright.

13. A particular type of tumor of the adrenal gland causes excessive secretion of sex hormones. This condition expresses itself in females as:
A. male pattern baldness
B. hirsutism
C. increase in growth of vellus hairs over the whole body
D. increase in length of terminal hairs

14. At the end of a follicle's growth cycle:
A. the follicle atrophies
B. the hair falls out
C. the hair elongates
D. the hair turns white

15. What is the major factor accounting for the waterproof nature of the skin?
A. Desmosomes in stratum corneum
B. Glycolipid between stratum corneum cells
C. The thick insulating fat of the hypodermis
D. The leathery nature of the dermis

16. In investigating the cause of thinning hair, which of the following questions needs to be asked?
A. Is the diet deficient in proteins?
B. Is the person taking megadoses of vitamin C?
C. Has the person been exposed to excessive radiation?
D. Has the person recently suffered severe emotional trauma?

17. Which structures are not associated with a nail?
A. Nail bed
B. Lunule
C. Nail folds
D. Nail follicle

18. One of the following is not associated with production of perspiration. Which one?
A. Sweat glands
B. Sweat pores
C. Holocrine gland
D. Eccrine gland
E. Apocrine gland

19. Components of sweat include:
A. water
B. sodium chloride
C. sebum
D. ammonia
E. vitamin D

20. Which of the following is *true* concerning oil production in the skin?
    A. Oil is produced by sudoriferous glands.
    B. Secretion of oil is via the holocrine mode.
    C. The secretion is called sebum.
    D. Oil is usually secreted into hair follicles.

21. A disorder *not* associated with oil glands is:
    A. seborrhea
    B. cystic fibrosis
    C. acne
    D. whiteheads

22. Contributing to the chemical barrier of the skin is (are):
    A. perspiration
    B. stratified squamous epithelium
    C. epidermal dendritic cells
    D. sebum
    E. melanin

23. Which of the following provide evidence of the skin's role in temperature regulation?
    A. Shivering
    B. Flushing
    C. Blue fingernails
    D. Sensible perspiration
    E. Acid mantle

24. Contraction of the arrector pili would be "sensed" by:
    A. tactile discs
    B. Meissner's corpuscles
    C. root hair plexuses
    D. Pacinian corpuscles

25. A dermatologist examines a patient with lesions on the face. Some of the lesions appear as a shiny, raised spot; others are ulcerated with a beaded edge. What is the diagnosis?
    A. Melanoma
    B. Squamous cell carcinoma
    C. Basal cell carcinoma
    D. Either squamous or basal cell carcinoma

26. A burn patient reports that the burns on her hands and face are not painful, but she has blisters on her neck and forearms and the skin on her arms is very red. This burn would be classified as:
    A. first-degree only
    B. second-degree only
    C. third-degree only
    D. critical

27. Which of the following is associated with vitamin synthesis in the body?
    A. Cholesterol
    B. Calcium metabolism
    C. Carotene
    D. Ultraviolet radiation
    E. Melanin

28. A patient has a small (about 1 mm), regular, round, pale mole. Is this likely to be melanoma?
    A. Yes
    B. No, because it is small
    C. No, because it is pale
    D. No, because it is round
    E. No, because it is regular

Word Dissection

For each of the following word roots, fill in the literal meaning and give an example, using a word found in this chapter.

Word root	Translation	Example
1. arrect		
2. carot		
3. case		
4. cere		
5. corn		
6. cort		
7. cutic		
8. cyan		
9. derm		
10. folli		
11. hemato		
12. hirsut		
13. jaune		
14. kera		
15. lanu		
16. lunul		
17. medull		
18. melan		
19. pall		
20. papilla		
21. pili		
22. plex		
23. rhea		
24. seb		
25. spin		
26. sudor		
27. tegm		
28. vell		

Bones and Skeletal Tissues

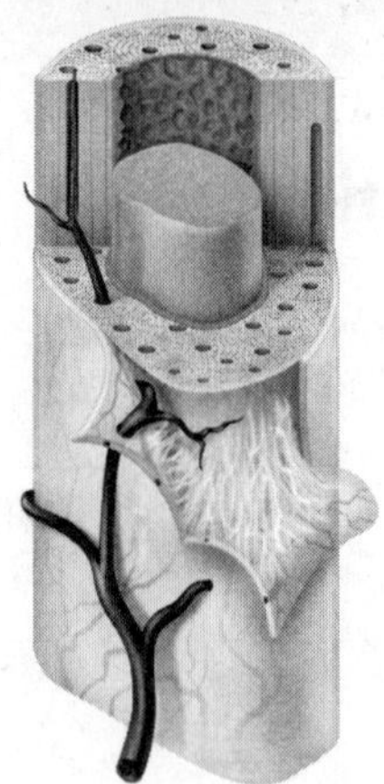

Student Objectives

When you have completed the exercises in this chapter, you will have accomplished the following objectives:

Skeletal Cartilages

1. Describe the functional properties of the three types of cartilage tissue.
2. Locate the major cartilages of the adult skeleton.
3. Explain how cartilage grows.

Classification of Bones

4. Name the major regions of the skeleton and describe their relative functions.
5. Compare and contrast the structure of the four bone classes and provide examples of each class.

Functions of Bones

6. List and describe five important functions of bones.

Bone Structure

7. Indicate the functional importance of bone markings.
8. Describe the gross anatomy of a typical long bone and flat bone. Indicate the locations and functions of red and yellow marrow, articular cartilage, periosteum, and endosteum.
9. Describe the histology of compact and spongy bone.
10. Discuss the chemical composition of bone and the advantages conferred by the organic and inorganic components.

Bone Development

11. Compare and contrast intramembranous ossification and endochondral ossification.
12. Describe the process of long bone growth that occurs at the epiphyseal plates.

Bone Homeostasis: Remodeling and Repair

13. Compare the locations and remodeling functions of the osteoblasts, osteocytes, and osteoclasts.
14. Explain how hormones and physical stress regulate bone remodeling.
15. Describe the steps of fracture repair.

Homeostatic Imbalances of Bone

16. Contrast the disorders of bone remodeling seen in osteoporosis, osteomalacia, and Paget's disease.

Bone and cartilage are the principal tissues that provide support for the body; together they make up the skeletal system. The early skeleton is formed largely from cartilage. Later, the cartilage is replaced almost entirely by bone. Bones provide attachments for muscles and act as levers for the movements of body parts. Bones perform additional essential functions. They are the sites for blood cell formation and are storage depots for many substances, such as fat, calcium, and phosphorus.

Bone tissue has an intricate architecture that is both stable and dynamic. Although it has great strength and rigidity, bone tissue is able to change structurally in response to a variety of chemical and mechanical factors.

Chapter 6 topics for student review include an overview of skeletal cartilages, the structures of long and flat bones, the remodeling and repair of bone, and bone development and growth.

BUILDING THE FRAMEWORK

Skeletal Cartilages

1. Use the key choices to identify the type of cartilage tissue found in the following body locations:

Key Choices

A. Elastic cartilage B. Fibrocartilage C. Hyaline cartilage

_____ 1. At the junction of a rib and the sternum

_____ 2. The skeleton of the external ear

_____ 3. Supporting the trachea walls

_____ 4. Forming the intervertebral discs

_____ 5. Forming the epiglottis

_____ 6. At the ends of long bones

_____ 7. Most of the fetal skeleton

_____ 8. Knee menisci

2. In comparing bone and cartilage tissue, indicate whether each of the following statements is true (T) or false (F).

_____ 1. Cartilage is more resilient than bone.

_____ 2. Cartilage is especially strong in resisting shear (bending and twisting) forces.

_____ 3. Cartilage can grow faster than bone in the growing skeleton.

_____ 4. In the adult skeleton, cartilage regenerates faster than bone when damaged.

_____ 5. Neither bone nor cartilage contains capillaries.

_____ 6. Bone tissue contains relatively little water compared with cartilage tissue, which contains a large amount of water.

_____ 7. Nutrients diffuse quickly through cartilage matrix but very poorly through solid bone matrix.

3. What single structural characteristic accounts for the resilience of cartilage and its ability to grow rapidly in the developing skeleton?

____

____

Classification of Bones

1. Identify each of the following bones as a member of one of the four major bone categories. Use L for long bone, S for short bone, F for flat bone, and I for irregular bone. Enter the appropriate letters in the answer blanks.

_____ 1. Calcaneus

_____ 2. Frontal

_____ 3. Femur

_____ 4. Humerus

_____ 5. Mandible

_____ 6. Metacarpal

_____ 7. Radius

_____ 8. Sternum

_____ 9. Vertebra

Functions of Bones

1. List and explain five important functions of bones. Write your answers in the answer blanks below.

1. ____

2. ____

3. ____

4. ____

5. ____

Bone Structure

1. Figure 6.1A is a drawing of a sagittal section of the femur. Do not color the articular cartilage. Leave it white. Select different colors for the bone regions listed at the coding circles below. Color the coding circles and the corresponding regions on the drawing. Complete Figure 6.1A by labeling compact bone and spongy bone.

   Figure 6.1B is a midlevel, cross-sectional view of the diaphysis of the femur. As in A, identify by color the area where yellow marrow is found. Label the membrane that lines the cavity and the membrane that covers the outside surface. Indicate by an asterisk (*) the membrane that contains both osteoblasts and osteoclasts.

   ◯ Diaphysis
   ◯ Epiphyseal plate
   ◯ Articular cartilage
   ◯ Area where red marrow is found
   ◯ Area where yellow marrow is found

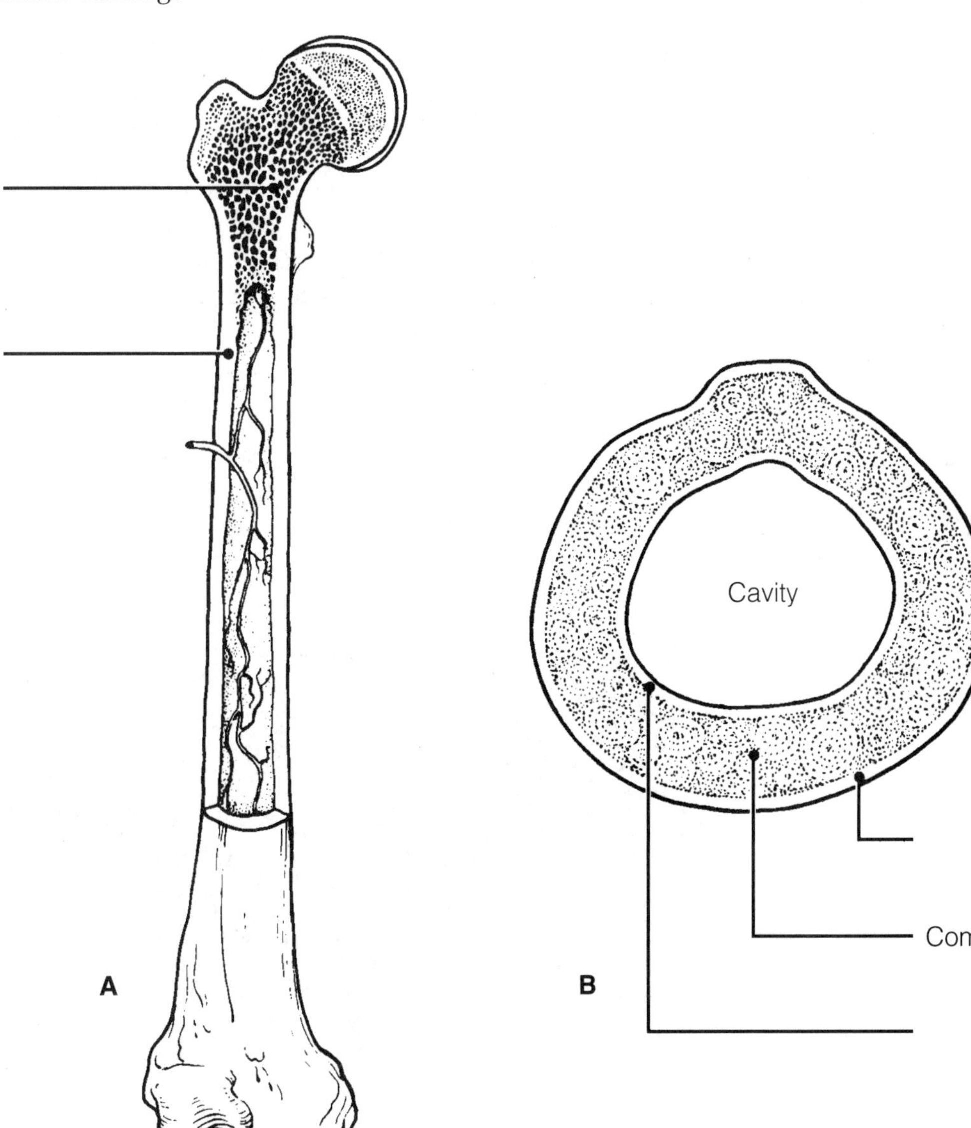

Figure 6.1

2. Using the key choices, characterize the following statements relating to the structure of a long bone. Enter the appropriate answers in the answer blanks.

Key Choices

A. Diaphysis C. Epiphysis E. Yellow marrow cavity

B. Epiphyseal plate D. Red marrow F. Periosteum

__________________ 1. Location of spongy bone in an adult's bone

__________________ 2. Location of compact bone in an adult's bone

__________________ 3. Site of hematopoiesis in an adult's bone

__________________ 4. Scientific name for bone shaft

__________________ 5. Site of fat storage

__________________ 6. Region of longitudinal growth in a child

__________________ 7. Composed of hyaline cartilage until the end of adolescence

__________________ 8. Inner layer consists primarily of osteoblasts and osteogenic cells

3. Figure 6.2 is a sectional diagram showing the five-layered structure of a typical flat bone. Select different colors for the layers below. Add labels and leaders to identify Sharpey's fibers and trabeculae. Then answer the questions that follow, referring to Figure 6.2 and inserting your answers in the answer blanks.

○ Spongy bone ○ Compact bone ○ Periosteum

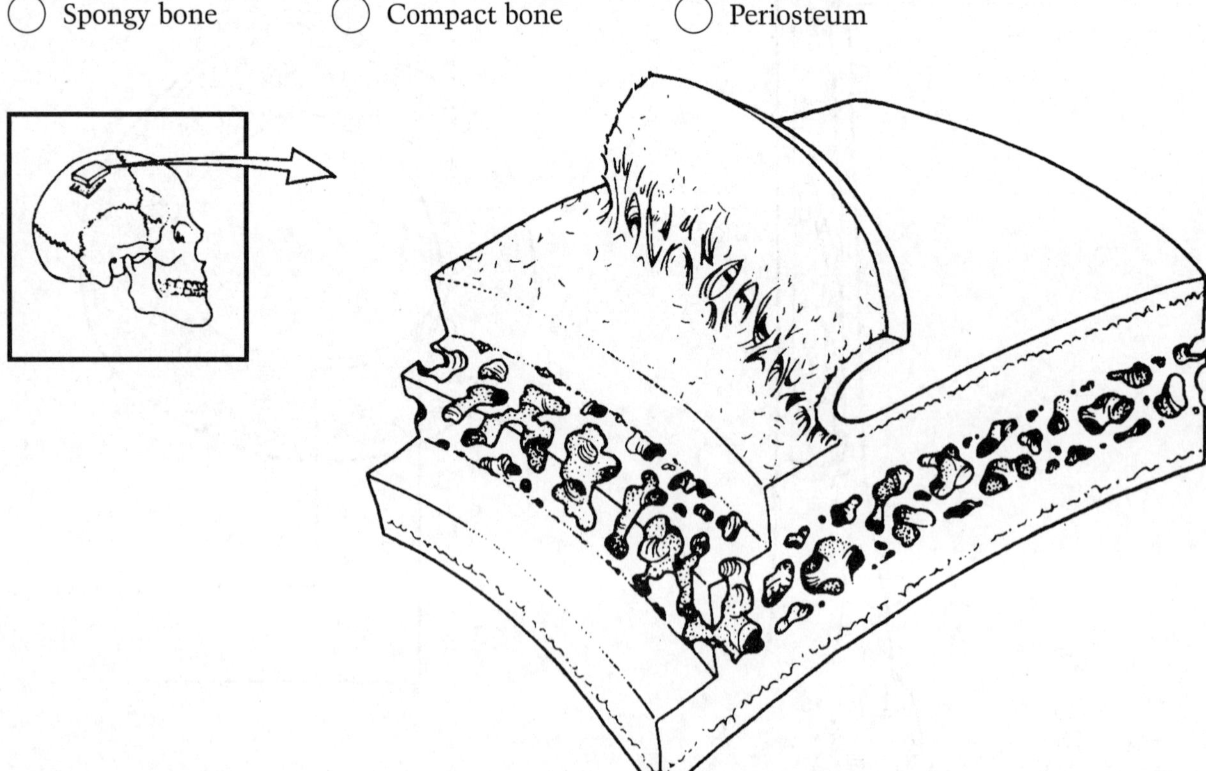

Figure 6.2

1. Which layer is called the diploë? ____________________

2. Name the membrane that lines internal bone cavities. ____________________

4. Five descriptions of bone structure are provided in Column A.

First, identify the structure by choosing the appropriate term from Column B and placing the corresponding answer in the answer blank.

Second, consider Figure 6.3A, a diagrammatic view of a cross section of bone, and Figure 6.3B, a higher-magnification view of compact bone tissue. Select different colors for the structures and bone areas in Column B and use them to color the coding circles and corresponding structures on the diagrams. As concentric lamellae would be difficult to color without confusing other elements, identify one lamella by using a bracket and label.

	Column A	**Column B**
____________	1. Layers of calcified matrix	A. Central (Haversian) canal ○
____________	2. "Residences" of osteocytes	B. Concentric lamellae ○
____________	3. Longitudinal canal, carrying blood vessels and nerves	C. Lacunae ○
		D. Canaliculi ○
____________	4. Nonliving, structural part of bone	E. Bone matrix ○
____________	5. Tiny canals connecting lacunae	F. Osteocyte ○

A

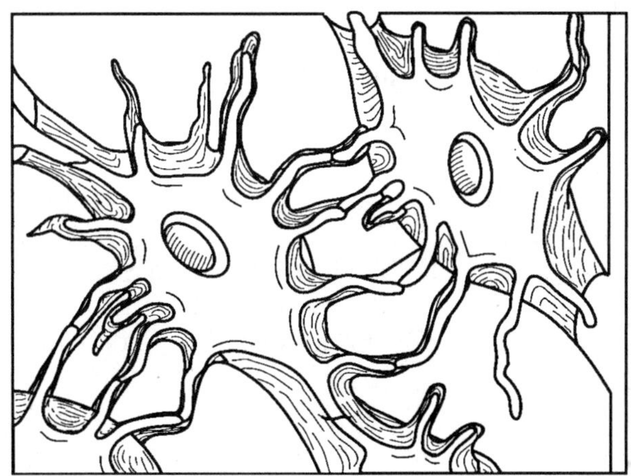

B

Figure 6.3

5. Classify each of the following terms as a projection (P), a depression (D), or an opening (O). Enter the appropriate letter in the answer blanks.

_____ 1. Condyle	_____ 4. Foramen	_____ 7. Ramus	_____ 10. Fossa
_____ 2. Crest	_____ 5. Head	_____ 8. Spine	_____ 11. Facet
_____ 3. Fissure	_____ 6. Meatus	_____ 9. Tuberosity	_____ 12. Sinus

6. Circle the term that does not belong in each of the following groupings.

1. Rigidity Calcium salts Hydroxyapatites Collagen Hardness
2. Hematopoiesis Red marrow Yellow marrow Spongy bone Diploë
3. Lamellae Cellular extensions Canaliculi Circulation Osteoclasts
4. Osteon Marrow cavity Volkmann's canals Haversian canal Canaliculi
5. Epiphysis Articular cartilage Periosteum Hyaline cartilage
6. Perichondrium Periosteum Appositional growth Osteoblasts
7. Spongy Cancellous Woven Lamellar Trabecular

Bone Development

1. The following events apply to the endochondral ossification process as it occurs in the primary ossification center. Put these events in their proper order by assigning each a number (1–6).

_____ 1. Cavity formation occurs within the hyaline cartilage.

_____ 2. Collar of bone is laid down around the hyaline cartilage model just beneath the periosteum.

_____ 3. Periosteal bud invades the marrow cavity.

_____ 4. Perichondrium becomes vascularized to a greater degree and becomes a periosteum.

_____ 5. Osteoblasts lay down bone around the cartilage spicules in the bone's interior.

_____ 6. Osteoclasts remove the cancellous bone from the shaft interior, leaving a marrow cavity that then houses fat.

2. For each statement that is true, insert T in the answer blank. For false statements, correct the underlined words by inserting the correct words in the answer blanks.

____________________ 1. When a bone forms from a fibrous membrane, the process is called <u>endochondral</u> ossification.

______________________ 2. Membrane bones develop from hyaline cartilage structures.

______________________ 3. The organic bone matrix is called the osteoid.

______________________ 4. The enzyme alkaline phosphatase encourages the deposit of collagen fibers within the matrix of developing bone.

______________________ 5. When trapped in lacunae, osteoblasts change into osteocytes.

______________________ 6. Large numbers of osteocytes are found in the inner periosteum layer.

______________________ 7. During endochondral ossification, the periosteal bud invades the deteriorating hyaline cartilage shaft.

______________________ 8. Primary ossification centers appear in the epiphyses.

______________________ 9. Epiphyseal plates are made of spongy bone.

______________________ 10. In appositional growth, bone reabsorption occurs on the periosteal surface.

______________________ 11. "Maturation" of newly formed (noncalcified) bone matrix takes about 10 days.

3. Follow the events of intramembranous ossification by writing the missing words in the answer blanks.

1. ______________________
2. ______________________
3. ______________________
4. ______________________
5. ______________________
6. ______________________
7. ______________________
8. ______________________
9. ______________________
10. ______________________

The initial supporting structure for this type of ossification is a fibrous membrane formed by __(1)__. The first recognizable event is a clustering of the __(1)__ cells to form a(n) __(2)__ in the fibrous membrane. These cells then differentiate into __(3)__, which begin secreting __(4)__ around the fibers of the membrane. Within a few days, calcium salt deposit or __(5)__ occurs, producing true __(6)__. The first network of trabeculae formed are arranged irregularly. This early membrane bone is referred to as __(7)__ bone. As it forms, a layer of vascular __(8)__ condenses on the external face of the bone structure, forming a __(9)__. Eventually lamellar bone replaces __(7)__, and the vascular tissue within the __(10)__ differentiates into red marrow. The final result is a flat bone.

4. Figure 6.4 is a diagram representing the histological changes in the epiphyseal plate of a growing long bone.

First, select different colors for the types of cells named below. Color the coding circles and the corresponding cells in the diagram.

○ Cells depositing osteoid ○ Dividing cartilage cells

○ Older, enlarging, vesiculating cells ○ Cells cutting themselves off from nutrients

Second, identify bracketed zones A–D on the diagram as: *growth, ossification, hypertrophic*, or *calcification.*

Third, complete the statements on page 115, referring to Figure 6.4 and the labeled regions on the diagram. Insert the correct words in the spaces provided.

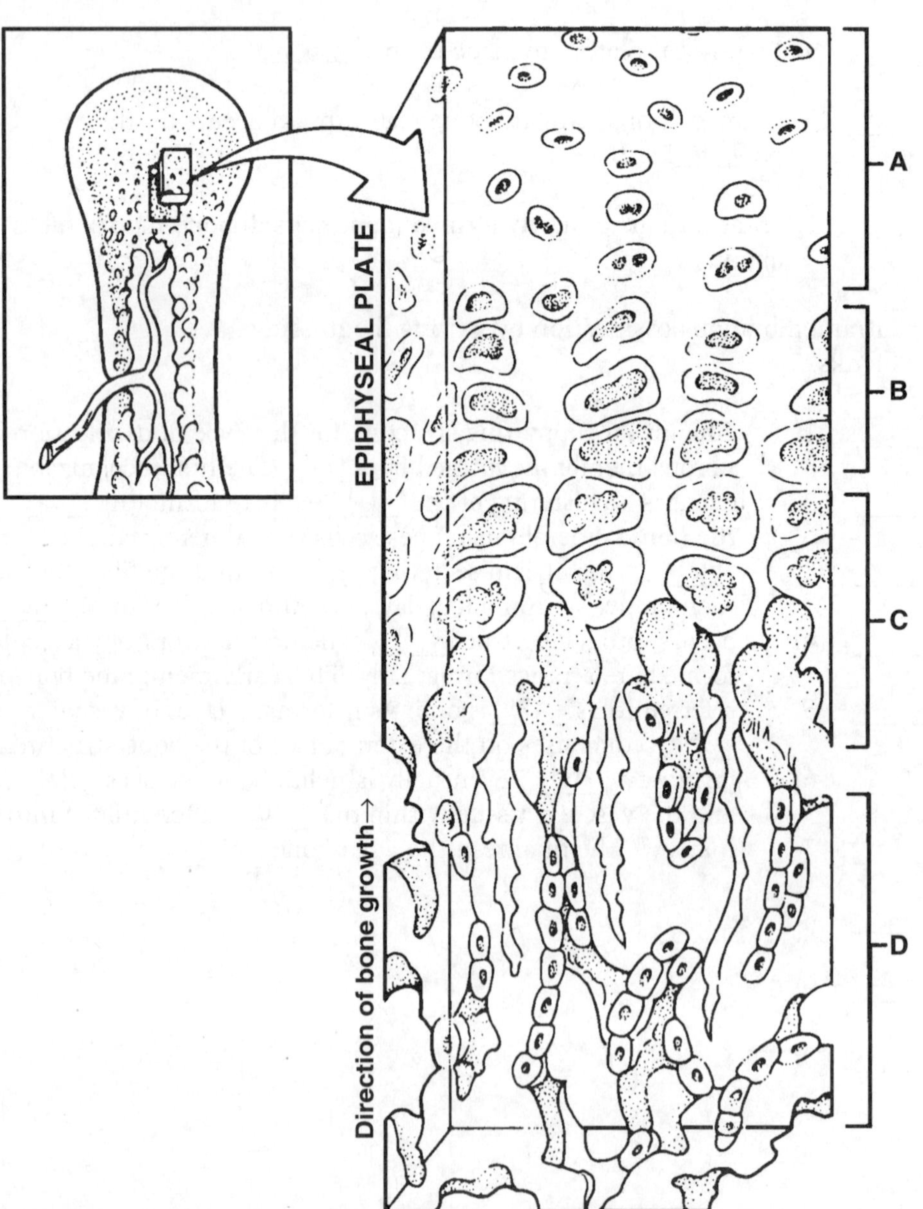

Figure 6.4

____________________ 1. The cell type in region A is __(1)__.

____________________ 2. The type of cell in region B is __(2)__.

____________________ 3. Calcification of the cartilage matrix begins in region __(3)__.

____________________ 4. The cartilaginous matrix begins to deteriorate in region __(4)__.

____________________ 5. The matrix being deposited in region D is __(5)__ matrix.

Bone Homeostasis: Remodeling and Repair

1. Using the key choices, insert the correct answers in the answer blanks below.

Key Choices

A. Atrophy	C. Gravity	E. Osteoclasts	G. Parathyroid hormone
B. Calcitonin	D. Osteoblasts	F. Osteocytes	H. Stress and/or tension

____________________ 1. When blood calcium levels begin to drop below homeostatic levels, __(1)__ is released, causing calcium to be released from bones.

____________________ 2. Mature bone cells, called __(2)__, maintain bone in a viable state.

____________________ 3. Disuse such as that caused by paralysis or severe lack of exercise results in muscle and bone __(3)__.

____________________ 4. Large tubercles and/or increased deposit of bony matrix occur at sites of __(4)__.

____________________ 5. Immature, or matrix-depositing, bone cells are referred to as __(5)__.

____________________ 6. __(6)__ causes blood calcium to be deposited in bones as calcium salts.

____________________ 7. Bone cells that liquefy bone matrix and release calcium to the blood are called __(7)__.

____________________ 8. Astronauts must perform isometric exercises when in outer space because bones atrophy under conditions of weightlessness or lack of __(8)__.

2. Circle the term that does not belong in each of the following groupings.

1. Bone deposit   Injury sites   Growth zones   Repair sites   Bone resorption

2. Osteoid   Organic matrix   Calcium salts   Osteoblasts   10 μm wide

3. Hypercalcemia   Hypocalcemia   Calcium   Salt deposit in soft tissue Ca^{2+} over 11 mg/100 ml blood

4. Mechanical forces   Gravity   Muscle pull   Blood calcium levels   Wolff's law

5. Growth in diameter   Growth in length   Appositional growth   Increase in thickness

3. According to Wolff's law, bones form according to the stresses placed upon them. Figure 6.5 is a simple diagram of the proximal end of a femur (thigh bone). There are two sets of double arrows and two single arrows. Your job is to decide which of the single or paired arrows represents each of the following conditions and indicate those conditions on the diagram. Color the arrows to agree with your coding circles.

○ Site of maximal compression   ○ Load (body weight) exertion site

○ Site of maximal tension   ○ Point of no stress

Explain why long bones can "hollow out" without jeopardy to their integrity (soundness of structure).

____

____

____

Now an application of Wolff's law: Rafael is a tennis pro. Explain why the bones of his serving arm are much thicker than the bones of the other arm.

____

____

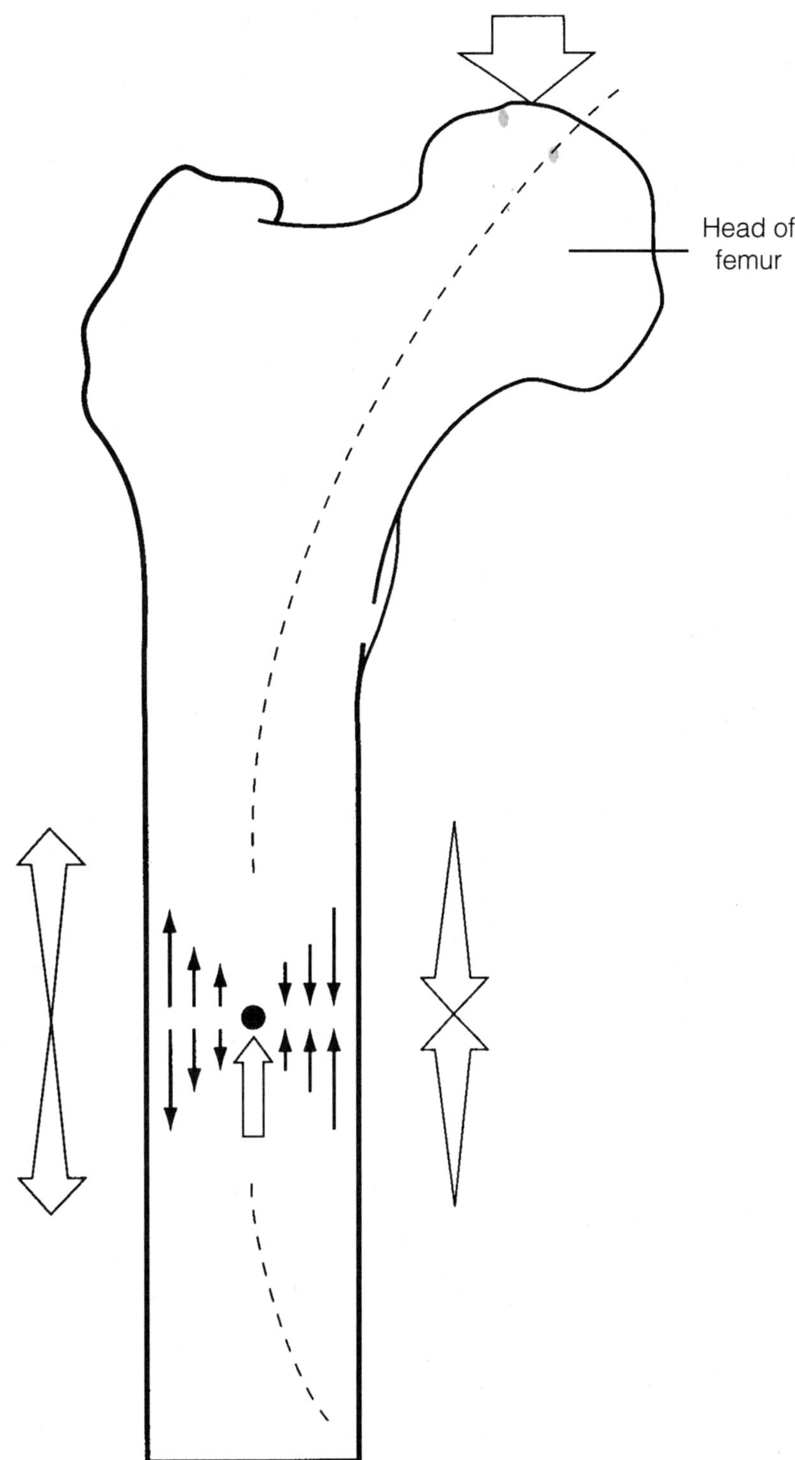

Figure 6.5

4. Use the terms in Column B to identify the fracture types described in Column A.

	Column A	Column B
________________	1. Break parallels the long axis of the bone	A. Complete fracture
		B. Displaced fracture
________________	2. Bone is broken through	C. Incomplete fracture
________________	3. Bone ends are aligned	D. Linear fracture
		E. Nondisplaced fracture
		F. Transverse fracture

5. Use the key choices to identify the fracture (fx) types shown in Figure 6.6 and the fracture types and treatments described below. Enter the appropriate answer in each answer blank.

Key Choices

A. Closed reduction	E. Depressed fracture	I. Spiral fracture
B. Comminuted fracture	F. Greenstick fracture	
C. Compression fracture	G. Open reduction	
D. Compound fracture	H. Simple fracture	

________________ 1. Bone is broken cleanly; the ends do not penetrate the skin

________________ 2. Nonsurgical realignment of broken bone ends and splinting of bone

________________ 3. Bone breaks from twisting forces

________________ 4. A break common in children; bone splinters, but break is incomplete

________________ 5. A fracture in which the bone is crushed; common in the vertebral column

________________ 6. A fracture in which the bone ends penetrate through the skin surface

________________ 7. Surgical realignment of broken bone ends

________________ 8. A common type of skull fracture

________________ 9. Also called a closed fracture

________________ 10. A common sports fracture

________________ 11. Often seen in the brittle bones of the elderly

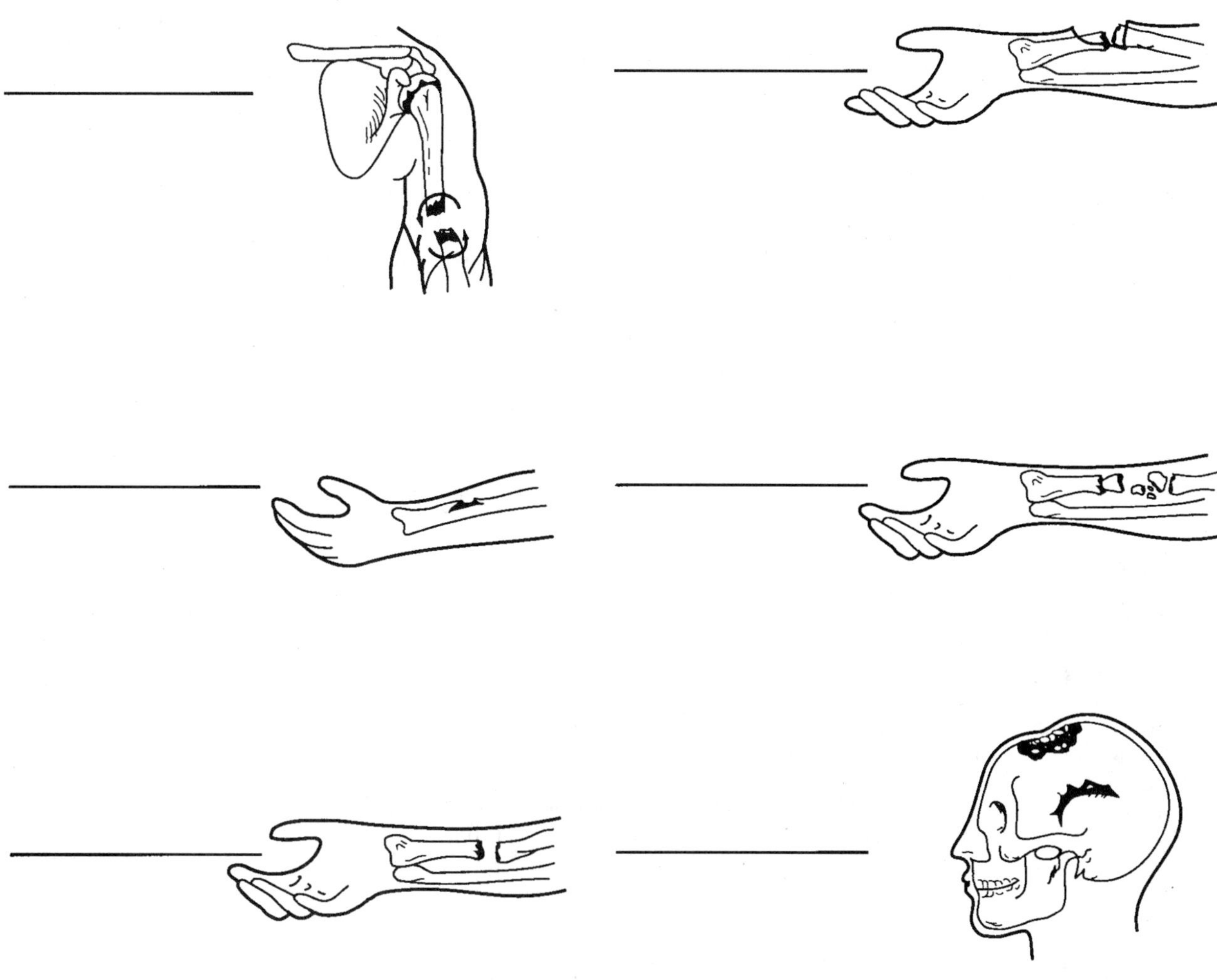

Figure 6.6

6. For each of the following statements about bone breakage and the repair process that is true, insert T in the answer blank. For false statements, correct the underlined words by inserting the correct words in the answer blanks.

________________ 1. A <u>hematoma</u> usually forms at a fracture site.

________________ 2. Deprived of nutrition, <u>osteocytes</u> at the fracture site die.

________________ 3. Nonbony debris at the fracture site is removed by <u>fibroblasts</u>.

________________ 4. <u>Osteocytes</u> produce collagen fibers that span the break.

________________ 5. Osteoblasts from the <u>medullary cavity</u> migrate to the fracture site.

________________ 6. The <u>fibrocartilaginous callus</u> is the first repair mass to splint the broken bone.

________________ 7. The bony callus is composed of <u>compact</u> bone.

Homeostatic Imbalances of Bone

1. Circle the term that does not belong in each of the following groupings.

1. Bacterial infection Osteoporosis Inflammation Osteomyelitis
2. Osteomalacia Elderly Vertebral compression fractures Osteoporosis
3. Increased reabsorption Decreased density Paget's disease Elderly
4. Soft bones Rickets Osteomalacia Porous bones
5. Bone-filled marrow cavity Pagetic bone Osteomalacia Bone thickenings
6. Rickets Calcified epiphyseal discs Vitamin D deficiency Children

THE INCREDIBLE JOURNEY:

A Visualization Exercise for the Skeletal System

. . . you look around and find yourself examining the stalagmite- and stalactite-like structures that surround you.

1. Complete the narrative by inserting the missing words in the answer blanks.

1. ____________
2. ____________
3. ____________
4. ____________
5. ____________
6. ____________
7. ____________
8. ____________

For this journey you are miniaturized and injected into the interior of the largest bone of your host's body, the __(1)__. Once inside this bone, you look around and find yourself examining the stalagmite- and stalactite-like structures that surround you. Although you feel as if you are in an underground cavern, you know that it has to be bone. Because the texture is so full of holes, it obviously is __(2)__ bone.

Although the arrangement of these bony spars seems to be haphazard, as if someone randomly had dropped straws, they are precisely arranged to resist points of __(3)__. All about you is frantic, hurried activity. Cells are dividing rapidly, nuclei are being ejected, and disc-like cells are appearing. You decide that these disc-like cells are __(4)__ and that this is the __(5)__ cavity. As you explore further, strolling along the edge of the cavity, you spot many tunnels leading into the solid bony area on which you are walking. Walking into one of these drainpipelike openings, you notice that it contains a glistening white ropelike structure, a __(6)__, and blood vessels running the length of the tube. You eventually come to a point in the channel where the horizontal passageway joins with a vertical passage that runs with the longitudinal axis of the bone. This is obviously a __(7)__ canal. You would like to see how nutrients are brought into __(8)__ bone, so you decide

______________________ 9. to follow this channel. Reasoning that there is no way you can possibly scale the slick walls of the channel, you leap up and

______________________ 10. grab onto a white cord hanging down its length. Because it is easier to slide down than to try to climb up the cord, you begin

______________________ 11. to lower yourself, hand over hand. During your descent, you notice small openings in the wall, which are barely large enough

______________________ 12. for you to wriggle through. You conclude that these are the

__(9)__ that connect all the __(10)__ to the nutrient supply in the central canal. You decide to investigate one of these tiny openings and begin to swing on your cord, trying to get a foothold on one of the openings. After managing to anchor yourself, and squeezing into an opening, you use a flashlight to illuminate the passageway in front of you. You are startled by a giant cell with many dark nuclei that appears to be plastered around the entire lumen directly ahead of you. As you watch this cell, the bony material beneath it, the __(11)__, begins to liquefy. The cell apparently is a bone-digesting cell, or __(12)__, and as you are unsure whether or not its enzymes can also liquefy you, you slither backward hurriedly and begin your trek back to your retrieval site.

CHALLENGING YOURSELF

At the Clinic

1. Mrs. Bruso, a woman in her 80s, stumbled slightly while walking and then felt a terrible pain in her hip. At the hospital, X rays reveal that her hip is broken and that she has compression fractures in her lower vertebral column and extremely low bone density in her vertebrae, hip bones, and femurs. What are the condition, cause, and treatment?

2. A child with a "red-hot" infected finger is brought into the clinic a week after the infection has set in. Why does the doctor order an X ray?

3. An X ray of the arm of an accident victim reveals a faint line curving around and down the shaft. What kind of fracture might this indicate?

4. Johnny, a child with unusually short stature, comes in for diagnostic tests. The tests reveal that he has abnormal collagen production. What name is given to this condition?

5. Rita's bone density scan revealed that she has osteoporosis. Her physician prescribed a drug that inhibits osteoclast activity. Explain this treatment.

6. A woman taking triple the recommended calcium supplement is suffering from lower back pains. X rays reveal that she has kidney stones. What is the name of the condition that causes this problem?

7. Larry, a 6-year-old boy, is already 5 feet tall. For what hormone will the pediatrician order a blood test?

8. Old Norse stories tell of a famous Viking named Egil who lived around 900 AD. His skull was greatly enlarged and misshapen, and the cranial bones were hardened and thickened (6 cm, or several inches, thick!). After he died, his skull was dug up and it withstood the blow of an ax without damage. In life, he had headaches from the pressure exerted by enlarged vertebrae on his spinal cord. What bone disorder did Egil probably have?

9. Ming posed the following question: "If the epiphyseal growth plates are growing so fast, why do they stay thin? Growing things are supposed to get larger or thicker, but these plates remain the same thickness." How would you answer her?

Stop and Think

1. Which appears first, the blood vessels within an osteon or the lamellae of bone matrix around the central canal?
2. Is there any form of interstitial growth in bone tissue? (Explain)
3. Contrast the arrangement and articulation of long bones and short bones.
4. Which type of cell, osteoclast or osteoblast, is more active in the medullary cavity of a growing long bone? On the internal surface of cranial bones?
5. Compare the vascularity of bone to that of the fibrous and cartilaginous tissues that precede it.
6. While walking home from a meeting of his adult singles support group, 52-year-old Ike broke a bone and damaged a knee cartilage in a fall. Assuming no special tissue grafts are made, which will probably heal faster, the bone or the cartilage? Why?
7. Eric, a radiologic tech student, was handed an X ray of the right femur of a 10-year-old boy. He noted that the area of the epiphyseal plate was beginning to show damage from osteomyelitis (the boy's diagnosis). Eric expressed his concern about the future growth of the boy's femur to the head radiologist. Need he have worried? Why or why not?
8. In several early cultures, bone deformation was a sign of social rank. For example, in the Mayan culture, head binding was done to give the skull a flattened forehead and a somewhat pointed top. Explain this in terms of Wolff's law.
9. Signs of successful (or at least nonfatal) brain surgery can be found on skulls many thousands of years old. How can an anthropologist tell that the patient didn't die from the surgery?
10. Explain (a) why cartilages are springy and (b) why cartilage can grow so quickly in the developing skeleton.
11. After splashdown in the Atlantic Ocean, the American astronaut team was brought to the Naval Hospital for checkups. X rays revealed decreased bone mass in all of them. Isn't this surprising in view of the fact that they exercised regularly in the space capsule? Explain.

COVERING ALL YOUR BASES

Multiple Choice

Select the best answer or answers from the choices given.

1. Important bone functions include:
   A. support of the pelvic organs
   B. protection of the brain
   C. providing levers for movement of the limbs
   D. protection of the skin and limb musculature
   E. storage of water
2. Which of the following are correctly matched?
   A. Short bone—wrist
   B. Long bone—leg
   C. Irregular bone—sternum
   D. Flat bone—cranium
3. Terms that can be associated with any type of bone include:
   A. periosteum
   B. diaphysis
   C. diploë
   D. cancellous bone
   E. medullary cavity
4. Which would be common locations of osteoblasts?
   A. Osteogenic layer of periosteum
   B. Lining of red marrow spaces
   C. Covering articular cartilage
   D. Lining central canals
   E. Aligned with Sharpey's fibers

5. Which of the listed bone markings are sites of muscle or ligament attachment?
   A. Trochanter
   B. Meatus
   C. Facet
   D. Spine
   E. Condyle

6. Which of the following are openings or depressions?
   A. Fissure
   B. Tuberosity
   C. Meatus
   D. Fossa
   E. Tubercle

7. A passageway connecting neighboring osteocytes in an osteon is a:
   A. central canal
   B. lamella
   C. lacuna
   D. canaliculus
   E. perforating canal

8. Between complete osteons are remnants of older, remodeled osteons known as:
   A. circumferential lamellae
   B. concentric lamellae
   C. interstitial lamellae
   D. lamellar bone
   E. woven bone

9. Which of these could be found in cancellous bone?
   A. Osteoid
   B. Trabeculae
   C. Canaliculi
   D. Central canals
   E. Osteoclasts

10. Elements prominent in osteoblasts include:
    A. rough ER
    B. secretory vesicles
    C. lysosomes
    D. smooth ER
    E. heterochromatin

11. Which of the following are prominent in osteoclasts?
    A. Golgi apparatus
    B. Lysosomes
    C. Microfilaments
    D. Exocytosis

12. Endosteum is in all these places, except:
    A. around the exterior of the femur
    B. on the trabeculae of spongy bone
    C. lining the central canal of an osteon
    D. often directly touching bone marrow

13. Which precede(s) intramembranous ossification?
    A. Chondroblast activity
    B. Mesenchymal cells
    C. Woven bone
    D. Collagen formation
    E. Osteoid formation

14. Which of the following is (are) part of the process of endochondral ossification and growth?
    A. Vascularization of the fibrous membrane surrounding the cartilage template
    B. Formation of diploë
    C. Destruction of cartilage matrix
    D. Appositional growth
    E. Mitosis of chondroblasts

15. What is the earliest event (of those listed) in endochondral ossification?
    A. Ossification of proximal epiphysis
    B. Appearance of the epiphyseal plate
    C. Invasion of the shaft by the periosteal bud
    D. Cavitation of the cartilage shaft
    E. Formation of secondary ossification centers

16. Which zone of the epiphyseal plate is most influenced by sex hormones?
    A. Zone of resting cartilage
    B. Zone of hypertrophic cartilage
    C. Zone of proliferating cartilage
    D. Zone of calcification

17. The region active in appositional growth is:
    A. osteogenic layer of periosteum
    B. within central canals
    C. endosteum of red marrow spaces
    D. internal callus
    E. epiphyseal plate

18. Deficiency of which of the following hormones will cause dwarfism?
    A. Growth hormone
    B. Sex hormones
    C. Thyroid hormones
    D. Calcitonin
    E. Parathyroid hormone

19. A remodeling unit consists of:
    A. osteoblasts
    B. osteoid
    C. osteocytes
    D. osteoclasts
    E. chondroblasts

20. The calcification front marks the location of:
    A. newly formed osteoid
    B. newly deposited hydroxyapatite
    C. actively mitotic osteoblasts
    D. active osteoclasts
    E. the activity of alkaline phosphatase

21. A deficiency of calcium in the diet would lead to:
- A. an increase of parathyroid hormone in the blood
- B. an increase in calcitonin secretion
- C. an increase in somatomedin levels in the blood
- D. increased secretion of growth hormone

22. Ionic calcium plays a role in:
- A. the transmission of nerve impulses
- B. blood clotting
- C. muscle contraction
- D. cytokinesis
- E. the activity of sudoriferous glands

23. Which of the following is not associated with Wolff's law?
- A. Compression
- B. Gravity
- C. Growth hormone
- D. Orientation of trabeculae
- E. Bone atrophy following paralysis

24. The initial event following a bone fracture is:
- A. formation of granulation tissue
- B. ossification of internal callus
- C. hemorrhage and hematoma formation
- D. remodeling
- E. endochondral ossification

25. Women suffering from osteoporosis are frequent victims of _______________ fractures of the vertebrae.
- A. compound
- B. spiral
- C. comminuted
- D. compression
- E. depression

26. Which of the listed bone disorders is (are) caused by hormonal imbalances?
- A. Osteomalacia
- B. Osteoporosis
- C. Gigantism
- D. Achondroplasia
- E. Paget's disease

27. At birth, ossification has progressed to the point where:
- A. only intramembranous ossification has begun
- B. endochondral ossification is complete
- C. some secondary ossification centers have appeared
- D. only major long bones have primary centers of ossification
- E. appositional growth has yet to begin

28. The growth spurt of puberty is triggered by:
- A. high levels of sex hormones
- B. the initial, low levels of sex hormones
- C. growth hormone
- D. parathyroid hormone
- E. calcitonin

Word Dissection

For each of the following word roots, fill in the literal meaning and give an example, using a word found in this chapter.

Word root	Translation	Example
1. call		
2. cancel		
3. clast		
4. fract		
5. lamell		
6. malac		
7. myel		
8. physis		
9. poie		
10. soma		
11. trab		

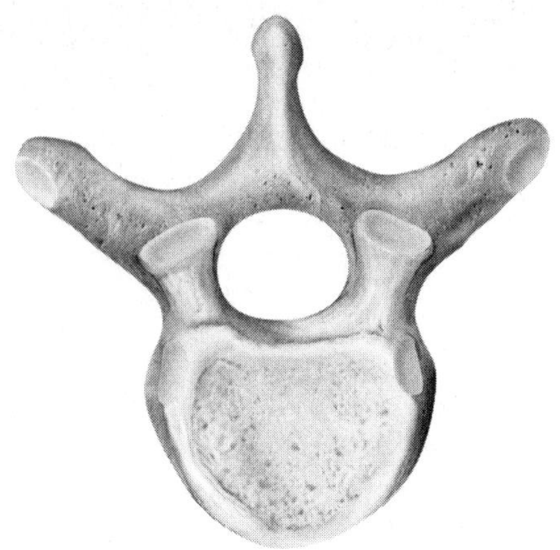

7

The Skeleton

Student Objectives

When you have completed the exercises in this chapter, you will have accomplished the following objectives:

PART 1: THE AXIAL SKELETON

1. Name the major parts of the axial and appendicular skeletons and describe their relative functions.

The Skull

2. Name, describe, and identify the skull bones. Identify their important markings.
3. Compare and contrast the major functions of the cranium and the facial skeleton.
4. Define the bony boundaries of the orbits, nasal cavity, and paranasal sinuses.

The Vertebral Column

5. Describe the structure of the vertebral column, list its components, and describe its curvatures.
6. Indicate a common function of the spinal curvatures and the intervertebral discs.
7. Discuss the structure of a typical vertebra and describe the regional features of cervical, thoracic, and lumbar vertebrae.

The Thoracic Cage

8. Name and describe the bones of the thoracic cage (bony thorax).
9. Differentiate true from false ribs.

PART 2: THE APPENDICULAR SKELETON

The Pectoral (Shoulder) Girdle

10. Identify bones forming the pectoral girdle and relate their structure and arrangement to the function of this girdle.
11. Identify important bone markings on the pectoral girdle.

The Upper Limb

12. Identify or name the bones of the upper limb and their important markings.

The Pelvic (Hip) Girdle

13. Name the bones contributing to the os coxae and relate the pelvic girdle's strength to its function.
14. Describe differences in the male and female pelves and relate these to functional differences.

The Lower Limb

15. Identify the lower limb bones and their important markings.
16. Name the arches of the foot and explain their importance.

The human skeleton is composed of 206 bones, which form a strong, flexible framework that supports the body and protects the vital organs. The bones articulate at joints, forming a complex system of levers stabilized by ligaments and activated by muscles. This system provides an extraordinary range of body movements.

The skeleton is, by evolution and appearance, constructed of two coordinated divisions. The axial skeleton is made up of the bones located along the body's longitudinal axis and center of gravity. The appendicular skeleton consists of the bones of the shoulders, hips, and limbs, appended to the axial division.

Topics for student study and review in Chapter 7 include the identification, location, and function of anatomically important bones, as well as the changes that occur in the bones of the skeleton throughout life.

BUILDING THE FRAMEWORK

1. Figure 7.1 on the opposite page illustrates an articulated skeleton. Identify all bones or groups of bones by writing the correct labels at the leader lines. Then, using two different colors, color the bones of the axial and appendicular skeletons.

○ Axial skeleton ○ Appendicular skeleton

2. Take a moment to go over some basic skull "geography" by matching the key choices to the skull features described below.

Key Choices

A. Base E. Fossae

B. Calvaria F. Orbits

C. Cranial cavity G. Paranasal sinuses

D. Facial bones H. Vault

__________ 1. The superolateral parts of the skull (2 choices)

__________ 2. The floor of the skull

__________ 3. Air-filled cavities that lighten the skull

__________ 4. Concavities in the skull floor that support parts of the brain

__________ 5. House the eyes

__________ 6. Present most special sense organs in the anterior position

__________ 7. Encloses the brain

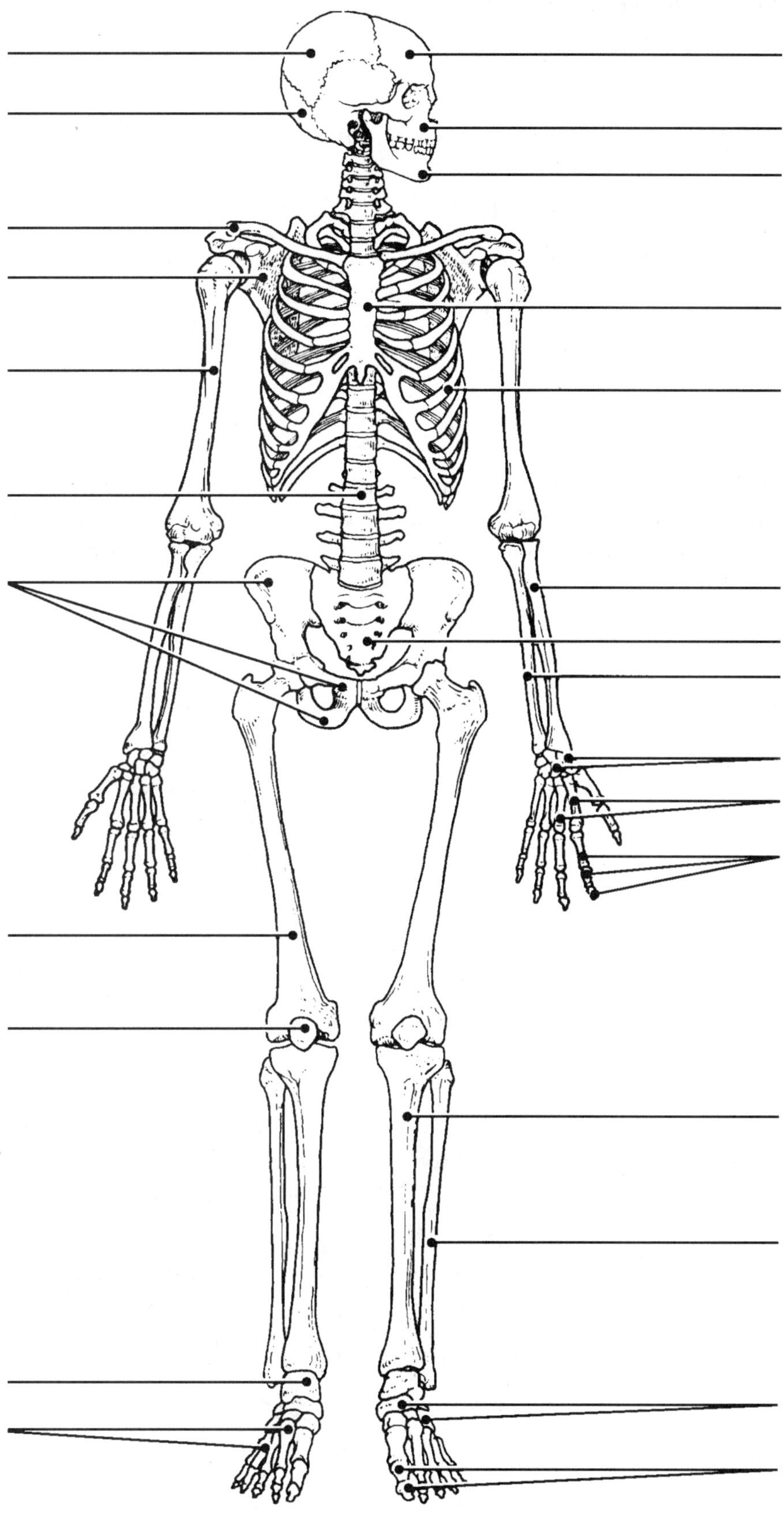

Figure 7.1

PART 1: THE AXIAL SKELETON

The Skull

1. Circle all of the bone names that represent *cranial* bones.

A. Ethmoid
B. Frontal
C. Hyoid
D. Inferior conchae
E. Lacrimal
F. Mandible
G. Maxillary
H. Nasal
I. Occipital
J. Palatine
K. Parietal
L. Sphenoid
M. Temporal
N. Vomer
O. Zygomatic

2. Figure 7.2 shows the lateral (A), inferior (B), and anterior (C) views of the skull. Select different colors for the bones below, and color the coding circles and corresponding bones. Complete views A, B, and C by labeling the bone markings indicated by leader lines.

○ Frontal
○ Parietal
○ Mandible
○ Maxilla
○ Sphenoid
○ Ethmoid
○ Temporal
○ Zygomatic
○ Palatine
○ Occipital
○ Nasal
○ Lacrimal
○ Vomer

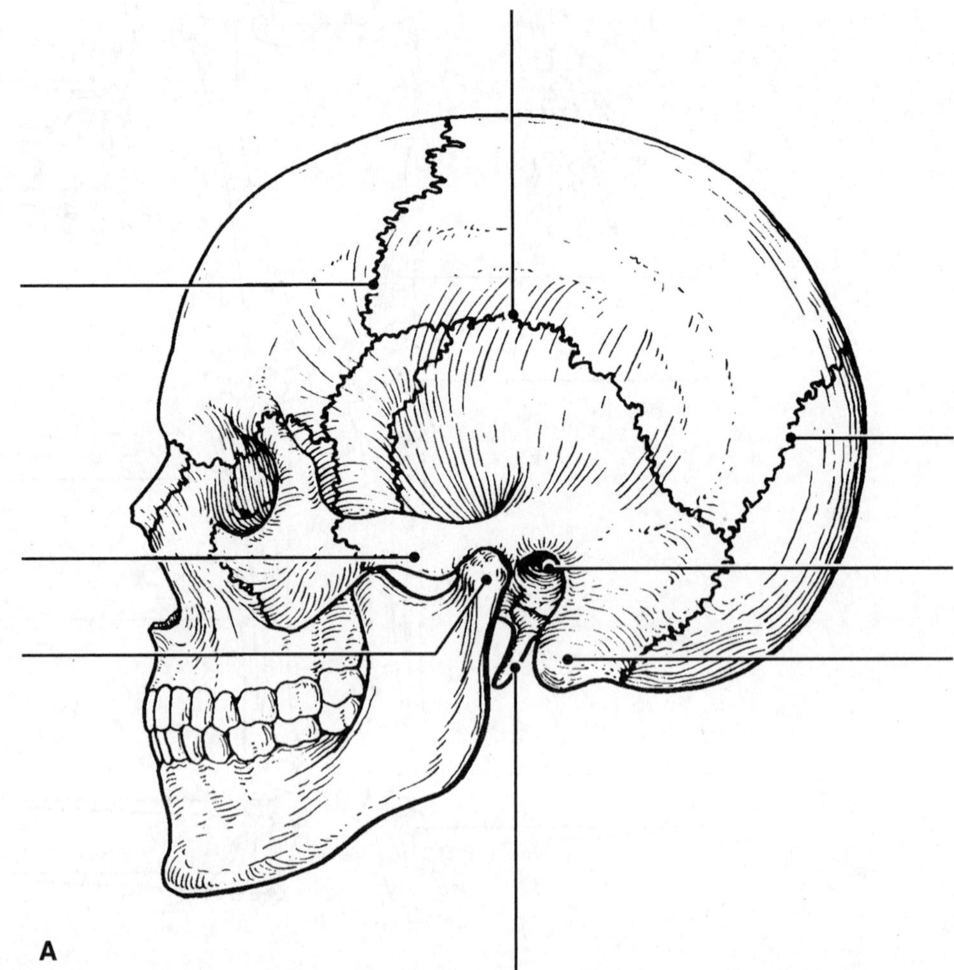

Figure 7.2

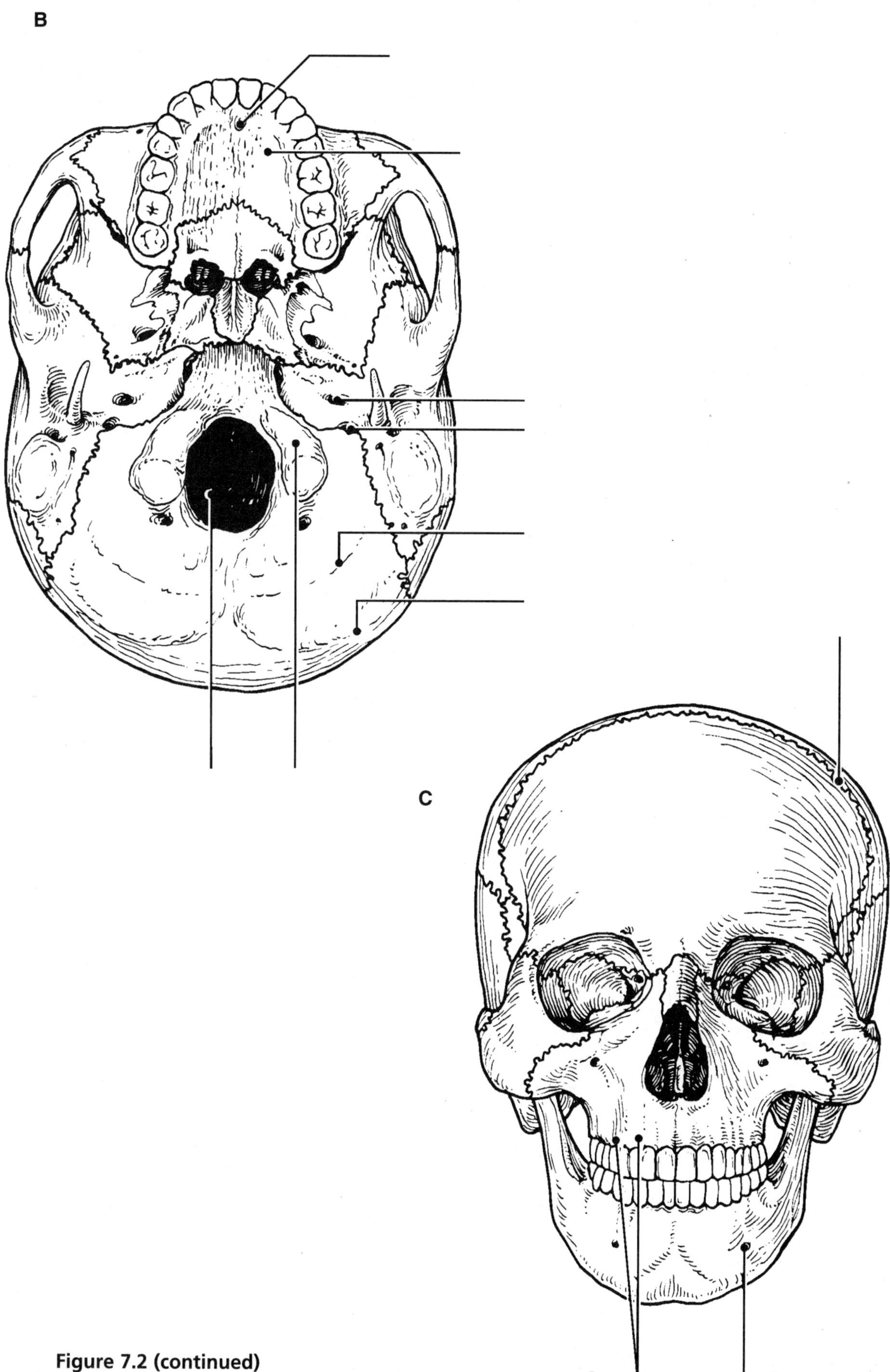

Figure 7.2 (continued)

3. Match the bone names listed in the key choices with the following bone descriptions. (Note that some descriptions apply to more than one bone.)

Key Choices

A. Ethmoid
B. Frontal
C. Hyoid
D. Inferior conchae
E. Lacrimal
F. Mandible
G. Maxillary
H. Nasal
I. Occipital
J. Palatine
K. Parietal
L. Sphenoid
M. Temporal
N. Vomer
O. Zygomatic

__________ 1. Connected by the frontal suture

__________ 2. Connected by the lambdoidal suture

__________ 3. Connected by the squamous suture

__________ 4. Connected by the sagittal suture

__________ 5. Cheekbone

__________ 6. Superolateral part of the cranium

__________ 7. Contains olfactory foramina

__________ 8. Posterior part of the hard palate

__________ 9. Posteriormost part of the cranium

__________ 10. Has two turbinates (conchae) as part of its structure; also contributes to the nasal septum

__________ 11. Foramen magnum contained here

__________ 12. Site of the sella turcica

__________ 13. Houses hearing and equilibrium receptors

__________ 14. Forms the bony eyebrow ridges and roofs of orbits

__________ 15. Forms the chin

__________ 16. The only bone connected to the skull by a freely movable joint

__________ 17. Site of the mastoid process

__________ 18. Contains the mental foramina

__________ 19. Neither a cranial nor a facial bone

__________ 20. __________ 21. __________ 22. __________ 23. Four bones containing paranasal sinuses

__________ 24. Bears an upward protrusion called the crista galli

__________ 25. Keystone bone of cranium

__________ 26. Tiny bones with openings for the tear ducts

__________ 27. Bony part of nasal septum

__________ 28. Site of external acoustic meatus

4. Anterior and lateral views of the skull, showing the positions of the sinuses, are shown in Figure 7.3. First, select different colors for each of the sinuses and use them to color the figure. Then, briefly answer the following questions concerning the sinuses.

1. What are sinuses? ______________________________

2. What purpose do they serve in the skull? ______________________________

____

3. Why are they so susceptible to infection? ______________________________

____

4. What is the function of the mucus-secreting mucosae?

____

○ Sphenoid sinus

○ Frontal sinus

○ Ethmoid sinuses

○ Maxillary sinus

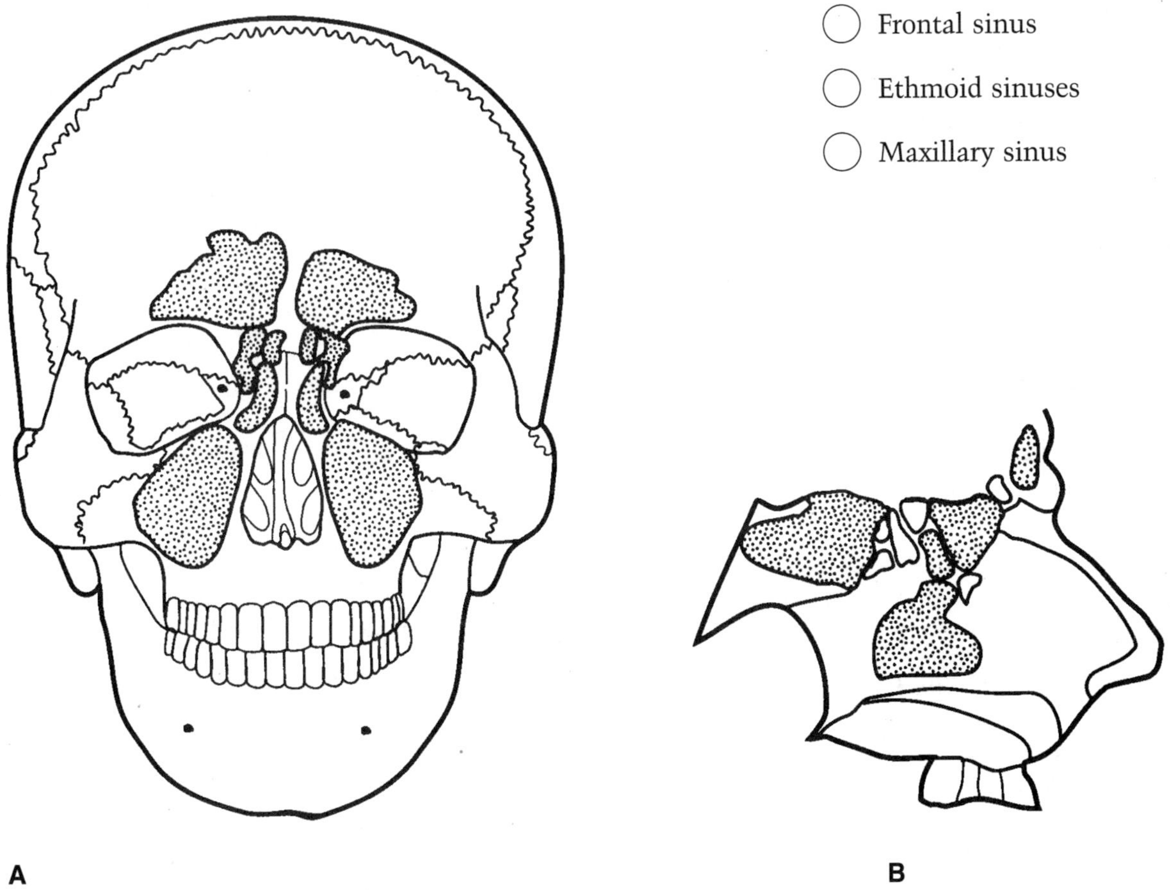

A

B

Figure 7.3

5. Complete the following table, which pertains to important bone markings on skull bones. Insert your answers in the spaces provided under each heading.

Bone marking(s)	Skull bone	Function of bone marking
	Mandible	Attachment for temporalis muscle
Carotid canal		
		Supports, partly encloses pituitary gland
Jugular foramen		
		Form anterior two-thirds of hard palate
		Attachment site for several neck muscles
Optic canals		
		Pass olfactory nerve fibers
		Attachment for ligamentum nuchae and neck muscles
Mandibular condyles	Mandibles	
Mandibular fossa		
Stylomastoid foramen		

The Vertebral Column

1. Using the key choices, correctly identify the vertebral parts and areas described below. Enter the appropriate answers in the answer blanks.

Key Choices

A. Body

B. Intervertebral foramina

C. Spinous process

D. Superior articular process

E. Transverse process

F. Vertebral arch

__________________ 1. Portion enclosing the nerve cord

__________________ 2. Weight-bearing part; also called the centrum

__________________ 3. Provide levers for the muscles to pull against

__________________ 4. Provide articulation points for the ribs

__________________ 5. Openings allowing spinal nerves to pass

2. Figure 7.4 is a lateral view of the vertebral column. Identify each numbered region of the column by listing in the answer blanks the region name first and then the specific vertebrae involved (for example, sacral region, S# to S#). Also identify the modified vertebrae indicated by numbers 6 and 7 in Figure 7.4. Select different colors for each vertebral region and use them to color the coding circles and the corresponding regions.

______________________ 1. ○

______________________ 2. ○

______________________ 3. ○

______________________ 4. ○

______________________ 5. ○

______________________ 6. ○

______________________ 7. ○

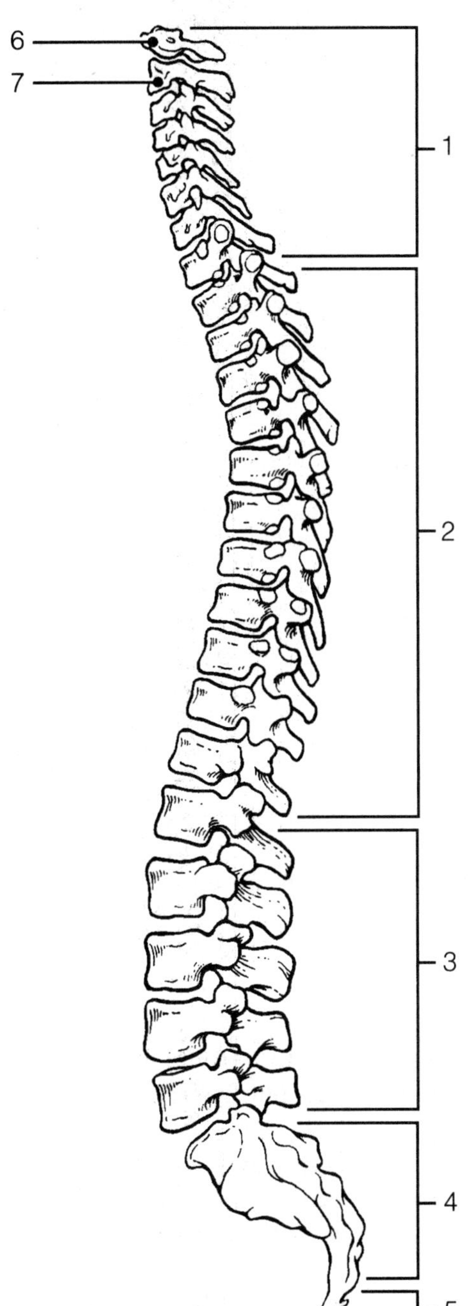

Figure 7.4

3. Match the vertebral column structures in Column B with the appropriate descriptions in Column A.

	Column A	Column B
________	1. Contain foramina in the transverse processes	A. Atlas (C_1)
________	2. Have articular facets for the ribs on their bodies and transverse processes	B. Axis (C_2)
________	3. Bifid spinous process	C. Cervical vertebrae (C_3–C_7)
________	4. A circle of bone; articulates superiorly with the occipital condyles	D. Coccyx
________	5. Shield-shaped composite bone; has alae	E. Lumbar vertebrae
________	6. Thickest bodies; short blunt spinous processes	F. Sacrum
________	7. Bears a peg-shaped dens that acts as a pivot	G. Thoracic vertebrae
________	8. Fused rudimentary vertebrae; tailbone	
________	9. Are 5 in number; not fused	
________	10. Are 12 in number; not fused	
________	11. No intervertebral disc between these two bones	

4. Figure 7.5 shows superior views of four types of vertebrae.

First, in the answer blank below each figure, indicate in which region of the spinal column it would be found. In addition, specifically identify Figure 7.5A.

Second, select four different colors and use them to color the coding circles and the corresponding structures, if present, for each type of vertebra.

Third, complete the statements relating to articulating surfaces by selecting the correct letters from the vertebral drawings in Figure 7.5 and inserting your responses in the answer blanks.

○ Body (centrum) ○ Transverse process ○ Transverse foramen

○ Spinous process ○ Vertebral foramen

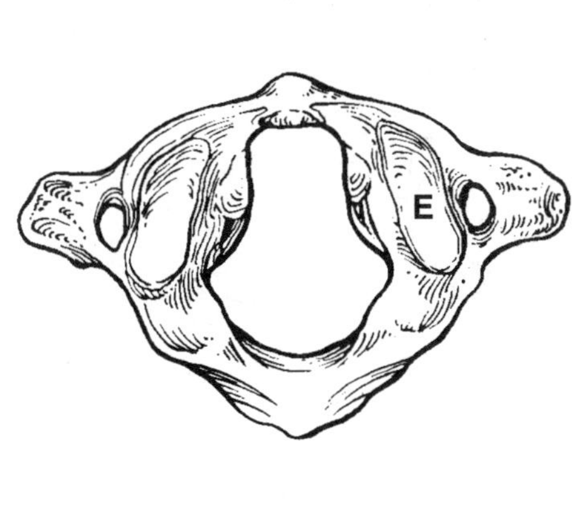

A ____________________

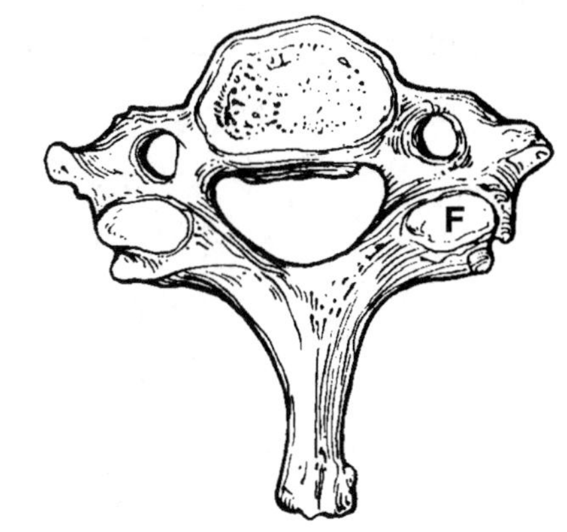

B ____________________

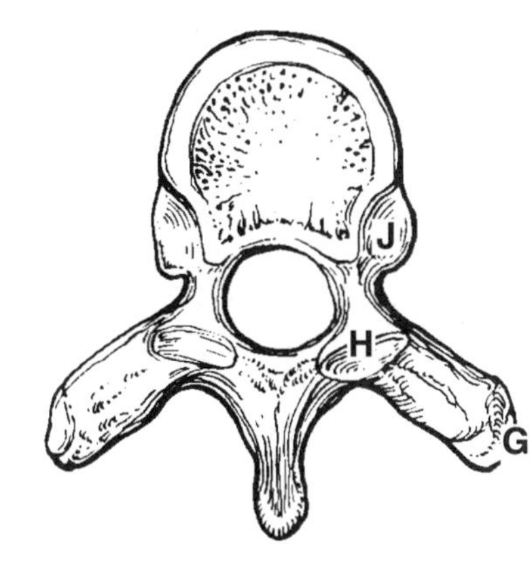

C ____________________

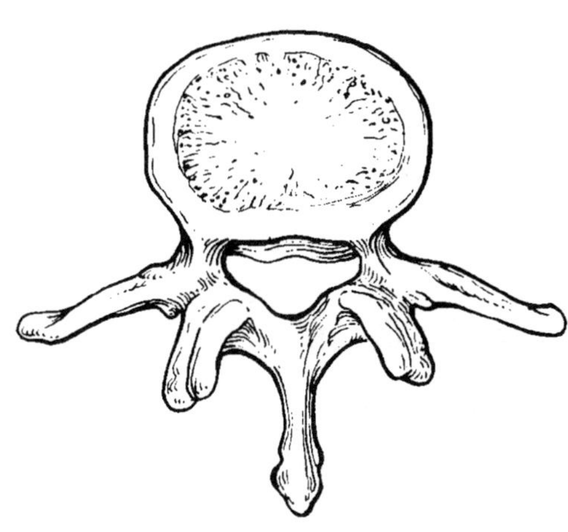

D ____________________

Figure 7.5

____________________ 1. Surface __(1)__ articulates with the head of a rib.

____________________ 2. Surface __(2)__ articulates with the inferior process of the superior cervical vertebra.

____________________ 3. Surface __(3)__ articulates with the tubercle of a rib.

____________________ 4. Surface __(4)__ receives the occipital condyle of the skull.

____________________ 5. Surface __(5)__ articulates with the inferior process of the superior thoracic vertebra.

5. Three views of spinal curvatures are shown in Figure 7.6. Identify each as being an example of one of the following conditions: lordosis, scoliosis, or kyphosis.

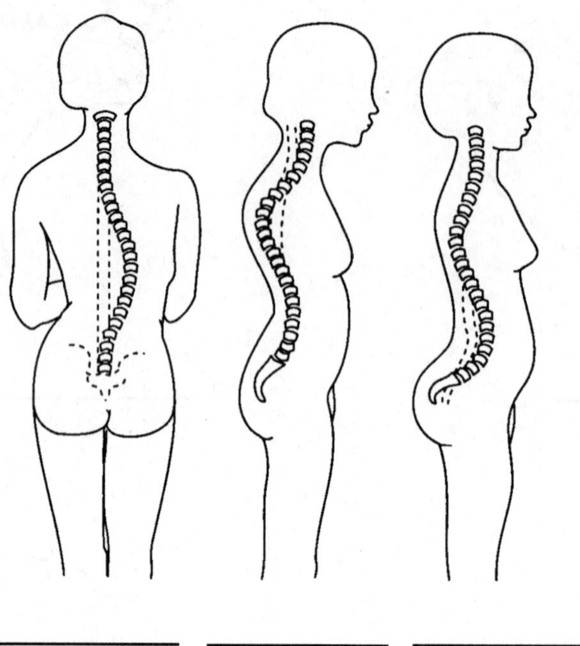

_______________ _______________ _______________

Figure 7.6

The Thoracic Cage

1. Complete the following statements referring to the thorax by inserting your responses in the answer blanks.

_______________ 1.
_______________ 2.
_______________ 3.
_______________ 4.
_______________ 5.
_______________ 6.
_______________ 7.
_______________ 8.

The organs protected by the thoracic cage include the __(1)__ and the __(2)__. Ribs 1–7 are commonly called __(3)__ ribs, whereas ribs 8–12 are called __(4)__ ribs. Ribs 11 and 12 are also called __(5)__ ribs. All ribs articulate posteriorly with the __(6)__ and most connect anteriorly to the __(7)__, either directly or indirectly.

The general shape of the thoracic cage is __(8)__.

2. Figure 7.7 is an anterior view of the thorax. Select different colors to identify the structures below and color the coding circles and corresponding structures. Then label the subdivisions of the sternum indicated by leader lines.

- ◯ Vertebrosternal ribs
- ◯ Vertebrochondral ribs
- ◯ Costal cartilages
- ◯ Sternum
- ◯ Vertebral ribs

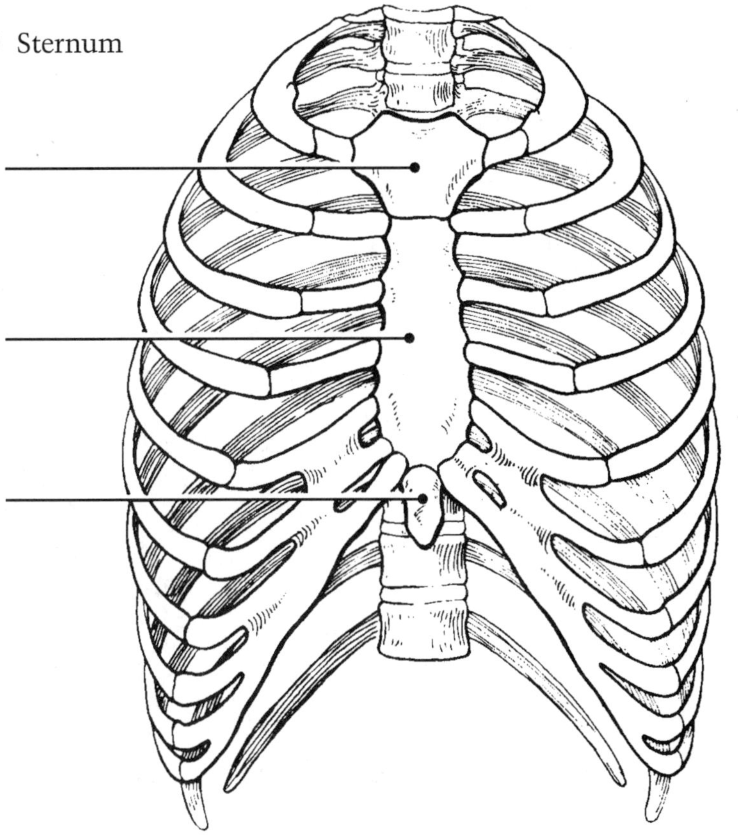

Figure 7.7

PART 2: THE APPENDICULAR SKELETON

The Pectoral (Shoulder) Girdle

1. Complete the following statements. Insert your answers in the answer blanks.

________________ 1.
________________ 2.
________________ 3.
________________ 4.
________________ 5.
________________ 6.

The lower limb bones are attached to the axial skeleton by the __(1)__. The upper limb bones are attached to the axial skeleton by the __(2)__. The two major functions of the axial skeleton are __(3)__ and __(4)__. The two major functions of the appendicular skeleton are __(5)__ and __(6)__.

2. Figure 7.8 is a posterior view of one bone of the pectoral girdle. Identify this bone by inserting its name on the line beside the figure. Select different colors for the structures listed below and use them to color the coding circles and the corresponding structures. Then, label the borders and angles indicated by leader lines.

   ◯ Spine

   ◯ Glenoid cavity

   ◯ Coracoid process

   ◯ Acromion

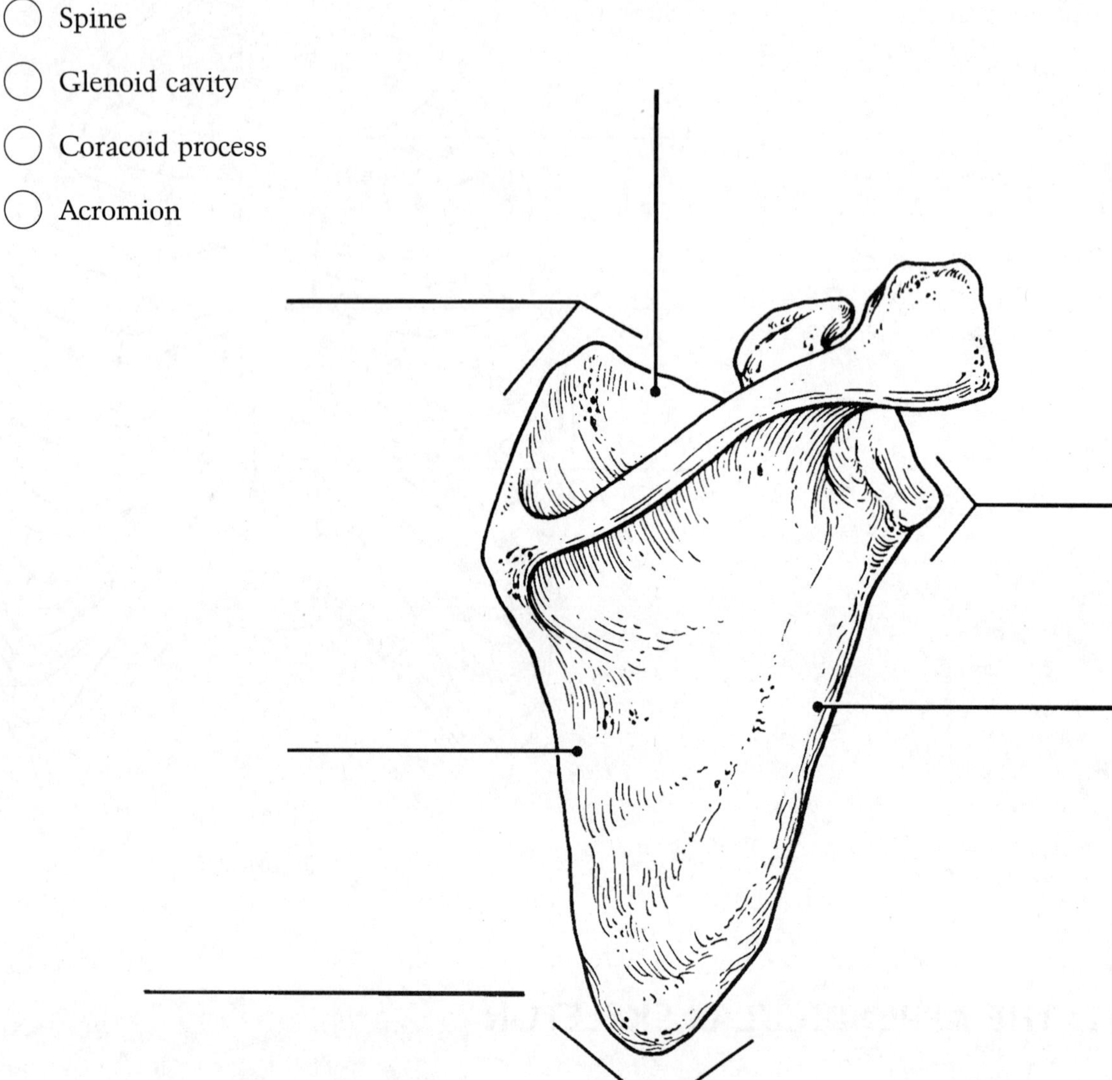

Figure 7.8

The Upper Limb

1. Figure 7.9 illustrates the anterior views of the three bones (A, B, C) of the left arm and forearm in the anatomical position. Identify each bone (A, B, and C) by writing its name at the leader line. Then label all the bone markings listed below by inserting leader lines and the names of the bone markings on Figure 7.9. Finally, color the coding circles and the bone markings.

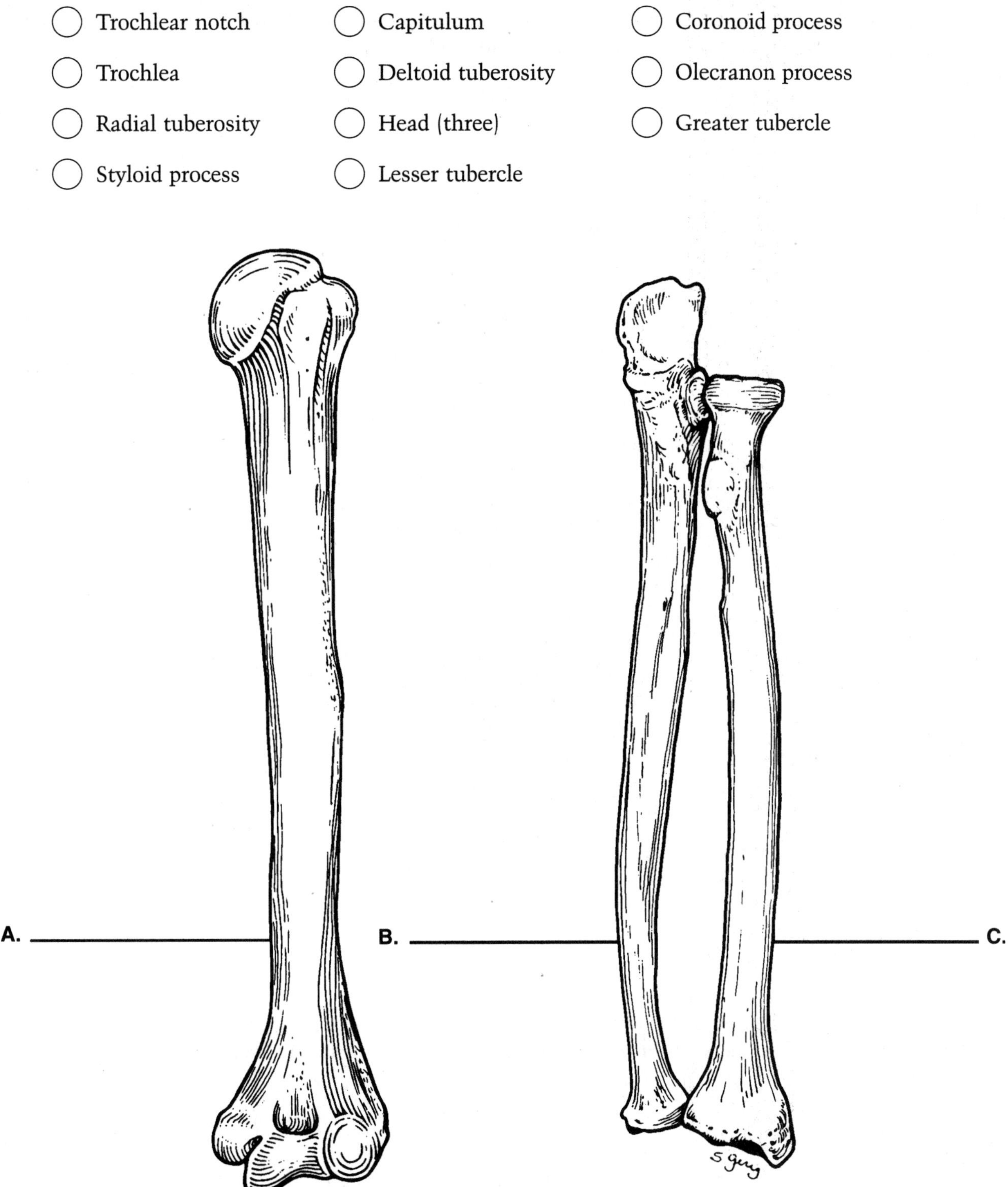
Trochlear notch
Capitulum
Coronoid process
Trochlea
Deltoid tuberosity
Olecranon process
Radial tuberosity
Head (three)
Greater tubercle
Styloid process
Lesser tubercle
A.
B.
C.

Figure 7.9

2. Figure 7.10 is a diagram of the hand. Select different colors for the following structures and use them to color the coding circles and diagram.

○ Carpals ○ Phalanges ○ Metacarpals

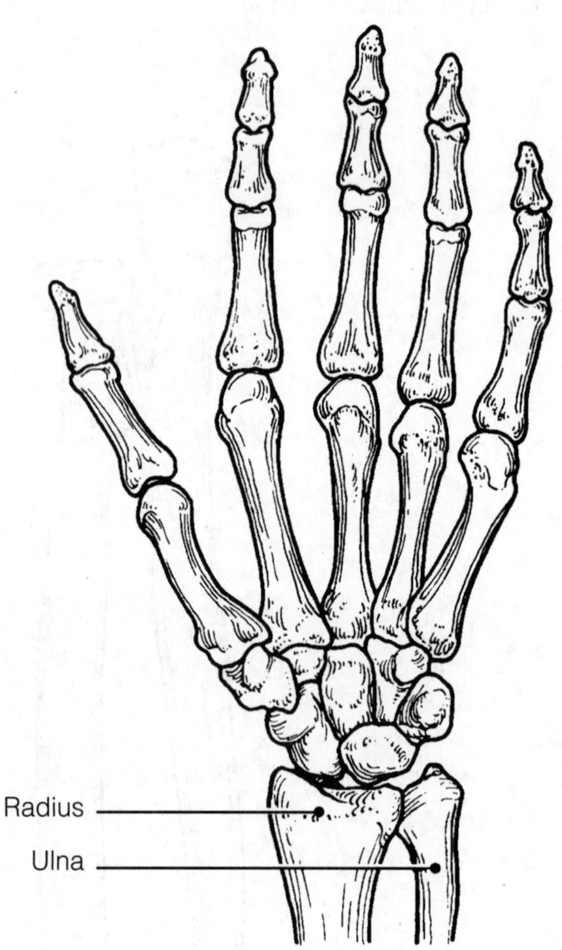

Figure 7.10

3. Compare the pectoral and pelvic girdles by choosing appropriate descriptive terms from the key choices. Insert the appropriate key letters in the answer blanks.

Key Choices

A. Flexibility D. Shallow socket for limb attachment

B. Massive E. Deep, secure socket for limb attachment

C. Lightweight F. Weight-bearing

1. Pectoral: _____ _____ _____ 2. Pelvic: _____ _____ _____

4. Using the key choices, identify the bones or bone markings according to the descriptions that follow. Insert the appropriate answers in the answer blanks.

Key Choices

A. Acromion	F. Coronoid fossa	K. Olecranon fossa	P. Scapula
B. Capitulum	G. Deltoid tuberosity	L. Olecranon process	Q. Sternum
C. Carpals	H. Glenoid cavity	M. Phalanges	R. Styloid process
D. Clavicle	I. Humerus	N. Radial tuberosity	S. Trochlea
E. Coracoid process	J. Metacarpals	O. Radius	T. Ulna

________________ 1. Projection on lateral surface of humerus to which the deltoid muscle attaches

________________ 2. Arm bone

________________ 3. ________________ 4. Bones of the shoulder girdle

________________ 5. ________________ 6. Bones of the forearm

________________ 7. Point where the scapula and clavicle connect

________________ 8. Pectoral girdle bone that is freely movable

________________ 9. Pectoral girdle bone that articulates anteriorly with the sternum

________________ 10. Socket in the scapula for the arm bone

________________ 11. Process above the glenoid fossa that permits muscle attachment

________________ 12. Commonly called the collarbone

________________ 13. Distal medial process of the humerus; joins the ulna

________________ 14. Medial bone of the forearm in anatomical position

________________ 15. Rounded knob on the humerus that articulates with the radius

________________ 16. Anterior depression, superior to the trochlea, that receives part of the ulna when the forearm is flexed

________________ 17. Forearm bone most involved in formation of the elbow joint

________________ 18. ________________ 19. Bones that articulate with the clavicle

________________ 20. Bones of the wrist

________________ 21. The fingers have three of these but the thumb has only two

________________ 22. Form the palm of the hand

The Pelvic (Hip) Girdle

1. Figure 7.11 is a diagram of the articulated pelvis.

First, identify the bones and bone markings indicated by leader lines on the figure. Also, label the dashed lines showing the dimensions of the true pelvis and the diameter of the false pelvis.

Second, select different colors for the structures listed at the right and use them to color the coding circles and the corresponding structures in the figure.

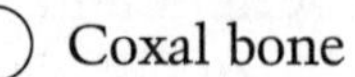

- ◯ Coxal bone
- ◯ Sacrum
- ◯ Pubic symphysis
- ◯ Acetabulum

Third, complete the illustration by labeling the following bone markings: obturator foramen, anterior superior iliac spine, iliac crest, ischial spine, pubic ramus, and pelvic brim.

Fourth, list three ways the female pelvis differs from the male pelvis in the answer blanks below the illustration.

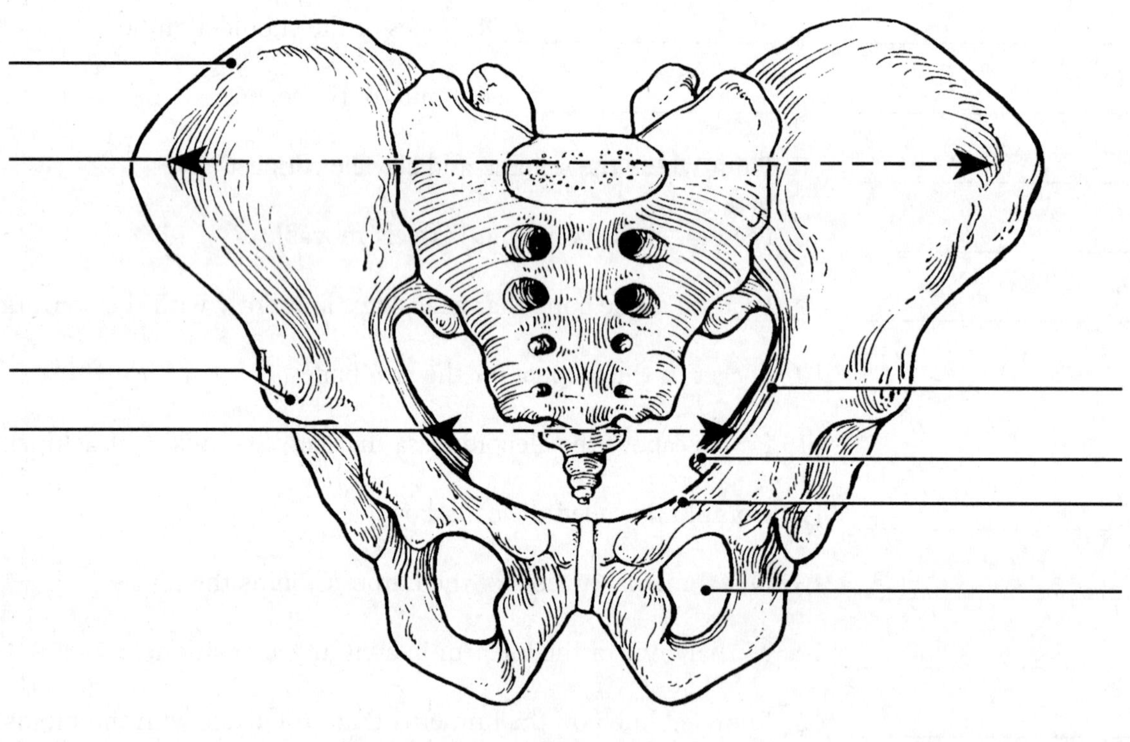

Figure 7.11

1. ______________________________

   ____

2. ______________________________

   ____

3. ______________________________

   ____

The Lower Limb

1. The bones of the thigh and leg are shown in Figure 7.12. Identify each and put your answers in the blanks below the diagrams. Select different colors for those structures below with a color-coding circle and use them to color the corresponding structures on the diagrams. Complete the illustration by inserting the following terms at the ends of the appropriate leader lines in the figure.

◯ Femur	◯ Tibia	◯ Fibula
Head of femur	Anterior border	Head of fibula
Intercondylar eminence	Lesser trochanter	Medial malleolus
Tibial tuberosity	Greater trochanter	Lateral malleolus

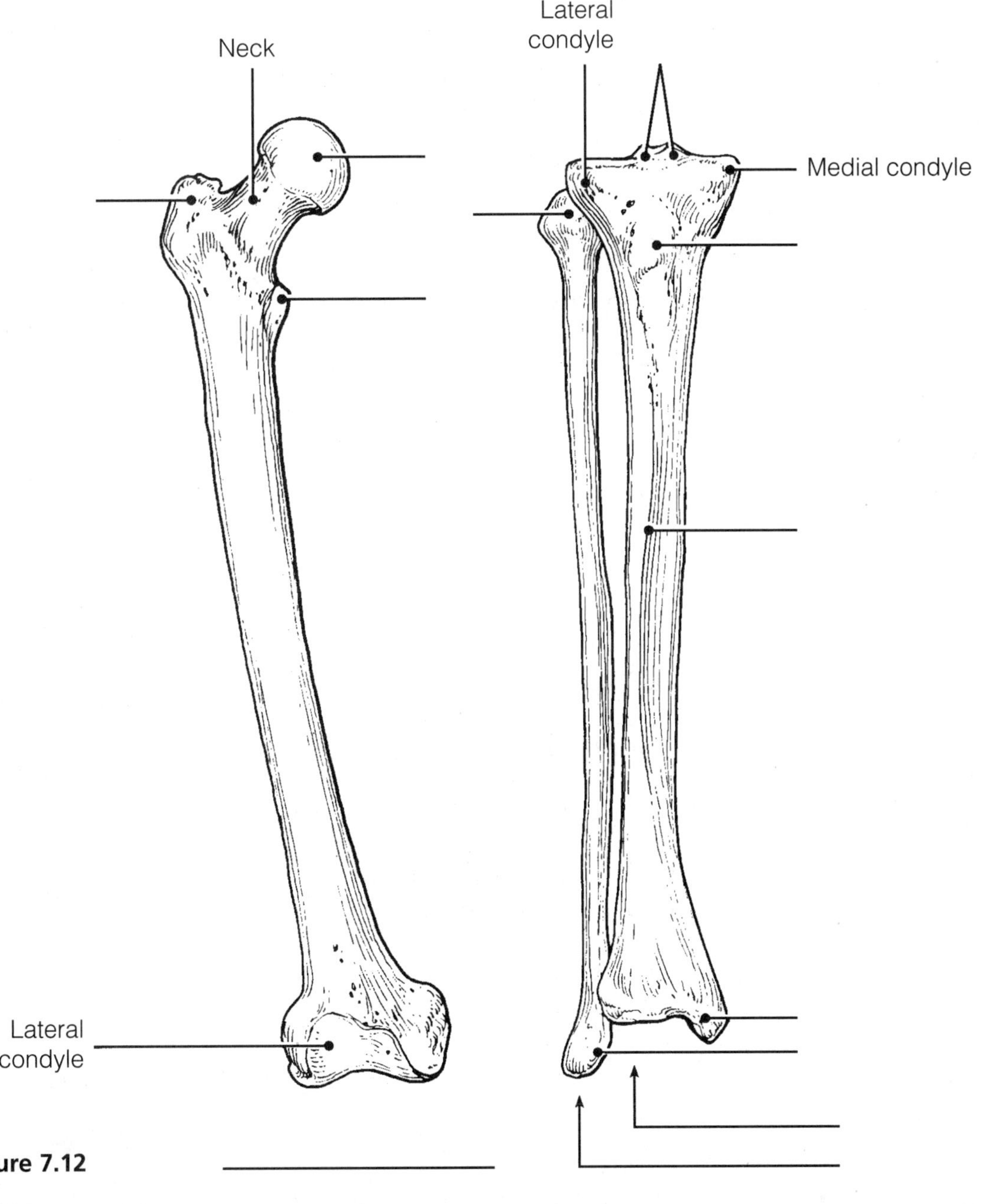

Figure 7.12

2. Match the bone names and markings in Column B to the descriptions in Column A.

	Column A	Column B
________	1. Fuse to form the coxal bone (hip bone)	A. Acetabulum
________	2. Receives the weight of the body when sitting	B. Calcaneus
________	3. Point where the coxal bones join anteriorly	C. Femur
________	4. Upper margin of the iliac bones	D. Fibula
________	5. Deep socket in the coxal bone that receives the head of the thigh bone	E. Gluteal tuberosity
________	6. Point where the axial skeleton attaches to the pelvic girdle	F. Greater sciatic notch
________	7. Longest bone in the body, articulates with the coxal bone	G. Greater and lesser trochanters
________	8. Lateral bone of the leg	H. Iliac crest
________	9. Medial bone of the leg	I. Ilium
________	10. Bones forming the knee joint	J. Ischial tuberosity
________	11. Point where the patellar ligament attaches	K. Ischium
________	12. Kneecap	L. Lateral malleolus
________	13. Shinbone	M. Lesser sciatic notch
________	14. Distal process on medial tibial surface	N. Medial malleolus
________	15. Process forming the outer ankle	O. Metatarsals
________	16. Heel bone	P. Obturator foramen
________	17. Bones of the ankle	Q. Patella
________	18. Bones forming the instep of the foot	R. Pubic symphysis
________	19. Opening in a coxal bone formed by the pubic and ischial rami	S. Pubis
________	20. Sites of muscle attachment on the proximal end of the femur	T. Sacroiliac joint
________	21. Tarsal bone that articulates with the tibia	U. Talus
		V. Tarsals
		W. Tibia
		X. Tibial tuberosity

_____________________ 22. Its tubercle serves as an attachment point for the inguinal ligament

_____________________ 23. Sports the greater and lesser trochanters and the linea aspera

CHALLENGING YOURSELF

At the Clinic

1. Jack, a young man, is treated at the clinic for an accident in which he hit his forehead. When he returns for a checkup, he complains that he can't smell anything. A hurried X ray of his head reveals a fracture. What part of which bone was fractured to cause his loss of the sense of smell?
2. The pediatrician at the clinic explains to parents of a newborn that their child suffers from cleft palate. She tells them that the normal palate fuses in an anterior-to-posterior pattern. The child's palatine processes have not fused. Have his palatine bones fused normally?
3. Janet, a 10-year-old girl, is brought to the clinic after falling out of a tree. An X ray shows she has small fractures of the transverse processes of T3 to T5 on the right side. Janet will be watched for what abnormal spinal curvature over the next several years?
4. A young woman is brought to the emergency room after falling on her outstretched arm. Her forearm is bruised both anteriorly and posteriorly along most of its length, but there is no fracture. The doctor explains that she dislocated the head of the radius, but it has relocated spontaneously. What persistent injury is responsible for the bruising?
5. Bart, a truck driver, tells his physician that he has a shooting pain in his right leg. What is a likely diagnosis?
6. Mr. Ogally, a heavy beer drinker with a large potbelly, complained of severe lower back pains. X rays showed displacement of his lumbar vertebrae. What is the condition called and what probably caused it in Mr. Ogally?
7. A man in his late 50s comes to the clinic, worried about a "hard spot" right below his breastbone. An X ray reveals nothing abnormal, although the man insists the spot wasn't there until the last few years. What is the explanation?
8. Antonio is hit in the face with a bad hop grounder during baseball practice. An X ray reveals multiple fractures of the bones around an orbit. Name the bones that form the margins of the orbit.
9. Malcolm was rushed to the emergency room after trying to break a fall with outstretched arms. The physician took one look at his shoulder and saw that he had a broken clavicle (and no other injury). Describe the position of Malcolm's shoulder. Malcolm was worried about injury to the main blood vessels to his arm (the subclavian vessels), but was told such injury was unlikely. Why could the doctor predict this with some assurance?
10. A 75-year-old woman and her 9-year-old granddaughter were in a train crash in which both sustained trauma to the chest. X rays showed that the grandmother had several fractured ribs but her granddaughter had none. Explain these surprisingly (?) different findings.
11. After having a severe cold accompanied by nasal congestion, Helen complained that she had a frontal headache and the right side of her face ached. What bony structures probably became infected by the bacteria or viruses causing the cold?

Stop and Think

1. Professor Ron Singer, who is a big kidder, pointed to the foramen magnum of the skull and said, "That is where the food goes. The food passes down through this hole when you swallow." Some of the students believed him, but others said this was a big mistake. Can you correct his statement?
2. What purpose is served by the unique structure of the turbinates (conchae)?
3. Name all the parts of the temporal bone that play a part in hearing.
4. Snakes can swallow prey with diameters larger than that of the snake's open mouth. What design modifications of the snake's jaw might allow this?
5. Does nodding the head "yes" involve movement at the same joint as shaking the head "no"? Explain your answer.
6. Look at a skeleton or a picture of the thoracic cage and try to figure out which ribs are most easily fractured.
7. The thoracic cage doesn't provide nearly as much protection for the heart as the cranium does for the brain. However, its more open structure is advantageous. How so?
8. The acetabulum is much deeper than the glenoid cavity. What is (are) the structure-function relationship(s) here?
9. Name the bones in order as prescribed by the mnemonic device "**S**ally **l**eft **t**he **p**arty **t**o **t**ake **C**athy **h**ome." Then identify the bone group, and indicate which four bones are more proximal.
10. List the bones in each of the three cranial fossae.
11. Bone X rays are sometimes used to determine if a person has reached his or her final height. What are the clinicians checking out?

COVERING ALL YOUR BASES

Multiple Choice

Select the best answer or answers from the choices given.

1. The type of tissue in vertebral ligaments is:
   A. fibrocartilage
   B. elastic cartilage
   C. elastic connective tissue
   D. dense regular connective tissue
   E. dense irregular connective tissue
2. Which of the following are classified as short bones?
   A. Distal phalanx
   B. Cuneiform
   C. Pisiform
   D. Patella
   E. Calcaneus
3. Flat bones include:
   A. ribs
   B. maxilla
   C. nasal
   D. sternum
   E. sphenoid
4. Which of the following bones are part of the axial skeleton?
   A. Vomer
   B. Clavicle
   C. Sternum
   D. Parietal
   E. Coxal bone
5. Bone pain behind the external auditory meatus probably involves the:
   A. maxilla
   B. ethmoid
   C. sphenoid
   D. temporal
   E. lacrimal

6. Bones that articulate with the sphenoid include:
   A. parietal
   B. vomer
   C. maxilla
   D. zygomatic
   E. ethmoid

7. Which parts of the temporal bone contribute to the calvaria?
   A. Squamous region
   B. Petrous region
   C. Mastoid region
   D. Tympanic region

8. Which of the following are bony landmarks of the anterior skull surface?
   A. Mental foramina
   B. Anterior cranial fossa
   C. Glabella
   D. Supraorbital foramen
   E. Frontal sinuses

9. Points of muscle, tendon, or ligament attachment include:
   A. superior nuchal lines
   B. greater wings of the sphenoid
   C. external occipital crest
   D. mastoid process
   E. coronoid process

10. Which of the following are part of the sphenoid?
    A. Crista galli
    B. Sella turcica
    C. Petrous portion
    D. Pterygoid process
    E. Lesser wings

11. Which of the following are found within the cranial cavity?
    A. Perpendicular plate of ethmoid
    B. Internal acoustic meatus
    C. Crista galli
    D. Cranial fossae
    E. Lacrimal fossa

12. Bones helping to form the oral cavity include:
    A. mandible
    B. vomer
    C. maxilla
    D. sphenoid
    E. palatine

13. Which of the following bones are at least partly covered by a mucous membrane?
    A. Zygomatic
    B. Temporal
    C. Ethmoid
    D. Occipital

14. Which of these start as separate bones in the fetus, but are fused in an adult skeleton?
    A. Coxal bone
    B. Vertebral ribs
    C. Mandible
    D. Coccyx
    E. Cuneiforms

15. The protective bony ring around the spinal cord is formed by several parts of the vertebra including:
    A. laminae
    B. centrum
    C. transverse process
    D. pedicles
    E. facet

16. Ribs articulate with the:
    A. transverse processes of lumbar vertebrae
    B. manubrium
    C. spinous processes of thoracic vertebrae
    D. xiphoid process
    E. centrum of cervical vertebrae

17. Structural characteristics of all cervical vertebrae are:
    A. small body
    B. bifid spinous process
    C. transverse foramina
    D. small vertebral foramen
    E. costal facets

18. What types of movement are possible between thoracic vertebrae?
    A. Flexion
    B. Rotation
    C. Some lateral flexion
    D. Circumduction
    E. Extension

19. Which humeral process articulates with the radius?
    A. Trochlea
    B. Greater tubercle
    C. Lesser tubercle
    D. Capitulum
    E. Olecranon fossa

20. A coronoid process is found on the:
    A. maxilla
    B. mandible
    C. scapula
    D. humerus
    E. ulna

21. Which of the following bones exhibit a styloid process?
    A. Hyoid
    B. Temporal
    C. Humerus
    D. Radius
    E. Ulna

22. Which bones articulate with metacarpals?
    A. Hamate
    B. Pisiform
    C. Proximal phalanges
    D. Capitate
    E. Cuneiform

23. Hip bone markings include:
    A. ala
    B. sacral hiatus
    C. gluteal surface
    D. pubic ramus
    E. fovea capitis

24. Which of the following bones or bone parts articulate with the femur?
    A. Ischial tuberosity
    B. Pubis
    C. Patella
    D. Fibula
    E. Tibia

25. The length ratio of the head and trunk to the lower limbs becomes 1 to 1:
    A. at birth
    B. by 10 years of age
    C. at puberty
    D. when the epiphyseal plates fuse
    E. never

26. Which pair of ribs inserts on the sternum at the sternal angle?
    A. First
    B. Second
    C. Third
    D. Fourth

27. The inferior angle of the scapula is at the same level as the spinous process of vertebra:
    A. C_7
    B. C_5
    C. T_3
    D. T_7
    E. L_4

28. An important bony landmark that can be recognized by a distinct dimple in the skin is:
    A. posterior superior iliac spine
    B. styloid process of the ulna
    C. shaft of the radius
    D. acromion of the scapula

29. A blow to the cheek is most likely to break what superficial bone or bone part?
    A. Superciliary arches
    B. Zygomatic process
    C. Mandibular ramus
    D. Styloid process
    C. Mandibular ramus
    D. Styloid process

30. The name of the first cervical vertebra is:
    A. atlas
    B. axis
    C. occiput
    D. vertebra prominens

31. All these bony features are near or in the shoulder joint, except:
    A. acromion
    B. greater tubercle
    C. glenoid cavity
    D. anatomical neck of humerus
    E. deltoid tuberosity

32. All these bony features are near or in the hip joint, except:
    A. acetabulum
    B. sacral promontory
    C. greater trochanter
    D. neck of femur

33. Which of the following bones are not paired?
    A. Parietal
    B. Frontal
    C. Sternum
    D. Pubis
    E. Calcaneus

Word Dissection

For each of the following word roots, fill in the literal meaning and give an example, using a word found in this chapter.

Word root	Translation	Example
1. acetabul		
2. alve		
3. append		
4. calv		
5. clavicul		
6. crib		
7. den		
8. ethm		
9. glab		
10. ham		
11. ment		
12. odon		
13. pal		
14. pect		
15. pelv		
16. pis		
17. pter		
18. scaph		
19. skeleto		
20. sphen		
21. styl		
22. sutur		
23. vert		
24. xiph		

8

Joints

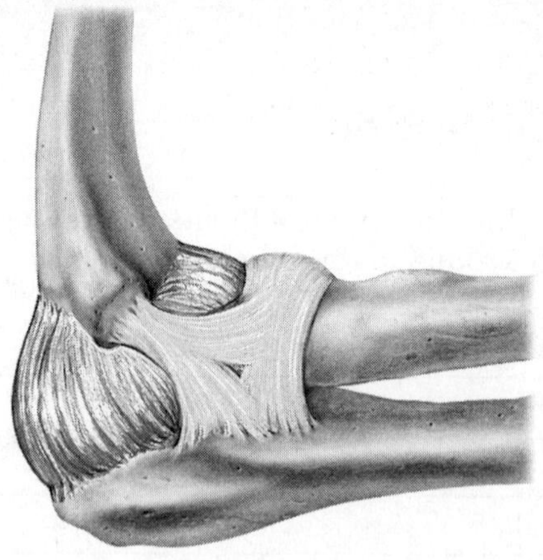

Student Objectives

When you have completed the exercises in this chapter, you will have accomplished the following objectives:

Classification of Joints

1. Define joint or articulation.
2. Classify joints structurally and functionally.

Fibrous Joints

3. Describe the general structure of fibrous joints. Name and give an example of each of the three common types of fibrous joints.

Cartilaginous Joints

4. Describe the general structure of cartilaginous joints. Name and give an example of each of the two common types of cartilaginous joints.

Synovial Joints

5. Describe the structural characteristics of synovial joints.
6. Compare the structures and functions of bursae and tendon sheaths.
7. List three natural factors that stabilize synovial joints.
8. Name and describe (or perform) the common body movements.
9. Name and provide examples of the six types of synovial joints based on the type of movement(s) allowed.
10. Describe the elbow, knee, hip, jaw, and shouder joints in terms of articulating bones, anatomical characteristics of the joint, movements allowed, and joint stability.

Homeostatic Imbalances of Joints

11. Name the most common joint injuries and dicuss the symptoms and problems associated with each.
12. Compare and contrast the common types of arthritis.
13. Describe the cause and consequences of Lyme disease.

Joints are structures that connect adjoining bones. Except for the hyoid bone, and sesamoid bones like the patella, each bone contacts at least one other bone at a joint. A typical joint includes the adjacent surfaces of the bones and the fibrous tissue or ligaments that bind the bones together. In addition to binding the bones together, joints provide the skeleton with the flexibility to permit body movements.

Because joints vary in structure and range of motion, it is convenient to classify them on the basis of their structure (fibrous, cartilaginous, or synovial) and function (the degree of joint movement allowed).

Topics for review in Chapter 8 include the classification of joints, the structures of selected synovial joints, joint impairment, and changes in the joints throughout life.

BUILDING THE FRAMEWORK

Classification of Joints

1. Write your answers to the following questions in the answer blanks.

1. What are the two major functions of joints? ________________________________

____

2. List three criteria used to classify joints. ________________________________

____

____

3. For each of the structural joint categories—fibrous, cartilaginous, and synovial—give the *most common* functional classification and describe the degree of movement.

____

____

____

Fibrous, Cartilaginous, and Synovial Joints

1. For each joint described below, select an answer from Key A. Then, if the Key A selection is *other than C* (synovial joint), classify the joint further by making a choice from Key B.

KEY A:	KEY B:	
A. Cartilaginous	1. Gomphosis	4. Syndesmosis
B. Fibrous	2. Suture	5. Synchondrosis
C. Synovial	3. Symphysis	6. Synostosis

_____ 1. Characterized by hyaline cartilage connecting the bony portions

_____ 2. All have a fibrous capsule lined with a synovial membrane surrounding a joint cavity

_____ 3. Bone regions united by fibrous connective tissue

_____ 4. Joints between skull bones

_____ 5. Joint between atlas and axis

_____ 6. Hip, elbow, knee, and intercarpal joints

_____ 7. Intervertebral joints (between vertebral bodies)

_____ 8. Pubic symphysis

_____ 9. All are reinforced by ligaments

_____ 10. Costosternal joints 2–7

_____ 11. Joint providing the most protection to underlying structures

_____ 12. Often contains a fluid-filled cushion

_____ 13. Child's epiphyseal plate made of hyaline cartilage

_____ 14. Most joints of the limbs

_____ 15. Teeth in body alveolar sockets

_____ 16. Joint between first rib and manubrium of sternum

_____ 17. Ossified sutures

_____ 18. Distal tibiofibular joint

2. Which structural joint type is *not* commonly found in the axial skeleton

and why not? ____

____

____

3. Figure 8.1 shows the structure of a typical synovial joint. Select different colors to identify and color the following areas. Then label the following: the more proximal epiphyseal line, the more distal epiphyseal line, spongy bone, periosteum.

◯ Articular cartilage of bone ends ◯ Synovial membrane

◯ Fibrous capsule ◯ Joint cavity

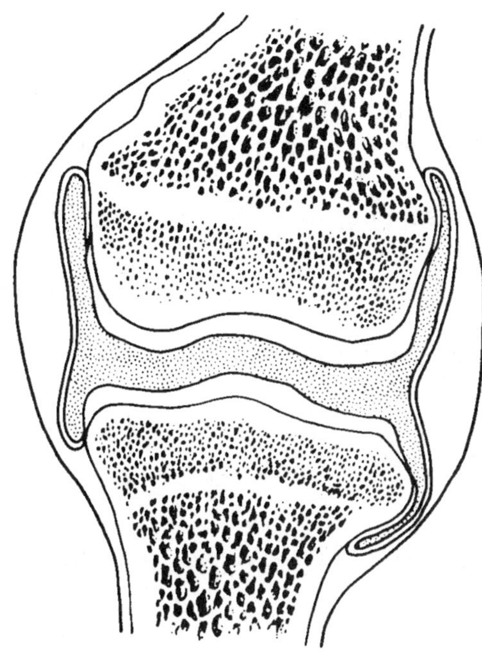

Figure 8.1

4. Match the key choices with the descriptive phrases below, to properly characterize selected aspects of synovial joints.

Key Choices

A. Articular cartilage C. Ligaments and fibrous capsule

B. Synovial fluid D. Muscle tendons

_____ 1. Keeps bone ends from crushing when compressed; resilient

_____ 2. Resists tension placed on joints

_____ 3. Lubricant that minimizes friction and abrasion of joint surfaces

_____ 4. Keeps joints from overheating

_____ 5. Helps prevent dislocation

5. Match the body movement terms in Column B with the appropriate description in Column A. (More than one choice may apply.)

	Column A	Column B
________	1. Movement along the sagittal plane that decreases the angle between two bones	A. Abduction
		B. Adduction
________	2. Movement along the frontal plane, away from the body midline; raising the arm laterally	C. Circumduction
		D. Depression
________	3. Circular movement around the longitudinal bone axis; shaking the head "no"	E. Dorsiflexion
________	4. Slight displacement or slipping of bones, as might occur between the carpals of the wrist	F. Elevation
		G. Eversion
________	5. Describing a cone-shaped pathway with the arm	H. Extension
		I. Flexion
________	6. Lifting or raising a body part; shrugging the shoulders	J. Gliding
________	7. Moving the hand into a palm-up (or forward) position	K. Inversion
		L. Plantar flexion
________	8. Movement of the superior aspect of the foot toward the leg; standing on the heels	M. Pronation
		N. Protraction
________	9. Turning the sole of the foot medially	O. Retraction
________	10. Movement of a body part anteriorly; jutting the lower jaw forward	P. Rotation
________	11. Common angular movements	Q. Supination

6. Figure 8.2 illustrates types of movements allowed by synovial joints. Match the letters on the figure with the types of movements listed below. Insert your answers in the answer blanks. Then color the drawing to suit your fancy.

________	1. Flexion	________	6. Protraction
________	2. Plantar flexion	________	7. Circumduction
________	3. Abduction	________	8. Adduction
________	4. Rotation	________	9. Extension
________	5. Pronation	________	10. Dorsiflexion

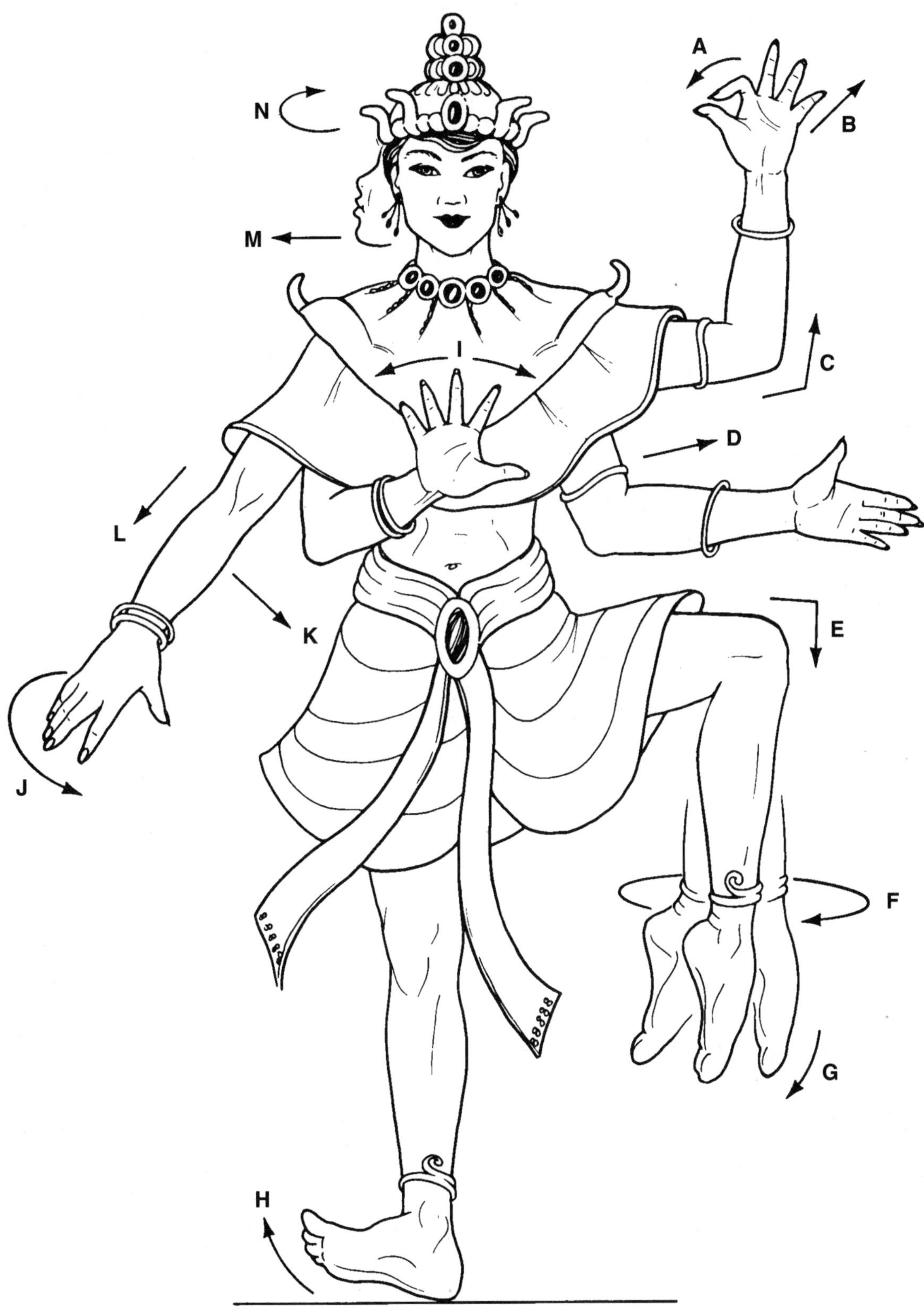

Figure 8.2

7. Match the joint types in Column B with the examples or descriptions of joints listed in Column A.

	Column A	Column B
________	1. Knuckles	A. Ball and socket
________	2. Sacroiliac joints	B. Condyloid
________	3. Intercarpal joints	C. Hinge
________	4. Femoropatellar joint of the knee	D. Pivot
________	5. Hip and shoulder joints	E. Plane
________	6. Radiocarpal	F. Saddle
________	7. Proximal tibiofibular joint	
________	8. Carpometacarpal joint of the thumb	
________	9. Elbow, knee, and interphalangeal joints	
________	10. Joint between C_1 and C_2	
________	11. Uniaxial joints	
________	12. Biaxial joints	
________	13. Multiaxial joints	
________	14. Nonaxial joints	

8. Circle the term that does not belong in each of the following groupings.

1. Pivot joint Uniaxial joint Multiaxial joint Atlas/axis joint
2. Ball-and-socket joint Multiaxial joint Hip joint Saddle joint
3. Tibiofibular joints Sutures Synostoses Gomphoses
4. Articular discs Multiaxial joint Largest joint in the body Knee joint
5. Saddle joint Carpometacarpal joint of thumb Elbow joint Biaxial joint
6. Amphiarthrotic Intercarpal joints Plane joints Nonaxial joints
7. Intervertebral joints Bursae Cartilaginous joints Symphyses
8. Hinge joint Condyloid joint Elbow joint Uniaxial joint
9. Muscle tendon reinforcement Shoulder joint Biaxial joint
   Most freely moving joint in the body

9. Several characteristics of specific synovial joints are described below. Identify each of the joints described by choosing a response from the key choices.

Key Choices

A. Elbow B. Hip C. Knee D. Shoulder

__________ 1. Rotator cuff muscles are important in stabilizing this joint; capsule reinforced only anteriorly by ligaments; articular surfaces shallow.

__________ 2. Three joints in one; capsule incomplete anteriorly; has menisci and intracapsular cruciate ligaments.

__________ 3. Capsule is loose; reinforced by medial and lateral collateral ligaments; articular surfaces most important in ensuring joint stability.

__________ 4. Articular surfaces deep and secure; capsule heavily reinforced by ligaments and muscle tendons; intracapsular ligamentum teres. Extremely stable joint.

10. Figure 8.3 shows diagrams of two synovial joints. Identify each joint by inserting the name of the joint in the blank below each diagram. Then select different colors and use them to color the coding circles and the structures in the diagrams. Finally, add labels and leader lines on the diagrams to identify the following structures: the ligamentum teres, the anterior cruciate ligament, the posterior cruciate ligament, the suprapatellar bursa, the subcutaneous prepatellar bursa, and the deep infrapatellar bursa.

○ Fibrous capsule (lined with synovial membrane)
○ Joint cavity
○ Articular cartilage of bone ends
○ Acetabular labrum

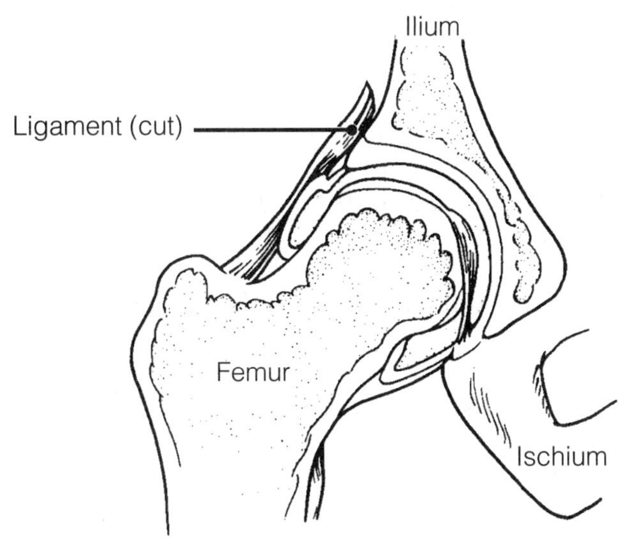

A. __________

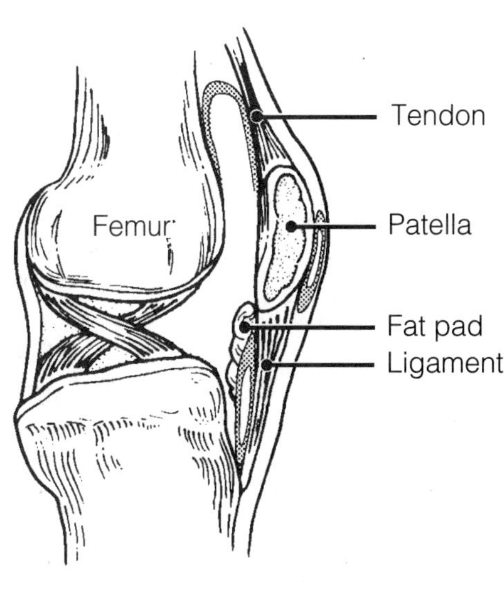

B. __________

Figure 8.3

Homeostatic Imbalances of Joints

1. For the following statements that are true, insert T in the answer blanks. For false statements, correct the underlined words and insert your corrections in the answer blanks.

_______________ 1. The term arthritis refers to bone ends that are forced out of their normal positions in a joint cavity.

_______________ 2. In a sprain, the ligaments reinforcing a joint are excessively stretched or torn.

_______________ 3. Erosion of articular cartilages and formation of painful bony spurs are characteristic of gouty arthritis.

_______________ 4. "Housemaid's knee" is an example of inflammation of the tendon sheaths.

_______________ 5. Chronic arthritis usually results from bacterial invasion.

_______________ 6. Healing of a partially torn ligament is slow because its hundreds of fibrous strands are poorly aligned.

_______________ 7. Acute arthritis is an autoimmune disease.

_______________ 8. Hyperuricemia may lead to rheumatoid arthritis.

_______________ 9. Torn menisci of the knee rarely heal because cartilage is avascular.

_______________ 10. In rheumatoid arthritis, the initial pathology is inflammation of the bursae.

_______________ 11. Gouty arthritis involves several joints that are affected in a bilateral manner.

_______________ 12. Most cartilage injuries of the knee involve torn ligaments.

_______________ 13. The function of the bursae is to absorb the shock between bony structures in a joint.

_______________ 14. The most common form of chronic arthritis is rheumatoid arthritis.

_______________ 15. The local swelling seen in bursitis is caused by excessive production of plasma fluid.

2. Explain why sprains and injuries to joint cartilages are particularly troublesome.

____

____

THE INCREDIBLE JOURNEY:

A Visualization Exercise for the Knee Joint

This place is like an underground cave, and you flip on your back to swim in the shallow pool.

1. Complete the narrative by inserting the missing words in the answer blanks.

1. ____________________
2. ____________________
3. ____________________
4. ____________________
5. ____________________
6. ____________________
7. ____________________
8. ____________________
9. ____________________
10. ____________________
11. ____________________
12. ____________________
13. ____________________
14. ____________________
15. ____________________

For this journey, you are miniaturized and injected into the knee joint of a football player who is resting before the game. You find yourself standing on a firm, white, flexible, C-shaped structure, a ___(1)___, which prevents side-to-side rocking of the ___(2)___ on the ___(3)___. The space you are in, the ___(4)___, is filled with a clear, viscous liquid, the ___(5)___, which serves to ___(6)___ the glassy-smooth articulating surfaces. You note that this place is like an underground cave, and you flip on your back to swim in the shallow pool. As you gaze upward, you see, almost touching your head, two enormous boulderlike structures. At first, you're afraid they might fall on you, but then, of course, you recognize them as the ___(7)___ of the femur. After becoming accustomed to this environment, you begin to wonder what this football player is doing with his knee joint. Because you see that his ___(8)___ ligament is taut, you surmise that his knee joint must be in extension.

Suddenly, the liquid around you becomes less viscous, and you are swept upward, past a rather small, flat, roundish bone, which you identify as the ___(9)___, into an enormous ocean of that same viscous liquid. You note that it is surrounded by a membrane. After some puzzlement, you realize that this must be the suprapatellar ___(10)___!

Knowing that your football player is no longer at rest, you swim, with great effort, back to the joint cavity to watch his knee joint in action. Within a moment, a hard, lateral blow to the knee sends him (and you) spinning. You examine his joint and note that the meniscus has been torn from its attachment at the ___(11)___ margins of the ___(12)___ and that the extracapsular lateral collateral ___(13)___ has also suffered some damage. Because you know that cartilage rarely heals because it has no ___(14)___ for nourishment, you predict that your football player will most likely undergo surgery for removal of his ___(15)___. You prefer not to wait until then to leave your host, so you signal headquarters that you are ready to be picked up by the syringe.

CHALLENGING YOURSELF

At the Clinic

1. Jenny's father loved to hold his 4-year-old daughter by her hand and swing her around in great circles. One day, Jenny's glee was suddenly replaced by tears and she screamed that her left elbow hurt. When examined, the little girl was seen to hold her elbow semiflexed and her forearm pronated. What is your diagnosis?
2. A patient complains of pain starting in the jaw and radiating down the neck. Upon questioning, he states that when he is under stress he grinds his teeth. What joint is causing his pain?
3. In the condition *cleidocranial dysostosis*, ossification of the skull is delayed and clavicles may be completely absent. What effect would you predict that this condition has on the function and appearance of the shoulders?
4. Mrs. Carlisle, an elderly woman, suffered sharp pain in her shoulder while trying to raise a stuck window. Examination reveals degeneration and rupture of the rotator cuff. What is the function of the rotator cuff, and how does its rupture affect the function of the shoulder joint?
5. A complete fracture of the femoral neck will not result in death of the femoral head, unless the ligamentum teres is ruptured. What does this ligament contribute to maintain the integrity of the femoral head?
6. At work, a heavy crate fell from a shelf onto Bertha's acromial region. In the emergency room, the physician could feel that the head of her humerus had moved into the axilla. What had happened to Bertha?
7. A surgeon has removed part of the shaft of the radius of a patient suffering from bone cancer. What bone will he likely use to obtain a graft to replace the excised bone?
8. What type of cartilage is damaged in a case of "torn cartilage"?
9. Marjorie, a middle-aged woman, comes to the clinic complaining of stiff, painful joints. A glance at her hands reveals knobby, deformed knuckles. What condition will she be tested for?

Stop and Think

1. Name all the diarthrotic joints in the axial skeleton.
2. What joints are involved in pronation and supination? How are these joints classified?
3. To what joint classification would the developing coxal bone be assigned? How is the adult coxal bone classified?
4. Explain the significance of the interdigitating and interlocking surfaces of sutures.
5. What microscopic feature marks the end of the periosteum and the beginning of an intrinsic ligament?
6. Is the lining of a joint cavity an epithelial membrane? If so, what type? If not, what is it?
7. What is the role of menisci in the jaw and sternoclavicular joints?
8. What is the effect of years of poor posture on the posterior vertebral ligaments?
9. Only one joint flexes posteriorly. Which one?
10. Is the head capable of circumduction? Why or why not?
11. How does the knee joint differ from a typical hinge joint?

COVERING ALL YOUR BASES

Multiple Choice

Select the best answer or answers from the choices given.

1. Which of the joints listed would be classified as synarthrotic?
   A. Vertebrocostal
   B. Sternocostal (most)
   C. Acromioclavicular
   D. Distal tibiofibular
   E. Epiphyseal plate

2. Cartilaginous joints include:
   A. syndesmoses
   B. symphyses
   C. synostoses
   D. synchondroses

3. Which of the following may be an amphiarthrotic fibrous joint?
   A. Syndesmosis
   B. Suture
   C. Sternocostal
   D. Intervertebral

4. Joints that eventually become synostoses include:
   A. fontanelles
   B. first rib—sternum
   C. ilium-ischium-pubis
   D. epiphyseal plates

5. Which of the following joints have a joint cavity?
   A. Glenohumeral
   B. Acromioclavicular
   C. Intervertebral (at bodies)
   D. Intervertebral (at articular processes)

6. Joints that *contain* fibrocartilage include:
   A. symphysis pubis
   B. knee
   C. acromioclavicular
   D. atlanto-occipital

7. Considered part of a synovial joint are:
   A. bursae
   B. articular cartilage
   C. tendon sheath
   D. capsular ligaments

8. Synovial fluid:
   A. is indistinguishable from fluid in cartilage matrix
   B. is formed by synovial membrane
   C. thickens as it warms
   D. nourishes and lubricates

9. Which joint type would be best if the joint receives a moderate to heavy stress load and requires moderate flexibility in the sagittal plane?
   A. Gliding
   B. Saddle
   C. Condyloid
   D. Hinge

10. In comparing two joints of the same type, what characteristic(s) would you use to determine strength and flexibility?
    A. Depth of the depression of the concave bone of the joint
    B. Snugness of fit of the bones
    C. Size of bone projections for muscle attachments
    D. Presence of menisci

11. Which of the following is a multiaxial joint?
    A. Atlanto-occipital
    B. Atlantoaxial
    C. Glenohumeral
    D. Metacarpophalangeal

12. Plane joints allow:
    A. pronation
    B. flexion
    C. rotation
    D. gliding

13. The temporomandibular joint is capable of which movements?
    A. Elevation
    B. Protraction
    C. Hyperextension
    D. Inversion

14. Abduction is:
    A. moving the right arm out to the right
    B. spreading out the fingers
    C. wiggling the toes
    D. moving the sole of the foot laterally

15. Which of the joints listed is classified as a hinge joint functionally but is actually a condyloid joint structurally?
    A. Elbow
    B. Interphalangeal
    C. Knee
    D. Temporomandibular

16. Which of the following joints has the greatest freedom of movement?
    A. Interphalangeal
    B. Saddle joint of thumb
    C. Distal tibiofibular
    D. Coxal

17. In what condition would you suspect inflammation of the superficial olecranal bursa?
    A. Student's elbow
    B. Housemaid's knee
    C. Target shooter's trigger finger
    D. Teacher's shoulder

18. Contributing substantially to the stability of the hip joint is the:
A. ligamentum teres
B. acetabular labrum
C. patellar ligament
D. iliofemoral ligament

19. The ligament located on the posterior aspect of the knee joint is the:
A. fibular collateral
B. tibial collateral
C. oblique popliteal
D. posterior cruciate

20. Which of the following apply to a sprain?
A. Dislocation
B. Slow healing
C. Surgery for ruptured ligaments
D. Grafting

21. Examining a person's skeleton would enable one to learn about his or her:
A. age
B. gender
C. muscular strength
D. height

22. Erosion of articular cartilage and subsequent formation of bone spurs occur in:
A. ankylosis
B. ankylosing spondylitis
C. torn cartilage
D. osteoarthritis

23. An autoimmune disease resulting in inflammation and eventual fusion of diarthrotic joints is:
A. gout
B. rheumatoid arthritis
C. degenerative joint disease
D. pannus

24. Which of the following is correlated with gouty arthritis?
A. High levels of antibodies
B. Hyperuricemia
C. Hypouricemia
D. Hammertoe

25. Synovial joints are richly innervated by nerve fibers that:
A. monitor how much the capsule is stretched
B. supply articular cartilages
C. cause the joint to move
D. monitor pain if the capsule is injured

26. Which specific joint does the following description identify? "Articular surfaces are deep and secure, multiaxial; capsule heavily reinforced by ligaments; labrum helps prevent dislocation; the first joint to be built artificially; very stable."
A. Elbow
B. Hip
C. Knee
D. Shoulder

27. How does the femur move with respect to the tibia when the knee is locking during extension?
A. Glides anteriorly
B. Flexes
C. Glides posteriorly
D. Rotates medially
E. Rotates laterally

28. Movements made in chewing food are:
A. flexion
B. extension
C. elevation
D. depression
E. opposition

Word Dissection

For each of the following word roots, fill in the literal meaning and give an example, using a word found in this chapter.

Word root	Translation	Example
1. ab		
2. ad		
3. amphi		
4. ankyl		
5. arthro		
6. artic		
7. burs		
8. cruci		
9. duct		
10. gompho		
11. labr		
12. luxa		
13. menisc		
14. ovi		
15. pron		
16. rheum		
17. spondyl		
18. supine		

9

Muscles and Muscle Tissue

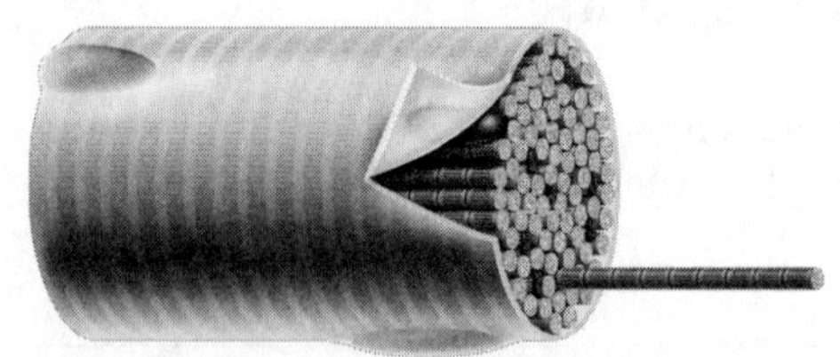

Student Objectives

When you have completed the exercises in this chapter, you will have accomplished the following objectives:

Overview of Muscle Tissues

1. Compare and contrast the basic types of muscle tissue.
2. List four important functions of muscle tissue.

Skeletal Muscle Anatomy

3. Describe the gross structure of a skeletal muscle.
4. Describe the microscopic structure and functional roles of the myofibrils, sarcoplasmic reticulum, and T tubules of skeletal mucle fibers.
5. Describe the sliding filament model of muscle contraction.

Physiology of Skeletal Muscle Fibers and Skeletal Muscles

6. Explain how muscle fibers are stimulated to contract by describing events that occur at the neuromuscular junction.
7. Describe how an action potential is generated.
8. Follow the events of excitation-contraction coupling that lead to cross bridge activity.
9. Define motor unit and muscle twitch, and describe the events occurring during the three phases of a muscle twitch.
10. Explain how smooth, graded contractions of a skeletal muscle are produced.
11. Differentiate between isometric and isotonic contractions.
12. Describe three ways in which ATP is regenerated during skeletal muscle contraction.
13. Define oxygen deficit and muscle fatigue. List possible causes of muscle fatigue.
14. Describe factors that influence the force, velocity, and duration of skeletal muscle contraction.
15. Describe three types of skeletal muscle fibers and explain the relative value of each type.
16. Compare and contrast the effects of aerobic and resistance exercise on skeletal muscles and on other body systems.

Smooth Muscle

17. Compare the gross and microscopic anatomy of smooth muscle cells to that of skeletal muscle cells.
18. Compare and contrast the contractile mechanisms and the means of activation of skeletal and smooth muscles.
19. Distinguish between single-unit and multiunit smooth muscle structurally and functionally.

The specialized muscle tissues are responsible for virtually all body movements, both internal and external. Most of the body's muscle is of the voluntary type and is called skeletal muscle because it is attached to the skeleton. Voluntary muscles allow us to manipulate the external environment, move our bodies, and express emotions on our faces. The balance of the body's muscle is smooth muscle and cardiac muscle, which form the bulk of the walls of hollow organs and the heart, respectively. Smooth and cardiac muscles are involved in the transport of materials within the body.

Study activities in Chapter 9 deal with the microscopic and gross structures of skeletal and smooth muscle. (Cardiac muscle is discussed more thoroughly in Chapter 17.) Important concepts of muscle physiology are also examined in this chapter.

BUILDING THE FRAMEWORK

Overview of Muscle Tissues

1. Nine characteristics of muscle tissue are listed below. Identify each muscle type by choosing the correct key choices and writing the letters in the answer blanks.

Key Choices

A. Cardiac B. Smooth C. Skeletal

_____ 1. Involuntary

_____ 2. Banded appearance

_____ 3. Longitudinally and circularly arranged layers

_____ 4. Dense connective tissue packaging

_____ 5. Gap junctions

_____ 6. Coordinated activity allows it to act as a pump

_____ 7. Moves bones and the facial skin

_____ 8. Referred to as the muscular system

_____ 9. Voluntary

_____ 10. Best at regenerating when injured

2. Identify the type of muscle in each of the illustrations in Figure 9.1. Color the diagrams as you wish.

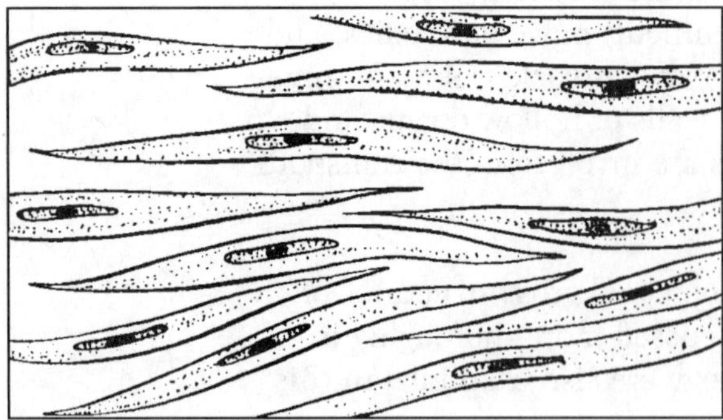

A. ____________________

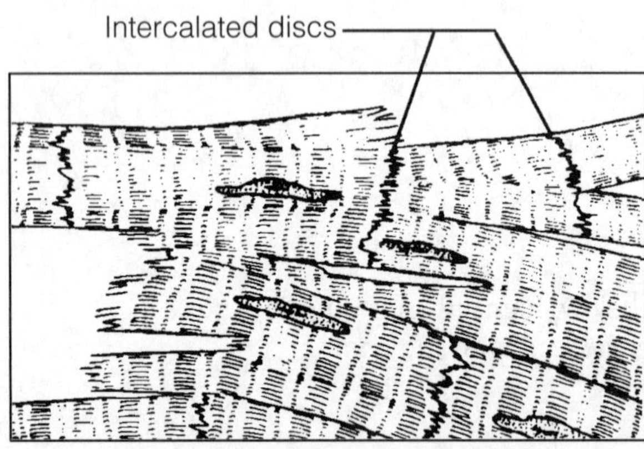

B. ____________________

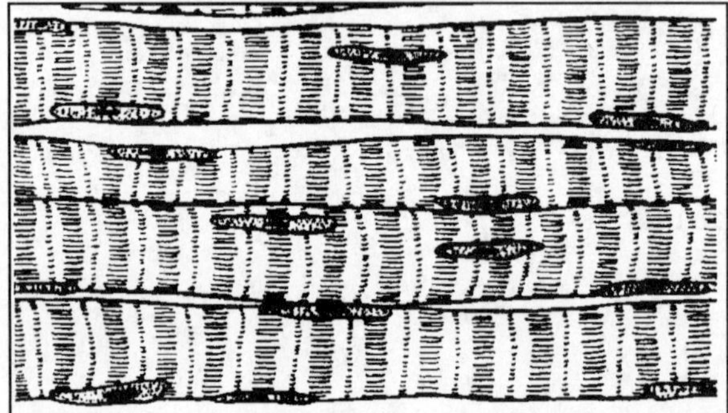

C. ____________________

Figure 9.1

3. Regarding the functions of muscle tissues, circle the term in each of the groupings that does not belong with the other terms.

1. Urine Foodstuffs Blood Bones Smooth muscle
2. Heart Cardiac muscle Blood pump Promotes labor during birth
3. Excitability Response to a stimulus Contractility Action potential
4. Elasticity Recoil properties Contractility Resumption of resting length
5. Ability to shorten Contractility Pulls on bones Stretchability
6. Posture maintenance Movement Promotes growth Generates heat

Skeletal Muscle Anatomy

1. Identify the structures described in Column A by matching them with the terms in Column B. Enter the correct letters (and terms if desired) in the answer blanks. Then, select a different color for each of the terms in Column B that has a color-coding circle and color the structures in Figure 9.2.

	Column A	**Column B**
________	1. Connective tissue surrounding a fascicle	A. Endomysium ○
________	2. Just deep to the deep fascia	B. Epimysium ○
________	3. Contractile unit of muscle	C. Fascicle ○
________	4. A muscle cell	D. Fiber ○
________	5. Thin connective tissue investing each muscle cell	E. Myofilament
		F. Myofibril ○
________	6. Plasma membrane of the muscle cell	G. Perimysium ○
________	7. A long filamentous organelle found within muscle cells that has a banded appearance	H. Sarcolemma
		I. Sarcomere
________	8. Actin-, myosin-, or titin-containing structure	J. Sarcoplasm
________	9. Cordlike extension of connective tissue beyond the muscle, serving to attach it to the bone	K. Tendon ○
________	10. A discrete bundle of muscle cells	

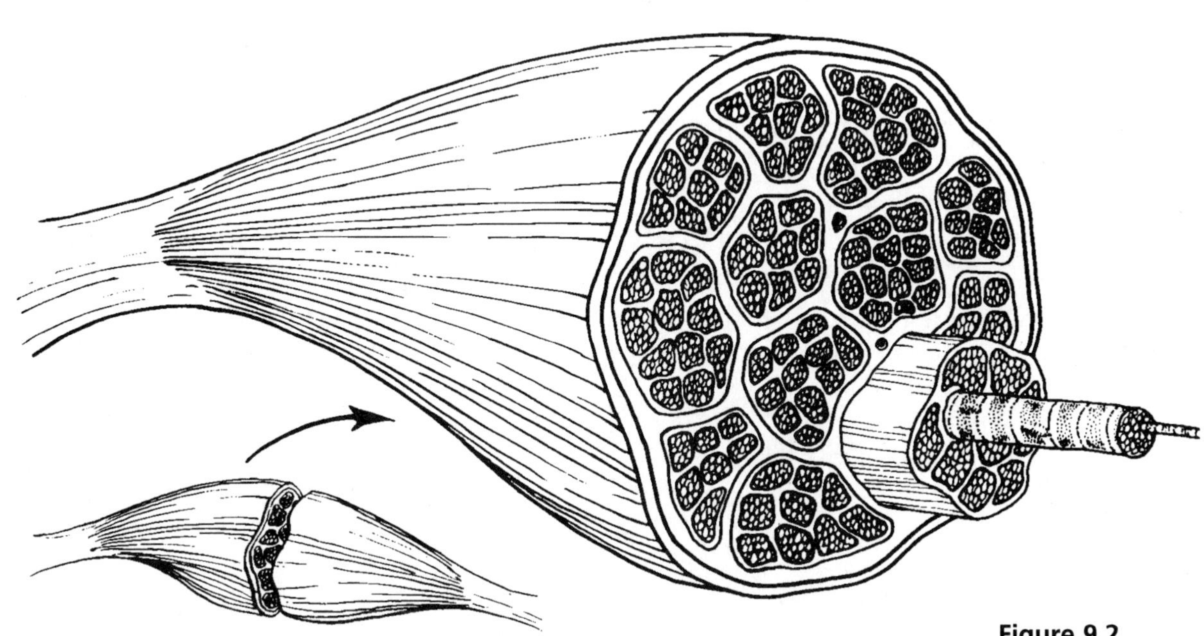

Figure 9.2

2. Figure 9.3 is a diagrammatic representation of a small portion of a relaxed muscle cell (the bracket indicates the portion that has been enlarged). First, select a different color for each of the structures with a coding circle. Color the coding circles and the corresponding structures on Figure 9.3. When you have finished, *bracket* and *label* an A band, an I band, and a sarcomere. Then, match the numbered lines (1, 2, and 3) in part B to the cross sections in part C.

○ Thin myofilaments ○ Thick myofilaments ○ Z discs

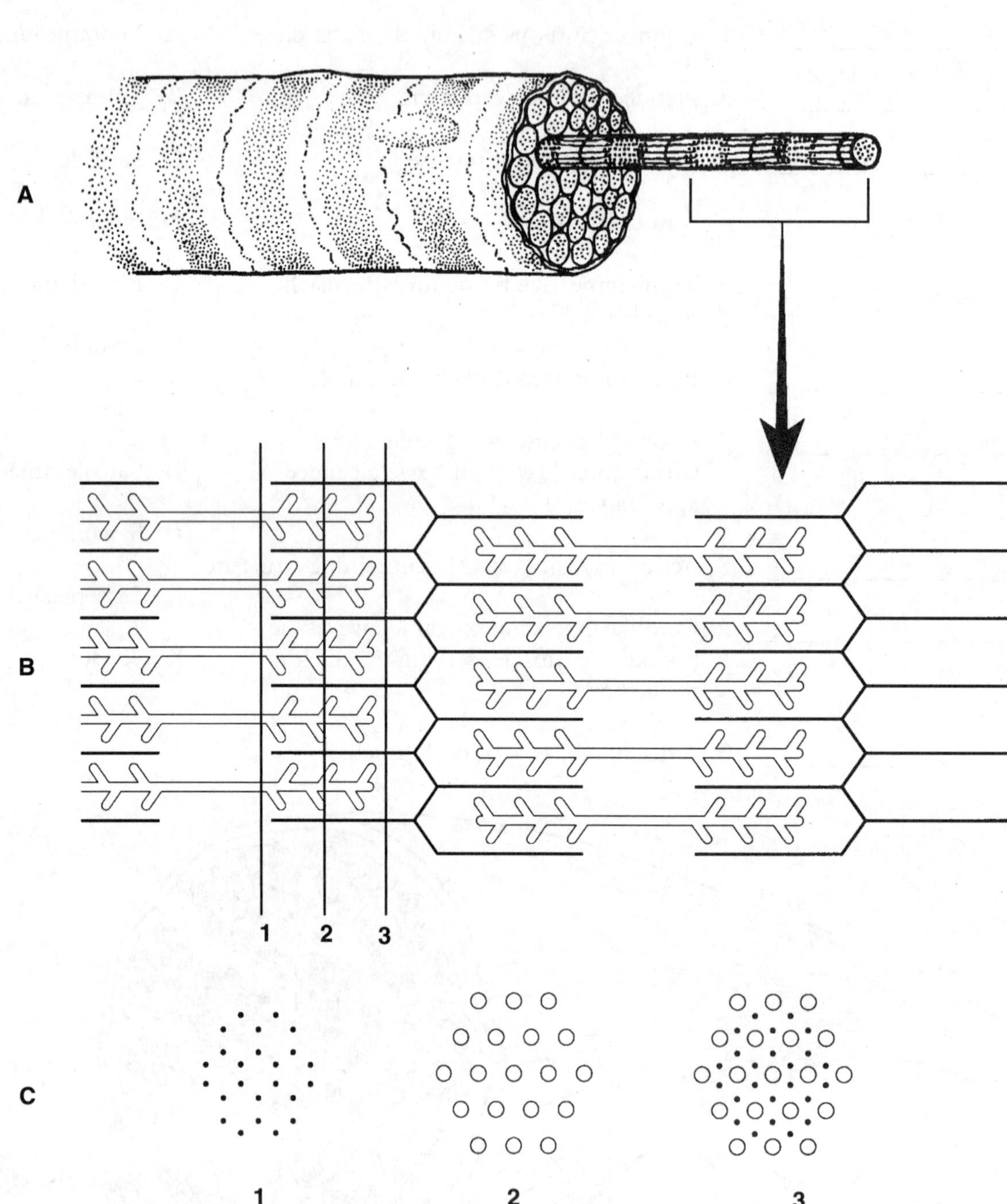

Figure 9.3

3. After referring to Figure 9.3 showing the configuration of a relaxed sarcomere, at the right, draw a contracted sarcomere. Label the thin myofilaments, the thick myofilaments, and the Z discs. Draw an arrow beside one myosin head on each end of the A band to indicate the direction of the working stroke.

4. Fill in the blanks at the right below, giving the location of calcium during the stated phases of muscle contraction. Use the key choices provided at the left.

Key Choices

Key Choices		
A. Becoming bound to SR pumps	________________	1. Unstimulated fiber
B. Attached to troponin	________________	2. Contracting fiber
C. In terminal cisternae	________________	3. Recovering fiber

5. Figure 9.4 shows the intimate relationship between the sarcolemma and two important muscle cell organelles—the sarcoplasmic reticulum (SR) and the myofibrils—in a small segment of a muscle cell. Identify each structure below by coloring the coding circles and the corresponding structures on the diagram. Then bracket and label the composite structure called the triad.

○ Mitochondrion ○ Myofibrils ○ Sarcolemma

○ SR ○ T tubule

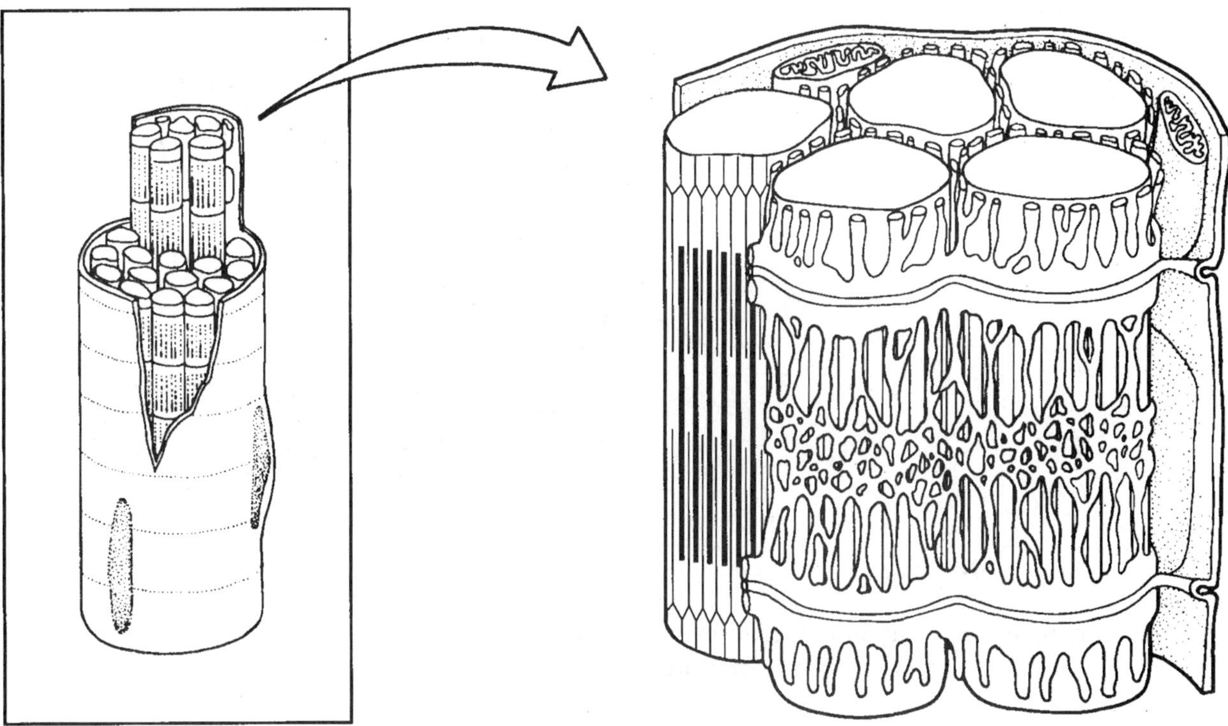

Figure 9.4

6. The integral proteins of the SR and T tubule membranes are in intimate contact at the triads, forming so-called double zippers. Complete the following statements relating to this relationship.

__________________ 1.
__________________ 2.
__________________ 3.
__________________ 4.

The name given to the integral proteins of the SR is __(1)__, and these proteins seem to function as __(2)__. The corresponding T tubule proteins are believed to function as __(3)__, which change their configuration in response to a(n) __(4)__ traveling along the T tubule.

7. Which of these bands or lines narrows when a skeletal muscle contracts? Check (✓) the appropriate answer.

____ 1. H band ____ 2. A band ____ 3. I band ____ 4. M line

Physiology of Skeletal Muscle Fibers and Skeletal Muscles

1. Relative to general terminology concerning muscle activity, first, label the following structures on Figure 9.5: insertion, origin, tendon, resting muscle, and contracting muscle. Next, identify the two structures named below by choosing different colors for the coding circles and the corresponding structures in the figure.

○ Movable bone

○ Immovable bone

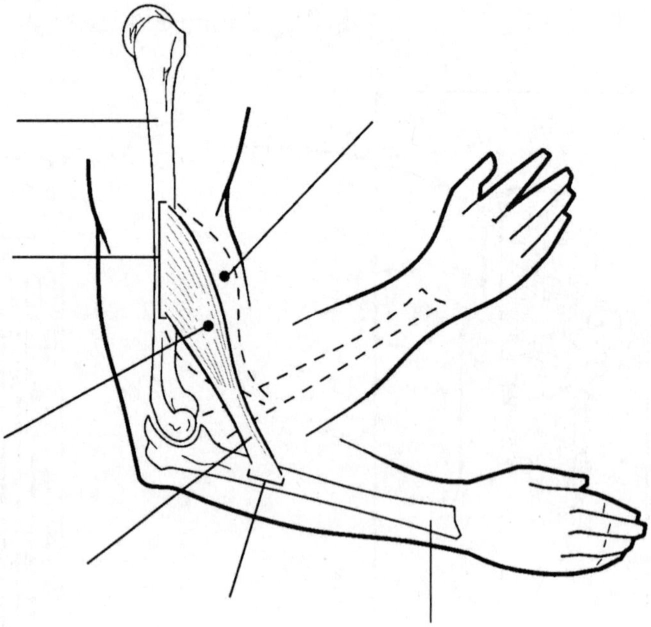

Figure 9.5

2. Correctly relate the story of contraction events in a muscle fiber by *numbering* each event below. The first step is indicated (number 1).

____ 1. Myosin heads bind to active sites on actin molecules.

____ 2. ATP is hydrolyzed.

_____ 3. Myosin heads return to their high-energy shape (cocked), ready for the next working stroke.

__1__ 4. Calcium ions bind to troponin.

_____ 5. Cycling continues until calcium ions are sequestered by the SR.

_____ 6. Myosin cross bridges detach from actin.

_____ 7. Troponin changes shape.

_____ 8. ADP and P_i (inorganic phosphate) are released from the thick filament.

_____ 9. Myosin heads pull on the thin filaments (working stroke) and slide them toward the center of the sarcomere.

_____ 10. ATP binds to the thick filament.

_____ 11. Tropomyosin is moved into the groove between the F-actin strands, exposing active sites on actin.

3. Figure 9.6 shows the components of a neuromuscular junction. Identify the parts by coloring the coding circles and the corresponding structures in the diagram. Add small arrows to indicate the location of the ACh receptors and label appropriately.

○ Mitochondrion

○ Synaptic vesicles

○ T tubule

○ Junctional folds

○ Synaptic cleft

○ Sarcomere

Figure 9.6

4. Figure 9.7 diagrams the elements involved in excitation-contraction coupling. Color the coding circles and the corresponding structures.

- ○ Axonal ending
- ○ Tropomyosin
- ○ T tubule
- ○ Actin
- ○ Myosin
- ○ Sarcolemma
- ○ Troponin
- ○ SR
- ○ Mitochondria
- ○ Synaptic vesicles containing acetylcholine
- ○ Calcium ions

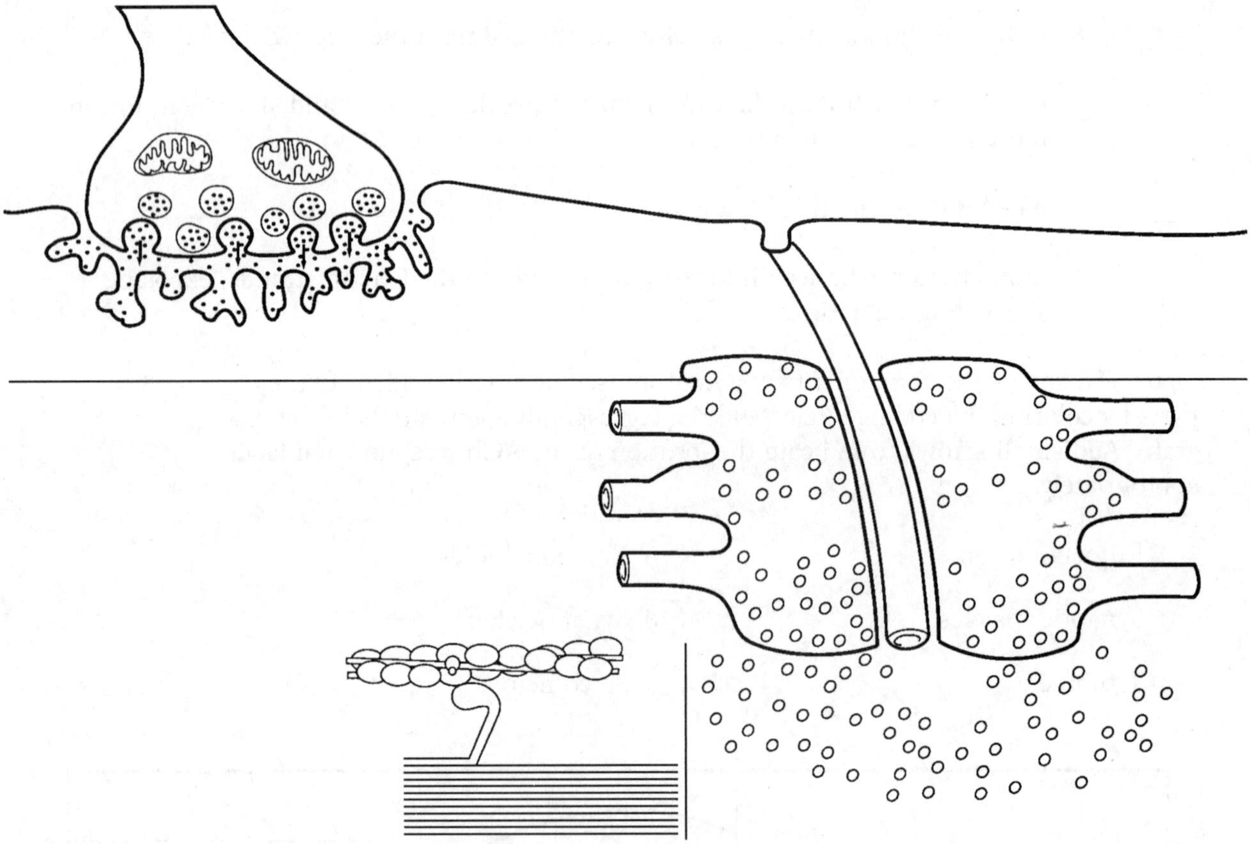

Figure 9.7

5. Number the following statements in their proper sequence to describe excitation-contraction coupling in a skeletal muscle cell. The first step has already been identified as number 1.

__1__ 1. Acetylcholine is released by the axonal ending, diffuses to the muscle cell, and attaches to ACh receptors on the sarcolemma.

_____ 2. The action potential, carried deep into the cell via the T tubules, causes the SR to release calcium ions.

_____ 3. AChE breaks down ACh, which separates from its receptors.

_____ 4. The muscle cell relaxes and lengthens.

_____ 5. The calcium ion concentration at the myofilaments increases; the myofilaments slide past one another, and the cell shortens.

_____ 6. Depolarization occurs, and the action potential is generated along the sarcolemma.

_____ 7. Within 30 ms after the action potential ends, Ca^{2+} concentration at the myofilaments decreases.

6. Define *motor unit.* ____

____

7. Circle the term in each grouping that does not belong with the others.

1. Tropomyosin Troponin Blocks myosin binding sites on actin
   Stabilizes actin
2. 10–15 seconds High-power muscle activity CP + ATP
   Relies on aerobic respiration Immediate transfer of PO_4^{2-} to ADP
3. Oxygen Mitochondria $CO_2 + H_2O$ Oxidative phosphorylation
   End product is lactic acid Jogging
4. Anaerobic metabolism Lactic acid Glycogen reserves
   Fuels extended periods of strenuous activity Energizes 100-m swim
5. Muscle fatigue Increased ADP Decreased Ca^{2+} release from SR
   K^+ accumulates in T tubules Psychological factors
6. Series elastic elements stretched Internal tension = external tension
   Twitch contraction Prolonged stimulation Increased force exerted
7. Aerobic exercise Increases muscle vascularization Muscle hypertrophy
   Cycling More efficient body metabolism Increased cardiovascular efficiency
8. Weight lifting Resistance exercise Increased muscle size Increased endurance
   Good muscle definition Nearly immovable loads
9. Disuse atrophy Immobilization Increased muscle strength
   Loss of nerve stimulation Increased connective tissue in muscles

8. In the appropriate graph spaces below in Figure 9.8, draw the indicated myograms. Be sure to include arrows at the bottom to indicate each stimulus. For the twitch myogram, label the latent, contraction, and relaxation periods.

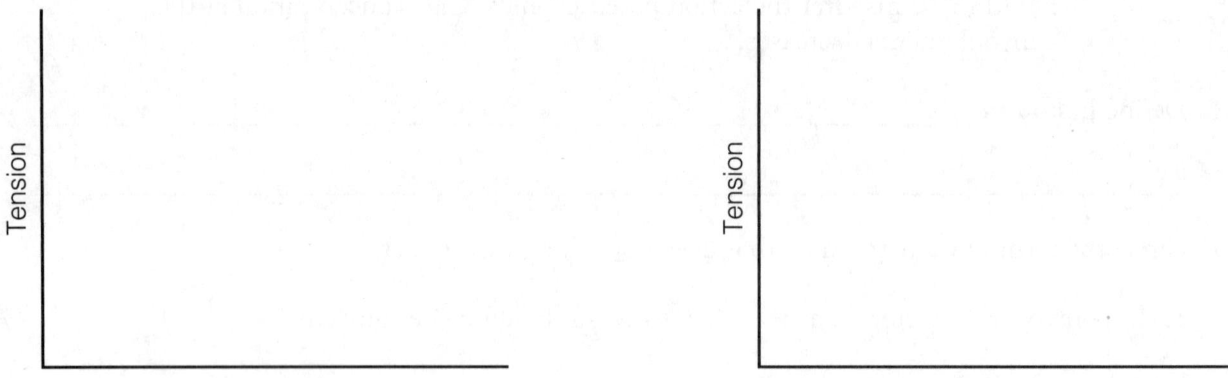

A. Twitch

B. Complete (fused) tetanus

Tension

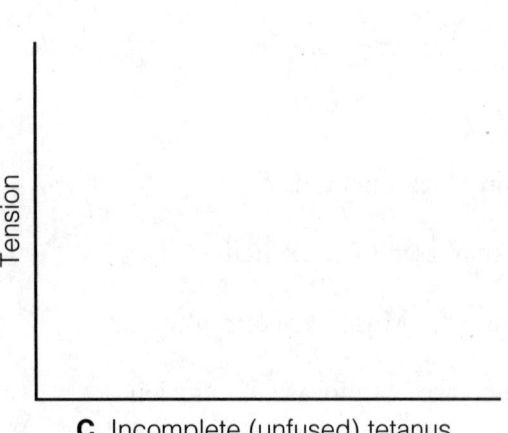

C. Incomplete (unfused) tetanus

Figure 9.8

9. Complete the following statements by choosing the correct responses from the key choices and entering the appropriate letters or terms in the answer blanks.

Key Choices

A. Fatigue	D. Isometric contraction	G. Many motor units
B. Isotonic contraction	E. Fused tetanus	H. Relax
C. Muscle tone	F. Few motor units	

________________ 1. __(1)__ is a continuous contraction that shows no evidence of relaxation.

________________ 2. A(n) __(2)__ is a contraction in which the muscle shortens and work is done.

________________ 3. To accomplish a strong contraction, __(3)__ are stimulated at a rapid rate.

________________ 4. When a weak but smooth muscle contraction is desired, __(4)__ are stimulated at a rapid rate.

________________ 5. When a muscle is being stimulated but is not able to respond, the condition is called __(5)__.

________________ 6. A(n) __(6)__ is a contraction in which the muscle does not shorten, but tension in the muscle keeps increasing.

________________ 7. Wave summation results in stronger contractions at the same stimulus strength because the muscle does not have time to completely __(7)__ between successive stimuli.

________________ 8. __(8)__ includes both concentric and eccentric contractions.

10. Using the key choices, identify the factors requested below.

Key Choices

A. Load	C. Relative fiber size	E. Degree of muscle stretch
B. Muscle fiber type	D. Number of fibers activated	F. Frequency of stimulation

________ 1. All factors that influence contractile force

________ 2. All factors that influence the velocity and duration of muscle contraction

11. Fill in the missing steps in the following flowchart of muscle cell metabolism by writing the appropriate terms in the blanks below it.

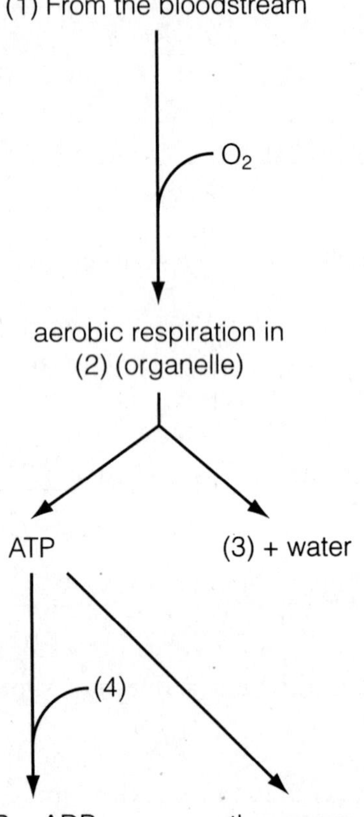

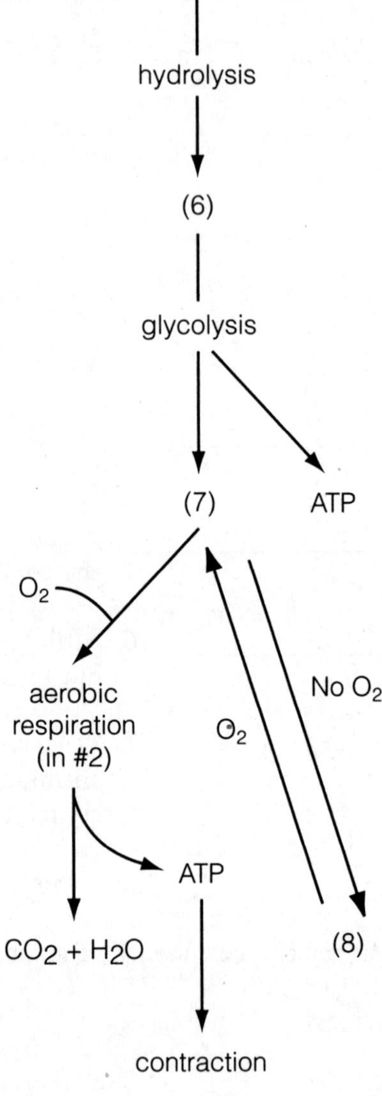

1. ______________________
2. ______________________
3. ______________________
4. ______________________
5. ______________________
6. ______________________
7. ______________________
8. ______________________

12. Check the appropriate column in the chart to characterize each type of skeletal muscle fiber.

Characteristics	Slow oxidative fibers	Fast glycolytic fibers	Fast oxidative fibers
Rapid twitch rate			
Fast myosin ATPases			
Use mostly aerobic metabolism			
Large myoglobin stores			
Large glycogen stores			
Fatigue slowly			
Fibers are white			
Fibers are small			
Fibers contain many capillaries and mitochondria			

13. Explain why muscles contract weakly or not at all when:

1. they are excessively stretched. ______________________________

____

2. the sarcomeres are strongly contracted. ______________________________

____

3. excessive levels of lactic acid and ADP have accumulated in the muscle and an excessive amount of K^+ is lost from the muscle fibers.

____

____

14. Which of the following occur within a muscle cell during oxygen deficit? Place a check mark (✓) by the correct choices.

_____ 1. Decreased ATP

_____ 2. Increased ATP

_____ 3. Increased lactic acid

_____ 4. Decreased oxygen

_____ 5. Increased oxygen

_____ 6. Decreased carbon dioxide

_____ 7. Increased carbon dioxide

_____ 8. Increased glucose

15. Briefly describe how you can tell you are repaying the oxygen deficit. ____________________

____

____

Smooth Muscle

1. Smooth muscle cells differ from skeletal muscle cells in several ways. Indicate which of the following elements seen in skeletal muscle are *absent* in smooth muscle by placing a check mark in the appropriate answer blanks. Circle elements seen in smooth muscle but not in skeletal muscle.

_____ 1. Dense bodies

_____ 2. Diffuse junctions

_____ 3. Gap junctions

_____ 4. Calveoli

_____ 5. Elaborate neuromuscular junctions

_____ 6. Myofibrils

_____ 7. Myofilaments

_____ 8. Sarcomeres

_____ 9. SR

_____ 10. Striations

_____ 11. Tropomyosin

_____ 12. Troponin

_____ 13. T tubules

_____ 14. Varicosities

_____ 15. Z discs

2. Complete the following statements concerning smooth muscle characteristics by inserting the correct terms in the answer blanks.

Whereas skeletal muscle exhibits elaborate connective tissue coverings, smooth muscle has a scant connective tissue (1), which is secreted by the (2) cells. The most common variety of smooth muscle, called (3) smooth muscle, is arranged in opposing layers in the walls of hollow organs. Most often, there are (4) such layers, one running (5) and the other running (6). As these layers alternately contract and relax, substances are squeezed along the internal passages. This phenomenon is called (7). The electrical coupling of the smooth muscle cells by (8) allows smooth muscle to contract as a unit because the (9) spread from cell to cell. Smooth muscle cells, referred to as (10) cells, set the pace for the entire muscle sheet in some cases. Nerve endings called (11) release neurotransmitter into wide synaptic clefts called (12). In smooth muscles, ionic calcium, which triggers contraction, enters the cytosol from both the SR and the (13). Ca^{2+} binds to the cytoplasmic protein (14), which in turn binds to (15) located on the myosin heads. (15) catalyzes the transfer of (16) from ATP to the myosin heads and cross bridge cycling begins.

____________________ 1.

____________________ 2.

____________________ 3.

____________________ 4.

____________________ 5.

____________________ 6.

____________________ 7.

____________________ 8.

____________________ 9.

____________________ 10.

____________________ 11.

____________________ 12. ____________________ 14. ____________________ 16.

____________________ 13. ____________________ 15.

3. For each of the following descriptions, indicate whether it is more typical of (A) skeletal muscle or (B) smooth muscle (single-unit variety).

_____ 1. Ionic calcium binds to the thick filaments (perhaps to calmodulin)

_____ 2. Myosin ATPases are very rapid acting

_____ 3. Total length change possible is 150%

_____ 4. Contraction is slow, and seemingly tireless, which saves energy

_____ 5. Contraction duration is always less

_____ 6. When stretched, contracts vigorously

_____ 7. ATP is *routinely* generated by anaerobic pathways

_____ 8. Excited by autonomic nerves

_____ 9. Excited by acetylcholine

_____ 10. Excited by norepinephrine

_____ 11. Excited by chemical stimuli, including many hormones

4. Describe the importance of the stress-relaxation response in single-unit smooth muscle. ____

____

____

____

5. 1. How is multiunit smooth muscle like skeletal muscle?

____

2. How is it different?

____

____

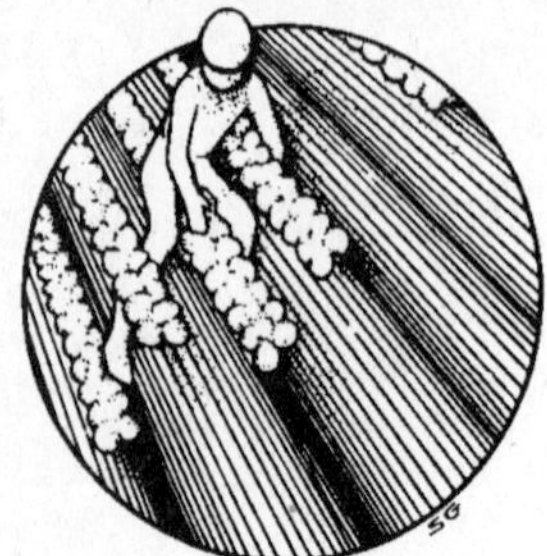

THE INCREDIBLE JOURNEY:

A Visualization Exercise for Skeletal Muscle Tissue

As you straddle this structure, you wonder what is happening.

1. Complete the narrative by inserting the missing words in the answer blanks.

______________ 1.

______________ 2.

______________ 3.

______________ 4.

______________ 5.

______________ 6.

______________ 7.

______________ 8.

______________ 9.

______________ 10.

On this incredible journey, you will enter a skeletal muscle cell to observe the events that occur during muscle contraction. You prepare yourself by donning a wetsuit and charging your ion detector. Then you climb into a syringe to prepare for injection. Your journey will begin when you see the gleaming connective tissue covering of a single muscle cell, the ___(1)___. Once injected, you monitor your descent through the epidermis and subcutaneous tissue. When you reach the muscle cell surface, you see that it is punctuated with pits at relatively regular intervals. Looking into the distant darkness, you can see a leash of fibers ending close to a number of muscle cells. Because all these fibers must be from the same motor neuron, this functional unit is obviously a ___(2)___. You approach the fiber ending on your muscle cell and scrutinize the ___(3)___ junction there. As you examine the junction, minute fluid droplets leave the nerve endings and attach to doughnut-shaped receptors on the muscle cell membrane. This substance released by the nerve ending must be ___(4)___. Then, as a glow falls over the landscape, your ion detector indicates that ions are disappearing from the muscle cell exterior and entering the muscle pits. The needle drops from high to low as the ___(5)___ ions enter the pits from the watery fluid outside. You should have expected this, as these ions must enter to depolarize the muscle cells and start the ___(6)___.

Next you begin to explore one of the surface pits. As the muscle jerks into action, you topple deep into the pit. Sparkling electricity lights up the wall on all sides. You grasp for a handhold. Finally successful, you pull yourself laterally into the interior of the muscle cell and walk carefully along what seems to be a log. Then, once again, you notice an eerie glow as your ion detector reports that ___(7)___ ions are entering the cytoplasm rapidly. The "log" you are walking on suddenly "comes to life" and begins to slide briskly in one direction. Unable to keep your balance, you fall. As you straddle this structure, you wonder what is happening. On all sides, cylindrical structures—such as the one you are astride—are moving past other similar but larger structures. Suddenly you remember, these are the myofilaments, ___(8)___ and ___(9)___, that slide past one another during muscle contraction.

Seconds later, the forward movement ends, and you begin to journey smoothly in the opposite direction. The ion detector now indicates low ___(10)___ ion levels. Because you cannot ascend the smooth walls of one of the entry pits, you climb from one myofilament to another to reach the underside of the sarcolemma. Then you travel laterally to enter a pit close to the surface and climb out onto the cell surface. Your journey is completed, and you prepare to leave your host once again.

CHALLENGING YOURSELF

At the Clinic

1. In preparation for tracheal intubation, curare (or a derivative) is given to relax the laryngeal muscles. What specific mechanism does curare inhibit?

2. Gregor, who works at a pesticide factory, comes to the clinic complaining of muscle spasms that interfere with his movement and breathing. A blood test shows that he has become contaminated with organophosphate pesticide. The doctor states that this type of pesticide is an acetylcholinesterase inhibitor. How would you explain to Gregor what this means?

3. Maureen Hamel, a young woman, comes to the clinic with symptoms of muscle weakness. The nurse notes on the record that her eyelids droop and her speech is slurred. What do you suspect is Maureen's problem, and what blood test would you run to support the diagnosis?

4. A young man appears for a routine physical before participating in his high school athletics program. A cursory examination shows striae on his thighs, which are much larger than his calves and arm muscles. What has caused the striae, and what would you suggest as a modification of his workout routine?

5. Fifty-seven-year-old Charlie Earl, who admits to a very sedentary lifestyle, has just completed a stress test at the clinic. The test results indicate that he has very little aerobic capacity, and his pulse and blood pressure are at the high end of the normal ranges. Taking into account the effects of aging on bones, joints, and muscles, what exercise program would you recommend for Charlie?

6. Samantha, a young gymnast, is brought to the clinic with a pulled calf muscle. What treatment will she receive?

7. Mrs. Sanchez says her 6-year-old son seems to be clumsy and tires easily. The doctor notices that his calf muscles do not appear atrophied; if anything, they seem enlarged. For what condition must the boy be checked? What is the prognosis?

8. Charlie, our patient from Question 5, has returned, very discouraged about his regimen of a 2-mile daily walk. He reports excruciating pain whenever he tries to walk at more than a "snail's pace." What condition is probably causing his pain? Should he stop his exercise program?

9. When a person dies, rigor mortis sets in as ATP synthesis ceases. Explain why the lack of ATP and the presence of Ca^{2+} in muscle cells would cause muscles to become rigid, rather than limp, after death.

Stop and Think

1. Do all types of muscle produce movement? Maintain posture? Are all three types important in generating heat?

2. Would movement be possible if muscles were not extensible? Why or why not?

3. Chickens are capable of only brief bursts of flight, and their flying muscles consist of white fibers. The breast muscles of ducks, by contrast, consist of red and intermediate fibers. What can you deduce about the flying abilities of ducks?

4. If you cut across a muscle fiber at the level of the T tubules, what would the cross section look like?

5. Which is a cross bridge attachment more similar to: a precision rowing team or a person pulling a rope and bucket out of a well?

6. How is calcium removed from the sarcomere when relaxation is initiated?

7. When a muscle fiber relaxes, do the thick and thin myofilaments remain in their contracted arrangement or slide back apart? If the latter, what makes them slide? What does this do to the length of the resting sarcomere?

8. All of the membranous structures involved in excitation-contraction coupling have a very large surface area. Explain how this is accomplished for each of these structures: synaptic vesicles, sarcolemma, T tubules, SR. Why is this membrane elaboration important in speeding up the processes involved?

9. Which of the following choices gives the best "reason" why the muscles that move the body and limbs are made of skeletal and not smooth muscle tissue? Explain your choice, and tell why each of the other answers is wrong. (a) Smooth muscle is much weaker than skeletal muscle tissue of the same size, so it could not move heavy limbs. (b) Smooth muscle cells do not contract by the sliding filament mechanism and therefore are inefficient. (c) Smooth muscle cells contract too slowly. If they moved the body, we could never move fast enough to survive dangerous situations. (d) Smooth muscle fatigues too easily.

10. What would happen if the refractory period of a skeletal muscle fiber were abnormally prolonged well into the period of relaxation?

11. Asynchronous motor unit summation reduces muscle fatigue. How so? How does this relate to muscle tone?

12. What factors decrease oxygen delivery in very active muscles? How do active muscles offset this effect?

13. Why would paralyzed muscles shorten as they atrophy? Why are physical therapy and braces essential to prevent tightening and contortion at joints?

14. Contrast the roles of calcium in initiating contraction in skeletal muscle and in smooth muscle.

15. How does the rudimentary nature of the SR in smooth muscle affect the contraction and relaxation periods of a smooth muscle twitch?

16. Heart transplant operations are becoming ever more common. What controls the rate of cardiac muscle contraction in a heart that has been transplanted?

COVERING ALL YOUR BASES

Multiple Choice

Select the best answer or answers from the choices given.

1. Select the type(s) of muscle tissue that fit the following description: self-excitable, pacemaker cells, gap junctions, extremely extensible, limited SR, calmodulin activated.
   A. Skeletal muscle
   B. Cardiac muscle
   C. Smooth muscle
   D. Involuntary muscle

2. Which of the following would be associated with a skeletal muscle fascicle?
   A. Perimysium
   B. Nerve
   C. Artery and at least one vein
   D. Epimysium

3. Skeletal muscle is *not* involved in:
   A. movement of skin
   B. propulsion of a substance through a body tube
   C. heat production
   D. inhibition of body movement

4. Which of the following are part of a thin myofilament?
   A. ATP-binding site
   B. TnI
   C. Globular actin
   D. Calcium

5. Factors involved in calcium release and uptake during contraction and relaxation include:
   A. calsequestrin
   B. T tubules
   C. SR
   D. terminal cisternae
   E. sarcolemma

6. In comparing electron micrographs of a relaxed skeletal muscle fiber and a fully contracted muscle fiber, which would be seen only in the *relaxed* fiber?
   A. Z discs
   B. Triads
   C. I bands
   D. A bands
   E. H zones

7. After ACh attaches to its receptors at the neuromuscular junction, the next step is:
   A. potassium-gated channels open
   B. calcium binds to troponin
   C. the T tubules depolarize
   D. cross bridges attach
   E. ATP is hydrolyzed

8. Detachment of the cross bridges is directly triggered by:
   A. hydrolysis of ATP
   B. repolarization of the T tubules
   C. the power stroke
   D. attachment of ATP to myosin heads

9. Transmission of the stimulus at the neuromuscular junction involves:
   A. synaptic vesicles
   B. TnT
   C. ACh
   D. junctional folds

10. Which of the following locations are important sites for calcium activity?
    A. SR
    B. T tubules
    C. Axonal ending
    D. Junctional folds of sarcolemma

11. Acetylcholinesterase hydrolyzes ACh during which phase(s) of a muscle twitch?
    A. Latent period
    B. Period of contraction
    C. Period of relaxation
    D. Refractory period

12–13. Use the following graph to answer Questions 12 and 13:

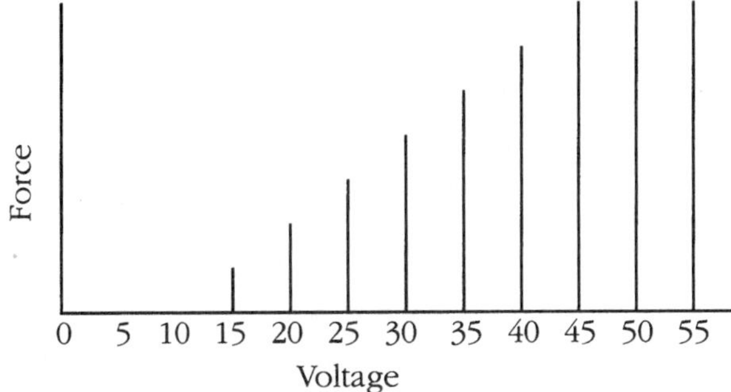

(The muscle is stimulated once at each voltage indicated)

12. What is the threshold stimulus for this muscle?
    A. 5 volts
    B. 10 volts
    C. 15 volts
    D. 40 volts
    E. 45 volts

13. What is the maximal stimulus for this muscle?
    A. 15 volts
    B. 40 volts
    C. 45 volts
    D. 55 volts

14. Small motor units are associated with:
    A. precision
    B. power
    C. fine control
    D. weak contraction

15. Holding up the corner of a heavy couch to vacuum under it involves which type(s) of contraction?
    A. Tetanic
    B. Isotonic
    C. Isometric
    D. Tone

16. Your ability to lift that heavy couch would be increased by which type of exercise?
    A. Aerobic
    B. Endurance
    C. Resistance
    D. Swimming

17. Which of the following activities depends most on anaerobic metabolism?
    A. Jogging
    B. Swimming a race
    C. Sprinting
    D. Running a marathon

18. The first energy source used to regenerate ATP when muscles are extremely active is:
    A. fatty acids
    B. glucose
    C. creatine phosphate
    D. pyruvic acid

19. ATP directly powers:
    A. exocytosis of synaptic vesicles
    B. sodium influx at the sarcolemma
    C. calcium release into the sarcoplasm
    D. recocking of the myosin head

20. Which statement is true about contractures?
    A. Depletion of ATP is absolute.
    B. Contraction is weak.
    C. Muscle cramping is due to depletion of electrolytes.
    D. The neuromuscular junction has been affected.

21. When paying back the oxygen deficit:
    A. lactic acid is formed
    B. lactic acid is reconverted to pyruvic acid
    C. ATP formation requires creatine phosphate
    D. muscle cells utilize glycogen reserves

22. Factors affecting force of contraction include:
    A. number of motor units stimulated
    B. cross-sectional area of the muscle/muscle fibers
    C. load
    D. degree of muscle stretch prior to the contraction

23. The ideal length-tension relationship is:
    A. maximal overlap of thin and thick myofilaments
    B. minimal overlap of thin and thick myofilaments
    C. no overlap of thin and thick myofilaments
    D. intermediate degree of overlap of thin and thick myofilaments

24. When the velocity of contraction is zero, the contraction is:
    A. strongest
    B. equal to the load
    C. greater than the load
    D. isometric

25. Which of these bands or lines does not narrow when a skeletal muscle fiber contracts?
    A. H zone
    B. A band
    C. I band
    D. M line

26. Characteristics of red, but not white, muscle are:
    A. abundant mitochondria
    B. richly vascularized
    C. fast contraction
    D. slow to fatigue

27. Which of the following is (are) unique to smooth muscle cells, compared with the other types of muscle cells?
    A. Produce endomysium
    B. Utilize calmodulin
    C. Can contract even when maximally stretched
    D. Self-excitable

28. Smooth muscle contraction is stimulated by:
    A. hormones
    B. inhibition of calmodulin
    C. neurotransmitters
    D. stretching

29. Which of the following is (are) true of single-unit but not multiunit smooth muscle?
    A. Presence of gap junctions
    B. Each muscle fiber has its own nerve ending
    C. Sheetlike arrangement in hollow organs
    D. Contracts in a corkscrewlike manner

30. What fiber types would be most useful in the leg muscles of a long-distance runner?
    A. White fast-twitch
    B. White slow-twitch
    C. Intermediate fast-twitch
    D. Red slow-twitch

31. Relative to differences between eccentric and concentric contractions, eccentric contractions:
    A. put the body in a position to contract concentrically
    B. occur as the muscle is actually increasing in length
    C. are more likely to cause muscle soreness the day after a strenuous activity
    D. involve unchanging muscle length and tension

Word Dissection

For each of the following word roots, fill in the literal meaning and give an example, using a word found in this chapter.

Word root	Translation	Example
1. fasci		
2. lemma		
3. metric		
4. myo		
5. penna		
6. raph		
7. sarco		
8. stalsis		
9. stria		
10. synap		
11. tetan		

10

The Muscular System

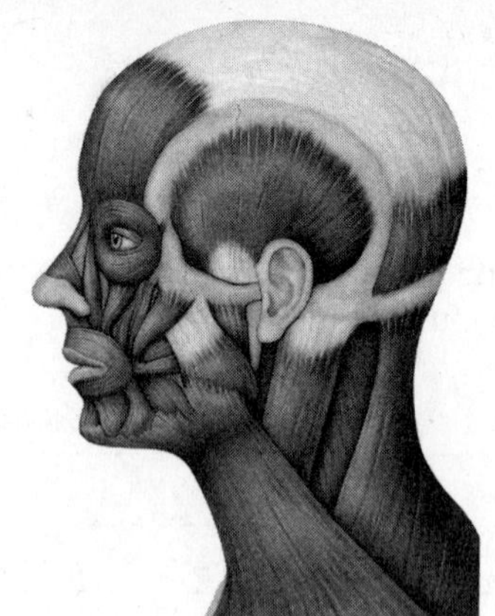

Student Objectives

When you have completed the exercises in this chapter, you will have accomplished the following objectives:

Interactions of Skeletal Muscles in the Body

1. Describe the function of prime movers, antagonists, synergists, and fixators.

Naming Skeletal Muscles

2. List the criteria used in naming muscles. Provide an example to illustrate the use of each criterion.

Muscle Mechanics: Importance of Fascicle Arrangement and Leverage

3. Name the common patterns of muscle fascicle arrangement and relate these to power generation.
4. Define lever, and explain how a lever operating at a mechanical advantage differs from one operating at a mechanical disadvantage.
5. Name the three types of lever systems and indicate the arrangement of effort, fulcrum, and load in each. Also note the advantages of each type of lever system.

Major Skeletal Muscles of the Body

6. Name and identify the muscles described in Tables 10.1 to 10.17. State the origin, insertion, and action of each.

All contractile tissues—smooth, cardiac, and skeletal muscle—are muscles, but the muscular system is concerned only with skeletal muscles, the organs composed of skeletal muscle tissue, and connective tissue wrappings. The skeletal muscles clothe the bony skeleton and make up about 40% of body mass. Skeletal muscles, also called voluntary muscles, help shape the body and enable you to smile, run, shake hands, grasp things, and otherwise manipulate your environment.

The topics of Chapter 10 include principles of leverage, classification of muscles on the basis of their major function, identification of voluntary muscles, and a description of their important actions and interactions.

BUILDING THE FRAMEWORK

Interactions of Skeletal Muscles in the Body

1. The terms in the key choices are often applied to the manner in which muscles interact with other muscles. Select the terms that apply to the following definitions and insert the correct answers in the answer blanks.

Key Choices

A. Antagonist B. Fixator C. Prime mover D. Synergist

____________________ 1. Holds parts of the body in proper position for the action of other muscles, primarily postural muscles

____________________ 2. Bears the major responsibility for producing a particular movement

____________________ 3. Acts to reverse or act against the action of another muscle

____________________ 4. Assists an agonist by a like contraction movement or by holding a joint over which an agonist acts immobile

Naming Skeletal Muscles

1. From the key choices, select the criteria that are used in naming the muscles listed below. In some cases, more than one criterion applies.

Key Choices

A. Action

B. Direction of fibers

C. Location of the muscle

D. Location of origin and insertion

E. Number of origins

F. Relative muscle size

G. Shape of the muscle

_____ 1. Adductor brevis

_____ 2. Brachioradialis

_____ 3. Deltoid

_____ 4. Extensor digitorum longus

_____ 5. Internal oblique

_____ 6. Levator scapulae

_____ 7. Orbicularis oculi

_____ 8. Pectoralis major

_____ 9. Quadriceps femoris

_____ 10. Sternohyoid

Muscle Mechanics: Importance of Fascicle Arrangement and Leverage

1. Complete the following short paragraph, which considers the operation of lever systems in the body. Insert your responses in the answer blanks.

__________________ 1.

__________________ 2.

__________________ 3.

__________________ 4.

In the body, lever systems involve the interaction of __(1)__, which provide the force, and __(2)__, which provide the lever arms. The force is exerted where __(3)__. The site at which movement occurs in each case is the intervening __(4)__, which acts as the fulcrum.

2. Circle all of the elements that characterize a lever that operates at a mechanical disadvantage:

Fast Moves a large load over a small distance Requires maximal shortening

Slow Moves a small load over a large distance Requires minimal shortening

Muscle force greater than the load Muscle force less than load moved

3. The brachialis is a short stout muscle inserting close to the fulcrum of the elbow joint; the brachioradialis is a slender muscle inserting distally, well away from the fulcrum of the elbow joint. Which muscle is associated with mechanical advantage and which with mechanical disadvantage?

4. Figure 10.1 illustrates the elements of three different lever systems on the left and provides examples of each lever system from the body on the right. Select different colors for the lever elements listed above the figure and use them to color the diagrams of each lever system. Complete this exercise by following the steps below for each muscle diagram.

1. Add a red arrow to indicate the point at which the effort is exerted.

2. Draw a circle enclosing the fulcrum.

3. Indicate whether the lever shown is operating at a mechanical advantage or disadvantage by filling in the line below the diagram.

4. Color the load yellow.

○ Effort ○ Fulcrum ○ Load (resistance)

A. First-class lever

____

B. Second-class lever

____

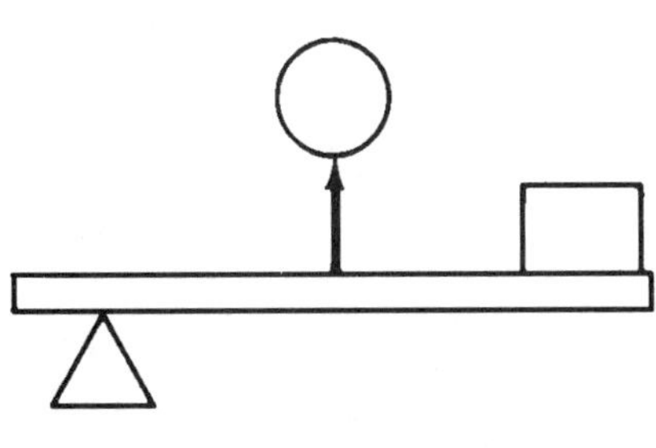

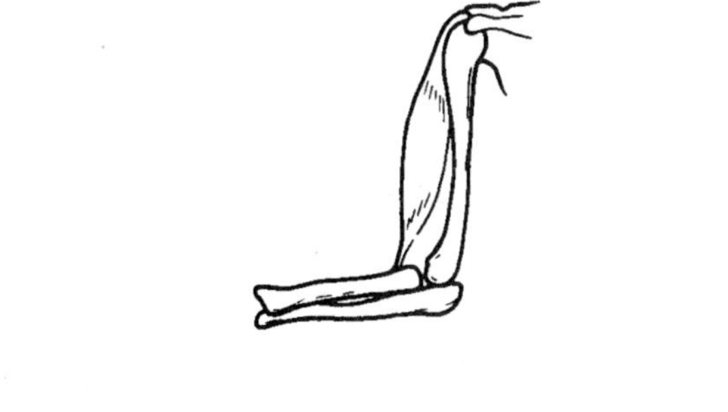

C. Third-class lever

____

Figure 10.1

5. Four fascicle arrangements are shown in Figure 10.2; these same arrangements are described below. First, correctly identify each fascicle arrangement by (1) writing the appropriate term in the answer blank and (2) writing the letter of the corresponding diagram in the answer blank. Second, on the remaining lines beside the diagrams, reidentify the fascicle arrangement and then name two muscles that have the fascicle arrangement shown.

________________ 1. Fascicles insert into a midline tendon from both sides.

________________ 2. Muscle fibers are arranged in concentric array around an opening.

________________ 3. Fascicles run with the long axis of the muscle.

________________ 4. Fascicles angle from a broad origin to a narrow insertion.

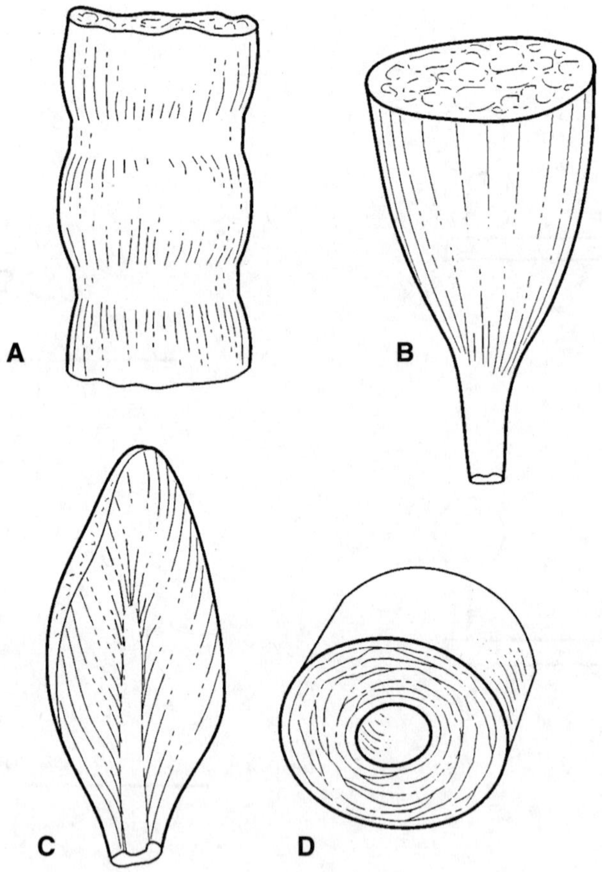

____
____
____

A

B

____
____
____

____
____
____

C

D

____
____
____

Figure 10.2

Major Skeletal Muscles of the Body

1. Match the names of the suprahyoid and infrahyoid muscles listed in Column B with their descriptions in Column A. (Note that not all muscles in Column B are described.) For each muscle that has a color-coding circle, color the coding circles and the corresponding muscles on Figure 10.3 as you like, but color the hyoid bone yellow and the thyroid cartilage blue.

	Column A	Column B
○ ________________	1. Most medial muscle of the neck; thin; depresses larynx	A. Digastric
○ ________________	2. Slender muscle that parallels the posterior belly of the digastric muscle; elevates and retracts hyoid bone	B. Geniohyoid
○ ________________	3. Consists of two bellies united by an intermediate tendon; prime mover to open the mouth; also depresses mandible	C. Mylohyoid
○ ________________	4. Lateral and deep to sternohyoid; pulls thyroid cartilage inferiorly	D. Omohyoid
○ ________________	5. Flat triangular muscle deep to digastric; elevates hyoid, enabling tongue to force food bolus into the pharynx	E. Sternohyoid
○ ________________	6. Narrow muscle in the midline running from chin to hyoid; widens pharynx for receiving food as it elevates the hyoid bone	F. Sternothyroid
○ ________________	7. Superior continuation of the sternothyroid	G. Stylohyoid
		H. Thyrohyoid

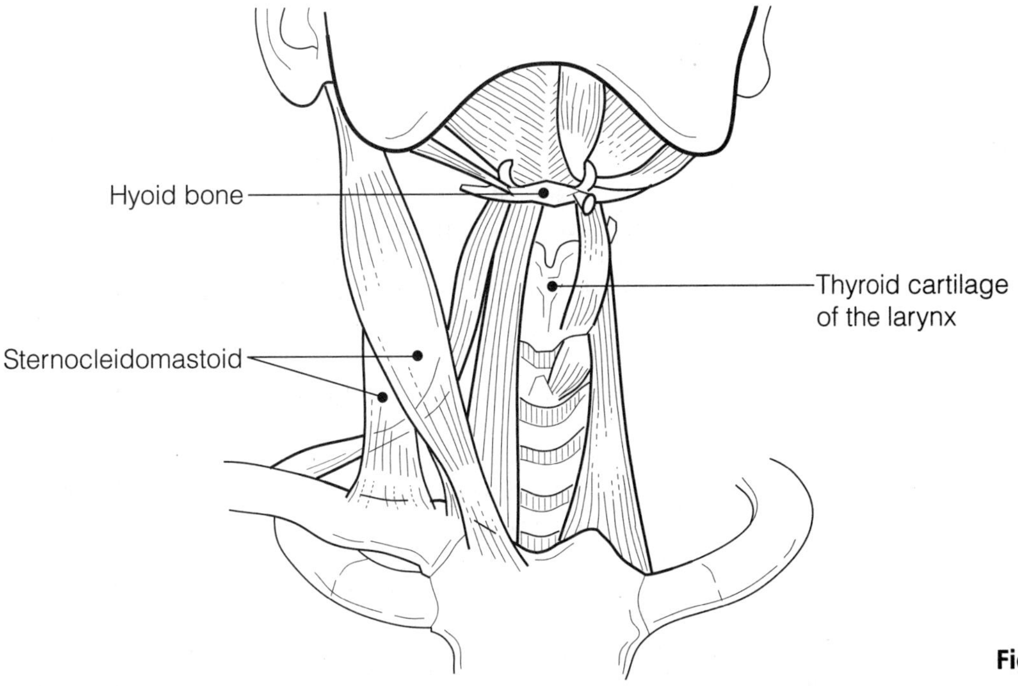

Figure 10.3

2. Match the muscle names in Column B to the facial muscles described in Column A.

	Column A	Column B
________	1. Squints the eyes	A. Corrugator supercilii
________	2. Furrows the forehead horizontally	B. Depressor anguli oris
________	3. Smiling muscle	C. Frontal belly of the epicranius
________	4. Puckers the lips	D. Occipital belly of the epicranius
________	5. Draws the corners of the lips laterally and downward	E. Orbicularis oculi
________	6. Pulls the scalp posteriorly	F. Orbicularis oris
________	7. Tensed during shaving of the chin and neck	G. Platysma
		H. Zygomaticus

3. Relative to muscles of the head and neck, name the major muscles described here. Select a different color for each muscle and color the coding circles and corresponding muscles on Figure 10.4. Notice that the last muscle to be identified lacks a coding circle.

○ ________ 1. Used to show you're happy

○ ________ 2. Well developed in a clarinet player; holds food between the teeth

○ ________ 3. Used in winking

○ ________ 4. Used to raise your eyebrows

○ ________ 5. The "kissing" muscle

○ ________ 6. Prime mover of jaw closure

○ ________ 7. Synergist muscle for jaw closure; elevates and retracts the mandible

○ ________ 8. Posterior neck muscle, called the "bandage" muscle

○ ________ 9. Prime mover of head flexion, a two-headed neck muscle

________ 10. Protrudes the mandible; side-to-side grinding movements

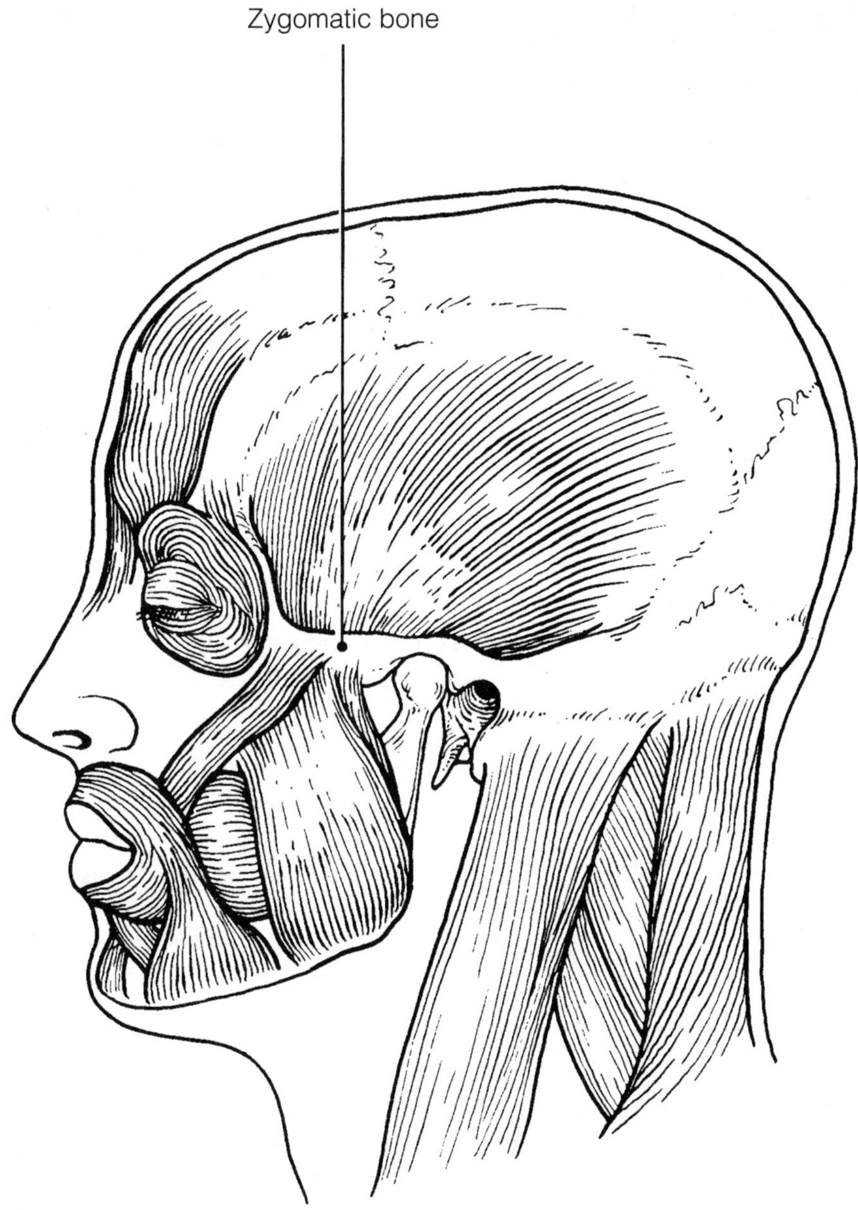

Figure 10.4

4. Describe briefly the function of the pharyngeal constrictor muscles.

____

____

5. Name the anterior trunk muscles described below. Then, for each muscle name that has a color-coding circle, select a different color to color the coding circle and the corresponding muscle on Figure 10.5.

○ ______________________ 1. A major spine flexor; the name means "straight muscle of the abdomen"

○ ______________________ 2. Prime mover for shoulder flexion and adduction

○ ______________________ 3. Prime mover for shoulder abduction

○ ______________________ 4. Forms the external lateral walls of the abdomen

○ ______________________ 5. Acting alone, each muscle of this pair turns the head toward the opposite shoulder

○ ______________________ 6. Prime mover to protract and hold the scapula against the thorax wall; the "boxer's muscle"

○ ______________________ 7. Four muscle pairs that together form the so-called abdominal girdle

______________________ 8. A tendinous "seam" running from the sternum to the pubic symphysis that indicates the midline point of fusion of the abdominal muscle sheaths

○ ______________________ 9. The deepest muscle of the abdominal wall

______________________ 10. Deep muscles of the thorax that promote the inspiratory phase of breathing

______________________ 11. An unpaired muscle that acts in concert with the muscles named immediately above to accomplish inspiration

______________________ 12. A flat, thoracic muscle deep to the pectoralis major that acts to draw the scapula inferiorly or to elevate the rib cage

6. What is the functional reason the muscle group on the dorsal leg (calf) is so much larger than the muscle group in the ventral region of the leg?

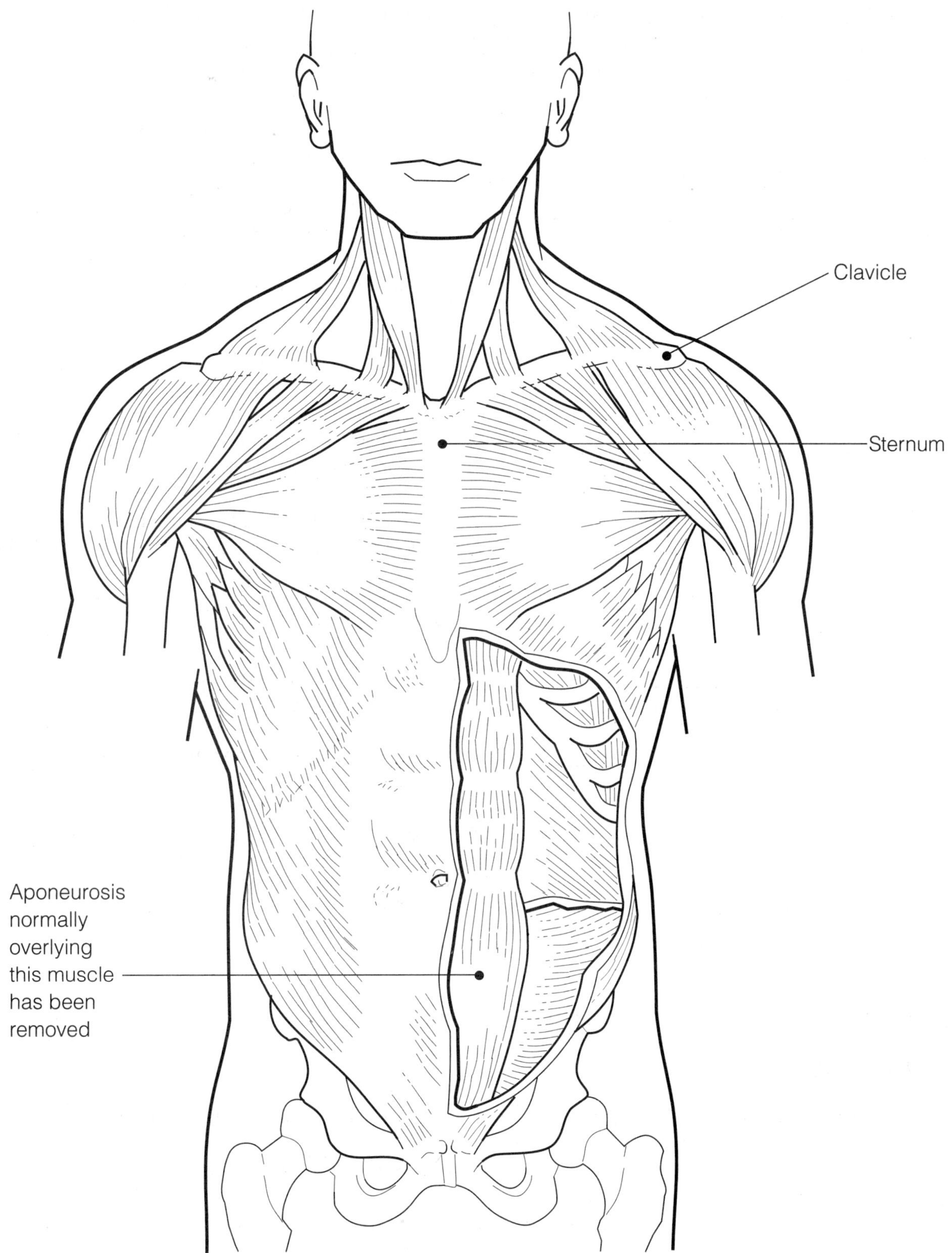
Clavicle
Sternum
Aponeurosis
normally
overlying
this muscle
has been
removed

Figure 10.5

7. Name the posterior trunk muscles described below. Select a different color for each muscle with a color-coding circle and color the coding circles and the corresponding muscles on Figure 10.6.

○ ____________________ 1. A muscle that enables you to shrug your shoulders or extend your head

○ ____________________ 2. A muscle that adducts the shoulder and causes extension of the shoulder joint

○ ____________________ 3. The shoulder muscle that is the antagonist of the muscle described in Question 2 above

____________________ 4. Prime mover of back extension; a composite muscle consisting of three columns

____________________ 5. A fleshy muscle forming part of the posterior abdominal wall that helps to maintain upright posture

○ ____________________ 6. Acting individually, small rectangular muscles that rotate the glenoid cavity of the scapulae inferiorly

○ ____________________ 7. Synergist of the trapezius in scapular elevation; acts to flex the head to the same side

○ ____________________ 8. Synergist of latissimus dorsi in extension and adduction of the humerus

○ ____________________ 9. A rotator cuff muscle; prevents downward dislocation of the humerus

○ ____________________ 10. A rotator cuff muscle; rotates the humerus laterally

○ ____________________ 11. A rotator cuff muscle; lies immediately inferior to the infraspinatus

8. Several muscles that act to move and/or stabilize the scapula are listed in the key choices. Match them to the appropriate descriptions below.

Key Choices

A. Levator scapulae B. Rhomboids C. Serratus anterior D. Trapezius

____________________ 1. Muscle that holds the scapula tightly against the thorax wall

____________________ 2. Kite-shaped muscle pair that elevates, stabilizes, and depresses the scapulae

____________________ 3. Small rectangular muscles that square the shoulders as they act together to retract the scapula

____________________ 4. Small muscle pair that elevates the scapulae

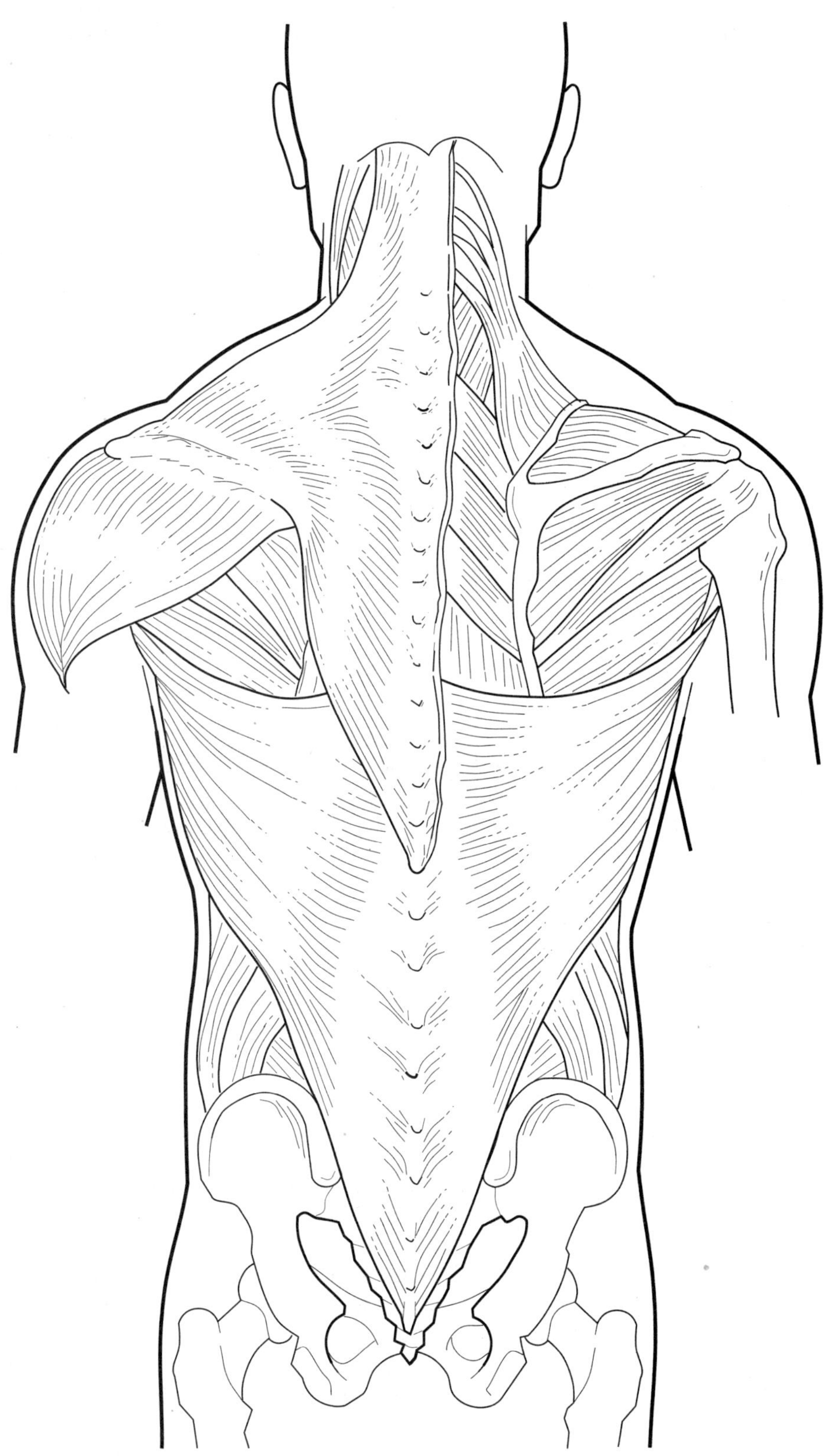

Figure 10.6

9. The muscles described in Column A comprise the pelvic floor and perineum and help support the abdominopelvic organs. Match the muscle names in Column B with the appropriate descriptions in Column A by filling in the answer blanks with the correct letters.

	Column A	Column B
______ ______	1. Two paired muscles forming the bulk of the pelvic diaphragm	A. Bulbospongiosus
		B. Coccygeus
______ ______	2. Muscles that form the urogenital diaphragm	C. Deep transverse perineal muscle
______ ______ ______	3. Muscles forming the superficial space	
______	4. A muscle that constricts the urethra	D. Ischiocavernosus
______	5. Empties the urethra; assists in penile erection	E. Levator ani
______	6. The most important muscle pair in supporting the pelvic viscera; forms sphincters at the anorectal junction	F. External urethral sphincter
______	7. Retards venous drainage and helps maintain penile erection	G. Superficial transverse perineal muscle

10. Identify the neck or vertebral column muscles described below by selecting answers from the key choices. Put each answer in the appropriate answer blank, then color the coding circles and the corresponding muscles on Figure 10.7.

Key Choices

A. Erector spinae C. Scalenes E. Splenius

B. Quadratus lumborum D. Semispinalis F. Sternocleidomastoid

______________________ 1. Elevates the first two ribs

○ ______________________ 2. Prime mover of back extension; consists of three muscle columns (iliocostalis, longissimus, and spinalis)

○ ______________________ 3. One flexes the vertebral column laterally; the pair extends the lumbar spine and fixes the 12th rib

○ ______________________ 4. Extends the vertebral column and head and rotates them to the opposite side

______________________ 5. Prime mover of head flexion; spasms of one causes torticolis

______________________ 6. Acting together, the pair extends the head; one rotates the head and bends it laterally

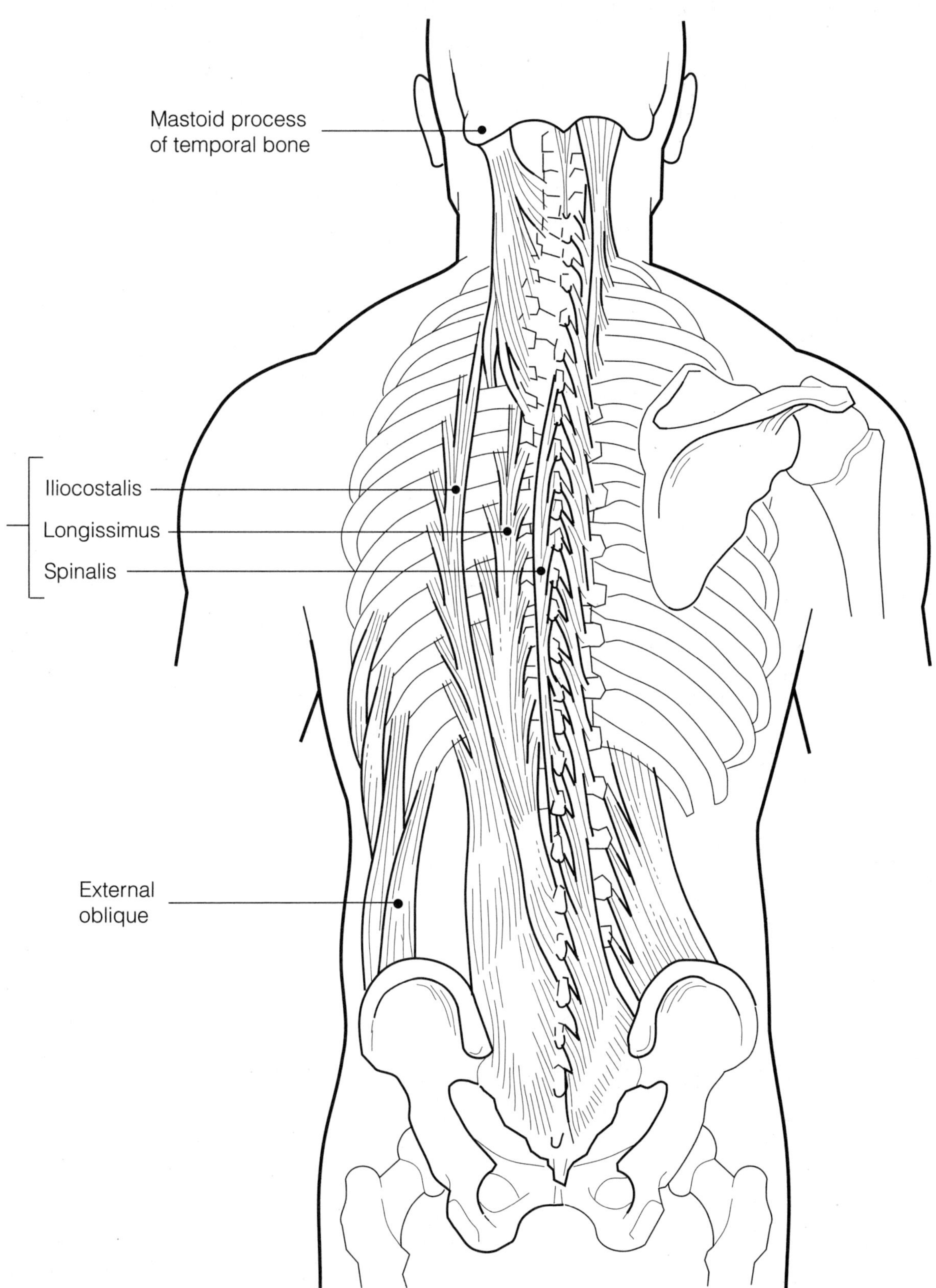
Mastoid process
of temporal bone
Iliocostalis
Longissimus
Spinalis
External
oblique

Figure 10.7

11. Match the forearm muscles listed in Column B with their major action in Column A. Then choose colors and color the muscles provided with coding circles on Figure 10.8. Also identify (by labeling)—two muscles that act on the elbow rather than the wrist and hand—the biceps brachii and the brachioradialis—and the thenar and lumbrical muscles of the hand. All of these muscles have leader lines.

Column A

_____ 1. Flexes the wrist and middle phalanges

_____ 2. Pronate the forearm

_____ 3. Flexes the distal interphalangeal joints

_____ _____ 4. Powerful wrist flexors

_____ 5. Thumb flexors

_____ 6. Extend and abduct the wrist

_____ 7. Prime mover of finger extension

_____ 8. Extend the thumb

_____ 9. Puts the forearm and hand in the anatomical position

_____ 10. Small muscle that provides a guide to locate the median nerve at the wrist

Column B

○ A. Extensor carpi radialis longus (and brevis)

B. Extensor carpi ulnaris

C. Extensor digitorum

D. Extensor pollicis longus and brevis

○ E. Flexor carpi radialis

○ F. Flexor carpi ulnaris

G. Flexor digitorum profundus

○ H. Flexor digitorum superficialis

○ I. Flexor pollicis longus and brevis

○ J. Palmaris longus

○ K. Pronators teres and quadratus

L. Supinator

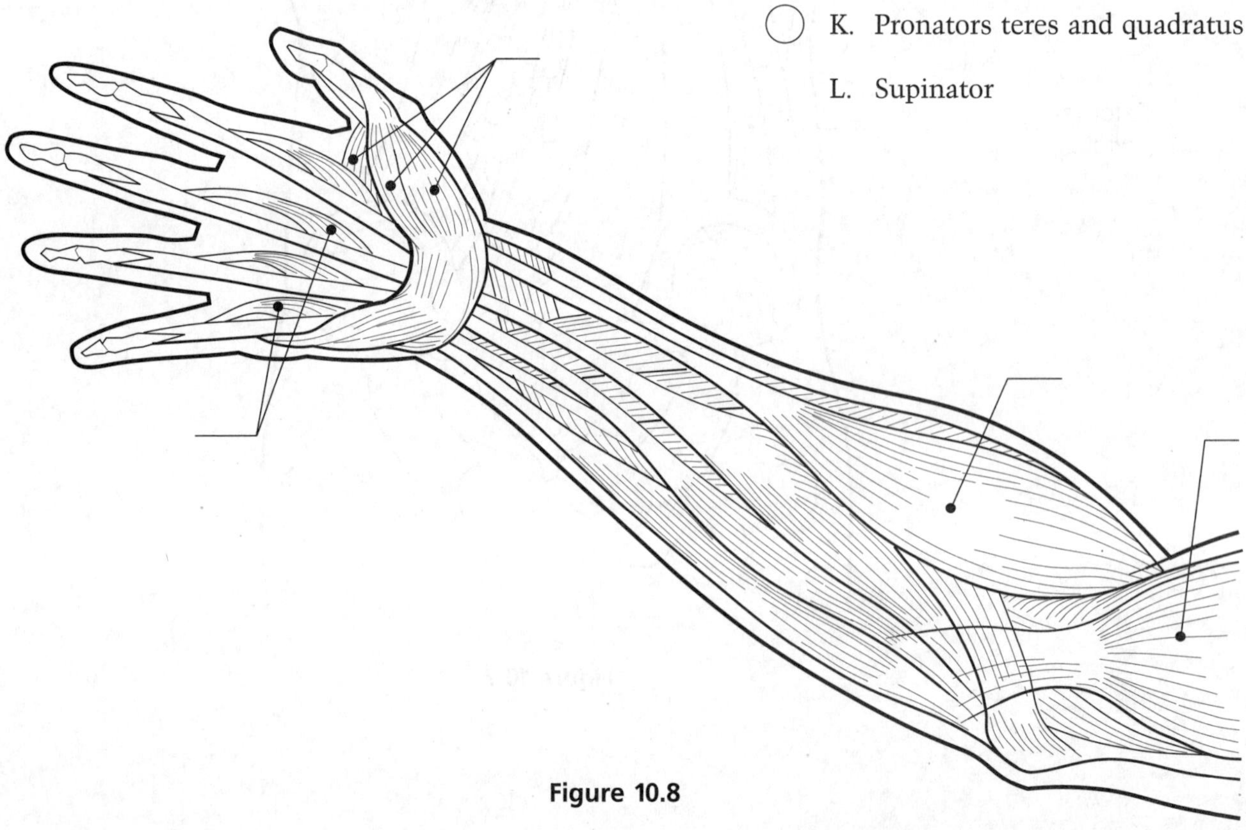

Figure 10.8

12. Name the upper limb muscles described below. Then select a different color for each muscle that has a color-coding circle and color the muscles on Figure 10.9.

1. Wrist flexor that follows the ulna

 ○ ______________________________

2. The muscle that extends the fingers

 ○ ______________________________

3. Two elbow flexor muscles: the first also supinates

 ○ ______________________________

4. The second is a synergist at best

 ○ ______________________________

5. The muscle that extends the elbow: prime mover

 ○ ______________________________

6. A short muscle; synergist of triceps brachii

 ○ ______________________________

7. A stocky muscle, deep to the biceps brachii; a prime mover of elbow flexion

 ____

8.–9. Two wrist extensors that follow the radius

 ○ ______________________________

 ○ ______________________________

10. The muscle that abducts the thumb

 ○ ______________________________

11. The muscle that extends the thumb

 ○ ______________________________

12. A powerful shoulder abductor; used to raise the arm overhead

 ○ ______________________________

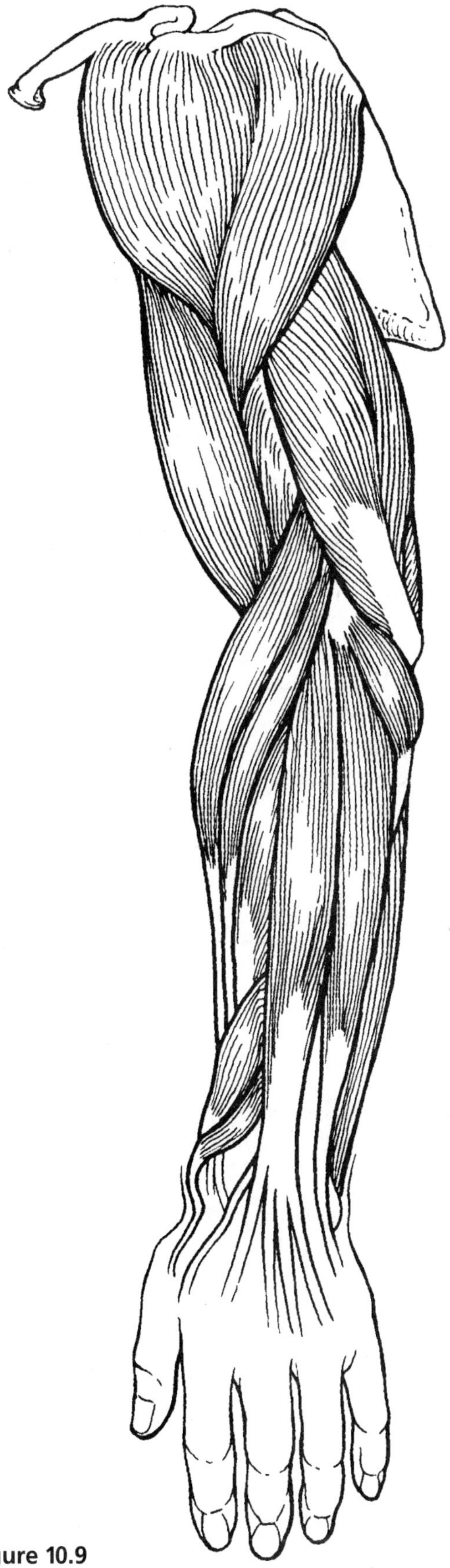

Figure 10.9

13. Name the muscles of the lower limb described below. Select a different color for each muscle that has a color-coding circle and color the circles and the muscles on Figure 10.10. Complete the illustration by labeling those muscles with leader lines.

________________ 1. Strong hip flexor, deep in pelvis; a composite of two muscles

○ ________________ 2. Power extensor of the hip; forms most of buttock mass

○ ________________ 3. Prime mover of plantar flexion; a composite of two muscles

○ ________________ 4. Inverts and dorsiflexes the foot

○ ________________ 5. The group that enables you to draw your legs to the midline of your body, as when standing at attention

○ ________________ 6. The muscle group that extends the knee

○ ________________ 7. The muscle group that extends the thigh and flexes the knee

○ ________________ 8. The smaller hip muscle commonly used as an injection site

○ ________________ 9. The thin superficial muscle of the medial thigh

○ ________________ 10. A muscle enclosed within fascia that blends into the iliotibial tract; a synergist of the iliopsoas

○ ________________ 11. Dorsiflexes and everts the foot; prime mover of toe extension

○ ________________ 12. A superficial muscle of the lateral leg; plantar flexes and everts the foot

________________ 13. A muscle deep to the soleus; prime mover of foot inversion; stabilizes the medial longitudinal arch of the foot and plantar flexes the ankle

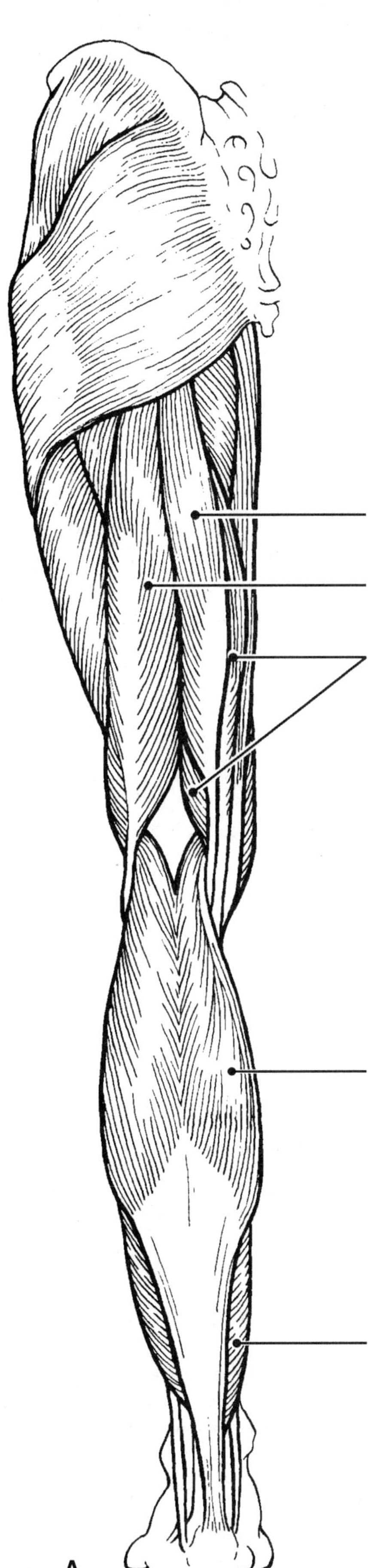

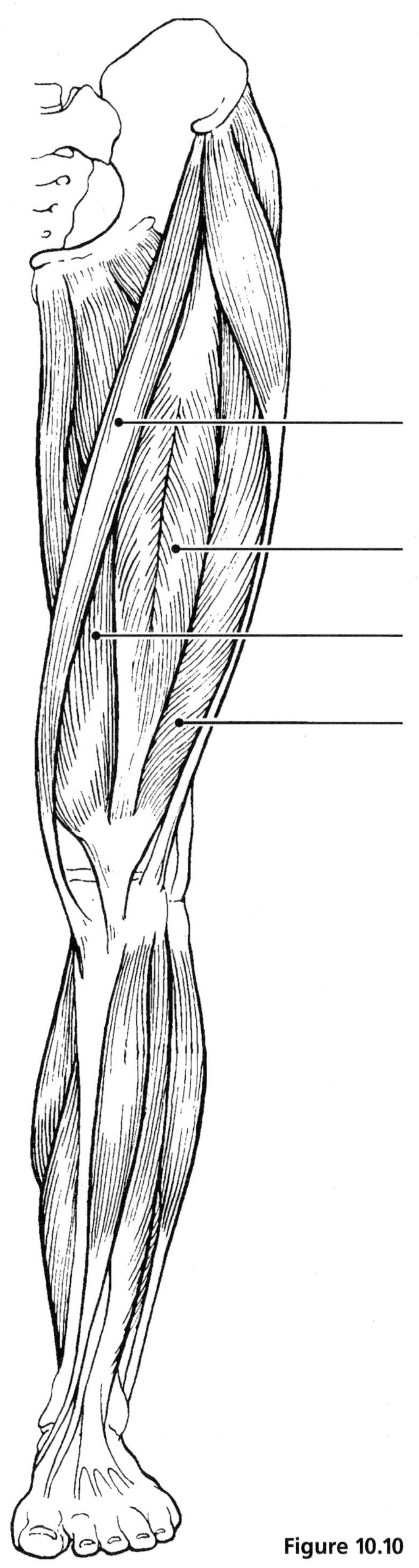

Figure 10.10

14. Complete the following statements describing muscles. Write the correct answers in the answer blanks.

______________ 1.	Three muscles, __(1)__, __(2)__, and __(3)__, are commonly used for intramuscular injections.
______________ 2.	
______________ 3.	The insertion tendon of the __(4)__ group contains a large sesamoid bone, the patella.
______________ 4.	The triceps surae insert in common into the __(5)__ tendon.
______________ 5.	The bulk of the tissue of a muscle tends to lie __(6)__ to the part of the body it causes to move.
______________ 6.	The extrinsic muscles of the hand originate on the __(7)__.
______________ 7.	
______________ 8.	Most flexor muscles are located on the __(8)__ aspect of the body; most extensors are located __(9)__. An exception to this generalization is the extensor-flexor musculature of the __(10)__.
______________ 9.	
______________ 10.	The pectoralis major and deltoid muscles act synergistically to __(11)__ the arm.
______________ 11.	

15. When kicking a football, at least three major actions of the lower limb are involved. Name the major muscles (or muscle groups) responsible for the following:

1. Flexing the hip joint: ______________________________

2. Extending the knee: ______________________________

3. Dorsiflexing the foot: ______________________________

____

16. Match the muscles of the foot (Column B) with their descriptions in column A.

	Column A	Column B
____	1. Small four-part muscle on the dorsal aspect of the foot	A. Abductor digiti minimi
		B. Abductor hallucis
____	2. Third-layer muscle; flexes metatarsophalangeal joint of the great toe	C. Adductor hallucis
		D. Extensor digitorum brevis
____	3. Important in maintaining the transverse arch of the foot	E. Flexor digitorum brevis
____	4. Abducts the little toe	F. Flexor hallucis brevis
____	5. "Worm"-like muscles of the second muscle layer	G. Interossei
		H. Lumbricals

_____ 6. Fourth-layer muscles; abduct and adduct the toes

_____ 7. Abducts the great toe

17. Now that you have begun your study of the skeletal muscles of the body, use what you have learned to match a specific muscle to its fascicle arrangement:

	Column A	**Column B**
____________	1. Deltoid	A. Convergent
____________	2. Rectus femoris	B. Fusiform
____________	3. Gracilis	C. Circular
____________	4. Gluteus maximus	D. Bipennate
____________	5. Vastus medialis	E. Multipennate
____________	6. External urethral sphincter	F. Parallel
		G. Unipennate
____________	7. Biceps brachii	

18. Circle the term that does not belong in each of the following groupings.

1. Vastus lateralis Vastus medialis Knee extension Biceps femoris
2. Latissimus dorsi Pectoralis major Synergists Adduction of shoulder Antagonists
3. Gluteus minimus Lateral rotation Gluteus maximus Piriformis Obturator externus
4. Adductor magnus Vastus medialis Rectus femoris Origin on os coxa Iliacus
5. Lateral rotation Teres minor Supraspinatus Infraspinatus
6. Tibialis posterior Flexor digitorum longus Fibularis longus Foot inversion
7. Supraspinatus Rotator cuff Teres major Teres minor Subscapularis
8. Biceps brachii/triceps brachii Pectoralis/latissimus dorsi Hamstrings/quadriceps Gastrocnemius/soleus

19. Identify the numbered muscles in Figure 10.11. Match each number with one of the following muscle names. Then select a different color for each muscle that has a color-coding circle and color each muscle group on Figure 10.11.

○ ______________________ 1. Orbicularis oris

○ ______________________ 2. Pectoralis major

○ ______________________ 3. External oblique

○ ______________________ 4. Sternocleidomastoid

○ ______________________ 5. Biceps brachii

○ ______________________ 6. Deltoid

○ ______________________ 7. Vastus lateralis

○ ______________________ 8. Frontal belly of epicranius

○ ______________________ 9. Rectus femoris

○ ______________________ 10. Sartorius

○ ______________________ 11. Gracilis

○ ______________________ 12. Adductor group

○ ______________________ 13. Fibularis longus

○ ______________________ 14. Temporalis

○ ______________________ 15. Orbicularis oculi

○ ______________________ 16. Zygomaticus

○ ______________________ 17. Masseter

○ ______________________ 18. Vastus medialis

○ ______________________ 19. Tibialis anterior

○ ______________________ 20. Transversus abdominus

○ ______________________ 21. Tensor fascia lata

○ ______________________ 22. Rectus abdominis

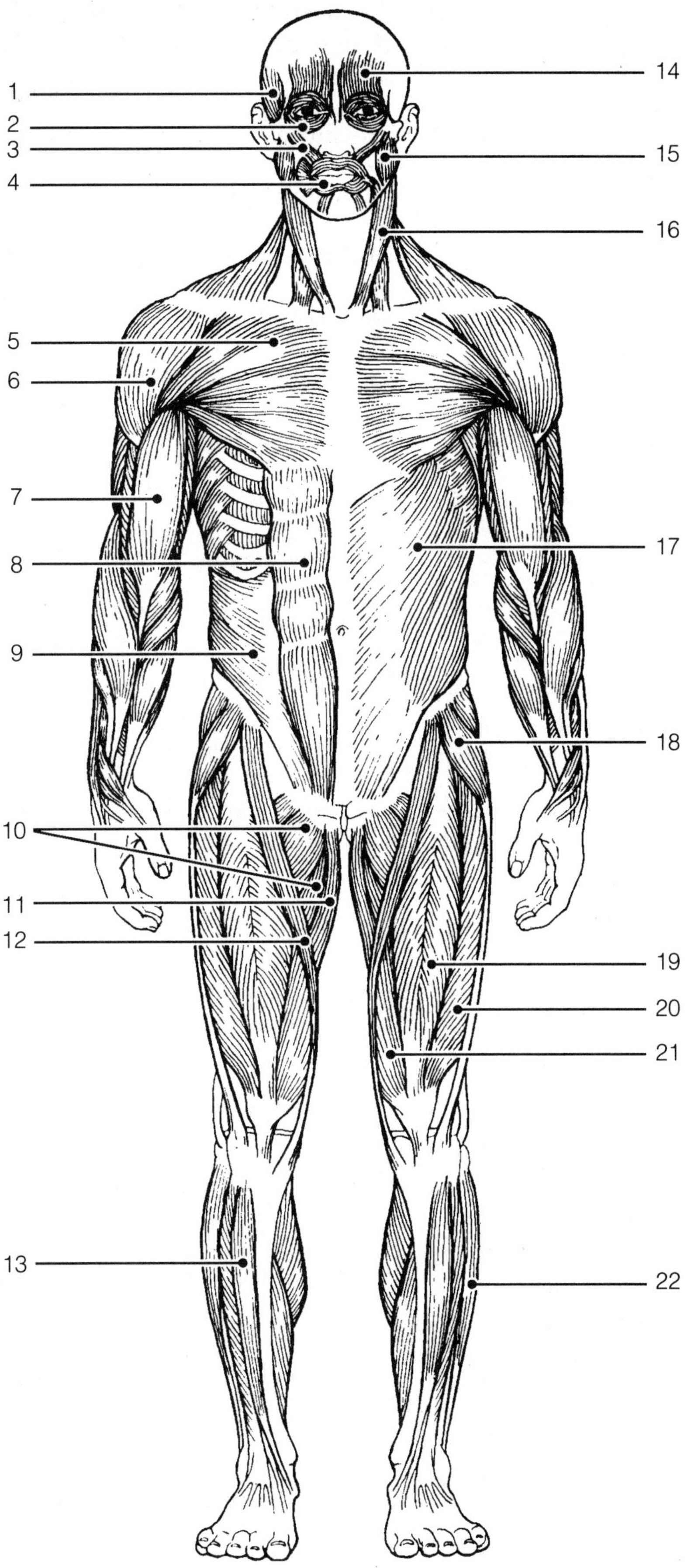

Figure 10.11

20. Identify each of the numbered muscles in Figure 10.12. Match each number with one of the following muscle names. Then select different colors for each muscle and color the coding circles and corresponding muscles on Figure 10.12.

○ ____________________ 1. Gluteus maximus

○ ____________________ 2. Adductor muscles

○ ____________________ 3. Gastrocnemius

○ ____________________ 4. Latissimus dorsi

○ ____________________ 5. Deltoid

○ ____________________ 6. Semitendinosus

○ ____________________ 7. Trapezius

○ ____________________ 8. Biceps femoris

○ ____________________ 9. Triceps brachii

○ ____________________ 10. External oblique

○ ____________________ 11. Gluteus medius

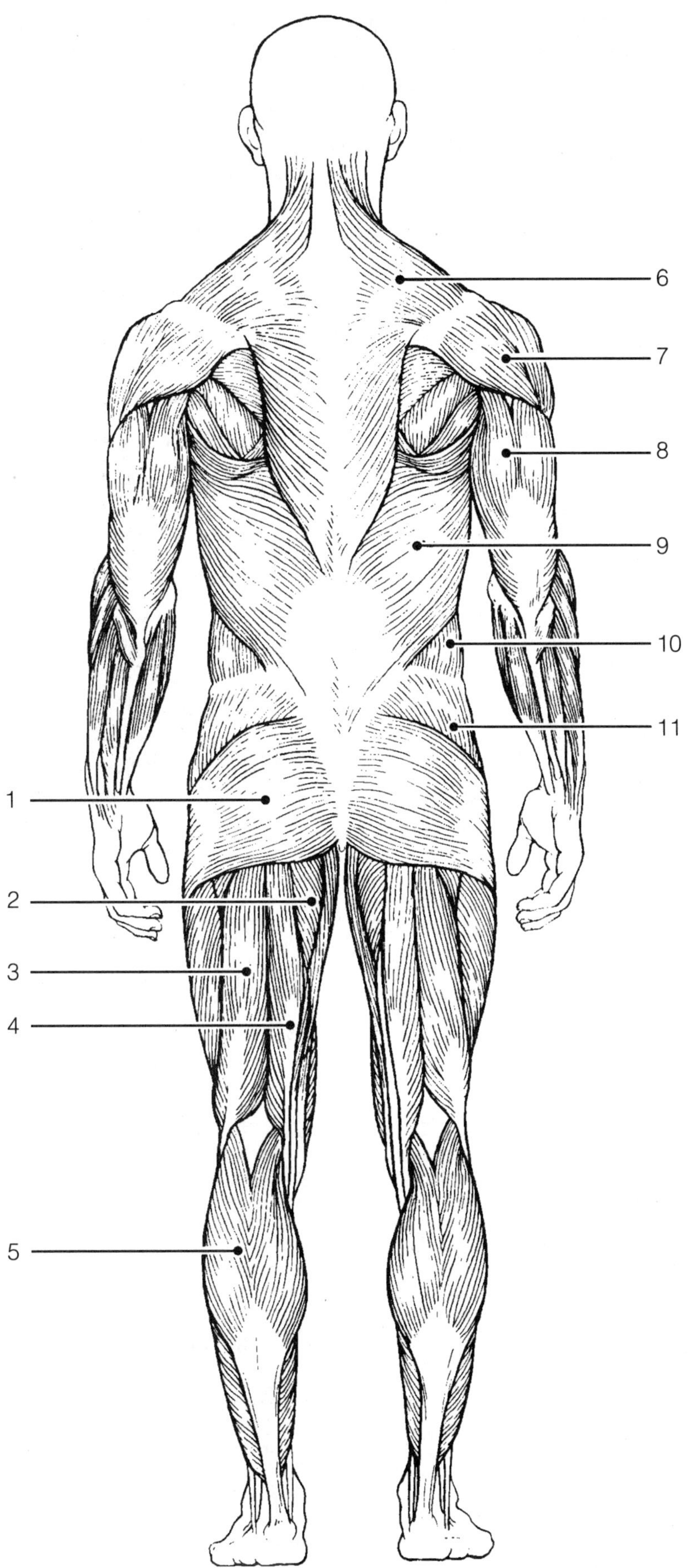

Figure 10.12

21. On the following two pages are skeleton diagrams, both anterior and posterior views (Figures 10.13 and 10.14). Listed below are muscles that should be drawn on the indicated sides of the skeleton diagrams. Using your muscle charts and, if necessary, referring back to bone landmarks in Chapter 7, draw the muscles as accurately as possible. Be sure to draw each muscle on the correct side of the proper figure; the muscles are grouped to minimize overlap.

Figure 10.13 (Anterior View)

Right side of body

1. Orbicularis oculi
2. Sternocleidomastoid
3. Deltoid
4. Brachialis
5. Vastus medialis
6. Vastus lateralis

Left side of body

1. Rectus femoris
2. Tibialis anterior
3. Quadratus lumborum
4. Frontal belly of the epicranius
5. Platysma

Figure 10.13 (Posterior View)

Right side of body

1. Deltoid
2. Gluteus minimus
3. Semimembranosus
4. Occipital belly of the epicranius

Left side of body

1. Trapezius
2. Biceps femoris

Figure 10.14 (Anterior View)

Right side of body

1. Pectoralis major
2. Psoas major
3. Iliacus
4. Adductor magnus
5. Gracilis

Left side of body

1. Pectoralis minor
2. Rectus abdominis
3. Biceps brachii
4. Sartorius
5. Vastus intermedius

Figure 10.14 (Posterior View)

Right side of body

1. Latissimus dorsi
2. Gluteus maximus
3. Gastrocnemius

Left side of body

1. Triceps brachii
2. Gluteus medius
3. Semitendinosus
4. Soleus

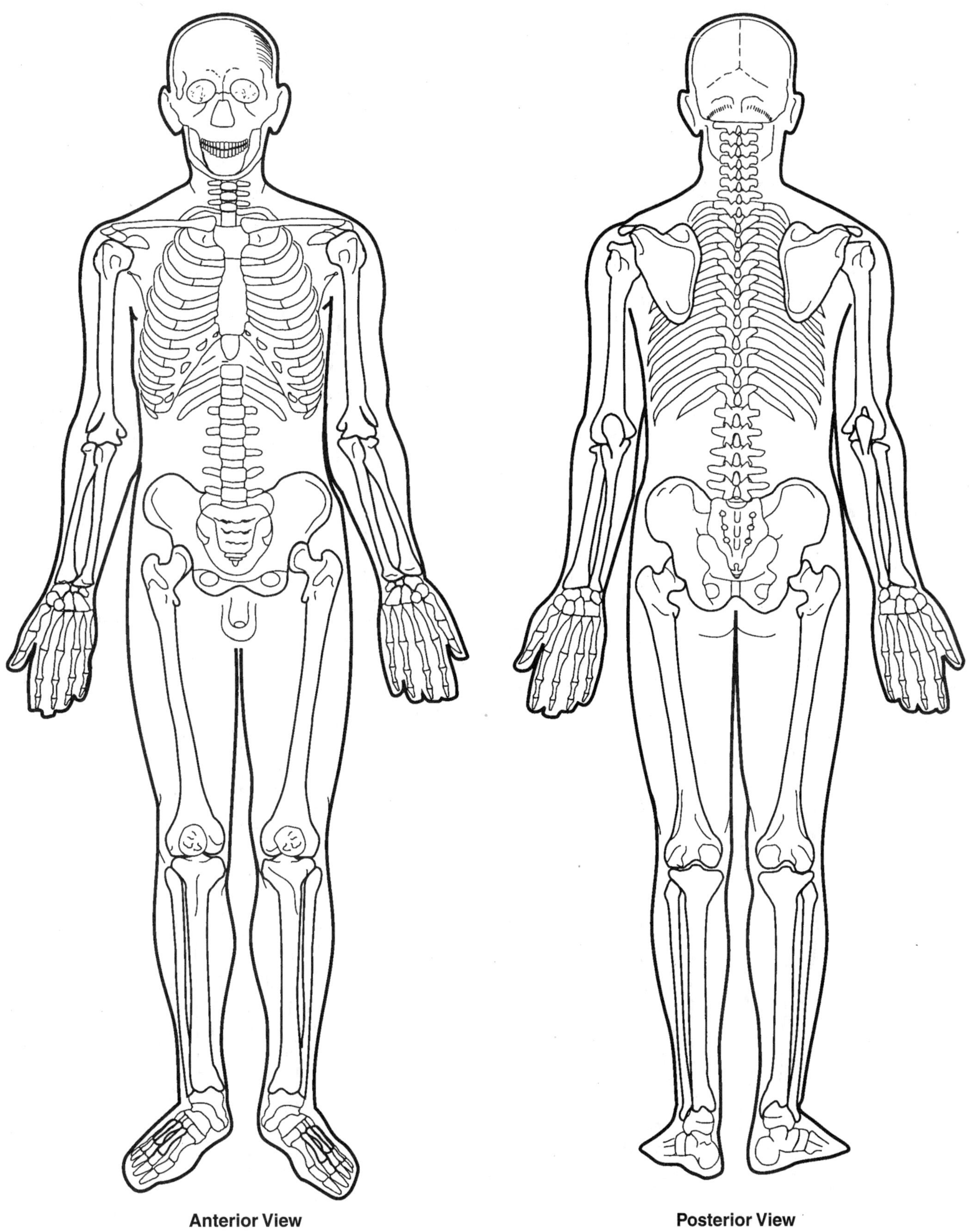
Anterior View
Posterior View

Figure 10.13

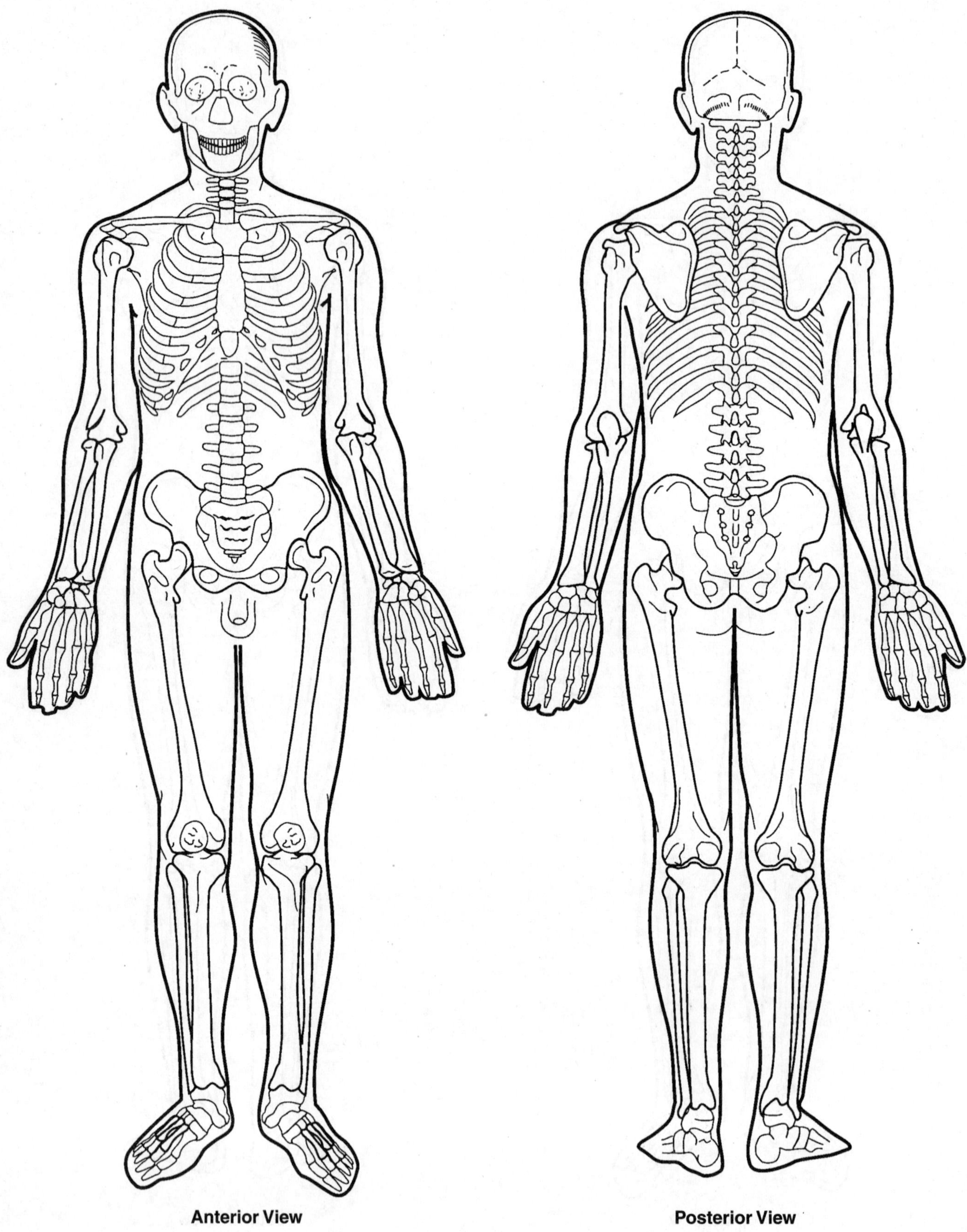

Figure 10.14

CHALLENGING YOURSELF

At the Clinic

1. An elderly man is brought to the clinic by his distraught wife. Among other signs, the nurse notices that the muscles on the right side of his face are slack. What nerve is not functioning properly?

2. Pete, who has been moving furniture all day, arrives at the clinic complaining of painful spasms in his back. He reports having picked up a heavy table by stooping over. What muscle group has Pete probably strained, and why are these muscles at risk when one lifts objects improperly?

3. An accident victim (who was not wearing a seat belt) was thrown from a vehicle and pronounced dead at the scene. The autopsy reveals the cause of death to be spinal cord injury resulting in paralysis of the phrenic and intercostal nerves. Why is this injury fatal?

4. Mr. Posibo has had gallbladder surgery. Now he is experiencing weakness of the muscles on his right side only, the side in which the incision was made through the abdominal musculature. Consequently, the abdominal muscles on his left side contract more strongly, throwing his torso into a lateral flexion. Mr. Posibo needs physical therapy. What abnormal spinal curvature will result if he doesn't get it and why?

5. An emergency appendectomy is performed with the incision made at the lateral edge of the right iliac abdominopelvic region. Was the rectus abdominis cut?

6. In some women who have borne several children, the uterus prolapses (everts through the weakened pelvic diaphragm). The weakening of what muscles allows this to happen?

7. What muscles must be immobilized to prevent movement of a broken clavicle?

8. Out of control during a temper tantrum, Malcolm smashed his fist through a glass door and severed several tendons at the anterior wrist. What movements are likely to be lost if tendon repair is not possible?

9. During an overambitious workout, a high school athlete pulls some muscles by forcing his knee into extension when his hip is already fully flexed. What muscles does he pull?

10. Susan, a massage therapist, was giving Mr. Graves a back rub. What two broad superficial muscles of the back were receiving the "bulk" of her attention?

11. An elderly woman with extensive osteoarthritis of her left hip joint entered the hospital to have total hip joint replacement (prosthesis implantation). After surgery, her left hip was maintained in adduction to prevent dislocation of the prosthesis while healing was occurring. Physical therapy was prescribed to prevent atrophy of the gluteal muscles during the interval of disuse. Name the gluteal muscles and describe which of their actions were being prevented while the hip was adducted.

12. In adolescents who are still growing, steroids act to speed up conversion of cartilage to bone. In regard to bone growth, what negative consequence might ensue from steroid use by young athletes?

13. Ten-year-old Billy is jumping up and down as he tells his mother he has to "go bad" (urinate). She tells him he will have to "hold it" until they get home. What muscles must he keep contracted until then to prevent urination?

Stop and Think

1. Do all skeletal muscles attach to a bone? If not, what else do they attach to, and how is the result different from muscle attachment to bone?

2. Are the *stated* origins and insertions always the *functional* origins and insertions? That is, is the stated origin always fixed, and does the stated insertion always move?

3. State, in general terms, how to tell by a muscle's location what its action will be.

4. How can the latissimus dorsi, which is on the back, aid in medial, instead of lateral, rotation of the humerus?

5. Where is the insertion of the tibialis anterior, which is able to invert the foot? Where is the insertion of the fibularis longus, a foot everter?

COVERING ALL YOUR BASES

Multiple Choice

Select the best answer or answers from the choices given.

1. Which one of the following muscles has the most motor units relative to its size?
   A. Flexor pollicis longus
   B. Rectus femoris
   C. Soleus
   D. Latissimus dorsi

2. Which muscle has the largest motor units?
   A. Fibularis brevis
   B. Gluteus maximus
   C. Frontalis
   D. Platysma

3. Head muscles that insert on a bone include the:
   A. zygomaticus
   B. masseter
   C. buccinator
   D. temporalis

4. Which muscles change the position or shape of the lips?
   A. Zygomaticus
   B. Orbicularis oris
   C. Platysma
   D. Mentalis

5. Chewing muscles include the:
   A. buccinator
   B. medial pterygoid
   C. lateral pterygoid
   D. sternothyroid

6. Which of the following plays a role in swallowing?
   A. Lateral pterygoid
   B. Styloglossus
   C. Digastric
   D. Geniohyoid

7. The hyoid bone provides an insertion for the:
   A. digastric
   B. pharyngeal constrictor muscles
   C. sternothyroid
   D. thyrohyoid

8. Muscles that function in head rotation include:
   A. scalenes
   B. sternocleidomastoid
   C. splenius
   D. semispinalis capitis

9. Lateral flexion of the torso involves:
   A. erector spinae
   B. rectus abdominis
   C. quadratus lumborum
   D. external oblique

10. Muscles attached to the vertebral column include:
    A. quadratus lumborum
    B. external oblique
    C. diaphragm
    D. latissimus dorsi

11. Which of the following muscles attach to the hip bones?
    A. Transversus abdominis
    B. Rectus femoris
    C. Vastus medialis
    D. Longissimus group of erector spinae

12. Muscles that attach to the rib cage include:
    A. scalenes
    B. internal oblique
    C. trapezius
    D. psoas major

13. Which muscles are part of the pelvic diaphragm?
A. Superficial transverse perineal muscle
B. Deep transverse perineal muscle
C. Levator ani
D. External urethral sphincter

14. Which of the following insert on the arm?
A. Biceps brachii C. Trapezius
B. Triceps brachii D. Coracobrachialis

15. Rotator cuff muscles include:
A. supraspinatus D. teres major
B. infraspinatus E. subscapularis
C. teres minor

16. Muscles that help stabilize the scapula and shoulder joint include:
A. triceps brachii C. trapezius
B. biceps brachii D. rhomboids

17. Which muscles insert on the radius?
A. Triceps brachii
B. Biceps brachii
C. Flexor carpi radialis
D. Pronator teres

18. Which muscles insert on the femur?
A. Quadratus lumborum
B. Iliacus
C. Gracilis
D. Tensor fasciae latae

19. Which of these thigh muscles causes movement at the hip joint?
A. Rectus femoris
B. Biceps femoris
C. Vastus lateralis
D. Semitendinosus

20. Leg muscles that can cause movement at the knee joint include:
A. tibialis anterior C. gastrocnemius
B. fibularis longus D. soleus

21. Muscles that function in lateral hip rotation include:
A. gluteus medius
B. gluteus maximus
C. obturator externus
D. tensor fasciae latae

22. Hip adductors include:
A. sartorius C. vastus medialis
B. gracilis D. pectineus

23. Which muscles are in the posterior compartment of the leg?
A. Flexor digitorum longus
B. Flexor hallucis longus
C. Extensor digitorum longus
D. Fibularis tertius

24. Which of the following insert distal to the tarsus?
A. Soleus
B. Tibialis anterior
C. Extensor hallucis longus
D. Fibularis longus

25. Which muscles are contracted while standing at attention?
A. Iliocostalis
B. Rhomboids
C. Tensor fascia latae
D. Flexor digitorum longus

26. The main muscles used when doing chin-ups are:
A. triceps brachii and pectoralis major
B. infraspinatus and biceps brachii
C. serratus anterior and external oblique
D. latissimus dorsi and brachialis

27. In walking, which two lower limb muscles keep the forward-swinging foot from dragging on the ground?
A. Pronator teres and popliteus
B. Flexor digitorum longus and popliteus
C. Adductor longus and abductor digiti minimi in foot
D. Gluteus medius and tibialis anterior

28. The major muscles used in doing push-ups are:
A. biceps brachii and brachialis
B. supraspinatus and subscapularis
C. coracobrachialis and latissimus dorsi
D. triceps brachii and pectoralis major

29. Someone who sticks out a thumb to hitch a ride is ______ the thumb.
A. extending C. adducting
B. abducting D. opposing

30. Which are ways in which muscle names have been derived?
A. Attachments C. Function
B. Size D. Location

Word Dissection

For each of the following word roots, fill in the literal meaning and give an example, using a word found in this chapter.

Word root	Translation	Example
1. agon		
2. brevis		
3. ceps		
4. cleido		
5. gaster		
6. glossus		
7. pectus		
8. perone		
9. rectus		

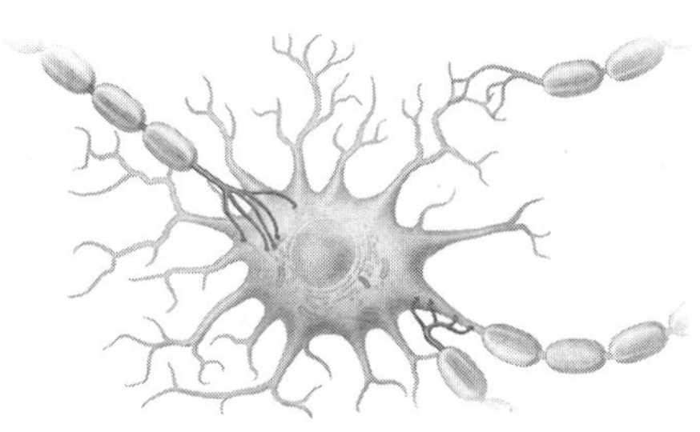

11 Fundamentals of the Nervous System and Nervous Tissue

Student Objectives

When you have completed the exercises in this chapter, you will have accomplished the following objectives:

Functions and Divisions of the Nervous System

1. List the basic functions of the nervous system.
2. Explain the structural and functional divisions of the nervous system.

Histology of Nervous Tissue

3. List the types of neuroglia and cite their functions.
4. Define neuron, describe its important structural components, and relate each to a functional role.
5. Differentiate between a nerve and a tract, and between a nucleus and a ganglion.
6. Explain the importance of the myelin sheath and describe how it is formed in the central and peripheral nervous systems.
7. Classify neurons structurally and functionally.

Membrane Potentials

8. Define resting membrane potential and describe its electrochemical basis.
9. Compare and contrast graded potentials and action potentials.
10. Explain how action potentials are generated and propagated along neurons.
11. Define absolute and relative refractory periods.
12. Define saltatory conduction and contrast it to conduction along unmyelinated fibers.

The Synapse and Neurotransmitters and Their Receptors

13. Define synapse. Distinguish between electrical and chemical synapses by structure and by the way they transmit information.
14. Distinguish between excitatory and inhibitory postsynaptic potentials.
15. Describe how synaptic events are integrated and modified.
16. Define neurotransmitter and name several classes of neurotransmitters.

Basic Concepts of Neural Integration

17. Describe common patterns of neuronal organization and processing.
18. Distinguish between serial and parallel processing.

The nervous system is the master coordinating system of the body. Every thought, action, and sensation reflects its activity. Because of its complexity, the anatomical structures of the nervous system are considered in terms of two principal divisions—the central nervous system (CNS), consisting of the brain and spinal cord, and the peripheral nervous system (PNS). The PNS, consisting of cranial nerves, spinal nerves, and ganglia, provides the communication lines between the CNS and the body's muscles, glands, and sensory receptors. It is most important to recognize that the nervous system acts in an integrated manner both structurally and functionally.

In Chapter 11, we discuss the organization of the nervous system and the histology of nervous tissue, but the primary focus is the structure and function of neurons. Because every body system is controlled, at least in part, by the nervous system, a basic comprehension of how it functions is essential to understanding overall body functioning.

BUILDING THE FRAMEWORK

Functions and Divisions of the Nervous System

1. List the three major functions of the nervous system.

1. ______________________________

____

2. ______________________________

____

3. ______________________________

____

2. Only one part of the structural classification of the nervous system is depicted in Figure 11.1. Identify that part by color coding it and coloring it on the diagram, and label its two parts. Then, draw in and color code the other structural subdivision of the nervous system, and add labels and leader lines to identify the main parts.

◯ Central nervous system (CNS) ◯ Peripheral nervous system (PNS)

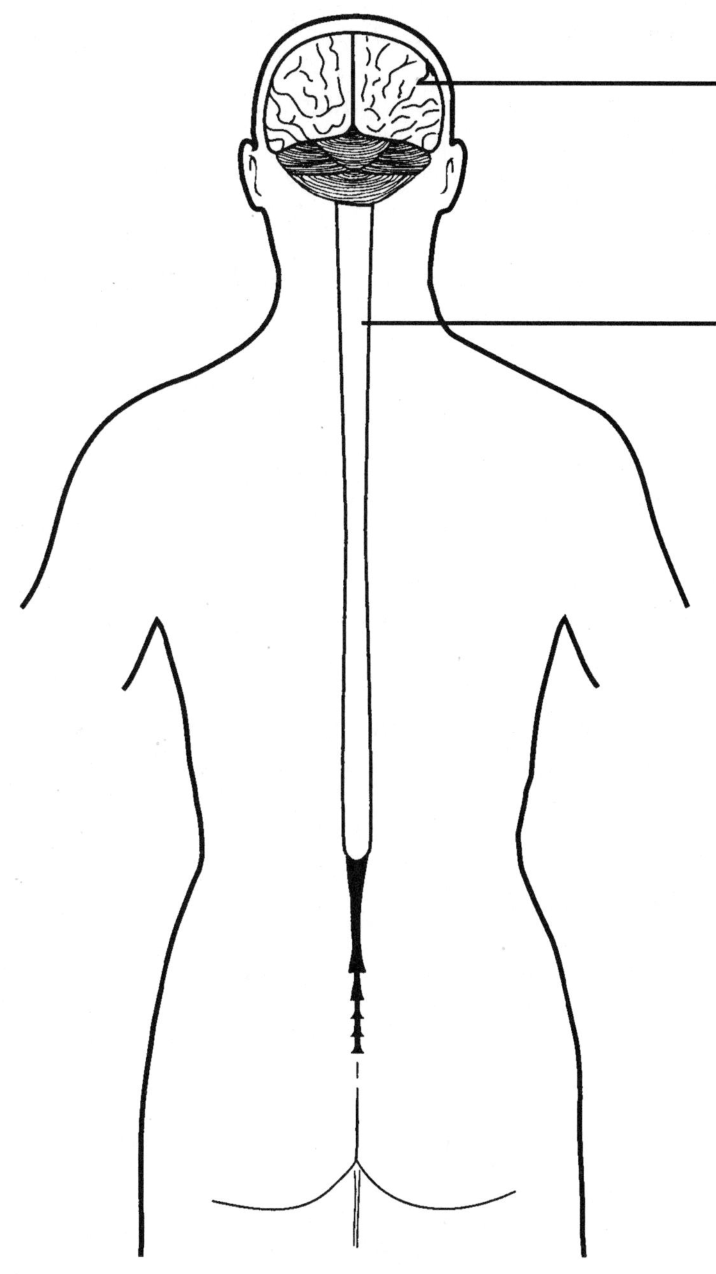

Figure 11.1

3. Choose the key choices that best correspond to the following descriptions. Insert the appropriate answers in the answer blanks.

Key Choices

A. Autonomic nervous system

B. Central nervous system

C. Peripheral nervous system

D. Somatic nervous system

__________ 1. Nervous system subdivision that is composed of the brain and spinal cord

__________ 2. Subdivision of the PNS that controls voluntary activities such as the activation of skeletal muscles

__________ 3. Nervous system subdivision that is composed of the cranial and spinal nerves and ganglia

__________ 4. Subdivision of the PNS that regulates the activity of the heart and smooth muscle and of glands; also called the involuntary nervous system

__________ 5. A major subdivision of the nervous system that interprets incoming information and issues orders

__________ 6. A major subdivision of the nervous system that serves as the communication lines, linking all parts of the body to the CNS

Histology of Nervous Tissue

1. This exercise emphasizes the difference between neurons and supporting cells. If a statement is true, write the letter T in the answer blank. If a statement is false, change the underlined word(s) and write the correct word(s) in the answer blank.

__________ 1. Supporting cells found in the CNS are called <u>neuroglia</u>.

__________ 2. <u>Neurons</u> are mitotic and therefore are responsible for most brain neoplasms.

__________ 3. Schwann cells and satellite cells are found only in the <u>CNS</u>.

__________ 4. <u>Ependymal cells</u> show irritability and conductivity.

__________ 5. Almost 50% of the volume of neural tissue in the CNS is made up of <u>neurons</u>.

__________ 6. In the CNS, <u>oligodendrocytes</u> engulf invading microorganisms and dead neural tissue.

__________ 7. <u>Astrocytes</u> line the central cavities of the brain.

____________________ 8. <u>Schwann cells</u> wrap their cytoplasmic extensions around thick neuron fibers in the CNS.

____________________ 9. The bulbous ends of <u>the axons of neurons</u> cling to capillaries.

2. Relative to neuron anatomy, match the anatomical terms in Column B with the appropriate descriptions of function in Column A. Place the correct answers in the answer blanks.

	Column A	Column B
____________	1. Releases neurotransmitters	A. Axon
____________	2. Conducts local currents toward the soma	B. Axon terminal
		C. Cell body
____________	3. Increases the speed of impulse transmission	D. Dendrite
____________	4. Location of the nucleus	E. Myelin sheath
____________	5. Conducts impulses away from the cell body	F. Nissl bodies
____________	6. Most are located and protected within the CNS	
____________	7. Short, tapering, diffusely branched extension from the cell body	
____________	8. The process called a nerve fiber	
____________	9. Formed by Schwann cells in the PNS	
____________	10. Clustered ribosomes and rough ER	
____________	11. Patchy disappearance in multiple sclerosis	
____________	12. Site of biosynthetic activities	

3. Circle the term that does not belong in each of the following groupings.

1. Nucleus Soma Centrioles Nucleolus
2. Mitochondria Rough ER Ribosomes Nissl bodies
3. Melanin Glycogen Lipofuscin Pigment
4. Dendritic spine Input Output Receptive
5. Axon terminal Synaptic knob Bouton Axon collateral

4. Figure 11.2 is a diagram of a neuron. First, label the parts with leader lines on the illustration. Then, choose different colors for each of the structures listed below and color the illustration. Finally, draw arrows on the figure to indicate the direction of impulse transmission along the neuron's membrane.

○ Axon ○ Nerve cell body ○ Neurilemma

○ Dendrites ○ Myelin sheath

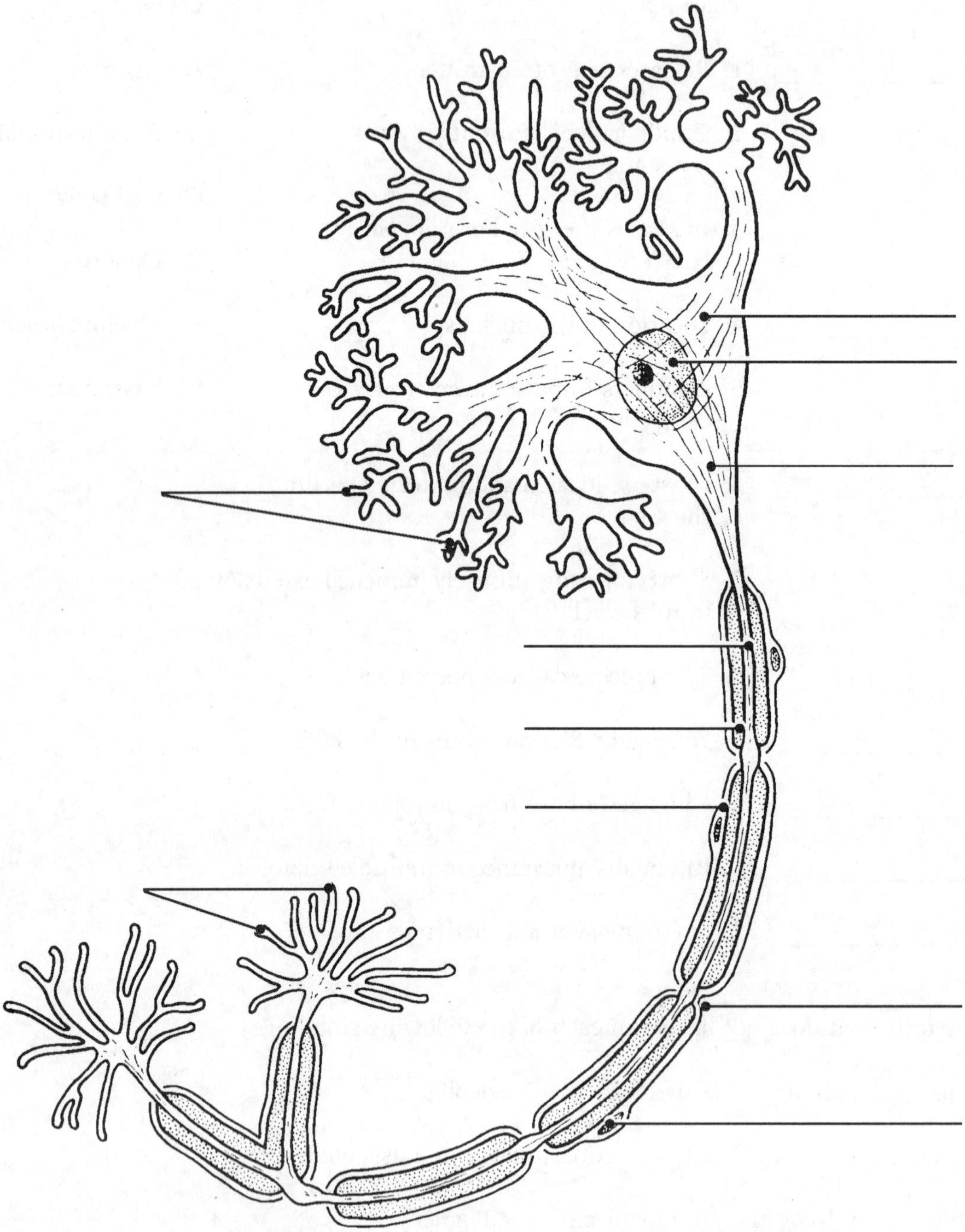

Figure 11.2

5. Match the key choices to their functional designation.

Key Choices

A. Axon B. Axon hillock C. Axon terminals D. Dendrite

__________ 1. Receptive region

__________ 2. Trigger zone

__________ 3. Conducting component

__________ 4. Secretory components

6. List the three special characteristics of neurons. __________

____

____

7. Using the key choices, select the terms that match the following descriptions. Insert the correct answers in the answer blanks.

Key Choices

A. Ganglion D. Nerve G. Synapse

B. Neuroglia E. Nodes of Ranvier H. Stimuli

C. Neurotransmitters F. Nucleus I. Tract

__________ 1. Junction or point of close contact between neurons

__________ 2. Chemicals released by neurons that stimulate other neurons, muscles, or glands

__________ 3. Gaps in a myelin sheath

__________ 4. Bundle of axons in the CNS

__________ 5. Collection of cell bodies found outside the CNS

__________ 6. Collection of cell bodies found within the CNS

__________ 7. Changes, occurring inside or outside the body, that activate the nervous system

__________ 8. Bundle of axons in the PNS

8. Circle the term that does not belong in each of the following groupings.

   1. Nodes of Ranvier   Myelin sheath   Cell body   Axon
   2. Cell body   Centrioles   Nucleolus   Ribosomes
   3. Ganglia   Clusters of cell bodies   PNS   Clusters of glial cells
   4. Dendrites   Neurotransmitters in vesicles   Telodendria   Synaptic knobs
   5. Gray matter   Myelin   Fiber tracts   White matter

9. Three diagrams of neurons are shown in Figure 11.3. On the lines below each diagram, indicate the neuron's structural and (most likely) functional classification. Color the diagram according to your fancy.

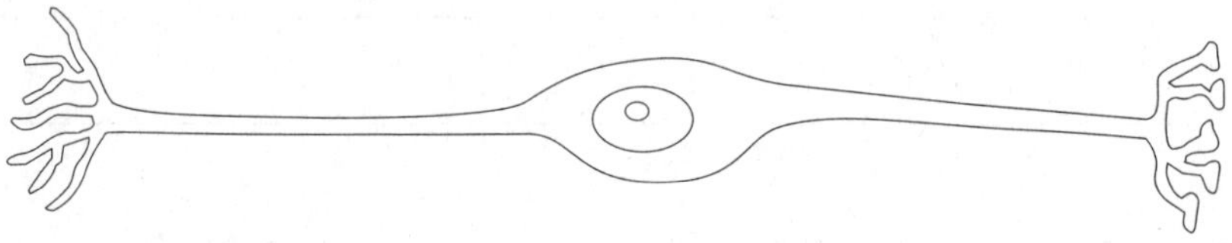

A. Classification: Structural ______________________________

Functional ______________________________

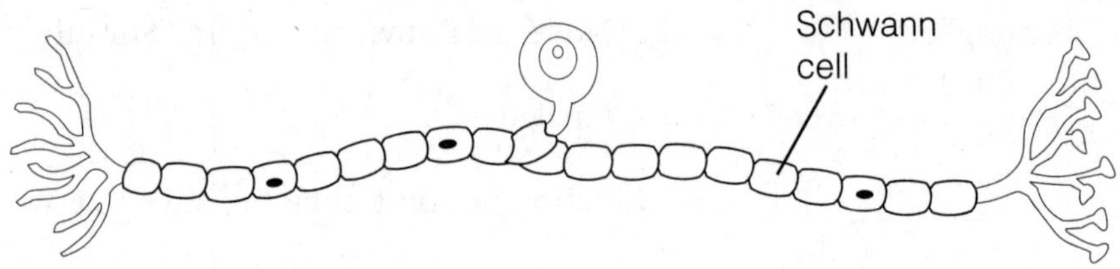

B. Classification: Structural ______________________________

Functional ______________________________

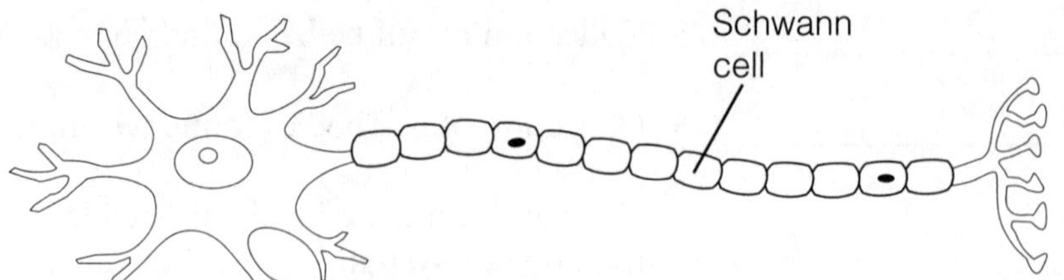

C. Classification: Structural ______________________________

Functional ______________________________

Figure 11.3

10. Several descriptions of neurons are given below. Two keys are provided. Key A lists terms that classify neurons structurally; Key B lists terms that classify neurons functionally. Choose responses from one or both keys and insert the terms in the answer blanks as needed to complete the descriptive statements that follow.

Key A:

Structural

Bipolar

Multipolar

Unipolar

Key B:

Functional

Sensory or afferent

Motor or efferent

Association

A neuron that excites skeletal muscle cells in your biceps muscle is functionally a(n) __(1)__ neuron and structurally a(n) __(2)__ neuron.

Neurons that reside entirely within the CNS are __(3)__ neurons. Structurally, most of these neurons belong to the __(4)__ classification.

A structural class of neurons that is rare in the body is the __(5)__ type.

Structurally, the most common neurons are __(6)__.

A neuron that transmits impulses from pain receptors in your skin to your spinal cord is classified as a(n) __(7)__ neuron. Structurally, this type of neuron is __(8)__.

____________	1.	____________	5.
____________	2.	____________	6.
____________	3.	____________	7.
____________	4.	____________	8.

11. Two diagrams showing a small section of PNS axon(s) are provided in Figure 11.4. First, identify which axons are *myelinated* and which are *unmyelinated* by writing one of these terms below the appropriate diagram. Then, color code and color the structures listed below. Finally, bracket the neurilemma.

○ Axon ○ Schwann cell cytoplasm ○ Schwann cell plasma membrane

○ Axolemma ○ Schwann cell nucleus

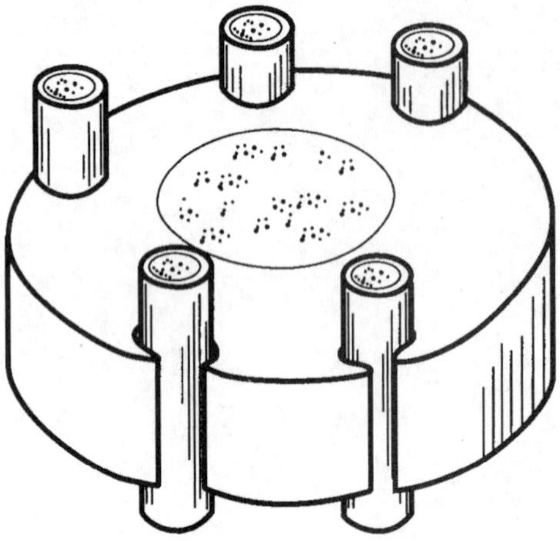

Figure 11.4

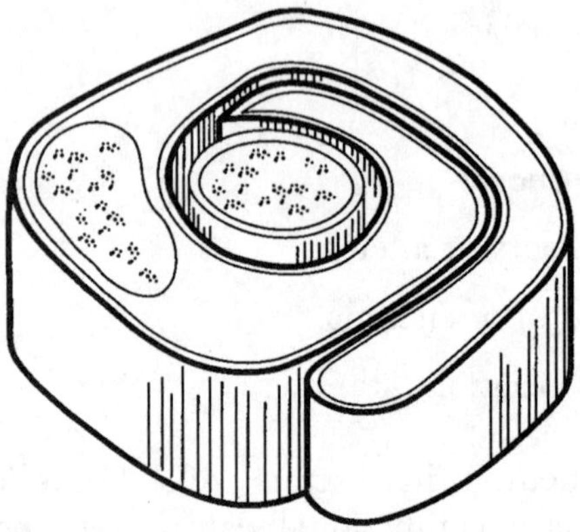

A. ____________________

B. ____________________

Membrane Potentials

1. Complete the following statements, which refer to the resting membrane potential. Write the missing terms in the answer blanks.

____________ 1.
____________ 2.
____________ 3.
____________ 4.
____________ 5.
____________ 6.
____________ 7.
____________ 8.
____________ 9.
____________ 10.
____________ 11.

The relationship of voltage (V), current (I), and resistance (R) is known as __(1)__ law and is expressed by the formula __(2)__. In the body, currents across cellular membranes are carried by __(3)__. The difference of charge on the two sides of the membrane is called the __(4)__. The resistance to current flow is provided by the __(5)__. Ion channels in plasma membranes are constructed of __(6)__. Channels that remain open are called __(7)__ channels. Channels that open and close in response to various stimuli are called __(8)__ channels. Ions move across the membrane in response to a difference in concentration, which is called the __(9)__ gradient, and also in response to a difference in charge, which is called the __(10)__ gradient. Channels that open in response to physical deformation of the receptor are called __(11)__.

2. Check (✓) all descriptions that apply to a resting neuron.

_____ 1. Its inside is negative relative to its outside.

_____ 2. Its outside is negative relative to its inside.

_____ 3. The cytoplasm contains more sodium and less potassium than does the extracellular fluid.

_____ 4. The cytoplasm contains more potassium and less sodium than does the extracellular fluid.

_____ 5. A charge separation exists at the membrane.

_____ 6. The electrochemical gradient for the movement of sodium across the membrane is greater than that for potassium.

_____ 7. The electrochemical gradient for the movement of potassium across the membrane is greater than that for sodium.

_____ 8. The membrane is more permeable to sodium than potassium.

_____ 9. The membrane is more permeable to potassium than sodium.

3. Graph the following set of data on the axes provided in Figure 11.5, then label the following: absolute refractory period, action potential (AP), depolarization, graded potential, hyperpolarization, relative refractory period, repolarization, resting membrane potential (RMP).

Voltage (mV)	–70	–70	–65	–70	–70	–60	–70	–70	–50	+30	–65	–75	–78	–71	–70	–70
Time (ms)	2.0	0.5	0.7	1.0	1.5	1.7	2.0	2.5	3.0	3.5	4.0	4.2	4.5	5.0	5.5	6.0

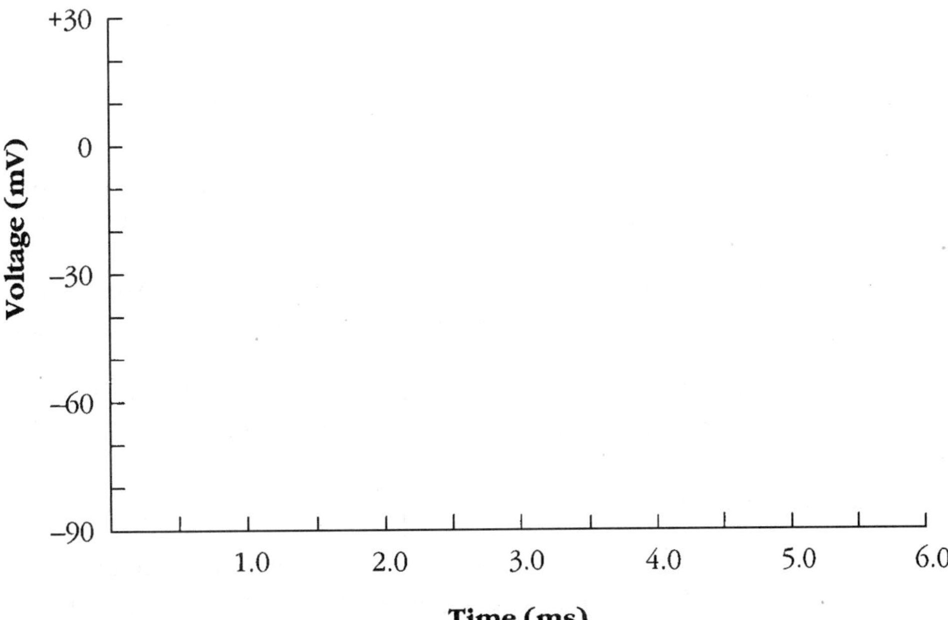

Figure 11.5

4. Using the key choices, select the terms defined in the following statements. Write the correct answers in the answer blanks.

Key Choices

A. Absolute refractory period	E. Graded potential	I. Repolarization
B. Action potential	F. Hyperpolarization	J. Sodium-potassium pump
C. Depolarization	G. Polarized	K. Subthreshold
D. Frequency of impulses	H. Relative refractory period	L. Threshold

________________ 1. Corresponds to the period of repolarization of the neuron

________________ 2. Process by which the resting potential is decreased as sodium ions move into the axon

________________ 3. State of an unstimulated neuron's membrane

________________ 4. Period (event) during which potassium ions move out of the axon

________________ 5. Also called the nerve impulse

________________ 6. Period when a neuron cannot be restimulated because its sodium gates are open and an AP is being generated

________________ 7. Mechanism by which ATP is used to move sodium ions out of the cell and potassium ions into the cell; completely restores and maintains the resting conditions of the neuron

________________ 8. Point at which an axon "fires"

________________ 9. Term for a weak stimulus

________________ 10. Self-propagated depolarization

________________ 11. Codes for intensity of the stimulus

________________ 12. Membrane potential at which the outward current carried by K^+ is exactly equal to the inward current carried by Na^+

________________ 13. A voltage change that reduces the ability of a neuron to conduct an impulse; the membrane potential becomes more negative

________________ 14. A local change in membrane potential in which current flow is quickly dissipated, that is, decremental

________________ 15. An all-or-none electrical event

________________ 16. A voltage change that brings a neuron closer to its threshold for firing; the membrane potential becomes less negative and moves toward 0

__________ 17. Results from the opening of voltage-regulated ionic gates

__________ 18. Results from the opening of chemically regulated gates or energetic stimuli

__________ 19. Characterized by a rapid polarity reversal

5. Circle the term that does not belong in each of the following groupings. (*Note:* Cl^- = chloride ion; K^+ = potassium ion; Na^+ = sodium ion.)

1. Inside resting cell   High Na^+ Low Na^+ High K^+

2. High Cl^- Protein anions   Low Cl^- Cytoplasm of resting cell

3. $Na^+ - K^+$ pump   3 Na^+ out / 2 K^+ in   3 Na^+ in / 2 K^+ out   ATP required

4. Nerve impulse   Graded potential   Short-distance signal   Short-lived signal

5. Depolarization   Inside membrane less negative   Hyperpolarization   –70 mV to –50 mV

6. Can generate action potentials   Axons   Peripheral processes   Cell bodies

6. For each of the following statements that is true, insert T in the answer blank. For false statements, correct the underlined word(s) by inserting the correct word(s) in the answer blanks.

__________ 1. The larger the diameter, the <u>slower</u> an axon conducts impulses.

__________ 2. Almost all the gated <u>calcium</u> channels are concentrated at the nodes of Ranvier of myelinated axons.

__________ 3. Saltatory conduction is much <u>faster</u> than impulse propagation along unmyelinated membranes.

__________ 4. Anesthetics block nerve impulses by blocking voltage-gated <u>potassium</u> channels.

__________ 5. A <u>postsynaptic</u> neuron conducts impulses toward the synapse.

__________ 6. Chemically gated channels open when <u>the membrane potential changes</u>.

__________ 7. Net current flows between two points that have <u>the same potential</u>.

__________ 8. The fatty interior of the neuronal membrane acts as a <u>capacitor</u> because it is a poor conductor of current.

__________ 9. G protein–linked receptors are also referred to as <u>ionotropic</u> receptors.

Synapses and Neurotransmitters and Their Receptors

1. Complete the following statements referring to chemical synapses by writing the correct terms in the answer blanks.

1. ________________
2. ________________
3. ________________
4. ________________
5. ________________
6. ________________

When the nerve impulse reaches the axon terminal, __(1)__ ions enter the terminal through voltage-regulated gates. The effect of the entry of these ions is to promote fusion of synaptic vesicles with the __(2)__. Exocytosis of __(3)__ molecules follows. These diffuse across the __(4)__ and bind to __(5)__ on the postsynaptic membrane, causing ion channels to open. The current flows that result cause local changes in the __(6)__ of the postsynaptic membrane. The neurotransmitter is inactivated enzymatically or otherwise removed from the synaptic cleft.

7. ________________
8. ________________
9. ________________
10. ________________
11. ________________
12. ________________

At __(7)__ synapses, neurotransmitter binding opens a single type of channel, allowing both __(8)__ ions to diffuse through the membrane. Because accumulation of __(9)__ charges inside the cell is prevented, local depolarization events, instead of __(10)__, occur. The function of these graded potentials is to initiate a(n) __(11)__ distally at the axon __(12)__ of the postsynaptic neuron.

13. ________________
14. ________________
15. ________________

At __(13)__ synapses, potassium and/or chloride permeability of the postsynaptic membrane is increased. Consequently, the charge inside the cell becomes relatively more __(14)__, and the resting membrane potential changes from –70 mV toward –90 mV, indicating __(15)__ of the postsynaptic membrane.

16. ________________
17. ________________
18. ________________
19. ________________

A single EPSP (excitatory postsynaptic potential) is insufficient to generate an action potential, but threshold depolarization can be achieved by __(16)__ of many EPSPs. Monitoring and integration of all excitatory and/or inhibitory neurotransmitters is accomplished by the postsynaptic membranes, which __(17)__ EPSPs with IPSPs (inhibitory postsynaptic potentials). Sometimes, the release of excitatory neurotransmitters is inhibited by the activity of another neuron via a(n) __(18)__ synapse or by the action of chemicals called __(19)__.

2. Use the key choices to identify the types of synapses described below. Write the correct answers in the answer blanks.

Key Choices

A. Chemical synapse B. Electrical synapse

________________ 1. Releases neurotransmitters from vesicles

________________ 2. Very rapid transmission

________________ 3. Protein channels connect cytoplasm of adjacent neurons

________________ 4. Postsynaptic neuron has receptor region

_______________ 5. Transmission may be uni- or bidirectional

_______________ 6. Presynaptic neuron has a knoblike axon terminal

_______________ 7. Synchronizes activities of all interconnected neurons

_______________ 8. Allows the flow of ions between neurons

_______________ 9. Fluid fills synaptic space between neurons

3. Using the key choices, first match the neurotransmitters listed in Column A with the numbered descriptions given below. Then, select the appropriate classes of these neurotransmitters from Column B. Finally, if applicable, select the correct effect of each neurotransmitter from Column C. Write the letters of all your choices in the answer blanks.

Key Choices

Column A	Column B	Column C
A. Acetylcholine	E. Excitatory	I. Affects emotions
B. Amino acids	F. Inhibitory	J. Painkillers
C. Biogenic amines	G. Direct acting	K. Reduce anxiety
D. Peptides	H. Indirect acting	L. Regulate biological clock

_______________ 1. An example is gamma-aminobutyric acid (GABA)

_______________ 2. Examples are beta-endorphins and enkephalins

_______________ 3. Released at neuromuscular junctions of skeletal muscles

_______________ 4. Typically are broadly distributed in the brain

4. Circle the term that does not belong in each of the following groupings.

1. Ion channels open Change in membrane potential Slow response Direct action neurotransmitter

2. Indirect action neurotransmitter Short-lived effect G protein linked Second messenger

3. Receptor potential Weak stimuli Threshold strength More nerve impulses per second

4. Pain receptors Receptor membrane response declines Adaptation Light pressure receptor

5. Nitric oxide ATP Novel neurotransmitter ACh

5. In Figure 11.6, identify by coloring the following structures, which are typically part of a chemical synapse. Also, bracket the synaptic cleft, and identify the arrows showing (1) the direction of the presynaptic impulse and (2) the direction of net neurotransmitter movements.

○ Axon terminal ○ Postsynaptic membrane ○ Presynaptic membrane

○ Mitochondria ○ Na^+ ions ○ Ca^{2+} ions

○ K^+ ions ○ Chemically gated channels ○ Synaptic vesicles

○ Postsynaptic neurotransmitter receptors ○ Neurotransmitter molecules

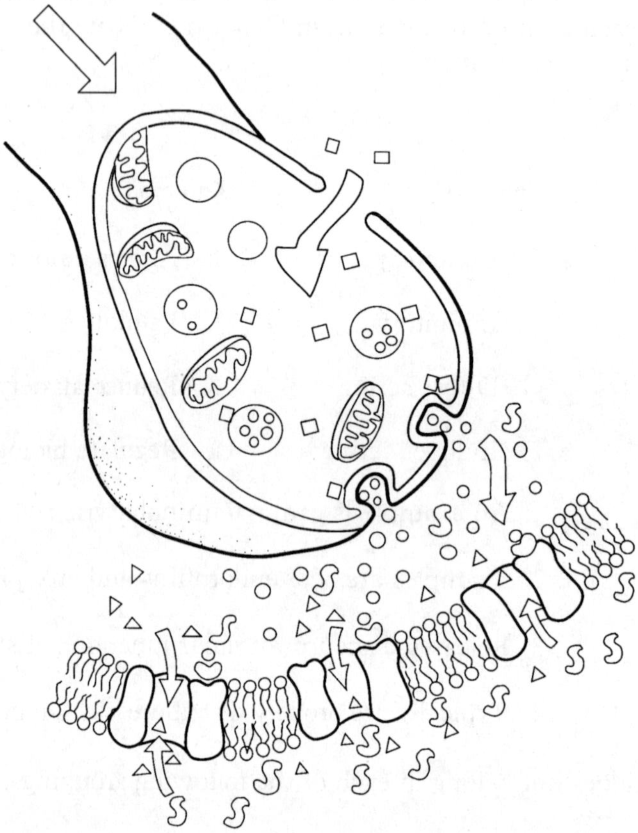

Figure 11.6

6. Using the key choices, select the phase of action potential generation described in the following statements. Write the correct key letter in the answer blanks.

Key Choices

A. Depolarizing phase C. Repolarization, ↓ in Na^+ permeability E. Resting state

B. Hyperpolarization D. Repolarization, ↑ in K^+ permeability

__________ 1. Voltage change caused by sodium influx opens more Na^+ channels

__________ 2. All voltage-gated Na^+ and K^+ channels are closed

__________ 3. Na^+ entry declines, voltage-gated K^+ channels open

__________ 4. Produces the undershoot

__________ 5. Only leakage channels are open

__________ 6. Na^+ inactivation gates closing

__________ 7. Na^+ voltage-sensitive activation gates open

__________ 8. AP spike reverses direction

__________ 9. Na^+ channels resetting to resting state position; K^+ entry continues

7. On Figure 11.7, several types of chemical synapses are illustrated. Identify each type, using the key choices. Color the diagram as you wish.

Key Choices

A. Axoaxonic B. Axodendritic C. Axosomatic D. Dendrodendritic

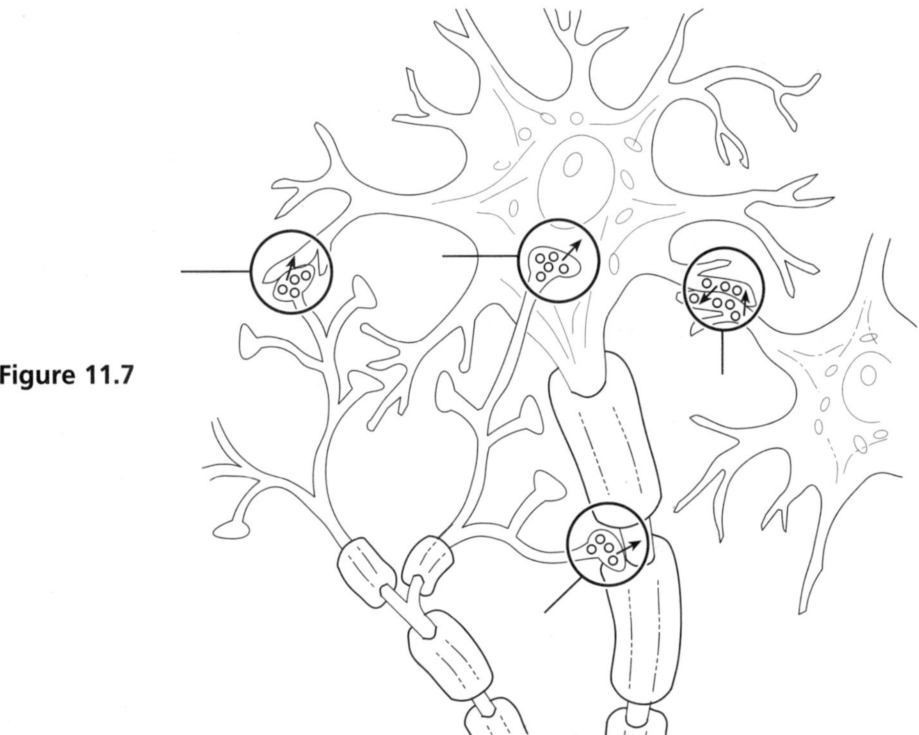

Figure 11.7

Basic Concepts of Neural Integration

1. Circle the term that does not belong in each of the following groupings.

1. PNS Neural integration Neuronal pools Circuits
2. Parallel processing Several unique responses CNS integration All-or-nothing response
3. Reflex arc Variety of stimuli Serial processing One anticipated response

2. Refer to Figure 11.8, showing a reflex arc, as you complete the following exercise. First, briefly answer the following questions by writing your answers in the answer blanks.

1. What is the stimulus? ______________________________

2. What tissue is the effector? ______________________________

3. How many synapses occur in this reflex arc? ______________________________

Next, select different colors for each of the following structures, and use them to color the diagram. Finally, draw arrows on the figure indicating the direction of impulse transmission through this reflex pathway.

◯ Receptor region ◯ Association neuron ◯ Effector

◯ Sensory neuron ◯ Motor neuron

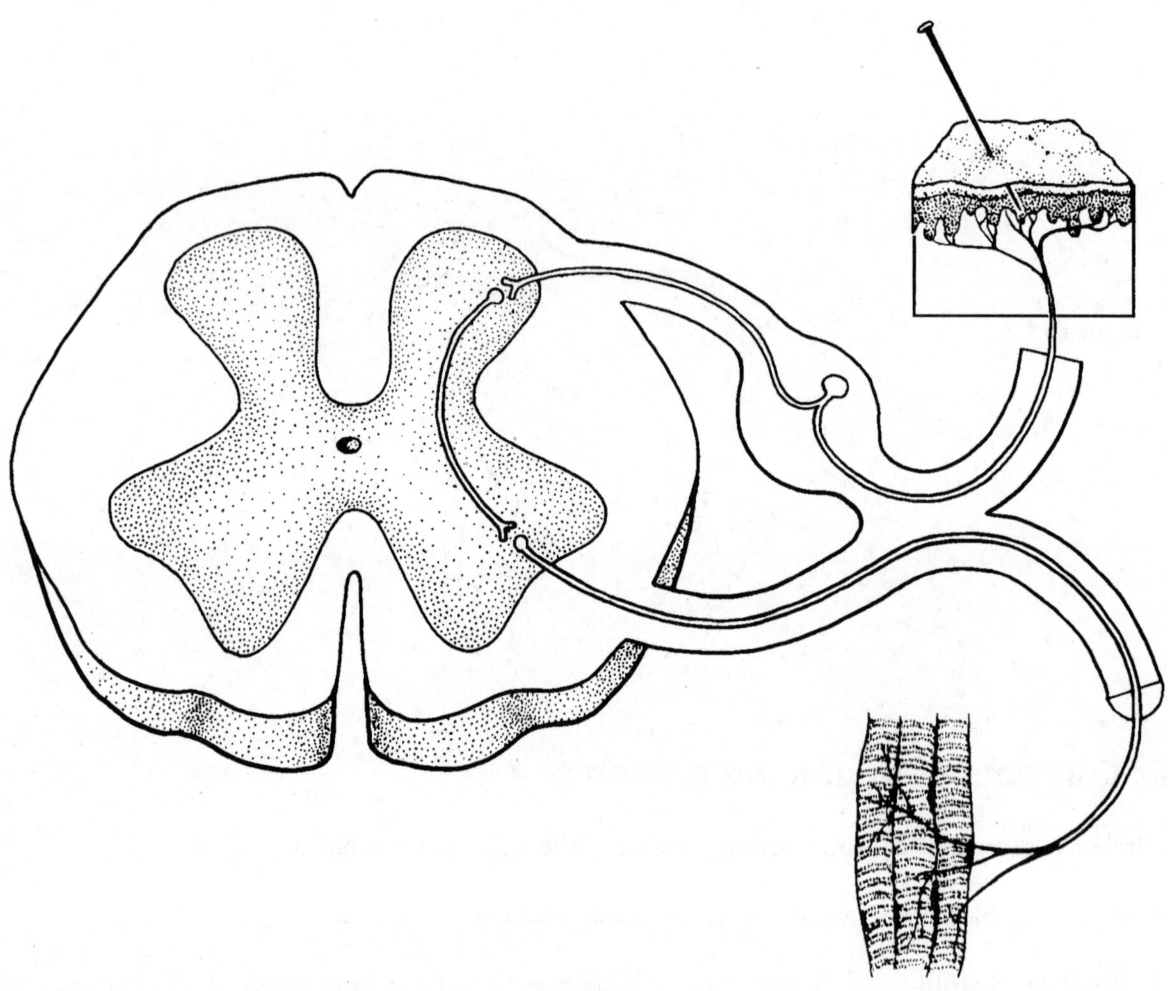

Figure 11.8

3. Using the key choices, match the types of circuits with their descriptions. Write the correct answers in the answer blanks.

Key Choices

A. Converging B. Diverging C. Parallel after-discharge D. Reverberating

_______________ 1. Large number of skeletal muscle fibers stimulated by a few motor neurons

_______________ 2. The sight, sound, and smell of popping popcorn all elicit the same feelings

_______________ 3. One fiber stimulates increasing numbers of neurons

_______________ 4. Impulses repeatedly sent through the same circuit, lasting perhaps a lifetime

_______________ 5. Many presynaptic neurons stimulate a few neurons

_______________ 6. Control of rhythmic activities, such as respiration and the sleep-wake cycle

_______________ 7. Impulses from many neurons reach a common output cell at different times

_______________ 8. Used to solve this conversion for the speed of impulse transmission: 100 m/s = ? mi/hr

CHALLENGING YOURSELF

At the Clinic

1. A patient taking diuretics comes to the clinic complaining of muscular weakness and fatigue. A deficiency of which ion will be suspected? (*Hint:* you might also want to do a little reading in Chapter 25 for this one.)

2. Being a long-time herpesvirus sufferer, Brad was not surprised to learn (at a herpes information session) that herpesviruses tend to "hide out" in nerve tissue to avoid attack by the hody's immune system. However, their "mode of travel" in the body did surprise him. How *do* herpesviruses travel to the neuron cell body?

3. A brain tumor is found in a CT scan of Mr. Childs' head. The physician is assuming that it is not a secondary tumor (i.e., it did not spread from another part of the body) because an exhaustive workup has revealed no signs of cancer elsewhere in Mr. Childs' body. Is the brain tumor more likely to have developed from nerve tissue or from neuroglia? Why?

4. With what specific process does the lack of myelination seen in multiple sclerosis interfere?

5. Sally has been diagnosed as having a certain type of epilepsy associated with a lack of GABA. Why would this deficiency lead to the increased and uncontrolled neuronal activity exhibited by Sally's seizures?

6. Mr. Jacobson, a tax accountant, comes to the clinic complaining of feeling very "stressed out" and anxious. He admits to drinking 10 to 12 cups of coffee daily. His doctor (knowing that caffeine alters the threshold of neurons) suggests he reduce his intake of coffee. What is caffeine's effect on the threshold of neurons, and how might this effect explain Mr. Jacobson's symptoms?

7. Mr. Marple staggered home after a "good night" at the local pub. While attempting to navigate the stairs, he passed out cold and lay (all night) with his right armpit straddling the staircase banister. When he awoke the next morning, he had a severe headache, but what bothered him more was that he had no sensation in his right arm and hand, which also appeared to be paralyzed. Explain.

Stop and Think

1. Histological examination during a brain autopsy revealed a superabundance of microglia in a certain brain area. What abnormal condition might this indicate?

2. What is the benefit of having highly branched cell processes such as those of neurons?

3. Axon diameter increases noticeably at the nodes of Ranvier. What purpose does this serve?

4. What elements (molecules or organelles) are involved in axon transport?

5. How might a neuron without an axon transmit impulses?

6. Describe ion flow through a neuronal membrane that is simultaneously being excited and inhibited by various presynaptic axons. Assuming an exact balance in excitatory and inhibitory stimuli, how would a graph of membrane potential look?

7. If two parallel pathways were equal in all regards (such as myelination, axon diameter, and total pathway length), except that the axons in one pathway were twice as long as the axons in the neighboring pathway, in which pathway would nerve impulses travel more quickly? To which type of neuronal pool does this type of arrangement apply?

8. Chloride moves fairly freely through the neuronal membrane, and yet the opening of chloride gates can cause inhibition. How can a chloride flux be inhibitory?

9. Neurotransmitter remains in the synaptic cleft only briefly because it is removed by enzymes or other mechanisms. If the mechanism for removal of neurotransmitter is impaired, will the postsynaptic cell remain depolarized and refractory?

10. Each neuron makes and releases one neurotransmitter—true or false? Elaborate on your response.

11. With learning, new synapses can form. How can this affect a neuron's facilitated and discharge zones?

12. Are neurons incorporated into more than one neuronal circuit type?

13. Two anatomists were arguing about the sensory neuron. One said its peripheral process is an axon and gave two good reasons. But the other anatomist called the peripheral process a dendrite and gave one good reason. Cite all three reasons given and state your own opinion.

14. Shortly after birth, essentially no new neurons are formed. This being the case, how can the enhancement of certain pathways, such as those promoting more acute hearing in blind people, develop?

15. During AP transmission, many ions cross the neuronal membrane. What is it that travels *along* the membrane and acts as a signal?

16. As Melanie woke up, she stretched and quickly did 20 sit-ups before getting out of bed. As she brushed her teeth, the aroma of coffee stimulated her smell receptors and her stomach began to gurgle. Indicate the division of the nervous system involved in each of these activities or events.

COVERING ALL YOUR BASES

Multiple Choice

Select the best answer or answers from the choices given.

1. An example of integration by the nervous system is:
   A. the feel of a cold breeze
   B. the shivering and goose bumps that result
   C. the sound of rain
   D. the decision to go back for an umbrella

2. Which of the following functions would utilize a visceral efferent cranial nerve?
   A. Sense of taste
   B. Sense of position of the eye
   C. Movement of the tongue in speech
   D. Secretion of salivary glands

3. The cell type *most immediately affected* by damage to a blood vessel in the brain is a(n):
   A. neuron C. astrocyte
   B. microglia D. oligodendrocyte

4. Which of the following neuroglial types would be classified as epithelium?
   A. Astrocyte
   B. Ependymal cells
   C. Oligodendrocyte
   D. Microglia

5. Cell types with an abundance of smooth ER include:
   A. neuron C. astrocyte
   B. Schwann cell D. oligodendrocyte

6. The region of a neuron with voltage-gated sodium channels is the:
   A. axon hillock C. dendrite
   B. soma D. perikaryon

7. Which of the following would be a direct result of destruction of a neuron's neurofilaments?
   A. Cessation of protein synthesis
   B. Loss of normal cell shape
   C. Inhibition of impulse transmission
   D. Interference with intracellular transport

8. Where might a gray matter nucleus be located?
   A. Alongside the vertebral column
   B. Within the brain
   C. Within the spinal cord
   D. In the sensory receptors

9. Histological examination of a slice of neural tissue reveals a bundle of nerve fibers held together by cells whose multiple processes wrap around several fibers and form a myelin sheath. The specimen is likely to be:
   A. a nucleus C. a nerve
   B. a ganglion D. a tract

10. Neurotransmitters:
    A. act directly to produce action potentials
    B. produce only graded potentials
    C. have a direct effect on the secretory component of a neuron
    D. can act only at voltage-gated channels

11. What do myelin sheath membranes lack which makes them good insulators?
    A. Lipids
    B. Carbohydrate groups
    C. Channel and carrier proteins
    D. Close contact with adjacent membranes

12. Bipolar neurons:
    A. are found in the head
    B. are always part of an afferent pathway
    C. have two dendrites
    D. have two axons

13. Which of the following would be true of the peripheral process of a unipolar neuron?
    A. Its membrane contains voltage-gated channels.
    B. It is never myelinated.
    C. It connects directly to the central process.
    D. It typically has a wrapping of Schwann cells.

14. Which of the following skin cells would form a junction with a motor neuron?
    A. Keratinocyte
    B. Sudoriferous glandular epithelial cell
    C. Arrector pili muscle cell
    D. Fibroblast

15. The term that refers to a measure of the potential energy of separated electrical charges is:
    A. voltage C. resistance
    B. current D. conductance

16. Current is inversely related to:
    A. voltage C. potential difference
    B. resistance D. capacitance

17. An unstimulated plasma membrane is characterized by:
    A. ions flowing along their electrochemical gradients
    B. cytoplasmic side slightly negatively charged
    C. all gated channels open
    D. low concentration of Na^+ in cytoplasm compared to extracellular fluid

18. Which would result from inhibition of a neuron's sodium-potassium pump?
    A. The membrane would lose its polarity.
    B. Potassium would undergo a net movement to the cell's interior.
    C. Cytoplasmic anions would diffuse out of the cell.
    D. Sodium would accumulate outside the cell.

19. Impulse *propagation* is associated with:
    A. graded potentials
    B. chemically gated ion channels
    C. hyperpolarization
    D. voltage-gated sodium channels

20. Hyperpolarization results from:
    A. sodium flux into the cell
    B. chloride flux into the cell
    C. potassium flux into the cell
    D. potassium flux out of the cell

21. Decremental current flow is associated with:
    A. chemically gated channels
    B. postsynaptic potential
    C. saltatory conduction
    D. axolemma

22. Which of these is involved with a positive feedback cycle?
    A. Sodium-potassium pump
    B. Chemically gated channels
    C. Voltage-gated sodium channels
    D. Membrane potential

23. In the graph below, which stimulus (1, 2, 3, or 4) is the strongest? (*Note:* The vertical lines represent action potential tracings.)

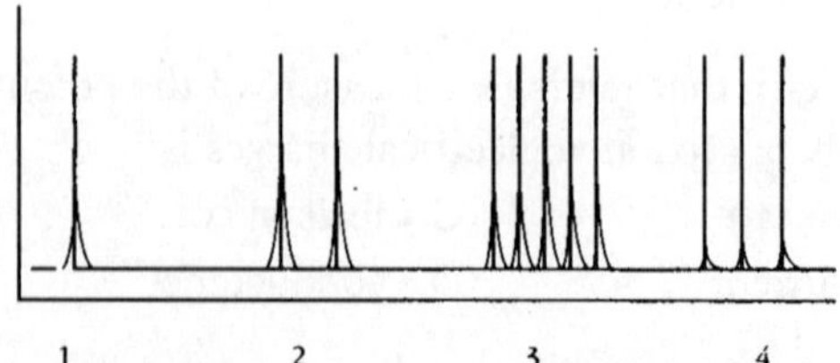

    A. Stimulus 1
    B. Stimulus 2
    C. Stimulus 3
    D. Stimulus 4

24. If the following receptors were stimulated simultaneously, which signal would reach the brain first?
    A. Pain receptors in the skin of the toe
    B. Sensory receptors in a nearby blood vessel wall
    C. Small touch receptors in the toe's skin
    D. Pressure receptors in the toe's musculature

25. Which is *not* characteristic of a chemical synapse?
    A. Abundant gap junctions
    B. Voltage-gated channels in the postsynaptic membrane
    C. Unidirectional communication
    D. Synaptic delay

26. An IPSP is associated with:
    A. opening of chemically gated channels for K^+
    B. hyperpolarization
    C. decremental conduction
    D. depolarization

27. Which of the following is true concerning tetanic potentiation?
    A. Calcium concentrations increase during stimulation.
    B. Membrane proteins can be altered.
    C. The postsynaptic cell is inhibited.
    D. The presynaptic cell is inhibited.

28. Which of the following are characteristic of the biogenic amine neurotransmitters?
    A. They are direct acting.
    B. Some are catecholamines.
    C. Endorphins are in this class.
    D. At least some are involved in regulation of mood.

29. A neuronal pool that might rely on fatigue to end its impulse propagation is a(n):
    A. converging circuit
    B. diverging circuit
    C. oscillating circuit
    D. parallel after-discharge circuit

30. A synapse between an axon terminal and a neuron cell body is called:
    A. axodendritic
    B. axoaxonic
    C. axosomatic
    D. axoneuronic

31. Myelin is most closely associated with which of the following cell parts introduced in Chapter 3?
    A. Cell nucleus
    B. Smooth ER
    C. Ribosomes
    D. The plasma membrane

32. Anesthetics reduce pain by inhibiting nerve transmission. Of the three chemicals described below, which might be effective anesthetics?

1. A chemical that prevents opening of Na^+ channels in the membrane
2. A chemical that binds with and blocks neurotransmitter receptors
3. A chemical that inhibits degradation of the neurotransmitter

A. 1 only
B. 1 and 2
C. 1 and 3
D. 3 only

Word Dissection

For each of the following word roots, fill in the literal meaning and give an example, using a word found in this chapter.

Word root	Translation	Example
1. dendr		
2. ependy		
3. gangli		
4. glia		
5. neur		
6. nom		
7. oligo		
8. salta		
9. syn		

12

The Central Nervous System

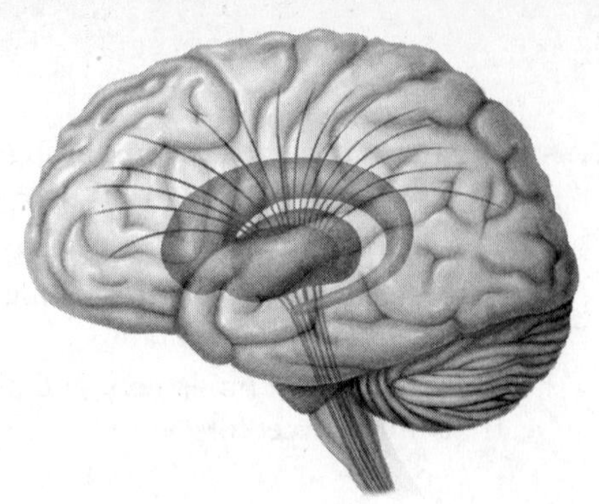

Student Objectives

When you have completed the exercises in this chapter, you will have accomplished the following objectives:

The Brain

1. Describe the process of brain development.
2. Name the major regions of the adult brain.
3. Name and locate the ventricles of the brain.
4. List the major lobes, fissures, and functional areas of the cerebral cortex.
5. Explain lateralization of hemisphere function.
6. Differentiate between commissures, association fibers, and projection fibers.
7. Describe the general function of the basal nuclei (basal ganglia).
8. Describe the location of the diencephalon, and name its subdivisions and functions.
9. Identify the three major regions of the brain stem, and note the functions of each area.
10. Describe the structure and function of the cerebellum.
11. Locate the limbic system and the reticular formation, and explain the role of each functional system.

Higher Mental Functions

12. Define EEG and distinguish between alpha, beta, theta, and delta brain waves.
13. Describe consciousness clinically.
14. Compare and contrast the events and importance of slow-wave and REM sleep, and indicate how their patterns change through life.
15. Compare and contrast the stages and categories of memory.
16. Describe the relative roles of the major brain structures believed to be involved in declarative and procedural memories.

Protection of the Brain

17. Describe how meninges, cerebrospinal fluid, and the blood-brain barrier protect the CNS.
18. Describe the formation of cerebrospinal fluid, and follow its circulatory pathway.
19. Indicate the cause (if known) and major signs and symptoms of cerebrovascular accidents, Alzheimer's disease, Parkinson's disease, and Huntington's disease.

The Spinal Cord

20. Describe the embryonic development of the spinal cord.
21. Describe the gross and microscopic structure of the spinal cord.
22. List the major spinal cord tracts, and classify each as a motor or sensory tract.
23. Distinguish between flaccid and spastic paralysis, and between paralysis and paresthesia.

Diagnostic Procedures for Assessing CNS Dysfunction

24. List and explain several techniques used to diagnose brain disorders.

The human brain is a marvelous biological computer. It can receive, store, retrieve, process, and dispense information with almost instantaneous speed. Together, the brain and spinal cord make up the central nervous system (CNS). By way of inputs from the peripheral nervous system, the CNS is advised of changes in both the external environment and internal body conditions. After perception, integration, and coordination of this knowledge, the CNS dispatches instructions to initiate appropriate responses.

Included in Chapter 12 are student exercises on the regions and associated functions of the brain and spinal cord. Traumatic and degenerative disorders of the central nervous system are also considered, as are topics such as EEGs, sleep, and selected higher mental functions.

BUILDING THE FRAMEWORK

The Brain

1. Figure 12.1 (which continues on page 242) shows diagrams of embryonic development. Arrange the diagrams in the correct order by numbering each one. Where possible, insert the time, in days or weeks, of each stage of development. Label all structures that have leader lines.

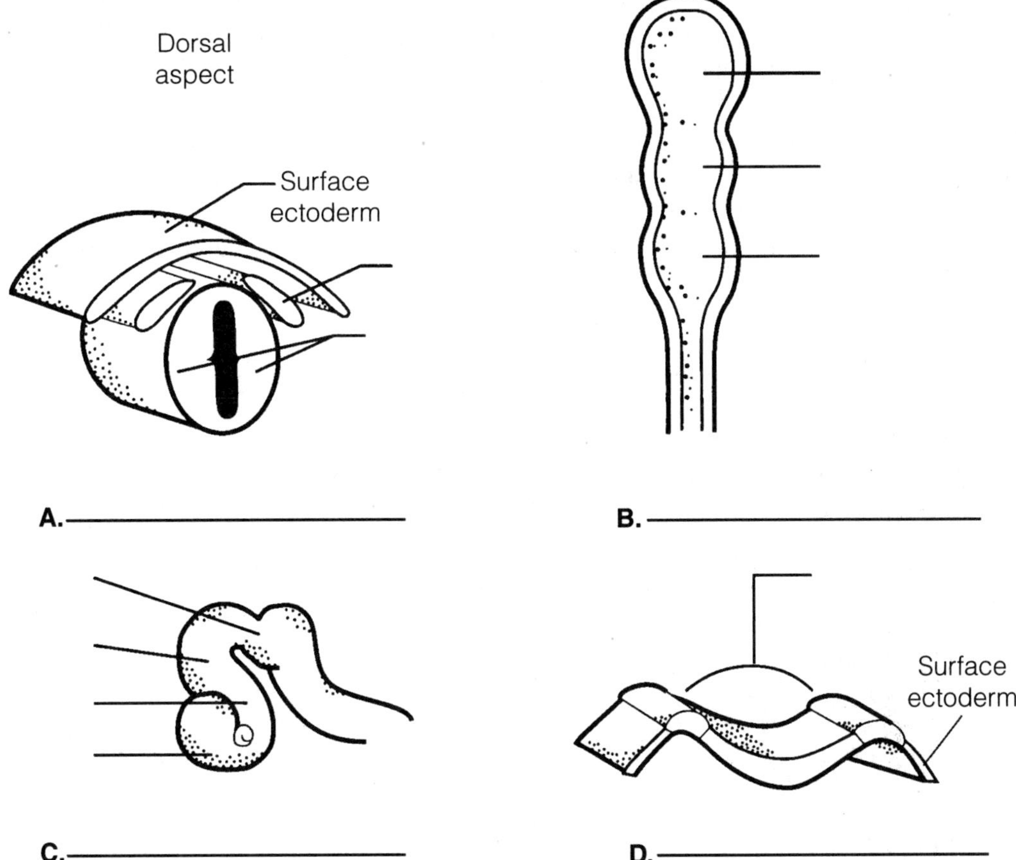

Figure 12.1

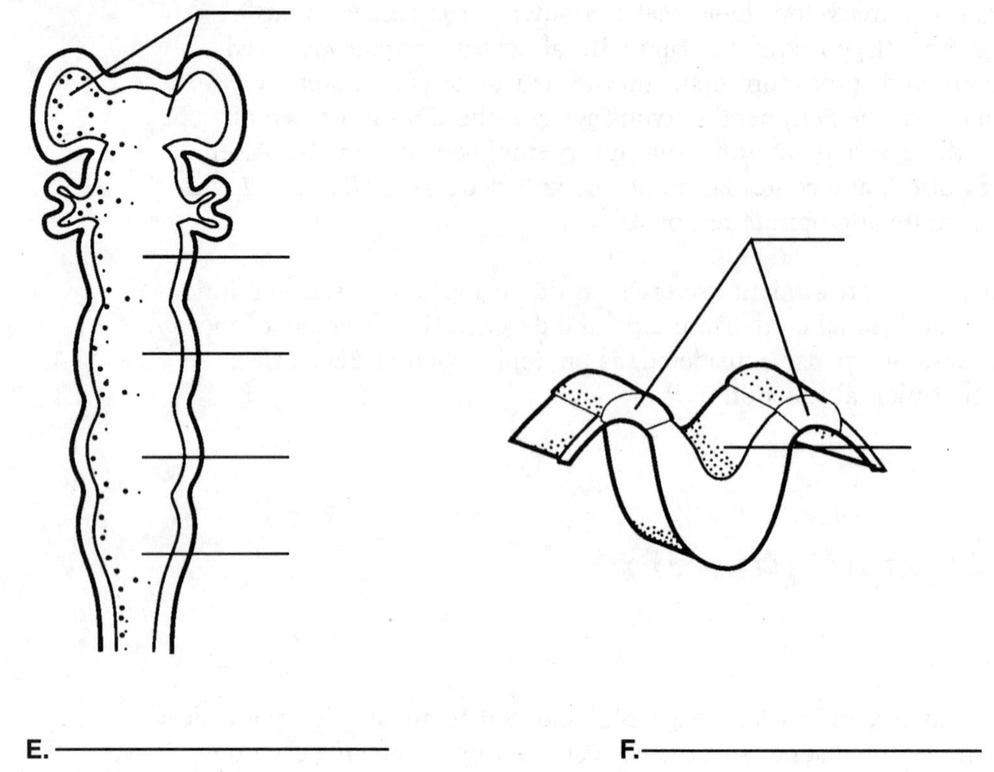

Figure 12.1

2. Fill in the table below by indicating the adult brain structures formed from each of the secondary brain vesicles and the adult neural canal regions. Some of that information has already been entered.

Secondary brain vesicle	Adult brain structures	Neural canal regions
Telencephalon		Lateral ventricles
Diencephalon	Diencephalon (thamus, hypothalamus, epithalamus)	
Mesencephalon		
Metencephalon	Brain stem: pons; cerebellum	
Myelencephalon		

3. Figure 12.2 is a diagram of the right lateral view of the human brain. First, match the letters on the diagram with the following list of terms and insert the appropriate letters in the answer blanks. Then, select different colors for each of the areas of the brain with a color coding circle and use them to color the diagram. If an identified area is part of a lobe, use the color you selected for the lobe but use stripes for that area.

○ _____ 1. Frontal lobe

○ _____ 2. Parietal lobe

○ _____ 3. Temporal lobe

○ _____ 4. Precentral gyrus

_____ 5. Parieto-occipital fissure

○ _____ 6. Postcentral gyrus

_____ 7. Lateral fissure

_____ 8. Central sulcus

○ _____ 9. Cerebellum

○ _____ 10. Medulla

○ _____ 11. Occipital lobe

○ _____ 12. Pons

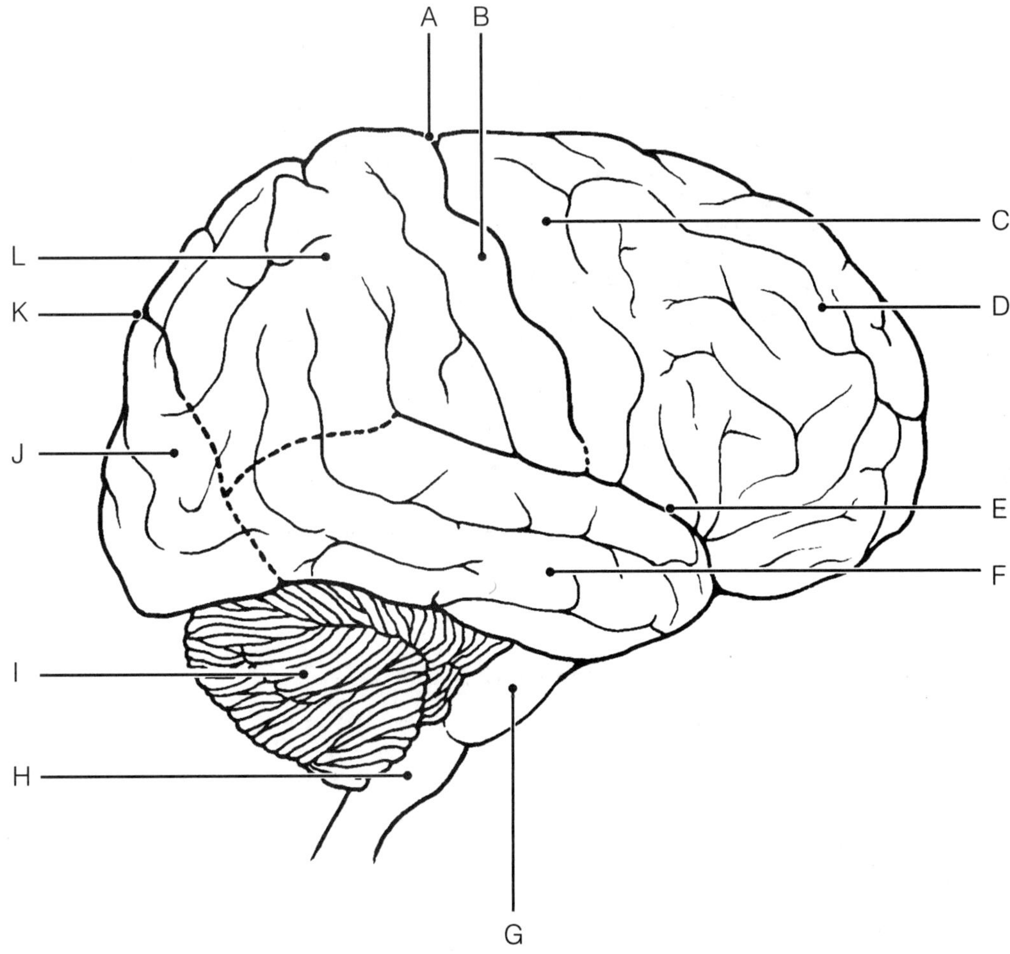

Figure 12.2

4. Figure 12.3 illustrates a "see-through" brain showing the positioning of the ventricles and connecting canals or apertures. Correctly identify all structures having leader lines by using the key choices provided below. One of the lateral ventricles has already been identified. Color the spaces filled with cerebrospinal fluid blue.

Key Choices

A. Anterior horn

B. Central canal

C. Cerebral aqueduct

D. Fourth ventricle

E. Inferior horn

F. Interventricular foramen

G. Lateral aperture

H. Third ventricle

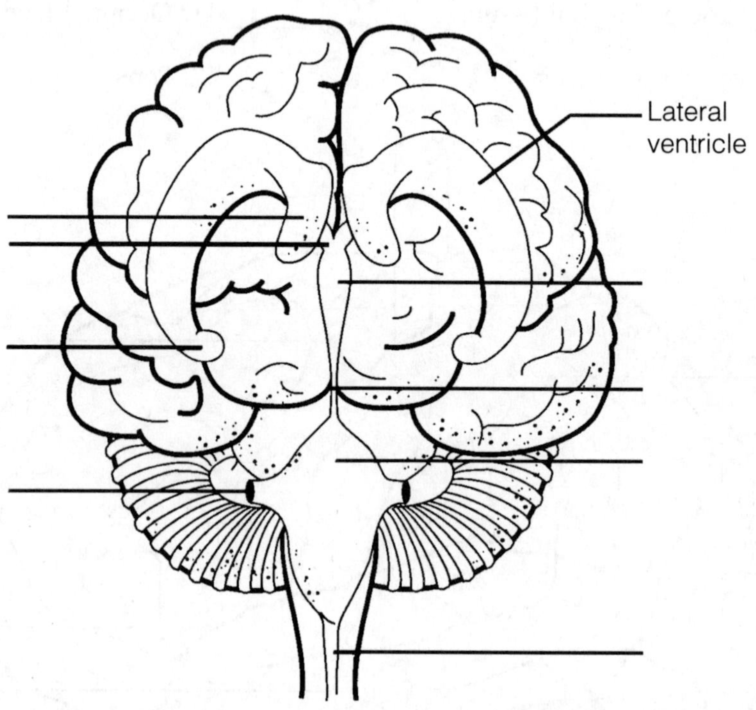

Figure 12.3

5. Figure 12.4 shows a left lateral view of the brain with some of its functional areas indicated by numbers. These areas are listed below. Identify each cortical area by its corresponding number on the diagram. Color the diagram as you wish.

______	Primary motor cortex	______	Primary somatosensory cortex
______	Premotor cortex	______	Somatosensory association cortex
______	Visual cortex	______	Auditory cortex
______	Prefrontal cortex	______	Broca's area
______	Frontal eye field	______	Wernicke's area
______	Posterior association area		

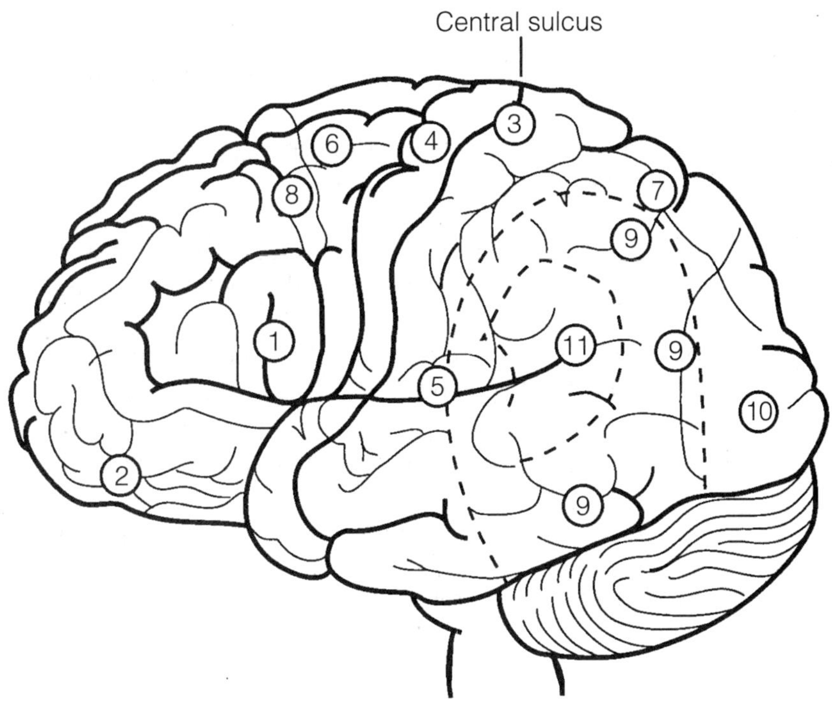

Figure 12.4

6. Some of the following brain structures consist of gray matter; others are white matter. Write G (for gray) or W (for white) as appropriate.

_____ 1. Cortex of cerebellum

_____ 2. Internal capsule

_____ 3. Anterior commisure

_____ 4. Medial lemniscus

_____ 5. Pyramids

_____ 6. Olives

_____ 7. Thalamic nuclei

_____ 8. Cerebellar peduncle

7. If a statement is true, write the letter T in the answer blank. If a statement is false, correct the underlined word(s) and write the correct word(s) in the answer blank.

__________________ 1. The primary somatosensory area of the cerebral hemisphere(s) is found in the <u>precentral</u> gyrus.

__________________ 2. Cortical areas involved in audition are found in the <u>occipital</u> lobe.

__________________ 3. The primary motor area in the <u>temporal</u> lobe is involved in the initiation of voluntary movements.

__________________ 4. The specialized motor speech area is located at the base of the precentral gyrus in an area called <u>Wernicke's</u> area.

__________________ 5. The right cerebral hemisphere receives sensory input from the <u>right</u> side of the body.

__________________ 6. The <u>pyramidal</u> tract is the major descending voluntary motor tract.

__________________ 7. The primary motor cortex is located in the <u>postcentral</u> gyrus.

__________________ 8. Centers for control of repetitious or stereotyped motor skills are found in the <u>primary motor</u> cortex.

__________________ 9. The largest parts of the motor homunculi are the lips, tongue, and <u>toes</u>.

__________________ 10. Sensations such as touch and pain are integrated in the <u>primary sensory cortex</u>.

__________________ 11. The primary visual cortex is in the <u>frontal</u> lobe of each cerebral hemisphere.

__________________ 12. In most humans, the area that controls the comprehension of language is located in the <u>left</u> cerebral hemisphere.

__________________ 13. Elaboration of the <u>visual</u> cortex sets humans apart from other animals.

_____________ 14. Complex sensory memory patterns are stored in an area called the general interpretation area.

_____________ 15. Areas in the cerebral hemisphere opposite the ones containing Broca's and Wernicke's areas are centers for cognitive language.

_____________ 16. Cerebral dominance designates the hemisphere that is dominant for memory.

_____________ 17. The right cerebral hemisphere of left-handed humans is usually involved with intuition, poetry, and creativity.

8. Referring as necessary to the brain areas listed in Exercise 9, match the appropriate brain structures with the following descriptions. Insert the terms selected in the answer blanks.

_____________ 1. Site of regulation of water balance, body temperature, rage, and pain centers; the main visceral (autonomic) center of brain

_____________ 2. Reflex centers involved in regulating respiratory rhythm in conjunction with lower brain stem centers

_____________ 3. Responsible for the regulation of posture and coordination of skeletal muscle movements

_____________ 4. Important relay station for afferent fibers, traveling to the sensory cortex for interpretation

_____________ 5. Contains autonomic centers that regulate blood pressure and respiratory rhythm, as well as coughing and sneezing centers

_____________ 6. Midbrain area consisting of large, descending motor tracts

_____________ 7. Influences body rhythms; interacts with the biological clock

_____________ 8. Location of middle cerebellar peduncles

_____________ 9. Locations of visual and auditory reflex centers

9. Figure 12.5 is a diagram of the sagittal view of the human brain. First, match the letters on the diagram with the following list of terms and insert the appropriate letters in the answer blanks. Then, color the brain stem areas blue and the areas where cerebrospinal fluid is found yellow.

_____ 1. Cerebellum

_____ 2. Cerebral aqueduct

_____ 3. Cerebral hemisphere

_____ 4. Cerebral peduncle

_____ 5. Choroid plexus

_____ 6. Corpora quadrigemina

_____ 7. Corpus callosum

_____ 8. Fornix

_____ 9. Fourth ventricle

_____ 10. Hypothalamus

_____ 11. Medulla oblongata

_____ 12. Optic chiasma

_____ 13. Pineal body

_____ 14. Pituitary gland

_____ 15. Pons

_____ 16. Thalamus (interthalamic adhesion)

_____ 17. Third ventricle

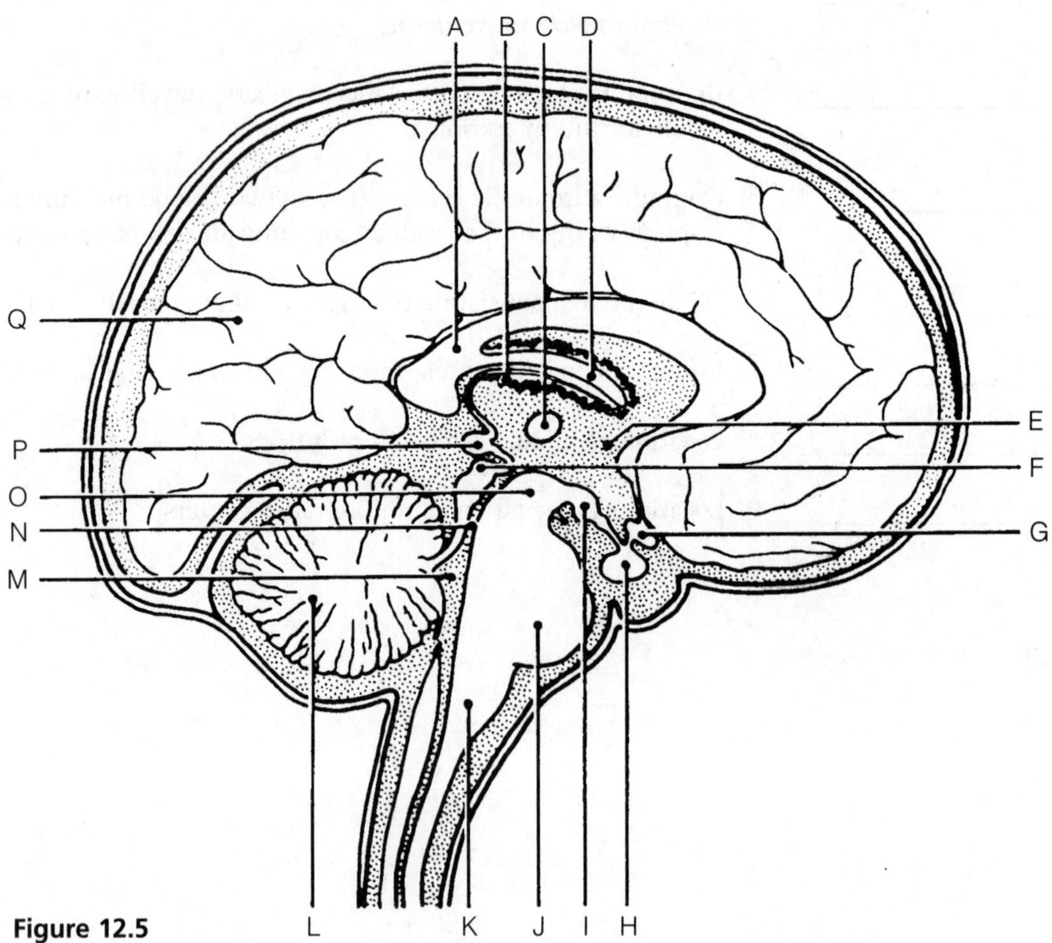

Figure 12.5

10. In the horizontal section shown in Figure 12.6 (see plane of cut in inset), identify, by color coding and coloring, the structures listed below.

○ Caudate nucleus ○ Choroid plexus ○ Putamen

○ Claustrum ○ Pineal body ○ Thalamus

○ Corpus callosum

Then, using the leader lines provided, correctly identify the fornix, inferior horn of the lateral ventricle, third ventricle, insula, internal capsule, and septum pellucidum.

11. Using the key choices, select the terms identified in the following descriptions by inserting the appropriate letters in the answer blanks.

Key Choices

A. Basal nuclei	D. Cerebral hemispheres	G. Septum pellucidum
B. Brain stem	E. Cortex	H. Ventricles
C. Cerebellum	F. Diencephalon	I. White matter

__________________ 1. The four major subdivisions of the adult brain

__________________ 2. Contain cerebrospinal fluid

__________________ 3. Masses of gray matter embedded deep within the cerebral white matter

__________________ 4. Myelinated fiber tracts

__________________ 5. Consists of the midbrain, pons, and medulla

__________________ 6. Separates the lateral ventricles

__________________ 7. Thin layer of gray matter on outer surface of cerebral hemispheres and cerebellum

__________________ 8. Account for more than 60% of the total brain weight

__________________ 9. Consists of the hypothalamus, thalamus, epithalamus, and retinas of the eyes

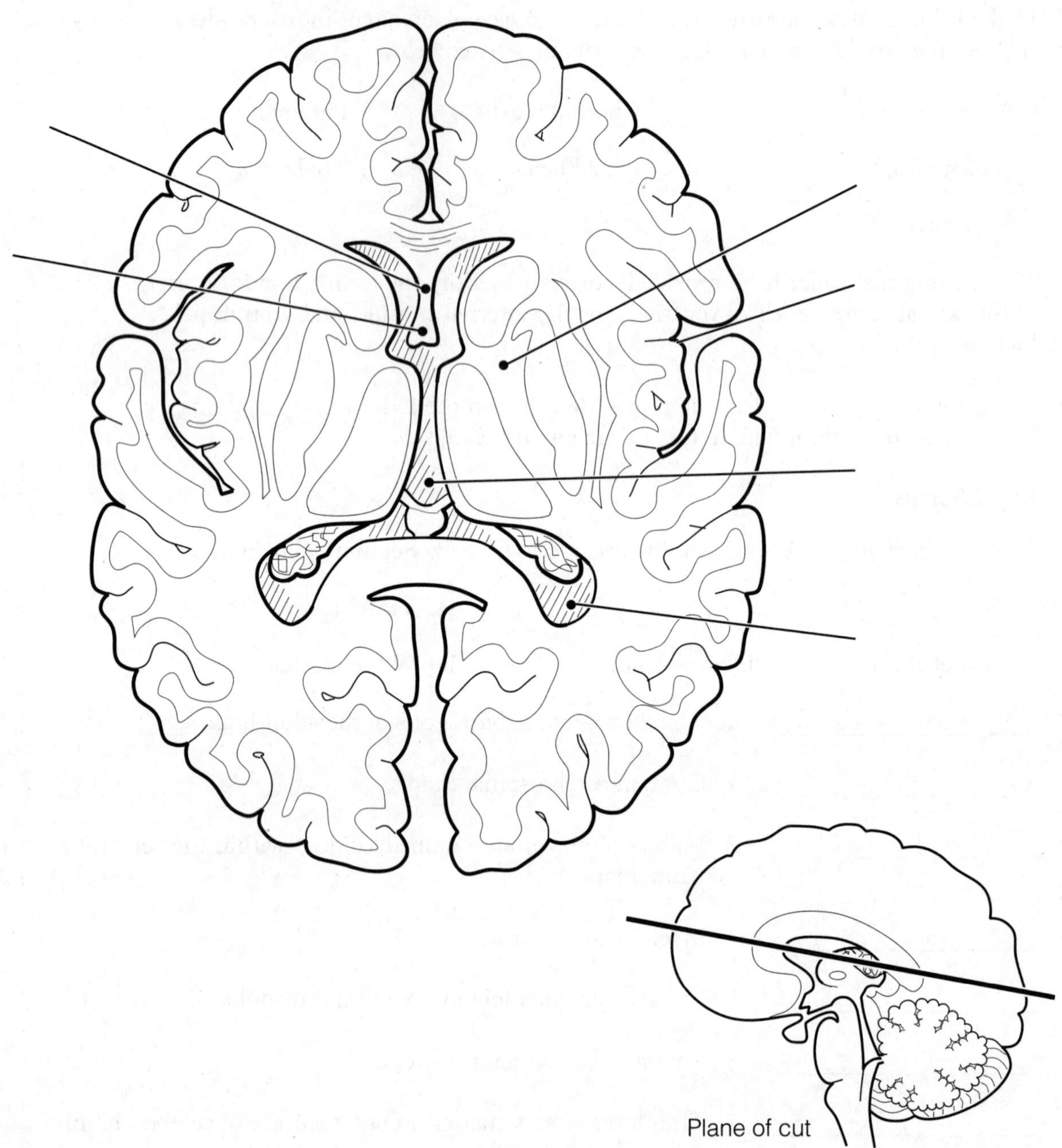

Figure 12.6

12. Figure 12.7 is a diagram of a frontal section through the brain, as indicated on the orientation diagram. Label the ventricles and the longitudinal fissure, both of which are indicated by leader lines. Color code and color the structures listed below.

- ◯ Cerebral cortex
- ◯ Basal nuclei
- ◯ Thalamus
- ◯ Cerebral white matter
- ◯ Corpus callosum
- ◯ Internal capsule
- ◯ Hypothalamus
- ◯ Septum pellucidum

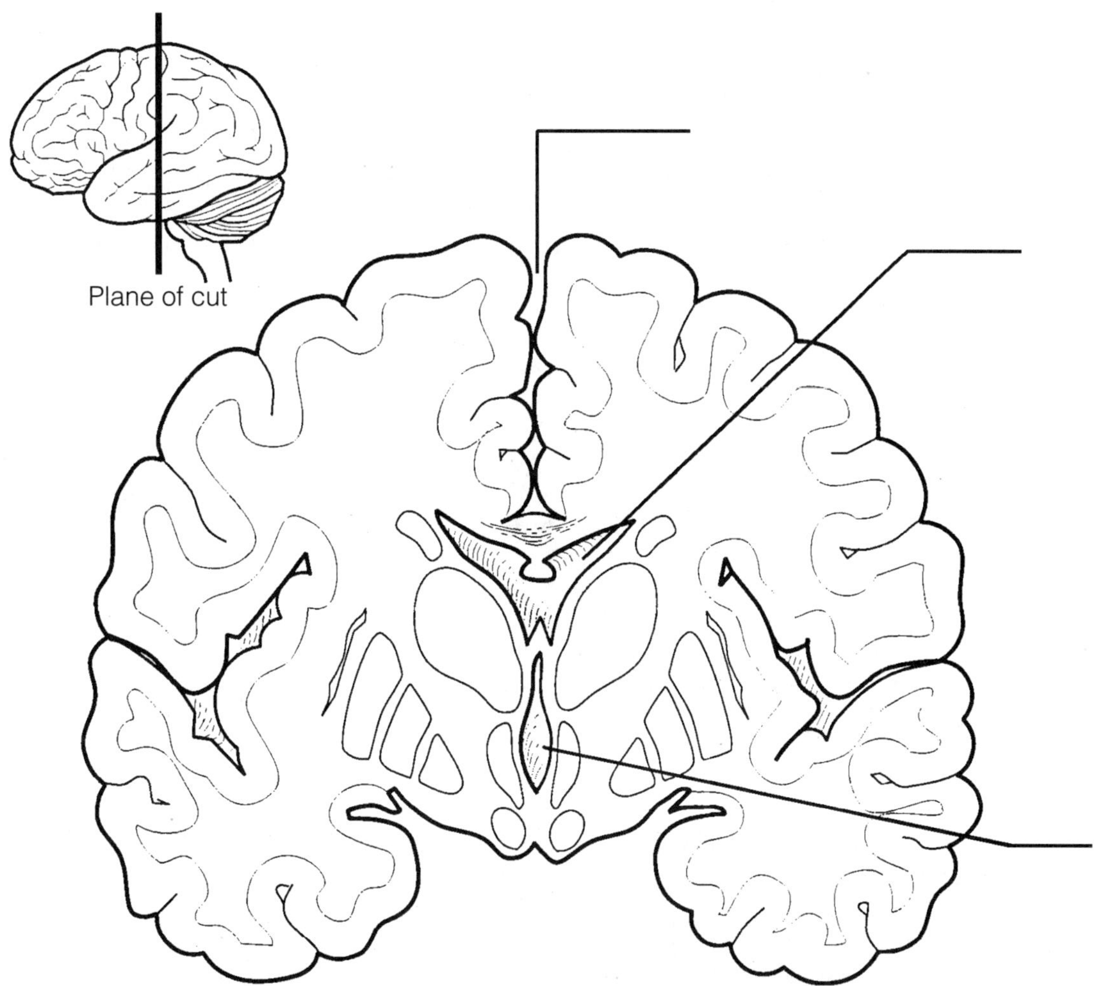

Figure 12.7

13. Figure 12.8 shows the brain stem and associated diencephalon. Using the key choices, label all structures provided with leader lines. Then, color code and color the following structures or groups of structures.

○ All cranial nerves ○ Medulla oblongata ○ Midbrain

○ Diencephalon ○ Pons ○ Infundibulum

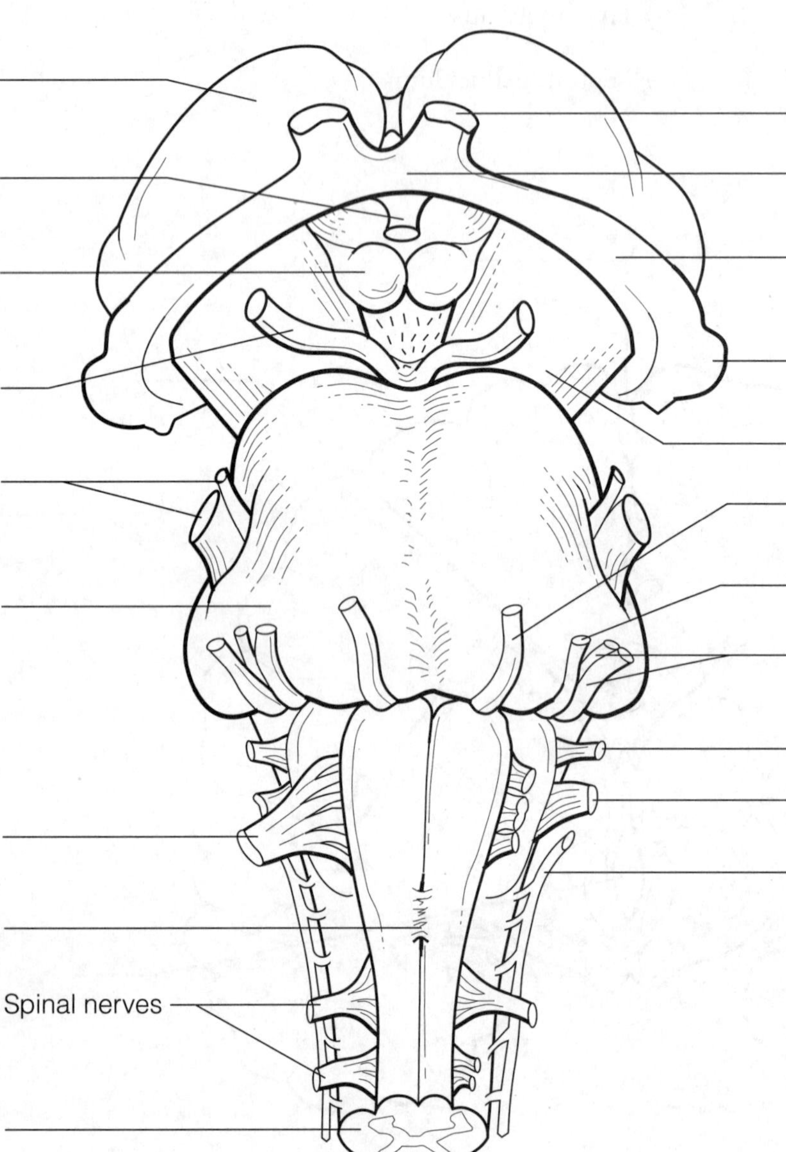

Figure 12.8

Key Choices

A. Abducens nerve (VI)

B. Accessory nerve (XI)

C. Cerebral peduncle

D. Decussation of the pyramids

E. Facial nerve (VII)

F. Glossopharyngeal nerve (IX)

G. Hypoglossal nerve (XII)

H. Infundibulum

I. Lateral geniculate body

J. Mammillary body

K. Oculomotor nerve (III)

L. Optic chiasma

M. Optic nerve (II)

N. Optic tract

O. Pons

P. Spinal chord

Q. Thalamus

R. Trigeminal nerve (V)

S. Vagus nerve (X)

T. Vestibulocochlear nerve (VIII)

14. Circle the term that does not belong in each of the following groupings.

1. Connects cerebral hemispheres   Corpus callosum   Cerebral peduncles   Commissure
2. Hypophysis cerebri   Pineal gland   Produces hormones   Pituitary gland
3. Medulla oblongata   Pons   Cardiac center   Decussation of pyramids
4. Prefrontal cortex   Distractability   Loss of initiative   Paralysis
5. Nonverbal components   Speech disorder   Affective language areas   Loss of vocal expression
6. Lack of cerebral dominance   Reading disorder   Motor impairment   Dyslexia
7. Projection fibers   Internal capsule   Association fibers   Corona radiata
8. Basal nuclei   Internal capsule   Subcortical motor nuclei   Striatum
9. Caudate nucleus   Putamen   Globus pallidus   Lentiform nucleus
10. Cerebellum   Vermis   Parietal lobe   Arbor vitae

15. Below is a listing of events involved in voluntary motor activity that also considers the role of the cerebellum. Put the event(s) in their correct temporal (time) sequence by numbering them 1 to 5. (*Note:* There are *two* number 5 events.)

__________ 1. Cerebellar output to brain stem nuclei initiates subconscious motor output.

__________ 2. Primary motor cortex sends impulses along pyramidal tracts.

__________ 3. Feedback from cerebellum to cerebrum.

__________ 4. Premotor cortex initiates motor activity.

__________ 5. Pyramidal collaterals signal cerebellum.

__________ 6. Visual, auditory, equilibrium, and body sensory input to cerebellum is assessed as the motor activity is initiated.

16. Relative to the limbic and reticular systems, identify the correct system for each characteristic described below. Insert L for the limbic system or R for the reticular system in the answer blanks.

__________ 1. Maintains cortex in a conscious state

__________ 2. Includes the RAS

__________ 3. Originates in primitive rhinencephalon

__________ 4. Its hypothalamus is the gatekeeper for visceral responses

__________ 5. Site of rage and anger; moderated by cerebral cortex

__________ 6. Has far-flung axonal connections

__________ 7. Functioning may be associated with psychosomatic illness

__________ 8. Helps distinguish and filter out unimportant stimuli

__________ 9. Depressed by alcohol and some drugs; severe injury may cause coma

__________ 10. Severe injury may result in personality changes

__________ 11. Extends through the brain stem

__________ 12. Includes cingulate gyrus, hippocampus, amygdala, and some thalamic nuclei

__________ 13. Located in the medial aspect of both hemispheres

Higher Mental Functions

1. Using the key choices, identify the brain wave patterns described below. Write your answers in the answer blanks. (*Note:* Hz = hertz, or cycles per second.)

Key Choices

A. Alpha B. Beta C. Delta D. Theta

__________ 1. Slow, synchronous waves with an average frequency of 10 Hz

__________ 2. More irregular than beta waves; normal in children but uncommon in awake adults

__________ 3. Recorded when a person is awake and relaxed, with eyes closed

__________ 4. High-amplitude waves with a very low frequency (4 Hz or less)

__________ 5. Recorded during sleep or anesthesia

__________ 6. Recorded when a person is awake and fully alert

2. Complete the following statements by writing the missing terms in the answer blanks.

1. ____________________
2. ____________________
3. ____________________
4. ____________________
5. ____________________
6. ____________________
7. ____________________
8. ____________________
9. ____________________

A recording of the electrical activity of the brain is called a(n) __(1)__. Brain wave patterns are identified by their __(2)__, which are measured in __(3)__. When large numbers of neurons fire synchronously, the __(4)__ of the waves increases, a condition typical of the normal adult in a state of __(5)__. Brain waves that are abnormally slow or fast are typical when __(6)__ is interfered with. Very fast brain waves with large spikes are common in a patient who suffers from uncontrollable seizures, a condition called __(7)__. The most severe form of this disease is called __(8)__, during which a sensory "aura" is followed by unconsciousness and intense convulsions. A record showing the complete absence of brain waves is clinical evidence of __(9)__.

3. For any of the following statements about sleep that are true, write T in the answer blanks. For any false statements, correct the underlined word by writing the correct word in the answer blanks. (*Note:* EEG = electroencephalogram; REM = rapid eye movements; RAS = reticular activating system; NREM = non–rapid eye movements.)

____________________ 1. Circadian rhythms recur every <u>12</u> hours.

____________________ 2. The unconscious state, when arousal is not possible, is called <u>sleep</u>.

____________________ 3. During sleep, vital <u>cortical</u> activities continue.

____________________ 4. The sleep-wake cycle is most likely timed by the <u>RAS</u>.

____________________ 5. During deep sleep, vital signs decline, and EEG waves <u>decrease</u> in amplitude.

____________________ 6. During REM sleep, vital signs increase, most muscle movement is inhibited, and the EEG shows <u>alpha</u> waves.

____________________ 7. In normal adults, NREM and REM sleep periods alternate, with <u>REM</u> sleep accounting for the greatest proportion of sleeping time.

____________________ 8. Most dreaming occurs during REM sleep, when certain brain neurons release more norepinephrine and <u>serotonin</u>.

____________________ 9. Deprivation of <u>NREM</u> sleep may lead to emotional instability.

____________________ 10. In the elderly, <u>slow wave (stage 4)</u> sleep declines and may disappear.

____________________ 11. Two important sleep disorders are insomnia and <u>coma</u>.

4. State the definition of *consciousness* as proposed by cognitive scientists, and list the three underlying suppositions.

____

____

____

____

5. List the four clinical states of consciousness, starting with the highest state of cortical activity.

__________, __________, __________, __________

6. Categorize the following descriptions as characteristic of either long-term memory (L) or short-term memory (S). Insert your answers in the answer blanks.

_____ 1. Very brief

_____ 2. Lasts from seconds to hours

_____ 3. Enormous capacity

_____ 4. May last a lifetime

_____ 5. Limited capacity

_____ 6. Association process required

7. List four factors that promote memory consolidation (the transfer of information from short-term to long-term memory).

1. ____

2. ____

3. ____

4. ____

8. Distinguish between declarative (fact) memory and nondeclarative memory by defining each term and by comparing how each type of memory is best remembered.

1. Declarative memory ____

____

____

2. Nondeclarative memory ______________________________

____

____

9. Several proposed pathways and regions, including the amygdala, hippocampus, thalamus, hypothalamus, prefrontal cortex, and basal forebrain, may be involved in the consolidation of declarative memory. In terms of these pathways, respond to the questions that follow by writing your answers in the answer blanks.

______________ 1. According to present assumptions, are the initial connections between old memories and the new perceptions made in the cortex or in the subcortical structures?

______________ 2. The medial temporal lobe plays a role in memory consolidation by communicating with the thalamus and what cortical area?

______________ 3. This feedback pathway seems to transform the initial perception into what?

______________ 4. What is the result if the hippocampus and surrounding temporal lobe structures are destroyed?

______________ 5. Can a person with amnesia still learn skills?

______________ 6. What neurotransmitter released by the basal forebrain prunes the thalamus and prefrontal cortex to allow formation of memories?

______________ 7. As opposed to declarative memory, nondeclarative memory is believed to be processed/mediated by which brain regions?

______________ 8. What neurotransmitter is needed for this circuit (7) to operate?

10. Broca's area, Wernicke's area, and the basal nuclei form a system involved with language. What is the general functional focus of this system?

____

____

Protection of the Brain

1. Figure 12.9 shows a frontal view of the meninges of the brain at the level of the superior sagittal (dural) sinus. First, label arachnoid villi and falx cerebri on the figure. Then, select different colors for each of the following structures and use them to color the diagram.

- ◯ Dura mater
- ◯ Arachnoid
- ◯ Pia mater
- ◯ Subarachnoid space

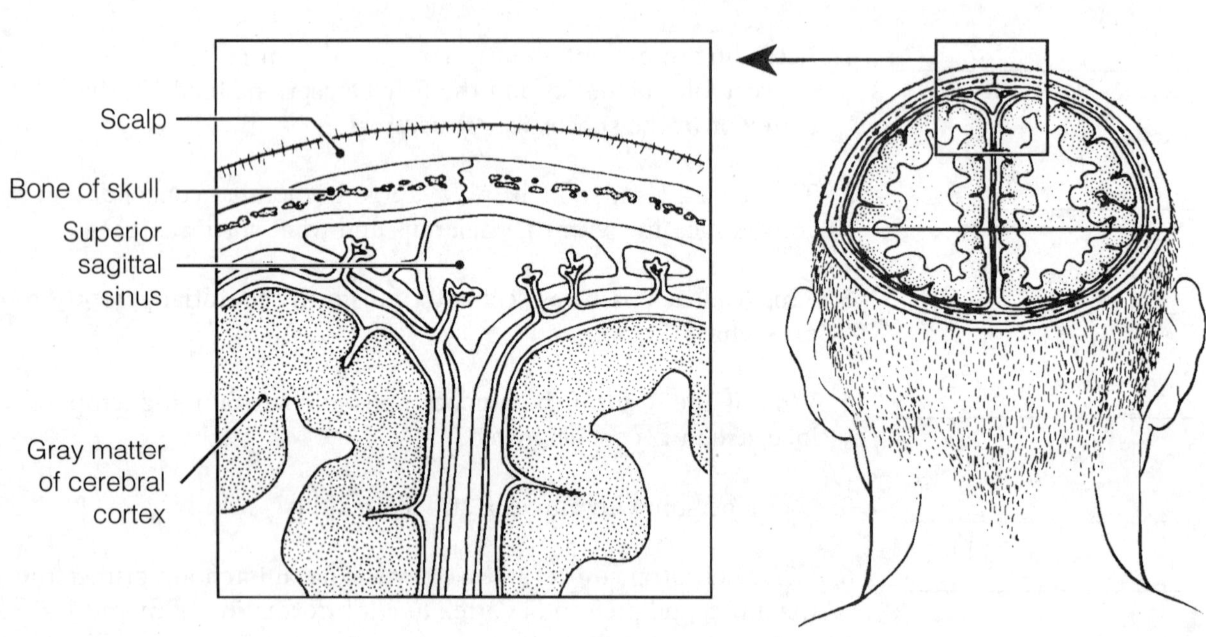

Figure 12.9

2. Referring again to the structures in Figure 12.9, identify the meningeal (or associated) structures described here. Write the correct terms in the answer blanks.

__________ 1. Innermost covering of the brain; delicate and vascular

__________ 2. Structures that return cerebrospinal fluid to the venous blood in the dural sinuses

__________ 3. Its outer layer forms the periosteum of the skull

__________ 4. Contains cerebrospinal fluid

__________ 5. Location of major arteries and veins

__________ 6. Contains venous blood

__________ 7. Attaches to crista galli of the ethmoid bone

3. Complete the following statements by inserting the missing terms in the answer blanks.

______ 1.
______ 2.
______ 3.
______ 4.
______ 5.
______ 6.
______ 7.
______ 8.
______ 9.
______ 10.
______ 11.
______ 12.

The composition of the cerebrospinal fluid (CSF) is similar to __(1)__, from which it arises. The concentration of __(2)__ in the CSF is important in the control of cerebral blood flow and breathing.

The CSF is formed by capillary knots called __(3)__, which hang from the roof of each __(4)__. Circulation of the CSF is aided by the beating cilia of the __(5)__ cells lining the ventricles.

The CSF ordinarily flows from the lateral ventricles to the third ventricle and then through the __(6)__ to the fourth ventricle. Some of the CSF continues down the __(7)__ of the spinal cord, but most of it circulates into the __(8)__ by passing through three tiny openings in the walls of the __(9)__. The CSF is returned to the blood via the __(10)__, located in the dural sinuses. If drainage of the CSF is obstructed, the fluid accumulates under pressure, causing a condition called __(11)__. In the newborn, the condition results in an enlarged head because the skull can expand, but in adults the increasing pressure may cause __(12)__.

4. Relative to the blood-brain barrier, determine whether the following statements are true or false. For true statements, insert T in the answer blanks. For false statements, correct the underlined word(s) by writing the correct word(s) in the answer blanks.

______ 1. <u>Desmosomes</u> are the epithelial structures that make brain capillaries relatively impermeable.

______ 2. Water, glucose, and <u>metabolic wastes</u> pass freely from the blood into the brain, but proteins and most drugs are normally prevented from entering the brain.

______ 3. <u>Calcium</u> ions, in particular, are actively pumped from the brain because they modify neural thresholds.

______ 4. The blood-brain barrier is absent in the <u>cerebrum</u>.

______ 5. Alcohol and other <u>water-soluble</u> molecules easily cross the blood-brain barrier.

______ 6. The blood-brain barrier is <u>incompletely developed</u> in newborns.

5. Match the brain disorders listed in Column B with the conditions described in Column A. Place the correct answers in the answer blanks.

	Column A	Column B
________	1. Slight and transient brain injury	A. Alzheimer's disease
________	2. Traumatic injury that destroys brain tissue	B. Cerebral edema
________	3. Total nonresponsiveness to stimulation	C. Cerebrovascular accident (CVA)
________	4. May cause medulla oblongata to be wedged into foramen magnum by pressure of blood	D. Coma
________	5. After head injury, retention of water by brain	E. Concussion
________	6. Results when a brain region is deprived of blood or exposed to prolonged ischemia; probably reflects excessive NO release	F. Contusion
________	7. Reversible CVA	G. Intracranial hemorrhage
________	8. Progressive degeneration of the brain with abnormal protein deposits	H. Multiple sclerosis
________	9. Autoimmune disorder with extensive demyelination	I. Transient ischemic attack (TIA)

The Spinal Cord

1. Using the key choices, select the correct terms as defined by the following descriptions. Write the correct letters in the answer blanks.

Key Choices

A. Decussation	C. Gray matter	E. Sensory
B. Funiculus	D. Motor	F. Somatotopy

________ 1. Its central location separates white matter into columns

________ 2. Crossing of fibers from one side to the other side of the cord

________ 3. Precise spatial relationship of most spinal cord pathways to an orderly mapping of the body

________ 4. Pathways from the brain to the spinal cord

________ 5. Ascending spinal cord pathways

________ 6. A white column of the cord containing several tracts

2. Figure 12.10 is a cross-sectional view of the spinal cord. First identify the areas listed in the key choices by inserting the correct letters next to the appropriate leader lines on parts A and B of the figure. Then, color the bones of the vertebral column in part B gold.

Key Choices

A. Central canal	E. Dorsal root	I. Ventral horn
B. Columns of white matter	F. Dorsal root ganglion	J. Ventral root
C. Conus medullaris	G. Filum terminale	
D. Dorsal horn	H. Spinal nerve	

On part A, color the butterfly-shaped gray matter gray, and color the spinal nerves and roots yellow. Finally, select different colors to identify the following structures and use them to color the figure.

○ Pia mater ○ Dura mater ○ Arachnoid mater

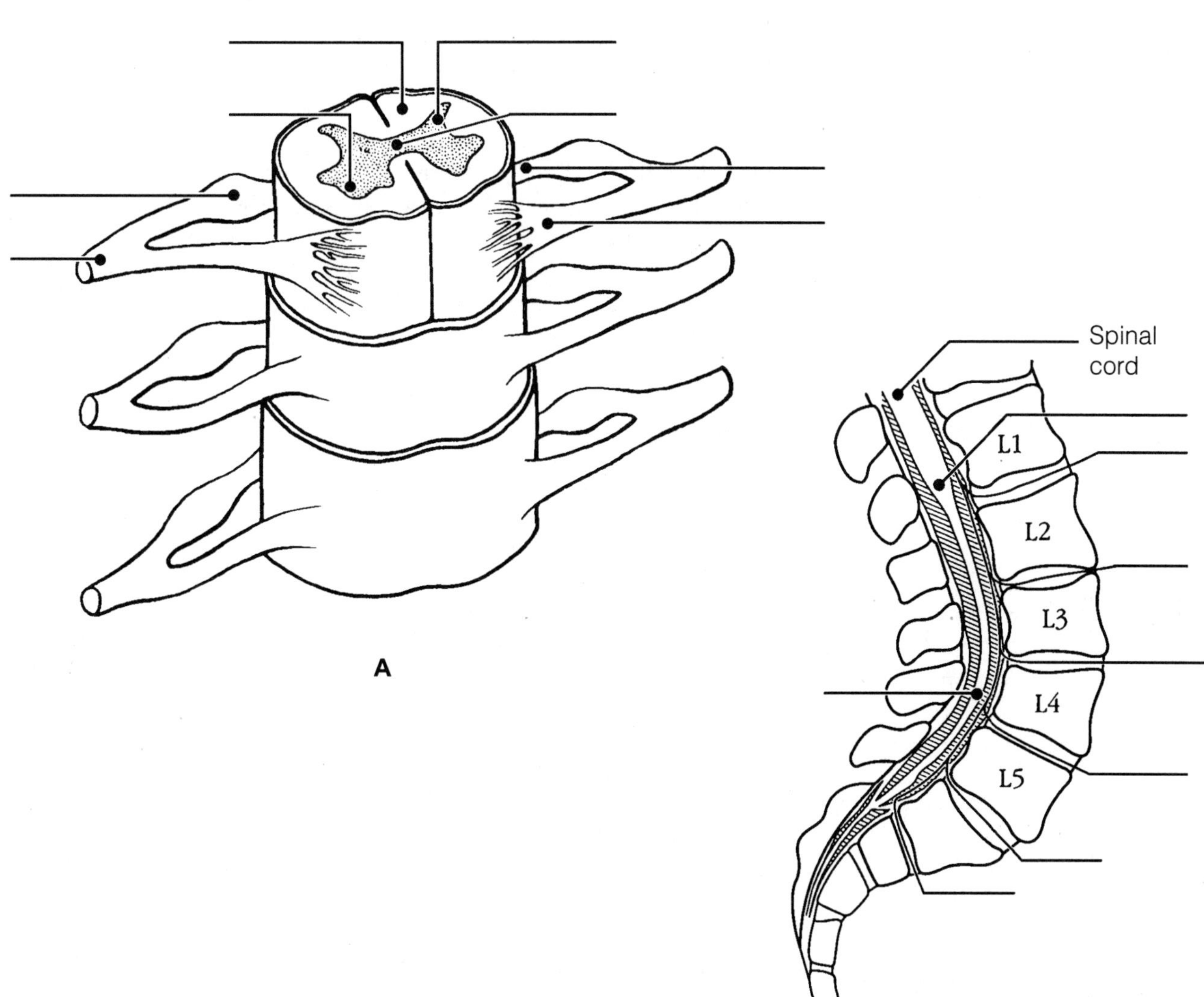

Figure 12.10

3. Complete the following statements by writing the missing terms in the answer blanks.

________ 1.
________ 2.
________ 3.
________ 4.
________ 5.
________ 6.
________ 7.
________ 8.
________ 9.

The spinal cord extends from the __(1)__ of the skull to the __(2)__ region of the vertebral column. The meninges, which cover the spinal cord, extend more inferiorly to form a sac, from which cerebrospinal fluid can be withdrawn without damage to the spinal cord. This procedure is called a(n) __(3)__. __(4)__ pairs of spinal nerves arise from the cord. Of these, __(5)__ pairs are cervical nerves, __(6)__ pairs are thoracic nerves, __(7)__ pairs are lumbar nerves, and __(8)__ pairs are sacral nerves. The tail-like collection of spinal nerves at the inferior end of the spinal cord is called the __(9)__.

4. Complete the following statements by writing the missing terms in the answer blanks.

________ 1.
________ 2.
________ 3.
________ 4.
________ 5.
________ 6.
________ 7.
________ 8.
________ 9.
________ 10.
________ 11.

The spinal cord sheath is composed of a single __(1)__ layer of the dura mater. Soft, protective padding around the spinal cord fills the __(2)__ space. The space between the arachnoid and pia mater meninges is filled with __(3)__. The anterior groove extending the length of the cord is called the __(4)__. The interior central gray matter mass is surrounded by __(5)__ matter. The two sides of the central gray mass are connected by the __(6)__. Neurons in the __(7)__ horn are somatic motor neurons that serve the skeletal muscles, whereas neurons in parts of the __(8)__ horn are autonomic motor neurons serving the visceral organs. If the dorsal root of a spinal nerve is severed, loss of __(9)__ function follows. If the ventral root is damaged, loss of __(10)__ function occurs. In the embryo, neurons of the dorsal root ganglion arise from the __(11)__.

5. Classify the following inputs and outputs as somatic sensory (SS), visceral sensory (VS), somatic motor (SM), or visceral motor (VM) relative to regions of the spinal cord involved in transmission.

_____ 1. Pain from skin

_____ 2. Proprioception

_____ 3. Efferent innervation of a gland

_____ 4. Efferent innervation of your gluteus maximus

_____ 5. A stomachache

_____ 6. A sound you hear

_____ 7. Efferent innervation of the muscle of the urinary bladder wall

6. First, examine the pathway in Figure 12.11 to determine whether this ascending pathway (spinal cord tract) is a *specific* or *nonspecific* pathway. Label accordingly on the left line beneath the pathway. Next, identify the pathway more precisely by selecting one of the choices below, and insert your response on the right line beneath the figure.

Spinocerebellar | Fasciculus cuneatus | Fasciculus gracilis | Spinothalamic

Then, color code the following structures and identify them by coloring them on the diagram.

○ First-order neuron

○ Second-order neuron

○ Third-order neuron

○ Sensory receptor

○ Sensory homunculus

○ Spinal cord

○ Thalamus

○ Postcentral gyrus

Finally, circle all sites of synapse, and indicate your reasons for identifying this pathway as you did. ____

____

____

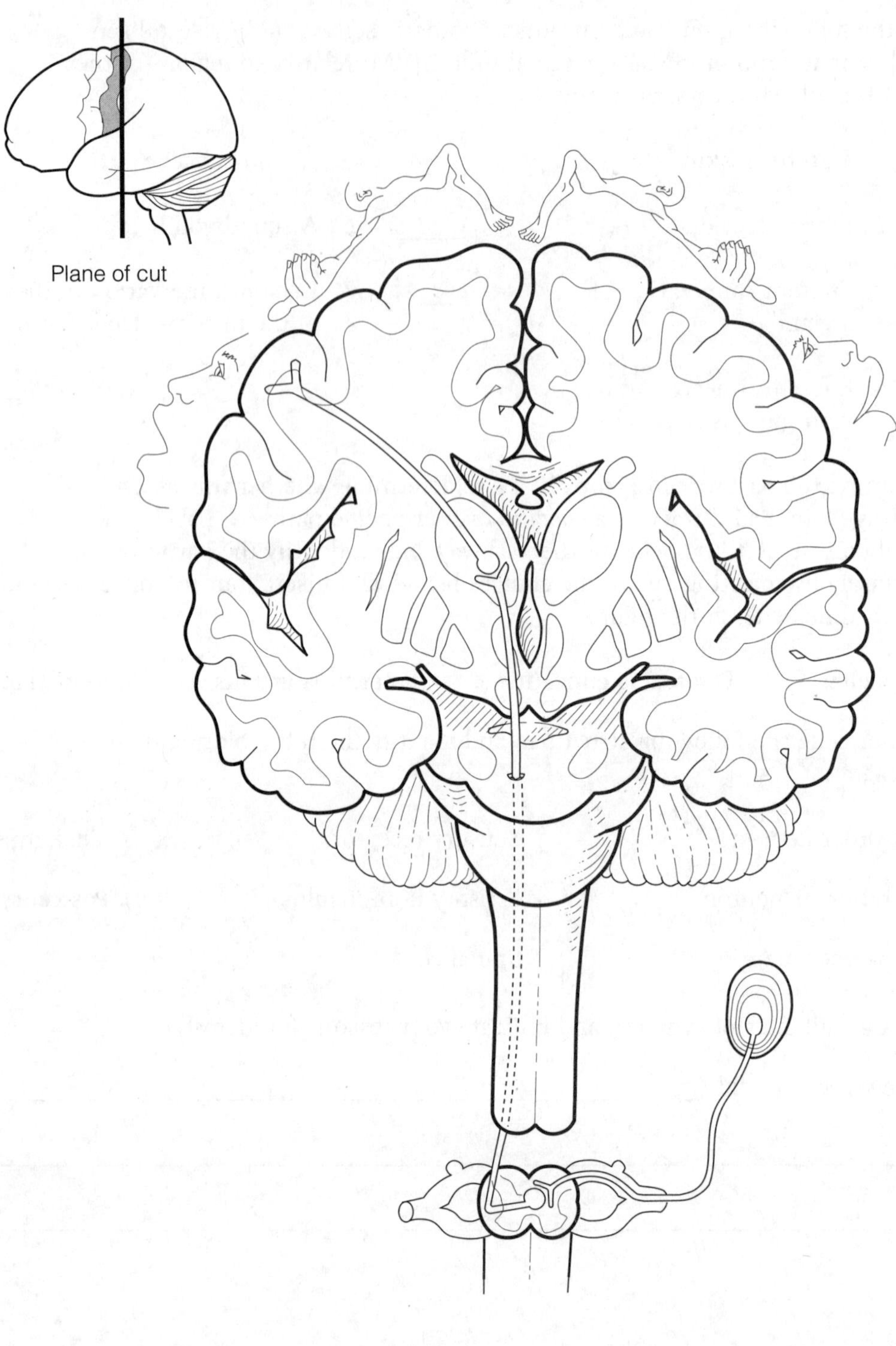

Figure 12.11

7. First, examine the diagram in Figure 12.12 to determine whether it is an example of a *direct* or *indirect* motor pathway. Write your response on the line below the figure. Next, determine what specific tracts are depicted by carefully examining the two locations where the fiber tract has been "lassoed" by a leader line, and label the pathways appropriately on the diagram. Then, color code the structures provided with coloring circles and identify them by coloring them on the diagram.

○ Basal nuclei	○ Brain stem motor nuclei	○ Cerebellum
○ Alpha motor neuron	○ Primary motor cortex	○ Effector
○ Motor homunculus	○ Medullary pyramid	○ Internal capsule
○ Thalamus		

Finally, indicate on the diagram which of these structures is the point of decussation.

8. Match the terms listed in Column B with the injuries described in Column A.

	Column A	Column B
________	1. Loss of sensation	A. Flaccid paralysis
________	2. Paralysis without atrophy	B. Hemiplegia
________	3. Loss of motor function	C. Paralysis
________	4. Traumatic flexion and/or extension of the neck	D. Paraplegia
________	5. Result of transection of the cord between T_{11} and L_1	E. Paresthesia
________	6. Transient period of functional loss induced by trauma to the cord	F. Quadriplegia
________	7. Result of permanent injury to the cervical region of the cord	G. Spastic paralysis
________	8. Paralysis of one side of the body that usually reflects brain injury rather than injury to the spinal cord	H. Spinal shock
		I. Whiplash

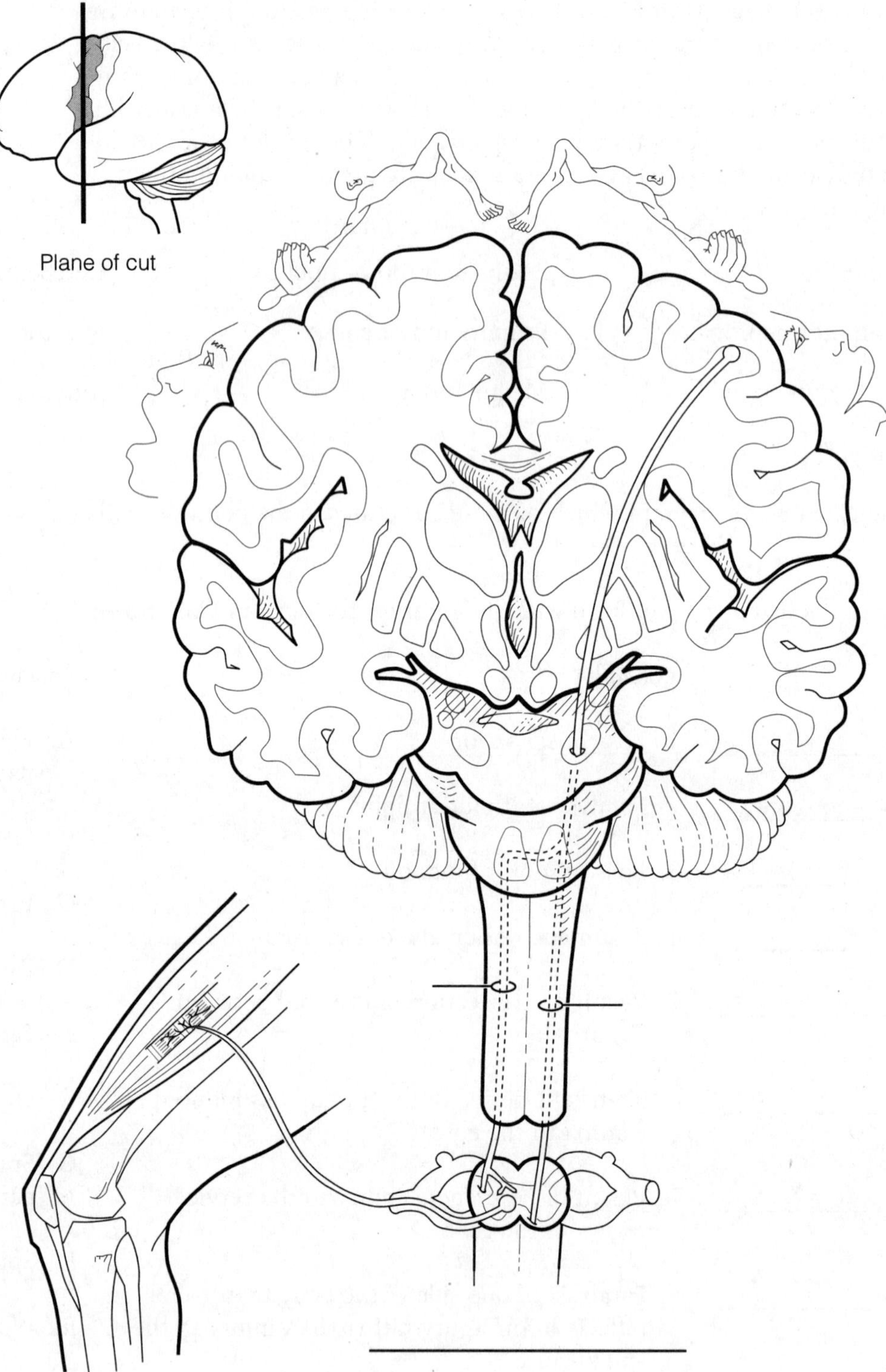

Figure 12.12

Diagnostic Procedures for Assessing CNS Dysfunction

1. If a statement is true, write the letter T in the answer blank. If a statement is false, change the underlined word(s) and write the correct word(s) in the answer blank.

__________________ 1. A flat electroencephalogram is clinical evidence of brain <u>death</u>.

__________________ 2. Damage to the cerebral arteries of TIA victims is usually assessed by a <u>PET scan</u>.

__________________ 3. The biochemical activity of the brain may be monitored by <u>MRI</u>.

__________________ 4. A lumbar puncture is done below L_1 because the spinal cord ends <u>below</u> that level.

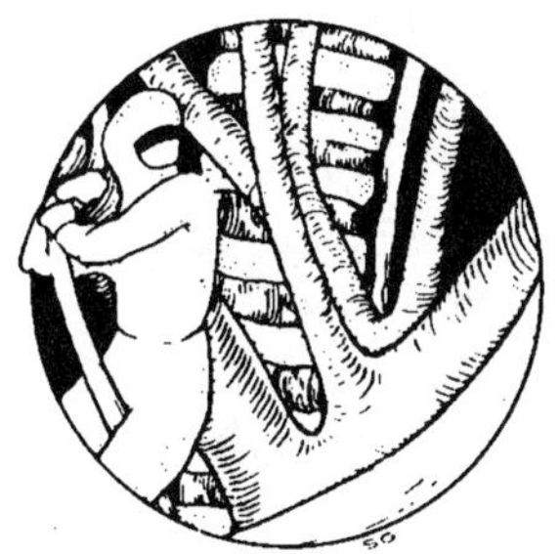

THE INCREDIBLE JOURNEY:

A Visualization Exercise for the Nervous System

You climb onto the first cranial nerve you see . . .

1. Complete the following narrative by inserting the missing words in the answer blanks.

__________________ 1.

__________________ 2.

Nervous tissue is quite densely packed, and it is difficult to envision strolling through its various regions. Imagine instead that each of the various functional regions of the brain has a computerized room in which you might observe what occurs in that particular area. Your assignment is to determine where you are at any given time during your journey through the nervous system.

You begin your journey after being injected into the warm pool of cerebrospinal fluid in your host's fourth ventricle. As you begin your stroll through the nervous tissue, you notice a huge area of branching white matter overhead. As you enter the first computer room, you hear an announcement through the loudspeaker: "The pelvis is tipping too far posteriorly—please correct—we are beginning to fall backward and will soon lose our balance." The computer responds immediately, decreasing impulses to the posterior hip muscles and increasing impulses to the anterior thigh muscles. "How is that, proprioceptor 1?" From this information, you determine that your first stop is the __(1)__.

At the next stop, you hear "Blood pressure to head is falling; increase sympathetic nervous system stimulation of the blood vessels." Then, as it becomes apparent that your host has not only stood up but is going to run, you hear "Increase rate of impulses to the heart and respiratory muscles—we are going to need more oxygen and a faster blood flow to the skeletal muscles of the legs." You recognize that this second stop must be the __(2)__.

____________ 3.
____________ 4.
____________ 5.
____________ 6.
____________ 7.

Computer room 3 presents a problem. There is no loudspeaker here; instead, incoming messages keep flashing across the wall, giving only bits and pieces of information. "Four hours since last meal—stimulate appetite center. Slight decrease in body temperature—initiate skin vasoconstriction. Mouth dry—stimulate thirst center. Um, a stroke on the arm—stimulate pleasure center." Looking at what has been recorded here—appetite, temperature, thirst, and pleasure—you conclude that this has to be the __(3)__.

____________ 8.
____________ 9.
____________ 10.
____________ 11.
____________ 12.

Continuing your journey upward toward the higher brain centers, finally you are certain you have reached the cerebral cortex. The first center you visit is quiet, like a library with millions of encyclopedias of facts and recordings of past input. You conclude that this must be the area where __(4)__ are stored and that you are probably in the __(5)__ lobe. The next stop is close by. As you enter the computer center, you once again hear a loudspeaker: "Let's have the motor instructions to say tintinnabulation—we don't want them to think we're tongue-tied." This area is obviously __(6)__. Your final stop in the cerebral cortex is a very hectic center. Electrical impulses are traveling back and forth between giant neurons, sometimes in different directions and sometimes back and forth between a small number of neurons. Watching intently, you try to make some sense out of these interactions, and you suddenly realize that this *is* what is happening here. The neurons are trying to make some sense out of something, which helps you decide that this must be the brain area where __(7)__ occurs in the __(8)__ lobe.

You hurry out of this center and retrace your steps back to the cerebrospinal fluid, deciding en route to observe a cranial nerve. You decide to pick one randomly and follow it to the organ it serves. You climb onto the first cranial nerve you see and slide down past the throat. Picking up speed, you quickly pass the heart and lungs and see the stomach and small intestine coming up fast. A moment later you land on the stomach, and now you know that this wandering nerve has to be the __(9)__. As you look upward, you see that the nerve is traveling almost straight up and decide you'll have to find an alternative route back to the cerebrospinal fluid. You begin to walk posteriorly until you find a spinal nerve, which you follow until you reach the vertebral column. You squeeze between two adjacent vertebrae to follow the nerve to the spinal cord. With your pocket knife, you cut away the tough connective tissue covering the cord. Thinking that the __(10)__ covering deserves its name, you finally manage to cut an opening large enough to get through, and you return to the warm bath of cerebrospinal fluid that these membranes enclose. At this point you are in the __(11)__, and from here you swim upward until you get to the lower brain stem. Once there, it should be an easy task to find the holes leading into the __(12)__ ventricle, where your journey began.

CHALLENGING YOURSELF

At the Clinic

1. Following a train accident, Sharon Money, a woman with an obvious head injury, is observed stumbling about the scene. An inability to walk properly and loss of balance are quite obvious. What brain region was injured?

2. A child is brought to the clinic with a high temperature. The doctor states that the child's meninges are inflamed. What name is given to this condition and how is it diagnosed?

3. Jemal, an elderly man with a history of TIAs, complained to his daughter that he had a severe headache. Shortly thereafter, he lapsed into a coma. At the hospital, he was diagnosed as having a brain hemorrhage. Which part of the brain was damaged by the hemorrhage?

4. A young man has just received serious burns resulting from standing with his back too close to a bonfire. He is muttering that he never felt the pain; otherwise, he would have smothered the flames by rolling on the ground. What part of his CNS might be malfunctioning?

5. Elderly Mrs. Barker has just suffered a stroke. She is able to understand verbal and written language, but when she tries to respond, her words are garbled. What cortical region has been damaged by the stroke?

6. An elderly man is brought to the clinic by his wife, who noticed that his speech is slurred, the right side of his face is slack, and he has difficulty swallowing. In which area of the brain would a stroke be suspected?

7. Sufferers of Tourette's syndrome have an excess of dopamine in the basal nuclei. They tend to exhibit explosive, uncontrollable verbal episodes of motor activity. The drug haloperidol reduces dopamine levels and returns voluntary control to the patient. What condition would an overdose of haloperidol mimic?

8. Huntington's disease results from a genetic defect that causes degeneration of the basal nuclei and, eventually, the cerebral cortex. Its initial symptoms involve involuntary motor activity, such as arm flapping. Would it be treated with a drug that increases or decreases dopamine levels?

9. Cindy's parents report that she has "spells," lasting 15 to 20 minutes, during which the 4-year-old is unresponsive although apparently awake. What is Cindy's probable diagnosis? Is improvement likely?

10. A young woman comes to the clinic complaining that she is plagued with seizures during which she loses consciousness. Afterwards, she is extremely disoriented for a while. From what type of epilepsy is she suffering?

11. Mrs. Tonegawa is brought to the hospital in a semiconscious state after falling from a roof. She did not lose consciousness immediately, and she was initially lucid. After a while, though, she became confused and then unresponsive. What is a likely explanation of her condition?

12. Beth is brought to the clinic by her frantic husband. He explains that she has gradually lost control of her right hand. No atrophy of the muscles is apparent. Should upper or lower motor neuron damage be investigated?

13. A woman brings her elderly father to the clinic. He has been having more and more difficulty caring for himself and now even dressing and eating are a problem. The nurse notes the man's lack of change of facial expression; slow, shuffling gait; and tremor in the arms. What are the likely diagnosis and treatment?

14. Mary, an elderly woman, complains that she cannot sleep at night. She says she sleeps no more than four hours a night, and she demands sleeping medication. Is Mary's sleep pattern normal? Will a sleeping aid help her?

15. An alcoholic in his 60s is brought to the clinic in a stupor. When he has regained consciousness, a PET scan is ordered. The PET scan shows no activity in parts of the limbic system. The man has no memory of what befell him; in fact, he has no memory of the last 20 years! What type of amnesia does he have? Will it affect his ability to learn new skills?

Stop and Think

1. Why do you think so much of the cerebral cortex is devoted to sensory and motor connections to the eyes?
2. Electroconvulsive therapy (ECT) can be used to treat severe clinical depression. After ECT, patients exhibit some memory loss. The electrical current apparently wipes out reverberating circuits, resulting in lifting of the depression as it causes loss of information. Is ECT more likely to affect STM or LTM?
3. Do all tracts from the cerebrum to the spinal cord go through the entire brain stem?
4. Contrast damage to the primary visual cortex with damage to the visual association cortex.
5. Any CSF accumulation will cause hydrocephalus, which in a fetus will lead to expansion of the skull. In a fetus, would blockage of CSF circulation in the cerebral aqueduct have a different effect than blockage of the arachnoid villi?
6. What does mannitol's ability to make capillary cells shrivel tell you about the cells' permeability to mannitol?
7. Suppose you cut the little finger of your left hand. Would you expect that the cut might interfere with motor function, sensory function, or both? Explain your answer.
8. A victim of a motorcycle accident fractured vertebra T_{12}, and there was concern that the spinal cord was crushed at the level of this vertebra. At what spinal cord segment was the damage expected? Choose and explain: (a) between C_4 and C_8, and the ability to move the arms was tested, (b) at C_1, and the respiratory movements of the diaphragm were tested, (c) spinal-cord level L_3, (d) there could be no deficits because the cord always ends above T_{12}.
9. Why is hemiplegia more likely to be a result of brain injury than of spinal cord injury?
10. Which would more likely result from injury exclusively to the dorsal side of the spinal cord—paresthesia or paralysis? Explain your answer.
11. Ralph had brain surgery to remove a small intracranial hematoma (blood mass). The operation was done under local anesthesia, and Ralph remained conscious while the surgeon removed a small part of the skull. The operation went well, and Ralph asked the surgeon to mildly stimulate his (unharmed) postcentral gyrus with an electrode. The surgeon did so. What happened? Choose and explain: (a) Ralph was seized with uncontrollable rge; (b) he saw things that were not there; (c) he asked to see what was touching his hand, but nothing was; (d) he started to kick; (e) he heard his mother's voice from 30 years ago.

COVERING ALL YOUR BASES

Multiple Choice

Select the best answer or answers from the choices given.

1. The secondary brain vesicle that is least developed in the adult brain is the:
   A. telencephalon
   B. mesencephalon
   C. metencephalon
   D. rhombencephalon
2. Which is an incorrect association of brain region and ventricle?
   A. Mesencephalon—third ventricle
   B. Cerebral hemispheres—lateral ventricles
   C. Pons—fourth ventricle
   D. Medulla—fourth ventricle

3. Which of the following is not part of the brain stem?

A. Medulla C. Pons
B. Cerebellum D. Midbrain

4. Connecting a ventricle to the subarachnoid space is the function of the:

A. interventricular foramen
B. cerebral aqueduct
C. lateral aperture
D. median aperture

5. The discrete correlation of body regions to CNS structures is:

A. a homunculus C. lateralization
B. somatotopy D. cephalization

6. Which is not associated with the frontal lobe?

A. Olfaction
B. Motor control of the eyes
C. Prefrontal cortex
D. Skilled motor programs

7. Regions involved with language include:

A. auditory association area
B. prefrontal cortex
C. Wernicke's area
D. Broca's area

8. When neurons in Wernicke's area send impulses to neurons in Broca's area, the white matter tracts utilized are:

A. commissural fibers
B. projection fibers
C. association fibers
D. anterior funiculus

9. The basal nuclei include:

A. hippocampus C. lentiform nucleus
B. caudate nucleus D. mammillary bodies

10. Which of the brain areas listed are involved in normal voluntary muscle activity?

A. Amygdala C. Medullary pyramids
B. Putamen D. Precentral gyrus

11. If impulses to the cerebral cortex from the thalamus were blocked, which of the following sensory inputs would get through to the appropriate sensory cortex?

A. Visual impulses
B. Auditory impulses
C. Olfactory impulses
D. General sensory impulses

12. Functions controlled by the hypothalamus include:

A. crude interpretation of pain
B. regulation of many homeostatic mechanisms
C. setting of some biological rhythms
D. secretion of melatonin

13. The pineal gland is located in the:

A. hypophysis cerebri C. epithalamus
B. mesencephalon D. corpus callosum

14. Which of the following are part of the midbrain?

A. Corpora quadrigemina
B. Cerebellar peduncles
C. Substantia nigra
D. Nucleus of cranial nerve V

15. Cranial nerves with their nuclei in the pons include:

A. facial C. trigeminal
B. vagus D. trochlear

16. Functions that are at least partially overseen by the medulla are:

A. regulation of the heart
B. maintaining equilibrium
C. regulation of respiration
D. visceral motor function

17. Which of the following are important in cerebellar processing?

A. Feedback to cerebral motor cortex
B. Input from body parts
C. Output for subconscious motor activity
D. Direct impulses from motor cortex via a separate cerebellar pathway

18. Parts of the limbic system include:

A. cingulate gyrus C. fornix
B. hippocampus D. septal nuclei

19. Inability to prevent expressing one's emotions most likely indicates damage to the:

A. affective brain
B. cognitive brain
C. affective brain's control of the cognitive brain
D. cognitive brain's control of the affective brain

20. Which statements concerning the reticular activating system are true?

A. It consists of neural columns extending through the brain stem.
B. Its connections reach as far as the cerebral cortex.
C. It is housed primarily in the thalamus.
D. It filters sensory input to the cerebrum.

21. Relative to the cranial meninges:
    A. the arachnoid produces CSF
    B. the dura contains several blood sinuses
    C. the pia mater has two layers
    D. three dural folds help support brain tissue

22. Which structures are directly involved with formation, circulation, and drainage of CSF?
    A. Ependymal cilia
    B. Ventricular choroid plexuses
    C. Arachnoid villi
    D. Serous layers of the dura mater

23. Which of the following are associated with the conus medullaris?
    A. Filum terminale
    B. The end of the dural sheath
    C. Cauda equina
    D. Location of lumbar puncture

24. The spinal cord feature associated with the leash of nerves supplying the upper limbs is the:
    A. brachial plexus
    B. brachial enlargement
    C. cervical enlargement
    D. lateral gray horns

25. Features associated with the anterior surface of the spinal cord include the:
    A. ventral horns of gray matter
    B. median fissure
    C. median sulcus
    D. root ganglia

26. Poliomyelitis affects the:
    A. dorsal white columns
    B. dorsal gray horns
    C. ventral white columns
    D. ventral gray horns

27. Which spinal cord tracts carry impulses for conscious sensations?
    A. Fasciculus gracilis
    B. Lateral spinothalamic
    C. Anterior spinocerebellar
    D. Fasciculus cuneatus

28. Damage to which descending tract would be suspected if a person had unilateral poor muscle tone and posture?
    A. Reticulospinal
    B. Rubrospinal
    C. Tectospinal
    D. Vestibulospinal

29. Damage to the lower motor neurons may result in:
    A. paresthesia
    B. flaccid paralysis
    C. spastic paralysis
    D. paraplegia

30. Which neuron parts occupy the gray matter in the spinal cord?
    A. Tracts of long axons
    B. Motor neuron cell bodies
    C. Sensory neuron cell bodies
    D. Nerves

31. A professor unexpectedly blew a loud horn in his anatomy classroom, and all his students looked up, startled. These reflexive movements of their neck and eye muscles were mediated by (the):
    A. cerebral cortex
    B. inferior olives
    C. raphe nuclei
    D. inferior colliculus

32. Which of the following is (are) associated with the medial lemniscal pathways?
    A. Fasciculus gracilis
    B. Fasciculus cuneatus
    C. Ventral posterior thalamic nucleus
    D. Trigeminal nerves

33. Both specific and nonspecific pathways:
    A. transmit impulses to the thalamus
    B. initiate interpretation of emotional aspects of sensation
    C. stimulate the RAS
    D. cross over within the CNS

34. Relaxation of muscles prior to and during sleep is due to:
    A. increased inhibition at the projection level
    B. decreased stimulation by the vestibular nuclei
    C. increased activity of the RAS
    D. inhibition of the reticular nuclei

35. An adult male is having an EEG. The recording shows irregular waves with a frequency of 4–7 Hz. This is:
    A. normal
    B. normal in REM sleep
    C. normal in deep sleep
    D. abnormal in an awake adult

36. During REM sleep:
    A. alpha waves appear
    B. oxygen utilization of the brain increases
    C. rapid eye movement occurs
    D. dreaming occurs

37. Consciousness involves:
A. discrete, localized stimulation of the cortex
B. activity concurrent with localized activities
C. simultaneous, interconnected activity
D. stimulation of the RAS

38. States of unconsciousness include:
A. sleep
B. tonic-clonic seizure
C. syncope
D. narcoleptic seizure

39. Fact memory involves:
A. the hippocampi
B. rehearsal
C. language areas of the cortex
D. consolidation

40. Sleep apnea:
A. is associated with obesity
B. is exacerbated by alcohol
C. is due to a lack of oxygen
D. tends to occur repeatedly throughout the night

Word Dissection

For each of the following word roots, fill in the literal meaning and give an example, using a word found in this chapter.

Word root	Translation	Example
1. campo		
2. collicul		
3. commis		
4. cope		
5. enceph		
6. epilep		
7. falx		
8. forn		
9. gyro		
10. hippo		
11. infundib		
12. isch		
13. lemnisc		
14. nigr		
15. rhin		
16. rostr		
17. uncul		
18. uncus		

13

The Peripheral Nervous System and Reflex Activity

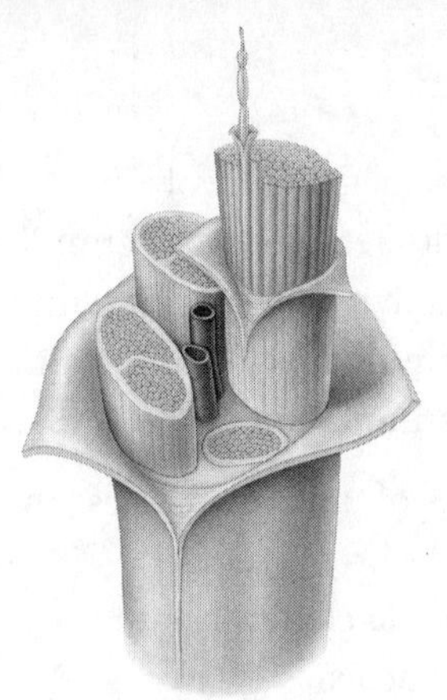

Student Objectives

When you have completed the exercises in this chapter you will have accomplished the following objectives:

1. Define peripheral nervous system and list its components.

PART 1: SENSORY RECEPTORS AND SENSATION

2. Classify general sensory receptors by structure, stimulus detected, and body location.
3. Outline the events that lead to sensation and perception.
4. Describe receptor and generator potentials and sensory adaptation.
5. Describe the main aspects of sensory perception.
6. Describe the structure and function of accessory eye structures, eye layers, the lens, and humors of the eye.
7. Outline the causes and consequences of cataracts and glaucoma.
8. Trace the pathway of light through the eye to the retina, and explain how light is focused for distant and close vision.
9. Outline the causes and consequences of astigmatism, myopia, hyperopia, and presbyopia.
10. Describe the events involved in the stimulation of photoreceptors by light, and compare and contrast the roles of rods and cones in vision.
11. Compare and contrast light and dark adaptation.
12. Trace the visual pathway to the visual cortex, and briefly describe the steps in visual processing.
13. Describe the location, structure, and afferent pathways of taste and smell receptors, and explain how these receptors are activated.
14. Describe the structure and general function of the outer, middle, and internal ears.
15. Describe the sound conduction pathway to the fluids of the internal ear, and follow the auditory pathway from the spiral organ (of Corti) to the temporal cortex.
16. Explain how one is able to differentiate pitch and loudness, and localize the source of sounds.
17. List possible causes and symptoms of otitis media, deafness, and Ménière's syndrome.
18. Explain how the balance organs of the semicircular canals and the vestibule help maintain dynamic and static equilibrium.

PART 2: TRANSMISSION LINES: NERVES AND THEIR STRUCTURE AND REPAIR

19. Define ganglion and indicate the general body location of ganglia.
20. Describe the general structure of a nerve.
21. Follow the process of nerve regeneration.
22. Name the 12 pairs of cranial nerves; indicate the body region and structures innervated by each.
23. Describe the formation of a spinal nerve and the general distribution of its rami.
24. Define plexus. Name the major plexuses and describe the distribution and function of the peripheral nerves arising from each plexus.

PART 3: MOTOR ENDINGS AND MOTOR ACTIVITY

25. Compare and contrast the motor endings of somatic and autonomic nerve fibers.
26. Outline the three levels of the motor hierarchy.
27. Compare the roles of the cerebellum and basal nuclei in controlling motor activity.

PART 4: REFLEX ACTIVITY

28. Name the components of a reflex arc and distinguish between autonomic and somatic reflexes.
29. Compare and contrast stretch, flexor, crossed extensor, and Golgi tendon reflexes.

The part of the nervous system that lies outside the brain and spinal cord is called the peripheral nervous system (PNS). The PNS consists of sensory receptors, all peripheral nerves (12 pairs of cranial nerves, 31 pairs of spinal nerves) and their associated ganglia, and the motor endings. The PNS serves as the two-way communication network between the environment (both inside and outside the body) and the central nervous system (CNS), where the sensory information generated by the PNS is received, interpreted, and used to maintain homeostasis.

The topics for study in Chapter 13 are the sensory receptors, the structure and function of the special sense organs, the structure and types of PNS nerves, their pathways, the body regions they serve, and motor endings and the basics of motor activity. Reflex activities are also studied in this chapter.

BUILDING THE FRAMEWORK

1. Complete the following statements by writing the missing terms in the answer blanks.

1. ____________
2. ____________
3. ____________
4. ____________
5. ____________
6. ____________
7. ____________
8. ____________

The two functional divisions of the PNS are the __(1)__ division, in which the nerve fiber type called __(2)__ carries impulses toward the CNS, and the __(3)__ division, in which the nerve fiber type called __(4)__ carries impulses away from the CNS. The motor division of the PNS includes two types of fibers, the __(5)__ and __(6)__ motor nerve fibers. Skeletal muscles are innervated by __(7)__ fibers, whereas visceral organs are innervated by __(8)__ fibers.

2. By assigning them numbers from 1 to 10, arrange the following elements in order of stimulation. (*Note:* One of the elements will be used twice.)

Spinal nerve   Ventral ramus of spinal nerve   Dorsal root of spinal nerve

Sensory nerve fiber   Motor nerve fiber   Receptor   Effector   Spinal Cord

PART 1: SENSORY RECEPTORS AND SENSATION

1. Complete the following statements by writing the missing terms in the answer blanks.

1. ____________________
2. ____________________
3. ____________________
4. ____________________
5. ____________________
6. ____________________
7. ____________________
8. ____________________
9. ____________________

__(1)__ is the awareness of internal and external stimuli. The conscious interpretation of such stimuli is called __(2)__.

Sensory receptors transduce stimulus energy into __(3)__. As stimulus energy is absorbed by the receptor, the __(4)__ of the receptor membrane changes. This allows ions to flow through the membrane and results in a(n) __(5)__ potential called the receptor potential. If the receptor potential reaches __(6)__ a generator potential is produced and a(n) __(7)__ potential is generated and transmitted.

The ascending pathways that end at the cerebellum are the __(8)__ pathways. The brain region that acts as a relay station and projects fibers to the sensory cortex is the __(9)__.

2. If a statement about the three levels of sensory integration is true, write T in the answer blank. If a statement is false, change the underlined word(s) and write the correct word(s) in the answer blank.

____________________ 1. Sensation occurs in the <u>sensory association cortex</u>.

____________________ 2. The stronger the stimulus, the greater is the <u>velocity</u> of impulse transmission.

____________________ 3. The <u>spinothalamic</u> ascending pathways transmit information about pain, coarse touch, and temperature.

____________________ 4. Second-order neurons of both specific and nonspecific ascending circuits terminate in the <u>medulla</u>.

____________________ 5. The fasciculus cuneatus, fasciculus gracilis, and median lemniscal tracts form the <u>ventral</u> ascending pathways.

____________________ 6. The cell bodies of <u>second-order</u> neurons are located in the dorsal root ganglia.

____________________ 7. To accommodate precise, discriminatory information, impulses are carried by <u>dorsal column-medial lemniscal</u> ascending pathways.

3. Classify the receptor types described below by *location* (make a choice from Key A) and then by *stimulus type detected* (make a choice from Key B).

Key A: A. Exteroceptor
B. Interoceptor
C. Proprioceptor

Key B: 1. Chemoreceptor
2. Mechanoreceptor
3. Nociceptor
4. Photoreceptor
5. Thermoreceptor

__________, ________ 1. In skeletal muscles; respond to muscle stretch

__________, ________ 2. In the walls of blood vessels; respond to oxygen content of surrounding interstitial fluid

__________, ________ 3. In the skin; respond to a hot surface

__________, ________ 4. In the eyes; respond to light

__________, ________ 5. In the inner ear; respond to sound

__________, ________ 6. In the skin; respond to acid splashing on the skin

__________, ________ 7. In the wall of blood vessels; respond to blood pressure

__________, ________ 8. In the stomach wall; respond to an overfull stomach

__________, ________ 9. In the nose; respond to a skunk's spray (Yuk!)

4. In relation to sensory receptors, circle the term that does not belong in each of the following groupings.

1. Cutaneous receptors   Free dendritic endings   Tendon stretch   Pain and touch
2. Meissner's corpuscle   Encapsulated dendritic endings   Dermal papillae   Numerous in muscle
3. Largest of corpuscular receptors   Light touch   Pacinian corpuscle   Multiple layers of supporting cells
4. Thermoreceptor   Hair movement   Light touch   Root hair plexus
5. Tactile discs   Meissner's corpuscle   Muscle spindle   Touch receptor
6. Proprioceptors   Nociceptors   Sensitive to tendon stretch   Muscle spindle
7. Tactile discs   Sensitive to joint orientation   Articular capsules   Joint kinesthetic receptors

8. Free dendritic endings   Root hair plexuses   Tactile discs

   Modified free dendritic endings

9. Phasic receptors   Fast adapting   Pacinian corpuscles   Tonic receptors

5. Identify the highly simplified cutaneous receptors shown in Figure 13.1 and color the figure as it strikes your fancy.

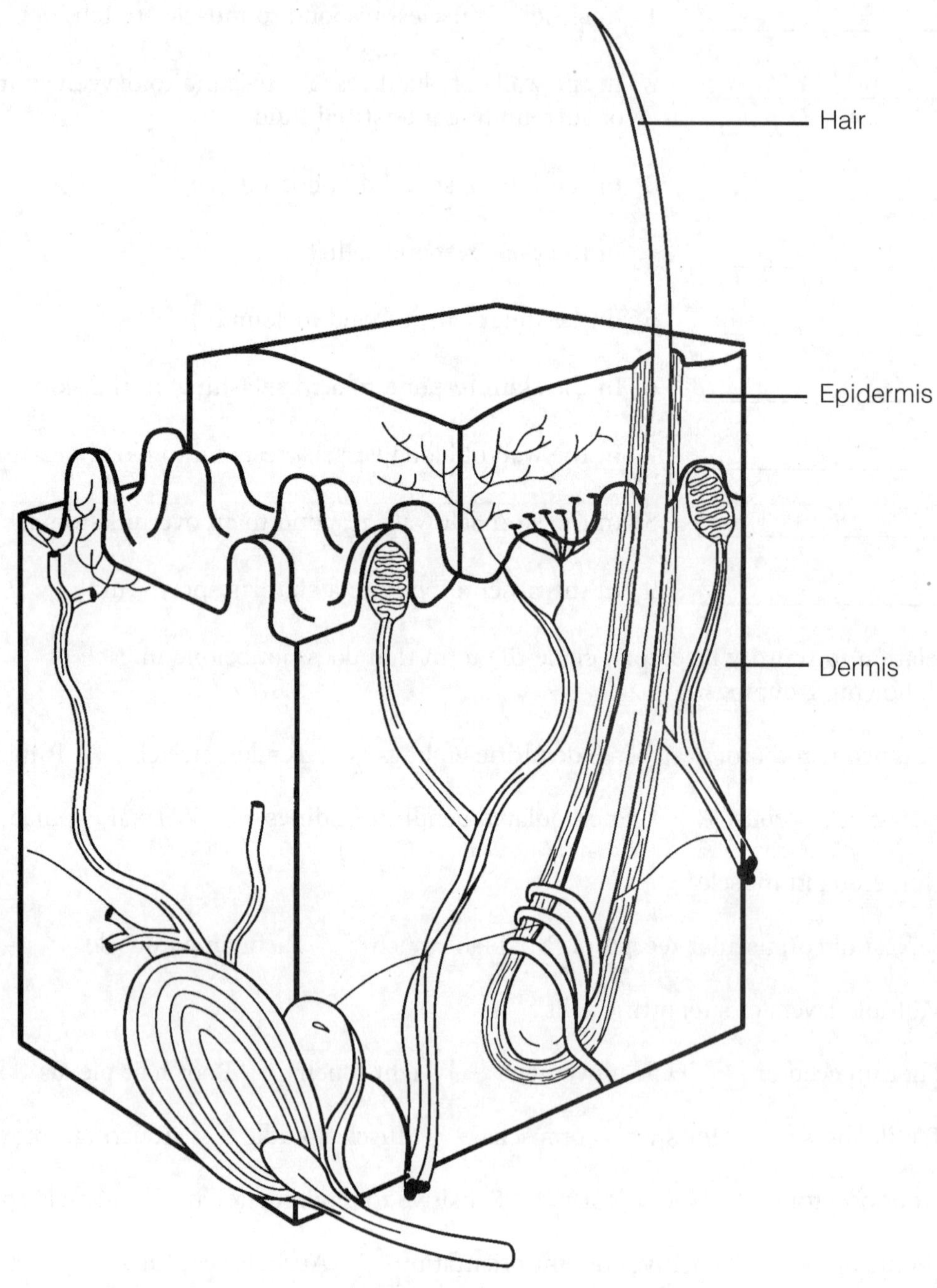

Figure 13.1

6. Certain activities and sensations are listed below. Using the key choices, select the specific receptor type that would be activated by the activity or sensation. Insert the correct letters in the answer blanks. Note that more than one receptor type may be activated in some cases.

Key Choices

A. Free nerve endings (pain)	D. Tactile discs	G. Joint kinesthetic receptor
B. Golgi tendon organ	E. Muscle spindle	H. Root hair plexuses
C. Meissner's corpuscle	F. Pacinian corpuscle	I. Ruffini endings

__________ 1. Standing on hot pavement (identify two)

__________ 2. Feeling a pinch (identify two)

__________ 3. Leaning on a shovel (identify four)

__________ 4. Muscle sensations when rowing a boat (identify three)

__________ 5. Feeling a caress (identify three)

__________ 6. Feeling the coldness of an iced-tea glass

__________ 7. Reading braille

The Eye and Vision

7. Complete the following statements by writing the missing terms in the answer blanks.

__________ 1.

__________ 2.

__________ 3.

__________ 4.

Attached to the eyes are the __(1)__ muscles, which enable us to direct our eyes toward a moving object. The anterior aspect of each eye is protected by the __(2)__, which have eyelashes projecting from their edges. Associated with the eyelids are both typical sebaceous glands and modified sebaceous glands called __(3)__ that help lubricate the eyes. An inflammation of one of these (latter) glands is called a(n) __(4)__.

8. Three accessory eye structures contribute to the formation of tears and/or aid in lubricating the eyeball. In the table, name each structure and its major secretory product. Indicate the various components of the secretion if applicable.

Accessory eye structures	**Secretory product**

9. Trace the pathway that the secretion of the lacrimal glands takes from the surface of the eye by assigning a number to each structure.

_____ 1. Lacrimal sac

_____ 2. Nasal cavity

_____ 3. Nasolacrimal duct

_____ 4. Lacrimal canaliculi

10. Identify each of the eye muscles indicated by leader lines in Figure 13.2. Color each muscle a different color. Then, in the blanks below, indicate the eye movement caused by each muscle.

1. Superior rectus ____________________
2. Inferior rectus ____________________
3. Superior oblique ____________________
4. Lateral rectus ____________________
5. Medial rectus ____________________
6. Inferior oblique ____________________

____

____

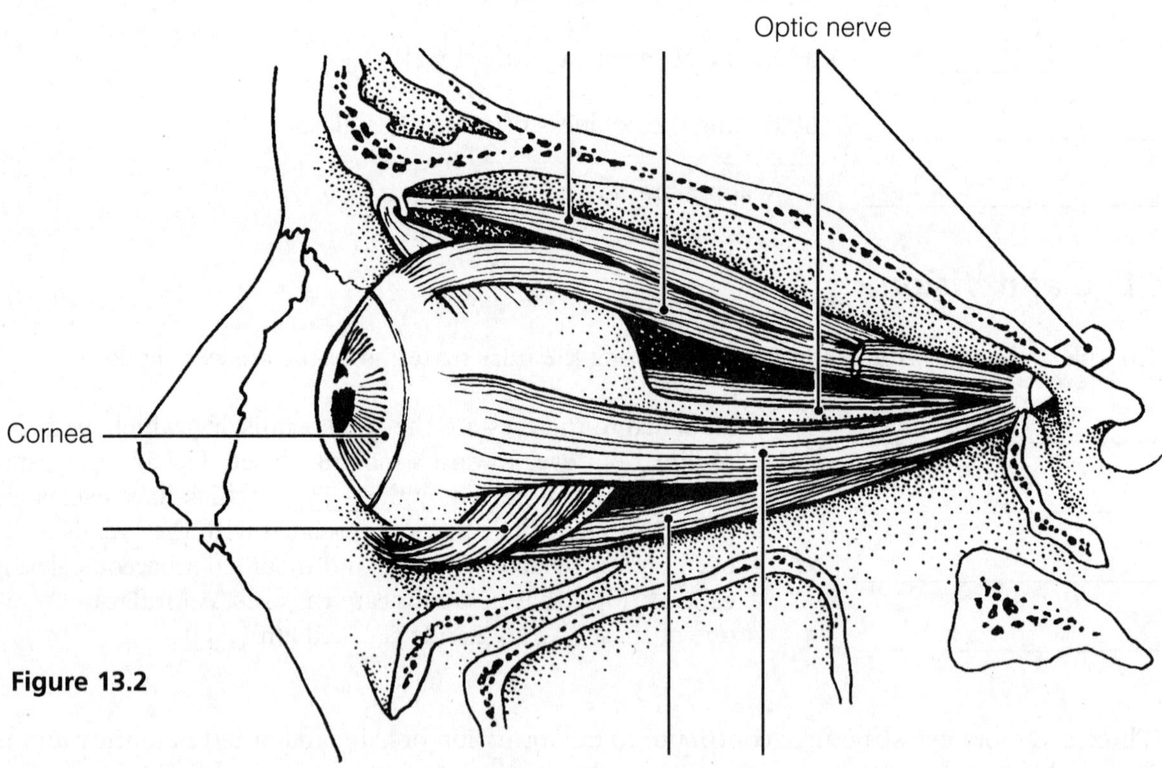

Figure 13.2

11. Choose terms from the key choices to match the following descriptions.

Key Choices

A. Commissure C. Ciliary glands E. Tarsal glands G. Tarsal plate

B. Caruncle D. Conjunctival sac F. Palpebra

____________________ 1. Modified sebaceous (oil) glands embedded in the tarsal plates

____________________ 2. Connective tissue sheet supporting the eyelid internally

____________________ 3. Fleshy elevation at the medial canthus

____________________ 4. Angular point where the eyelids meet

____________________ 5. Recess at the junction of palpebral and bulbar conjunctivae

____________________ 6. Another name for the eyelid

12. Circle the term that does not belong in each of the following groupings.

1. Choroid Sclera Vitreous humor Retina
2. Inferior oblique Iris Superior rectus Inferior rectus
3. Pupil constriction Far vision Accommodation Bright light
4. Mechanoreceptors Rods Cones Photoreceptors
5. Ciliary body Iris Ciliary zonule Lens
6. Uvea Choroid Ciliary body Pigmented retina Iris
7. Retina Neural layer Pigmented layer Transparent Contains photoreceptors
8. Optic disc Blind spot Lacks photoreceptors Macula lutea
9. Cornea Active sodium ion pump Transparent Richly vascular Many pain fibers
10. Inner segment Outer segment Photoreceptor pigments Receptor region
11. Conjunctiva Tarsal glands Tarsal plate Lacrimal gland

13. Using the key choices in Exercise 8, identify the structures indicated by leader lines on the diagram of the eye in Figure 13.3. Select different colors for all the structures with a color coding circle in Exercise 8 and color the structures on the figure.

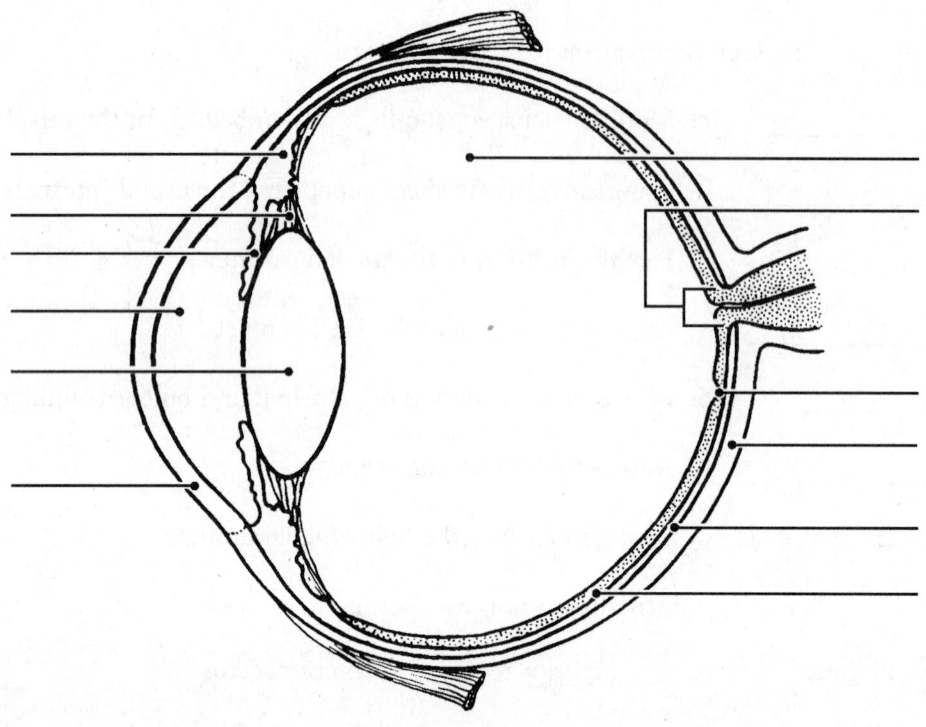

Figure 13.3

14. Using the key choices, identify the parts of the eye described in the following statements. Insert the correct letters in the answer blanks.

Key Choices

○ A. Aqueous humor	○ F. Fovea centralis	○ K. Sclera
○ B. Canal of Schlemm	○ G. Iris	○ L. Ciliary zonule
○ C. Choroid	○ H. Lens	○ M. Vitreous humor
○ D. Ciliary body	○ I. Optic disc	
○ E. Cornea	○ J. Retina	

_____ 1. Secures the lens to the ciliary body

_____ 2. Fluid that fills the anterior segment of the eye; provides nutrients to the lens and cornea

_____ 3. White, opaque portion of the fibrous layer

_____ 4. Area of retina that lacks photoreceptors

_____ 5. Muscular structure that manipulates the lens

_____ 6. Nutritive (vascular) region of the middle layer of eyeball

_____ 7. Drains the aqueous humor of the eye

_____ 8. Portion of the inner layer concerned with image formation

_____ 9. Gel-like substance, filling the posterior segment of the eyeball; helps reinforce the eyeball

_____ 10. Drains into the scleral venous sinus (canal of Schlemm)

_____ 11. _____ 12. Smooth muscle structures (sites of intrinsic eye muscles)

_____ 13. Area of acute or discriminatory vision

_____ 14. _____ 15. _____ 16. _____ 17. Refractory media of the eye

_____ 18. Anteriormost clear part of the fibrous layer

_____ 19. Pigmented "diaphragm" of the eye

15. Match the functionally corresponding parts of the eye (Column B) to those of a camera (Column A).

	Column A	Column B
__________	1. Lenses	A. Retina
__________	2. Diaphragm	B. Sclera
__________	3. Light-sensitive layer of film	C. Iris
__________	4. Focusing by changing the focal distance	D. Accommodation
__________	5. Case	E. Lens
__________	6. Reducing light entry by selecting a smaller aperture	F. Pupillary constriction

16. In the following table, circle each word that correctly describes events occurring within the eye during close and distant vision.

Vision	Ciliary muscle		Lens convexity		Degree of light refraction	
1. Distant	Relaxed	Contracted	Increased	Decreased	Increased	Decreased
2. Close	Relaxed	Contracted	Increased	Decreased	Increased	Decreased

17. Explain why it helps to look up and gaze into space after reading for a prolonged period.

___

___

___

18. Match the key choices with the following descriptions concerning optics.

Key Choices

A. Concave lens   C. Far point of vision   E. Near point of vision

B. Convex lens   D. Focal point   F. Visible light

__________ 1. The closest point at which clear focus is possible

__________ 2. The electromagnetic waves to which the photoreceptors of the eyes respond

__________ 3. The point at which light rays are converged by a convex lens

__________ 4. A lens that is thickest at the edges; diverges the light rays

__________ 5. The point beyond which accommodation is unnecessary

19. Match the terms in Column B with the appropriate descriptions in Column A. Insert the correct answers in the answer blanks.

	Column A	Column B
__________	1. Light bending	A. Accommodation
__________	2. Ability to focus for close vision (under 20 feet)	B. Accommodation pupillary reflex
__________	3. Normal vision	C. Astigmatism
__________	4. Inability to focus well on close objects; farsightedness	D. Cataract
		E. Convergence
__________	5. Reflex constriction of pupils when they are exposed to bright light	F. Emmetropia
		G. Glaucoma
__________	6. Clouding of lens, resulting in loss of sight	H. Hyperopia
__________	7. Nearsightedness	I. Myopia
__________	8. Blurred vision, resulting from unequal curvatures of the lens or cornea	J. Night blindness
__________	9. Condition of increasing pressure inside the eye, resulting from blocked drainage of aqueous humor	K. Photopupillary reflex
		L. Refraction
__________	10. Medial movement of the eyes during focusing on close objects	
__________	11. Reflex constriction of the pupils when viewing close objects	
__________	12. Inability to see well in the dark; often a result of vitamin A deficiency	

20. Complete the following statements by writing the missing terms in the answer blanks.

1. ____________________
2. ____________________
3. ____________________
4. ____________________
5. ____________________
6. ____________________
7. ____________________
8. ____________________ 9. ____________________ 10. ____________________

There are (1) varieties of cones. One type responds most vigorously to (2) light, another to (3) light, and still another to (4) light. The ability to see intermediate colors such as purple results from the fact that more than one cone type is being stimulated (5). Lack of all color receptors results in (6). Because this condition is sex linked, it occurs most commonly in (7). Black and white or dim light vision is a function of the (8). The density of cones is greatest in the (9), whereas rods are densest in the (10).

21. Answer the following questions concerning rod photopigment and physiology, or complete the statements as indicated.

1. The bent 11-*cis* form of retinal is combined with a protein called ____________________ to form the photoreceptor pigment called ____________________.

2. The light-induced event during which the photoreceptor pigment breaks down to its two components and retinal assumes its straighter ____________________ shape is called ____________________.

3. Retinal is produced from vitamin ____________________, which is ordinarily stored in large amounts by the ____________________.

4. What ionic and electrical events occur in the photoreceptors (rods) when it is dark?

____

____

5. How is this changed in the light? ____________________

____

____

22. Name in sequence the neural elements of the visual pathway, beginning with the retina and ending with the optic cortex.

Retina → ____________________ → ____________________ → ____________________

synapse in thalamus → ____________________ → optic cortex

23. Check (✓) all of the following conditions that pertain to dark adaptation.

_____	1. Retinal sensitivity increases	_____	7. Retinal sensitivity decreases
_____	2. Rhodopsin accumulates	_____	8. Rhodopsin is broken down rapidly
_____	3. The cones are inactive	_____	9. The cones are active
_____	4. Rods are activated	_____	10. Rods are inactivated
_____	5. Visual acuity increases	_____	11. Visual acuity decreases
_____	6. Pupils constrict	_____	12. Pupils dilate

24. Literally, *binocular vision* means "two-eyed vision." However, in common practice this term means that:

____

____

____

25 Complete the following statements pertaining to visual processing.

________________ 1.
________________ 2.
________________ 3.
________________ 4.
________________ 5.
________________ 6.
________________ 7.
________________ 8.
________________ 9.
________________ 10.
________________ 11.
________________ 12.
________________ 13.
________________ 14.
________________ 15.

The only cells of the retina that are capable of generating action potentials are the __(1)__. These cells receive inputs from the __(2)__ cells, which in turn receive inputs from the __(3)__. Ganglion cells that depolarize when the center of their receptive field is illuminated are said to have __(4)__ fields, whereas those that hyperpolarize when the center of their receptive field is illuminated are said to have __(5)__ fields. Retinal cells outside the direct pathway to the ganglion cells are the __(6)__ cells and the __(7)__ cells. These cells are involved in retinal processing. As a result of retinal processing, the information flowing from the retina is more concerned with reporting areas of light and dark __(8)__ and edges than with transmitting information about individual light points.

After being segregated and processed in the __(9)__, visual inputs flow to the visual cortex. __(10)__ cortex neurons have a receptive field that corresponds to several ganglion cell fields with the same orientation and type of center. Two common types of ganglion cells are the __(11)__ cells that receive input from __(12)__ and relay information about the color of nonmoving objects. The __(13)__ cells respond best to large moving objects at the image edge. The __(14)__ visual processing stream of the striate cortex specializes in identification of objects and extends through the temporal lobe. Spatial localization of objects is the role of the __(15)__ stream that extends through the parietal cortex.

26. Label the structures having leader lines in Figure 13.4, which shows the visual pathways to the brain. Then use red to color the left visual field and green to color the right visual field. Use red and green stripes to mark the area of overlap of the two visual fields. Color the entire pathway to the visual cortex in the occipital lobe on both sides: red for the left visual field (not for the left eye) and green for the right visual field.

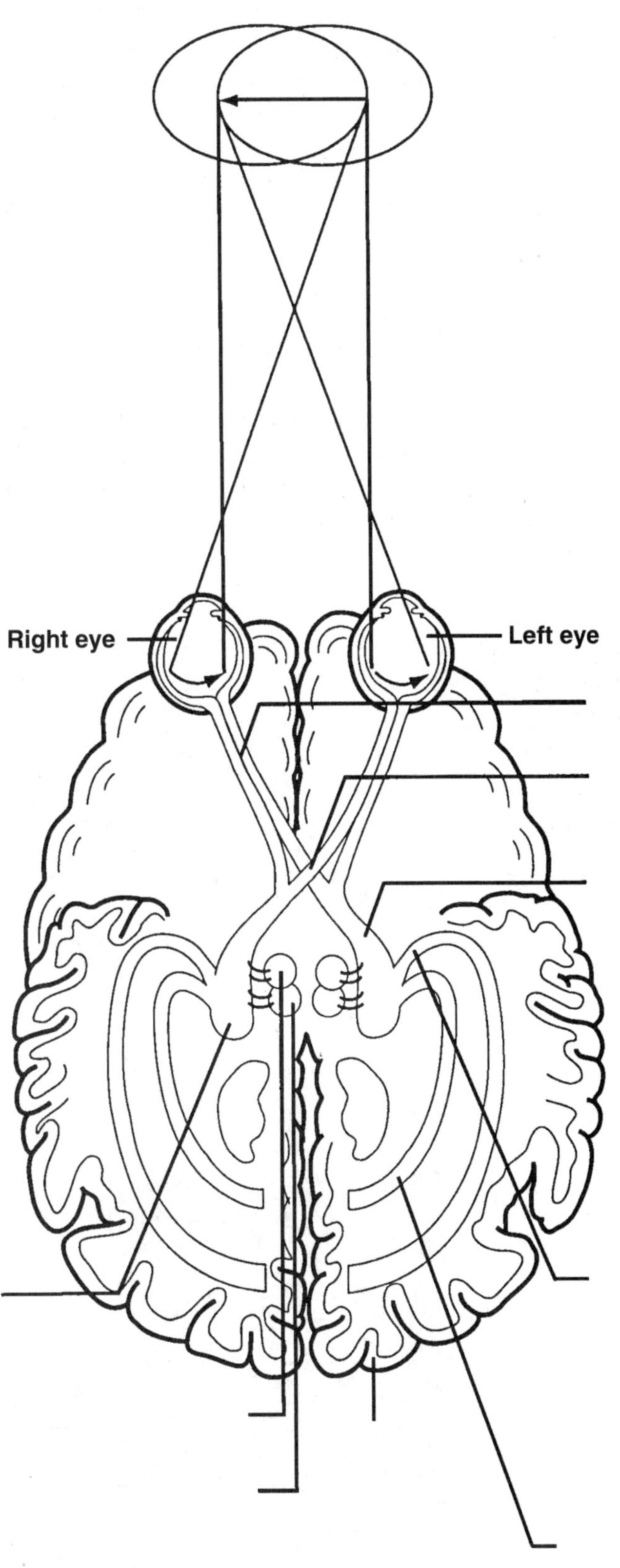

Figure 13.4

The Chemical Senses: Taste and Smell

27. Complete the following statements by writing the missing terms in the answer blanks.

________ 1.	The gustatory and olfactory senses rely on __(1)__, meaning that chemicals dissolved in fluid will stimulate these receptors.
________ 2.	The receptors for __(2)__ taste, mostly in the back of the tongue, are thought to be protective because many poisons
________ 3.	stimulate these receptors. The sensation we call taste is reduced when the nasal passages are swollen; this indicates
________ 4.	that "taste" relies heavily on the __(3)__ sense. The loss of smell that accompanies a cold is the result of the less-than-
________ 5.	optimal position of the olfactory epithelium, on the __(4)__ of the nasal cavity. The pathway for the sense of smell also runs
________ 6.	to the __(5)__ system and thus has emotional ties. It is the only sensory input that does not pass through the __(6)__ in the diencephalon to reach its destination in the cerebral cortex.

28. On Figure 13.5A, label the two types of tongue papillae containing taste buds. On Figure 13.5B, color the taste buds green. On Figure 13.5C, color the gustatory cells red, the supporting cells blue, the basal cells purple, and the sensory fibers yellow. Add appropriate labels at the leader lines to identify the *taste pore* and *microvilli* of the gustatory cells.

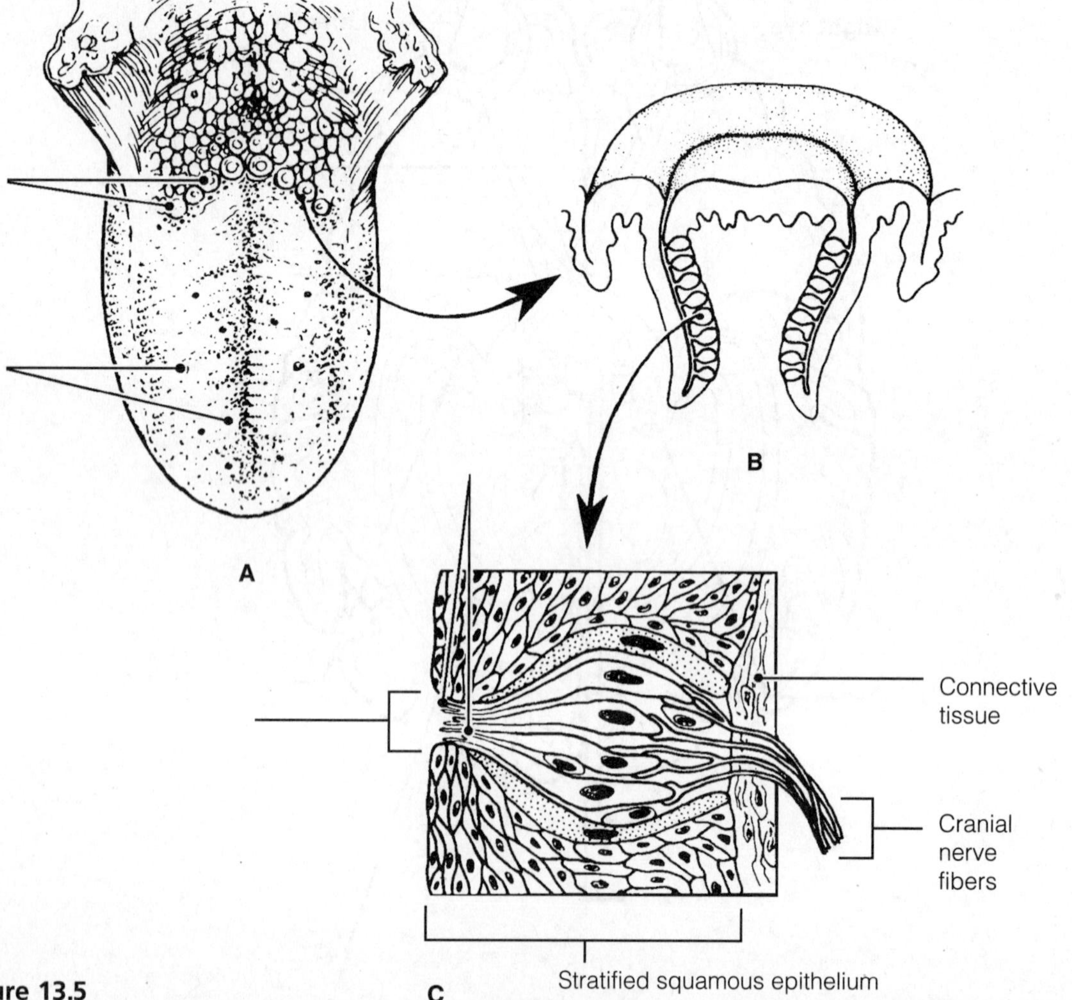

Figure 13.5

29. Figure 13.6 illustrates the site of the olfactory epithelium in the nasal cavity (part A is an enlarged view of the olfactory receptor area). Select different colors to identify the structures listed below and use them to color the illustration. Then add labels and leader lines to identify the filaments of the olfactory nerve, the fibers of the olfactory tract, the mitral cells, and the olfactory cilia, and add arrows to indicate the direction of impulse transmission. Finally, respond to the questions following the diagrams.

- ◯ Olfactory neurons (receptor cells)
- ◯ Supporting cells
- ◯ Glomeruli
- ◯ Olfactory bulb
- ◯ Cribriform plate of the ethmoid bone

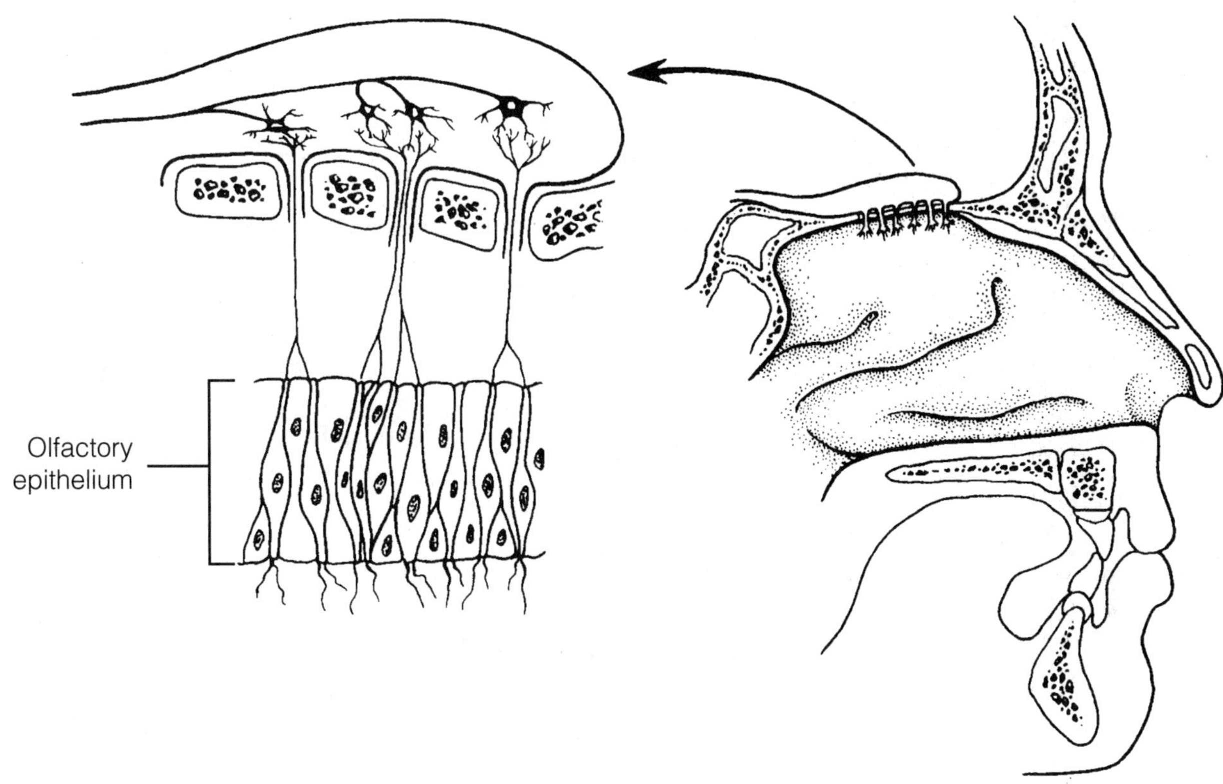

Figure 13.6 **A** **B**

1. What substance "captures" airborne odorants (that is, acts as a solvent)? ____________

2. What name is given to the complex spherical structures that serve as the first relay station for olfaction? ____________

3. Olfactory neurons are structurally ____________ neurons.

4. List two brain regions to which olfactory tract fibers project, other than the olfactory cortex in the temporal lobes. ____________

____

30. Circle the term that does not belong in each of the following groupings.

1. Sweet Musky Sour Bitter Salty
2. Bipolar neuron Epithelial cell Olfactory receptor Ciliated
3. Gustatory cell Taste pore Microvilli Yellow-tinged epithelium
4. Vagus nerve Facial nerve Glossopharyngeal nerve Olfactory nerve
5. Olfactory receptor Low specificity Variety of stimuli Five receptor types
6. Sugars Sweet Saccharine Metal ions Amino acids
7. Alkaloids H^+ Nicotine Quinine Bitter
8. Olfactory aura Epileptic seizure Anosmia Uncinate fit

31. Complete the following statements by writing the missing terms in the answer blanks.

____________ 1.
____________ 2.
____________ 3.
____________ 4.
____________ 5.

Olfactory transduction involves binding to a(n) __(1)__ linked to a G protein. Activation of the G proteins leads to formation of __(2)__, a second messenger that acts on a plasma membrane channel so that __(3)__ and __(4)__ ions can enter. This ion entry leads to depolarization and __(5)__.

The Ear: Hearing and Balance

32. Using the key choices, select the terms that apply to the following descriptions. Place the correct letters in the answer blanks.

Key Choices

A. Anvil (incus)	F. Hammer (malleus)	K. Semicircular canals
B. Pharyngotympanic tube	G. Oval window	L. Stirrup (stapes)
C. Cochlea	H. Perilymph	M. Tympanic membrane
D. Endolymph	I. Pinna (auricle)	N. Vestibule
E. External acoustic meatus	J. Round window	

_____ 1. _____ 2. _____ 3. Structures composing the external, or outer, ear

_____ 4. _____ 5. _____ 6. Elements of the bony or osseous labyrinth

_____ 7. _____ 8. _____ 9. The ossicles

_____ 10. Ear structures not involved with hearing

_____ 11. Allows middle ear pressure to be equalized with atmospheric pressure

_____ 12. Transmits sound vibrations to the ossicles

_____ 13. Contains the organ of Corti

_____ 14. Passage from the nasopharynx to the middle ear

_____ 15. _____ 16. House receptors for the sense of equilibrium

_____ 17. Transfers vibrations from the stirrup to the fluid in the internal ear

_____ 18. Fluid inside the membranous labyrinth

_____ 19. Fluid within the osseous labyrinth, and surrounding the membranous labyrinth

_____ 20. Fits into the oval window

33. Figure 13.7 is a diagram of the ear. Use anatomical terms (as needed) from the key choices in Exercise 1 to correctly identify all structures in the figure with leader lines. Color all external ear structures yellow, color the ossicles red, color the equilibrium areas of the internal ear green, and color the internal ear structures involved with hearing blue.

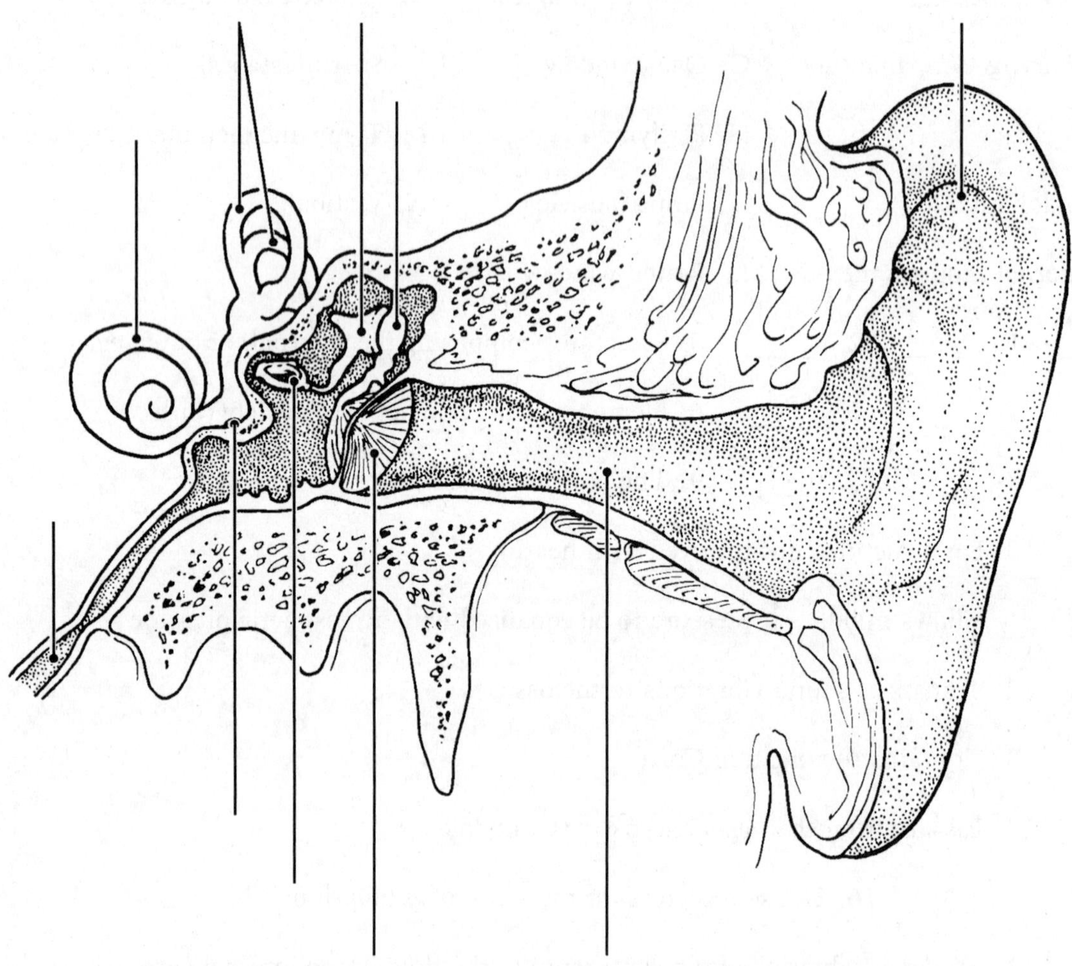

Figure 13.7

34. Sound waves hitting the eardrum set it into vibration. Trace the pathway through which vibrations and fluid currents travel to finally stimulate the hair cells in the organ of Corti. Name the appropriate ear structures in their correct sequence.

Eardrum → ______________ → ______________ → ______________ →

oval window → ______________ → ______________ → ______________

→ hair cells.

35. Match the terms in Column B concerning cochlear structures with the appropriate descriptions in Column A.

	Column A	Column B
______________	1. Bony pillar supporting the coiled cochlea	A. Basilar membrane
		B. Helicotrema
______________	2. Superior cavity of the cochlea	C. Modiolus
______________	3. Inferior cavity of the cochlea	D. Scala media
______________	4. The cochlear duct	E. Scala tympani
______________	5. The apex of the cochlea; where the perilymph-containing chambers meet	F. Scala vestibuli
______________	6. Shelflike extension of the modiolus	G. Spiral lamina
______________	7. Forms the endolymph	H. Spiral organ of Corti
______________	8. Roof of the cochlear duct	I. Stria vascularis
______________	9. Membrane that supports the organ of Corti	J. Vestibular membrane

36. Two tiny skeletal muscles are associated with the bones of the middle ear. Name these two muscles. Note their bony points of attachment and describe their common function.

____

____

____

37. Figure 13.8 is a view of the structures of the membranous labyrinth. Correctly identify and label the following major areas of the labyrinth on the figure: membranous semicircular canals, saccule and utricle, and the cochlear duct. Next, correctly identify and label each of the receptor types shown in enlarged views (organ of Corti, crista ampullaris, and macula). Finally, using terms from the key choices, identify all receptor structures with leader lines. (Some of these terms may need to be used more than once.) Color the diagram as you wish.

Key Choices

A. Basilar membrane

B. Cochlear nerve fibers

C. Cupula

D. Otolithic membrane

E. Hair cells

F. Otoliths

G. Tectorial membrane

H. Vestibular nerve fibers

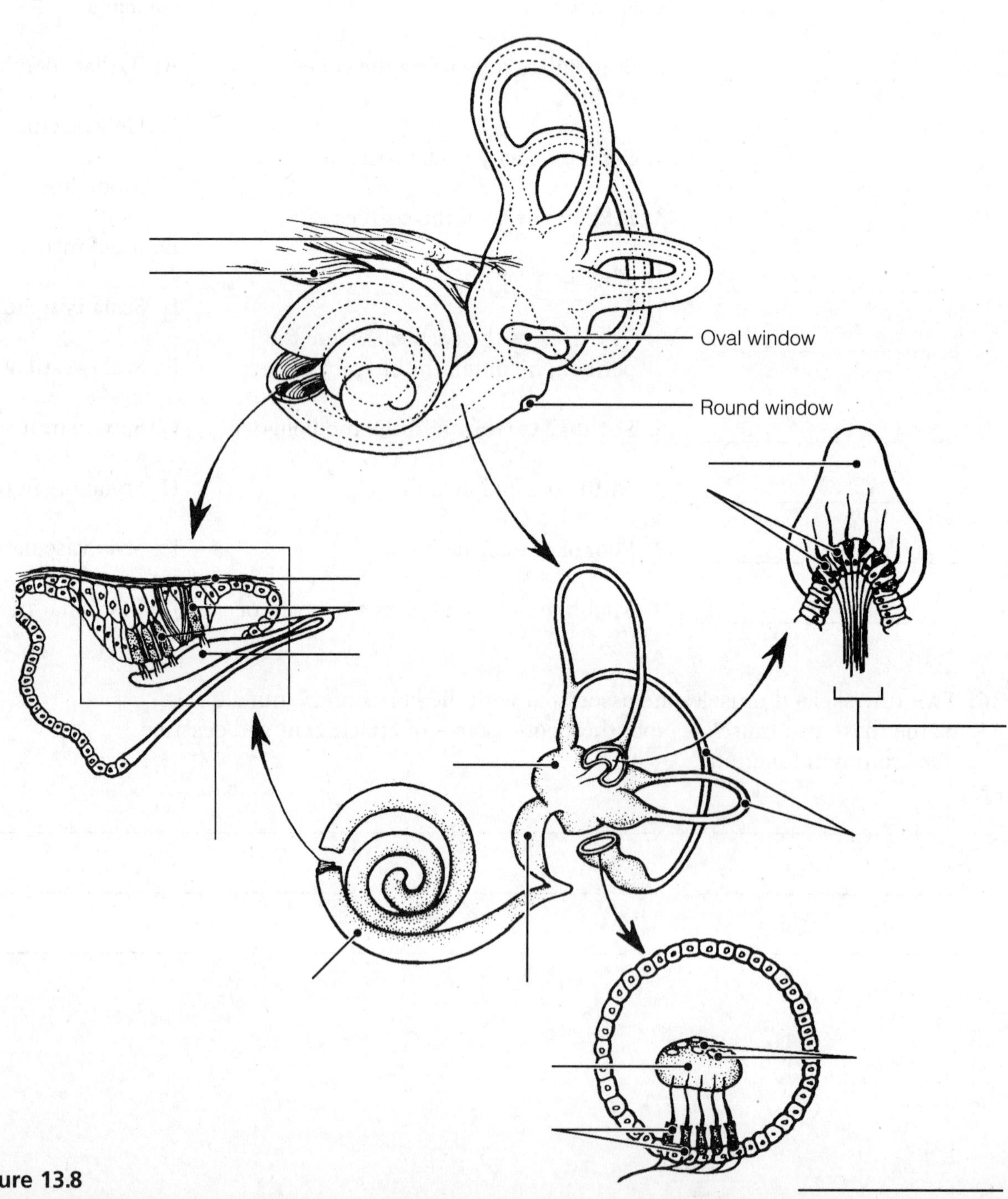

Figure 13.8

38. Why is the lever system constructed by the ear ossicles sometimes compared with a hydraulic press?

____

____

____

39. Complete the following statements referring to the physics of sound.

The source of sound is a(n) __(1)__. In order for sound to be propagated, there must be a medium, and it must be __(2)__ so that alternating regions of __(3)__ and __(4)__ can be produced. Sound waves are periodic, and the distance between successive peaks of a sound sine wave is referred to as the __(5)__ of sound. As a rule, sounds of high frequency have __(6)__ wavelengths, whereas sounds of low frequency have __(7)__ wavelengths. Different frequencies are perceived as differences in __(8)__, whereas differences in amplitude or intensity of a sound are perceived as differences in __(9)__. Sound intensities are measured in units called __(10)__.

________________ 1.

________________ 2.

________________ 3.

________________ 4.

________________ 5.

________________ 6.

________________ 7. ________________ 9.

________________ 8. ________________ 10.

40. The fibers of the basilar membrane vary in length and thickness. Those nearest the oval window are short and stiff, whereas those near the cochlear apex are long and floppy. How is the structure of the basilar membrane related to the excitation of cochlear hair cells?

____

____

____

____

41. Indicate whether the following conditions relate to conduction deafness (C) or sensorineural (central) deafness (S). Place the correct letters in the answer blanks.

__________ 1. Can result from a bug wedged in the external auditory canal

__________ 2. Can result from a stroke (CVA)

__________ 3. Sound is heard in both ears during bone conduction, but only in one ear during air conduction

__________ 4. Not improved much by a hearing aid

__________ 5. Can result from otitis externa

__________ 6. Can result from otosclerosis, excessive earwax, or a perforated eardrum

__________ 7. Can result from a lesion on the auditory nerve

42. Complete the following description of the functioning of the static and dynamic equilibrium receptors by writing the key choices in the answer blanks.

Key Choices

A. Angular/rotatory	E. Gravity	I. Semicircular canals
B. Cupula	F. Perilymph	J. Static
C. Dynamic	G. Proprioception	K. Utricle
D. Endolymph	H. Saccule	L. Vision

__________ 1.
__________ 2.
__________ 3.
__________ 4.
__________ 5.
__________ 6.
__________ 7.
__________ 8.

The receptors for __(1)__ equilibrium are found in the crista ampullaris of the __(2)__. These receptors respond to changes in __(3)__ motion. When motion begins, the __(4)__ fluid lags behind and the __(5)__ is bent, which excites the hair cells. When the motion stops suddenly, the fluid flows in the opposite direction and again stimulates the hair cells. The receptors for __(6)__ equilibrium are found in the maculae of the __(7)__ and __(8)__. These receptors report on the position of the head in space. Tiny stones found in a gel overlying the hair cells roll in response to the pull of __(9)__. As they roll, the gel moves and tugs on the hair cells, exciting them. Besides the equilibrium receptors of the internal ear, the senses of __(10)__ and __(11)__ are also important in helping maintain equilibrium.

__________ 9. __________ 10. __________ 11.

43. List three problems about which a person with equilibrium imbalance might complain.

1. ____________________ 2. ____________________ 3. ____________________

44. Circle the term that does not belong in each of the following groupings.

1. Malleus Incus Pinna Stapes
2. Tectorial membrane Crista ampullaris Semicircular canals Cupula
3. Gravity Angular motion Sound waves Rotation
4. Utricle Saccule Pharyngotympanic tube Vestibule
5. Vestibular nerve Optic nerve Cochlear nerve Vestibulocochlear nerve
6. Crista ampullaris Maculae Retina Proprioceptors Pressure receptors
7. Fast motility Inner hair cells Ear sounds Outer hair cells

PART 2: TRANSMISSION LINES: NERVES AND THEIR STRUCTURE AND REPAIR

1. Name the four functional types of nerve fibers.

____

____

____

____

2. Figure 13.9 is a diagrammatic view of a section of a nerve wrapped in its connective tissue coverings. Select different colors to identify the following structures and use them to color the figure. Then, label these sheaths, indicated by leader lines on the figure.

- ○ Endoneurium
- ○ Perineurium
- ○ Epineurium

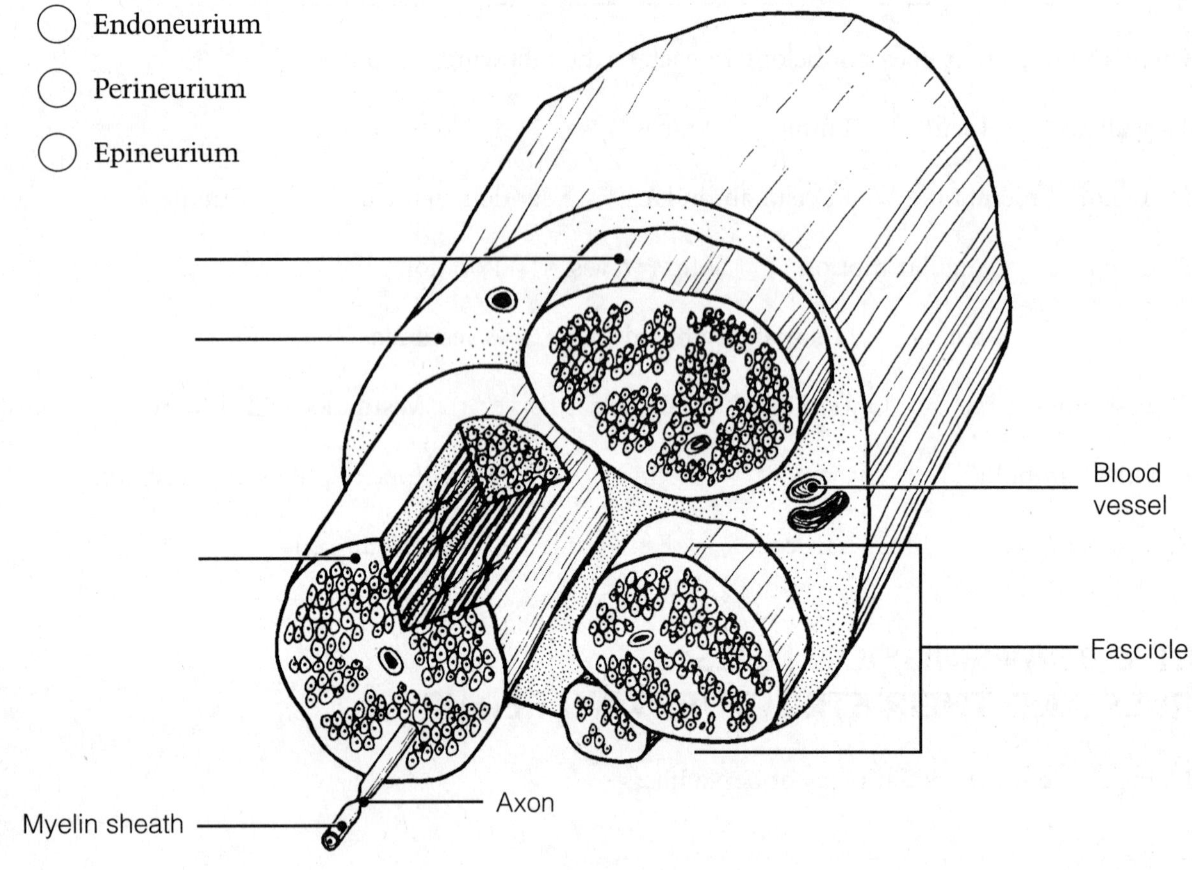

Figure 13.9

3. Complete the following statements relating to regeneration of nervous tissue. Write the missing terms in the answer blanks.

1. ____________
2. ____________
3. ____________
4. ____________
5. ____________
6. ____________
7. ____________
8. ____________
9. ____________
10. ____________
11. ____________
12. ____________

Lost neurons cannot be replaced because mature neurons usually do not __(1)__. An axon in the __(2)__ nervous system can regenerate successfully if the injury is distal to the __(3)__, which must remain intact. After the injured axon ends seal off, the axon and its myelin sheath distal to the injury __(4)__. This debris is disposed of by __(5)__. Regeneration tubes, a system of cellular __(6)__, formed by __(7)__, guide growing axonal processes toward their original contacts.

Nerve fibers located in the __(8)__ nervous system essentially never regenerate. Although the __(9)__ cells clean up the degenerated neural debris, __(10)__ are notably absent from the CNS, thus cleanup is slower. Additionally, the __(11)__ that form the myelin sheath __(12)__. Consequently no guiding channels are formed.

4. Match the names of the cranial nerves listed in Column B with the appropriate descriptions in Column A by inserting the correct letters in the answer blanks.

	Column A	Column B
________	1. The only nerve that originates from the forebrain	A. Accessory
		B. Abducens
________	2. The only cranial nerve that extends beyond the head and neck region	C. Facial
________	3. The largest cranial nerve	D. Glossopharyngeal
________	4. Fibers arise from the sensory apparatus within the inner ear	E. Hypoglossal
		F. Oculomotor
________	5. Supplies somatic motor fibers to the lateral rectus muscle of the eye	G. Olfactory
________	6. Has five major branches; transmits sensory, motor, and autonomic impulses	H. Optic
		I. Trigeminal
________	7. Cell bodies are located within their associated sense organs	J. Trochlear
________	8. Transmits sensory impulses from pressure receptors of the carotid artery	K. Vagus
		L. Vestibulocochlear
________	9. Arises from ventral rootlets of the spinal cord	
________	10. Innervates four of the muscles that move the eye and the iris	
________	11. Supplies the superior oblique muscle of the eye	
________	12. Two nerves that supply the tongue muscles	
________	13. Its *tract* is frequently misidentified as this nerve	
________	14. Serves muscles covering the facial skeleton	

5. Provide the name and number of the cranial nerves involved in each of the following activities, sensations, or disorders. Write your answers in the answer blanks.

________________ 1. Hyperextending and flexing the neck

________________ 2. Smelling freshly baked bread

________________ 3. Constricting the pupils for reading

________________ 4. Stimulates the mobility and secretory activity of the digestive tract

________________ 5. Involved in frowning and puzzled looks

________________ 6. Crunching an apple and chewing gum

________________ 7. Tightrope walking and listening to music

________________ 8. Gagging and swallowing; tasting bitter foods

________________ 9. Involved in "rolling" the eyes (three nerves; provide numbers only)

________________ 10. Feeling a toothache

________________ 11. Watching tennis on TV

________________ 12. If this nerve is damaged, deafness results

________________ 13. Stick out your tongue!

________________ 14. Inflammation of this nerve may cause Bell's palsy

________________ 15. Damage to this nerve may cause anosmia

6. Match each of the cranial nerves listed in Column A to its terminal connection in the cerebral cortex (an item from Column B).

	Column A	Column B
__________	1. Oculomotor	A. Frontal eye field
__________	2. Olfactory	B. Hypothalamus
__________	3. Facial	C. Parietal lobe
__________	4. Trigeminal	D. Primary motor cortex
__________	5. Optic	E. Occipital lobe
__________	6. Vagus	F. Temporal lobe

7. The 12 pairs of cranial nerves are indicated by leader lines on Figure 13.10. First, label each by name and Roman numeral on the figure. Then, color each nerve with a different color.

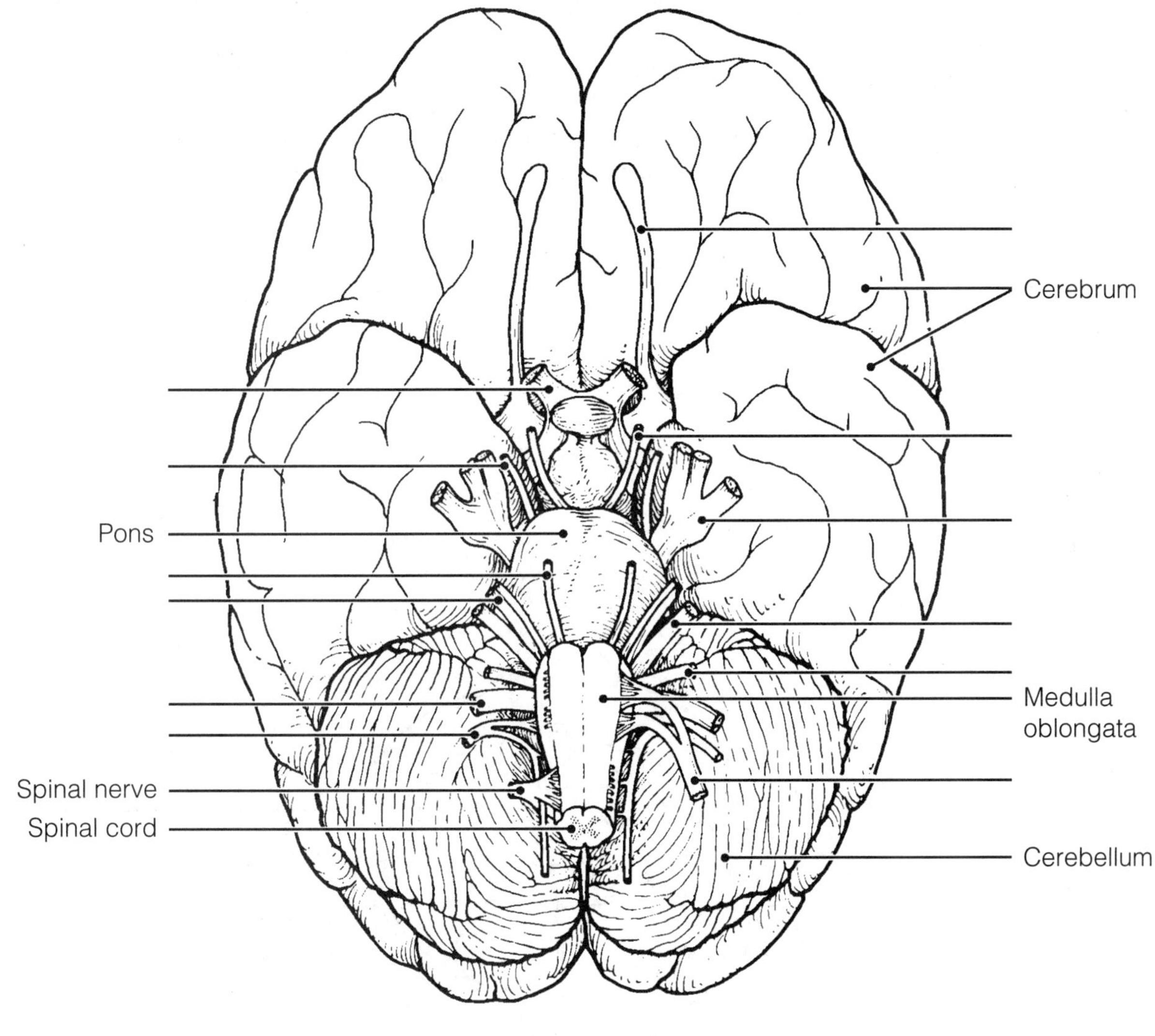

Figure 13.10

8. Name the five major motor branches of the facial nerve.

____________________, ____________________, ____________________,

____________________, and ____________________.

9. Identify all structures provided with leader lines in Figure 13.11. Color the diagram as you wish and then complete the following statements by inserting your responses in the answer blanks.

________________ 1.
________________ 2.
________________ 3.
________________ 4.
________________ 5.
________________ 6.
________________ 7.

Each spinal nerve is formed from the union of __(1)__ and __(2)__. After its formation, the spinal nerve splits into __(3)__. The ventral rami of spinal nerves C_1–T_1 and L_1–S_4 take part in forming __(4)__, which serve the __(5)__ of the body. The ventral rami of T_1–T_{12} run between the ribs to serve the __(6)__. The posterior rami of the spinal nerves serve the __(7)__.

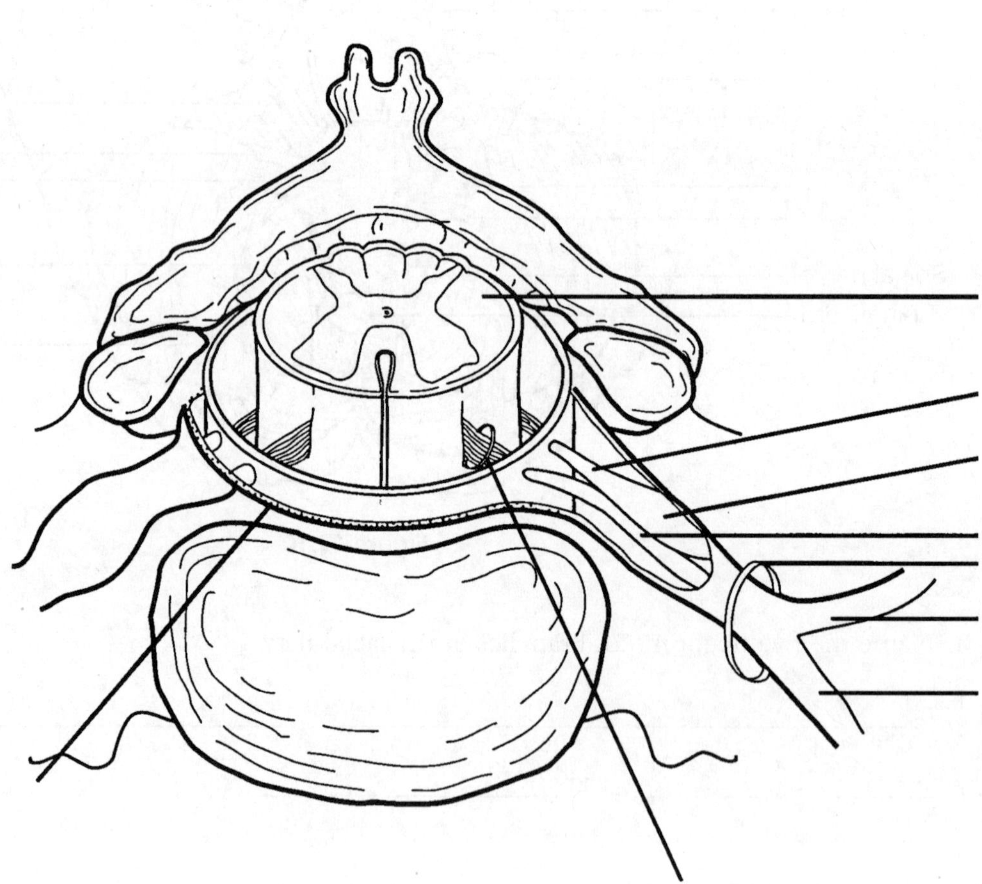

Figure 13.11

10. Referring to the distribution of spinal nerves, circle the term that does not belong in each of the following groupings.

1. Division of spinal nerve   Two rami   Meningeal branch   Two roots
2. Thoracic roots   Lumbar roots   Sacral roots   Cauda equina
3. Activation of skeletal muscles   Ventral root   Motor root   Afferent fibers
4. Anterior ramus   Motor function   Sensory function   Ventral root
5. Ventral ramus   Dorsal ramus   Division of spinal nerve   Two spinal roots

11. Name the major nerves that serve the following body parts. Insert your responses in the answer blanks.

____________________ 1. Head, neck, shoulders (name plexus only)

____________________ 2. Diaphragm

____________________ 3. Hamstrings

____________________ 4. Leg and foot (name two)

____________________ 5. Most anterior forearm muscles

____________________ 6. Flexor muscles of the arm

____________________ 7. Abdominal wall (name plexus only)

____________________ 8. Thigh flexors and knee extensors

____________________ 9. Medial side of the hand

____________________ 10. Plexus serving the upper limb

____________________ 11. Large nerve composed of two nerves in a common sheath

____________________ 12. Buttock, pelvis, lower limb (name plexus only)

____________________ 13. Largest branch of the brachial plexus

____________________ 14. Largest nerve of the lumbar plexus

____________________ 15. Adductor muscles of the thigh

____________________ 16. Gluteus maximus

12. Figure 13.12 is an anterior view of the principal nerves arising from the cords of the brachial plexus. Select five different colors and color the coding circles and the nerves listed below. Also, label each nerve by inserting its name at the appropriate leader line.

- ◯ Axillary nerve
- ◯ Musculocutaneous nerve
- ◯ Median nerve
- ◯ Radial nerve
- ◯ Ulnar nerve

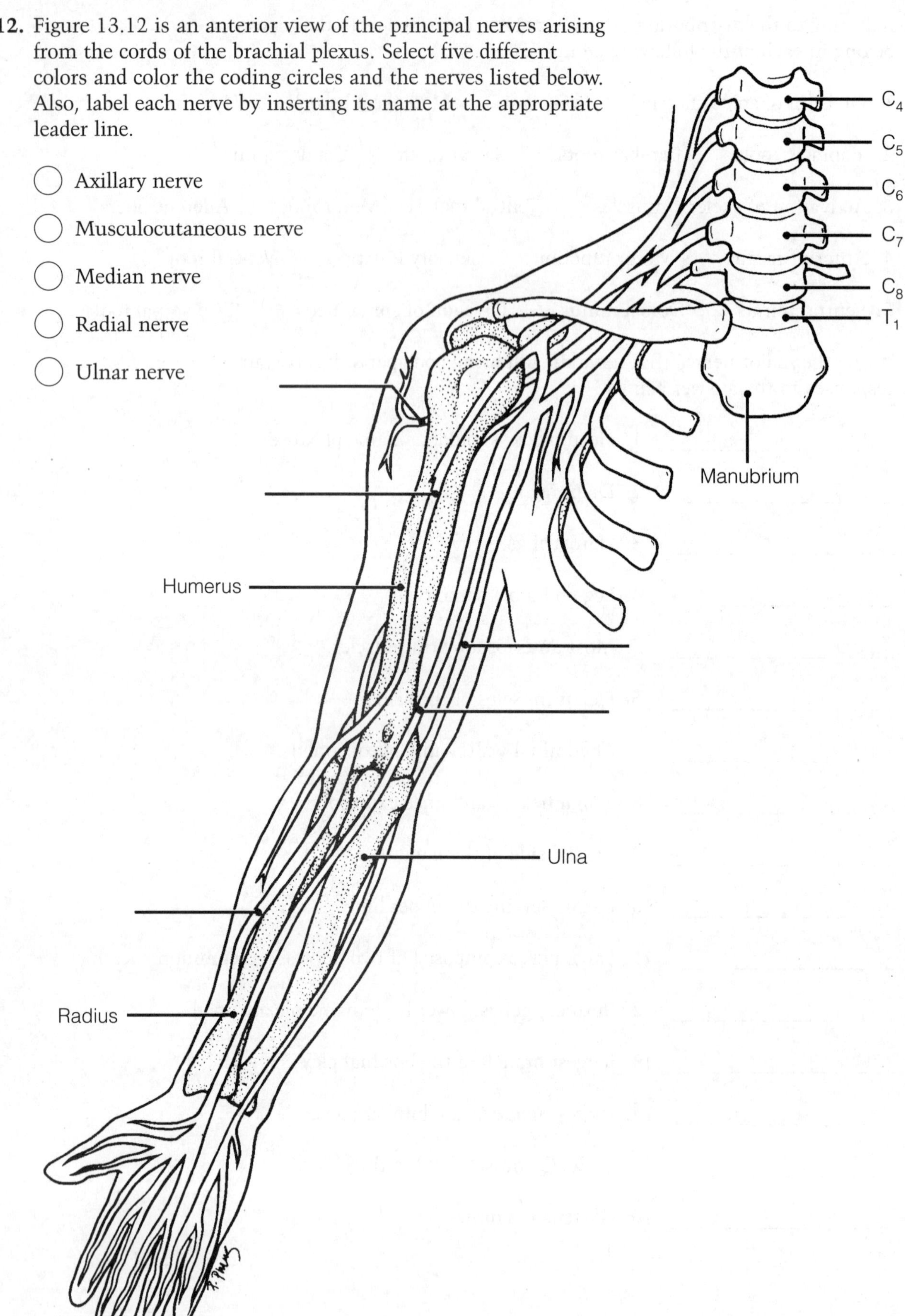

Figure 13.12

13. Figure 13.13 is a posterior view of the major nerves arising from the cords of the sacral plexus. Choose seven colors and color the coding circles and the nerves listed below. Then, identify each nerve by inserting its name at the appropriate leader line.

- ◯ Common fibular nerve
- ◯ Inferior gluteal nerve
- ◯ Plantar nerve branches
- ◯ Sciatic nerve
- ◯ Superior gluteal nerve
- ◯ Sural nerve
- ◯ Tibial nerve

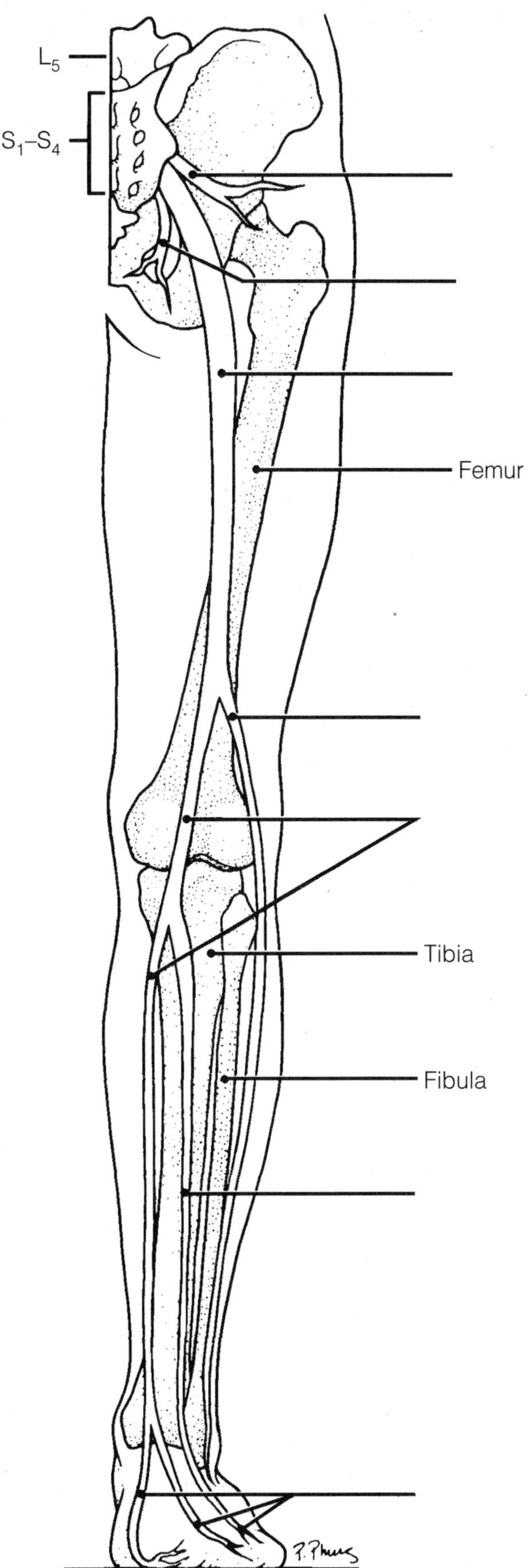

Figure 13.13

14. If a statement is true, write the letter T in the answer blank. If a statement is false, change the underlined word(s) and write the correct word(s) in the answer blank.

__________________ 1. Inability to extend the hand at the wrist (wristdrop) can be caused by injury to the median nerve.

__________________ 2. Injury to the common fibular nerve causes loss of foot and ankle movements (footdrop).

__________________ 3. Thumb and little finger opposition is lost if the axillary nerve is injured.

__________________ 4. Injury to the radial nerve causes loss of flexion of the fingers at the distal interphalangeal joints (clawhand).

__________________ 5. Irritation of the phrenic nerve may cause hiccups.

PART 3: MOTOR ENDINGS AND MOTOR ACTIVITY

1. Several characteristics of motor endings (both autonomic and somatic) are listed below. Differentiate between the autonomic and somatic motor endings by checking (✓) all those that distinguish the synapses en passant of autonomic fibers.

____ 1. Wide synaptic cleft

____ 2. Narrow synaptic cleft

____ 3. May contain NE

____ 4. Beadlike varicosities

____ 5. Clustered boutons

____ 6. Glycoprotein-rich basal lamina in synaptic cleft

____ 7. Innervate skeletal muscle fibers

____ 8. Innervate smooth muscle fibers and glands

2. Using the key choices, identify the structures, or associated structures, in the following descriptions. Write the correct letters in the answer blanks.

Key Choices

A. Basal nuclei | D. Cerebellum | G. Indirect (multineuronal) system

B. Brain stem motor areas | E. Command neurons | H. Direct (pyramidal) system

C. Central pattern generators | F. Cortical motor area | I. Spinal cord

__________________ 1. Activates anterior horn neurons of a spinal cord segment during locomotion

__________________ 2. Includes the reticular, red, and vestibular nuclei of the brain stem

__________________ 3. Able to start, stop, or modify the central pattern generators

__________________ 4. The crucial center for sensory-motor integration and control

__________________ 5. Neurons located in the precentral gyri of the frontal lobes

_______________ 6. May involve networks of spinal cord neurons arranged in reverberating circuits

_______________ 7. Function as a liaison between various areas of the cerebral cortex and interact with brain stem "motor programs" areas

_______________ 8. Precommand areas

3. Several neural structures play important roles in the hierarchy of motor control. In the flowchart shown in Figure 13.14, there are several blanks. First, fill in the names for the lowest, middle, and highest levels of the motor hierarchy (put terms inside the appropriate boxes). Then, next to the level terms (lowest, etc.), write the names of the brain regions that contribute to that level of the hierarchy. Also, identify the boxes representing motor output and sensory input. Finally, color the diagram as you like.

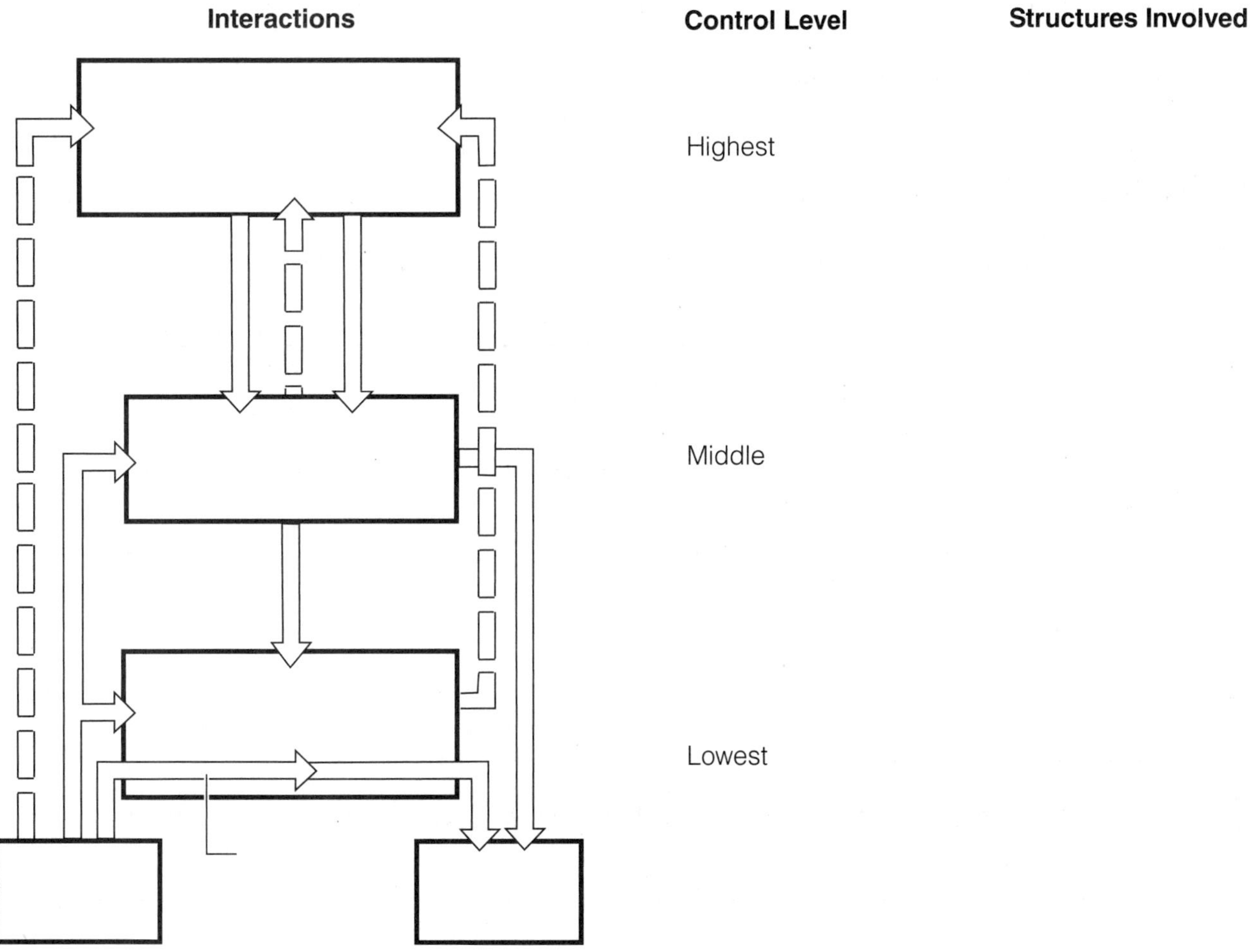

Figure 13.14

4. Circle the term that does not belong in each of the following groupings.

1. Acquired reflex Consciously considered actions Complex motor behavior
All-or-none sequence of motor actions
2. Cerebral cortex Pyramidal neurons Conscious level Command neurons
3. Cerebellum Rhythmic bursts of motor impulses Inherited
Central pattern generators
4. Command neurons Breathing rhythms Interneurons Precommand areas
5. Reticular nuclei Vestibular nuclei Basal nuclei Red nuclei
6. Anterior horn neurons Multineuronal tracts Direct tracts
Voluntary muscle contractions

PART 4: REFLEX ACTIVITY

1. Answer the following questions by writing your answers in the answer blanks.

1. Define the term *reflex*. ______________________________

____

2. List in order the five essential elements of a reflex arc.

A. ____________ C. ____________ E. ____________

B. ____________ D. ____________

3. List the two functional classifications of reflexes. ______________________________

____

4. Name the type of reflex that can occur without the involvement of higher centers.

____

2. Using the key choices, identify the types of reflexes involved in each of the following situations.

Key Choices

A. Somatic reflex(es) B. Autonomic reflex(es)

_____ 1. Patellar (knee-jerk) reflex

_____ 2. Pupillary light reflex

_____ 3. Effectors are skeletal muscles

_____ 4. Effectors are smooth muscle and glands

_____ 5. Flexor reflex

_____ 6. Regulation of blood pressure

_____ 7. Salivary reflex

_____ 8. Clinical testing for spinal cord assessment

3. Complete the following statements by writing the missing terms in the answer blanks:

Stretch reflexes are initiated by __(1)__, which monitor changes in muscle length. Each consists of small __(2)__ enclosed in a capsule. The nerve endings wrapped around these cells are of two types, responding to two different aspects of stretch: the __(3)__ endings of the larger __(4)__ are stimulated by amount and rate of stretch. The __(5)__ endings of the small __(6)__ respond only to degree of stretch. Motor innervation to these special muscle cells is via __(7)__, which arise from small motor neurons. The large, contractile __(8)__ are innervated by __(9)__. These nerve fibers' cell bodies, known as __(10)__, are excited in the stretch reflex; antagonists are inhibited via __(11)__.

_______________ 1.

_______________ 2.

_______________ 3.

_______________ 4.

_______________ 5.

_______________ 6.

_______________ 7.

_______________ 8.

_______________ 9.

_______________ 10.

_______________ 11.

4. Some of the structural elements described in Exercise 3 are illustrated in Figure 13.15. Identify each of the numbered structures by matching it to the terms listed below the diagram.

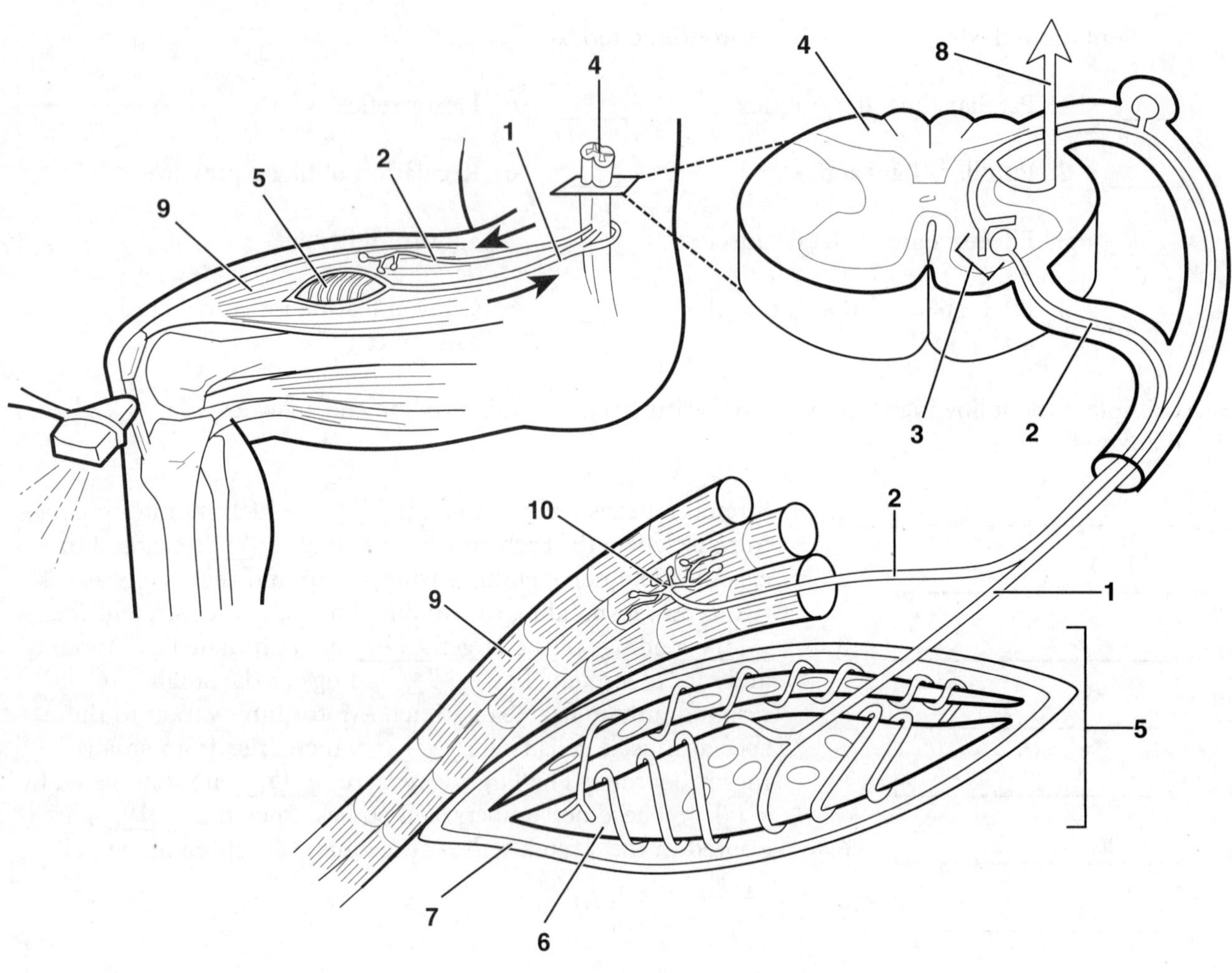

Figure 13.15

_____ Primary sensory endings

_____ Alpha (α) motor neuron

_____ Spinal cord

_____ Synapse

_____ Muscle spindle

_____ Skeletal muscle (extrafusal fibers)

_____ Motor ending to extrafusal fiber

_____ Intrafusal fiber

_____ Receptor capsule

_____ Ascending fibers

5. The table below lists several reflex characteristics. Check (✓) each characteristic that relates to a stretch reflex, to a Golgi tendon reflex, or to both reflexes.

Characteristic	Stretch	Golgi tendon
1. Sensor is the muscle spindle		
2. Sensor is the Golgi tendon organ		
3. Alpha motor neuron innervates agonist muscle		
4. Contraction of agonist muscle is inhibited		
5. Agonist muscle contracts		
6. Contraction of antagonist muscle is inhibited		
7. Antagonist muscle is stimulated		
8. Neuron to antagonist muscle is inhibitory		
9. Type of innervation is reciprocal		
10. Helps ensure smooth onset and termination of muscle contraction		

6. Using the key choices, characterize the types of somatic spinal reflexes according to the descriptions listed below. Insert the letter that best matches each description in the corresponding answer blank.

Key Choices

A. Abdominal   C. Deep tendon   E. Stretch

B. Crossed extensor   D. Flexor   F. Superficial

__________ 1. Maintain normal muscle tone and body posture

__________ 2. Initiated by painful stimuli

__________ 3. Particularly important in maintaining balance

__________ 4. Polysynaptic

__________ 5. Initiated by proprioceptors

__________ 6. Monosynaptic and ipsilateral

__________ 7. Polysynaptic and both ipsilateral and contralateral

__________ 8. Initiated by cutaneous stimulation

__________ 9. Polysynaptic and ipsilateral

_________________ 10. Functioning of corticospinal tracts is required

_________________ 11. Protective and overrides the spinal pathways

_________________ 12. The patellar reflex (knee-jerk reflex)

_________________ 13. Plantar reflex

_________________ 14. Stimulation causes the umbilicus to move toward the stimulated site

THE INCREDIBLE JOURNEY:

A Visualization Exercise for the Special Senses

You . . . see a discontinuous sea of glistening, white rock slabs . . .

1. Complete the narrative by inserting the missing words in the answer blanks.

_________________ 1.

_________________ 2.

_________________ 3.

_________________ 4.

_________________ 5.

_________________ 6.

_________________ 7.

_________________ 8.

Your present journey will take you through your host's internal ear to observe and record events documenting what you have learned about how hearing and equilibrium receptors work.

This is a very tightly planned excursion. Your host has been instructed to move his head at specific intervals and will be exposed to various sounds so that you can make specific observations. For this journey you are injected into the bony cavity of the internal ear, the __(1)__, and are to make your way through its various chambers in a limited amount of time.

Your first observation is that you are in a warm sea of __(2)__ in the vestibule. To your right are two large sacs, the __(3)__ and __(4)__. You swim over to one of these membranous sacs, cut a small semicircular opening in the wall, and wiggle through. Because you are able to see very little in the dim light, you set out to explore this area more fully. As you try to move, however, you find that your feet are embedded in a thick, gluelike substance. The best you manage is slow-motion movements through this __(5)__.

It is now time for your host's first scheduled head movement. Suddenly your world tips sharply sideways. You hear a roar (rather like an avalanche) and look up to see a discontinuous sea of glistening, white rock slabs sliding toward you. You protect yourself from these __(6)__ by ducking down between the hair cells that are bending vigorously with the motion of the rocks. Now that you have seen and can document the operation of a(n) __(7)__, a sense organ of __(8)__ equilibrium, you quickly back out through the hole you made.

_______________ 9.
_______________ 10.
_______________ 11.
_______________ 12.
_______________ 13.
_______________ 14.
_______________ 15.
_______________ 16.

Keeping in mind the schedule and the fact that it is nearly time for your host to be exposed to tuning forks, you swim quickly to the right, where you see what looks like the opening of a cave with tall seaweed waving gently in the current. Abruptly, as you enter the cave, you find that you are no longer in control of your movements, but instead are swept along in a smooth undulating pattern through the winding passageway of the cave, which you now know is the cavity of the (9). As you move up and down with the waves, you see hair cells of the (10), the sense organ for (11), being vigorously disturbed below you. Flattening yourself against the chamber wall to prevent being carried further by the waves, you wait for the stimulus to stop. Meanwhile you are delighted by the electrical activity of the hair cells below you. As they depolarize and send impulses along the (12) nerve, the landscape appears to be alive with fireflies.

Now that you have witnessed the events for this particular sense receptor, you swim back through the vestibule toward your final observation area at the other end of the bony chambers. You recognize that your host is being stimulated again because of the change in fluid currents, but because you are not close to any of the sensory receptors, you are not sure just what the stimulus is. Then, just before you, three dark openings appear, the (13). You swim into the middle opening and see a strange structure that looks like the bristles of an artist's paintbrush; you swim upward and establish yourself on the soft brushy top portion. This must be the (14) of the (15), the sensory receptor for (16) equilibrium. As you rock back and forth in the gentle currents, a sudden wave of fluid hits you. Clinging to the hairs as the fluid thunders past you, you realize that there will soon be another such wave in the opposite direction. You decide that you have seen enough of the special senses and head back for the vestibule to leave your host once again.

CHALLENGING YOURSELF

At the Clinic

1. Brian is brought to the clinic by his parents, who noticed that his right eye does not rotate laterally very well. The doctor explains that the nerve serving the lateral rectus muscle is not functioning correctly. What nerve is involved?
2. After brain surgery, Ray complains that the hospital food tastes unbearably repugnant. His ability to taste salt and sugar are normal, but he firmly maintains that the food is rotten and inedible. From what condition is he suffering, and why?
3. Nine children attending the same Happy-Time Day-care Center developed red, inflamed eyes and eyelids. What is the most likely cause and name of the condition? Would antibiotics be an appropriate treatment?
4. An infant girl is brought to the clinic with strabismus, and tests show that she can control both eyes independently. What noninvasive procedure will be tried before surgery?
5. Mrs. Krikbaum comes to the clinic because of excruciating pain in the left side of her face whenever she drinks cold fluids. What nerve is likely to be inflamed and what name is given to this condition?
6. A teenage boy has been diagnosed with Bell's palsy. What is the cause of Bell's palsy and what are some of its symptoms?

7. Larry Robinson, a patient with a CNS tumor, has lost his ability to taste sour and bitter substances. What cranial nerve nucleus has been impaired by the tumor?

8. A blow to the neck has caused Yolanda to lose her voice. What nerve is affected and what would happen if the entire nerve were nonfunctional?

9. In John's checkup, one year after an accident severed his right accessory nerve, severe muscle atrophy was noted. What two prominent muscles have been affected?

10. Three-year-old Samantha is sobbing that her right arm is "gone" and the examination shows her to have little muscle strength in that limb. Questioning the parents reveals that her father had been swinging her by the arms. What part of the PNS has been damaged?

11. A man comes to the clinic for physical therapy after injuring his right elbow. The flexors of the forearm are weak, and he has great difficulty picking up objects with thumb and forefinger. What nerve has been damaged?

12. Kathleen suffered extensive damage to her sciatic nerve, requiring nerve resection (suturing the nerve back together). She now reports that the sensations from the medial and lateral sides of the knee seem to be reversed. Explain how this could have happened.

13. Harry fell off a tall ladder and fractured the anterior cranial fossa of his skull. On arrival at the hospital, a watery, blood-tinged fluid was dripping from his right nostril. Several days later, Harry complained that he could no longer smell. What nerve was damaged in the fall?

14. Frita, a woman in her early 70s, was having problems chewing. She was asked to stick out her tongue. It deviated to the right and its right side was quite wasted. What nerve was injured?

15. Ted is a war veteran who was hit in the back by fragments of an exploding bomb. His skin is numb in the center of his buttocks and along the entire posterior side of a lower limb, but there is no motor problem at all. One of the following choices is the most likely site of his nerve injury. Choose and explain: (a) a few dorsal roots of the cauda equina; (b) spinal cord transection at C_6; (c) spinal cord transection at L_5; (d) femoral nerve transected in the lumbar region.

16. A 67-year-old woman was diagnosed as having advanced nasopharyngeal cancer with cranial infiltration. How would you test her for integrity of cranial nerves IX–XI?

17. Clancy is the star of his hometown hockey team. During a game he is hit so hard with a hockey stick that he hits the ice. He is brought to the hospital complaining of pain on his left side and later he is unable to flex his hip or extend his knee on that side. Which nerve has been damaged?

18. Mr. Daysh is experiencing night blindness and seeks help at the clinic. What is the technical name for this disorder? What dietary supplement will be recommended? If the condition has progressed too far, what retinal structure will be degenerated?

19. Jan, a senior citizen, is experiencing disturbances in equilibrium: dizziness, vertigo, and nausea. What part of the ear is probably malfunctioning?

20. When Mrs. House visits her ophthalmologist, she complains of pain in her right eye. The intraocular pressure of that eye is found to be abnormally elevated. What is the name of Mrs. House's probable condition? What causes it? What might be the outcome if the problem is not rectified?

Stop and Think

1. When a PNS neuron's axon is cut, several structural changes occur in the cell body (particularly concerning the Nissl bodies [chromatophilic substance] and nucleoli). What are these changes and what is their significance to the repair process?

2. Follow the pathway from pricking your finger with a needle to saying "ouch."
3. Name exteroceptors that are not cutaneous receptors.
4. You wake up in the middle of the night from a sound sleep. Still groggy, you can't feel your body. As you become more awake, you realize this is not a dream. What characteristic of sensation are you experiencing and how could this have happened?
5. Name, in order, the cranial nerves involved in rolling your eyes to follow the second hand on a watch through one complete rotation. Do this separately for each eye.
6. Why are the cerebellum and basal nuclei called "precommand" areas?
7. Are the cords of nerve fibers forming the cauda equina truly spinal nerves?
8. What is the benefit of having the nerve supply of the diaphragm, which is located in the thoracic-lumbar area of the spinal cord, arise from cervical nerves?
9. How does accommodation of muscle spindles figure in the importance of stretch routines as a warm-up for exercise?
10. Use Hilton's law to deduce the nerves that innervate the ankle joint.
11. Dharma, a psychiatric nurse, prepares a bath for a patient and as the tub fills she continuously mixes the water with her hand. As she helps the patient into the tub, he cries out, "What's the matter with you—you're scalding me!" What important aspect of sensory receptor behavior has the nurse forgotten? Explain this phenomenon.
12. Dr. Omata noticed that her anatomy students always confused the glossopharyngeal (IX) and hypoglossal (XII) nerves because the two names have certain similarities. Thus, every year she made her class explain some basic differences between these two nerves. What would you tell her?
13. Adrian and Abdul, two anatomy students, were arguing about the facial nerve. Adrian said it innervates all the skin of the face, and that is why it is called the facial nerve. Abdul said the facial nerve does not innervate face skin at all. Who was more correct? Explain your choice.
14. As Craig stepped onto the sailboat, he smelled the salty sea air and felt the boat rocking beneath his feet. After a few minutes, the smell faded but he was aware of the rocking motion for the entire sail—four hours. What type of receptors are involved in smell and detection of motion? Why did the sensation of smell fade but the rather unpleasant sensation of rocking persist?
15. Can reflex arcs (such as the one triggered when touching something hot) involve (incorporate) pathways for the perception of pain?

COVERING ALL YOUR BASES

Multiple Choice

Select the best answer or answers from the choices given.

1. In an earthquake, which type of sensory receptor is most likely to sound the *first* alarm?
   A. Exteroceptor
   B. Visceroceptor
   C. Mechanoreceptor
   D. Proprioceptor
2. Which of the following can be either somatic or visceral?
   A. Mechanoreceptor C. Chemoreceptor
   B. Photoreceptor D. Nociceptor
3. A cerebral cortical area *not* associated with a special sense is the:
   A. postcentral gyrus C. temporal lobe
   B. occipital lobe D. precentral gyrus

4. Examples of encapsulated nerve endings are:
   A. nociceptors
   B. tactile discs
   C. Pacinian corpuscles
   D. muscle spindles

5. Structures that can be either exteroceptors or proprioceptors include:
   A. Meissner's corpuscle
   B. tactile disc
   C. Pacinian corpuscle
   D. Ruffini ending

6. Which of the following are examples of graded potentials?
   A. Generator potential
   B. Action potential
   C. Receptor potential
   D. Synaptic potential

7. The trochlear nerve contains which class of nerve fibers?
   A. Somatic sensory only
   B. Somatic motor and proprioceptor
   C. Somatic sensory, visceral sensory, and visceral motor
   D. Somatic sensory, visceral sensory, somatic motor, and visceral motor

8. Which sheath contains the most collagen?
   A. Endoneurium
   B. Epineurium
   C. Perineurium
   D. Neurilemma

9. Which of the following could be found in a ganglion?
   A. Cell body of somatic afferent neuron
   B. Autonomic synapse
   C. Visceral efferent neuron
   D. Somatic interneuron

10. Axonal regeneration involves:
    A. disintegration of the proximal portion of the cut nerve fibers
    B. loss of the oligodendrocyte's myelin sheath
    C. proliferation of the oligodendrocytes
    D. maintenance of the neurilemma

11. Chances of CNS nerve fiber regeneration would (might) be increased by a(n):
    A. oligodendrocyte inhibitor
    B. astrocyte inhibitor
    C. secondary demyelination inhibitor
    D. nerve growth–inhibitor blocking agent

12. Cranial nerves that have some function in vision include the:
    A. trochlear
    B. trigeminal
    C. abducens
    D. facial

13. Eating difficulties would result from damage to the:
    A. mandibular division of trigeminal nerve
    B. facial nerve
    C. glossopharyngeal nerve
    D. vagus nerve

14. If the right trapezius and sternocleidomastoid muscles were atrophied, you would suspect damage to the:
    A. vagus nerve
    B. motor branches of the cervical plexus
    C. facial nerve
    D. accessory nerve

15. Which nerve has cutaneous branches?
    A. Facial
    B. Trochlear
    C. Trigeminal
    D. Accessory

16. Which nerve is tested by the corneal reflex?
    A. Optic
    B. Oculomotor
    C. Ophthalmic division of the trigeminal
    D. Abducens

17. Dependence on a respirator (artificial breathing machine) would result from spinal cord transection:
    A. between C_1 and C_2
    B. between C_2 and C_3
    C. between C_6 and C_7
    D. between C_7 and T_1

18. Which nerve stimulates muscles that flex the forearm?
    A. Ulnar
    B. Musculocutaneous
    C. Radial
    D. Median

19. Motor functions of arm, forearm, and fingers would be affected by damage to which one of these nerves?
    A. Radial
    B. Axillary
    C. Ulnar
    D. Median

20. An inability to extend the leg would result from a loss of function of the:
    A. lateral femoral cutaneous nerve
    B. ilioinguinal nerve
    C. saphenous branch of femoral nerve
    D. femoral nerve

21. The gastrocnemius muscle is served by the:
    A. branches of the tibial nerve
    B. inferior gluteal
    C. deep branch of the common peroneal nerve
    D. superficial branch of common peroneal nerve

22. Which contains only motor fibers?
    A. Dorsal root
    B. Dorsal ramus
    C. Ventral root
    D. Ventral ramus

23. Which nerve would be blocked by anesthetics administered during childbirth?
A. Superior gluteal
B. Inferior gluteal
C. Obturator
D. Pudendal

24. Dermatomes on the posterior body surface are supplied by:
A. cutaneous branches of dorsal rami of spinal nerves
B. cutaneous branches of ventral rami of spinal nerves
C. cutaneous branches of all cranial nerves
D. cutaneous branch of cranial nerve V

25. After staying up too late the previous night (studying, no doubt), a college student dozed off in his eight o'clock anatomy and physiology lecture. As his head slowly drifted forward to his chest, he snapped it erect again. What type of reflex does this exemplify?
A. Acquired reflex
B. Autonomic reflex
C. Deep tendon reflex
D. Stretch reflex

26. Gustatory cells are:
A. bipolar neurons
B. multipolar neurons
C. unipolar neurons
D. specialized epithelial cells

27. Alkaloids excite gustatory hairs at the:
A. tip of the tongue
B. back of the tongue
C. circumvallate papillae
D. fungiform papillae

28. Cranial nerves that are part of the gustatory pathway include:
A. trigeminal
B. facial
C. hypoglossal
D. glossopharyngeal

29. Which of the following parasympathetic responses can be triggered reflexively by the activation of taste receptors?
A. Coughing
B. Gagging
C. Secretion of saliva
D. Secretion of gastric juice

30. The receptors for olfaction are:
A. the ends of dendrites of bipolar neurons
B. cilia
C. specialized nonneural receptor cells
D. in papilla

31. Which cranial nerve controls contraction of the circular smooth muscle of the iris?
A. Trigeminal
B. Facial
C. Oculomotor
D. Abducens

32. Second-order neurons for olfaction are located in the:
A. olfactory bulb
B. olfactory epithelium
C. uncus of the limbic system
D. primary olfactory cortex

33. Nociceptors in the nasal mucosa stimulate the:
A. olfactory nerve
B. trigeminal nerve
C. facial nerve
D. glossopharyngeal nerve

34. Nonfunctional granule cells would result in:
A. excessive GABA secretion in the olfactory bulb
B. extremely rapid olfactory adaptation
C. diminished olfactory adaptation
D. anosmia

35. Removal of a tumor in a woman has necessitated destruction of the right facial nerve's branch to the lacrimal gland. Which of the following right eye problems will result from the loss of innervation?
A. Lack of lysozyme
B. Chalazion
C. Sty
D. Reduced lubrication of the conjunctiva

36. The eye movements involved in reading are:
A. saccades
B. scanning movements
C. nystagmus
D. strabismus

37 Which of the following would be found in the fovea centralis?
A. Ganglion neurons
B. Bipolar neurons
C. Cones
D. Rhodopsin

38. The vitreous humor:
A. helps support the lens
B. holds the retina in place
C. contributes to intraocular pressure
D. is constantly replenished

39. Blockage of which of the following is suspected in glaucoma?
A. Ciliary processes
B. Retinal blood vessels
C. Choroid vessels
D. Scleral venous sinus

40. The cornea is nourished by:
A. corneal blood vessels
B. aqueous humor
C. vitreous humor
D. scleral blood vessels

41. In an emmetropic eye:
A. the real image is right side up
B. the focal point is on the fovea centralis
C. convex lenses are required for correction
D. light refraction occurs

42. Refraction can be altered for near or far vision by the:
A. cornea
B. ciliary muscles
C. vitreous humor
D. neural layer of the retina

43. Convergence:
A. requires contraction of the medial rectus muscles of both eyes
B. is needed for near vision
C. involves transmission of impulses along the abducens nerves
D. can promote eye strain

44. The near point of vision:
A. occurs when the lens is at its maximum thickness
B. gets closer with advancing age
C. changes due to loss of lens elasticity
D. occurs when the ciliary muscles are totally relaxed

45. In far vision:
A. the lens is at its thinnest
B. the ciliary muscles contract
C. the light rays are nearly parallel
D. suspensory fibers are slack

46. An unequal corneal curvature:
A. can be corrected with concave lenses
B. is a type of astigmatism
C. is treated by transplant
D. results in cataracts

47. Objects in the periphery of the visual field:
A. stimulate cones
B. cannot have their color determined
C. can be seen in low light intensity
D. appear fuzzy

48. Vitamin A deficiency leads to depletion of:
A. retinal
B. rhodopsin
C. scotopsin
D. photopsins

49. Which of the following statements apply to rhodopsin?
A. Rhodopsin consists of opsin and retinal.
B. Rhodopsin is identical to the visual pigment in the cones.
C. The concentration of rhodopsin increases in the rods during dark adaptation.
D. A solution of rhodopsin looks reddish-purple but becomes colorless after illumination.

50. When moving from darkness to bright light:
A. rhodopsin breakdown accelerates
B. adaptation inhibits cones
C. retinal sensitivity declines
D. visual acuity increases

51. When light strikes the lateral aspect of the left retina, activity increases in the:
A. left optic tract
B. superior colliculus
C. pretectal nucleus
D. right primary visual cortex

52. Receptive fields of retinal ganglion cells:
A. are smaller in the fovea than in the periphery of the retina
B. are equal in size, over the entire retina
C. can be subdivided into functionally distinct center and surround regions
D. are found only in the fovea

53. Excitation of a retinal bipolar cell:
A. can result from excitation of its photoreceptor cells
B. causes excitation of its ganglion cell
C. always results from light striking its photoreceptor cells
D. is modified by lateral inhibition

54. Depth perception is due to all of the following factors except which one(s)?
A. The eyes are frontally located.
B. There is total crossover of the optic nerve fibers at the optic chiasma.
C. There is partial crossover of the optic nerve fibers at the optic chiasma.
D. Each visual cortex receives input from both eyes.

55. Which structures are contained within the petrous portion of the temporal bone?
A. Tympanic cavity
B. Mastoid air cells
C. External auditory meatus
D. Stapedius muscle

56. Which of the following is a space within the membranous labyrinth?
A. Scala vestibuli
B. Scala tympani
C. Scala media
D. Helicotrema

57. Movement of the ______________ membrane triggers bending of hairs of the hair cells in the spiral organ of Corti.
A. tympanic
B. tectorial
C. basilar
D. vestibular

58. Sounds entering the external auditory canal are eventually converted to nerve impulses via a chain of events including:
A. vibration of the eardrum
B. vibratory motion of the ossicles against the round window
C. stimulation of hair cells in the organ of Corti
D. resonance of the basilar membrane

59. In the cochlea:
A. high-frequency sounds resonate close to the helicotrema
B. low-frequency sounds resonate farther from the oval window than high-frequency sounds
C. amplitude determines the intensity of movements of the basilar membrane
D. sound signals are mechanically processed by the basilar membrane before reaching the receptor cells

60. Transmission of impulses from the sound receptors along the cochlear nerve includes which of the following "way stations"?
A. Spiral ganglion
B. Superior olivary nucleus
C. Cochlear nuclei of the medulla
D. Auditory cortex

61. Auditory processing is:
A. analytic
B. synthetic
C. complex and involves local processing
D. interpreted in subjective terms such as pitch and loudness

62. Sound localization is possible if:
A. the sound source is in front of the head or slightly to the side; otherwise not
B. there is a slight difference in the amplitude of the sound entering the two ears
C. there is a slight difference in the time sound reaches the two ears
D. the sound is originating at a point exactly equidistant between the ears

63. According to the place theory:
A. a sound stimulus excites hair cells at a single site on the basilar membrane
B. the fibers of the cochlear nerve all arise from the same site in the organ of Corti
C. each frequency component of a sound excites hair cells at particular (and different) sites along the basilar membrane
D. each sound is interpreted at a specific "place" in the auditory cortex

64. A singer's high C, in the physical sense, is:
A. noise
B. a pure tone like the sound of a tuning fork
C. a sound with a fundamental tone and harmonics
D. music

65. Sound:
A. loses energy as it travels
B. is an example of a longitudinal wave phenomenon
C. is periodic
D. requires an elastic medium

66. If the loudness of a (20-dB) sound is doubled, the resulting sound's pressure level is:
A. 0 dB
B. 22 dB
C. 40 dB
D. 10 dB

67. Which of the following structures is (are) involved in static equilibrium?
A. Maculae
B. Saccule
C. Crista ampullaris
D. Otoliths

68. Which lies closest to the posterior pole of the eye?
A. Cornea
B. Optic disc
C. Macula lutea
D. Central artery

69. Which of the following are paired incorrectly?
A. Cochlear duct—cupula
B. Saccule—macula
C. Ampulla—otoliths
D. Semicircular duct—ampulla

70. Elements associated with static equilibrium include:
A. sensing the position of the head in space
B. response of calcium carbonate crystals to the pull of gravity
C. hair cells, stereocilia, and kinocilia
D. rotary motion

71. The vestibular apparatus:
A. responds to changes in linear acceleration
B. responds to unchanging acceleratory stimuli
C. does not usually contribute to conscious awareness of its activity
D. is helpful, but not crucial, to maintaining balance

72. Taste receptor cells are stimulated by:
   A. chemicals binding to the nerve fibers supplying them
   B. chemicals binding to their microvilli
   C. stretching of their microvilli
   D. impulses from the sensory nerves supplying them

73. The equilibrium pathway to the brain ends at which of the following brain centers?
   A. Vestibular nuclei of brain stem
   B. Flocculonodular node of cerebellum
   C. Inferior colliculi of midbrain
   D. Cerebral cortex

Word Dissection

For each of the following word roots, fill in the literal meaning and give an example, using a word found in this chapter.

Word root	Translation	Example
1. ampulla		
2. branchi		
3. caruncl		
4. cer		
5. cochlea		
6. esthesi		
7. fove		
8. glauc		
9. glosso		
10. gust		
11. kines		
12. lut		
13. macula		
14. metr		
12. modiol		
15. noci		
16. olfact		
17. palpebra		
18. papill		
19. pinn		
20. presby		
21. propri		
22. puden		
23. sacc		
24. scler		
25. tars		
26. trema		
27. tympan		
28. vagus		

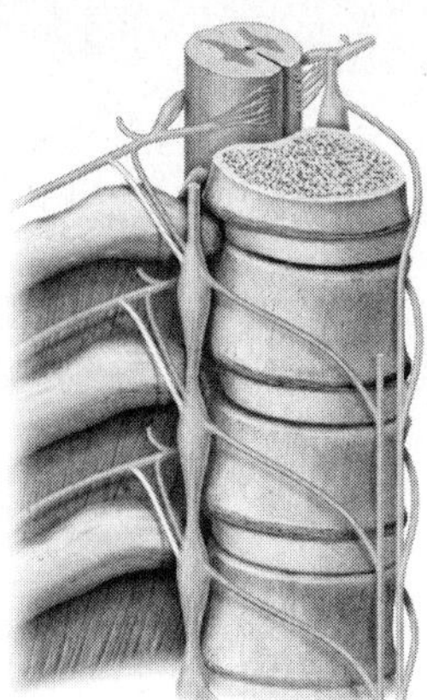

14

The Autonomic Nervous System

Student Objectives

When you have completed the exercises in this chapter, you will have accomplished the following objectives:

Introduction

1. Define autonomic nervous system and explain its relationship to the peripheral nervous system.
2. Compare the somatic and autonomic nervous systems relative to effectors, efferent pathways, and neurotransmitters released.
3. Compare and contrast the functions of the parasympathetic and sympathetic divisions.

ANS Anatomy

4. For the parasympathetic and sympathetic divisions, describe the site of CNS origin, locations of ganglia, and general fiber pathways.

ANS Physiology

5. Define cholinergic and adrenergic fibers, and list the different types of their receptors.
6. Describe the clinical importance of drugs that mimic or inhibit adrenergic or cholinergic effects.
7. State the effects of the parasympathetic and sympathetic divisions on the following organs: heart, blood vessels, gastrointestinal tract, lungs, adrenal medulla, and external genitalia.
8. Describe autonomic nervous system controls.

Homeostatic Imbalances of the ANS

9. Explain the relationship of some types of hypertension, Raynaud's disease, and autonomic dysreflexia to disorders of autonomic functioning.

The autonomic nervous system (ANS) is the involuntary part of the efferent motor division of the peripheral nervous system (PNS). The ANS is structurally and functionally subdivided into sympathetic and parasympathetic divisions. The two divisions innervate cardiac muscle, smooth muscle, and glands in a coordinated and reciprocal manner to maintain homeostasis of the internal environment.

In Chapter 14, topics for study include structural and functional comparisons between the somatic and autonomic nervous systems and between the sympathetic and parasympathetic divisions. Also included are exercises on the anatomy and physiology of the ANS and impairments and developmental aspects of the ANS.

BUILDING THE FRAMEWORK

Introduction

1. Identify the following descriptions as characteristic of the somatic nervous system (use S), of the autonomic nervous system (use A), or of both systems (use S, A). Insert your letter responses in the answer blanks.

_____ 1. Has efferent motor fibers

_____ 2. Single axonal pathway extends from the CNS to each effector

_____ 3. Slower conduction of nerve impulses

_____ 4. Motor neuron cell bodies are located in the CNS

_____ 5. Typically thick, heavily myelinated motor fibers

_____ 6. Target organs are skeletal muscles

_____ 7. Has preganglionic and ganglionic neurons

_____ 8. Effectors are glands and involuntary muscles

_____ 9. Neural pathways are found in cranial and spinal nerves

_____ 10. One ganglion of each motor unit is located outside the CNS

_____ 11. Typically thin fibers with little or no myelination

_____ 12. Effect of neurotransmitter on target organ is always excitatory

_______ 13. Neurotransmitter at effector site is always acetylcholine

_______ 14. No ganglia are present

_______ 15. Neurotransmitter effect on target organ may be excitation or inhibition

_______ 16. Motor activities under control of higher brain centers

2. Identify, by color coding and coloring, the following structures in Figure 14.1, which depicts the major anatomical differences between the somatic and autonomic motor divisions of the PNS. Also identify by labeling all structures provided with leader lines.

- ◯ Somatic motor neuron
- ◯ ANS preganglionic neuron
- ◯ ANS postganglionic axon
- ◯ ANS ganglionic neuron
- ◯ Autonomic ganglion
- ◯ Gray matter of spinal cord (CNS)
- ◯ Effector of the somatic motor neuron
- ◯ Effector of the autonomic motor neuron
- ◯ Intrinsic ganglionic cell
- ◯ Myelin sheath
- ◯ White matter of spinal cord (CNS)

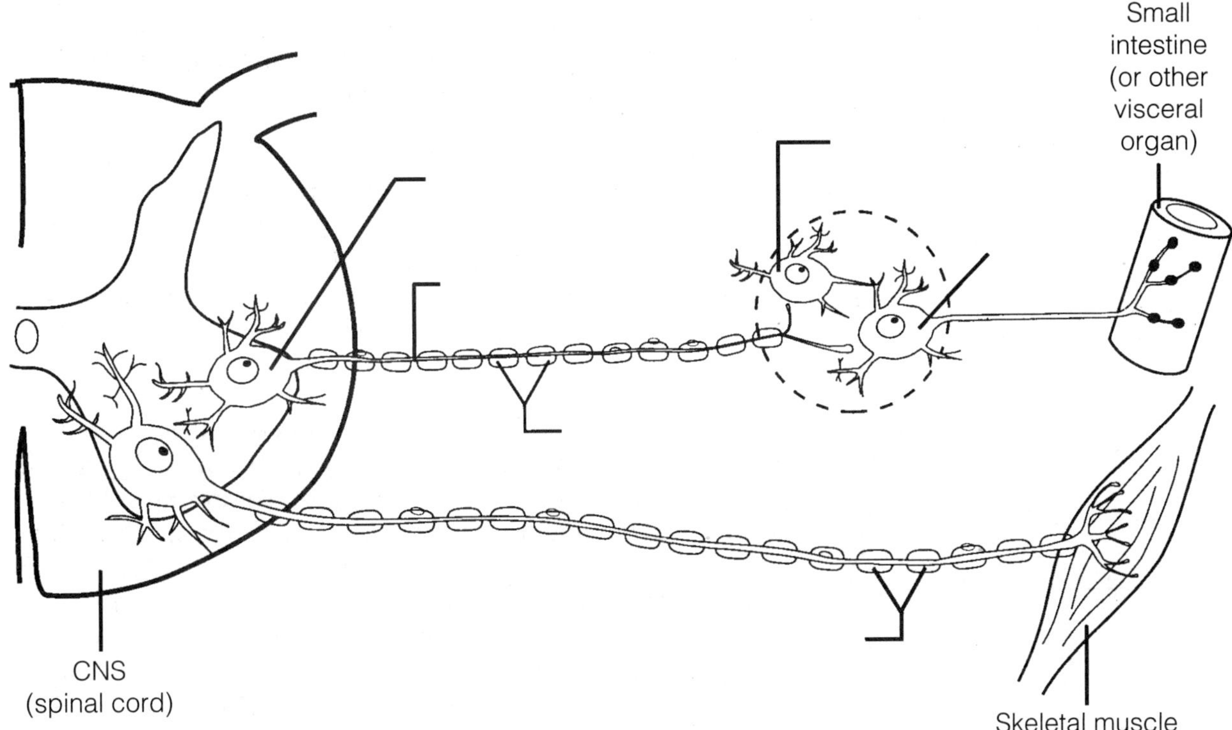

Figure 14.1

ANS Anatomy

1. Figure 14.2 is a highly simplified diagram of the anatomy of the two ANS divisions. Only certain target structures are indicated; for clarity, the spinal cord is depicted twice. Circles represent ganglia, and neural pathways are shown as dotted lines for use as guidelines.

Select eight colors; color the coding circles, structures, and specified boxes listed on the facing page. Use solid lines for preganglionic fibers and leave the dotted lines to represent postganglionic fibers. Next, label each division in the answer blanks below the figure. Then, insert leader lines and label the four cranial nerves by number.

2. The following paragraphs trace the sympathetic and parasympathetic pathways involved in the innervation of selected organs or structures. Complete the statements by inserting the missing terms in the answer blanks. Referring to Figure 14.2 should help you complete this exercise.

Sympathetic preganglionic fibers arise from cell bodies located in the __(1)__ horns of segments __(2)__ of the __(3)__. These fibers leave the ventral root by passing through the __(4)__ rami communicantes and enter adjoining __(5)__, which are aligned to form the __(6)__ trunk or chain. Synapses of sympathetic fibers may or may not occur in the adjoining ganglia. In cases where synapses are made within the adjoining ganglia, the __(7)__ axons enter the spinal nerves by way of the gray __(8)__. The term *gray* means that the axons are __(9)__. Examples of structures with this type of sympathetic innervation are the sweat glands and the __(10)__ in blood vessel walls.

Some preganglionic axons may travel within the sympathetic trunk to synapse in other than adjoining ganglia. For example, synapses within the superior __(11)__ ganglion contribute fibers innervating the __(12)__ of the eye, the __(13)__ glands, the heart, and the respiratory tract organs. Though some fibers serve the heart, most of the postganglionic fibers issuing from the other two cervical ganglia innervate the __(14)__.

Some fibers enter and leave the sympathetic chain without synapsing. Preganglionic fibers T_5–L_2 synapse in __(15)__ ganglia located anterior to the vertebral column. For example, fibers T_5–T_{12} contribute to the __(16)__ nerves, which synapse mainly in the __(17)__ ganglia. From there, postganglionic fibers distribute to serve most of the __(18)__ organs. A few fibers pass through the celiac ganglion without synapsing. These fibers synapse within the medulla of the __(19)__ gland. The __(20)__ (L_1 and L_2) splanchnic nerves synapse in collateral ganglia, from which postganglionic fibers innervate the __(21)__ and reproductive organs.

(continues on page 326)

1. ____________________
2. ____________________
3. ____________________
4. ____________________
5. ____________________
6. ____________________
7. ____________________
8. ____________________
9. ____________________
10. ____________________
11. ____________________
12. ____________________
13. ____________________
14. ____________________
15. ____________________
16. ____________________
17. ____________________
18. ____________________
19. ____________________
20. ____________________
21. ____________________

- ◯ Sites of parasympathetic neurons in the CNS
- ◯ Sites of sympathetic neurons in the CNS
- ◯ Sympathetic fibers
- ◯ Sympathetic trunk of paravertebral ganglia
- ◯ Terminal ganglia
- ◯ Boxes specifying organs that are provided only with sympathetic fibers
- ◯ Parasympathetic fibers
- ◯ Collateral ganglia

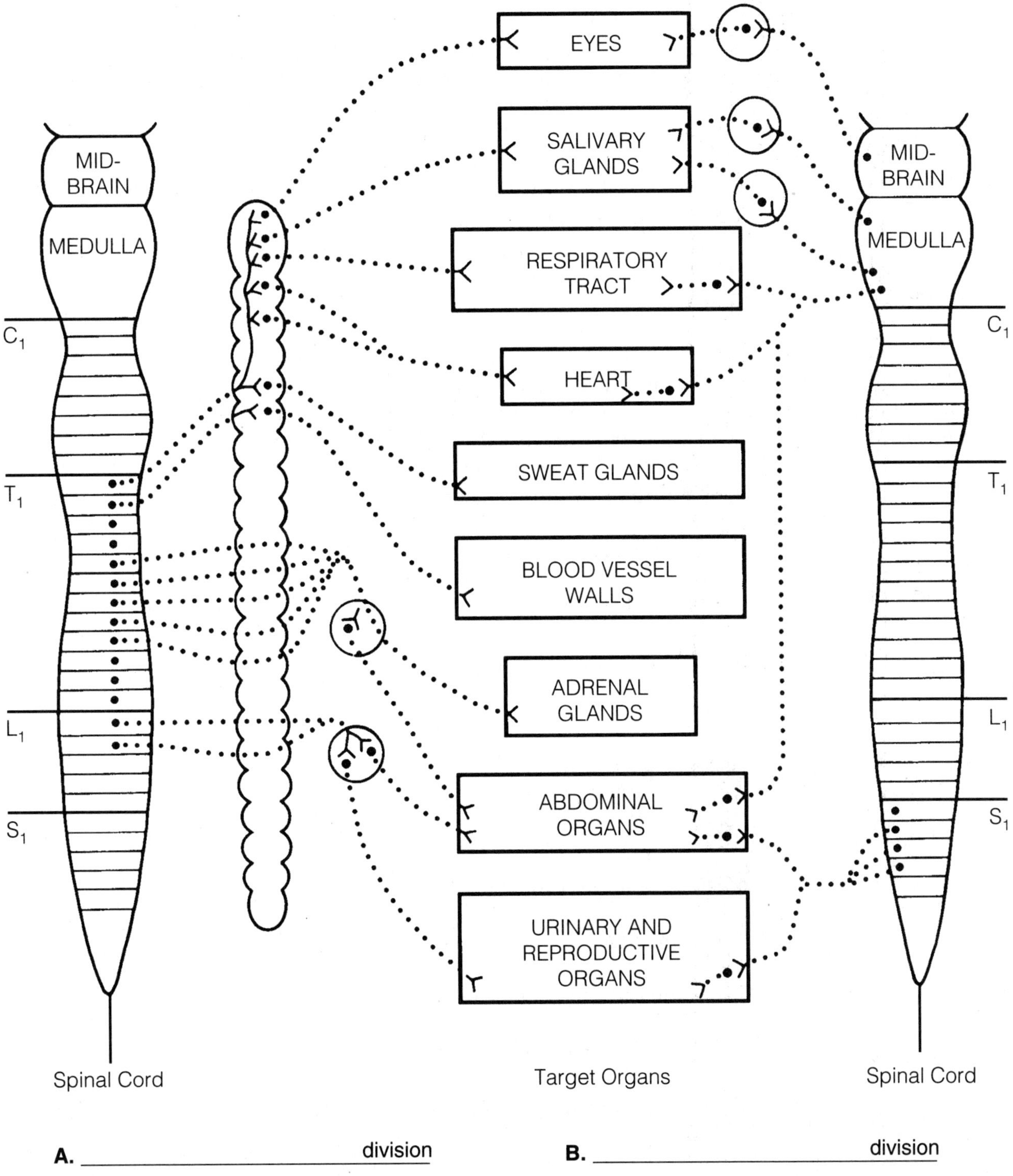

A. ______________________ division **B.** ______________________ division

Figure 14.2

______ 22.
______ 23.
______ 24.
______ 25.
______ 26.
______ 27.
______ 28.
______ 29.
______ 30.
______ 31.
______ 32.

Parasympathetic preganglionic neurons are located in the __(22)__ and in the __(23)__ region of the spinal cord. Preganglionic axons extend from the CNS to synapse in __(24)__ ganglia close to or within target organs. Some examples of cranial outflow are the following. In the ciliary ganglion, preganglionic fibers traveling with the __(25)__ nerve synapse with ganglionic neurons within the orbits. The salivary glands are innervated by parasympathetic fibers that travel with cranial nerves __(26)__ and __(27)__.

__(28)__ nerve fibers account for about 90% of all preganglionic parasympathetic fibers in the body. These fibers enter networks of interlacing nerve fibers called __(29)__, from which arise several branches that serve the organs located in the __(30)__ and in the __(31)__. Preganglionic fibers from the __(32)__ of the spinal cord synapse with ganglionic neurons in ganglia within the walls of the urinary and reproductive organs and the distal half of the large intestine.

3. Characterize each of the following anatomical descriptions as it relates to the sympathetic division (use S), the parasympathetic division (use P), or to both divisions (use S, P). Write the correct letter answers in the answer blanks.

______ 1. Short preganglionic and long postganglionic axons

______ 2. Travel within cranial and sacral nerves

______ 3. Called the thoracolumbar division

______ 4. Long preganglionic and short postganglionic axons

______ 5. No fibers in rami communicantes

______ 6. Called the craniosacral division

______ 7. Ganglia are close to the CNS

______ 8. Minimal branching of preganglionic fibers

______ 9. Terminal ganglia are close to or in the visceral organs served

______ 10. Gray and white rami communicantes utilized

______ 11. Extensive branching of preganglionic fibers

______ 12. Nearly all fibers are accompanied by sensory afferent fibers

ANS Physiology

1. The following table lists general functions of the ANS. Use a check mark (✓) to show which division of the ANS is involved in each function.

Function	Sympathetic	Parasympathetic
1. Normally in control		
2. "Fight-or-flight" system		
3. More specific local control		
4. Causes a dry mouth, dilates bronchioles		
5. Constricts eye pupils, decreases heart rate		
6. Conserves body energy		
7. Causes increased blood glucose levels		
8. Causes increase in digestive tract mobility		
9. Arrector pili muscles contract ("goose bumps")		

2. Define *cholinergic* fibers, and name the two types of cholinergic receptors below.

1. Cholinergic fibers ______________________________

2. Cholinergic receptors ______________________________

3. Define *adrenergic* fibers, and name the two major classes of adrenergic receptors.

1. Adrenergic fibers ______________________________

2. Adrenergic receptors ______________________________

4. Which structures innervated by the ANS would be affected after applying a nicotine patch?

____

5. Using the key choices, select which CNS centers control the autonomic activities listed below. Write the correct letters in the answer blanks.

Key Choices

A. Brain stem B. Cerebral cortex C. Hypothalamus D. Spinal cord

__________ 1. The main integration center of the ANS

__________ 2. Exerts the most direct control over ANS functioning

__________ 3. Coordinates blood pressure, water balance, and endocrine activity

__________ 4. Integrates defecation and micturition reflexes

__________ 5. Controls some autonomic functioning through meditation

__________ 6. Regulates heart and respiration rates and gastrointestinal reflexes

__________ 7. Awareness of autonomic functioning through biofeedback training

__________ 8. Influences autonomic functioning via limbic system connections

6. The following table lists several physiological conditions. Use a check mark (✓) to show which autonomic division is involved for each condition.

Function	Sympathetic	Parasympathetic
1. All neurons secrete acetylcholine		
2. Controls secretions of catecholamines		
3. Control of reflexes that act in thermoregulation		
4. Accelerates metabolism		
5. Short-lived control of effectors		
6. Ganglionic neurons secrete NE		
7. Localized control of effectors; not diffuse		

7. Circle the term that does not belong in each of the following groupings. (*Note:* ACh = acetylcholine; NE = norepinephrine.)

1. Skeletal muscle contraction Cholinergic receptors ACh NE

2. NE Usually stimulatory Nicotinic receptor Adrenergic receptor

3. Beta receptor Cardiac muscle Heart rate slows NE

4. Blood pressure control   Sympathetic tone   Parasympathetic tone

   Blood vessels partially constricted

5. Parasympathetic tone   Urinary tract   Digestive tract   Rapid heart rate

6. External genitalia   Sympathetic tone   Blood vessels dilate   Penis erection

Homeostatic Imbalances of the ANS

1. If a statement is true, write the letter T in the answer blank. If a statement is false, change the underlined word(s) and write the correct word(s) in the answer blank.

______________________ 1. Most autonomic disorders reflect the abnormal control of <u>adrenal medulla</u> activity.

______________________ 2. The cause of Raynaud's disease is thought to be intense <u>vasoconstriction</u> in response to exposure to cold or emotional stress.

______________________ 3. To promote vasodilation in the patient with a severe case of Raynaud's disease, a <u>parasympathectomy</u> is performed.

______________________ 4. Hypertension in the overly stressed patient can be controlled with <u>cholinergic</u> blocker medication.

CHALLENGING YOURSELF

At the Clinic

1. After surgery, patients are often temporarily unable to urinate, and bowel sounds are absent. What division of the ANS is affected by anesthesia?

2. Stress-induced stomach ulcers are due to excessive sympathetic stimulation. For example, one suspected cause of the ulcers is almost total lack of blood flow to the stomach wall. How is this related to sympathetic function?

3. Barium swallow tests on Mr. Bronson indicate gross distension of the inferior end of his esophagus. What condition does this indicate?

4. A young child has been diagnosed with Hirschsprung's disease. How would you explain the cause of the condition and the surgical treatment to the child's parents? (You will need to check a pathophysiology reference for this.)

5. Mrs. Griswold has been receiving treatment at the clinic for Raynaud's disease. Lately her symptoms have been getting more pronounced and severe, and her doctor determines that surgery is her only recourse. What surgical procedure will be performed, and what effect is it designed to produce?

6. Which is the more likely side effect of Mrs. Griswold's surgery—anhidrosis or hyperhidrosis—in the area of concern? (hidrosis = sweating.)

7. Brian, a young man who was paralyzed from the waist down in an automobile accident, has seen considerable progress in the return of spinal cord reflexes. However, quite unexpectedly, he is brought to the ER because of profuse sweating and involuntary voiding. From what condition is Brian suffering? For what life-threatening complication will the clinical staff be alert?

8. Bertha appears at the clinic complaining of intense lower back pain, although she claims she has not done anything to "knock her back out." Questioning reveals painful, frequent urination as well. What is a likely cause of the back pain?

9. John Ryder, a black male in his mid-40s, has hypertension for which no organic cause can be pinpointed. Which class of autonomic nervous system drugs will most likely be prescribed to manage his condition and why?

10. Roweena Gibson, a high-powered marketing executive, develops a stomach ulcer. She complains of a deep abdominal pain that she cannot quite locate, plus a pain in her abdominal wall. Exactly where on the abdominal wall is the superficial pain most likely to be located? (Doing a little research in Chapter 22 may be helpful here.)

11. Imagine that a mad scientist is seeking to invent a death ray that destroys a person's ciliary, pterygopalatine, and submandibular ganglia (and nothing else). List all the symptoms that would be apparent in the victim. Would the victim die, or would the scientist have to go back to the laboratory to try again?

12. How would a drug that inhibits the parasympathetic nervous system affect a person's pulse?

Stop and Think

1. Can the autonomic nervous system function properly in the absence of visceral afferent input? Explain your response.

   Use the following information for questions 2–6: Stretch receptors in the bladder send impulses along visceral afferent nerve fibers to the sacral region of the spinal cord. Synapses with motor neurons there initiate impulses along visceral efferent fibers that stimulate contraction of the smooth muscle of the bladder wall and relaxation of the smooth muscle (involuntary) sphincter to allow urination.

2. What division of the autonomic nervous system is involved?

3. What sort of neuronal circuit does this exemplify?

4. What nerves carry the impulses?

5. How is urination inhibited consciously?

6. How would this pathway be affected by spinal cord transection above the sacral region?

7. Migraine headaches are caused by constriction followed by dilation of the vessels supplying the brain. The pain is associated with the dilation phase and is apparently related to the high rate of blood flow through these vessels. Some migraine sufferers seek relief through biofeedback, learning to trigger dilation of the vessels in the hand(s). How does this provide relief?

8. Trace the sympathetic pathway from the spinal cord to the iris of the eye, naming all associated structures, fiber types, and neurotransmitters.

9. Trace the parasympathetic pathway from the brain to the heart, naming all associated structures, fiber types, and neurotransmitters.

COVERING ALL YOUR BASES

Multiple Choice

Select the best answer or answers from the choices given.

1. Which of the following is true of the autonomic, but not the somatic, nervous system?
   A. Neurotransmitter is acetylcholine
   B. Axons are myelinated
   C. Effectors are muscle cells
   D. Has motor neurons located in ganglia

2. Examination of nerve fibers supplying the rectus femoris muscle reveals an assortment of group A fibers and group C fibers. The group A fibers are known to be somatic motor fibers supplying the skeletal muscle. Which of these is likely to be the function of the group C fibers?
   A. Parasympathetic innervation to the muscle spindles
   B. Sympathetic innervation to the muscle spindles
   C. Parasympathetic innervation to the blood vessels
   D. Sympathetic innervation to the blood vessels

3. Which type of ganglia are located in or very near the enteric plexus of the GI tract?
   A. Collateral C. Chain
   B. Terminal D. Dorsal root

4. Which best describes ANS control?
   A. Completely under control of the cerebral cortex
   B. Completely under control of the brain stem
   C. Entirely controls itself
   D. Major control by the hypothalamus and spinal reflexes

5. Which of the following disorders is (are) related specifically to sympathetic functions?
   A. Achalasia
   B. Raynaud's disease
   C. Orthostatic hypotension
   D. Hirschsprung's disease

6. Adrenal medulla development involves:
   A. neural crest cells
   B. nerve growth factor
   C. sympathetic preganglionic axons
   D. parasympathetic preganglionic axons

7. Orville said he had a heartache because he broke up with his girlfriend and put his hand over his heart on his anterior chest. Staci told him that if his heart really hurt, he could also be pointing somewhere else. Where?
   A. His left arm C. His gluteal region
   B. His head D. His abdomen

Use the following choices to respond to questions 8–26:

A. sympathetic division
B. parasympathetic division
C. both sympathetic and parasympathetic
D. neither sympathetic nor parasympathetic

_____ 8. Typically has long preganglionic and short postganglionic fibers

_____ 9. Some fibers utilize gray rami communicantes

_____ 10. Courses through spinal nerves

_____ 11. Has nicotinic receptors on its ganglionic neurons

_____ 12. Has nicotinic receptors on its target cells

_____ 13. Has splanchnic nerves

_____ 14. Courses through cranial nerves

_____ 15. Originates in cranial nerves

_____ 16. Effects enhanced by direct stimulation of a hormonal mechanism

_____ 17. Includes otic ganglion

_____ 18. Includes celiac ganglion

_____ 19. Contains cholinergic fibers

_____ 20. Stimulatory impulses from hypothalamus and/or medulla pass through the thoracic spinal cord to connect to preganglionic neurons

_____ 21. Hyperactivity of this division canlead to ischemia (loss of circulation) to various body parts and hypertension

_____ 22. Affected by beta blockers

_____ 23. Hypoactivity of this division would lead to decrease in metabolic rate

_____ 24. Stimulated by the RAS

_____ 25. Has widespread, long-lasting effects

_____ 26. Sets the tone for the heart

Word Dissection

For each of the following word roots, fill in the literal meaning and give an example, using a word found in this chapter.

Word root	Translation	Example
1. adren		
2. chales		
3. epinephr		
4. mural		
5. ortho		
6. para		
7. pathos		
8. splanchn		

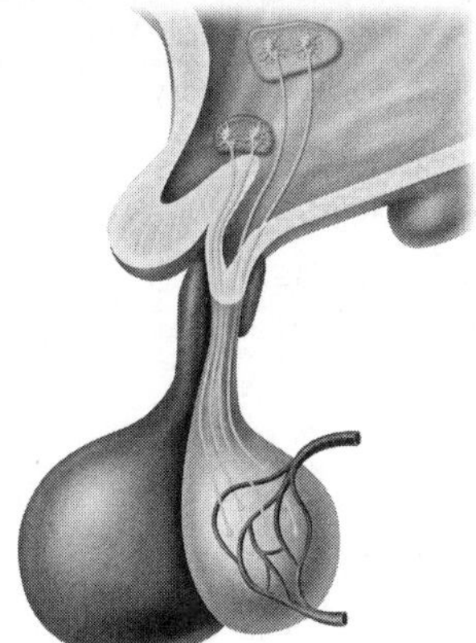

15

The Endocrine System

Student Objectives

When you have completed the exercises in this chapter, you will have accomplished the following objectives:

The Endocrine System: An Overview

1. Indicate important differences between hormonal and neural controls of body functioning.
2. List the major endocrine organs, and describe their body locations.
3. Distinguish between hormones, paracrines, and autocrines.

Hormones

4. Describe how hormones are classified chemically.
5. Describe the two major mechanisms by which hormones bring about their effects on their target tissues.
6. List three kinds of interaction of different hormones acting on the same target cell.
7. Explain how hormone release is regulated.

The Pituitary Gland and Hypothalamus

8. Describe structural and functional relationships between the hypothalamus and the pituitary gland.
9. List and describe the chief effects of anterior pituitary hormones.
10. Discuss the structure of the posterior pituitary, and describe the effects of the two hormones it releases.

Major Endocrine Glands

11. Describe important effects of the two groups of hormones produced by the thyroid gland.
12. Follow the process of thyroxine formation and release.
13. Indicate general functions of parathyroid hormone.
14. List hormones produced by the adrenal gland, and cite their physiological effects.
15. Briefly describe the importance of melatonin.
16. Compare and contrast the effects of the two major pancreatic hormones.
17. Describe the functional roles of hormones of the testes, ovaries, and placenta.

Hormone Secretion by Other Organs

18. Name a hormone produced by the heart.
19. State the location of enteroendocrine cells.
20. Briefly explain the hormonal functions of the kidney, skin, adipose tissue, bones, and thymus.

The endocrine system is vital to homeostasis and plays an important role in regulating the activity of body cells. Acting through bloodborne chemical messengers called hormones, the endocrine system organs orchestrate cellular changes that lead to growth and development, reproductive capability, and the physiological homeostasis of many body systems.

Activities in this chapter concern the localization of the various endocrine organs in the body and explain the general function of the various hormones and the results of their hypersecretion or hyposecretion.

BUILDING THE FRAMEWORK

The Endocrine System: An Overview

1. Complete the following statements by choosing answers from the key choices. Record the answers in the answer blanks.

Key Choices

A. Circulatory system	E. Metabolism	I. Nutrient
B. Electrolyte	F. More rapid	J. Reproduction
C. Growth and development	G. Nerve impulses	K. Slower and more prolonged
D. Hormones	H. Nervous system	L. Water

__________ 1. The endocrine system is a major controlling system in the body. Its means of control, however, is much __(1)__ than that of the
__________ 2. __(2)__, the other major body system that acts to maintain homeostasis. Perhaps the reason for this is that the endocrine
__________ 3. system uses chemical messengers, called __(3)__, instead of __(4)__. These chemical messengers enter the blood and are car-
__________ 4. ried throughout the body by the activity of the __(5)__.

__________ 5. The endocrine system has several important functions: It helps maintain __(6)__, __(7)__, and __(8)__ balance; regulates energy
__________ 6. balance and __(9)__; and prepares the body for childbearing or __(10)__.

__________ 7.

__________ 8. __________ 9. __________ 10.

2. Figure 15.1 is a diagram of the various endocrine organs of the body. Next to each letter on the diagram, write the name of the endocrine-producing organ (or area). Then select different colors for each and color the illustration. To complete your identification of the hormone-producing organs, name the organs (not illustrated) described in J, K, and L.

J. Small glands that ride "horseback" on the thyroid

____

K. Endocrine-producing organ present only in pregnant women

____

L. B and C hang from the floor of this neuroendocrine organ

____

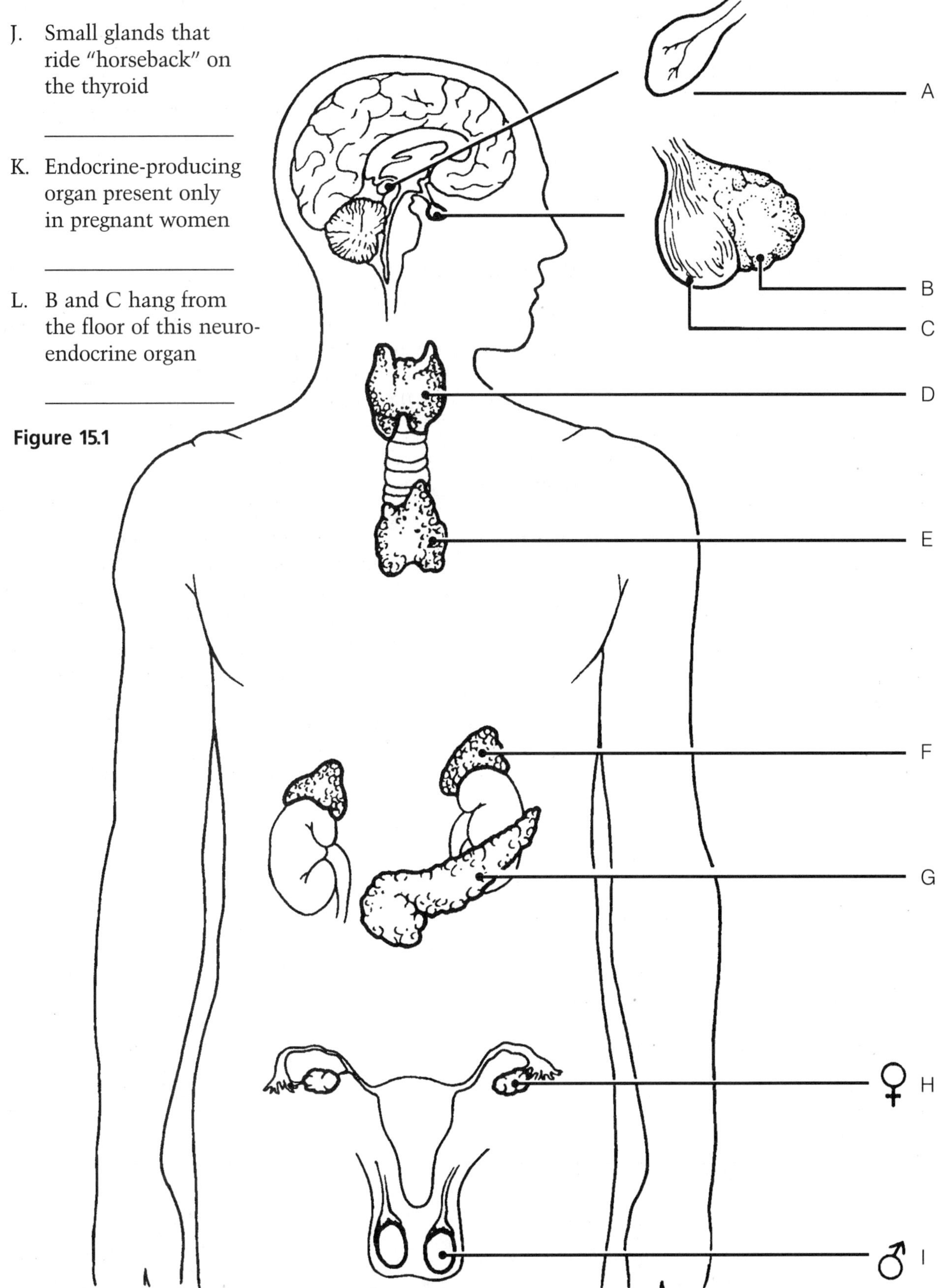

Figure 15.1

Hormones

1. Complete the following statements by choosing answers from the key choices. Record the answers in the answer blanks.

Key Choices

A. Altering activity	F. Negative feedback	K. Steroid- or amino acid-based
B. Anterior pituitary	G. Neural	L. Stimulating new or unusual activities
C. Hormonal	H. Neuroendocrine	M. Sugar or protein
D. Humoral	I. Receptors	N. Target cell(s)
E. Hypothalamus	J. Releasing and inhibiting factors (hormones)	

______________ 1.

______________ 2.

______________ 3.

______________ 4.

______________ 5.

______________ 6.

______________ 7.

______________ 8.

______________ 9.

______________ 10.

______________ 11.

______________ 12.

______________ 13.

All cells do not respond to endocrine system stimulation. Only those that have the proper __(1)__ on their cell membranes are activated by the chemical messengers. These responsive cells are called the __(2)__ of the various endocrine glands.

Hormones promote homeostasis by __(3)__ of body cells rather than by __(4)__. Most hormones are __(5)__ molecules. The various endocrine glands are prodded to release their hormones by nerve fibers (a __(6)__ stimulus), by other hormones (a __(7)__ stimulus), or by the presence of increased or decreased levels of various other substances in the blood (a __(8)__ stimulus). The secretion of most hormones is regulated by a __(9)__ system, in which increasing levels of that particular hormone "turn off" its stimulus. The __(10)__ is called the master endocrine gland because it regulates so many other endocrine organs. However, it is in turn controlled by __(11)__ secreted by the __(12)__. The structure identified in item 12 above is also part of the brain, so it is appropriately called a __(13)__ organ.

2. Differentiate clearly between a circulating hormone and an autocrine or paracrine.

____

____

3. Indicate the major stimulus for release of each of the hormones listed below. Choose your response from the key choices.

Key Choices

A. Hormonal B. Humoral C. Neural

_______________ 1. Adrenocorticotropic hormone

_______________ 2. Parathyroid hormone

_______________ 3. Insulin

_______________ 4. Thyroxine and triiodothyronine

_______________ 5. Epinephrine

_______________ 6. Oxytocin and antidiuretic hormone

_______________ 7. Estrogen and progesterone

_______________ 8. Calcitonin

4. Complete the following description of a second-messenger system by writing the missing words in the answer blanks.

_______________ 1.
_______________ 2.
_______________ 3.
_______________ 4.
_______________ 5.
_______________ 6.
_______________ 7.
_______________ 8.
_______________ 9.

The cyclic AMP mechanism is a good example of a second-messenger system. In this mechanism, the hormone, acting as the __(1)__ messenger, binds to target cell membrane receptors coupled to a signal transducer called __(2)__. This signal transducer molecule, in turn, acts as an intermediary to activate the enzyme __(3)__, which catalyzes the conversion of intracellular __(4)__ to cyclic AMP. Cyclic AMP then acts as the __(5)__ messenger to initiate a cascade of reactions in the target cell. Most of the subsequent events are mediated by the activation of enzymes called __(6)__, which in turn activate or inactivate other enzymes by adding a __(7)__ group to them. The sequence of events initiated by the second messenger depends on the __(8)__, __(9)__, and __(10)__. In addition to cyclic AMP, many other molecules are known to act as second messengers, including __(11)__, __(12)__, and __(13)__. Additionally, the ion __(14)__ can act intracellularly as a second messenger in certain cases.

_______________ 10. _______________ 12. _______________ 14.

_______________ 11. _______________ 13.

5. Explain why the persistence of a hormone in the blood is so limited.

____

6. Match the terms or phrases in Column B with the descriptions in Column A.

Column A

______ 1. ______ 2. ______ 3. The extent of target cell activation by hormone-receptor binding depends equally on these three factors

______ 4. The mechanism by which most steroid-based hormones influence their target cells

______ 5. ______ 6. ______ 7. ______ 8. Four ways hormones may alter cellular activity (depending on target cell type)

______ 9. Target cell responds to continued high hormone levels by forming more receptors capable of binding the hormone

______ 10. Reduced target cell response to continued high hormone levels in the blood

______ 11. Period of persistence of a hormone in the bloodstream

Column B

A. Affinity of the receptor for the hormone

B. Change in membrane permeability and/or voltage

C. Direct gene activation

D. Down-regulation

E. Enzyme activation/inactivation

F. Half-life

G. Hormone blood levels

H. Initiation of secretory activity

I. Relative number of hormone receptors on the target cells

J. Second-messenger system

K. Synthesis of regulatory molecules such as enzymes

L. Up-regulation

7. Using the key choices listed, respond to the following questions concerning hormone interaction at target cells.

Key Choices

A. Antagonism B. Permissiveness C. Synergism

______ 1. Another hormone must be present for a given hormone to exert its effects.

______ 2. Combined effects of several different hormones acting on a target cell are amplified.

______ 3. A hormone opposes or prevents the action of another hormone on a target cell.

8. For each of the following hormones, indicate specifically its chemical nature by choosing from the key choices.

Key Choices

A. Steroid B. Catecholamine C. Peptide D. Iodinated amino acid derivative

__________________ 1. Thyroxine

__________________ 2. Epinephrine

__________________ 3. Aldosterone

__________________ 4. Insulin

The Pituitary Gland and Hypothalamus

1. Figure 15.2 depicts the anatomical relationships between the hypothalamus and the anterior and posterior parts of the pituitary in a highly simplified way. First, identify each of the structures listed below by color coding and coloring them on the diagram. Then, on the appropriate lines write in the names of the hormones that influence each of the target organs shown at the bottom of the diagram. Color the target organ diagrams as you like.

◯ Hypothalamus

◯ Anterior pituitary

◯ Sella turcica of the sphenoid bone

◯ Posterior pituitary

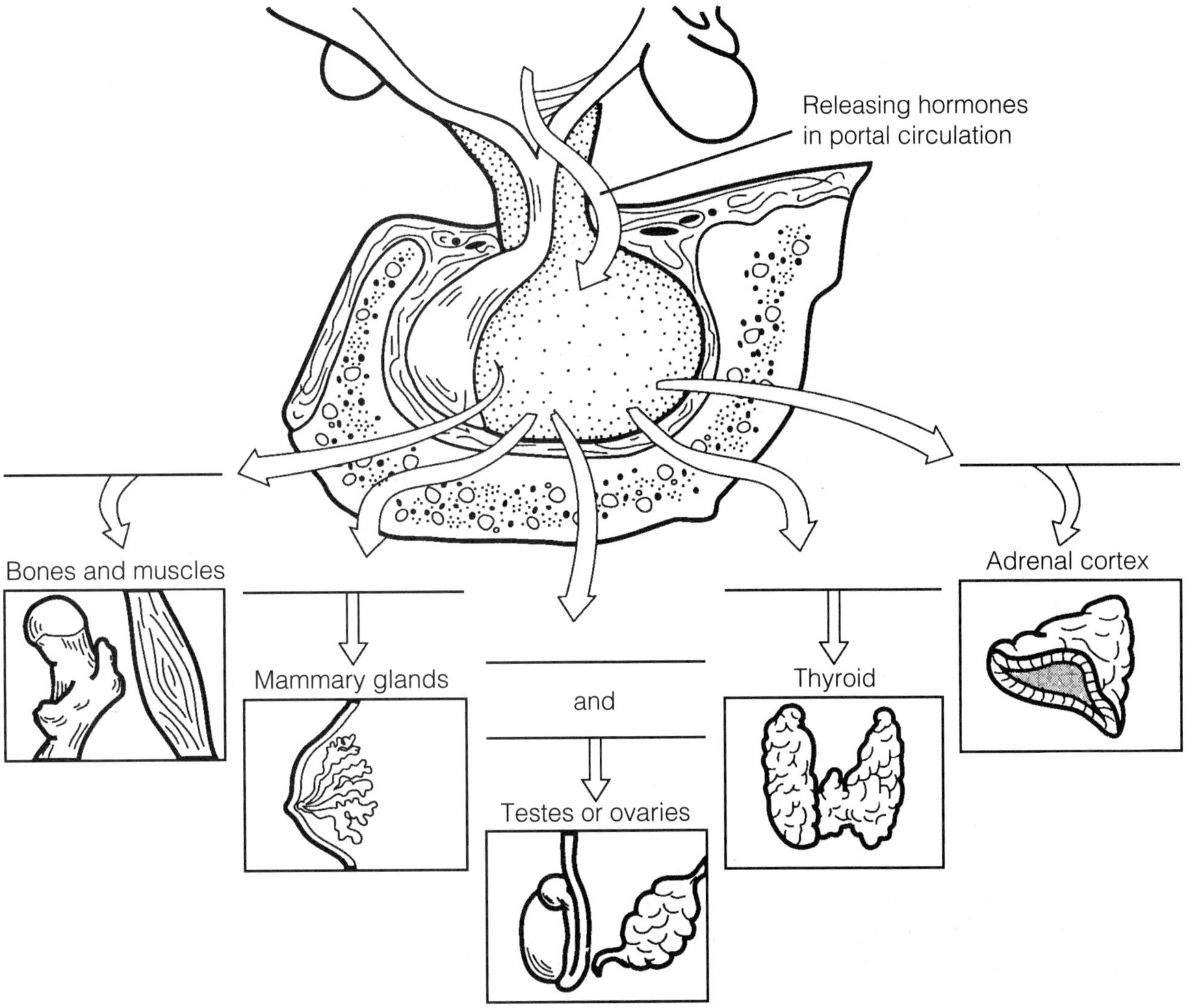

Figure 15.2

2. Circle the term that does not belong in each of the following groupings.

1. Posterior lobe   Neurohypophysis   Nervous tissue   Anterior lobe
2. Posterior lobe   Adenohypophysis   Glandular tissue   Anterior lobe
3. Growth hormone   Prolactin   Oxytocin   ACTH   Pro-opiomelanocortin
4. Neurohormones   Hypophyseal portal system   Axonal transport   Hypothalamic-hypophyseal tract

3. Explain why antidiuretic hormone is also called vasopressin.

____

____

4. Growth hormone has numerous effects, both direct and indirect. Indicate the direct effects of GH by checking (✓) the appropriate choices. Circle the choices that indicate effects mediated indirectly by somatomedins.

_____ 1. Encourages lipolysis

_____ 2. Promotes amino acid uptake from the blood and protein synthesis

_____ 3. Promotes an increase in size and strength of the skeleton

_____ 4. Promotes elevated blood sugar levels

_____ 5. Stimulates cartilage synthesis

_____ 6. Prompts the liver to produce somatomedins

_____ 7. Promotes glucose sparing

_____ 8. Promotes cell division

_____ 9. Causes muscle cell growth

_____ 10. Inhibits glucose metabolism

_____ 11. Increases blood levels of fatty acids

Major Endocrine Organs

1. Indicate the organ (or organ part) producing or releasing each of the hormones listed below by inserting the appropriate answers from the key choices in the answer blanks.

Key Choices

A. Adrenal gland (cortex)	E. Ovaries	I. Placenta
B. Adrenal gland (medulla)	F. Pancreas	J. Testes
C. Anterior pituitary	G. Parathyroids	K. Thymus
D. Hypothalamus	H. Pineal	L. Thyroid

________	1. ACTH	________	12. LH
________	2. ADH	________	13. Melatonin
________	3. Aldosterone	________	14. MSH
________	4. Calcitonin	________	15. Oxytocin
________	5. Cortisone	________	16. Progesterone
________	6. Epinephrine	________	17. Prolactin
________	7. Estrogens	________	18. PTH
________	8. FSH	________	19. Testosterone
________	9. Glucagon	________	20. Thymosin
________	10. Growth hormone	________	21. Thyroxine
		________	22. TSH
________	11. Insulin		

2. Name the hormone that best fits each of the following descriptions.

________ 1. Basal metabolic hormone

________ 2. "Programs" T lymphocytes

________ 3. Most important hormone regulating the amount of calcium circulating in the blood; released when blood calcium levels drop

________ 4. Helps protect the body during long-term stressful situations, such as extended illness and surgery

________ 5. Short-term stress hormone; aids in the fight-or-flight response; increases blood pressure and heart rate, for example

________ 6. Necessary if glucose is to be taken up by most body cells

________ 7. ________ 8. Regulate the function of another endocrine organ; four

________ 9. ________ 10. tropic hormones

________ 11. Acts antagonistically to insulin; produced by the same endocrine organ

________ 12. Hypothalamic hormone important in regulating water balance

________ 13. ________ 14. Anterior pituitary hormones that regulate the ovarian cycle

________ 15. ________ 16. Directly regulate the menstrual or uterine cycle

__________________ 17. Adrenal cortex hormone involved in regulating salt levels of body fluids

__________________ 18. __________________ 19. Necessary for milk production and ejection

3. Circle the term that does not belong in each of the following groupings.

1. Calcitonin — Increases blood Ca^{2+} levels — Thyroid gland — Enhances Ca^{2+} deposit
2. Thyroxine — Increases BMR — Calorigenic effect — Depresses glucose uptake — Enhances sympathetic nervous system activity
3. Glucocorticoids — Steroids — Aldosterone — Sex hormones — Thyroxine
4. Sympathomimetic amines — Norepinephrine — Epinephrine — Cortisol

4. Parathyroid hormone has multiorgan effects. Indicate its effects on the organs listed below.

1. Kidneys __________________
2. Intestine __________________
3. Bones __________________

5. Relative to thyroxine synthesis and release, put the following events in their correct time sequence by numbering them from 1 to 9. Event number 1 is already designated.

_____ 1. Within the Golgi apparatus, sugar molecules are attached to the thyroglobulin protein, and the glycolated molecules are packaged in membranous sacs.

_____ 2. Iodinated thyroglobulin is taken up into the follicle cells by endocytosis and combined with lysosomes.

_____ 3. T_3 and T_4 are cleaved out of the colloid by lysosomal enzymes.

_____ 4. Iodine is attached to tyrosine residues of the colloid, forming DIT and MIT.

_____ 5. T_4 and T_3 diffuse into the bloodstream.

__1__ 6. Thyroglobulin protein is synthesized on the ribosomes of the follicle cell.

_____ 7. Iodides are oxidized and transformed into iodine at the follicle lumen membrane.

_____ 8. Thyroglobulin is discharged into the lumen of the follicle.

_____ 9. Enzymes in the colloid link DITs and MITs together to form T_4 and T_3.

6. List the cardinal symptoms of diabetes mellitus and provide the rationale for the occurrence of each symptom.

1. ____

____

2. ____

____

3. ____

____

7. Compare the magnitude of the conditions in Column A with those immediately opposite in Column B. Circle the phrase indicating the condition that is always or usually greater. If the conditions are essentially always of the same magnitude, put an S on the dotted line between them.

Column A		Column B
1. Plasma levels of gonadotropins during childhood	....	Plasma levels of gonadotropins during puberty
2. Activity of T_3 at target cells	....	Activity of T_4 at target cells
3. Protein wasting and depressed immunity with high levels of glucocorticoids	..	Protein wasting and depressed immunity with low levels of glucocorticoids
4. Stimulation of aldosterone release by elevated Na^+ levels in plasma	.....	Stimulation of aldosterone release by elevated K^+ levels in plasma
5. Urine volume in a diabetes mellitus patient	......	Urine volume in a diabetes insipidus patient
6. Renin release by the kidneys when blood pressure is low	......	Renin release by the kidneys when blood pressure is elevated
7. Iodine atoms in a thyroxine molecule	....	Iodine atoms in a triiodothyronine molecule
8. Mineralocorticoid release by zona reticularis cells	....	Mineralocorticoid release by zona glomerulosa cells
9. PTH release when blood calcium levels are high (above 11 mg/100 ml)	...	PTH release when blood calcium levels are low (below 9 mg/100 ml)
10. Hyperglycemia with high levels of insulin in plasma	....	Hyperglycemia with high levels of glucagon in plasma
11. ACTH release with high plasma levels of cortisol	....	ACTH release with low plasma levels of cortisol
12. Synthesis of insulin by beta cells of pancreatic islets	......	Synthesis of insulin by alpha cells of pancreatic islets

8. For each of the hormones listed below, indicate its effect on blood glucose, blood calcium, and/or blood pressure by using the key choices.

Key Choices

A. Increases blood glucose

B. Decreases blood glucose

C. Increases blood calcium

D. Decreases blood calcium

E. Increases blood pressure

F. Decreases blood pressure

________ 1. Cortisol

________ 2. Insulin

________ 3. Parathyroid hormone

________ 4. Aldosterone

________ 5. Growth hormone

________ 6. Antidiuretic hormone

________ 7. Glucagon

________ 8. Thyroxine

________ 9. Epinephrine

________ 10. Calcitonin

9. The structure of endocrine cells often allows recognition of the type of hormone product they secrete. Use the key choices to characterize the endocrine cells listed below.

Key Choices

A. Has a well-developed rough ER and secretory granules

B. Has a well-developed smooth ER; prominent lipid droplets

_____ 1. Interstitial cell of the testis

_____ 2. Chief cell in the parathyroid gland

_____ 3. Zona fasciculata cell

_____ 4. Parafollicular cells of the thyroid

_____ 5. Beta cell of a pancreatic islet

_____ 6. Any endocrine cell in the anterior pituitary

10. Concerning the histology of the pure endocrine glands, match each endocrine gland in Column B with the best approximation of its histology in Column A.

Column A		Column B
_____	1. Spherical clusters of cells	A. Posterior pituitary
_____	2. Parallel cords of cells	B. Zona glomerulosa of the adrenal cortex
_____	3. Follicles	C. Thyroid gland
_____	4. Nervous tissue	D. Zona fasciculata of the adrenal cortex

11. Name the hormone that would be produced in *inadequate* amounts in each of the following conditions.

____________________ 1. Maturation failure of reproductive organs

____________________ 2. Tetany (death due to respiratory paralysis)

____________________ 3. Polyuria without high blood glucose levels; causes dehydration and tremendous thirst

____________________ 4. Goiter

____________________ 5. Cretinism, a type of dwarfism in which the individual retains childlike proportions and is mentally retarded

____________________ 6. Excessive thirst, high blood glucose levels, acidosis

____________________ 7. Abnormally small stature, normal proportions, a "Tom Thumb"

____________________ 8. Spontaneous abortion

____________________ 9. Myxedema in the adult

12. Name the hormone that would be produced in *excessive* amounts in each of the following conditions.

____________________ 1. Acromegaly in the adult

____________________ 2. Bulging eyeballs, nervousness, increased pulse rate, weight loss (Graves' disease)

____________________ 3. Demineralization of bones; spontaneous fractures

____________________ 4. Cushing's syndrome: moon face, hypertension, edema

____________________ 5. Abnormally large stature, relatively normal body proportions

____________________ 6. Abnormal hairiness; masculinization

Hormone Secretion by Other Organs

1. Besides the major endocrine organs, isolated clusters of cells produce hormones within body organs that are usually not associated with the endocrine system. A number of these hormones are listed in the table below. Complete the missing information on these hormones by filling in the blank spaces in the table.

Hormone	Chemical makeup	Source	Effects
Gastrin	Peptide		
Secretin		Duodenum	
Cholecystokinin	Peptide		
Erythropoietin		Kidney in response to hypoxia	
Cholecalciferol (vitamin D_3)		Skin; activated by kidneys	
Atrial natriuretic peptide (ANP)		Peptide	
	Peptide		Suppresses appetite, increases energy expenditure

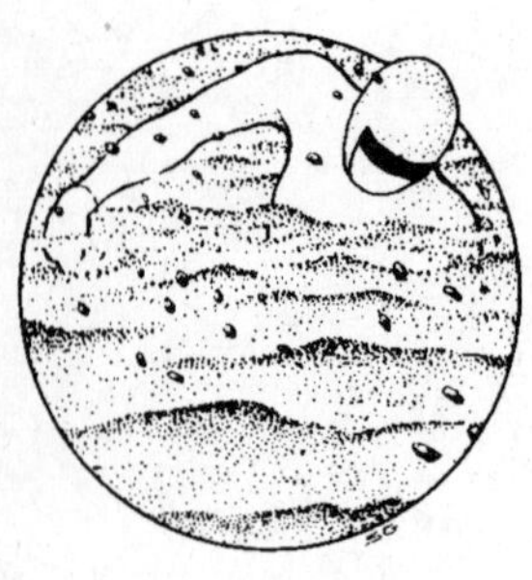

THE INCREDIBLE JOURNEY:

A Visualization Exercise for the Endocrine System

. . . you notice charged particles, shooting pell-mell out of the bone matrix. . .

1. Complete the following narrative by writing the missing words in the answer blanks.

For this journey, you will be miniaturized and injected into a vein of your host. Throughout the journey, you will be traveling in the bloodstream. Your instructions are to record changes in blood composition as you float along and to form some conclusions as to why these changes are occurring (that is, which hormone is being released).

__________________ 1.
__________________ 2.
__________________ 3.
__________________ 4.
__________________ 5.
__________________ 6.
__________________ 7.
__________________ 8.
__________________ 9.

Bobbing gently along in the slowly moving blood, you realize that there is a sugary taste to your environment; however, the sweetness begins to decrease quite rapidly. As the glucose levels of the blood have just decreased, obviously __(1)__ has been released by the __(2)__ so that the cells can take up glucose.

A short while later, you notice that the depth of the blood in the vein you are traveling in has diminished substantially. To remedy this potentially serious situation, the __(3)__ will have to release more __(4)__, so the kidney tubules will reabsorb more water. Within a few minutes the blood becomes much deeper; you wonder if the body is psychic as well as wise.

As you circulate past the bones, you notice charged particles, shooting pell-mell out of the bone matrix and jumping into the blood. You conclude that the __(5)__ glands have just released PTH because the __(6)__ levels have increased in the blood. As you continue to move in the bloodstream, the blood suddenly becomes sticky sweet, indicating that your host must be nervous about something. Obviously, his __(7)__ has released __(8)__ to cause this sudden increase in blood glucose.

Sometime later, you become conscious of a humming activity around you, and you sense that the cells are very busy. Obviously your host's __(9)__ levels are sufficient, as his cells are certainly not sluggish in their metabolic activities. You record this observation and prepare to end this journey.

CHALLENGING YOURSELF

At the Clinic

1. Pete is very short for his chronological age of 8. What physical features will allow you to determine quickly whether to check GH or thyroxine levels?
2. A young girl is brought to the clinic by her father. He complains that his daughter fatigues easily and seems mentally sluggish. You notice a slight swelling in the anterior neck. What condition do you suspect? What are some possible causes and their treatments?
3. Lauralee, a middle-aged woman, comes to the clinic and explains in an agitated way that she is "very troubled" by excessive urine output and consequent thirst. What two hormones might be causing the problem and what urine tests will be done to identify the problem?
4. A 2-year-old boy is brought to the clinic by his anguished parents. He is developing sexually and shows an obsessive craving for salt. Blood tests reveal hyperglycemia. What endocrine gland is hypersecreting?
5. Lester, a 10-year-old, has been complaining of severe lower back pains. The nurse notices that he seems weak and a reflex check shows abnormal response. Kidney stones are soon diagnosed. What abnormality is causing these problems?
6. Bertha Wise, age 40, is troubled by swelling in her face and unusual fat deposition on her back and abdomen. She reports that she bruises easily. Blood tests show elevated glucose levels. What is your diagnosis and what glands might be causing the problem?
7. A middle-aged man comes to the clinic, complaining of extreme nervousness, insomnia, and weight loss. The nurse notices that his eyes bulge and his thyroid is enlarged. What is the man's hormonal imbalance and what are two likely causes?

8. Mr. Holdt brings his wife to the clinic, concerned about her nervousness, heart palpitations, and excessive sweating. Tests show hyperglycemia and hypertension. What hormones are probably being hypersecreted? What is the cause? What physical factors allow you to rule out thyroid problems?
9. Phyllis, a type 1 diabetic, is rushed to the hospital. She had been regulating her diabetes extremely well, with no chronic problems, when her mother found her unconscious. Will blood tests reveal hypoglycemia or hyperglycemia? What probably happened?
10. A woman calls for an appointment at the clinic because she is not menstruating. She also reports that her breasts are producing milk, although she has never been pregnant. What hormone is being hypersecreted and what is the likely cause?
11. An accident victim who had not been wearing a seat belt received severe trauma to his forehead when he was thrown against the windshield. The physicians in the emergency room worried that his brain stem may have been driven inferiorly through the foramen magnum. To help assess this possibility, they quickly took a standard X ray of his head and searched for the position of the pineal gland. How could anyone expect to find a tiny, boneless gland like the pineal in an X ray?
12. Maryanne, a street person, is pregnant. She has not received prenatal care and her diet consists of what she has been able to scavenge from trash cans. What could you surmise about her PTH blood level?

Stop and Think

1. Compare and contrast protein and steroid hormones with regard to the following: (a) organelles involved in their manufacture; (b) ability to store the hormones within the cell; (c) rate of manufacture; (d) method of secretion; (e) means of transport in the bloodstream; (f) location of receptors in/on target cell; (g) use of second messenger; (h) relative time from attachment to receptor until effects appear; and (i) whether effects persist after the hormone is metabolized.
2. (a) Hypothalamic factors act both to stimulate and to inhibit the release of certain hormones. Name these hormones.

   (b) Blood levels of certain humoral factors are regulated both on the "up" and the "down" side by hormones. Name some such humoral factors.
3. What are enteroendocrine cells and why are they sometimes called paraneurons?
4. Would drug tolerance be due to up-regulation or down-regulation?
5. Why do the chemical structures of thyroxine and triiodothyronine require complexing with a large protein to allow long-term storage?
6. The brain is "informed" when we are in a stressful situation, and the hypothalamus responds to stressors by secreting a releasing hormone called corticotropin-releasing hormone. This hormone helps the body deal with the stress through a long sequence of events. Outline this entire sequence, starting with corticotropin-releasing hormone and ending with the release of cortisol. (Be sure to trace the hormones through the hypophyseal portal system and out of the pituitary gland.)
7. Joshua explained to his classmate Jennifer that the thyroid gland contains parathyroid cells in its follicles and the parathyroid cells secrete parathyroid hormone and calcitonin. Jennifer told him he was all mixed up. Can you correct Josh's mistakes?
8. When the carnival was scheduled to come to a small town, health professionals who felt that the sideshows were cruel and exploitive joined with the local consumer groups to enforce truth-in-advertising laws. They demanded that the fat man, the dwarf, the giant, and the bearded lady be billed as "people with endocrine system problems"

(which of course removed all the sensationalism usually associated with these attractions). Identify the endocrine disorder in each case and explain how (or why) the disorder produced the characteristic features of these four showpeople.

9. What would be the consequence of administering a phosphodiesterase-inhibiting drug on the action of a hormone that prompts production of the second messenger cAMP?

COVERING ALL YOUR BASES

Multiple Choice

Select the best answer or answers from the choices given.

1. Relative to the cyclic AMP second-messenger system, which of the following is not accurate?
   A. The activating hormone interacts with a receptor site on the plasma membrane.
   B. Binding of the galvanizing hormone directly activates adenylate cyclase.
   C. Activated adenylate cyclase catalyzes the transformation of AMP to cyclic AMP.
   D. Cyclic AMP acts within the cell to alter cell function as is characteristic for that specific hormone.

2. Which of the following hormones is (are) released by neurons?
   A. Oxytocin
   B. Insulin
   C. ADH
   D. Cortisol

3. The paraventricular nucleus of the hypothalamus is named for its proximity to the:
   A. lateral ventricles
   B. cerebral aqueduct
   C. third ventricle
   D. fourth ventricle

4. Which of the following might be associated with a second-messenger system?
   A. Phosphatidyl inositol
   B. Corticosteroids
   C. Glucagon
   D. Calmodulin

5. ANP, the hormone secreted by the heart, has exactly the opposite function to this hormone secreted by the zona glomerulosa:
   A. epinephrine
   B. cortisol
   C. aldosterone
   D. testosterone

6. Hormones that act directly or indirectly to elevate blood glucose include:
   A. GH
   B. cortisol
   C. insulin
   D. CRH

7. The release of which of the following hormones will be stimulated via the hypothalamic-hypophyseal tract?
   A. ACTH
   B. TSH
   C. ADH
   D. GH

8. Cells sensitive to the osmotic concentration of the blood include:
   A. chromaffin cells
   B. paraventricular neurons
   C. supraoptic neurons
   D. parafollicular cells

9. The gland derived from embryonic throat tissue known as Rathke's pouch is the:
   A. posterior pituitary
   B. anterior pituitary
   C. thyroid
   D. thymus

10. The primary capillary plexus is located in the:
    A. hypothalamus
    B. anterior pituitary
    C. posterior pituitary
    D. infundibulum

11. Pro-opiomelanocortin is the precursor of:
    A. cortisol
    B. corticotropin
    C. melatonin
    D. opium

12. Which of the following are direct or indirect effects of growth hormone?
    A. Stimulates cells to take in amino acids
    B. Increases synthesis of chondroitin sulfate
    C. Increases blood levels of fatty acids
    D. Decreases utilization of glucose by most body cells

13. Which of the following are tropic hormones secreted by the anterior pituitary gland?
    A. LH
    B. ACTH
    C. TSH
    D. FSH

14. Hormones secreted by females include:
    A. estrogens
    B. progesterone
    C. prolactin
    D. testosterone

15. Smooth muscle contractions are stimulated by:
    A. testosterone
    B. FSH
    C. prolactin
    D. oxytocin

16. Nerve input regulates the release of:
    A. oxytocin
    B. epinephrine
    C. melatonin
    D. cortisol

17. Hypertension may result from hypersecretion of:
    A. thyroxine
    B. cortisol
    C. aldosterone
    D. antidiuretic hormone

18. In initiating the secretion of stored thyroxine, which of these events occurs first?
    A. Production of thyroglobulin
    B. Discharge of thyroglobulin into follicle
    C. Attachment of iodine to thyroglobulin
    D. Lysosomal activity to cleave hormone from thyroglobulin

19. Hypothyroidism can cause:
    A. myxedema
    B. Cushing's syndrome
    C. cretinism
    D. exophthalmos

20. Calcitonin targets the:
    A. kidneys
    B. liver
    C. bone
    D. small intestine

21. Imbalances of which hormones will affect neural function?
    A. Thyroxine
    B. Parathormone
    C. Insulin
    D. Aldosterone

22. Hormones that regulate mineral levels include:
    A. calcitonin
    B. aldosterone
    C. atrial natriuretic peptide
    D. glucagon

23. Which of the following is given as a drug to reduce inflammation?
    A. Epinephrine
    B. Cortisol
    C. Aldosterone
    D. ADH

24. After menopause, steroids that maintain anabolism come from the:
    A. ovaries
    B. adrenal cortex
    C. anterior pituitary
    D. thyroid

25. Which is generally true of hormones?
    A. Exocrine glands produce them.
    B. They travel throughout the body in the blood.
    C. They affect only nonhormone-producing organs.
    D. All steroid hormones produce very similar physiological effects in the body.

26. The major endocrine organs of the body:
    A. tend to be very large organs
    B. are closely connected with each other
    C. all contribute to the same function (digestion)
    D. tend to lie near the midline of the body

27. Which type of cell secretes releasing hormones?
    A. Neuron
    B. Parafollicular cell
    C. Chromaffin cell
    D. Adenohypophysis cell

28. Of the following endocrine structures, which develops from the brain?
A. Neurohypophysis
B. Adenohypophysis
C. Thyroid gland
D. Thymus gland

Word Dissection

For each of the following word roots, fill in the literal meaning and give an example, using a word found in this chapter.

Word root	Translation	Example
1. adeno		
2. crine		
3. dips		
4. diuresis		
5. gon		
6. hormon		
7. humor		
8. mell		
9. toci		
10. trop		

16

Blood

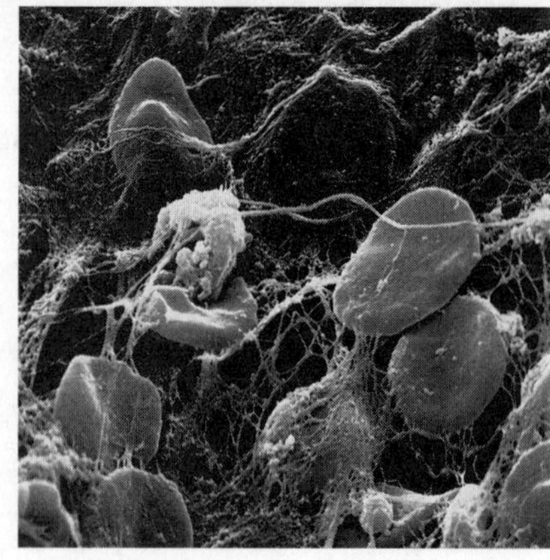

Student Objectives

When you have completed the exercises in this chapter, you will have accomplished the following objectives:

Overview: Blood Composition and Functions

1. Describe the composition and physical characteristics of whole blood. Explain why it is classified as a connective tissue.
2. List eight functions of blood.

Blood Plasma

3. Discuss the composition and functions of plasma.

Formed Elements

4. Describe the structure, function, and production of erythrocytes.
5. Describe the chemical makeup of hemoglobin.
6. Give examples of disorders caused by abnormalities of erythrocytes. Explain what goes wrong in each disorder.
7. List the classes, structural characteristics, and functions of leukocytes.
8. Describe how leukocytes are produced.
9. Give examples of leukocyte disorders, and explain what goes wrong in each disorder.
10. Describe the structure and function of platelets.

Hemostasis

11. Describe the processes of hemostasis. List factors that limit clot formation and prevent undesirable clotting.
12. Give examples of hemostatic disorders. Indicate the cause of each condition.

Transfusion and Blood Replacement

13. Describe the ABO and Rh blood groups. Explain the basis of transfusion reactions.
14. Describe fluids used to replace blood volume and the circumstances for their use.

Diagnostic Blood Tests

15. Explain the diagnostic importance of blood testing.

Blood, the indispensable "life fluid" that courses through the body's blood vessels, provides the means by which the body's cells receive essential nutrients and oxygen and dispose of their metabolic wastes. As blood flows past the tissue cells, exchanges continually occur between the blood and the cells so that vital functions are maintained.

This chapter provides an opportunity to review the general characteristics of whole blood and plasma, to identify the various formed elements (blood cells), and to recall their functions. Blood groups, transfusion reactions, clotting, and various types of blood abnormalities are also reviewed.

BUILDING THE FRAMEWORK

Overview: Blood Composition and Functions

1. Complete the following description of the components of blood by writing the missing words in the answer blanks.

________________ 1.
________________ 2.
________________ 3.
________________ 4.
________________ 5.
________________ 6.
________________ 7.
________________ 8.
________________ 9.
________________ 10.
________________ 11.

In terms of its tissue classification, blood is classified as a __(1)__ because it has living blood cells, called __(2)__, suspended in a nonliving fluid matrix called __(3)__. The "fibers" of blood only become visible during __(4)__.

If a blood sample is centrifuged, the heavier blood cells become packed at the bottom of the tube. Most of this compacted cell mass is composed of __(5)__, and the volume of blood accounted for by these cells is referred to as the __(6)__. The less dense __(7)__ rises to the top and constitutes about 45% of the blood volume. The so-called "buffy coat" composed of __(8)__ and __(9)__ is found at the junction between the other two blood elements. The buffy coat accounts for less than __(10)__ % of blood volume.

Blood is scarlet red in color when it is loaded with __(11)__; otherwise, it tends to be dark red.

2. List four delivery functions of blood, two regulatory functions, and two protection functions.

1. Delivery (distribution) functions ______________________________

____

____

2. Regulatory functions ______________________________

____

____

3. Protection functions ______________________________

____

____

Blood Plasma

1. List three classes of substances normally found dissolved in plasma.

1. ______________ 2. ______________ 3. ______________

2. Complete the following table relating to the proteins found in plasma.

Constituent	Description/importance
______________	60% of plasma proteins; important for osmotic balance
Fibrinogen	____% of plasma proteins; important in ______________
______________ ______________ ______________	36% of plasma proteins: transport proteins antibodies
Nonprotein nitrogenous substances	______________ ______________ (list 5)
______________	Organic chemicals absorbed from the digestive tract
Respiratory gases	______________, ______________

Formed Elements

1. Number the following cell types to indicate the sequence of erythrocyte maturation in the red bone marrow. Circle the cell type released to the blood; underline the cell type that ejects its nucleus and most organelles.

_____ 1. Reticulocyte

_____ 2. Hematopoietic stem cell

_____ 3. Late erythroblast

_____ 4. Normoblast

_____ 5. Erythrocyte

_____ 6. Early erythroblast

_____ 7. Proerythroblast

2. Figure 16.1 on the opposite page depicts in incomplete form the erythropoietin mechanism for regulating the rate of erythropoiesis. Several statements are incomplete. Complete the statements that have answer blanks and then choose colors (other than yellow) to identify the structures with color coding circles. Color all arrows on the diagram yellow. Finally, respond to the following questions to complete this exercise.

○ The kidney ○ Red bone marrow ○ RBCs

1. What is the normal life span of erythrocytes? ____________________ days

2. What three food nutrients (other than the normally required proteins and carbohydrates) are essential for erythropoiesis?

____

3. What is the fate of aged or damaged red blood cells? ____________________

____

4. What is the fate of the released hemoglobin? ____________________

____

____

3. Check (✓) all the factors that would serve as stimuli for erythropoiesis.

_____ 1. Hemorrhage

_____ 2. Insufficient hemoglobin per RBC

_____ 3. Living at a high altitude

_____ 4. Breathing pure oxygen

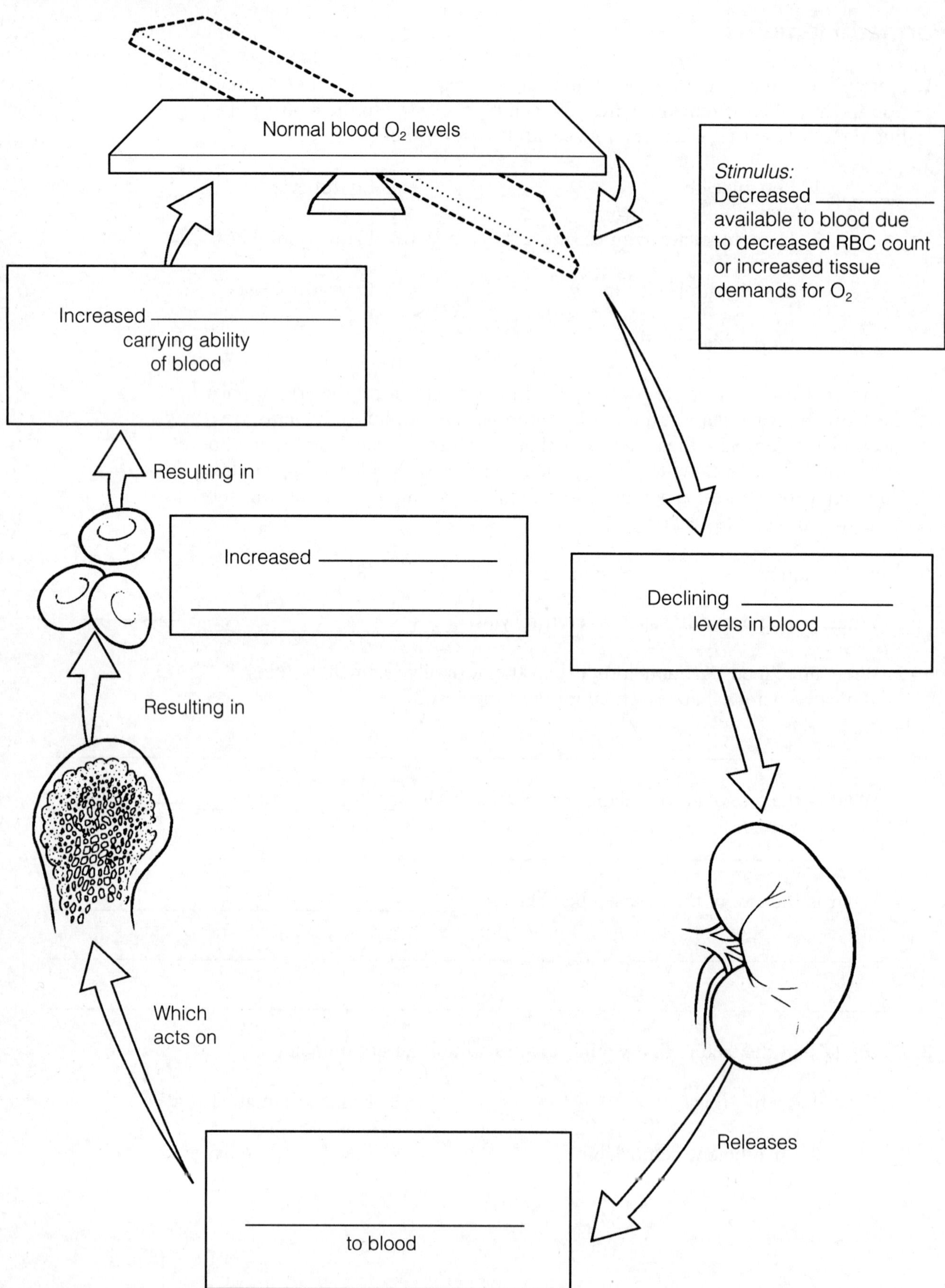
Normal blood O2 levels
Stimulus:
Decreased ____________ available to blood due to decreased RBC count or increased tissue demands for O2
Declining ____________ levels in blood
Releases
____________ to blood
Which acts on
Resulting in
Increased ____________
Resulting in
Increased ____________ carrying ability of blood

Figure 16.1

4. The following questions ask for information concerning structural characteristics of red blood cells. Provide the information requested to fully characterize each structural aspect.

1. Cell shape ______________________________
2. Nucleate or anucleate? ______________________________
3. Organelles present? ______________________________
4. Major molecular content of the cytoplasm ______________________________
5. Substance that accounts for the flexibility of the membrane ______________________________

The next section deals with functional characteristics of red blood cells. Provide the information requested to fully describe each functional characteristic.

6. Type of metabolism, aerobic or anaerobic? ______________________________
7. Explain why the RBC metabolism is of this type. ______________________________
8. Molecular makeup of hemoglobin ______________________________

   ____
9. The portion of hemoglobin that binds oxygen ______________________________
10. The portion of hemoglobin that binds carbon dioxide ______________________________
11. Site of oxygen loading ______________________________
12. Site of carbon dioxide loading ______________________________
13. Name given to oxygen-loaded hemoglobin ______________________________
14. Name given to oxygen-depleted hemoglobin ______________________________

5. Match the terms indicating specific types of anemias in Column B with the appropriate descriptions in Column A.

	Column A	Column B
____________	1. A genetic disorder in which abnormal hemoglobin is produced and becomes spiky under hypoxic conditions	A. Aplastic anemia B. Hemolytic anemia
____________	2. A common occurrence after transfusion error	C. Hemorrhagic anemia
____________	3. The bone marrow is destroyed or severely inhibited	D. Iron-deficiency anemia
____________	4. A consequence of acute blood loss	E. Pernicious anemia
____________	5. A possible consequence of sickle-cell anemia	F. Sickle-cell anemia

____________ 6. Results from inadequate intake of iron-rich foods or conditions involving chronic types of bleeding, such as gastric ulcers

____________ 7. A common problem of individuals who have a portion of their stomach removed to manage bleeding ulcers

6. If a statement is true, write the letter T in the answer blank. If a statement is false, change the underlined word(s) and write the correct word(s) in the answer blank.

____________ 1. White blood cells (WBCs) move into and out of blood vessels by the process of <u>positive chemotaxis</u>.

____________ 2. A count of white blood cells that provides information on the relative number of each WBC type is the <u>total</u> WBC count.

____________ 3. When blood becomes too acid or too basic, both the respiratory system and the <u>liver</u> may be called into action to restore it to its normal pH range.

____________ 4. Carbaminohemoglobin is formed when carbon dioxide binds to the <u>heme groups</u> of hemoglobin.

____________ 5. The cardiovascular system of an average adult contains approximately <u>4</u> liters of blood.

____________ 6. Blood is circulated through the blood vessels by the pumping action of the <u>heart</u>.

____________ 7. The only leukocyte type to arise from the lymphoid stem cells is the <u>lymphocyte</u>.

____________ 8. Normal <u>hemoglobin</u> values are in the range of 42% to 47% of the volume of whole blood.

____________ 9. An anemia resulting from a decreased RBC number causes the blood to become <u>more</u> viscous.

____________ 10. The leukocytes particularly important in the immune response are <u>monocytes</u>.

____________ 11. B-complex vitamins are necessary for <u>DNA</u> synthesis in RBCs.

____________ 12. In plasma, iron is transported bound to <u>hemosiderin</u>.

7. Rank the following lymphocytes in order of their relative abundance (in the blood of a healthy person) from 1 (most abundant) to 5 (least abundant).

____ 1. lymphocytes ____ 2. basophils ____ 3. neutrophils

____ 4. eosinophils ____ 5. monocytes

8. Using the key choices, identify the cell types or blood elements that fit the following descriptions.

Key Choices

A. Basophil	D. Lymphocyte	G. Neutrophil
B. Eosinophil	E. Megakaryocyte	H. Platelets
C. Formed elements	F. Monocyte	I. Red blood cell

__________ 1. Granulocyte with the smallest granules

__________ 2. Granular leukocytes (#2–#4)

__________ 3. __________ 4.

__________ 5. Also called an erythrocyte, anucleate

__________ 6. __________ 7. Phagocytic leukocytes that avidly engulf bacteria

__________ 8. __________ 9. Agranular leukocytes

__________ 10. Fragments to form platelets

__________ 11. All the others are examples of these

__________ 12. Increases in number during allergy attacks

__________ 13. Releases histamine during inflammatory reactions

__________ 14. Descendants may be formed in lymphoid tissue

__________ 15. Contains hemoglobin; therefore involved in oxygen transport

__________ 16. Does not use diapedesis

__________ 17. Increases in number during chronic infections; the largest WBC

__________ 18. __________ 19. The only formed elements that are not spherical

__________ 20. __________ 21. __________ 22.

__________ 23. __________ 24. Also called white blood cells (#20–#24)

__________ 25. Granulocyte with two types of granules

__________ 26. A T cell or a B cell

__________ 27. __________ 28. Kills by the respiratory burst

9. Four leukocytes are diagrammed in Figure 16.2. First, follow the directions for coloring each leukocyte as it appears when stained with Wright's stain. Then, identify each leukocyte type by writing the correct name in the blank below each illustration.

   A. Color the granules pale violet, the cytoplasm pink, and the nucleus dark purple.

   B. Color the nucleus deep blue and the cytoplasm pale blue.

   C. Color the granules bright red, the cytoplasm pale pink, and the nucleus red-purple.

   D. For this smallest white blood cell, color the nucleus deep purple-blue and the sparse cytoplasm pale blue-green.

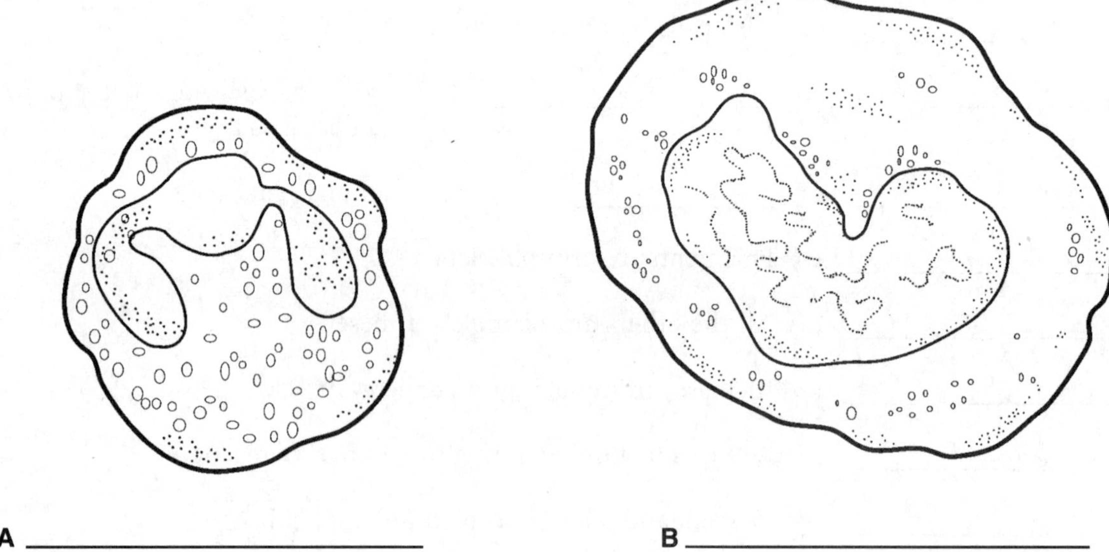

A ______________________ B ______________________

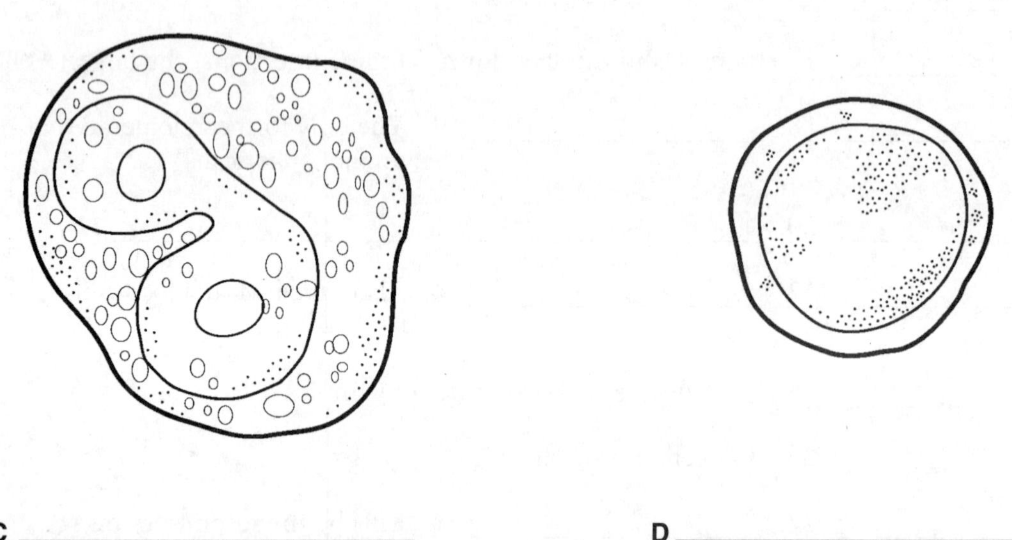

C ______________________ D ______________________

Figure 16.2

10. Circle the term that does not belong in each of the following groupings.

1. Erythrocytes Lymphocytes Monocytes Eosinophils
2. Neutrophils Monocytes Basophils Eosinophils
3. Histamine Heparin Basophil Antibodies
4. Hemoglobin Lymphocyte Oxygen transport Erythrocytes
5. Platelets Monocytes Phagocytosis Neutrophils
6. Thrombus Aneurysm Embolus Clot
7. Increased hemoglobin Decreased hemoglobin Red blood cell lysis Anemia
8. Plasma Nutrients Hemoglobin Wastes Clotting proteins
9. Myeloid stem cell Lymphocyte Monocyte Basophil Eosinophil

11. What are *colony-stimulating factors*? ____________________

____

Name two cells that are sources of CSFs: ____________________

____

12. Give the full names for the two cytokines abbreviated below:

1. G-CSF ____________________
2. IL-3 ____________________
3. These cytokines perform the same relative function in WBC production as __________ does in RBC production.

Hemostasis

1. Using the key choices, correctly complete the following brief description of the blood clotting process. Insert the answers in the answer blanks.

Key Choices

A. Break	D. Fibrinogen	G. Prothrombin	J. Thromboxane
B. Erythrocytes	E. Phosphatidylserine	H. Serotonin	
C. Fibrin	F. Platelets	I. Thrombin	

Clotting begins when a __(1)__ occurs in a blood vessel wall. Almost immediately, __(2)__ cling to the blood vessel wall and release __(3)__ and __(4)__, which help decrease blood loss by constricting the vessel. __(5)__ on the platelet surface promotes the pathway leading to formation of prothrombin activator, which causes __(6)__ to be converted to __(7)__. Once present, thrombin acts as an enzyme to attach __(8)__ molecules together to form long, threadlike strands of __(9)__, which then traps __(10)__ flowing by in the blood.

1. ____________________
2. ____________________
3. ____________________
4. ____________________
5. ____________________
6. ____________________
7. ____________________
8. ____________________
9. ____________________
10. ____________________

2. Several characteristics of the coagulation process are listed below. If the factor or effect is more typical of the intrinsic pathway, write I in the answer blank; if it is more typical of the extrinsic pathway, write E in the answer blank. If it applies equally to both pathways, write B in the answer blank.

_____ 1. Can occur outside the body in a test tube

_____ 2. Involves many more procoagulants

_____ 3. Entails the conversion of prothrombin to thrombin

_____ 4. Requires calcium ions

_____ 5. Occurs much more rapidly

_____ 6. A pivotal factor is phosphatidylserine (PF_3),a phospholipid associated with the external surfaces of aggregated platelets

_____ 7. Involves factors III and VII

_____ 8. Involves factors IX, XI, and XII

_____ 9. Tissue factor (TF) is exposed on injured cells extrinsic to the bloodstream

_____ 10. Clot retraction occurs

_____ 11. Prothrombin activator is a pivotal intermediate

3. Match the terms in Column B with the descriptions in Column A.

Column A

_____ 1. _____ 2. Two circumstances that prevent unnecessary enlargement of blood clots

_____ 3. Natural "clot-buster" chemical; causes fibrinolysis

_____ 4. Hereditary bleeder's disorder resulting from a lack of factor VIII

_____ 5. _____ 6. Two factors that prevent undesirable clotting in an unbroken blood vessel

_____ 7. Free-floating blood clot

_____ 8. _____ 9. Plasmin activators (2)

_____ 10. Clot formed in an unbroken blood vessel

_____ 11. Bleeder's disease that is a consequence of too few platelets

_____ 12. Hereditary disease resulting from a deficiency of factor IX

Column B

A. Activated clotting factors inhibited by heparin and antithrombin III

B. Activated factor XII

C. Coagulation factors washed away by flowing blood

D. Embolus

E. Hemophilia A

F. Hemophilia B

G. Intact endothelium

H. Plasmin

I. Endothelial cell secretions (prostacyclin and others)

J. Thrombin

K. Thrombocytopenia

L. Thrombus

Transfusion and Blood Replacement

1. Complete the following table concerning ABO blood groups.

Blood type	Agglutinogens or antigens	Agglutinins or antibodies in plasma	Can donate blood to type	Can receive blood from type
Type A	A			
Type B		Anti-A		
Type AB			AB	
Type O	None			

2. Ms. Pratt is claiming that Mr. X is the father of her child. Ms. Pratt's blood type is O negative. Her baby boy has type A positive blood. Mr. X's blood is typed and found to be type B positive. Could he be the father of her child? If not, what blood type would the father be expected to have?

____

3. When a person is given a transfusion of mismatched blood, a transfusion reaction occurs. Define the term *transfusion reaction*.

____

____

Diagnostic Blood Tests

1. Fill in the table of normal blood values; if appropriate, include units for both males and females. Also rank the leukocytes (listed under the heading "Differential WBC count") in order of their relative abundance (1 to 5, least to most) in the blood of a healthy person.

Characteristic	Normal value or normal range
% of body weight	
Blood volume	
Arterial pH	
Blood temperature	
RBC count	
Hematocrit	
Hemoglobin	
WBC count	
Differential WBC count	
neutrophils	
eosinophils	
basophils	
lymphocytes	
monocytes	
Platelet count	

2. A laboratory examination of blood yields valuable information that can be used to assess a patient's health status. Provide the names of diagnostic tests that fit the categories listed below.

Tests for certain nutrients

1. ______________________

2. ______________________

Tests for clotting

1. ______________________

2. ______________________

Tests for mineral levels	Tests for anemia
1. ____________________	1. ____________________
2. ____________________	2. ____________________
3. ____________________	3. ____________________

THE INCREDIBLE JOURNEY:

A Visualization Exercise for the Blood

Once inside, you quickly make a slash in the vessel lining . . .

1. Complete the following narrative by inserting the missing words in the answer blanks.

________________ 1.
________________ 2.
________________ 3.
________________ 4.
________________ 5.
________________ 6.
________________ 7.
________________ 8.
________________ 9.
________________ 10.
________________ 11.

For this journey, you will be injected into the external iliac artery and will be guided by a fluorescent monitor into the bone marrow of the iliac bone. You will observe and report events of blood cell formation, also called __(1)__, seen there and then move out of the bone into the circulation to initiate and observe the process of blood clotting, also called __(2)__. Once in the bone marrow, you observe several large dark-nucleated stem cells, or __(3)__, as they begin to divide and produce daughter cells. The daughter cells eventually formed have tiny cytoplasmic granules and very peculiarly shaped nuclei that look like small masses of nuclear material connected by thin strands of nucleoplasm. You note that you have just witnessed the formation of a type of white blood cell, called the __(4)__. You describe its appearance and make a mental note to try to observe its activity later. Meanwhile you can tentatively report that this cell type functions as a __(5)__ to protect the body.

At another site, daughter cells arising from the division of a stem cell are difficult to identify initially. As you continue to observe the cells, you see that they, in turn, divide. Eventually some of their daughter cells eject their nuclei and flatten out to assume a disk shape. You conclude that the kidneys must have released the hormone __(6)__ because those cells are __(7)__. That dark material filling their interior must be __(8)__ because those cells function to transport __(9)__ in the blood.

Now you turn your attention to the daughter cells being formed by the division of another stem cell. They are small round cells with relatively large round nuclei. In fact, their cytoplasm is very sparse. You record your observation of the formation of __(10)__. They do not remain in the marrow very long after formation but seem to enter the circulation almost as soon as they are produced. Some of those cells will produce __(11)__ that are released to the blood; others will act in other ways in the immune response. At this point, although you have yet to see the formation

______________ 12.

______________ 13.

______________ 14.

______________ 15.

______________ 16.

______________ 17.

______________ 18.

______________ 19.

______________ 20.

______________ 21.

______________ 22.

______________ 23.

______________ 24.

______________ 25.

of __(12)__, __(13)__, __(14)__, or __(15)__, you decide to proceed into the circulation to observe blood clotting.

You maneuver yourself into a small venule to enter the general circulation. Once inside, you quickly make a slash in the vessel lining, or __(16)__. Almost immediately what appear to be hundreds of jagged cell fragments swoop into the area and plaster themselves over the freshly made incision. You record that __(17)__ have just adhered to the damaged site. As you are writing, your chemical monitor flashes the message, "Vasoconstrictor substance released." You record that __(18)__ has been released, based on your observation that the vessel wall seems to be closing in. Peering out at the damaged site, you see that long ropelike strands are being formed at a rapid rate and are clinging to the site. You report that the __(19)__ mesh is forming and is beginning to trap RBCs to form the basis of the __(20)__. Even though you do not have the equipment to monitor the intermediate steps of this process, you know that the platelets must have also acted as activation sites for the formation of __(21)__, which then catalyzes the conversion of __(22)__ to __(23)__. This second enzyme then joined the soluble __(24)__ molecules together to form the network of strands you can see.

You carefully back away from the newly formed clot. You do *not* want to disturb the area because you realize that if the clot detaches, it might become a life-threatening __(25)__. Your mission here is completed, and you return to the entrance site.

CHALLENGING YOURSELF

At the Clinic

1. A patient on dialysis has a low RBC count. What hormone, secreted by the kidney, can be assumed to be deficient?
2. A bone marrow biopsy of Mr. Bongalonga, a man on a long-term drug therapy, shows an abnormally high percentage of nonhemopoietic connective tissue. What condition does this indicate? If the symptoms are critical, what short-term and long-term treatments may be necessary? Will infusion of whole blood or packed red cells be more likely?
3. Mrs. Francis comes to the clinic complaining of fatigue, shortness of breath, and chills. Blood tests show anemia and a bleeding ulcer is diagnosed. What type of anemia is this?
4. A patient is diagnosed with bone marrow cancer and has a hematocrit of 70%. What is this condition called?
5. The very concerned parents of a young boy named Teddy bring him to the clinic because he has chronic fatigue, bleeding, fever, and weight loss. A blood test reveals abnormally high numbers of lymphoblasts, anemia, and thrombocytopenia. What is his diagnosis?
6. What vitamin must be supplemented in cases of steatorrhea (inability to absorb fat) to avoid clotting problems?

7. After multiple transfusions, a weak transfusion reaction occurs. Is this likely to be caused by the ABO antigens or other antigens? What should be done to avoid this type of reaction?

8. Mrs. Carlyle is pregnant for the first time. Her blood type is Rh negative, her husband is Rh positive, and their first child has been determined to be Rh positive. Ordinarily, the first such pregnancy causes no major problems, but baby Carlyle is born blue and cyanotic.

    1. What is this condition, a result of Rh incompatibility, called?

    2. Why is the baby cyanotic?

    3. Because this is Mrs. Carlyle's first pregnancy, how can you account for the baby's problem?

    4. Assume that baby Carlyle was born pink and healthy. What measures should be taken to prevent the previously described situation from happening in the second pregnancy with an Rh positive baby?

    5. Mrs. Carlyle's sister has had two miscarriages before seeking medical help with her third pregnancy. Blood typing shows that she, like her sister, is Rh negative; her husband is Rh positive. What course of treatment will be followed?

9. A red marrow biopsy is ordered for two patients—one a child and the other an adult. The specimen is taken from the tibia of the child but from the iliac crest of the adult. Explain why different sites are used to obtain marrow samples in adults and children.

10. The Jones family let their dog Rooter lick their faces and kissed it on the mouth, not realizing that it had just explored the neighborhood dump and trash cans. Later the same day the veterinarian diagnosed Rooter as having pinworms. Three weeks later, blood tests ordered for routine physicals for camp indicated that both the family's daughters had blood eosinophil levels of over 3000 per cubic millimeter. Might there be a connection between these events?

11. An elderly man has been receiving weekly injections of vitamin B_{12} ever since nearly all of his stomach was removed 6 months previously (he had stomach cancer). Why is he receiving the vitamin injections? Why can't the vitamin be delivered in tablet form? What would be the result if he refuses the B_{12} injections?

12. List two blood tests that might be ordered if infectious mononucleosis is suspected and three that might be done if a person has bleeding problems.

13. Jenny, a young woman of 38, has had several episodes of excessive bleeding. Her physician orders tests to determine whether her problem is due to vitamin K deficiency or thrombocytopenia. What test results will be important in making this diagnosis?

14. A man of Greek ancestry goes to his doctor with the following symptoms. He is tired all the time and has difficulty catching his breath after even mild exercise. His doctor orders the following tests: complete blood count, hematocrit, differential WBC count. The tests show small, pale, immature erythrocytes, fragile erythrocytes, and around 2 million erythrocytes per cubic millimeter. What is the tentative diagnosis?

Stop and Think

1. Why is someone more likely to bleed to death when an artery is cleanly severed than when an artery is crushed and torn?

2. Knowing the pathways of hemopoiesis, explain why leukemia reduces RBC and platelet counts.

3. Dissect these words: polycythemia, diapedesis.

4. What are the functional advantages of the "shortcut" of the extrinsic pathway *and* of the multiple steps of the intrinsic pathway?

5. Sodium EDTA removes calcium ions from solution. What would be the effect of adding sodium EDTA to blood?

6. What is the functional importance of the fact that hemoglobin F has a higher oxygen affinity than hemoglobin A?

7. Explain the phenomenon called "athlete's anemia."

8. Relative to human blood groups, what are "private antigens"?

9. John, a novice to cigarette smoking, is trying to impress his buddies by smoking two packs a day—inhaling every drag. What do you think will happen to his reticulocyte count? Explain your reasoning.

COVERING ALL YOUR BASES

Multiple Choice

Select the best answer or answers from the choices given.

1. Which of the following are true concerning erythrocytes?
   A. They rely strictly on anaerobic respiration.
   B. About one-third of their volume is hemoglobin.
   C. Their precursor is called a megakaryoblast.
   D. Their shape increases membrane surface area.

2. A serious bacterial infection leads to more of these cells in the blood.
   A. Erythrocytes and platelets
   B. Mature and band neutrophils
   C. Erythrocytes andmonocytes
   D. All formed elements

3. Each hemoglobin molecule:
   A. contains either alpha or beta polypeptides
   B. can reversibly bind up to four oxygen molecules
   C. has one heme group
   D. has four iron atoms

4. If hemoglobin were carried loose in the plasma, it would:
   A. increase viscosity of blood plasma
   B. decrease blood osmotic pressure
   C. be unable to bind oxygen
   D. leak out of the bloodstream quite easily

5. Which of the following is (are) true of hemopoiesis?
   A. All blood-formed elements are derived from the hematopoietic stem cell.
   B. Hemopoiesis occurs in myeloid tissue.
   C. The processes of hemopoiesis include production of plasma proteins by the liver.
   D. Irradiation of the distal limbs of an adult probably will not affect hemopoiesis.

6. Which cell stage directly precedes the reticulocyte stage in erythropoiesis?
   A. Basophilic erythroblast
   B. Proerythroblast
   C. Late erythroblast
   D. Normoblast

7. Which would lead to increased erythropoiesis?
   A. Chronic bleeding ulcer
   B. Reduction in respiratory ventilation
   C. Decreased level of physical activity
   D. Reduced blood flow to the kidneys

8. Deficiency of which of the following will have a direct negative impact on RBC production?
   A. Transferrin
   B. Iron
   C. Folic acid
   D. Calcium

9. One possible test of liver function would measure the level in the bloodstream of:
   A. erythropoietin
   B. vitamin K
   C. bilirubin
   D. intrinsic factor

10. A possible cause of aplastic anemia is:
A. deficient production of intrinsic factor
B. ionizing radiation
C. vitamin B_{12} deficiency
D. thalassemia

11. A child is diagnosed with sickle-cell anemia. This means that:
A. one parent had sickle-cell anemia
B. one parent carried the sickle-cell gene
C. both parents had sickle-cell anemia
D. both parents carried the sickle-cell gene

12. Polycythemia vera will result in:
A. overproduction of WBCs
B. exceptionally high blood volume
C. abnormally high blood viscosity
D. abnormally low hematocrit

13. Which of the following does not characterize leukocytes?
A. Ameboid
B. Phagocytic (some)
C. Nucleated
D. Cells found in largest numbers in the bloodstream

14. The blood cell that can attack a specific antigen is a(n):
A. monocyte C. lymphocyte
B. neutrophil D. eosinophil

15. The first type of WBC whose development diverges from the common leukopoietic line is the:
A. monocyte C. granulocyte
B. lymphocyte D. basophil

16. The WBC type with the longest (possible) life span is the:
A. neutrophil C. lymphocyte
B. monocyte D. eosinophil

17. In leukemia:
A. the cancerous WBCs function normally
B. the cancerous WBCs fail to specialize
C. production of RBCs and platelets is decreased
D. infection and bleeding can be life threatening

18. Platelet formation:
A. involves mitosis without cytokinesis
B. includes specialization from the myeloid stem cell
C. requires stimulation by thrombopoietin
D. results in small, nucleated thrombocytes

19. A deficiency of albumin would result in:
A. increased blood volume
B. increased blood osmotic pressure
C. loss of water by osmosis from the bloodstream
D. a pH imbalance

20. All leukocytes share the following features except:
A. diapedesis
B. disease-fighting
C. distorted, lobed nuclei
D. more active in connective tissues than in blood

21. Which platelet factors attract more platelets?
A. Serotonin C. ADP
B. Thromboxane A_2 D. Prostacyclin

22. After activation of factor X, the next step in the coagulation sequence is:
A. activation of factor XI
B. activation of factor IX
C. formation of prothrombin activator
D. formation of thrombin

23. Fibrinolysis is increased by:
A. activation of thrombin
B. activation of plasminogen
C. t-PA
D. heparin

24. A condition resulting from thrombocytopenia is:
A. thrombus formation
B. embolus formation
C. petechiae
D. hemophilia

25. Which of the following can cause problems in a transfusion reaction?
A. Donor antibodies attacking recipient RBCs
B. Clogging of small vessels by agglutinated clumps of RBCs
C. Lysis of donated RBCs
D. Blockage of kidney tubules

26. The erythrocyte count increases when an individual moves from a low to a high altitude because:
A. the concentration of oxygen and/or total atmospheric pressure is lower at high altitudes
B. the basal metabolic rate is higher at high altitudes
C. the concentration of oxygen and/or total atmospheric pressure is higher at high altitudes
D. the temperature is lower at high altitudes

27. A hematocrit value that would be normal for females but not males is:

A. 70% C. 45%
B. 39% D. 51%

28. Sickling of RBCs can be induced in those with sickle-cell anemia by:

A. blood loss C. stress
B. vigorous exercise D. fever

29. If an Rh^- mother becomes pregnant, when can erythroblastosis fetalis *not possibly* occur in the child?

A. If the child is Rh^-
B. If the child is Rh^+
C. If the father is Rh^+
D. If the father is Rh^-

30. What is the difference between a thrombus and an embolus?

A. One occurs in the bloodstream, whereas the other occurs outside the bloodstream.
B. One occurs in arteries, the other in veins.
C. One is a blood clot, while the other is a parasitic worm.
D. A thrombus must travel to become an embolus.

31. The plasma component that forms the fibrous skeleton of a clot is:

A. platelets C. thromboplastin
B. fibrinogen D. thrombin

32. The leukocyte that releases histamine and other inflammatory chemicals is the:

A. basophil C. monocyte
B. eosinophil D. neutrophil

33. Which of the following formed elements are phagocytic?

A. Erythrocytes C. Monocytes
B. Neutrophils D. Lymphocytes

34. A pathological condition of widespread clotting in intact vessels is:

A. DIC C. thrombocytopenia
B. septicemia D. anemia

Word Dissection

For each of the following word roots, fill in the literal meaning and give an example, using a word found in this chapter.

Word root	Translation	Example
1. agglutin		
2. album		
3. bili		
4. embol		
5. emia		
6. erythro		
7. ferr		
8. hem		
9. karyo		
10. leuko		
11. lymph		
12. phil		
13. poiesis		
14. rhage		
15. thromb		

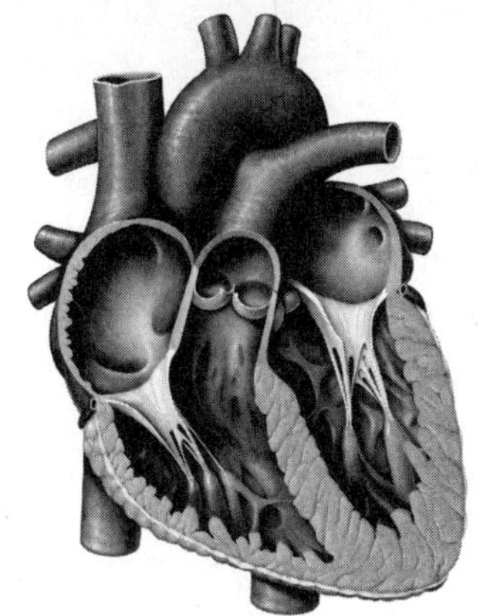

17 The Cardiovascular System: The Heart

Student Objectives

When you have completed the exercises in this chapter, you will have accomplished the following objectives:

Heart Anatomy

1. Describe the size, shape, location, and orientation of the heart in the thorax.
2. Name the coverings of the heart.
3. Describe the structure and function of each of the three layers of the heart wall.
4. Describe the structure and functions of the four heart chambers. Name each chamber and provide the name and general route of its associated great vessel(s).
5. Trace the pathway of blood through the heart.
6. Name the major branches and describe the distribution of the coronary arteries.
7. Name the heart valves and describe their location, function, and mechanism of operation.

Cardiac Muscle Fibers

8. Describe the structural and functional properties of cardiac muscle, and explain how it differs from skeletal muscle.
9. Briefly describe the events of cardiac muscle cell contraction.

Heart Physiology

10. Name the components of the conduction system of the heart, and trace the conduction pathway.
11. Draw a diagram of a normal electrocardiogram tracing. Name the individual waves and intervals, and indicate what each represents.
12. Name some of the abnormalities that can be detected on an ECG tracing.
13. Describe normal heart sounds, and explain how heart murmurs differ.
14. Describe the timing and events of the cardiac cycle.
15. Name and explain the effects of various factors regulating stroke volume and heart rate.
16. Explain the role of the autonomic nervous system in regulating cardiac output.

As part of the cardiovascular system, the heart has a single function—to pump the blood into the blood vessels so that it reaches the trillions of tissue cells of the body. Survival of tissue cells depends on constant access to oxygen and nutrients and removal of carbon dioxide and wastes. This critical homeostatic requirement is met by a continuous flow of blood. The heart is designed to maintain this continuous flow by pumping blood simultaneously through two circuits: the pulmonary circulation to the lungs and the systemic circulation to all body regions.

Topics for study in Chapter 17 include the microscopic and gross anatomy of the heart, the related events of the cardiac cycle, the regulation of cardiac output, malfunctions of the heart, and changes in the heart throughout life.

BUILDING THE FRAMEWORK

Heart Anatomy

1. Complete the following statements by writing the missing terms in the answer blanks.

The heart is a cone-shaped muscular organ located within the __(1)__ of the thorax. Its apex rests on the __(2)__ and its superior margin lies at the level of the __(3)__ rib. Approximately two-thirds of the heart mass is seen to the left of the __(4)__.

The heart is enclosed in a serosal sac called the pericardium. The loosely fitting double outer layer consists of the outermost __(5)__ pericardium, lined by the parietal layer of the serous pericardium. The inner __(6)__ pericardium, also called the __(7)__, is the outermost layer of the heart wall. The function of the fluid that fills the pericardial sac is to decrease __(8)__ during heart activity. The middle layer of the heart wall, called the __(9)__, is composed of __(10)__; it forms the bulk of the heart.

Connective tissue fibers that ramify throughout this layer construct the so-called __(11)__ of the heart. The membrane that lines the heart and also forms the valve flaps is the __(12)__. This layer is continuous with the __(13)__ linings of the blood vessels that enter and leave the heart. The heart has __(14)__ chambers. Relative to the roles of these chambers, the __(15)__ are the receiving chambers, whereas the __(16)__ are the discharging chambers. Teeth-shaped bundles of smooth muscle, found only in the auricles, are called __(17)__, while irregular ridges of muscle called __(18)__ are found in the ventricles.

______ 1.

______ 2.

______ 3.

______ 4.

______ 5.

______ 6.

______ 7.

______ 8.

______ 9.

______ 10.

______ 11.

______ 12.

______ 13.

______ 14.

______ 15.

______ 16.

______ 17.

______ 18.

2. Figure 17.1A is a transverse section through the thorax. Label the structures that have leader lines. Figure 17.1B is a highly schematic longitudinal section through the heart wall and pericardium. Select colors for each structure listed below; color the corresponding coding circles and the structure on the figure.

○ Fibrous pericardium ○ Parietal layer of serous pericardium ○ Endocardium

○ Myocardium ○ Visceral layer of serous pericardium (epicardium) ○ Diaphragm

○ Pericardial cavity

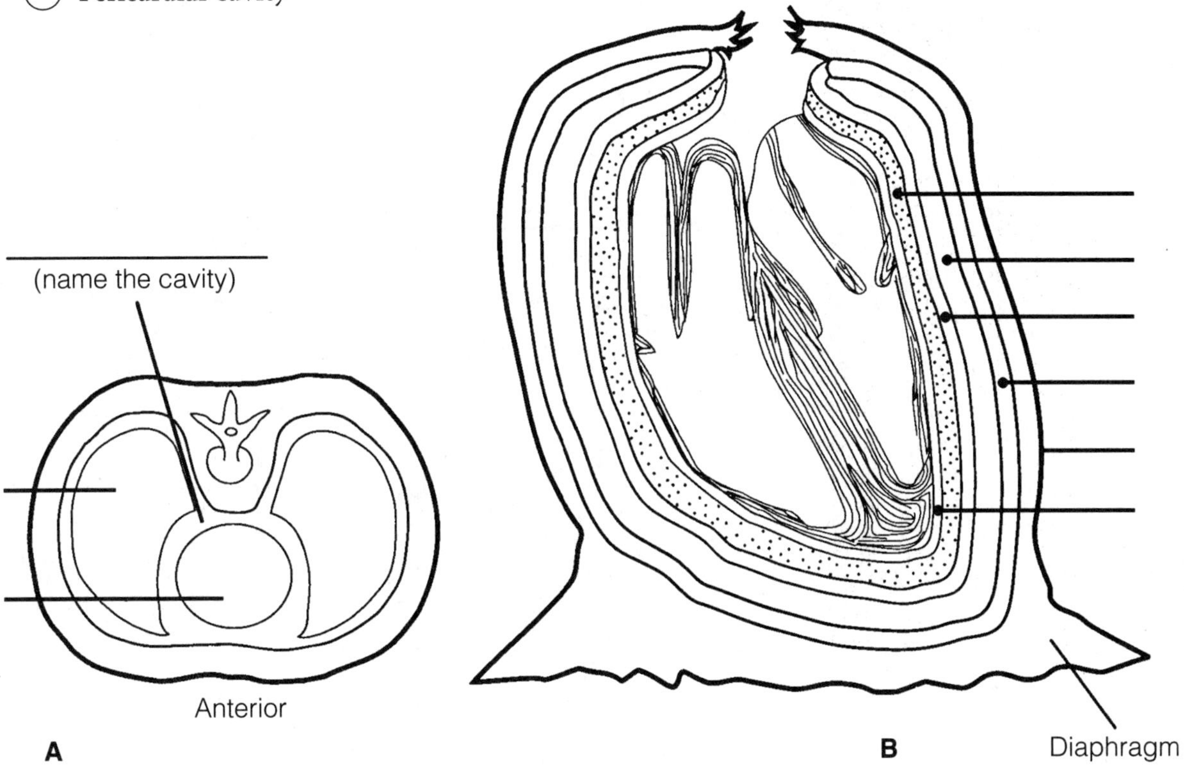

(name the cavity)

Anterior

A

B

Diaphragm

Figure 17.1

3. The heart is called a double pump because it serves two circulations. Trace the flow of blood through both pulmonary and systemic circulations by writing the missing terms in the answer blanks below. Then, identify the various regions of the circulation shown in Figure 17.2 by labeling them using the key choices. Color regions transporting O_2-poor blood blue and regions transporting O_2-rich blood red.

From the atrium through the tricuspid valve to the (1), through the (2) to the pulmonary trunk to the right and left (3), to the capillary beds of the (4), to the (5), to the (6) of the heart through the (7) valve, to the (8) through the (9) valve to the (10), to the systemic arteries, to the (11) of the body tissues, to the systemic veins, to the (12) and (13), which enter the right atrium of the heart.

1. ____________________
2. ____________________
3. ____________________
4. ____________________
5. ____________________
6. ____________________
7. ____________________
8. ____________________
9. ____________________
10. ____________________
11. ____________________
12. ____________________
13. ____________________

Key Choices

A. Vessels serving head and upper limbs

B. Vessels serving body trunk and lower limbs

C. Vessels serving the viscera

D. Pulmonary circulation

E. Pulmonary "pump"

F. Systemic "pump"

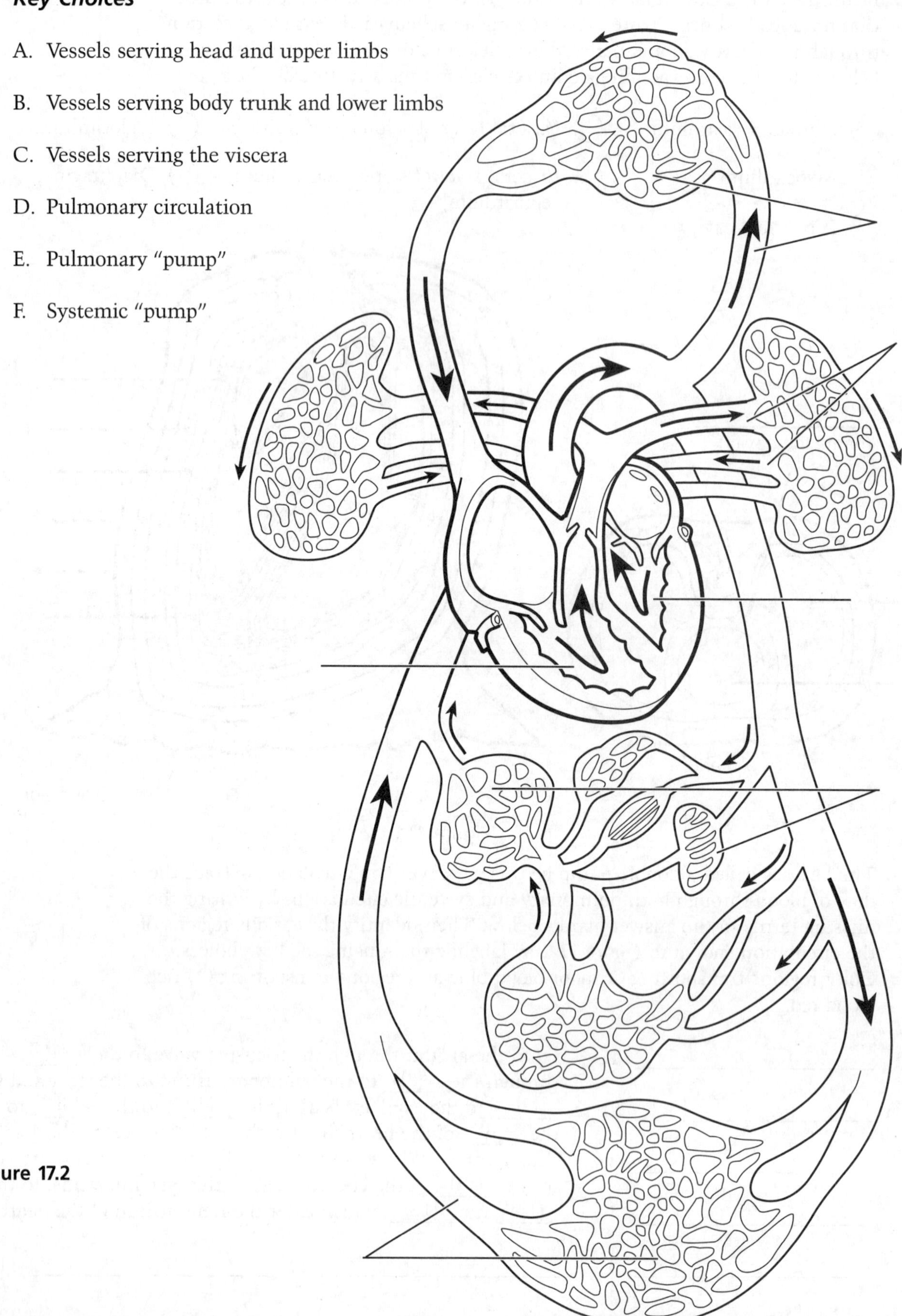

Figure 17.2

4. Figure 17.3 is an anterior view of the heart. Identify each numbered structure and write its name in the corresponding numbered answer blank. Then, select different colors for each structure with a coding circle and color the structures on the figure.

○	_______________ 1.	○	_______________ 6.	○	_______________ 11.
○	_______________ 2.	○	_______________ 7.		_______________ 12.
○	_______________ 3.	○	_______________ 8.		_______________ 13.
○	_______________ 4.		_______________ 9.		_______________ 14.
○	_______________ 5.		_______________ 10.	○	_______________ 15.

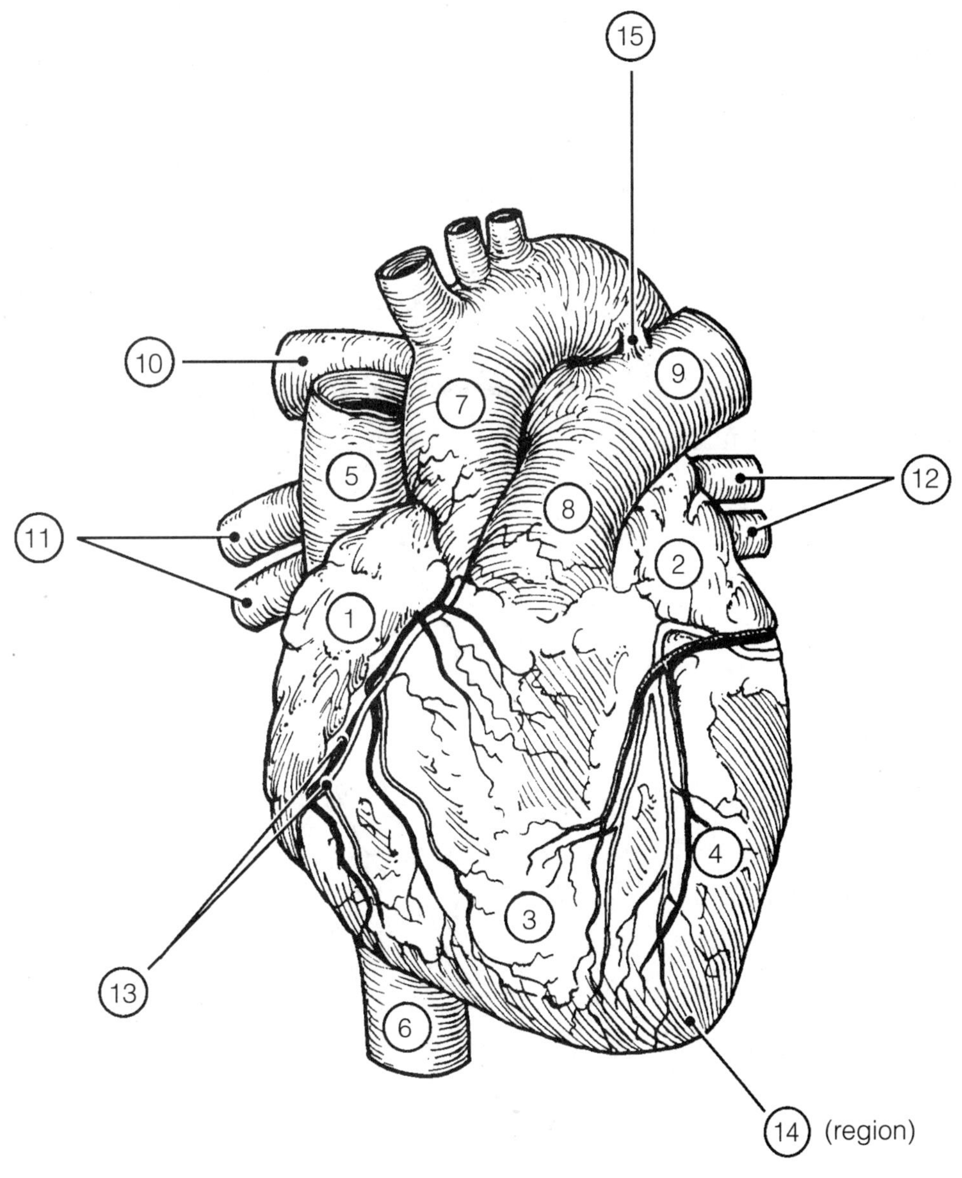

Figure 17.3

5. Figure 17.4 shows the vascular supply of the myocardium. Part A shows the arterial supply; part B shows the venous drainage. In each case, the anterior view is shown and the vessels located posteriorly are depicted as dashed lines. Use a light color to color in the *right atrium* and the *left ventricle* in each diagram. (Leave the left atrium and right ventricle uncolored.) Color the aorta red and the pulmonary trunk and arteries blue. Then color code and color the vessels listed below.

Part A

- ◯ Anterior interventricular artery
- ◯ Circumflex artery
- ◯ Left coronary artery
- ◯ Marginal artery
- ◯ Posterior interventricular artery
- ◯ Right coronary artery

Part B

- ◯ Anterior cardiac vein
- ◯ Coronary sinus
- ◯ Great cardiac vein
- ◯ Middle cardiac vein
- ◯ Small cardiac vein

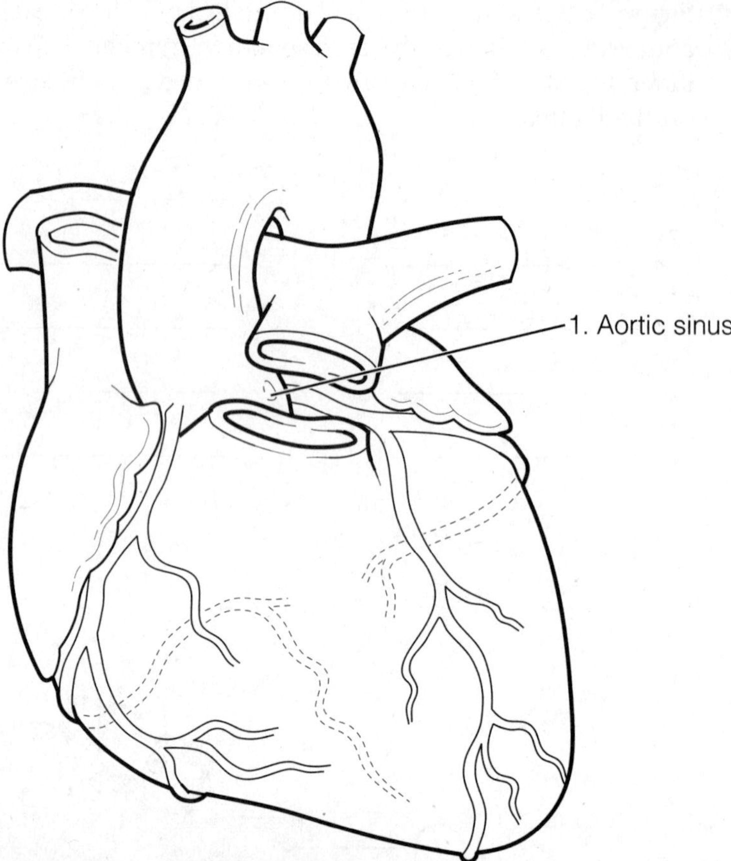

A. Anterior View: Arterial Supply

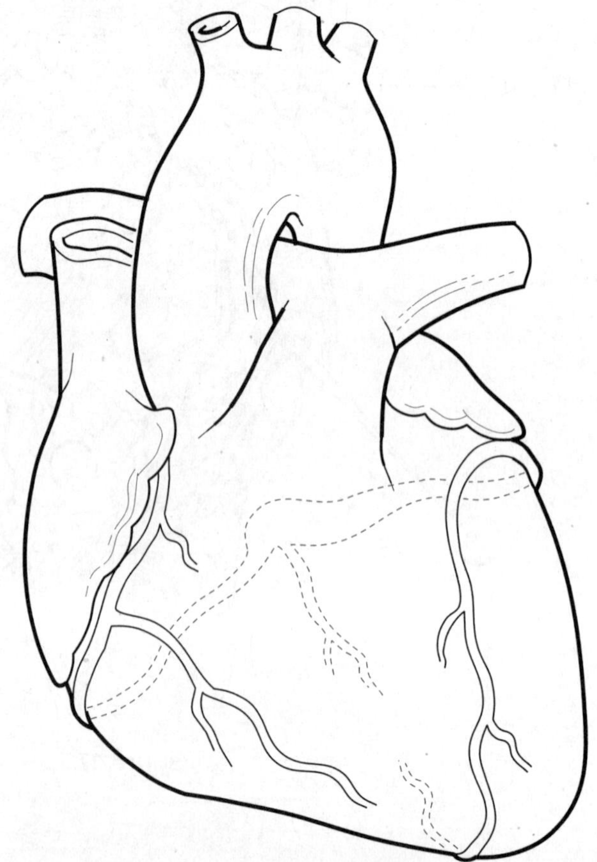

B. Anterior View: Venous Drainage

Figure 17.4

Cardiac Muscle Fibers

1. A number of characteristics peculiar to skeletal or cardiac muscle are listed below. Select which ones refer to each muscle type by writing S in the answer blanks preceding descriptions that apply to skeletal muscle and C in the answer blanks of those that identify characteristics of cardiac muscle.

_____ 1. One centrally located nucleus

_____ 2. More mitochondria per cell

_____ 3. T tubules line up at the Z lines

_____ 4. Uses fatty acids more effectively for ATP harvest

_____ 5. Requires stimulation by the nervous system to contract

_____ 6. Has a shorter refractory period

_____ 7. Contains self-excitable cells

_____ 8. Has well-developed terminal cisternae

_____ 9. Wide T tubules

_____ 10. More distinct myofibrils

_____ 11. All-or-none law applies at the organ level

_____ 12. Intercalated discs enhance intercellular electrical communication

2. Figure 17.5 is a schematic drawing of the microscopic structure of cardiac muscle. Using different colors, color the coding circles of the structures listed below and the corresponding structures on the figure. Then answer the questions that follow the figure. Write your answers in the answer blanks.

○ Nuclei (with nucleoli)

○ Muscle fibers

○ Intercalated discs

○ Striations

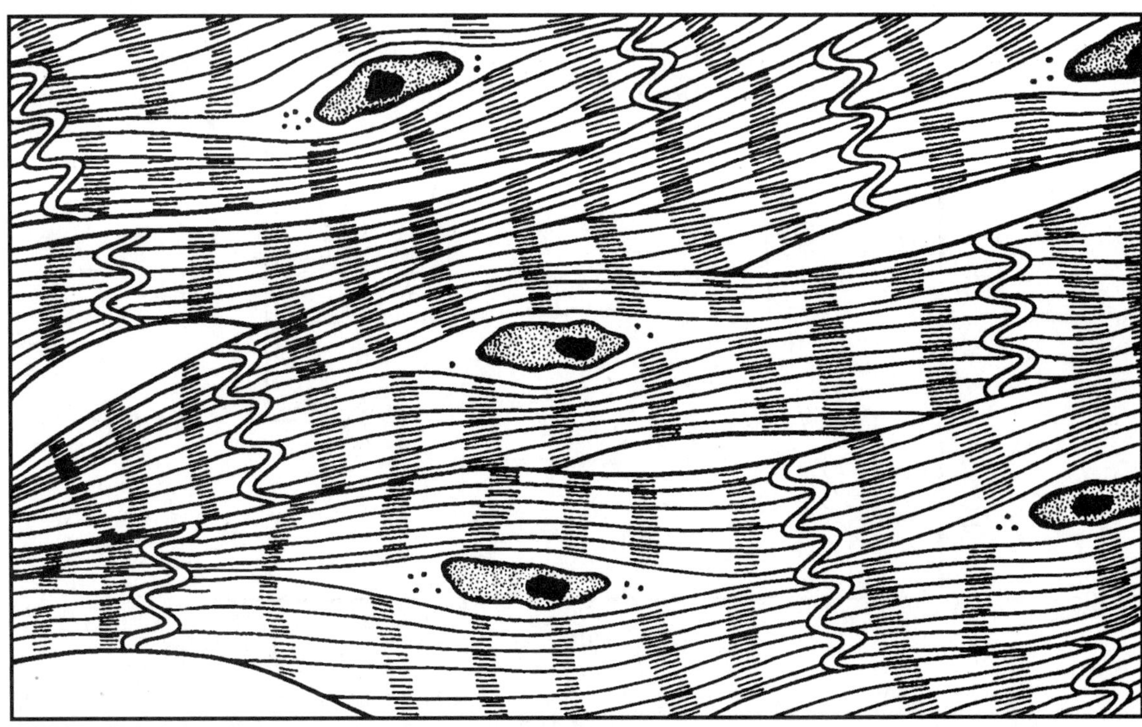

Figure 17.5

______________ 1. Name the loose connective tissue that fills the intercellular spaces.

______________ 2. What is the function during contraction of the desmosomes present in the intercalated discs? (*Note:* The desmosomes are not illustrated.)

______________ 3. What is the function of the gap junctions (not illustrated) also present in the intercalated discs?

______________ 4. What term describes the interdependent, interconnecting cardiac cells?

______________ 5. Which structures provide electrical coupling of cardiac cells?

3. On Figure 17.6, indicate (by adding labels) the following changes in membrane permeability and events of an action potential at their approximate point of occurrence. Also indicate by labeling a bracket: (1) the plateau and (2) the period when Ca^{2+} is pumped out of the cell. Also color code and color the arrows and brackets.

- ○ ↑ Na^+ (fast Na^+ channels open)
- ○ ↓ Na^+ (Na^+ channels close)
- ○ ↑ Ca^{2+} (slow Ca^{2+} channels open)
- ○ Ca^{2+} pumped from cell
- ○ ↓ Ca^{2+} (Ca^{2+} channels close)
- ○ ↑ K^+ (K^+ channels open)
- ○ Plateau

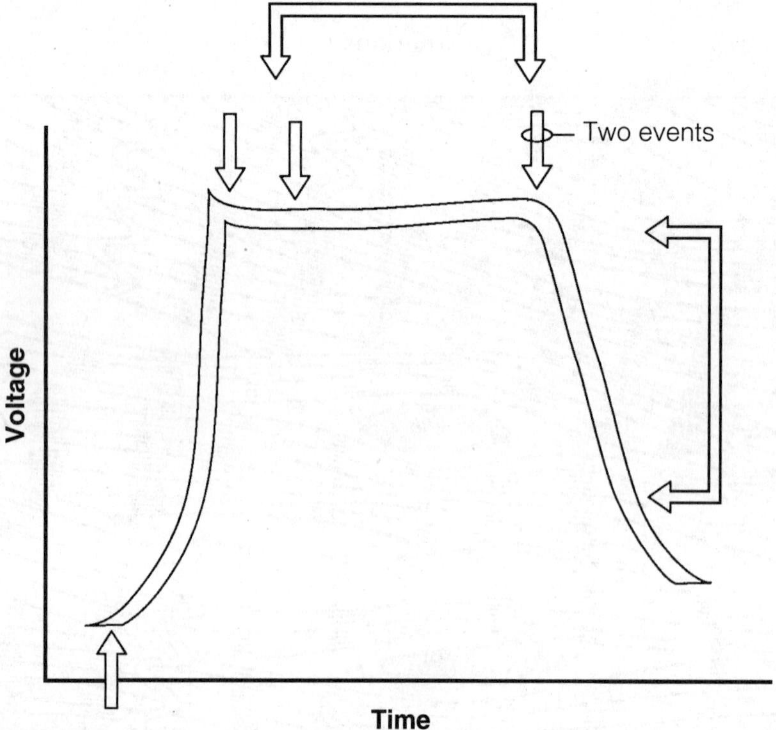

Figure 17.6

Heart Physiology

1. Complete the following statements concerning the cells of the nodal system of the heart. Write the missing terms in the answer blanks.

____________________ 1.
____________________ 2.
____________________ 3.
____________________ 4.
____________________ 5.
____________________ 6.
____________________ 7.
____________________ 8.

The nodal cells of the heart, unlike cardiac contractile muscle fibers, have an intrinsic ability to depolarize __(1)__. This reflects their unstable __(2)__, which drifts slowly toward the threshold for firing, that is, __(3)__. These spontaneously changing membrane potentials, called __(4)__, are probably due to reduced membrane permeability to __(5)__, while __(6)__ continues to diffuse __(7)__ the cell at a slow rate. Ultimately, when threshold is reached, gated __(8)__ channels open, allowing that ion to rush into the cells and reverse the membrane potential.

2. Figure 17.7 is a diagram of the frontal section of the heart. Follow the instructions below to complete this exercise, which considers both anatomical and physiological aspects of the heart.

1. Draw arrows to indicate the direction of blood flow through the heart. Draw the pathway of the oxygen-rich blood with red arrows and trace the pathway of oxygen-poor blood with blue arrows.

2. Identify each of the elements of the intrinsic conduction system (numbers 1–5 on the figure) by writing the appropriate terms in the numbered answer blanks. Then, indicate with green arrows the pathway that impulses take through this system.

3. Identify each of the heart valves (numbers 6–9 on the figure) by writing the appropriate terms in the numbered answer blanks. Draw and identify by name the cordlike structures that anchor the flaps of the atrioventricular (AV) valves.

4. Use the numbers from the figure to identify structures (A–H).

_____ A. _____ B. Prevent backflow into the ventricles when the heart is relaxed

_____ C. _____ D. Prevent backflow into the atria when the ventricles are contracting

_____ E. AV valve with three flaps

_____ F. AV valve with two flaps

_____ G. The pacemaker of the Purkinje system

_____ H. The point in the Purkinje system where the impulse is temporarily delayed

1. ____________
2. ____________
3. ____________
4. ____________
5. ____________
6. ____________
7. ____________
8. ____________
9. ____________

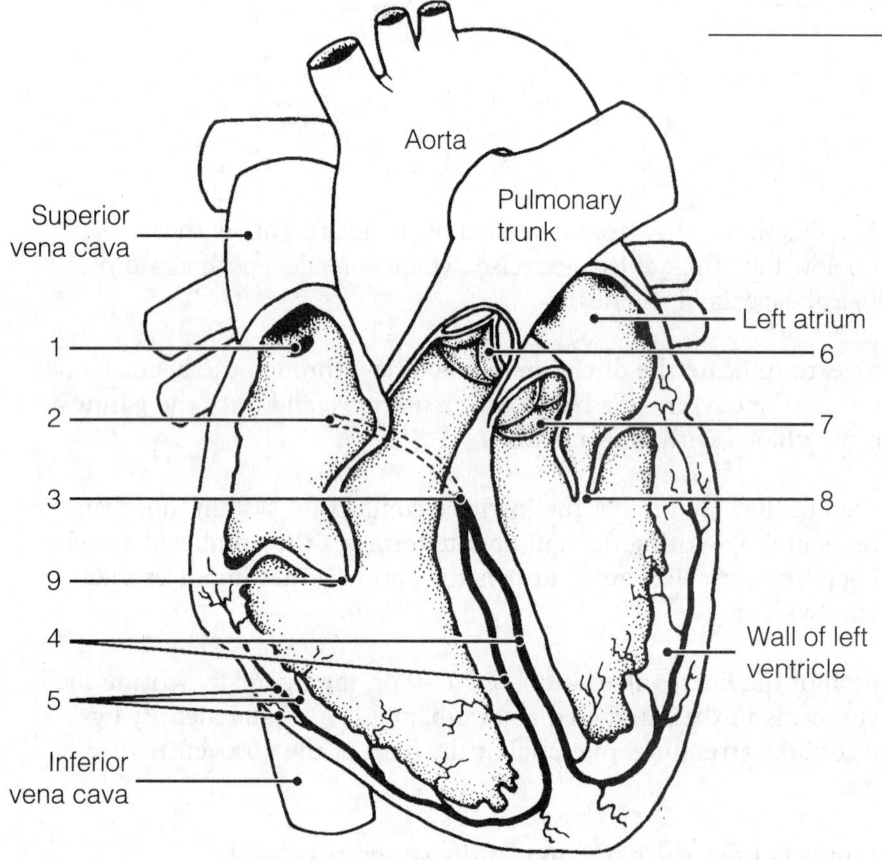

Figure 17.7

3. Respond to the questions below concerning the nodal system.

1. What name is given to the rate set by the heart's pacemaker? ________________________

2. What are the observed contraction rates of the different components of the intrinsic conduction system?

   SA node ________ beats/min AV node ________ beats/min

   AV bundle ________ beats/min Purkinje fibers ________ beats/min

3. The intrinsic conduction system enforces a faster rate of impulse conduction across the heart—at the rate of several meters per second in most parts of the conduction system. What would be the natural speed of impulse transmission across the heart in the absence of such a system? ________ m/s

4. What is the total time for impulse conduction across the healthy heart, on average? ________ seconds

4. Part of an electrocardiogram is shown in Figure 17.8. On the figure, identify the QRS complex, the P wave, and the T wave. Using a green pencil, bracket the P-Q interval and the Q-T interval. Then, using a red pencil, bracket a portion of the recording equivalent to the length of one cardiac cycle. Using a blue pencil, bracket a portion of the recording in which the *ventricles* would be in diastole.

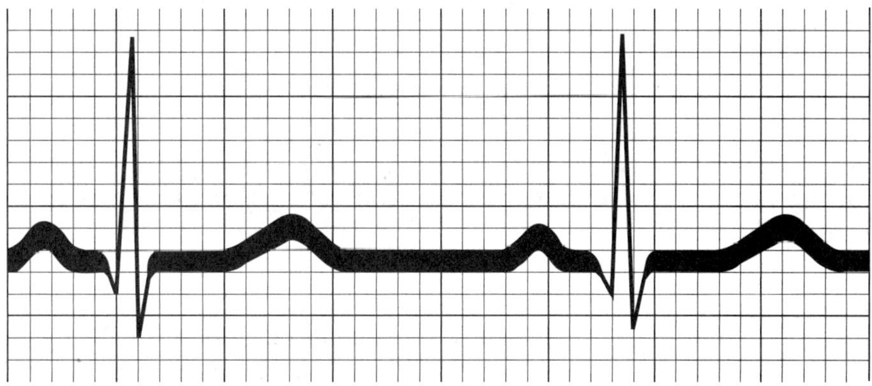

Figure 17.8

5. Examine the abnormal EGG tracings shown in Figure 17.9.

________________________ 1. Which shows extra P waves?

________________________ 2. Which shows tachycardia?

________________________ 3. Which has an abnormal QRS complex?

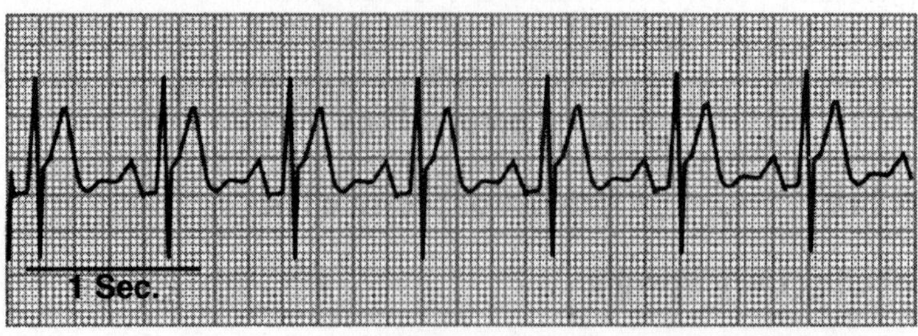

A

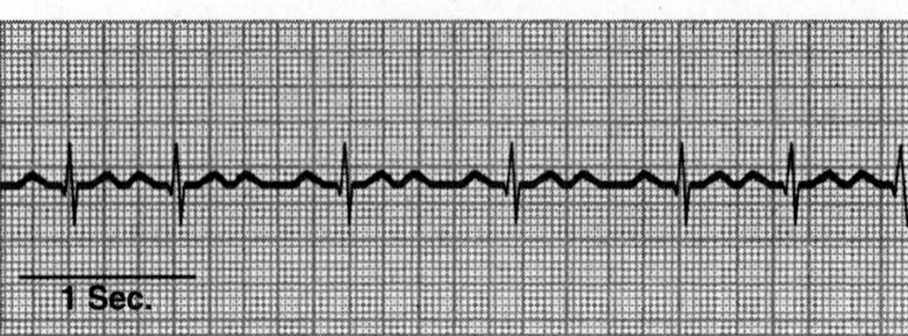

B

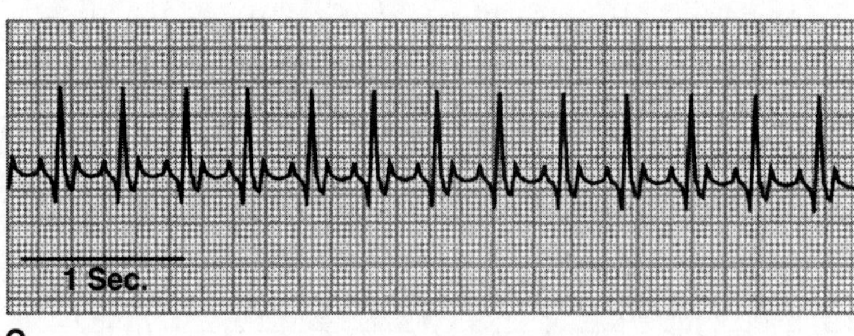

C

Figure 17.9

6. The events of one complete heartbeat are referred to as the cardiac cycle. Complete the following statements that describe these events by writing the missing terms in the answer blanks.

The contraction of the ventricles is referred to as ___(1)___ and the period of ventricular relaxation is called ___(2)___. The monosyllables describing heart sounds during the cardiac cycle are ___(3)___. The first heart sound is a result of closure of the ___(4)___ valves; closure of the ___(5)___ valves causes the second heart sound. The heart chambers that have just been filled when you hear the first heart sound are the ___(6)___ and the chambers that have just emptied are the ___(7)___. Immediately after the second heart sound, the ___(8)___ are filling with blood and the ___(9)___ are empty. Abnormal heart sounds, or ___(10)___, usually indicate valve problems.

1. ____________________
2. ____________________
3. ____________________
4. ____________________
5. ____________________
6. ____________________
7. ____________________
8. ____________________
9. ____________________
10. ____________________

7. The events of one cardiac cycle are graphed in Figure 17.10. First, identify the following by color:

◯ ECG tracing ◯ Atrial pressure line

◯ Aortic pressure line ◯ Ventricular pressure line

Then, identify by labeling the following:

- The P, QRS, and T waves of the ECG
- Points of opening and closing of the AV and semilunar valves
- Elastic recoil of the aorta

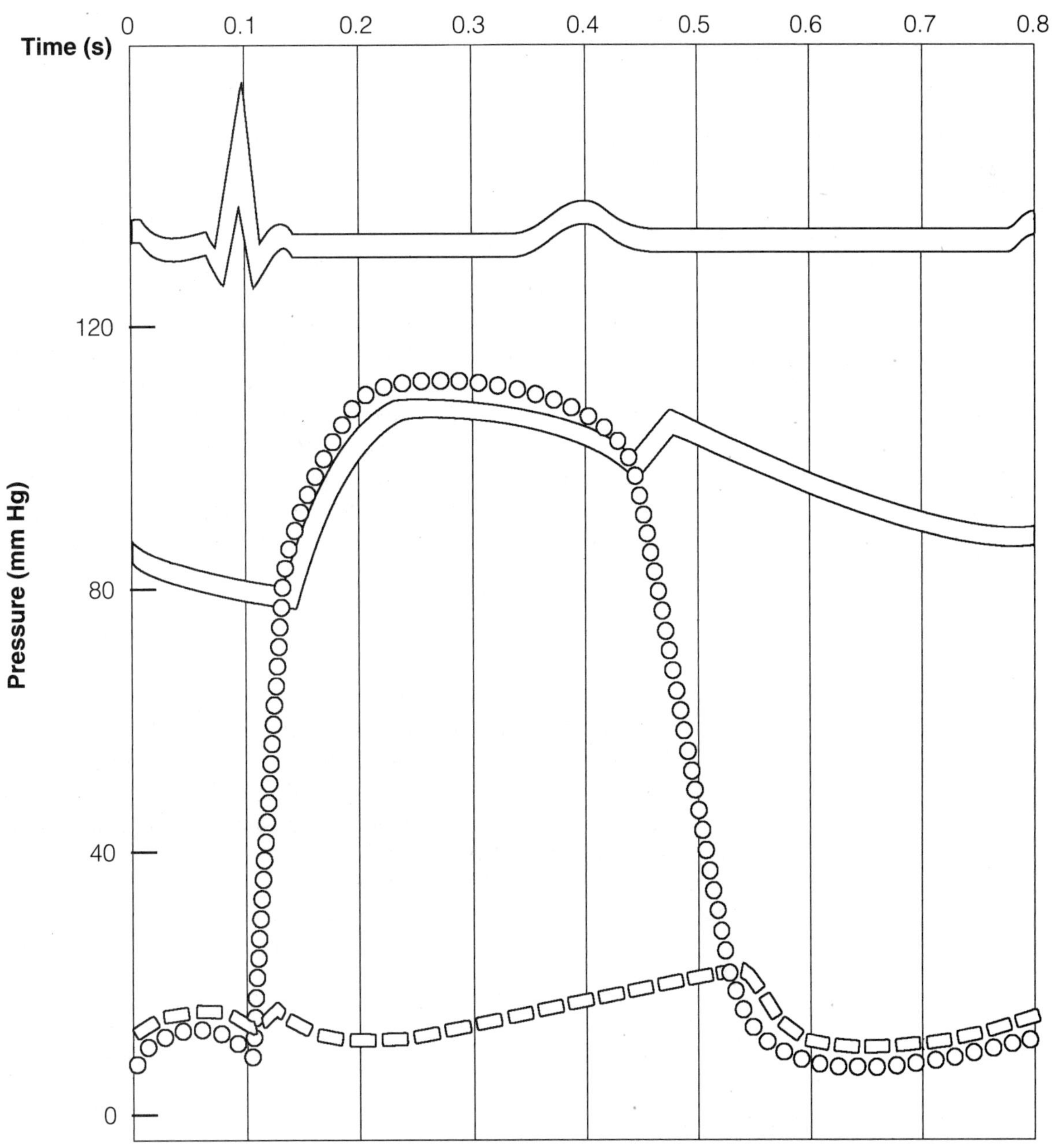

Figure 17.10

8. Circle the term that does not belong in each of the following groupings.

1. AV valves closed   AV valves open   Ventricular systole   Semilunar valves open
2. No P wave   SA node is pacemaker   AV node is pacemaker   Junctional rhythm
3. Ventricles fill   Ventricular systole   AV valves open   Late diastole
4. Early diastole   Semilunar valves open   Isovolumetric relaxation   Ventricular pressure drops
5. Stenotic valve   Restricted blood flow   High-pitched heart sound   Heart sound after valve closes
6. Isovolumetric ventricular systole   Blood volume unchanging   AV valve open   Semilunar valve closed

9. Complete the following statements relating to cardiac output by writing the missing terms in the answer blanks.

In the relationship CO = HR × SV, CO stands for __(1)__, HR stands for __(2)__, and SV stands for __(3)__. For the normal resting heart, the value of HR is __(4)__ and the value of SV is __(5)__. The normal average adult cardiac output, therefore, is __(6)__. The time for the entire blood supply to pass through the body is once each __(7)__. The normal ventricle holds a volume of blood equal to about __(8)__, of which about __(9)__ of blood remain after each contraction. The heart, however, can exceed its normal cardiac output, a property of the heart called __(10)__. In response to sudden demands, such as __(11)__, cardiac reserve is about four times the normal cardiac output.

SV represents the difference between the __(12)__, the amount of blood that collects in the ventricle during diastole, and the __(13)__, the volume of blood remaining in the ventricle after contraction. SV is critically related to __(14)__, or the degree of stretch of cardiac muscle before it contracts. Stretching increases the number of __(15)__ interactions between the actin and myosin filaments in the cardiac cells and hence the __(16)__ of heart contraction. Stretching cardiac muscle is accomplished by increasing the amount of __(17)__ returned to the heart, which distends the __(18)__.

1. ____________________
2. ____________________
3. ____________________
4. ____________________
5. ____________________
6. ____________________
7. ____________________
8. ____________________
9. ____________________
10. ____________________
11. ____________________
12. ____________________
13. ____________________
14. ____________________
15. ____________________
16. ____________________
17. ____________________
18. ____________________

10. Stroke volume may be enhanced in two major ways: by increasing venous return and by enhancing contractility of heart muscle.

1. Explain *contractility*. ______________________________

____

2. Name four factors, one neural and three hormonal, that enhance contractility.

____

3. Explain how exercise increases venous return. ______________________________

____

11. Check (✓) all factors that lead to an *increase* in cardiac output by influencing either heart rate or stroke volume.

_____ 1. Epinephrine

_____ 2. Thyroxine

_____ 3. Hemorrhage

_____ 4. Fear

_____ 5. Exercise

_____ 6. Fever

_____ 7. Depression

_____ 8. Anxiety

_____ 9. Activation of the vasomotor center

_____ 10. Activation of the cardio-acceleratory center

_____ 11. Activation of the cardioinhibitory center

_____ 12. Activation of the vagus nerves

_____ 13. Low blood pressure

_____ 14. High blood pressure

_____ 15. Increased end diastolic volume

_____ 16. Prolonged grief

_____ 17. Acidosis

_____ 18. Calcium channel blockers

_____ 19. Atrial reflex

_____ 20. Hyperkalemia

_____ 21. Increased afterload

12. If a statement is true, write the letter T in the answer blank. If a statement is false, change the underlined word(s) and write the correct word(s) in the answer blank.

______________ 1. Norepinephrine, released by <u>parasympathetic</u> fibers, stimulates the SA and AV nodes and the myocardium itself.

______________ 2. The resting heart is said to exhibit "vagal tone," meaning that the heart rate slows under the influence of <u>acetylcholine</u>.

______________ 3. Epinephrine secreted by the adrenal medulla <u>decreases</u> heart rate.

______________ 4. <u>Low</u> levels of ionic calcium cause prolonged cardiac contractions.

____________ 5. The resting heart rate is fastest in adult life.

____________ 6. Because the heart of the highly trained athlete hypertrophies, its stroke volume decreases.

____________ 7. In congestive heart failure, there is a marked rise in the end diastolic volume.

____________ 8. If the right side of the heart fails, pulmonary congestion occurs.

____________ 9. In peripheral congestion, the feet, ankles, and fingers swell.

____________ 10. The pumping action of the healthy heart ordinarily maintains a balance between cardiac output and venous return.

____________ 11. The cardioacceleratory center in the medulla gives rise to sympathetic nerves supplying the heart.

13. Match the terms in Column B with the statements in Column A. Place the correct letters in the answer blanks.

	Column A	Column B
____________	1. Results from prolonged coronary blockage	A. Angina pectoris
		B. Bradycardia
____________	2. Abnormal pacemaker	C. Congestive heart failure
____________	3. Allows backflow of blood	D. Ectopic focus
____________	4. Because of cardiac decompensation, circulation is inadequate to meet tissue needs	E. Fibrillation
		F. Heart block
____________	5. A slow heartbeat, that is, below 60 beats per minute	G. Incompetent valve
____________	6. A condition in which the heart is uncoordinated and useless as a pump	H. Myocardial infarction
		I. Pulmonary congestion
____________	7. A rapid heart rate, that is, over 100 beats per minute	J. Tachycardia
____________	8. Damage to the AV node, totally or partially releasing the ventricles from the control of the SA node	
____________	9. Chest pain, resulting from ischemia of the myocardium	
____________	10. Result of initial failure of the left side of the heart	

14. Match the ionic imbalances in Column B to the descriptions of their effects in Column A.

	Column A	Column B
________________	1. Depresses heart	A. Hypercalcemia
________________	2. Spastic contractions	B. Hyperkalemia
________________	3. Lowers resting potential	C. Hypocalcemia
________________	4. Feeble contractions, abnormal rhythms	D. Hypokalemia

CHALLENGING YOURSELF

At the Clinic

1. Homer Fox, a patient with acute severe pericarditis, has a critically low stroke volume. What is the name for the condition causing the low stroke volume and how does it cause it?
2. Mr. Greco, a patient with clotting problems, has been hospitalized with right-sided heart failure. What is his condition? Knowing Mr. Greco's history, what is the probable cause?
3. An elderly man is brought to the clinic because he fatigues extremely easily. An examination reveals a heart murmur associated with the bicuspid valve during ventricular systole. What is the diagnosis? What is a possible treatment?
4. Jimmy is brought to the clinic complaining of a sore throat. His mother says he has had the sore throat for about a week. Culture of a throat swab is positive for strep, and the boy is put on antibiotics. A week later, the boy is admitted to the hospital after fainting several times. He is cyanotic. What is a likely diagnosis?
5. After a bout with bacterial endocarditis, scar tissue often stiffens the edges of the heart valves. What condition will this cause? How would this be picked up in a routine examination?
6. Mary Ghareeb, a woman in her late 50s, has come to the clinic because of chest pains whenever she begins to exert herself. What is her condition called?
7. After a fairly severe heart attack, the ECG reveals normal sinus rhythm, but only for the P wave. The QRS and T waves are no longer in synchrony with the P wave, and the ventricular contraction corresponds to junctional rhythm. What is the problem? What part of the heart is damaged?
8. Ms. Hamad, who is 73 years old, is admitted to the coronary care unit of a hospital with a diagnosis of left ventricular failure resulting from a myocardial infarction. Her heart rhythm is abnormal. Explain what a myocardial infarction is, how it is likely to have been caused, and why the heart rhythm is affected.
9. Mr. Trump, en route to the hospital ER by ambulance, is in fibrillation. What is his cardiac output likely to be? He arrives at the emergency entrance DOA (dead on arrival). His autopsy reveals a blockage of the posterior interventricular artery. What is the cause of death?
10. Excessive vagal stimulation can be caused by severe depression. How would this be reflected in a routine physical examination?

11. Mrs. Suffriti has swollen ankles and signs of degenerating organ functions. What is a likely diagnosis?

12. You are a young nursing student and are called upon to explain how each of the following tests or procedures might be helpful in evaluating a patient with heart disease: blood pressure measurement, determination of blood lipid and cholesterol levels, electrocardiogram, chest X ray. How would you respond?

13. Mr. Langley is telling his friend about his recent visit to his doctor for a checkup. During his story, he mentions that the ECG revealed that he had a defective mitral valve and a heart murmur. Mr. Langley apparently misunderstood some of what the doctor explained to him about his diagnosis process. What has he misunderstood?

Stop and Think

1. The major coronary vessels are on the surface of the heart. What is the advantage of that location?

2. Because the SA node is at the top of the atrial mass, the atria contract from the top down. How does this increase the efficiency of atrial contraction? Do the ventricles have a similar arrangement? If so, how does it work?

3. What parts of the body feel referred pain when there is a major myocardial infarction?

4. What is the purpose of prolonged contraction of the myocardium?

5. How long does it take for a hormone released from the anterior pituitary gland to reach its target organ in the body?

6. What is the functional difference between ventricular hypertrophy due to exercise and hypertrophy due to congestive heart failure?

7. How does hypothyroidism affect heart rate?

8. The foramen ovale has a flap that partially covers it, and a groove leads from the opening of the inferior vena cava (which carries freshly oxygenated blood from the umbilical vein) to the foramen ovale. Explain these structural features in terms of directing blood flow.

9 A less-than-respectable news tabloid announced that "Doctors show exercise shortens life. Life expectancy is programmed into a set number of heartbeats; the faster your heart beats, the sooner you die!" Even if this "theory" were true, what is wrong with the conclusion concerning exercise?

COVERING ALL YOUR BASES

Multiple Choice

Select the best answer or answers from the choices given.

1. A knife plunged into the sixth intercostal space will:
   A. pierce the aorta
   B. pierce the atria
   C. pierce the ventricles
   D. miss the heart completely

2. The innermost layer of the pericardial sac is the:
   A. epicardium
   B. fibrous pericardium
   C. parietal layer of the serous pericardium
   D. visceral layer of the serous pericardium

3. Which of the heart layers are vascularized?
   A. Myocardium
   B. Endothelium of endocardial layer
   C. Mesothelium of epicardium
   D. Connective tissue layers of both endocardium and epicardium

4. The fibrous skeleton of the heart:
   A. supports valves
   B. anchors vessels
   C. provides electrical insulation to separate the atrial mass from the ventricular mass
   D. anchors cardiac muscle fibers

5. Which of the following is associated with the right atrium?
   A. Fossa ovalis
   B. Tricuspid valve
   C. Chordae tendineae
   D. Pectinate muscle

6. A surface feature associated with the circumflex artery is the:
   A. right coronary sulcus
   B. anterior interventricular sulcus
   C. left coronary sulcus
   D. posterior interventricular sulcus

7. Atrioventricular valves are held closed by:
   A. papillary muscles
   B. trabeculae carneae
   C. pectinate muscles
   D. chordae tendineae

8. During atrial systole:
   A. the atrial pressure exceeds ventricular pressure
   B. 70% of ventricular filling occurs
   C. the AV valves are open
   D. valves prevent backflow into the great veins

9. Occlusion of which of the following arteries would damage primarily the right ventricle?
   A. Marginal artery
   B. Posterior interventricular artery
   C. Circumflex artery
   D. Anterior interventricular artery

10. A choking sensation in the chest is the result of:
    A. myocardial infarction
    B. death of cardiac cells
    C. temporary lack of oxygen to cardiac cells
    D. possibly spasms of the coronary arteries

11. Which feature(s) is (are) greater in number/value/size or more developed in cardiac than in skeletal muscle?
    A. T tubule diameter
    B. Terminal cisternae
    C. Mitochondria
    D. Absolute refractory period

12. Which is (are) characteristic of cardiac but not of skeletal muscle?
    A. Gap junctions
    B. All-or-none law
    C. Endomysium
    D. Branching fibers

13. Ca^{2+} slow channels are open in cardiac cells' plasma membranes during:
    A. depolarization
    B. repolarization
    C. plateau
    D. resting membrane potential

14. Which of the following depolarizes next after the AV node?
    A. Atrial myocardium
    B. Ventricular myocardium
    C. Bundle branches
    D. Purkinje fibers

15. Threshold in pacemaker cells is marked by:
    A. opening of Na^+ channels
    B. opening of Ca^{2+} slow channels
    C. opening of Ca^{2+} fast channels
    D. opening of K^+ channels

16. Freshly oxygenated blood is first received by the:
    A. right ventricle
    B. left ventricle
    C. right atrium
    D. left atrium

17. A heart rate of 30 bpm indicates that the ____________________ is functioning as an ectopic focus.
    A. AV node
    B. bundle branches
    C. Purkinje fibers
    D. AV bundle

18. Stimulation of the cardiac plexus involves stimulation of the:
    A. baroreceptors in the great arteries
    B. cardioacceleratory center
    C. vagus nerve
    D. sympathetic chain

19. Atrial repolarization coincides in time with the:
    A. P wave
    B. T wave
    C. QRS wave
    D. P-Q interval

20. Soon after the onset of ventricular systole the:
    A. AV valves close
    B. semilunar valves open
    C. first heart sound is heard
    D. aortic pressure increases

21. Which of the following will occur first after the T wave?
   A. Increase in ventricular volume
   B. Decrease in atrial volume
   C. Closure of the semilunar valves
   D. Decrease in ventricular pressure

22. Given an end-diastolic volume of 150 ml, an end-systolic volume of 50 ml, and a heart rate of 60 bpm, the cardiac output is:
   A. 600 ml/min
   B. 6 liters/min
   C. 1200 ml/min
   D. 3 liters/min

23. The statement "strength of contraction increases intrinsically due to increased stretching of the heart wall" is best attributed to:
   A. contractility
   B. the Frank-Starling law of the heart
   C. the Bainbridge reflex
   D. the aortic sinus reflex

24. An increase in cardiac output is triggered by:
   A. the Bainbridge reflex
   B. stimulation of carotid sinus baroreceptors
   C. an increase in vagal tone
   D. epinephrine

25. Which of the following ionic imbalances inhibits Ca^{2+} transport into cardiac cells and blocks contraction?
   A. Hyponatremia
   B. Hypokalemia
   C. Hypernatremia
   D. Hyperkalemia
   E. Hypercalcemia

26. Cardiovascular conditioning results in:
   A. ventricular hypertrophy
   B. bradycardia
   C. increase in SV
   D. increase in CO

27. Conditions known to be associated with congestive heart failure include:
   A. coronary atherosclerosis
   B. diastolic pressure chronically elevated
   C. successive sublethal infarcts
   D. decompensated heart

28. In heart failure, venous return is slowed. Edema results because:
   A. blood volume is increased
   B. venous pressure is increased
   C. osmotic pressure is increased
   D. none of the above

29. Which structure(s) is (are) fetal remnant(s) associated with a healthy adult heart?
   A. Foramen ovale
   B. Ligamentum arteriosum
   C. Interventricular septal opening
   D. Patent ductus arteriosus

30. Which of the following terms refers to an unusually strong heartbeat, so that the person is aware of it?
   A. Extrasystole
   B. Tamponade
   C. Flutter
   D. Palpitation

31. The base of the heart is its ___________ surface.
   A. diaphragmatic
   B. posterior
   C. arterier
   D. superior

32. The thickest layer of the heart wall is:
   A. endocardium
   B. myocardium
   C. epicardium
   D. fibrous pericardium

Word Dissection

For each of the following word roots, fill in the literal meaning and give an example, using a word found in this chapter.

Word root	Translation	Example
1. angina		
2. baro		
3. brady		
4. carneo		
5. cusp		
6. diastol		
7. dicro		
8. ectop		
9. intercal		
10. pectin		
11. sino		
12. stenos		
13. systol		
14. tachy		

18

The Cardiovascular System: Blood Vessels

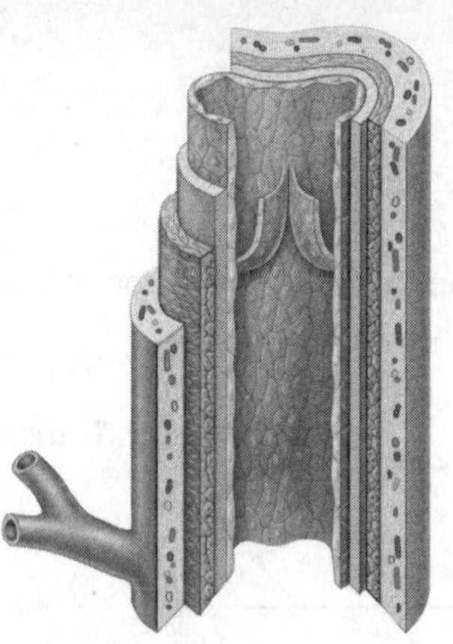

Student Objectives

When you have completed the exercises in this chapter, you will have accomplished the following objectives:

PART 1: OVERVIEW OF BLOOD VESSEL STRUCTURE AND FUNCTION

1. Describe the three layers that typically form the wall of a blood vessel, and state the function of each.
2. Define vasoconstriction and vasodilation.
3. Compare and contrast the structure and function of the three types of arteries.
4. Describe the structure and function of a capillary bed.
5. Describe the structure and function of veins, and explain how veins differ from arteries.

PART 2: PHYSIOLOGY OF CIRCULATION

6. Define blood flow, blood pressure, and resistance, and explain the relationships between these factors.
7. List and explain the factors that influence blood pressure, and describe how blood pressure is regulated.
8. Define hypertension. Describe its manifestations and consequences.
9. Explain how blood flow is regulated in the body in general and in its specific organs.
10. Outline factors involved in capillary dynamics, and explain the significance of each.
11. Define circulatory shock. List several possible causes.

PART 3: CIRCULATORY PATHWAYS: BLOOD VESSELS OF THE BODY

12. Trace the pathway of blood through the pulmonary circuit, and state the importance of this special circulation.
13. Describe the general functions of the systemic circuit.
14. Name and give the location of the major arteries and veins in the systemic circulation.
15. Describe the structure and special function of the hepatic portal system.

After leaving the heart, the blood is sent through a vast network of vessels that ultimately reach each cell, supplying oxygen and nutrients and removing wastes. Blood flows through an equally extensive vascular network back to the heart. In the adult human, the length of these circulatory pathways is about 60,000 miles. These are not passive tubular structures but dynamic structures that rapidly alter the blood flow in response to changing internal and external conditions.

The focus of Chapter 18 is on the vascular portion of the cardiovascular system. Topics include blood vessel structure and function, the physiology of circulation and blood pressure, the dynamics of the capillary bed, the anatomy of the pulmonary and systemic vasculature, and aspects of the development and diseases of blood vessels.

BUILDING THE FRAMEWORK

PART 1: OVERVIEW OF BLOOD VESSEL STRUCTURE AND FUNCTION

1. Assume someone has been injured in an automobile accident and is bleeding profusely. What pressure points could you compress to help stop the bleeding from the following areas?

____________________ 1. Thigh

____________________ 2. Forearm

____________________ 3. Calf

____________________ 4. Lower jaw

____________________ 5. Thumb

____________________ 6. Plantar surface of foot

____________________ 7. Temple

____________________ 8. Ankle

2. Select different colors for each of the three blood vessel tunics listed in the key choices and illustrated in Figure 18.1. (*Note:* Elastic laminae are *not* illustrated.) Color each tunic on the three diagrams and correctly identify each vessel type. Identify valves if appropriate. Write your answers in the blanks beneath the illustrations. In the additional blanks, list the structural details that allowed you to make the identifications. Then, using the key choices, identify the blood vessel tunics described in each of the following cases by writing the correct answers in the answer blanks.

Key Choices

○ A. Tunica intima ○ B. Tunica media ○ C. Tunica externa

__________ 1. Single thin layer of endothelium associated with scant connective tissue

__________ 2. Bulky middle coat, containing smooth muscle and elastin

__________ 3. Provides a smooth surface to decrease resistance to blood flow

__________ 4. The only tunic of capillaries

__________ 5. Called the adventitia in some cases

__________ 6. The only tunic that plays an active role in blood pressure regulation

__________ 7. Supporting, protective coat

__________ 8. Forms venous valves

Figure 18.1

A. __________ B. __________ C. __________

__________ __________ __________

__________ __________ __________

3. Match the terms pertaining to blood vessels in Column B to the appropriate descriptions in Column A. Some letters are used more than once.

	Column A	Column B
________	1. Transport blood away from heart	A. Arteries
________	2. Largest arteries, low resistance	B. Arterioles
________	3. Arteries with thickest tunica media; active in vasoconstriction	C. Capillaries
		D. Continuous
________	4. Control blood flow into individual capillary beds	E. Elastic
________	5. Lumen is the size of red blood cells	F. Fenestrated
________	6. Capillary type with an uninterrupted lining	G. Muscular
________	7. Capillary type with numerous pores and gap junctions	H. Sinusoids
		I. Veins
________	8. Capillaries with large intercellular clefts and irregular lumen	J. Venous sinuses
________	9. Vessels formed when capillaries unite	K. Venules
________	10. Vessels with thin walls and large lumens that often appear collapsed in histologic preparations	
________	11. Veins with only a tunica intima; supported by surrounding tissues	
________	12. Sometimes referred to as "conducting" arteries	
________	13. Artery type that accounts for most of the named arteries studied in an A&P lab.	

4. What is the importance of arterial anastomoses?

____

____

____

5. Figure 18.2 shows diagrammatic views of the three types of capillaries. Below each diagram, write in the type of capillary shown. Then, identify all elements and structures that have leader lines. Finally, color code and color the following structures:

○ Basement membrane ○ Endothelial cell(s) ○ Macrophage location (if present)

○ Capillary lumen ○ Erythrocyte(s)

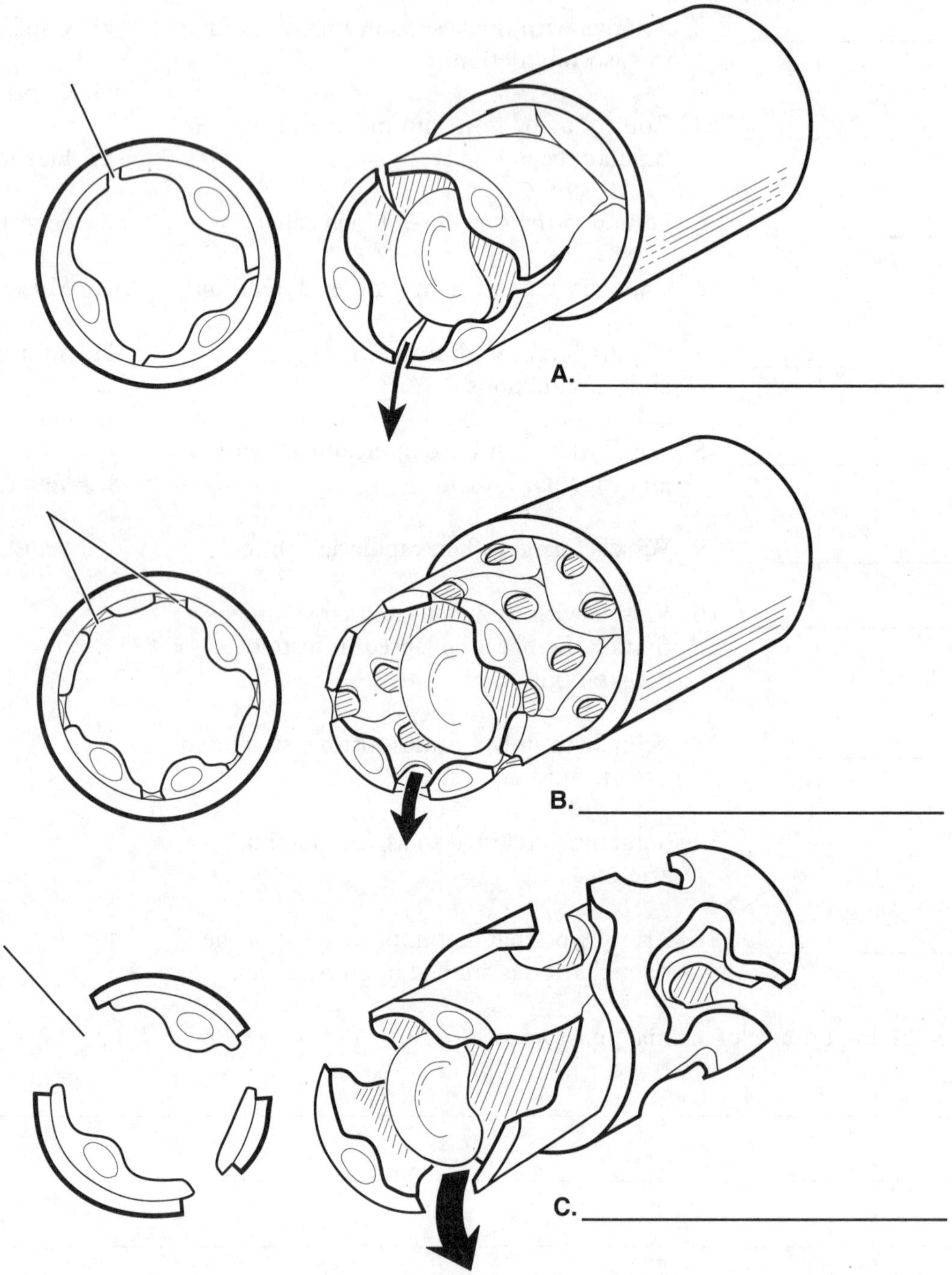

A. ______________________

B. ______________________

C. ______________________

Figure 18.2

6. Circle the term that does not belong in each of the following groupings.

1. Elastic Fenestrated Continuous Sinusoidal Capillaries
2. Distributing Vasoconstriction Muscular Most arteries Pressure points
3. Cornea Heart Cartilage Epithelium Tendons
4. Cytoplasmic vesicles Intercellular clefts Gap junctions Tight junctions Continuous capillaries
5. Liver Bone marrow Kidney Lymphoid tissue Sinusoids
6. High pressure Veins Capacitance vessels Valves Thick tunica externa Blood reservoirs
7. Anastomoses End arteries Collateral channels Capillary beds Venous Arterial
8. Valves Large lumen Thick media Collapsed lumen Veins
9. Metarteriole Anastomosis Thoroughfare channel Shunt True capillaries

7. Blood flows from high to low pressure. Hence, it flows from the high-pressure arteries through the capillaries and then through the low-pressure veins. Because blood pressure contributes less to blood propulsion in veins, special measures are required to ensure that venous return equals cardiac output. What role do the venous valves play?

____

8. Briefly explain why veins are called blood reservoirs and state where in the body venous blood reservoirs are most abundant.

____

____

____

____

PART 2: PHYSIOLOGY OF CIRCULATION

1. Fill in the blanks in the flowcharts below indicating the relationships between the terms provided. The position of some elements in these schemes has already been indicated.

1. difference in blood pressure, resistance, atherosclerosis, blood viscosity, vessel length, hematocrit, vasoconstriction

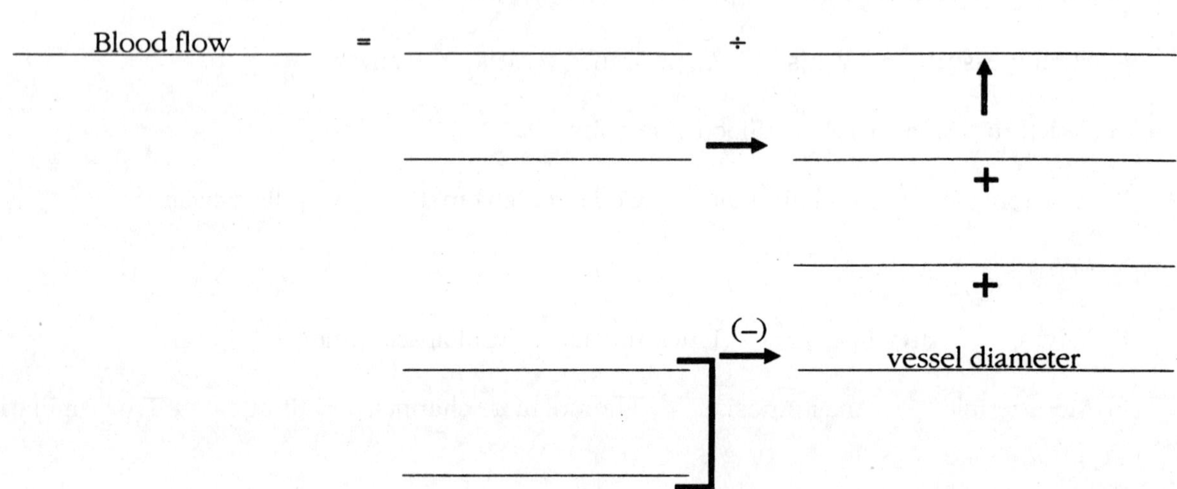

2. cardiac output, peripheral resistance, polycythemia, blood volume, heartbeat, excessive salt intake, vasoconstriction

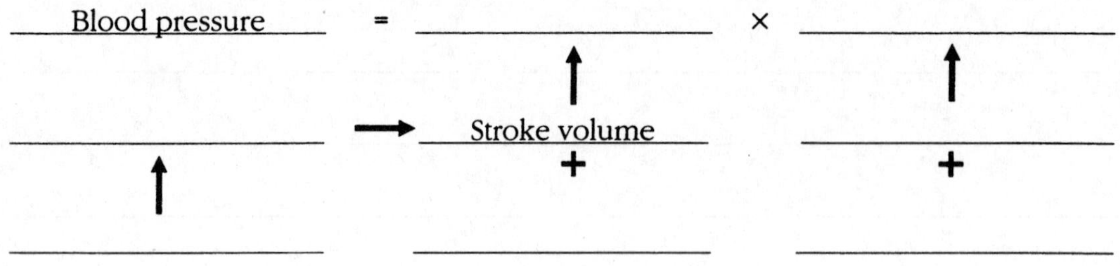

2. Briefly explain why blood flow in large, thick-walled arteries, such as the aorta and its branches, is fairly continuous and does not stop when the heart relaxes.

____

____

____

____

3. This exercise concerns blood pressure and pulse. Match the items in Column B with the appropriate descriptions in Column A.

	Column A	Column B
________	1. Expansion and recoil of an artery during heart activity	A. Over arteries
		B. Blood pressure
________	2. Pressure exerted by the blood against the blood vessel walls	C. Cardiac output
________	3. The product of these factors yields blood pressure (#3–#4)	D. Constriction of arterioles
________	4.	E. Diastolic blood pressure
________	5. Event primarily responsible for peripheral resistance	F. Peripheral resistance
		G. Pressure points
________	6. Blood pressure during heart contraction	H. Pulse
________	7. Blood pressure during heart relaxation	I. Sounds of Korotkoff
		J. Systolic blood pressure
________	8. Site where blood pressure determinations are normally made	K. Over veins
________	9. Points at the body surface where the pulse may be felt	
________	10. Sounds heard over a blood vessel when the vessel is partially compressed	

4. What effects do the following factors have on blood pressure? Use I to indicate an increase in pressure and D to indicate a decrease in pressure.

_____ 1. Increased diameter of the arterioles

_____ 2. Increased blood viscosity

_____ 3. Increased cardiac output

_____ 4. Increased pulse rate

_____ 5. Anxiety, fear

_____ 6. Increased urine output

_____ 7. Sudden change in position from reclining to standing

_____ 8. Physical exercise

_____ 9. Physical training

_____ 10. Alcohol

_____ 11. Hemorrhage

_____ 12. Nicotine

_____ 13. Arteriosclerosis

_____ 14. Stimulation of arterial baroreceptors

_____ 15. Stimulation of carotid body chemoreceptors

_____ 16. Release of epinephrine from adrenal medulla

_____ 17. Secretion of atrial natriuretic peptide

_____ 18. Secretion of antidiuretic hormone

_____ 19. Release of endothelin

_____ 20. Secretion of NO

_____ 21. Renin/angiotensin mechanism

_____ 22. Secretion of aldosterone

5. Circle the term that does not belong in each of the following groupings.

1. Blood pressure mm Hg Cardiac output Force of blood on vessel wall
2. Peripheral resistance Friction Arterioles Blood flow
3. Low viscosity High viscosity Blood Resistance to flow
4. Blood viscosity Blood pressure Vessel length Vessel diameter
5. High blood pressure Hemorrhage Weak pulse Low cardiac output
6. Resistance Friction Vasodilation Vasoconstriction
7. High pressure Vein Artery Spurting blood
8. Elastic arteries Diastolic pressure 120 mm Hg Auxiliary pumps
9. Cardiac cycle Intermittent blood flow Capillary bed Precapillary sphincters
10. Muscular pump Respiratory pump Inactivity Venous return
11. Hypotension Sympathetic activity Poor nutrition Training

6. If a statement is true, write the letter T in the answer blank. If a statement is false, change the underlined word(s) and write the correct word(s) in the answer blank.

________________ 1. Renin, released by the kidneys, causes a <u>decrease</u> in blood pressure.

________________ 2. The decreasing efficiency of the sympathetic nervous system vasoconstrictor functioning due to aging leads to a type of hypotension called <u>sympathetic</u> hypotension.

________________ 3. Two body organs in which vasoconstriction rarely occurs are the heart and the <u>kidneys</u>.

________________ 4. A <u>sphygmomanometer</u> is used to take the apical pulse.

________________ 5. The pulmonary circuit is a <u>high</u>-pressure circulation.

________________ 6. Cold has a <u>vasodilating</u> effect.

________________ 7. <u>Thrombophlebitis</u> is called the silent killer.

________________ 8. The nervous system controls blood pressure and blood distribution by altering the diameters of the <u>venules</u>.

________________ 9. The vasomotor center for blood pressure control is located in the <u>left atrium</u>.

________________ 10. Pressoreceptors in the <u>large arteries</u> of the neck and thorax detect changes in blood pressure.

________________ 11. ANP, the hormone produced by the atria, causes a <u>rise</u> in blood volume and blood pressure.

________________ 12. The hormone vasopressin causes intense <u>vasodilation</u>.

________________ 13. Hypertension in obese people can be promoted by increased <u>viscosity of the blood</u>.

________________ 14. The velocity of blood flow is slowest in the <u>veins</u>.

________________ 15. The total cross-sectional area of the vascular bed is least in the <u>capillary bed</u>.

________________ 16. <u>Hypertension</u> is the local adjustment of blood flow to a given tissue at any particular time.

________________ 17. In active skeletal muscles, autoregulated vasodilation is promoted by an increase of <u>acetylcholine</u> in the area.

7. Figure 18.3 is a diagram of a capillary bed. Arrows indicate the direction of blood flow. Select six different colors and color the coding circles and their structures on the figure. Then answer the questions that follow by referring to Figure 18.3. Notice that questions 6–14 concern fluid flows at capillary beds and the forces (hydrostatic and osmotic pressures) that promote such fluid shifts.

○ Arteriole ○ Thoroughfare channel ○ Postcapillary venule

○ Precapillary sphincters ○ True capillaries ○ Metarteriole

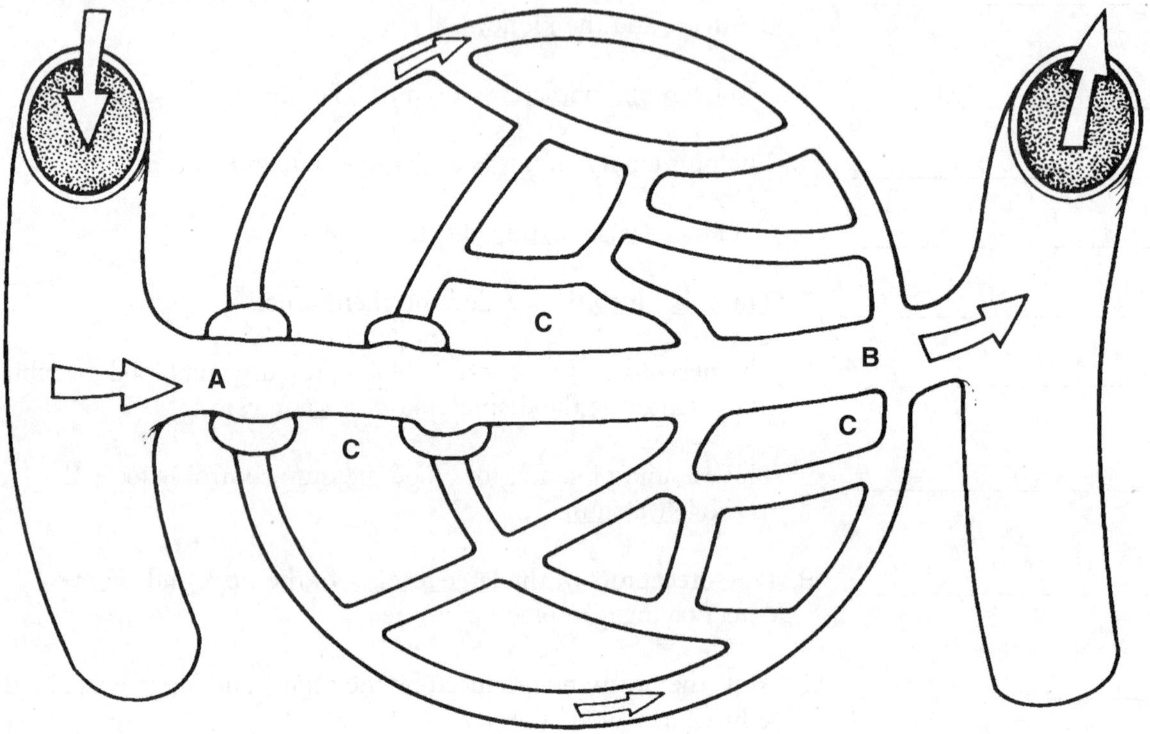

A

B

C

C

C

Figure 18.3

1. What is the liquid that surrounds tissue cells called?

____

2. What drives the movement of soluble substances between tissue fluids and the blood?

____

3. Which substances pass readily through the endothelial cell plasma membrane?

____

4. Which substances move through fluid-filled capillary clefts?

____

5. If the precapillary sphincters are contracted, by which route will the blood flow?

___

6. Under normal conditions, in which area does hydrostatic pressure predominate, A, B, or C?

___

7. Which area has the highest osmotic pressure? ___

8. Which pressure is in excess and causes fluids to move from A to C? (Be specific as to whether the force exists in the capillary or the interstitial space.)

___

9. Which pressure causes fluid to move from A to B? ___

10. Which pressure causes fluid to move from C to B? ___

11. Which blood protein is most responsible for osmotic pressure? ___

12. Where does the greater net flow of water out of the capillary occur? ___

13. If excess fluid does not return to the capillary, where does it go?

___

8. Briefly compare the characteristics, causes, and effects of hypovolemic shock and vascular shock.

___

___

___

___

___

___

9. Choose the vessel type (arteries, capillaries, veins) with the indicated characteristic.

____________________ 1. Highest total cross-sectional area

____________________ 2. Highest velocity of blood flow

____________________ 3. Lowest velocity of blood flow

____________________ 4. Pulse pressure

____________________ 5. Lowest blood pressure

10. Using the key choices, identify the special circulations described below.

Key Choices

A. Cerebral | C. Hepatic | E. Skeletal muscle

B. Coronary | D. Pulmonary | F. Skin

____________________ 1. Blood flow increases markedly when the body temperature rises

____________________ 2. The major autoregulatory stimulus is a drop in pH

____________________ 3. Arteries characteristically have thin walls and large lumens

____________________ 4. Vessels do not constrict but are compressed during systole

____________________ 5. Receives constant blood flow whether the body is at rest or strenuously exercising

____________________ 6. Vasodilation promoted by high oxygen levels

____________________ 7. Capillary flow markedly sluggish; phagocytes present

____________________ 8. Prolonged activity places extreme demands on cardiovascular system

____________________ 9. Additional oxygen can be supplied only by increased blood flow

____________________ 10. Much lower arterial pressure than that in systemic circulation

____________________ 11. Large, atypical capillaries with fenestrations

____________________ 12. Impermeable tight junctions in capillary endothelium

____________________ 13. One of the most precise autoregulatory systems in the body

____________________ 14. Venous blood empties into large dural sinuses rather than into veins

____________________ 15. Abundance of superficial veins

____________________ 16. During vigorous physical activity, receives up to two-thirds of total blood flow

PART 3: CIRCULATORY PATHWAYS: BLOOD VESSELS OF THE BODY

1. Figure 18.4 shows the pulmonary circuit. Identify all vessels that have leader lines. Color the vessels (and heart chambers) transporting oxygen-rich blood *red;* color those transporting carbon dioxide-rich blood *blue.*

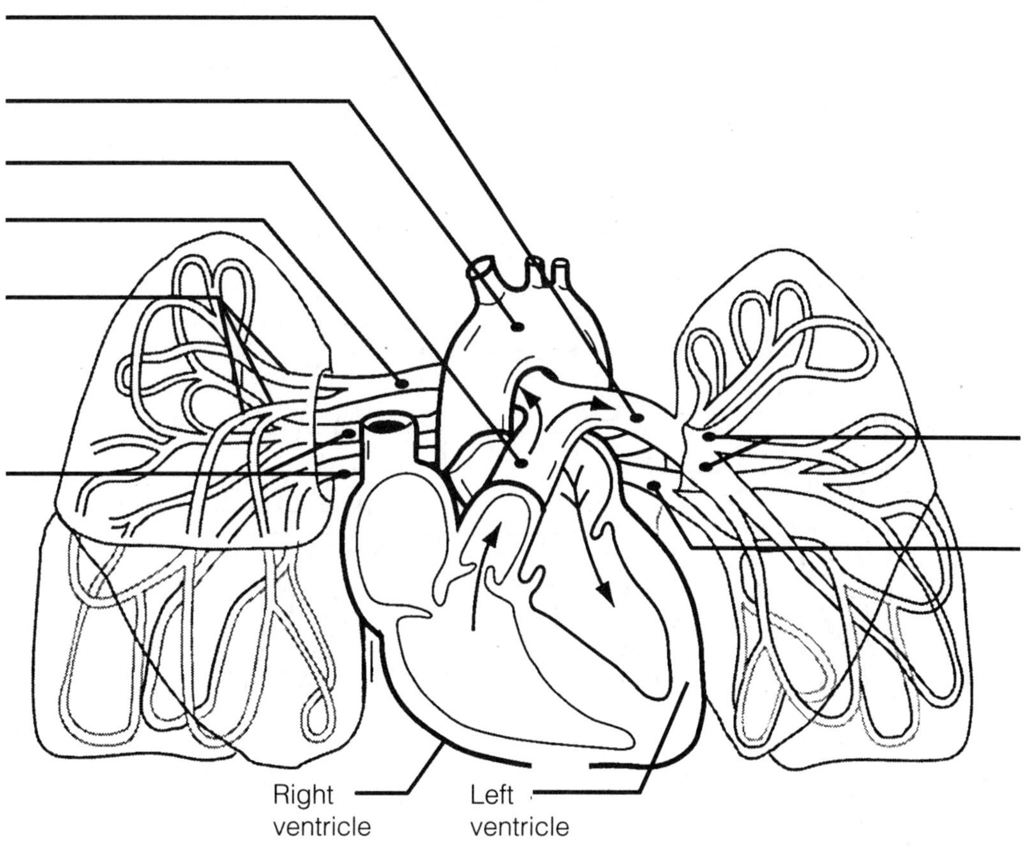

Figure 18.4

2. Figures 18.5 and 18.6 illustrate the locations of the major systemic arteries and veins of the body. These figures are highly simplified and will serve as a "warm-up" for the more detailed vascular diagrams to come. The arteries are shown in Figure 18.5. Color the arteries red, then identify those indicated by leader lines on the figure. The veins are shown in Figure 18.6. Color the veins blue, then identify each vein that has a leader line on the figure. Or, if you wish, color the individual vessels with different colors to help you to identify their extent.

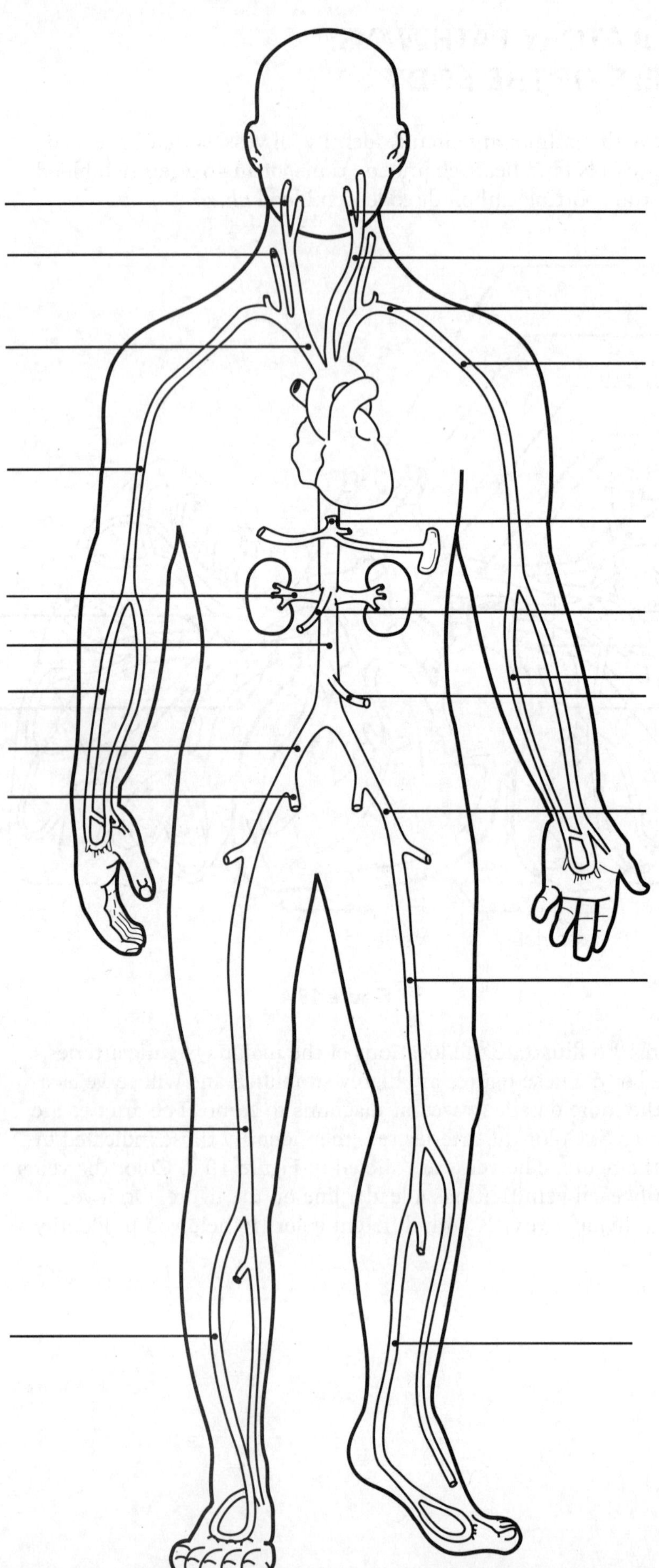

Figure 18.5
Arteries

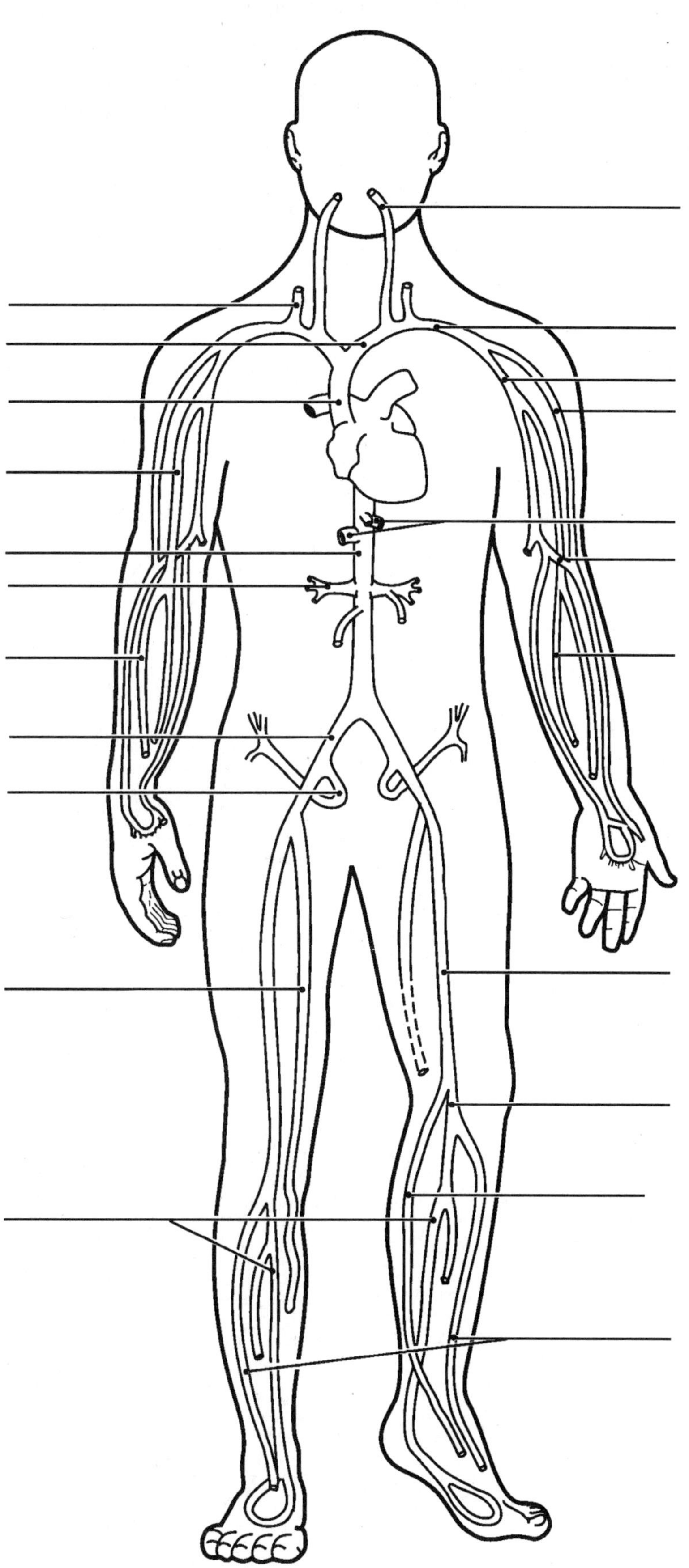

Figure 18.6
Veins

3. Figure 18.7 shows the major arteries of the head and neck. Note that the clavicle is omitted and that dashed lines represent deeper vessels. Color code and color the following vessels.

- ◯ Brachiocephalic
- ◯ Costocervical trunk
- ◯ External carotid
- ◯ Facial
- ◯ Internal carotid
- ◯ Internal thoracic
- ◯ Lingual
- ◯ Maxillary
- ◯ Occipital
- ◯ Ophthalmic
- ◯ Right common carotid
- ◯ Right subclavian
- ◯ Superficial temporal
- ◯ Superior thyroid
- ◯ Thyrocervical trunk
- ◯ Vertebral

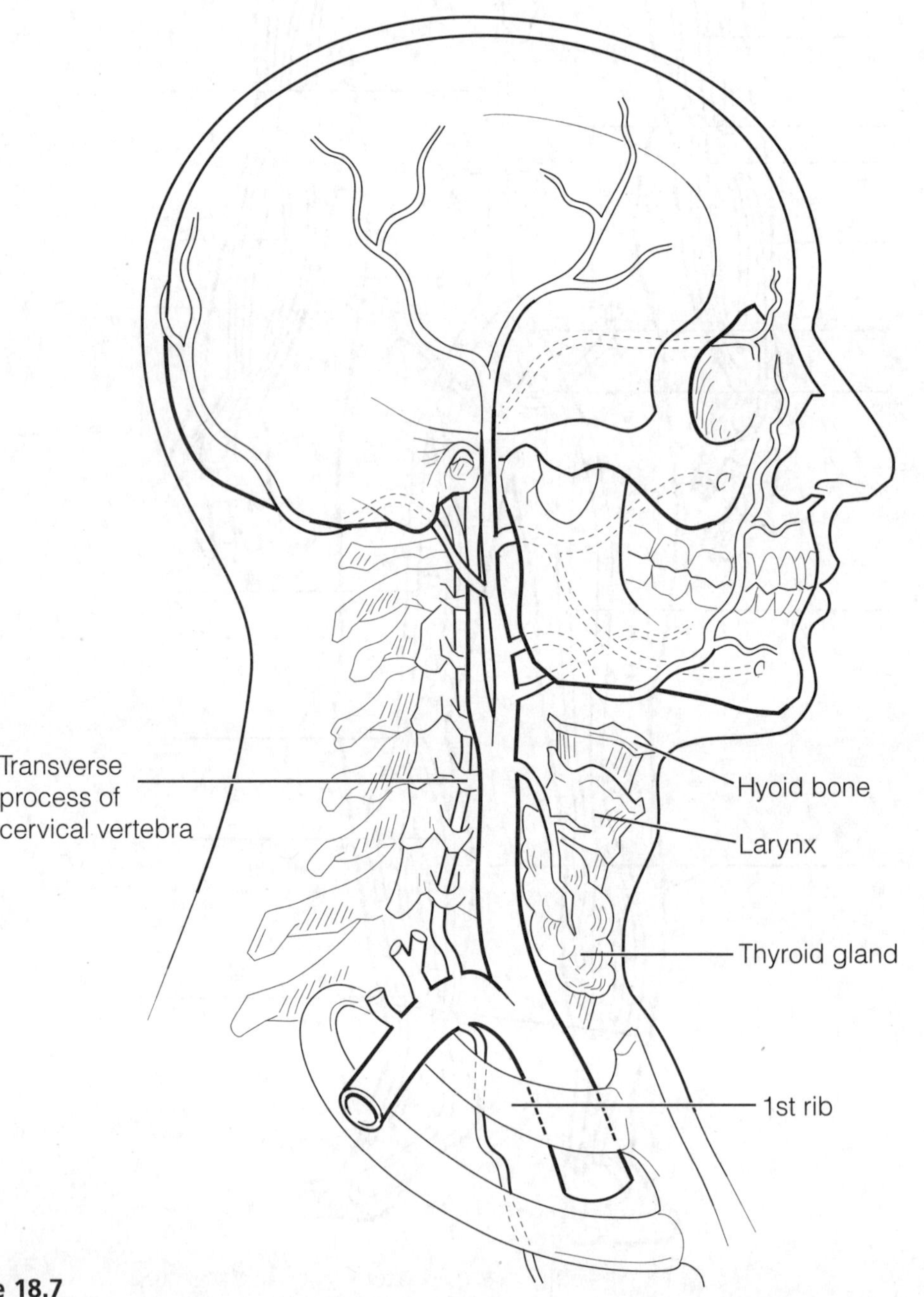

Figure 18.7
Arteries of the head and neck

4. Figure 18.8 illustrates the arterial circulation of the brain. Select different colors for the following structures, and color the diagram. Then respond to the questions following the diagram by choosing responses from the structures to be identified.

○ A. Basilar artery

○ B. Communicating arteries

○ C. Anterior cerebral arteries

○ D. Middle cerebral arteries

○ E. Posterior cerebral arteries

○ F. Cerebral arterial circle (Circle of Willis)

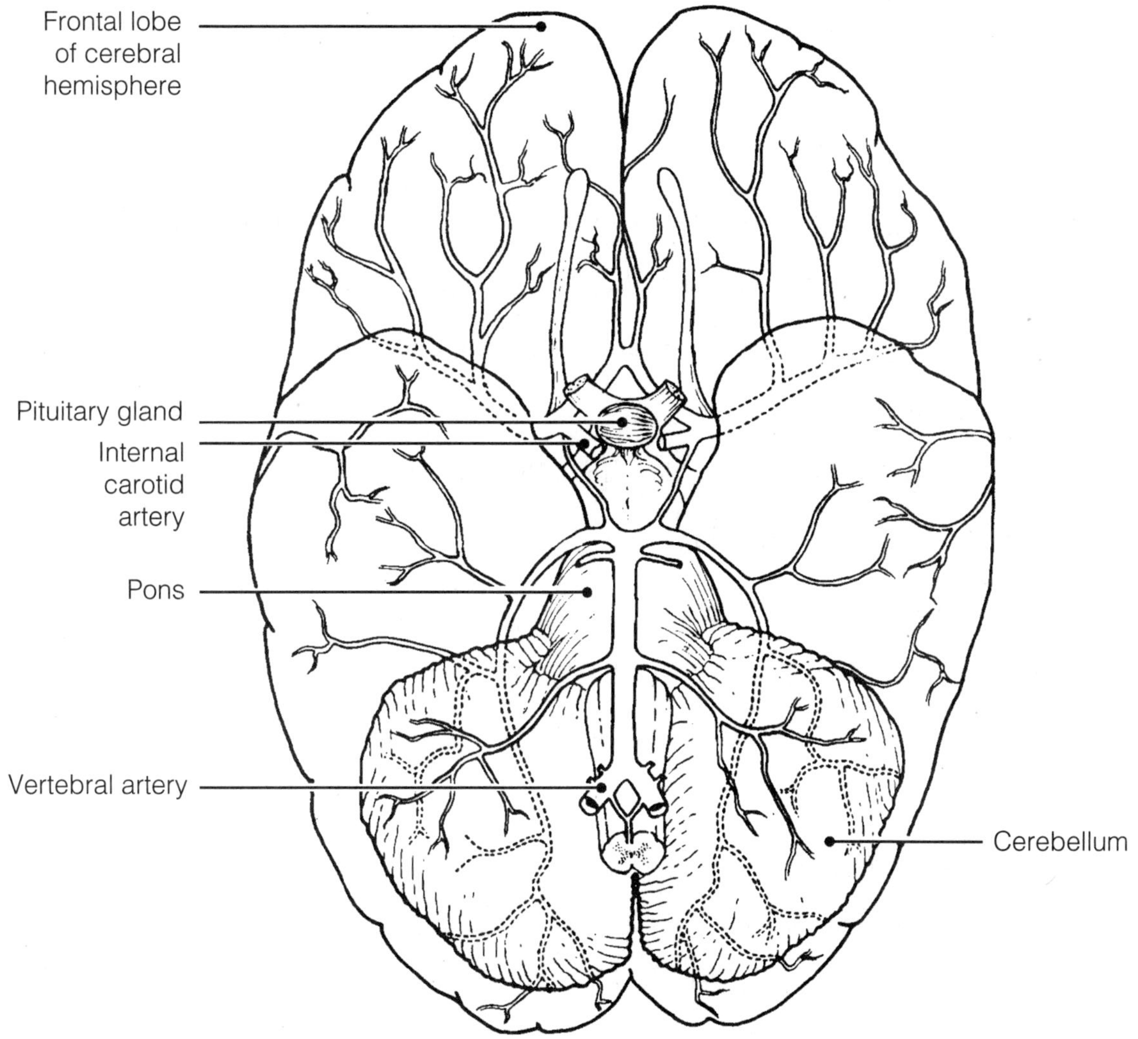

Frontal lobe of cerebral hemisphere
Pituitary gland
Internal carotid artery
Pons
Vertebral artery
Cerebellum

Figure 18.8

______________ 1. What is the name of the anastomosis that allows communication between the posterior and anterior blood supplies of the brain?

______________ 2. What two pairs of arteries arise from the internal carotid artery?

______________ 3. The posterior cerebral arteries serving the brain arise from what artery?

5. Using the key choices, identify the *arteries* in the following descriptions.

Key Choices

A. Anterior tibial	H. Coronary	O. Intercostals	V. Renal
B. Aorta	I. Deep artery of thigh	P. Internal carotid	W. Subclavian
C. Brachial	J. Dorsalis pedis	Q. Internal iliac	X. Superior mesenteric
D. Brachiocephalic	K. External carotid	R. Fibular	Y. Vertebral
E. Celiac trunk	L. Femoral	S. Phrenic	Z. Ulnar
F. Common carotid	M. Hepatic	T. Posterior tibial	
G. Common iliac	N. Inferior mesenteric	U. Radial	

_______ 1. _______ 2. Two arteries formed by the division of the brachiocephalic artery

_______ 3. First branches off the ascending aorta; serve the heart

_______ 4. _______ 5. Two paired arteries serving the brain

_______ 6. Largest artery of the body

_______ 7. Arterial network on the dorsum of the foot

_______ 8. Serves the posterior thigh

_______ 9. Supplies the diaphragm

_______ 10. Splits to form the radial and ulnar arteries

_______ 11. Auscultated to determine blood pressure in the arm

_______ 12. Supplies the last half of the large intestine

_______ 13. Serves the pelvis

_______ 14. External iliac becomes this artery on entering the thigh

_______ 15. Major artery serving the arm

_______ 16. Supplies the small intestine and part of the large intestine

_______ 17. Terminal branches of the dorsal, or descending, aorta

_______ 18. Arterial trunk that has three major branches, which serve the liver, spleen, and stomach

_______ 19. Major artery serving the tissues external to the skull

_______ 20. _______ 21. _______ 22. Three arteries serving the leg

_______ 23. Artery generally used to feel the pulse at the wrist

6. Figure 18.9 shows the venous drainage of the head. Color code and color each of the drainage veins individually. Label each of the dural venous sinuses that has a leader line, but color all the dural sinuses yellow. Note that the clavicle has been omitted.

Dural venous sinuses

Cavernous sinus
Inferior sagittal sinus
Straight sinus
Superior sagittal sinus
Transverse sinus

Drainage veins

- ○ Brachiocephalic
- ○ External jugular
- ○ Facial
- ○ Internal jugular
- ○ Middle thyroid
- ○ Ophthalmic
- ○ Subclavian
- ○ Superficial temporal
- ○ Superior thyroid
- ○ Vertebral

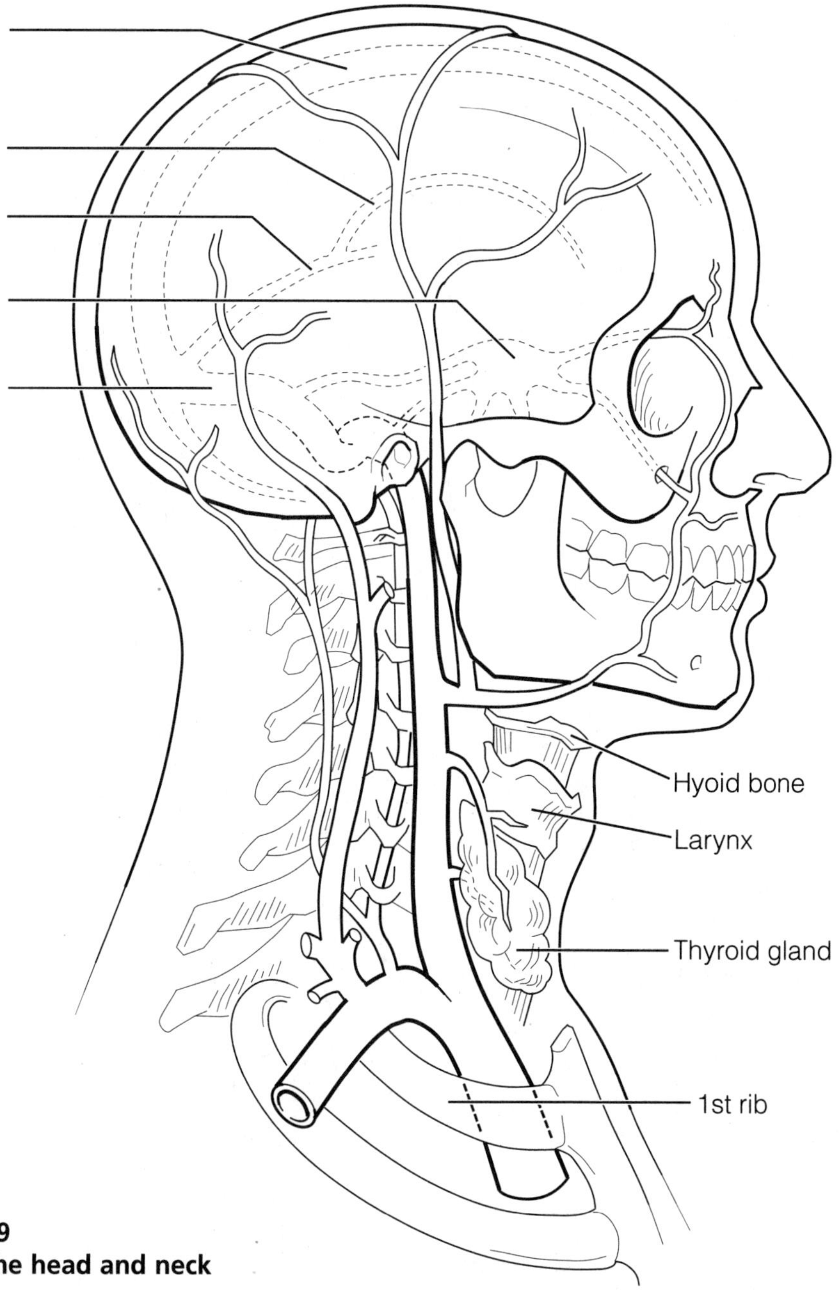

Figure 18.9
Veins of the head and neck

7. Figure 18.10 shows the arteries and veins of the upper limb. Using the key choices, identify each vessel provided with a leader line. (*Note:* In many cases, the same term will be used to identify both an artery and a deep vein.)

Key Choices

A. Axillary	H. Costocervical trunk	O. Subclavian
B. Basilic	I. Deep artery of arm	P. Subscapular
C. Brachial	J. Deep palmar arch	Q. Superficial palmar arch
D. Brachiocephalic	K. Digital	R. Thoracoacromial trunk
E. Cephalic	L. Median cubital	S. Thoracocervical trunk
F. Circumflex (anterior humeral)	M. Lateral thoracic	T. Thoracodorsal
G. Common interosseous	N. Radial	U. Ulnar

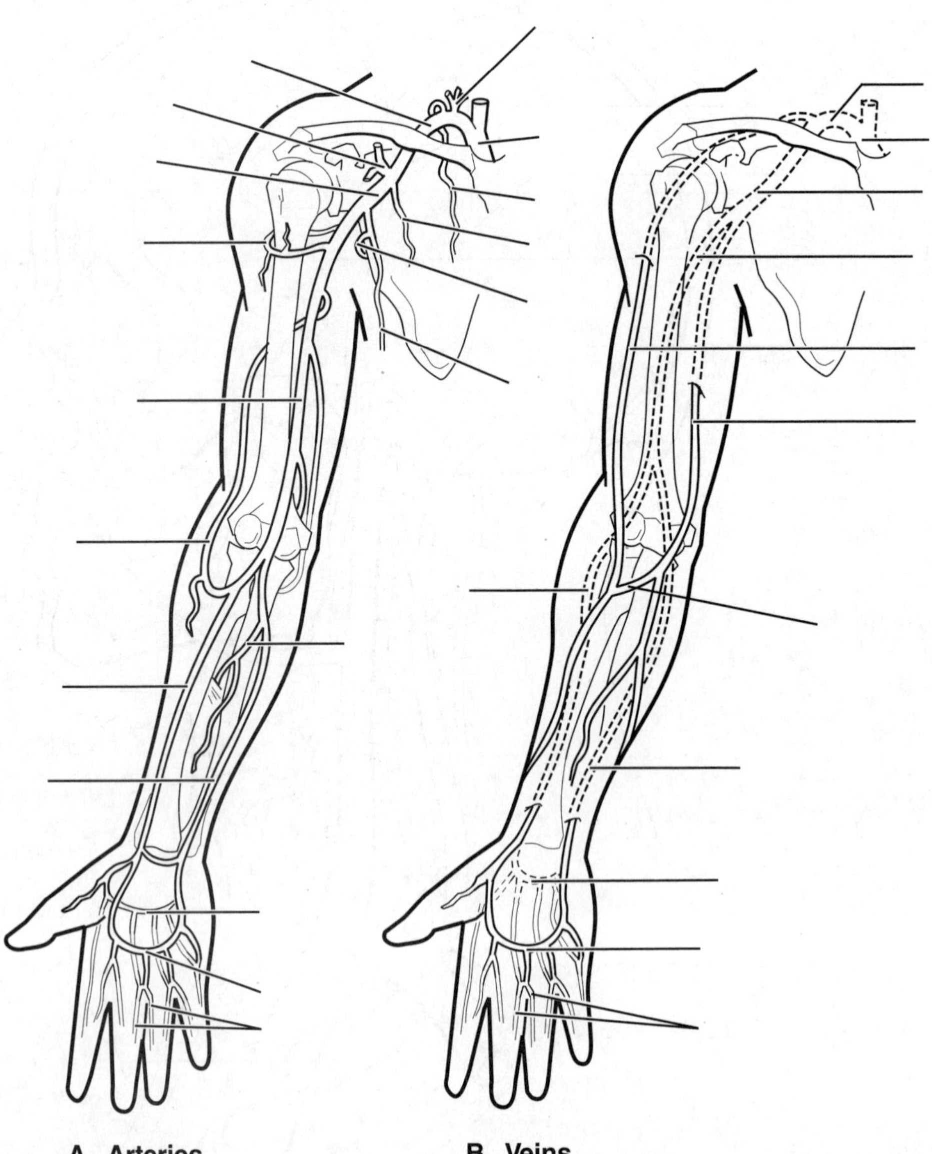

Figure 18.10

8. The abdominal vasculature is depicted in Figure 18.11. Using the key choices, identify the following vessels by selecting the correct letters. Color the diagram as you wish.

Key Choices

A. Aorta	F. Inferior mesenteric artery	K. Renal arteries
B. Celiac trunk	G. Inferior vena cava	L. Renal veins
C. Common iliac arteries	H. Lumbar arteries	M. Left gonadal vein
D. Gonadal arteries	I. Median sacral artery	N. Right gonadal vein
E. Hepatic veins	J. Superior mesenteric artery	

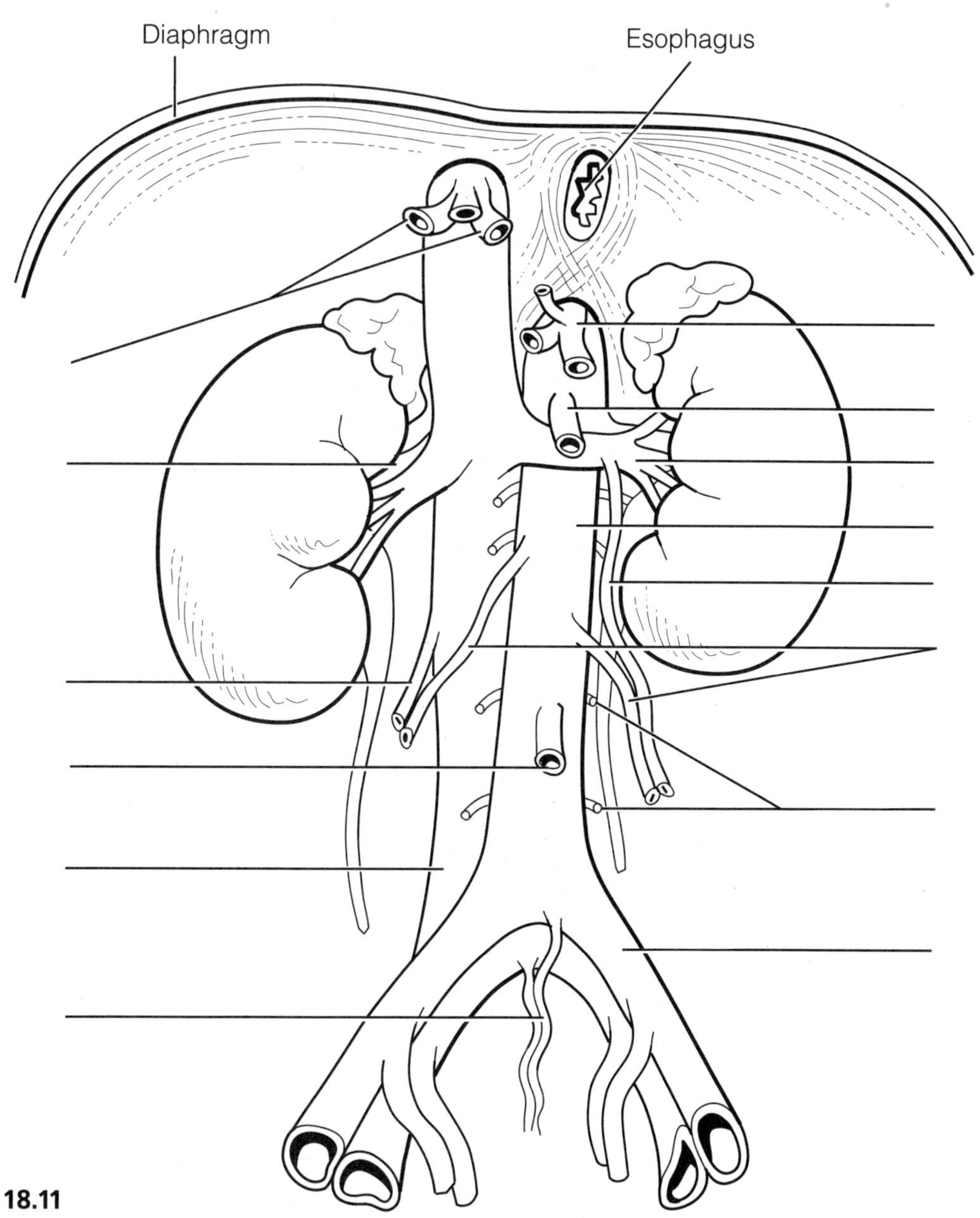

Figure 18.11

9. On Figure 18.12, identify the thoracic veins by choosing a letter from the key choices.

Key Choices

A. Accessory hemiazygos	E. External jugular	I. Left subclavian
B. Axillary	F. Inferior vena cava	J. Posterior intercostals
C. Azygos	G. Hemiazygos	K. Right subclavian
D. Brachiocephalic	H. Internal jugular	L. Superior vena cava

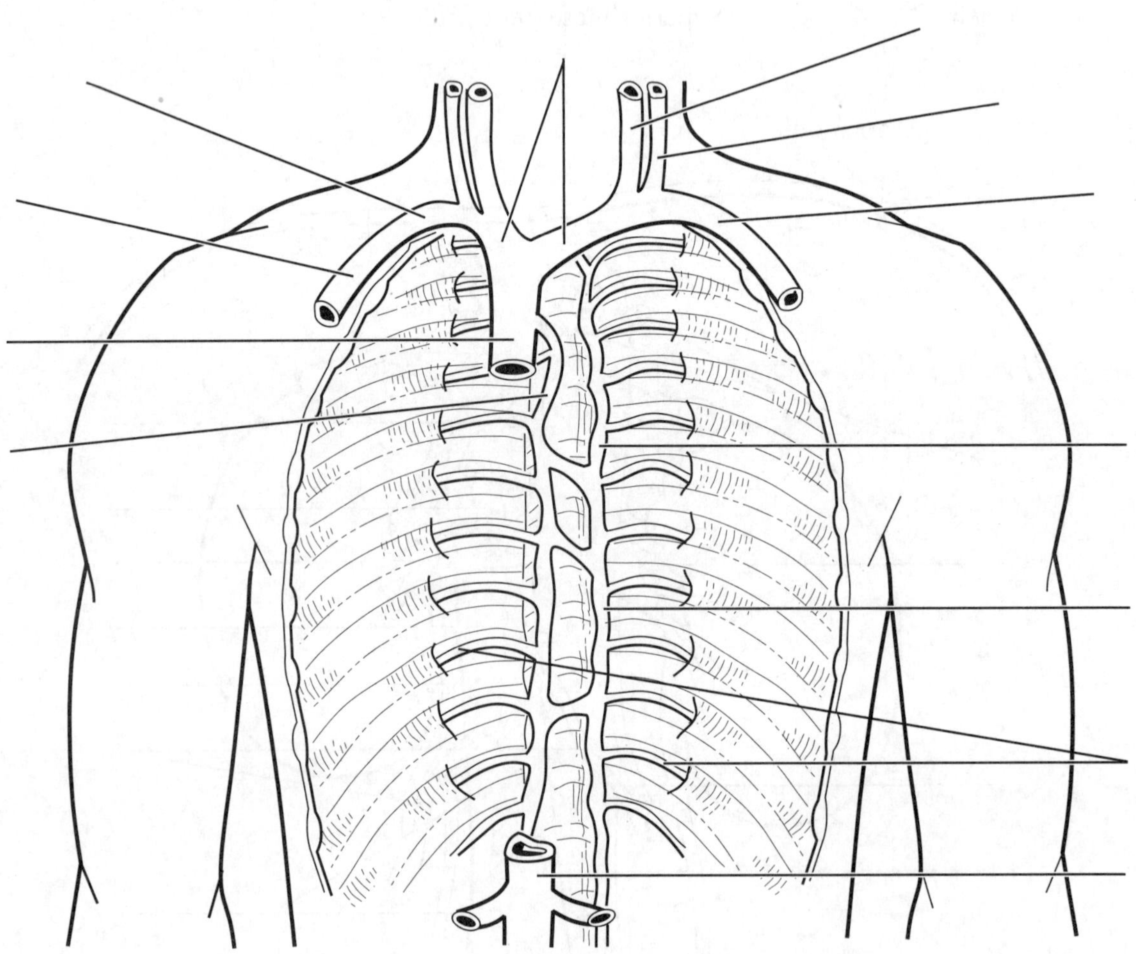

Figure 18.12

10. Figure 18.13, on the opposite page, illustrates the arterial supply of the right lower limb. Correctly identify all arteries that have leader lines by using letters from the key choices. *Notice that the key choices continue onto the opposite page.*

Key Choices

A. Abdominal aorta	C. Common iliac	E. Digital	G. External iliac
B. Anterior tibial	D. Deep artery of thigh	F. Dorsalis pedis	H. Femoral

18.14 is a diagram of the hepatic portal circulation. Select different for the structures listed below and color the structures on the ration.

- ○ Inferior mesenteric vein
- ○ Splenic vein
- ○ Hepatic portal vein
- ○ Superior mesenteric vein
- ○ Gastric vein

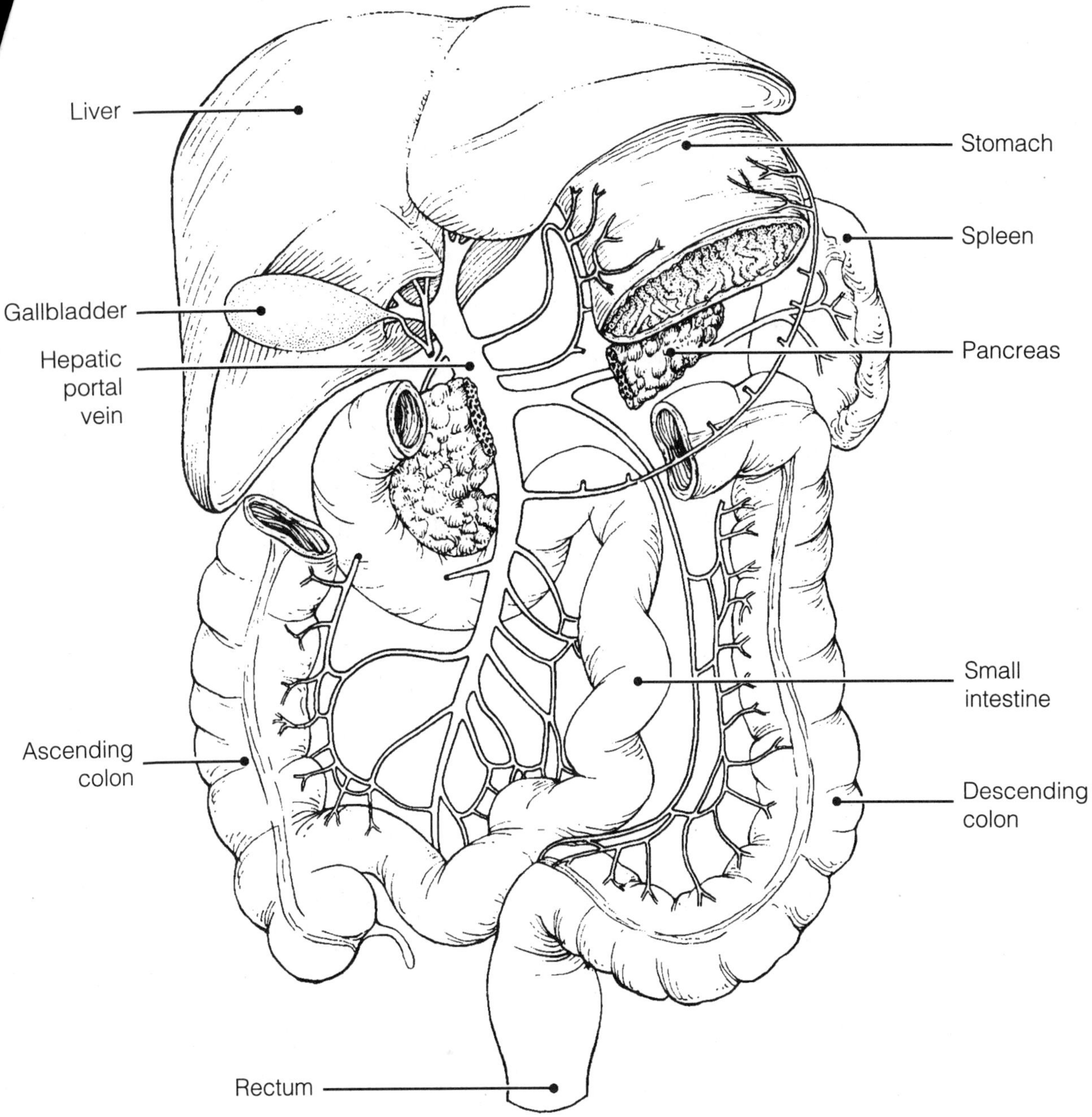

Figure 18.14

11. Figure
colors
illust

L.

M. Fib

N. Plantai

O. Popliteal

P. Posterior tibia

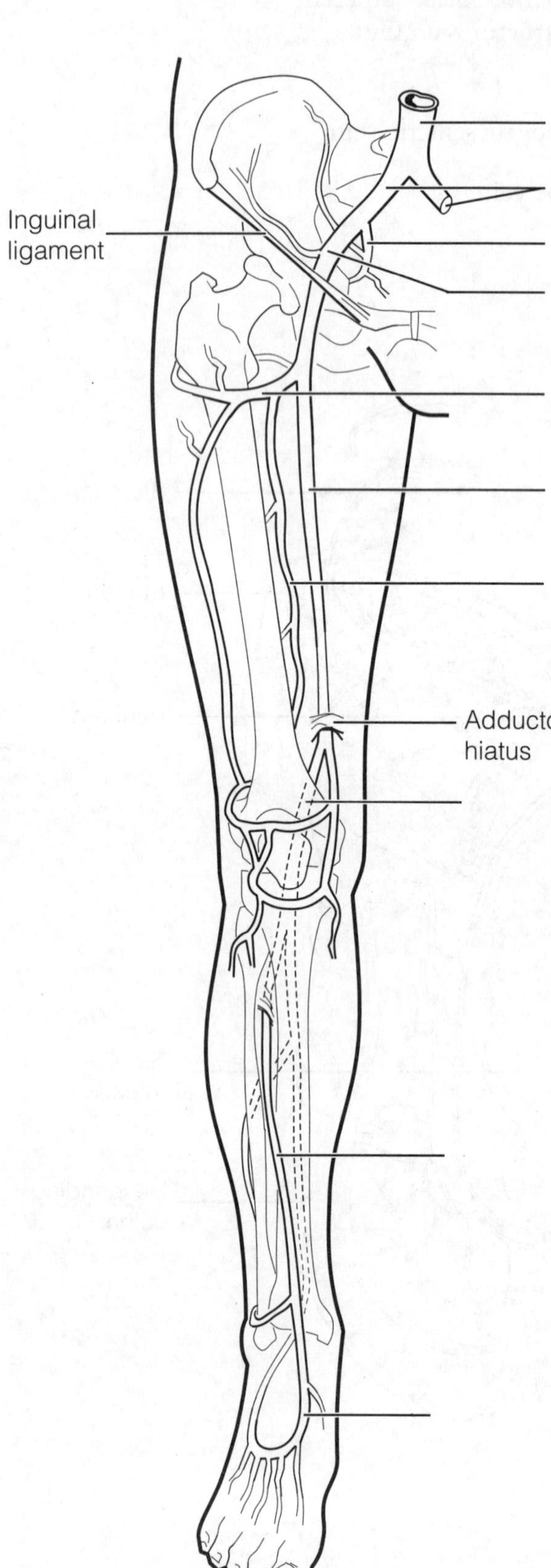

A. **Arteries of the pelvis, thigh, and leg (anterior view)**

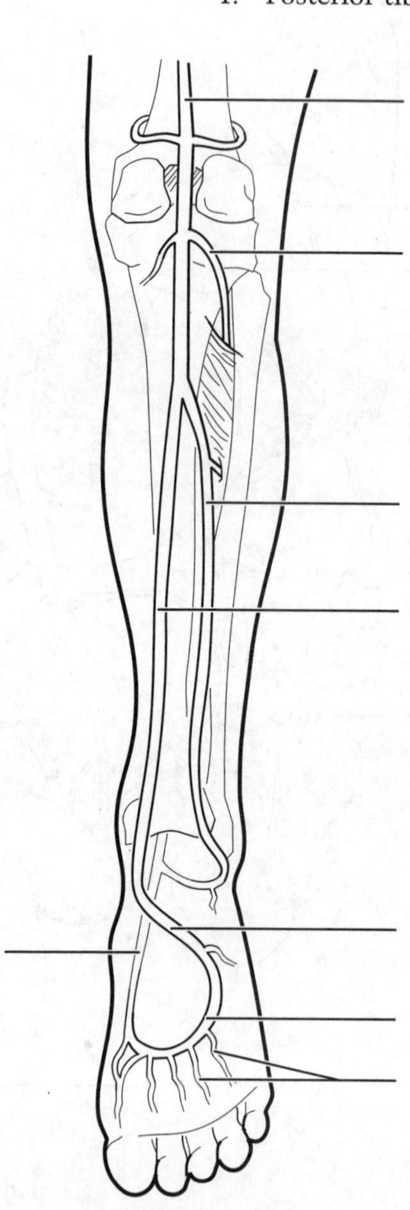

B. **Arteries of the leg (posterior view)**

Figure 18.13

12. Using the key choices, identify the veins in the following descriptions.

Key Choices

A. Anterior tibial	I. Gastric	Q. Internal jugular
B. Azygos	J. Gonadal	R. Posterior tibial
C. Basilic	K. Great saphenous	S. Radial
D. Brachiocephalic	L. Hepatic	T. Renal
E. Cardiac	M. Hepatic portal	U. Subclavian
F. Cephalic	N. Inferior mesenteric	V. Superior mesenteric
G. Common iliac	O. Inferior vena cava	W. Superior vena cava
H. Deep femoral	P. Internal iliac	X. Ulnar

_______ 1. _______ 2. Deep veins; drain the forearm

_______ 3. Receives blood from the arm via the axillary vein

_______ 4. Drains venous blood from the myocardium of the heart into the coronary sinus

_______ 5. Drains the kidney

_______ 6. Drains the dural sinuses of the brain

_______ 7. Join to become the superior vena cava (2)

_______ 8. _______ 9. Drain the leg and foot

_______ 10. Carries nutrient-rich blood from the digestive organs to the liver for processing

_______ 11. Superficial vein that drains the lateral aspect of the arm

_______ 12. Drains the ovaries or testes

_______ 13. Drains the thorax, empties into the superior vena cava

_______ 14. Largest vein inferior to the thorax

_______ 15. Drains the liver

_______ 16. _______ 17. _______ 18. Three veins that form/empty into the hepatic portal vein

_______ 19. Longest superficial vein of the body

_______ 20. Formed by the union of the external and internal iliac veins

_______ 21. Deep vein of the thigh

THE INCREDIBLE JOURNEY:

A Visualization Exercise for the Cardiovascular System

All about you are huge white cords, hanging limply from two flaps of endothelial tissue . . .

1. Complete the following narrative by writing the missing terms in the answer blanks.

1. ____________________
2. ____________________
3. ____________________
4. ____________________
5. ____________________
6. ____________________
7. ____________________
8. ____________________
9. ____________________
10. ____________________
11. ____________________
12. ____________________
13. ____________________

Your journey starts in the pulmonary vein and includes a trip to part of the systemic circulation and a special circulation. You ready your equipment and prepare to be injected into your host.

Almost immediately after injection, you find yourself swept into a good-sized chamber, the __(1)__. However, you do not stop in this chamber but continue to plunge downward into a larger chamber below. You land with a large splash and examine your surroundings. All about you are huge white cords, hanging limply from two flaps of endothelial tissue far above you. You report that you are sitting in the __(2)__ chamber of the heart, seeing the flaps of the __(3)__ valve above you. The valve is open, and its anchoring cords, the __(4)__, are lax. Since this valve is open, you conclude that the heart is in the __(5)__ phase of the cardiac cycle.

Suddenly you notice that the chamber walls seem to be closing in. You hear a thundering boom, and the whole chamber vibrates as the valve slams shut above you. The cords, now rigid and strained, form a cage about you, and you feel extreme external pressure. Obviously, the heart is in a full-fledged __(6)__. Then, high above on the right, the "roof" opens, and you are forced through this __(7)__ valve. A fraction of a second later, you hear another tremendous boom that sends shock waves through the whole area. Out of the corner of your eye, you see that the valve below you is closed, and it looks rather like a pie cut into three wedges.

As you are swept along in a huge artery, the __(8)__, you pass several branch-off points, but continue to careen along straight down at a dizzying speed until you approach the __(9)__ artery feeding the small intestine. After entering this artery and passing through successively smaller and smaller subdivisions of it, you finally reach the capillary bed of the small intestine. You watch with fascination as nutrient molecules move into the blood through the single layer of __(10)__ cells forming the capillary wall. As you move to the opposite shore of the capillary bed, you enter a venule and begin to move superiorly once again. The venules draining the small intestine combine to form the __(11)__ vein, which in turn combines with the __(12)__ vein to form the hepatic portal vein that carries you into the liver. As you enter the liver, you are amazed at the activity there. Six-sided hepatic cells, responsible for storing glucose and making blood proteins, are literally grabbing __(13)__ out of the blood as it percolates slowly past them.

________________ 14.
________________ 15.
________________ 16.
________________ 17.

Protective __(14)__ cells are removing bacteria from the slowly moving blood. Leaving the liver through the __(15)__ vein, you almost immediately enter the huge __(16)__, which returns blood from the lower part of the body to the __(17)__ of the heart. From here, you move consecutively through the right chambers of the heart into the __(18)__ artery, which carries you to the __(19)__.

________________ 18.
________________ 19.
________________ 20.
________________ 21.
________________ 22.

You report that your surroundings are now relatively peaceful. You catch hold of a red blood cell and scramble aboard. As you sit cross-legged on its concave surface, you muse that it is just like riding in a big, soft rubber boat. Riding smoothly along, you enter vessels that become narrower and narrower. Your red blood cell slows down as it squeezes through the windings of tiny blood vessels. You conclude that you are located in the pulmonary __(20)__. You report hearing great whooshes of sound as your host breathes, and you can actually see large air-filled spaces! But your little red cell "boat" seems strangely agitated. Its contours change so often that you are almost thrown out. You report that these contortions must be the result of the exchange of __(21)__ molecules because your "boat" is assuming a brilliant red color. This experience has left you breathless, so you poke your head between two thin endothelial cells for some fresh air. You then continue your journey, through increasingly wider vessels. After traveling through the left side of the heart again, you leave your host when you are aspirated out of the __(22)__ artery, which extends from the aorta to the axillary artery of the armpit.

CHALLENGING YOURSELF

At the Clinic

1. Mrs. Gray, age 50, is complaining of dull aching pains in her legs, which she claims have been getting progressively worse since the birth of her last child. (She is the mother of seven children.) During her physical examination, numerous varicosities are seen in both legs. How are varicosities recognized? What veins are most likely involved? What pathologic changes have occurred in these veins and what is the most likely causative factor in this patient's case? What instructions might be helpful to Mrs. Gray?

2. A routine scan on an elderly man reveals partial occlusion of the internal carotid artery, yet blood supply to his cerebrum is unimpaired. What are two possible causes of the occlusion? What compensatory mechanism is maintaining blood supply to the brain?

3. A patient with a bone marrow cancer is polycythemic. Will his blood pressure be high or low? Why?

4. Mrs. Baevich, an elderly, bedridden woman, has several "angry-looking" decubitus ulcers. During the past week, one of these has begun to hemorrhage very slowly. What will occur eventually if this situation is not properly attended to?

5. A blow to the base of the skull has sent an accident victim to the emergency room. Although there are no signs of hemorrhage, his blood pressure is critically low. What is a probable cause of his hypotension?

6. Mr. Grimaldi was previously diagnosed as having a posterior pituitary tumor that causes hypersecretion of ADH. He comes to the clinic regularly to have his blood pressure checked. Would you expect his BP to be chronically elevated or depressed? Why?

7. Renal (kidney) disease in a young man has resulted in blockage of the smaller arteries within the kidneys. Explain why this will lead to secondary hypertension.
8. A worried mother has brought Terry, her 11-year-old son, to the clinic after he complained of feeling dizzy and nauseous. No abnormalities are found during the examination. Upon questioning, however, Terry sheepishly admits that he was smoking with a friend, who had dared him to "take deep drags." Explain how this might have caused his dizziness.
9. An angiogram reveals coronary artery occlusion requiring medical intervention. What measure might be taken to reduce the occlusion in order to avoid bypass surgery?
10. Mrs. Shultz has died suddenly after what seemed to be a low-grade infection. A blood culture reveals that the infection had spread to the blood. What is this condition called and why was it fatal?
11. Progressive heart failure has resulted in insufficient circulation to sustain life. What type of circulatory shock does this exemplify?
12. Mr. Connors has just had a right femoral arterial graft and is resting quietly. Orders are given to palpate his popliteal and pedal (dorsalis pedis) pulses and to assess the color and temperature of his right leg four times daily. Why were these orders given?
13. Len, an elderly man, is bedridden after a hip fracture. He complains of pain in his legs, and phlebitis is diagnosed. What is phlebitis and what life-threatening complication can develop?
14. Examination of Mr. Cummings, a man in his 60s, reveals a blood pressure of 140/120. What is his pulse pressure? Is it normal, high, or low? What does this indicate about the state of his elastic arteries?
15. A man in his 40s was diagnosed as hypertensive. Dietary changes and exercise have helped, but his blood pressure is still too high. Explain to him why his doctor recommended beta blocker and diuretic drugs to treat his condition.
16. Sidney received a small but deep cut from broken glass in the exact midline of the anterior side of his distal forearm. He worried that he would bleed to death because he had heard stories about people committing suicide by slashing their wrist. Judge if Sid's fear of death is justified and explain your reasoning.

Stop and Think

1. Why is the cardiovascular lining a simple squamous epithelium rather than a thicker layer?
2. Does vasoconstriction increase or decrease the supply of blood to the tissues beyond the vasoconstricted area? Explain your answer.
3. When an entire capillary bed is closed off due to arteriolar vasoconstriction, are the precapillary sphincters open or closed? What is the pattern of blood flow into the capillaries when the arteriole dilates? From this, explain why the face flushes when coming inside on a cold day.
4. Standing up quickly after being in a horizontal position can cause dizziness. Why is this more likely in a warm room than in a cool room?
5. Why shouldn't a pregnant woman sleep on her back late in pregnancy?
6. The dura mater reinforces the walls of the intracranial sinuses. What reinforces the wall of the coronary sinus?
7. Why is necrosis more common in areas supplied by end arteries?
8. What would trigger a change in blood vessel length in the body?

9. Assume there is a clot blocking the right internal carotid artery. Briefly describe another route that might be used to provide blood to the areas served by its branches.

10. Kwashiorkor results from a dietary deficiency of protein. Explain why this causes edema.

11. Describe the positive feedback mechanism between the heart and vasomotor center that would be involved in untreated or irreversible circulatory shock.

12. Your friend, who knows very little about science, is reading a magazine article about a patient who had an "aneurysm at the base of his brain that suddenly grew much larger." The surgeons' first goal was to "keep it from rupturing" and the second goal was to "relieve the pressure on the brain stem and cranial nerves." The surgeons were able to "replace the aneurysm with a section of plastic tubing," so the patient recovered. Your friend asks you what all this means, and why the condition is life-threatening. What would you tell him?

13. Mr. Brown was distracted while trying to fell a large tree. His power saw whipped around and severed his right arm at the shoulder. Without a limb stump, applying a tourniquet is impossible. Where would you apply pressure to save Mr. Brown from fatal hemorrhage?

14. Freddy was looking for capillaries in his microscope slides of body organs. Although Freddy disagreed, Karl, his lab partner, kept insisting that a venule was a capillary. Finally, the teaching assistant said, "Look, here you see five erythrocytes lined up across the width of the lumen of this vessel so it cannot be a capillary." Explain the logic behind this statement and tell exactly how wide the vessel was (in micrometers).

COVERING ALL YOUR BASES

Multiple Choice

Select the best answer or answers from the choices given.

1. Which of the following is (are) part of the tunica intima?
   A. Simple squamous epithelium
   B. Basement membrane
   C. Loose connective tissue
   D. Smooth muscle

2. The tunica externa contains:
   A. nerves
   B. blood vessels
   C. lymphatics
   D. collagen

3. In comparing a parallel artery and vein, you would find that:
   A. the artery wall is thicker
   B. the artery diameter is greater
   C. the artery lumen is smaller
   D. the artery endothelium is thicker

4. Which vessels are conducting arteries?
   A. Brachiocephalic artery
   B. Common iliac artery
   C. Digital artery
   D. Arcuate artery (foot)

5. A pulse would be palpable in the:
   A. anterior cerebral artery
   B. hepatic portal vein
   C. muscular arteries
   D. inferior vena cava

6. Pulse pressure is increased by:
   A. an increase in diastolic pressure
   B. an increase in systolic pressure
   C. an equivalent increase in both systolic and diastolic pressures
   D. vasoconstriction

7. Vessels that do not have all three tunics include the:
   A. smallest arterioles
   B. smallest venules
   C. capillaries
   D. venous sinuses

8. Structures that are totally avascular are the:
   A. skin
   B. cornea
   C. articular surfaces of long bones
   D. joint capsules

9. Fenestrated capillaries occur in the:
   A. liver
   B. kidney
   C. cerebrum
   D. intestinal mucosa

10. Kupffer cells are:
   A. true capillary endothelium
   B. found throughout the body
   C. macrophages in the liver
   D. macrophages in the spleen

11. Which of the following is (are) part of a capillary bed?
   A. Precapillary sphincter
   B. Metarteriole
   C. Thoroughfare channel
   D. Terminal arteriole

12. Which vessels actively supply blood to associated capillary networks when the body is primarily controlled by the parasympathetic division?
   A. Femoral artery
   B. Superior mesenteric artery
   C. Cerebral arteries
   D. Coronary arteries

13. Which of the following can function as a blood reservoir?
   A. Brachiocephalic artery
   B. Cerebral capillaries
   C. Dural sinuses
   D. Inferior vena cava

14. Collateral circulation occurs in the:
   A. joint capsules
   B. cerebrum
   C. kidney
   D. myocardium

15. Venous return to the right atrium is increased by:
   A. increasing depth of respiration
   B. vigorous walking
   C. an increase in ventricular contraction strength
   D. activation of angiotensin II

16. The single most important factor in blood pressure regulation is:
   A. feedback by baroreceptors
   B. renal compensatory mechanisms
   C. regulation of cardiac output
   D. short-term changes in blood vessel diameter

17. Blood volume is altered by:
   A. vasodilation
   B. edema
   C. increased salt intake
   D. polycythemia

18. Vasomotor fibers that secrete acetylcholine are (probably) found in:
   A. cerebral arteries
   B. skeletal muscle
   C. skin
   D. digestive organs

19. The vasomotor center is inhibited by:
   A. impulses from the carotid sinus baroreceptors
   B. impulses from the aortic sinus baroreceptors
   C. stimulation by carotid body chemoreceptors
   D. alcohol

20. Chemical factors that increase blood pressure include:
   A. endothelin
   B. NO
   C. ADH
   D. ANP

21. Which of the regulatory chemicals listed involve or target the kidneys?
   A. Angiotensin
   B. ADH
   C. Aldosterone
   D. ANP

22. Mechanisms to decrease systemic arterial blood pressure are invoked by:
   A. standing up
   B. public speaking
   C. losing weight
   D. moving to higher altitude

23. Which of the following interfere with contraction of the smooth muscle of precapillary sphincters?
   A. Adenosine
   B. Oxygen
   C. Lactic acid
   D. Potassium ions

24. Increase in blood flow to a capillary bed after its supply has been blocked is called:
   A. a myogenic response
   B. reactive hyperemia
   C. active hyperemia
   D. a cardiogenic response

25. An increase in which of the following results in increased filtration from capillaries to the interstitial space?
   A. Capillary hydrostatic pressure
   B. Interstitial fluid hydrostatic pressure
   C. Capillary osmotic pressure
   D. Duration of precapillary sphincter contraction

26. Vessels involved in the circulatory pathway to and from the brain are the:
   A. brachiocephalic artery
   B. subclavian artery
   C. internal jugular vein
   D. internal carotid artery

27. The fight-or-flight response results in the constriction of:
A. renal arteries
B. celiac trunk
C. internal iliac arteries
D. external iliac arteries

28. Prominent venous anastomoses are formed by thc:
A. azygos vein
B. median cubital vein
C. cerebral arterial circle
D. hepatic portal vein

29. An abnormal thickening of the capillary basement membrane associated with diabetes mellitus is called:
A. phlebitis
B. aneurysm
C. microangiopathic lesion
D. atheroma

30. Which layer of the artery wall thickens most in atherosclerosis?
A. Tunica media
B. Tunica intima
C. Tunica adventitia
D. Tunica externa

31. Based on the vessels named pulmonary trunk, thyrocervical trunk, and celiac trunk, the term *trunk* must refer to:
A. a vessel in the heart wall
B. a vein
C. a capillary
D. a large artery from which other arteries branch

32. Which of these vessels is bilaterally symmetrical (i.e., one vessel of the pair occurs on each side of the body)?
A. Internal carotid artery
B. Brachiocephalic artery
C. Azygos vein
D. Superior mesenteric vein

33. The deep femoral and deep brachial veins drain the:
A. biceps brachii and hamstring muscles
B. flexor muscles in the hand and leg
C. quadriceps femoris and triceps brachii muscles
D. intercostal muscles of the thorax

34. A stroke that occludes a posterior cerebral artery will most likely affect:
A. hearing
B. vision
C. smell
D. higher thought processes

35. Tracing the drainage of the *superficial* venous blood from the leg, we find that blood enters the greater saphenous vein, femoral vein, inferior vena cava, and right atrium. Which veins are missing from that sequence?
A. Coronary sinus and superior vena cava
B. Posterior tibial and popliteal
C. Fibular (peroneal) and popliteal
D. External and common iliacs

36. Tracing the drainage of venous blood from the small intestine, we find that blood enters the superior mesenteric vein, hepatic vein, inferior vena cava, and right atrium. Which vessels are missing from that sequence?
A. Coronary sinus and left atrium
B. Celiac and common hepatic veins
C. Internal and common iliac veins
D. Hepatic portal vein and liver sinusoids

Word Dissection

For each of the following word roots, fill in the literal meaning and give an example, using a word found in this chapter.

Word root	Translation	Example
1. anastomos		
2. angio		
3. aort		
4. athera		
5. auscult		

Word root	Translation	Example
6. azyg		
7. capill		
8. carot		
9. celia		
10. entero		
11. epiplo		
12. fenestr		
13. jugul		
14. ortho		
15. phleb		
16. saphen		
17. septi		
18. tunic		
19. vaso		
20. viscos		

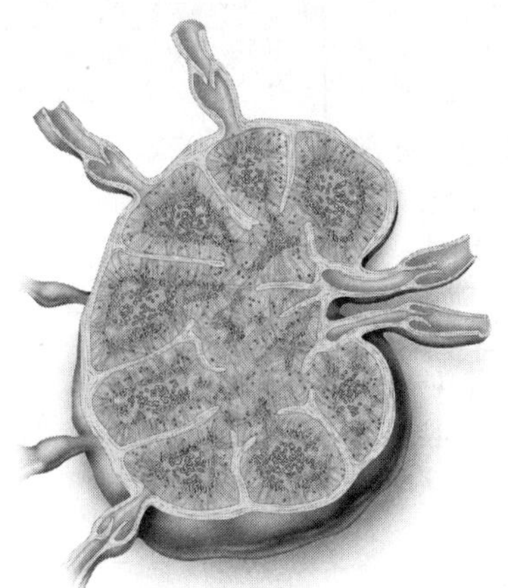

19

The Lymphatic System and Lymphoid Organs and Tissues

Student Objectives

When you have completed the exercises in this chapter, you will have accomplished the following objectives:

Lymphatic Vessels

1. List the functions of the lymphatic vessels.
2. Describe the structure and distribution of lymphatic vessels.
3. Describe the source of lymph and mechanism(s) of lymph transport.

Lymphoid Cells and Tissues

4. Describe the basic structure and cellular population of lymphoid tissue. Differentiate between diffuse and follicular lymphoid tissues.

Lymph Nodes

5. Describe the general location, histological structure, and functions of lymph nodes.

Other Lymphoid Organs

6. Name and describe the other lymphoid organs of the body. Compare and contrast them with lymph nodes, structurally and functionally.

The lymphatic system is a rather strange system, with its myriad lymphoid organs and its vessels derived from veins of the cardiovascular system. Although both elements help to maintain homeostasis, these two elements of the lymphatic system have substantially different roles. The lymphatic vessels help keep the cardiovascular system functional by maintaining blood volume. The lymphoid organs and tissues help defend the body from pathogens by providing operating sites for phagocytes and cells of the immune system. Chapter 19 tests your understanding of the functional roles of the various lymphatic system elements.

BUILDING THE FRAMEWORK

Lymphatic Vessels

1. Complete the following statements by writing the missing terms in the answer blanks.

____________ 1.
____________ 2.
____________ 3.
____________ 4.
____________ 5.
____________ 6.

The lymphatic system is a specialized subdivision of the circulatory system. Although the cardiovascular system has a pump (the heart) and arteries, veins, and capillaries, the lymphatic system lacks two of these structures: the __(1)__ and __(2)__. Like the __(3)__ of the cardiovascular system, the vessels of the lymphatic system are equipped with __(4)__ to prevent backflow. The lymphatic vessels act primarily to pick up leaked fluid, now called __(5)__, and return it to the bloodstream. About __(6)__ of fluid is returned every 24 hours.

2. Lymphatic capillaries exhibit unique junctions between their endothelial cells, and they are anchored to surrounding tissues by fine filaments. How do these structural modifications aid in their function?

____

____

____

____

____

3. Figure 19.1 provides an overview of the lymphatic vessels. In part A, the relationship between lymphatic vessels and the blood vessels of the cardiovascular system is depicted schematically. Part B shows the different types of lymphatic vessels in a simple way. First, color code and color the following structures in Figure 19.1.

○ Heart ○ Veins ○ Lymphatic vessels/lymph node

○ Arteries ○ Blood capillaries ○ Loose connective tissue around the blood and lymph capillaries

Then, identify by labeling these specific structures in part B:

A. Lymph capillaries C. Lymphatic collecting vessels E. Lymph node

B. Lymph duct D. Lymph trunks F. Valves

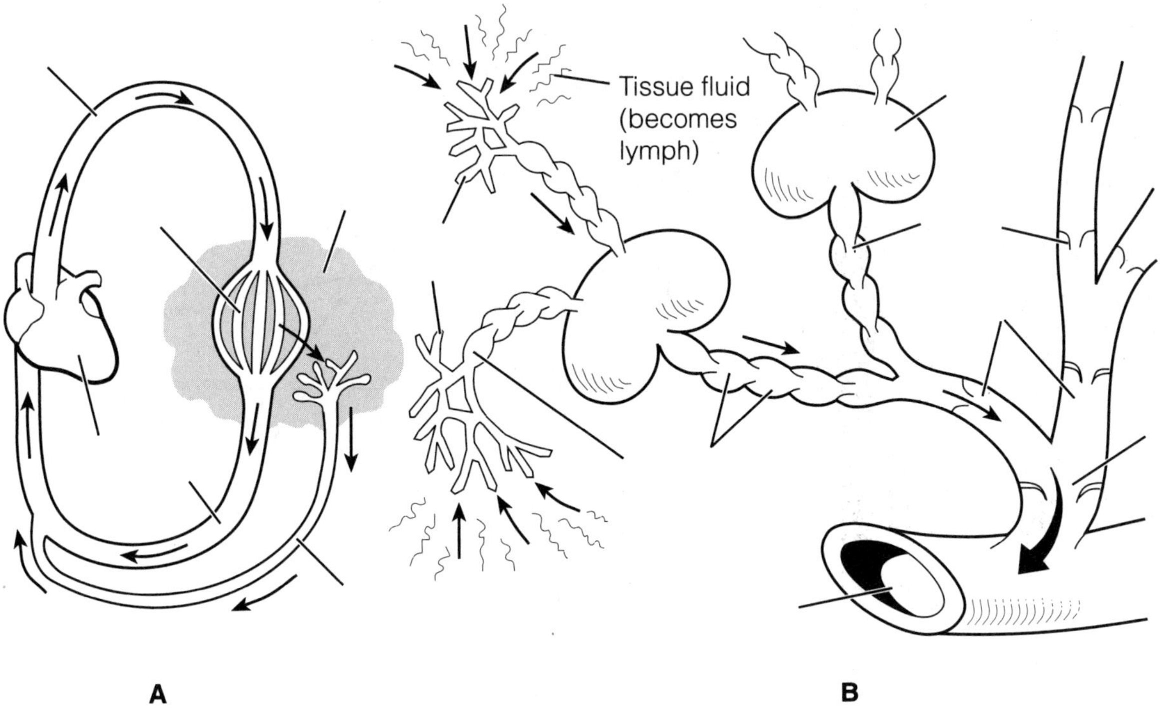

Figure 19.1

4. Circle the term that does not belong in each of the following groupings.

1. Lacteal Lymphatic capillary Chyle Fat-free lymph Intestinal mucosa
2. Blood capillary Lymph capillary Blind-ended Permeable to proteins
3. Edema Blockage of lymphatics Elephantiasis Inflammation Abundant supply of lymphatics
4. Skeletal muscle pump Flow of lymph Respiratory pump High-pressure gradient Action of smooth muscle cells in walls of lymph vessels
5. Minivalves Endothelial cell overlap Impermeable Anchoring filaments Lymphatic capillaries

5. On the more detailed diagram in Figure 19.2, identify all structures provided with leader lines and color all of the lymphatic vessels (trunks, etc.) green. The terms you will need are listed below. Then color code and color the following structures:

◯ All veins ◯ Ribs ◯ Lymph nodes

◯ Trachea ◯ Muscles of the thorax

Bronchomediastinal trunk(s)	Jugular trunk(s)	Lymphatic collecting vessels
Cisterna chyli	Lumbar trunk(s)	Right lymphatic duct
Intestinal trunk	Subclavian trunk(s)	Thoracic duct

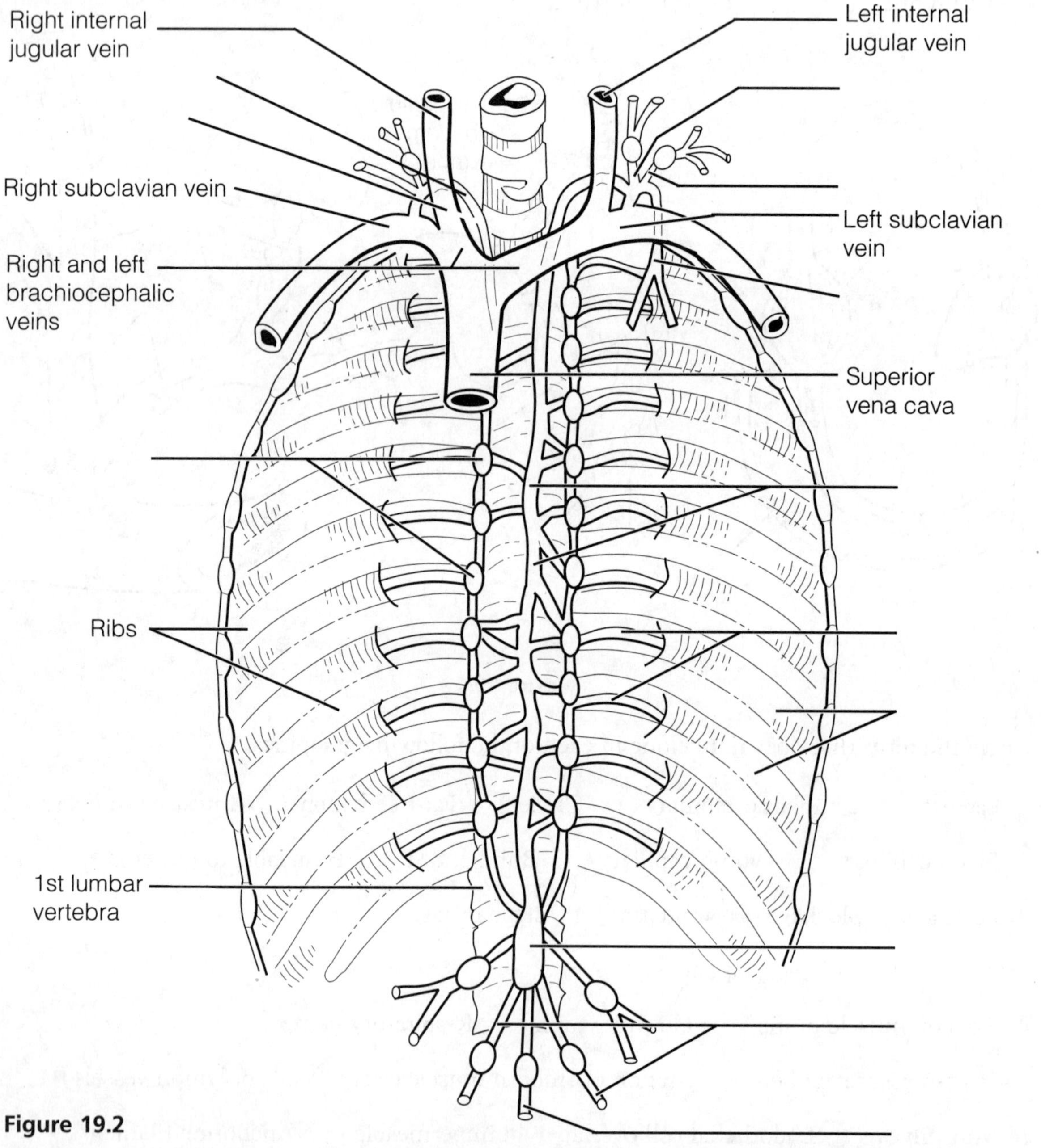

Figure 19.2

Lymphoid Cells and Tissues

1. Complete the following statements by writing the missing terms in the answer blanks.

________________ 1.

________________ 2.

________________ 3.

________________ 4.

________________ 5.

________________ 6.

________________ 7.

________________ 8.

________________ 9. ________________ 10.

There are three major types of cells in the lymphoid tissues of the body. One category, the __(1)__, contains immunocompetent cells that are able to destroy or immobilize foreign substances in the body; these foreign substances are called __(2)__. The two varieties in this group are __(3)__ and __(4)__. The other class of protective cells are called __(5)__. These defend the body by engulfing, or __(6)__, foreign substances. The third cellular group are the __(7)__, fibroblast-like cells that form the stroma or framework of the lymphoid organs. Lymphoid tissue is specifically classified as a type of connective tissue called __(8)__. When lymphoid tissue is loosely packed, it is called __(9)__; when it is tightly packed into spherical masses, the arrangement is said to be __(10)__.

2. Figure 19.3 depicts several different lymphoid organs. Label all lymphoid organs indicated by leader lines and add labels as necessary to identify the sites where the axillary, cervical, and inguinal lymph nodes would be located. Color the lymphoid organs as you like, and then shade in light green the portion of the body that is drained by the right lymphatic duct.

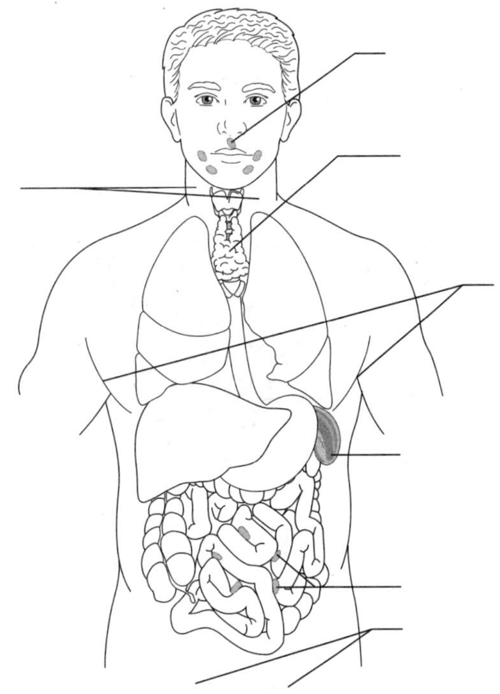

Figure 19.3

Lymph Nodes

1. Figure 19.4 is a diagram of a lymph node. First, using the terms with color coding circles, label all structures on the diagram that have leader lines. Color those structures as well. Then, add arrows to the diagram to show the direction of lymph flow through the organ. Circle the region that would approximately correspond to the medulla of the organ. Finally, answer the questions that follow.

- ○ Germinal centers of follicles
- ○ Cortex (other than germinal centers)
- ○ Medullary cords
- ○ Capsule and trabeculae
- ○ Hilum
- ○ Afferent lymphatics
- ○ Efferent lymphatics
- ○ Sinuses (subcapsular and medullary)

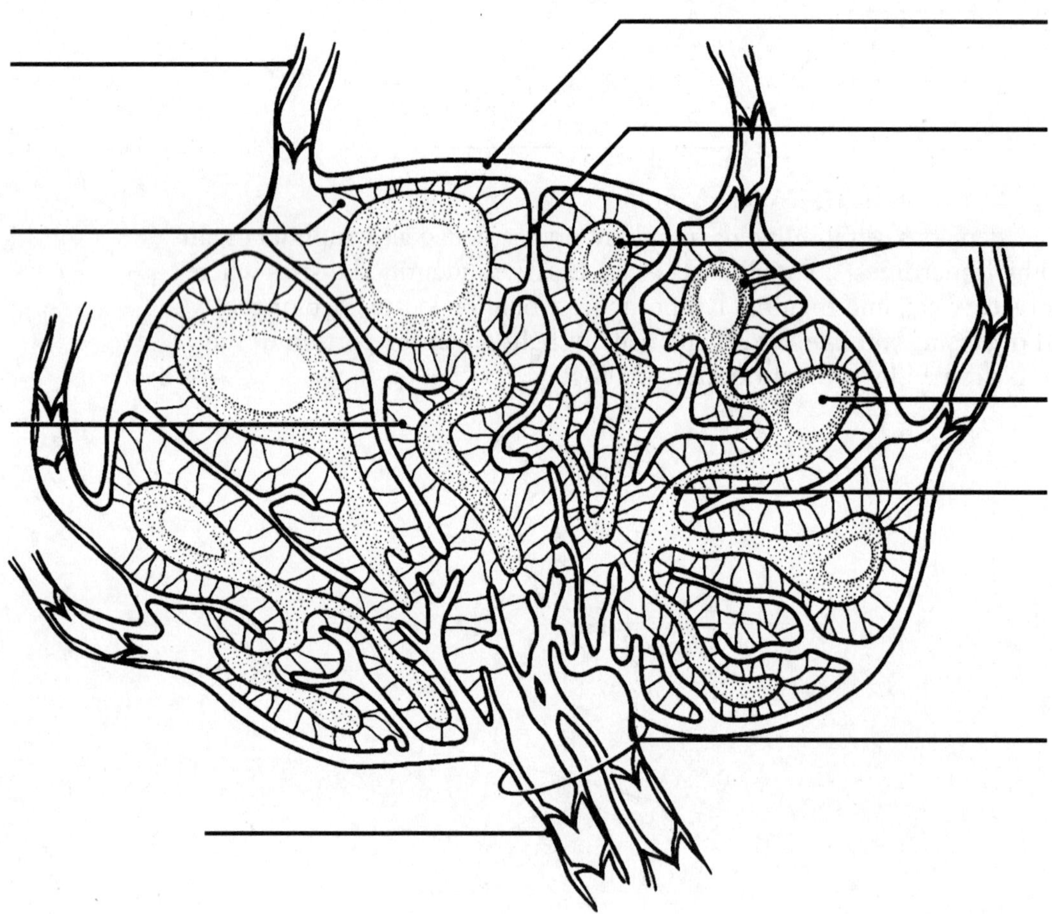

Figure 19.4

1. Which cell type is found in greatest abundance in the germinal centers?

   ____

2. What is the function of their daughter cells, the plasma cells? ______________________________

3. What is the predominant cell type in the cortical regions other than the germinal centers?

   ____

4. What cells cluster around the medullary sinuses? ______________________________

5. Functionally, the cells named in question 4 act as ______________________________.

6. What type of soft connective tissue forms the internal skeleton of the lymph node?

   ____

7. Of what importance is the fact that there are fewer efferent than afferent lymphatics associated with lymph nodes?

   ____

   ____

   ____

8. What structures assure the one-way flow of lymph through the node?

   ____

9. The largest collections of lymph nodes are found in what three body regions?

   ____

10. What is the general function of lymph nodes?

    ____

    ____

Other Lymphoid Organs

1. Match the terms in Column B with the appropriate descriptions in Column A. More than one choice may apply in some cases.

	Column A	Column B
________________	1. The largest lymphatic organ	A. Lymph nodes
________________	2. Filter lymph	B. Peyer's patches
________________	3. Particularly large and important during youth; helps to program T cells of the immune system	C. Spleen D. Thymus
________________	4. Collectively called MALT	E. Tonsils
________________	5. Removes aged and defective red blood cells	
________________	6. Contains red and white pulp	
________________	7. Exhibits Hassall's corpuscles: lacks follicles	
________________	8. Includes the adenoids	
________________	9. Acts against bacteria breaching the intestinal wall	
________________	10. Lack a complete capsule and have crypts in which bacteria can become trapped	

2. Complete the following "story" by inserting the missing terms in the answer blanks.

________________ 1.
________________ 2.
________________ 3.
________________ 4.
________________ 5.
________________ 6.
________________ 7.
________________ 8.
________________ 9.

The spleen is a versatile organ with many functions. For example, it temporarily stores __(1)__ released during hemoglobin breakdown as well as formed elements called __(2)__ that play a role in blood clotting. It cleanses the blood of both aging __(3)__ and foreign debris. In the fetus, the spleen also carries out __(4)__. The spleen's blood-processing functions are performed by its __(5)__ pulp regions, which contain venous __(6)__ and __(7)__. The spleen's lymphocytes and macrophages, which cluster around the central arteries in areas called __(8)__ pulp, are important in mounting a(n) __(9)__ response against antigens present in the blood percolating through the spleen.

3. Figure 19.5 depicts an enlarged view of a section from the spleen. Identify the following structures or regions by labeling the leader lines with the appropriate terms from the list below.

Hilum	Splenic artery	Splenic vein
Central artery	Splenic capsule	Trabecula
Red pulp	Splenic cords	Venous sinus

Once you have labeled the figure, color it as follows: Color the splenic vein and venous sinuses blue, the splenic artery and central arteries red, the white pulp lavender, and the splenic capsule and trabeculae green.

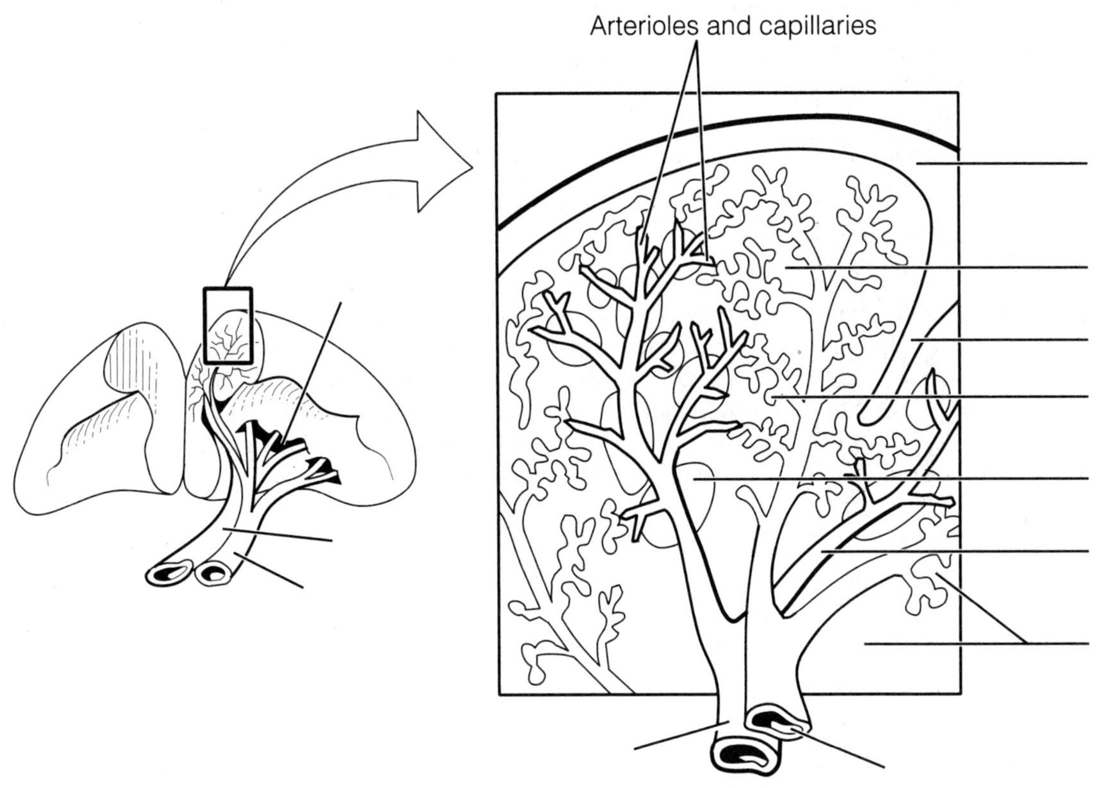

Figure 19.5

4. MALT means ____

Describe the important function of MALT. ________________________________

____

5. Name the three major types of tonsils. ________________________________

____

CHALLENGING YOURSELF

At the Clinic

1. Mrs. Burner comes to the clinic with red lines on her left arm, which are painful when touched. She reports being scratched by her cat recently. What is the diagnosis?
2. After surgery to remove lymphatic vessels associated with the removal of a melanoma, what condition can be expected relative to lymph drainage? Is this a permanent problem?
3. A young man appears at the clinic with swollen, tender inguinal lymph nodes. What term is used to describe an infected lymph node? What contagious disease is named for this term?
4. Lynn Oligopoulos, a young woman, has just been told her diagnosis is Hodgkin's disease. Describe, as best you can, the condition of her lymph nodes.
5. Young Joe Chang went sledding; the runner of a sled hit him in the left side and ruptured his spleen. Joe almost died because he did not get to the hospital fast enough. Upon arrival, a splenectomy was performed. What would you deduce is the immediate danger of a spleen rupture? Will Joe require a transplant for spleen replacement?
6. Gary has almost continuous earaches. The pediatrician explains to his parents that his adenoids should be removed to facilitate drainage of the pharyngotympanic tubes. To which lymphoid organs is she referring?
7. Angelica contracted both malaria and tuberculosis (diseases whose causative micro-organisms travel in the bloodstream) on a recent trip to the tropics. A week after returning home, she developed mononucleosis. During her clinic examination, the doctor felt the superior border of her spleen projecting far anterior to her left costal margin. He told Angelica that her spleen was enlarged, a condition called splenomegaly. Was something wrong with her spleen? What is the cause and effect here?

Stop and Think

1. Which lymphoid organs actually filter lymph?
2. Why are metastases common in lymph nodes?
3. Why are macrophages referred to as reticuloendothelial cells?
4. What would result from inability of lymphatic capillaries to pick up proteins from the interstitial fluid?
5. As Charlie read Chapter 19 for the second time, he suddenly had an insight, exclaiming to his roommate, "Lymph comes from blood and then it returns to the blood!" Is his insight correct? Explain.
6. Becky Sue is looking at slides of spleen tissue through the microscope and tells her instructor that she sees lymphoid follicles in the white pulp. Is she correct? Why or why not?

COVERING ALL YOUR BASES

Multiple Choice

Select the best answer or answers from the choices given.

1. Statements that apply to lymphatic capillaries include:
   A. the endothelial cells have continuous tight junctions
   B. the arterial end has a higher pressure than the venous end
   C. minivalves prevent the backflow of fluid into the interstitial spaces
   D. the endothelial cells are anchored by filaments to the surrounding structures

2. Chyle flows into the:
   A. lacteals
   B. intestinal lymph nodes
   C. intestinal trunk
   D. cisterna chyli

3. Metastasis of a melanoma to lymph nodes can result in:
   A. lymphedema
   B. lymphadenopathy
   C. Hodgkin's disease
   D. lymphangitis

4. Which region in a lymph node is involved in removing the coal dust inhaled by a miner?
   A. Subcapsular sinus C. Medullary cords
   B. TrabeculaeD. Follicles

5. Structurally, lympathic vessels are like:
   A. elastic arteries C. venules
   B. muscular arteries D. medium veins

6. Which parts of the lymph node show increased activity when antibody production is high?
   A. Germinal centers C. Medullary cords
   B. Outer follicle D. Sinuses

7. Which of the following connect to the lymph node at the hilum?
   A. Afferent lymphatic vessels
   B. Efferent lymphatic vessels
   C. Trabeculae
   D. Anchoring filaments

8. A cancerous lymph node would be:
   A. swollen C. painful
   B. firm D. fibrosed

9. The classification lymphoid tissues includes:
   A. the adenoids C. bone marrow
   B. the spleen D. the thyroid gland

10. There is no lymphatic supply to the:
    A. brain C. bone
    B. teeth D. bone marrow

11. The spleen functions to:
    A. remove aged RBCs
    B. house lymphocytes
    C. filter lymph
    D. store some blood components

12. Which characteristics are associated with the thymus?
    A. Providing immunocompetence
    B. Hormone secretion
    C. Hypertrophy in later life
    D. Cauliflower-like structural organization

13. The tonsils:
    A. have a complete epithelial capsule
    B. have crypts to trap bacteria
    C. filter lymph
    D. contain germinal centers

14. Lymphoid tissue is crowded with both macrophages and lymphocytes in close proximity. This design:
    A. enhances the ability of macrophages to phagocytize worn-out erythrocytes
    B. provides many opportunities for antigen challenge
    C. promotes effective bacterial killing
    D. leads to the production of many memory cells

15. Relative to distinguishing different lymphoid organs from each other in histological sections, how would you tell the thymus from a lymph node?
    A. Only the thymus has a cortex and medulla.
    B. Lymphocytes are far less densely packed in the thymus than in the lymph node.
    C. The thymus contains no blood vessels.
    D. Only the thymus has Hassall's corpuscles.

16. Which of the following are part of MALT?
    A. Tonsils
    B. Thymus
    C. Peyer's patches
    D. Any lymphoid tissue along the digestive tract

17. Which of the following result from infection?
A. Lymphangitis
B. Elephantiasis
C. Tonsillitis
D. Lymphedema

18. Cells of lymphatic tissue that play a crucial role in helping to activate T cells include:
A. dendritic cells
B. reticular cells
C. macrophages
D. plasma cells

19. Which of the following terms accurately applies to tonsils?
A. Pharyngeal
B. Crypts
C. Palatine
D. Lingual

20. Lymph capillaries:
A. are open ended like drinking straws
B. have continuous tight junctions like the capillaries of the brain
C. have endothelial cells separated by flaplike valves that open wide
D. have special barriers that stop cancer cells from entering

Word Dissection

For each of the following word roots, fill in the literal meaning and give an example, using a word found in this chapter.

Word root	Translation	Example
1. adeno		
2. angi		
3. chyle		
4. lact		
5. lymph		

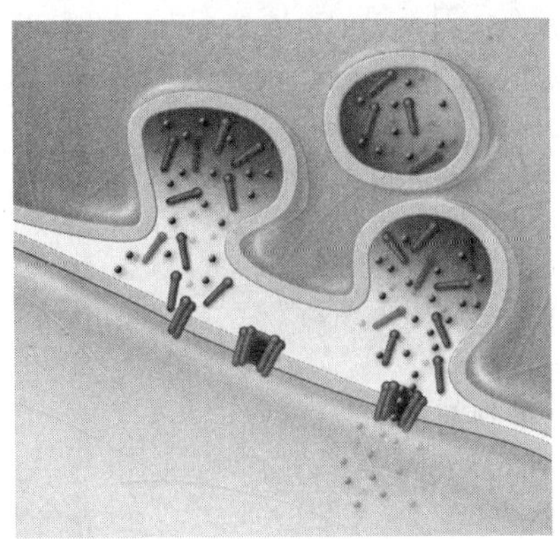

20

The Immune System: Innate and Adaptive Body Defenses

Student Objectives

When you have completed the exercises in this chapter, you will have accomplished the following objectives:

PART 1: INNATE DEFENSES

Surface Barriers: Skin and Mucosae

1. Describe surface membrane barriers and their protective functions.

Internal Defenses: Cells and Chemicals

2. Explain the importance of phagocytosis and natural killer cells in innate body defense.
3. Describe the inflammatory process. Identify several inflammatory chemicals and indicate their specific roles.
4. Name the body's antimicrobial substances and describe their function.
5. Explain how fever helps protect the body.

PART 2: ADAPTIVE DEFENSES

Antigens

6. Define antigen and describe how antigens affect the adaptive defenses.
7. Define complete antigen, hapten, and antigenic determinant.

Cells of the Adaptive Immune System: An Overview

8. Compare and contrast the origin, maturation process, and general function of B and T lymphocytes.
9. Define immunocompetence and self-tolerance, and describe their development in B and T lymphocytes.
10. Name several antigen-presenting cells and describe their roles in adaptive defenses.

Humoral Immune Response

11. Define humoral immunity.
12. Describe the process of clonal selection of a B cell.
13. Recount the roles of plasma cells and memory cells in humoral immunity.
14. Compare and contrast active and passive humoral immunity.
15. Describe the structure of an antibody monomer, and name the five classes of antibodies.
16. Explain the function(s) of antibodies and describe clinical uses of monoclonal antibodies.

Cell-Mediated Immune Response

17. Follow antigen processing in the body.
18. Define cell-mediated immunity and describe the process of activation and clonal selection of T cells.
19. Describe T cell functions in the body.
20. Indicate the tests ordered before an organ transplant is done, and methods used to prevent transplant rejection.

Homeostatic Imbalances of Immunity

21. Give examples of immune deficiency diseases and of hypersensitivity states.
22. Cite factors involved in autoimmune disease.

The immune system is a unique functional system made up of billions of individual cells, the bulk of which are lymphocytes. The sole function of this highly specialized defensive system is to protect the body against an incredible array of pathogens. In general, these "enemies" fall into three major camps: (1) microorganisms (bacteria, viruses, and fungi) that have gained entry into the body; (2) foreign tissue cells that have been transplanted (or, in the case of red blood cells, infused) into the body; and (3) the body's own cells that have become cancerous. The result of the immune system's activities is immunity, or specific resistance to disease.

The body is also protected by a number of innate defenses provided by intact surface membranes such as skin and mucosae, and a variety of cells and chemicals that can quickly mount the attack against foreign substances. The adaptive and innate defenses enhance each other's effectiveness.

This chapter first focuses on the innate body defenses. It then follows the development of the adaptive system and describes the mode of functioning of its effector cells—lymphocytes and macrophages. The chemicals that act to mediate or amplify the immune response are also discussed.

BUILDING THE FRAMEWORK

PART 1: INNATE DEFENSES

1. The three major elements of the body's innate defense system are:

   the (1) ____________________, consisting of the skin and ____________________;

   defensive cells, such as (2) ____________________ and phagocytes; and a profusion

   of (3) ____________________.

Surface Barriers: Skin and Mucosae

1. Indicate the sites of activity of the secretions of the surface membranes by writing the correct terms in the answer blanks.

   1. Lysozyme is found in the body secretions called ____________________ and ____________________.

   2. Fluids with an acid pH are found in the ____________________ and ____________________.

   3. Sebum is a product of the ____________________ glands and acts at the surface of the ____________________.

4. Mucus is produced by large mucus-secreting glands and the thousands of ____________________ cells found in the respiratory and ____________________ system mucosae.

2. Match the terms in Column B with the descriptions of the innate defenses of the body in Column A. More than one choice may apply.

	Column A	Column B
____________	1. Have antimicrobial activity	A. Acids
____________	2. Provide mechanical barriers	B. Lysozyme
____________	3. Constitute chemical barriers	C. Mucosae
____________	4. Entraps microorganisms entering the respiratory passages	D. Mucus
		E. Protein-digesting enzymes
____________	5. Part of the first line of defense	F. Sebum
		G. Skin

3. Describe the protective role of cilia.

____

____

____

Internal Defenses: Cells and Chemicals

1. Define *phagocytosis* and indicate why efforts at phagocytosis might not succeed.

____

____

____

____

2. Explain why many neutrophils are killed in their protective efforts.

____

____

____

3. Match the terms in Column B with the descriptions in Column A concerning events of the inflammatory response.

	Column A	Column B
________	1. Increased blood flow to an area	A. Chemotaxis
		B. Diapedesis
________	2. Inflammatory chemical released by degranulating mast cells	C. Exudate
________	3. Promotes release of white blood cells from the bone marrow	D. Histamine
		E. Leukocytosis-inducing factor
________	4. Cellular migration directed by a chemical gradient	F. Local hyperemia
		G. Macrophages
________	5. Fluid leaked from the bloodstream	H. Margination
________	6. Phagocytic progeny of monocytes	I. Neutrophils
________	7. Leukocytes pass through the wall of a capillary	
________	8. First leukocytes to migrate into the injured area	
________	9. White blood cells cling to capillary walls as blood flow slows due to fluid loss from the bloodstream	

4. Circle the term that does not belong in each of the following groupings.

1. Redness Pain Swelling Nausea Heat
2. Neutrophils Macrophages Phagocytes Natural killer cells
3. Hepatic macrophages Neutrophils Dendritic cells Alveolar macrophages
4. Inflammatory chemicals Histamine Kinins Complement Interferon

5. Interferons Antiviral Antimicrobial Host specific Anticancer

6. Natural killer cells Unique lymphocytes Macrophages NK cells

7. Defensins Free radicals Respiratory burst NO

5. Figure 20.1 diagrams the events involved in the inflammatory response. Events depicted in A precede those shown in B. Each event is represented by a square with one or more arrows. From the list below, write the correct number in each event square in the figure. Then, color code and color the structures that appear below the numbered list.

1. WBCs are drawn to the injured area by the release of inflammatory chemicals.
2. Tissue repair occurs.
3. Tissue injury occurs and microbes enter the injured tissues.
4. Local blood vessels dilate and the capillaries become engorged with blood.
5. Phagocytosis of microbes occurs.
6. Fluid containing clotting proteins is lost from the bloodstream and enters the injured tissue area.
7. Mast cells degranulate, releasing inflammatory chemicals.
8. Margination and diapedesis occur.

○ Mast cells ○ Erythrocyte(s) ○ Endothelium of capillary

○ Mast cell granules ○ Neutrophil(s) ○ Microorganisms

○ Monocyte ○ Macrophage ○ Fibrous repair tissue

○ Epithelium ○ Subcutaneous tissue

6. Explain the role of CAMs (selectins and integrins) on the process of margination by neutrophils.

____

____

____

____

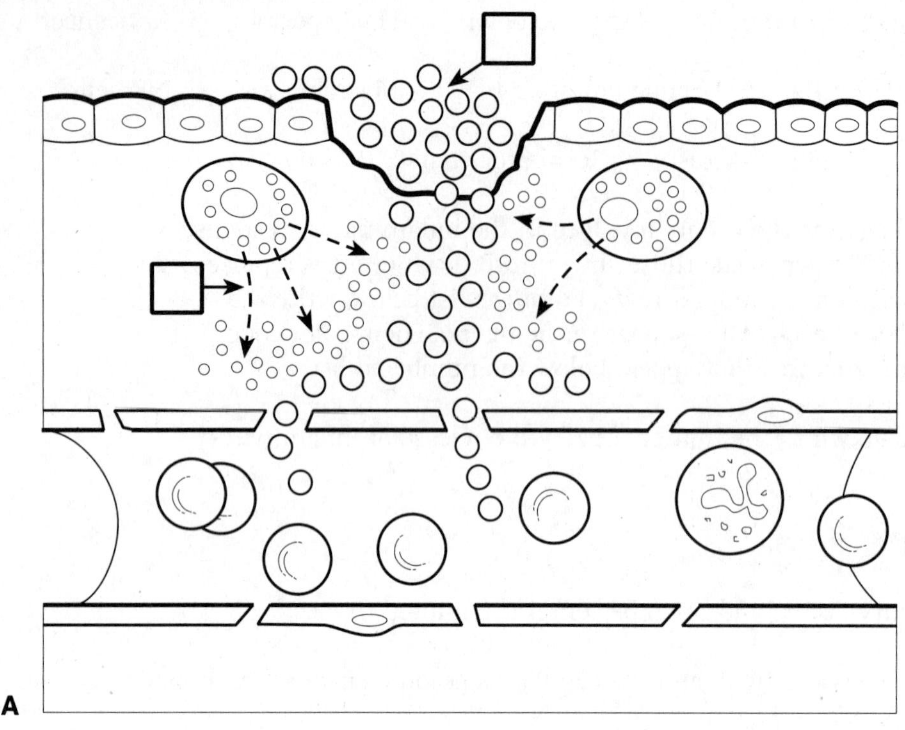

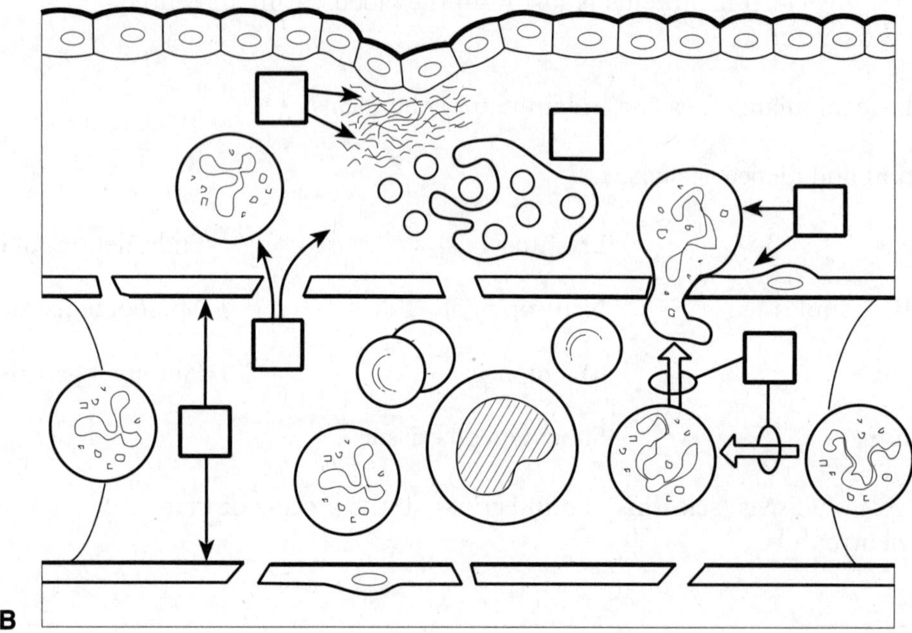

Figure 20.1

7. List three advantages of inflammation-induced edema.

1. ______________________________

2. ______________________________

3. ______________________________

8. Complete the following description of the activation and activity of complement by writing the missing terms in the answer blanks.

1. ____________________
2. ____________________
3. ____________________
4. ____________________
5. ____________________
6. ____________________
7. ____________________
8. ____________________
9. ____________________
10. ____________________
11. ____________________
12. ____________________
13. ____________________
14. ____________________

Complement is a system of at least __(1)__ proteins that circulate in the blood in an inactive form. Complement may be activated by either of two pathways. The __(2)__ pathway involves 11 proteins, designated C1–C9, and depends on the binding of antibodies to invading microorganisms and subsequent binding of complement to the __(3)__ complexes. The __(4)__ pathway is triggered by the interaction of three plasma proteins, factors B, D, and __(5)__, with __(6)__ molecules present on the surface of certain bacteria and fungi. Both pathways lead to the cleavage of C3 into C3a and C3b. Binding of __(7)__ to the target cell's surface results in the incorporation of C5–C9 into the target cell membrane. This incorporated complex is called the __(8)__, and its completed insertion results in target cell __(9)__. The complement fragments bound to the target cell surface also enable it to be phagocytized more readily, a phenomenon called __(10)__. The __(11)__ fragment of C3 amplifies the inflammatory response by attracting neutrophils and other inflammatory cells into the area.

__(12)__, produced by the liver in response to inflammation, is used as a clinical marker. Its binding to C1 of the __(13)__ complement pathway results in __(14)__ of bacteria.

9. Describe the event that leads to the synthesis of interferon and indicate the result of its synthesis.

____

____

____

10. Check (✓) all phrases that correctly describe the role of fever in body protection.

_____ 1. Is a normal response to pyrogens

_____ 2. Protects by denaturing tissue proteins

_____ 3. Reduces the availability of iron and zinc required for bacterial proliferation

_____ 4. Increases metabolic rate

PART 2: ADAPTIVE DEFENSES

Antigens

1. Complete the following statements relating to antigens by writing the missing terms in the answer blanks.

______________ 1.
______________ 2.
______________ 3.
______________ 4.
______________ 5.
______________ 6.
______________ 7.
______________ 8.
______________ 9.

Antigens are substances capable of mobilizing the __(1)__. Complete antigens can stimulate proliferation of specific lymphocytes and antibodies, a characteristic called __(2)__, and can react with the products of those reactions, a property called __(3)__. Of all the foreign molecules that act as complete antigens, __(4)__ are the most potent. Small molecules are not usually antigenic, but when they bind to self cell-surface proteins they may act as __(5)__, and then the complex is recognized as foreign. The ability of various molecules to act as antigens depends both on their __(6)__ and on the complexity of their structure. Large, complex molecules may have hundreds of recognizable 3-D structures called __(7)__, to which antibodies or activated lymphocytes can attach. Conversely, large molecules with many regularly repeating units are chemically __(8)__ and thus are poor antigens. Because such substances tend not to be __(9)__ by the body, they are useful for making artificial implants.

2. Complete this section dealing with self-antigens by writing the missing terms in the answer blanks.

______________ 1.
______________ 2.
______________ 3.

Self-antigens, the unique cell-surface glycoproteins called __(1)__, are coded for by genes of the __(2)__. These self-antigens have a deep groove that allows them to display protein fragments. Protein fragments derived from foreign antigens mobilize the __(3)__.

Cells of the Adaptive Immune System: An Overview

1. What are three important characteristics of the adaptive immune response?

____

2. Define *immunocompetence*. ____

____

____

____

3. A schematic drawing of the life cycle of the lymphocytes involved in immunity is shown in Figure 20.2. Select different colors for the areas listed below, and color the coding circles and the corresponding regions on the figure. If there is overlap, use stripes of a second color to indicate the second identification. Then respond to the questions following the figure, which relate to the two-phase differentiation process of B cells and T cells.

◯ Area where immature lymphocytes arise

◯ Area seeded by immunocompetent but naive B cells and T cells

◯ Area where T cells become immunocompetent

◯ Area where the antigen challenge and clonal selection are likely to occur

◯ Area where B cells become immunocompetent

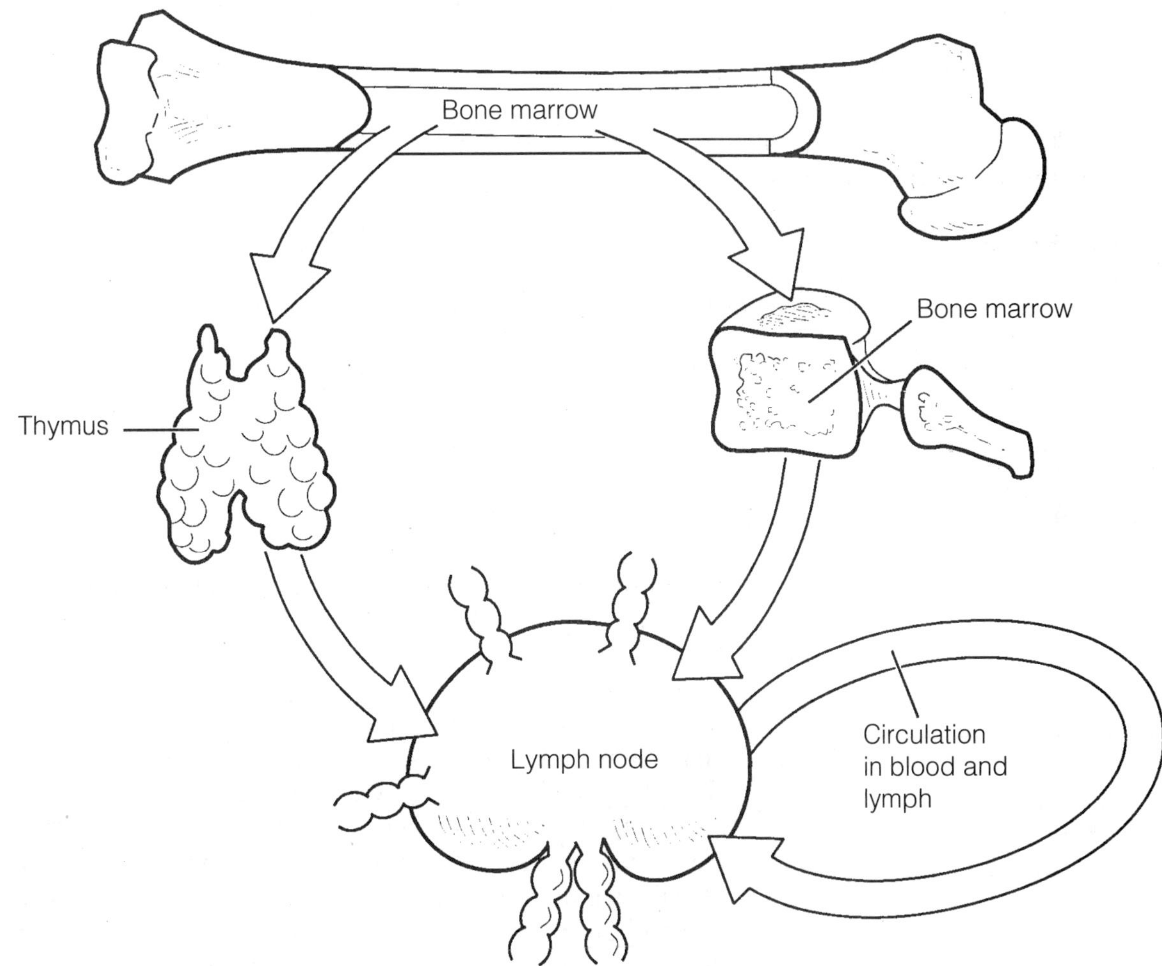

Figure 20.2

1. What signifies that a lymphocyte has become immunocompetent?

____

2. During what period of life does the development of immunocompetence occur? ______________________________

3. What determines which antigen a particular T cell or B cell will be able to recognize, its genes or "its" antigen? ______________________________

4. What triggers the process of clonal selection in a T cell or B cell, its genes or binding to "its" antigen? ______________________________

5. During the development of immunocompetence (during which B cells or T cells develop the ability to recognize a particular antigen), the ability to tolerate ____________________ must also occur if the immune system is to function normally.

4. Using the key choices, select the term that correctly completes each statement. Insert the appropriate answers in the answer blanks.

Key Choices

A. Antigen(s)	E. Dendritic cells	H. Lymph nodes
B. B cells	F. Humoral immunity	I. Macrophages
C. Blood	G. Lymph	J. T cells
D. Cellular immunity		

____________________ 1.

____________________ 2.

____________________ 3.

____________________ 4.

____________________ 5.

____________________ 6.

____________________ 7.

____________________ 8.

____________________ 9.

Immunity is resistance to disease resulting from the presence of foreign substances or __(1)__ in the body. When this resistance is provided by antibodies released to body fluids, the immunity is called __(2)__. When living cells provide the protection, the immunity is called __(3)__. The major actors in the adaptive immune response are two lymphocyte populations, the __(4)__ and the __(5)__. Phagocytic cells that act as accessory cells in the immune response are __(6)__. Because pathogens are likely to use both __(7)__ and __(8)__ as a means of getting around the body, __(9)__ and other lymphoid tissues (which house the immune cells) are in an excellent position to detect their presence.

5. T cells and B cells exhibit certain similarities and differences. Check (✓) the appropriate boxes in the table to indicate the lymphocyte type that exhibits each characteristic.

Characteristic	T cell	B cell
1. Originates in bone marrow from hemocytoblasts		
2. Progeny are plasma cells		
3. Progeny include helpers, regulatory cells, or killers		
4. Progeny include memory cells		
5. Is responsible for directly attacking foreign cells or virus-infected cells in the body		
6. Produces antibodies that are released to body fluids		
7. Bears a cell-surface receptor capable of recognizing a specific antigen		
8. Forms clones upon stimulation		
9. Accounts for most of the lymphocytes in the circulation		

Humoral Immune Response

1. The basic structure of an antibody molecule is diagrammed in Figure 20.3. Select two different colors and color the heavy chains and light chains. Add labels to the diagram to identify the type of bonds holding the polypeptide chains together. Also label the constant (C) and variable (V) regions of the antibody, and add "polka dots" to the variable portions. Then answer the questions following the figure.

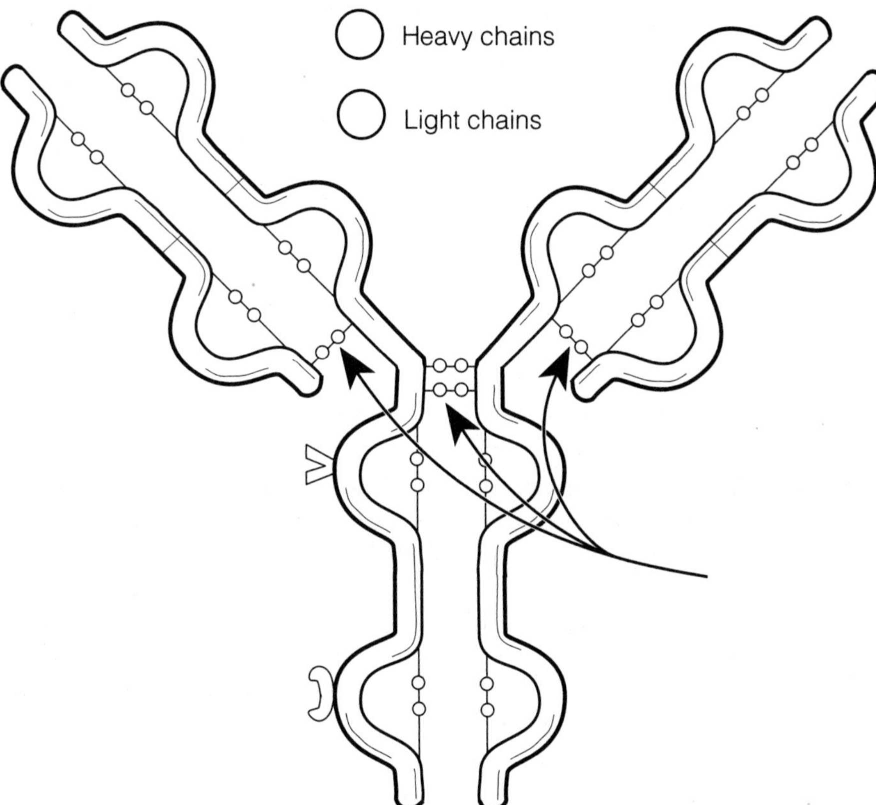

Figure 20.3

1. Which part of the antibody, V or C, is its antigen-binding site? ______________________

2. Which part determines antibody class and specific function? ______________________

2. Match the antibody classes in Column B with their descriptions in Column A.

Column A	Column B
_______ 1. Bound to the surface of a B cell; its antigen receptor	A. IgA
_______ 2. Crosses the placenta	B. IgD
_______ 3. The first antibody released during the primary response; indicates a current infection	C. IgE
_______ 4. Two classes that fix complement	D. IgG
_______ 5. The most abundant antibody found in blood plasma and the chief antibody released during secondary responses	E. IgM
_______ 6. Binds to the surface of mast cells and mediates an allergic response	
_______ 7. Predominant antibody in mucus, saliva, tears, and milk	
_______ 8. Is a pentamer	

3. Determine whether each of the following situations provides, or is an example of, active or passive immunity. If passive, write P in the blank; if active, write A in the blank.

_______ 1. An individual receives Sabin polio vaccine

_______ 2. Antibodies migrate through a pregnant woman's placenta into the vascular system of her fetus

_______ 3. A student nurse receives an injection of gamma globulin (containing antibodies to the hepatitis virus) after she has been exposed to viral hepatitis

_______ 4. "Borrowed" immunity

_______ 5. Immunologic memory is provided

_______ 6. An individual suffers through chicken pox

4. There are several important differences between primary and secondary immune responses to antigens. If a statement below best describes a primary response, write P in the blank; if a secondary response, write S in the blank.

_______ 1. The initial response to an antigen; "gearing up" stage

_______ 2. A lag period of several days occurs before antibodies specific to the antigen appear in the bloodstream

_______ 3. Antibody levels increase rapidly and remain high for an extended period

_______ 4. Immunologic memory is established

_______ 5. The second, third, and subsequent responses to the same antigen

5. Complete the following description of antibody function by writing the missing terms in the answer blanks.

______________ 1.
______________ 2.
______________ 3.
______________ 4.
______________ 5.
______________ 6.
______________ 7.

Antibodies can inactivate antigens in various ways, depending on the nature of the __(1)__. __(2)__ is the chief ammunition used against cellular antigens such as bacteria and mismatched red blood cells. The binding of antibodies to sites on bacterial exotoxins or viruses that can cause cell injury is called __(3)__. The cross-linking of cellular antigens into large lattices by antibodies is called __(4)__; Ig __(5)__, with its 10 antigen binding sites, is particularly efficient in this mechanism. When molecules are cross-linked into lattices by antibodies, the mechanism is more properly called __(6)__. In virtually all these cases, the protective mechanism mounted by the antibodies serves to disarm and/or immobilize the antigens until they can be disposed of by __(7)__.

Cell-Mediated Immune Response

1. Several populations of T cells exist. Match the type of cell in Column B with the descriptions in Column A.

	Column A	Column B
______________	1. Binds with and releases chemicals that activate B cells	A. Cytotoxic T cell
		B. Helper T cell
______________	2. Releases chemicals that activate cytotoxic T cells, macrophages, and a variety of nonimmune cells, but must be activated itself by recognizing both its antigen and a self-protein presented on the surface of an APC	C. Memory T cell
		D. Regulatory T cell
______________	3. Turns off the immune response when the "enemy" has been routed	
______________	4. Directly attacks and causes lysis of cellular pathogens	
______________	5. A clone member of any T cell variety	
______________	6. The major cell type in the CD4 population	
______________	7. The major cell type in the CD8 population	

2. Using the key choices, select the terms that correspond to the following descriptions of substances or events by inserting the appropriate answers in the answer blanks.

Key Choices

A. Antibodies

B. Antigen

C. Chemotactic factors

D. Complement

E. Cytokines

F. Inflammation

G. Interferon

H. Macrophage migration inhibiting factor

I. Perforin

J. Tumor necrosis factor

__________________ 1. A protein released by leukocytes and other cells that helps protect other body cells from viral multiplication

__________________ 2. Molecules that attract neutrophils and other protective cells into a region where an immune response is ongoing

__________________ 3. Any substance capable of eliciting an immune response and of binding with the product of that response

__________________ 4. Slow-acting protein that causes cell death; released in large amounts by macrophages

__________________ 5. A consequence of the release of histamine from mast cells and of complement activation by binding to antigen-antibody complexes

__________________ 6. C, G, H, and J are examples of this class of molecules

__________________ 7. A plasma protein complex that amplifies the immune response by causing lysis of cellular pathogens once it is "fixed" to their surface

__________________ 8. Chemicals released by activated macrophages; interleukin-1 and interferon, for example

__________________ 9. Proteins released by plasma cells that mark antigens for destruction by phagocytes or complement

__________________ 10. Chemical released by T_C cells that creates pores in the membranes of cellular antigens

3. Circle the term that does not belong in each of the following groupings.

1. Antibodies Gamma globulin Cytokines Immunoglobulins

2. Lymph nodes Liver Spleen Thymus Bone marrow

3. Clustering of receptors T cell activation B cell activation

   Endocytosis of antigen-receptor complexes

4. Autograft Isograft Allograft Xenograft Human

5. Cytotoxic T cells Cytolytic T cells Natural killer cells Killer T cells

6. Perforin Na^+-induced polymerization Membrane lesions Lethal hit Lysis

7. Dendritic cell presenters Helper T cells Class I MHC proteins

   Class II MHC proteins Interleukin-1

8. Anergy B cell activation Clonal deletion Receptor editing

4. Figure 20.4 is a flowchart of the immune response that tests your understanding of the interrelationships of that process. Several terms have been omitted from this flowchart. First, complete the figure by writing in the boxes the appropriate terms selected from the key choices below. (*Note:* Some terms are used more than once. Oval boxes represent cell types and rectangular boxes represent molecules. Also note that solid lines represent stimulatory or enhancing effects, whereas broken lines indicate inhibition.) Then color the coding circles of the cell types listed below and the corresponding oval boxes on the flowchart.

Key Choices

Cell Types	**Molecules**
○ B cell	Antibodies
○ Helper T cell	Chemotactic factors
○ Cytotoxic T cell	Complement
○ Macrophage	Cytokines
○ Memory B cell	Interferon
○ Memory T cell	Interleukin
○ Neutrophils	Perforin (and/or other cytotoxins)
○ Plasma cell	Suppressor factors
○ Suppressor T cell	

5. Complete the following paragraph about co-stimulation by inserting the appropriate terms in the answer blanks.

1. ____________________
2. ____________________
3. ____________________
4. ____________________

Not emphasized in Figure 20.4 is the importance of co-stimulatory signals, which may be __(1)__ or __(2)__. When the required co-stimulatory signals are absent, T cell activation is incomplete and the T cell tolerates the antigen, a situation called __(3)__. The two-step signaling sequence (antigen-MHC recognition, then co-stimulation) is thought to be a safeguard that prevents __(4)__ of our own healthy cells.

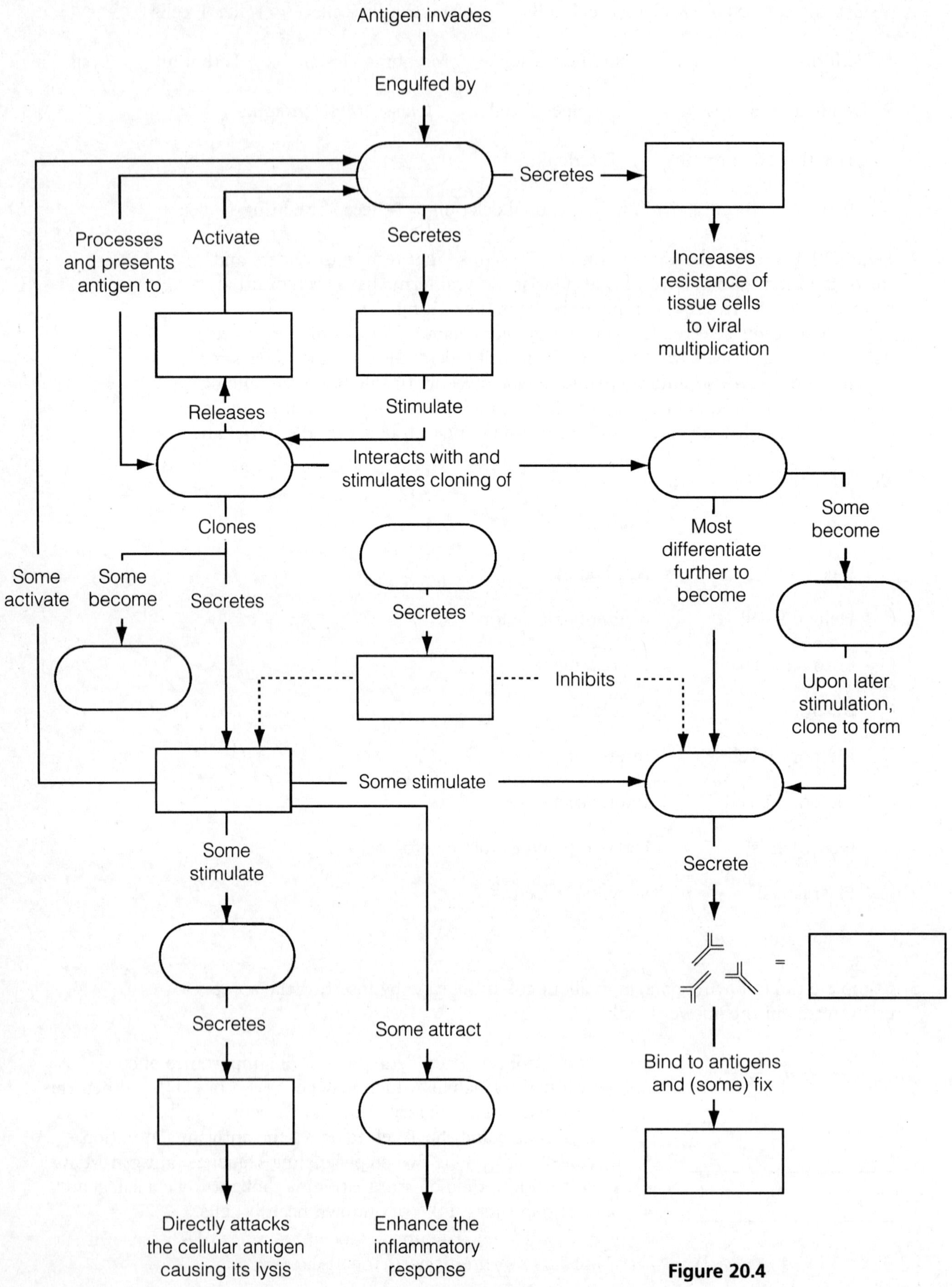

Antigen invades
Engulfed by
Secretes
Increases resistance of tissue cells to viral multiplication
Processes and presents antigen to
Activate
Secretes
Releases
Stimulate
Interacts with and stimulates cloning of
Some become
Most differentiate further to become
Clones
Some activate
Some become
Secretes
Secretes
Inhibits
Upon later stimulation, clone to form
Some stimulate
Some stimulate
Secrete
=
Secretes
Some attract
Bind to antigens and (some) fix
Directly attacks the cellular antigen causing its lysis
Enhance the inflammatory response

Figure 20.4

6. Organ transplants are often unsuccessful because MHC-encoded proteins vary in different individuals. However, chances of success increase if certain important procedures are followed. The following questions refer to this important area of clinical medicine.

1. Assuming that autografts and isografts are not possible, what is the next most successful graft type? ______________________

2. What is the source of the graft tissue or organ? ______________________

3. What two cell types are important in rejection phenomena?

____

4. Why are immunosuppressive drugs (or therapy) provided after transplant surgery?

____

____

5. What is the major shortcoming of immunosuppressive therapy?

____

____

Homeostatic Imbalances of Immunity

1. Using the key choices, identify the types of immunity disorders described below.

Key Choices

A. Hypersensitivity B. Autoimmune disease C. Immunodeficiency

______________ 1. AIDS and SCID

______________ 2. The immune system mounts an extraordinarily vigorous response to an otherwise harmless antigen

______________ 3. An allergic response such as hay fever or contact dermatitis

______________ 4. Occurs when the production or activity of immune cells or complement is abnormal

______________ 5. The body's own immune system produces the disorder; a breakdown of self-tolerance

______________ 6. Affected individuals are unable to combat infections that would present no problem for normally healthy people

_________________ 7. Multiple sclerosis, rheumatoid arthritis, and myasthenia gravis

_________________ 8. May be treated by providing normal stem cells from bone marrow or umbilical cord blood transplants in some cases

_________________ 9. Caused when self-antigens are changed by disease, or when antibodies produced against foreign antigens react with self-antigens

2. Match the terms relating to hypersensitivities in Column B with their descriptions in Column A.

	Column A	Column B
_________________	1. Allergies mediated by antibodies	A. Anaphylaxis (type I)
_________________	2. Occurs when the allergen cross-links to IgE molecules attached to mast cells or basophils, causing them to degranulate; reactions may be local or systemic	B. Type I hypersensitivity
		C. Cytotoxic (type II) hypersensitivity
_________________	3. Transfusion reaction	D. Delayed hypersensitivity (type IV)
_________________	4. Complement activation by circulating or attached immune complexes resulting in intense inflammatory responses	E. Immune complex (type III) hypersensitivity
_________________	5. Effective against facultative intracellular pathogens (salmonella and some yeasts)	
_________________	6. Can result in death due to circulatory shock in susceptible individuals when the allergen is bee venom	
_________________	7. Its major mediator is histamine	
_________________	8. Collectively make up the immediate hypersensitivities	
_________________	9. Can be passively transferred to others in whole blood transfusions	
_________________	10. Serum sickness, farmer's lung, and mushroom grower's lung, for example	
_________________	11. Redness, hives, local edema, and wheezing can all be symptoms	

_____________________ 12. Cytokines released by T cells are the major mediators

_____________________ 13. Poison ivy, contact dermatitis, and allergic reactions to detergents are examples

_____________________ 14. Together called subacute hypersensitivities

THE INCREDIBLE JOURNEY:

A Visualization Exercise for the Immune System

Something quite enormous and looking much like an octopus is nearly blocking the narrow tunnel just ahead.

1. Complete the following narrative by writing the missing words in the answer blanks.

_____________________ 1.

_____________________ 2.

_____________________ 3.

_____________________ 4.

_____________________ 5.

For this journey, you are equipped with scuba gear before you are miniaturized and injected into one of your host's lymphatic vessels. He has been suffering with a red, raw "strep throat." Your assignment is to travel into a swollen cervical lymph node and observe the activities going on there that reveal how your host's immune system is combating the infection.

On injection, you enter the lymph with a "whoosh" and then bob gently in the warm yellow fluid. As you travel along, you see thousands of spherical streptococcus bacteria and a few large globular __(1)__ molecules that, no doubt, have been picked up by the tiny lymphatic capillaries. Shortly thereafter, a large dark mass looms just ahead. This has to be a __(2)__, you conclude, and you dig in your wetsuit pocket to find the waterproof pen and recording tablet.

As you enter the gloomy mass, the lymphatic stream becomes shallow and begins to flow sluggishly. So that you can explore this little organ fully, you haul yourself to your feet and begin to wade through the slowly moving stream. On each bank you see a huge ball of cells that have large nuclei and such a scant amount of cytoplasm that you can barely make it out. You write, "Sighted the spherical germinal centers composed of __(3)__." As you again study one of the cell masses, you spot one cell that looks quite different and reminds you of a nest of angry hornets because it is furiously spewing out what seems to be a horde of tiny Y-shaped "bees." "Ah-ha," you think, "another valuable piece of information." You record, "Spotted a __(4)__ making and releasing __(5)__."

______________ 6. That done, you turn your attention to scanning the rest of the landscape. Suddenly you let out an involuntary yelp.
______________ 7. Something quite enormous and looking much like an octopus is nearly blocking the narrow tunnel just ahead. Your mind
______________ 8. whirls as it tries to figure out the nature of this cellular "beast" that appears to be guarding the channel. Then it hits you—this
______________ 9. has to be a __(6)__ on the alert for foreign invaders (more properly called __(7)__), which it "eats" when it catches them. The
______________ 10. giant cell roars, "Halt, stranger, and be recognized," and you dig frantically in your pocket for your identification pass. As you
______________ 11. drift toward the huge cell, you hold the pass in front of you, hands trembling because you know this cell could liquefy you as quick as the blink of an eye. Again the cell bellows at you, "Is this some kind of a security check? I'm on the job, as you can see!" Frantically you shake your head, no, and the cell lifts one long tentacle and allows you to pass. As you squeeze by, the cell says, "Being inside, I've never seen my body's outside. I must say, humans are a rather strange-looking lot!" Still shaking, you decide that you are in no mood for a chat and hurry along to put some distance between yourself and this guard cell.

Immediately ahead are what appear to be hundreds of the same type of cell sitting on every ledge and in every nook and cranny. Some are busily snagging and engulfing unfortunate strep bacteria that float too close. The slurping sound is nearly deafening. Then something grabs your attention—the surface of one of these cells is becoming dotted with parts of the same doughnut-shaped chemicals that you see on the strep bacteria membranes; and a round cell, similar but not identical to those you saw earlier in the germinal centers, is starting to bind to one of these "doorknobs." You smile smugly because you know you have properly identified the octopus-like cells. You then record your observations as follows: "Cells like the giant cell just identified act as __(8)__. I have just observed one in this role during its interaction with a __(9)__ cell."

You decide to linger a bit to see if the round cell becomes activated. You lean against the tunnel walls and watch quietly, but your wait is brief. Within minutes, the cell that was binding to the octopus-like cell begins to divide, and then its daughter cells divide again and again at a head-spinning pace. You write, "I have just witnessed the formation of a __(10)__ of like cells." Most of the daughter cells enter the lymph stream, but a few of them settle back and seem to go into a light sleep. You decide that the "napping cells" don't have any role to play in helping get rid of your host's present strep infection, but instead will provide for __(11)__ and become active at a later date.

You glance at your watch and wince as you realize that it is already five minutes past the time for your retrieval. You have already concluded that this is a dangerous place, and you are not sure how long your pass is good—so you swim hurriedly from the organ into the lymphatic stream to reach your pickup spot.

CHALLENGING YOURSELF

At the Clinic

1. As an infant receives her first dose of oral polio vaccine, the nurse explains to her parents that the vaccine is a preparation of weakened virus. What type of immunity will the infant develop?

2. After two late-term miscarriages, Mrs. Lyons, who is pregnant again, is having her Rh antibody level checked. What are the Rh types of Mrs. Lyons and her husband and what class of antibodies will be the chief culprit of any problems for the fetus?
3. What xenograft is performed fairly routinely in heart surgery?
4. Assuming their immunity is intact, why do some people get colds or the flu year after year?
5. A young man is rushed to the emergency room after fainting. His blood pressure is alarmingly low and his companion reports the man collapsed shortly after being stung by a wasp. What has caused his hypotension? What treatment will be given immediately?
6. Patty Hourihan is a strict environmentalist and a new mother. Although she is very much against using disposable diapers, she is frustrated by the fact that her infant breaks out in a diaper rash when she uses cloth diapers. New cloth diapers do not cause the rash, but washed ones do; what do you think the problem is?
7. A week after starting on aspirin to head off blood clots, Mr. Russell, a scholarly older man, reports to the clinic with large bruises over much of his body. What is his diagnosis, and in what class of immune disorders should it be placed?
8. About six months after an automobile accident in which her neck was severely lacerated, a young woman comes to the clinic for a routine checkup. Visual examination shows a slight swelling beneath the larynx. Upon questioning, the woman reports that she fatigues easily, has been gaining weight, and her hair is falling out. What is your diagnosis?
9. Anna is complaining of cramping, nausea, and diarrhea, which she thinks is a reaction to the dust stirred up by the construction near her home. What is a more likely explanation of her symptoms, assuming an allergy is involved?
10. The mother of a 7-year-old boy with a bad case of mumps is worried that he might become sterile. What will you tell her to allay her fears?
11. David's lymphatic stream contains a large number of plasma cells. Is the number of antibodies in his bloodstream increasing or decreasing at this time? Explain.

Stop and Think

1. Use of birth control pills decreases the acidity of the vaginal tract. Why might this increase the incidence of vaginitis?
2. Are all pathogens killed by stomach acid?
3. Why does the risk of infection increase when areas such as the auditory tube and sinus openings are swollen?
4. How does chewing sugar-free gum decrease the likelihood of cavities?
5. Dissect the words histamine, diapedesis, and interleukin.
6. The genetic "scrambling" of DNA segments that occurs during the process of gaining immunocompetence provides the best immunological response to the uncountable number of antigenic possibilities. What other part of the body must also respond to large numbers of variable chemicals, and so might also involve genetic "scrambling"? (Think!)
7. Is the allergen in poison ivy sap a water-soluble or lipid-soluble molecule? Explain your reasoning.
8. What is lymphocyte recirculation? What role does it play?

COVERING ALL YOUR BASES

Multiple Choice

Select the best answer or answers from the choices given.

1. Specialized macrophages include:
   A. dust cells of the lungs
   B. Kupffer cells
   C. epidermal dendritic cells
   D. natural killer cells

2. Which of the following are cardinal signs of inflammation?
   A. Phagocytosis C. Leukocytosis
   B. Edema D. Pain

3. Chemical mediators of inflammation include:
   A. pyrogens C. histamine
   B. prostaglandins D. cytokines

4. Aspirin inhibits:
   A. kinins C. chemotactic agents
   B. histamine D. prostaglandins

5. Which term indicates the WBCs' ability to squeeze their way through the capillary endothelium to get into the tissue spaces?
   A. Diapedesis C. Margination
   B. Degranulation D. Positive chemotaxis

6. Neutrophils die in the line of duty because:
   A. they ingest infectious organisms
   B. their membranes become sticky and they are attacked by macrophages
   C. they secrete cellular toxins, which affect them in the same way they affect pathogens
   D. the buildup of tissue fluid pressure causes them to lyse

7. Which of the following is a component of the alternative, but not the classical, pathway of complement activation?
   A. Activation of C1 and C2
   B. Exposure to polysaccharides of certain pathogens
   C. Complement fixation by antibodies
   D. Insertion of MAC in cell membranes of pathogenic organisms

8. Against which of the following will interferon do some good?
   A. Infection of body cells by a virus
   B. Circulating population of free virus
   C. Some types of cancer
   D. Bacterial infection

9. Characteristics of the immune response include:
   A. antigen specificity
   B. memory
   C. systemic protection
   D. species specificity

10. Antigen recognition is due to:
   A. genetic programming
   B. antigen exposure
   C. somatic recombination
   D. immune surveillance

11. Macrophages:
   A. form exudate
   B. present antigens
   C. secrete interleukin 1
   D. activate helper T cells

12. Immunogenicity is a property of:
   A. antibodies C. antigens
   B. effector T cells D. macrophages

13. The binding and clustering of B cell receptors by a multivalent antigen is required for:
   A. clonal selection
   B. antigen endocytosis
   C. activation
   D. degranulation

14. In the primary immune response:
   A. B cell cloning occurs
   B. plasma cell exocytosis is at a high level beginning about 6 days after exposure
   C. IgM antibodies initially outnumber IgG antibodies
   D. plasma cell life span is about 5 days

15. Antibodies secreted in mother's milk:
   A. are IgG antibodies
   B. are IgA antibodies
   C. provide natural active immunity
   D. provide natural passive immunity

16. Conditions for which passive artificial immunity is the treatment of choice include:
A. measles
B. botulism
C. rabies
D. venomous snake bite

17. Which parts of an antibody molecule are different for an IgG antibody than for an IgM antibody that attacks the same antigen?
A. Heavy chain constant region
B. Heavy chain variable region
C. Light chain constant region
D. Light chain variable region

18. Which of the following is important for activation of a B cell during the antigen challenge?
A. The antigen
B. A helper T cell
C. Chemicals that stimulate the lymphocytes to divide
D. Interleukin 2

19. Which of these antibody classes is usually arranged as a pentamer?
A. IgG
B. IgM
C. IgA
D. IgD

20. Which of the following antibody capabilities causes a transfusion reaction with A or B antigens?
A. Neutralization
B. Precipitation
C. Complement fixation
D. Agglutination

21. Class II MHC proteins are found on the membranes of:
A. all body cells
B. macrophages
C. activated B lymphocytes
D. some T cells

22. Cytotoxic T cells secrete:
A. tumor necrosis factor
B. histamine
C. perforin
D. gamma interferon

23. Which of the following terms is applicable to the use of part of the great saphenous vein in coronary bypass surgery?
A. Isograft
B. Xenograft
C. Allograft
D. Autograft

24. Which of the following is (are) examples of autoimmune disease?
A. Juvenile diabetes
B. Multiple sclerosis
C. Graves' disease
D. Rheumatoid arthritis

25. The predominant cell type in the germinal centers of the lymph nodes are:
A. proliferating B cells
B. circulating T cells
C. macrophages
D. plasma cells

26. The CD4 and CD8 proteins that identify the two major T cell populations are:
A. adhesion molecules that help maintain coupling during antigen recognition
B. associated with kinase enzymes
C. part of the TCR (T cell receptor)
D. cytokines

27. Possible APCs include:
A. dendritic cells
B. Langerhans cells
C. macrophages
D. B lymphocytes

28. Peptides derived from viral proteins are:
A. exogenous antigens
B. endogenous antigens
C. displayed by class I MHC
D. displayed by class II MHC

29. The main cellular target of the HIV virus that causes AIDS is:
A. T_H cells
B. T_C cells
C. macrophages
D. B cells

30. Toll-like receptors:
A. trigger release of cytokines
B. are found on macrophages
C. recognize general categories of microbes
D. play a central role in immune responses

Word Dissection

For each of the following word roots, fill in the literal meaning and give an example, using a word found in this chapter.

Word root	Translation	Example
1. hapt		
2. humor		
3. macro		
4. opso		
5. penta		
6. phylax		
7. phago		
8. pyro		
9. vacc		

21 The Respiratory System

Student Objectives

When you have completed the exercises in this chapter, you will have accomplished the following objectives:

Functional Anatomy of the Respiratory System

1. Identify the organs forming the respiratory passageway(s) in descending order until the alveoli are reached.
2. Describe the location, structure, and function of each of the following: nose, paranasal sinuses, pharynx, and larynx.
3. List and describe several protective mechanisms of the respiratory system.
4. Distinguish between conducting and respiratory zone structures.
5. Describe the makeup of the respiratory membrane, and relate structure to function.
6. Describe the gross structure of the lungs and pleurae.

Mechanics of Breathing

7. Explain the functional importance of the partial vacuum that exists in the intrapleural space.
8. Relate Boyle's law to the events of inspiration and expiration.
9. Explain the relative roles of the respiratory muscles and lung elasticity in producing the volume changes that cause air to flow into and out of the lungs.
10. List several physical factors that influence pulmonary ventilation.
11. Explain and compare the various lung volumes and capacities.
12. Define dead space.
13. Indicate types of information that can be gained from pulmonary function tests.

Gas Exchanges between the Blood, Lungs, and Tissues

14. State Dalton's law of partial pressures and Henry's law.
15. Describe how atmospheric and alveolar air differ in composition, and explain these differences.
16. Relate Dalton's and Henry's laws to events of external and internal respiration.

Transport of Respiratory Gases by Blood

17. Describe how oxygen is transported in the blood, and explain how oxygen loading and unloading is affected by temperature, pH, BPG, and P_{CO_2}.
18. Describe carbon dioxide transport in the blood.

Control of Respiration

19. Describe the neural controls of respiration.
20. Compare and contrast the influences of arterial pH, arterial partial pressures of oxygen and carbon dioxide, lung reflexes, volition, and emotions on respiratory rate and depth.

Respiratory Adjustments

21. Compare and contrast the hyperpnea of exercise with hyperventilation.
22. Describe the process and effects of acclimatization to high altitude.

Homeostatic Imbalances of the Respiratory System

23. Compare the causes and consequences of chronic bronchitis, emphysema, asthma, tuberculosis, and lung cancer.

Body cells require an abundant and continuous supply of oxygen to carry out their activities. As cells use oxygen, they release carbon dioxide, a waste product the body must get rid of. The circulatory and respiratory systems obtain and deliver oxygen to body cells and eliminate carbon dioxide from the body. The respiratory system structures are responsible for gas exchange between the blood and the external environment (that is, external respiration). The respiratory system also plays an important role in maintaining the acid-base balance of the blood.

Questions and activities in Chapter 21 consider both the anatomy and the physiology of the respiratory system structures.

BUILDING THE FRAMEWORK

Functional Anatomy of the Respiratory System

1. What are the four main events of respiration?

____

____

____

____

2. The respiratory system is divisible into *conducting zone* and *respiratory zone* structures.

1. Name the respiratory zone structures. ______________________________

____

2. What is their common function? ______________________________

3. Name the conducting zone structures. ______________________________

____

____

3. Use the terms provided to complete the pathway of air to the primary bronchi. Some of the terms have already been filled in. The main pathway—through the upper respiratory organs—is the vertical pathway at the far left. To the right of the name of most of those organs are horizontal lines for listing the subdivisions of that organ or the structures over which air passes as it moves through that organ.

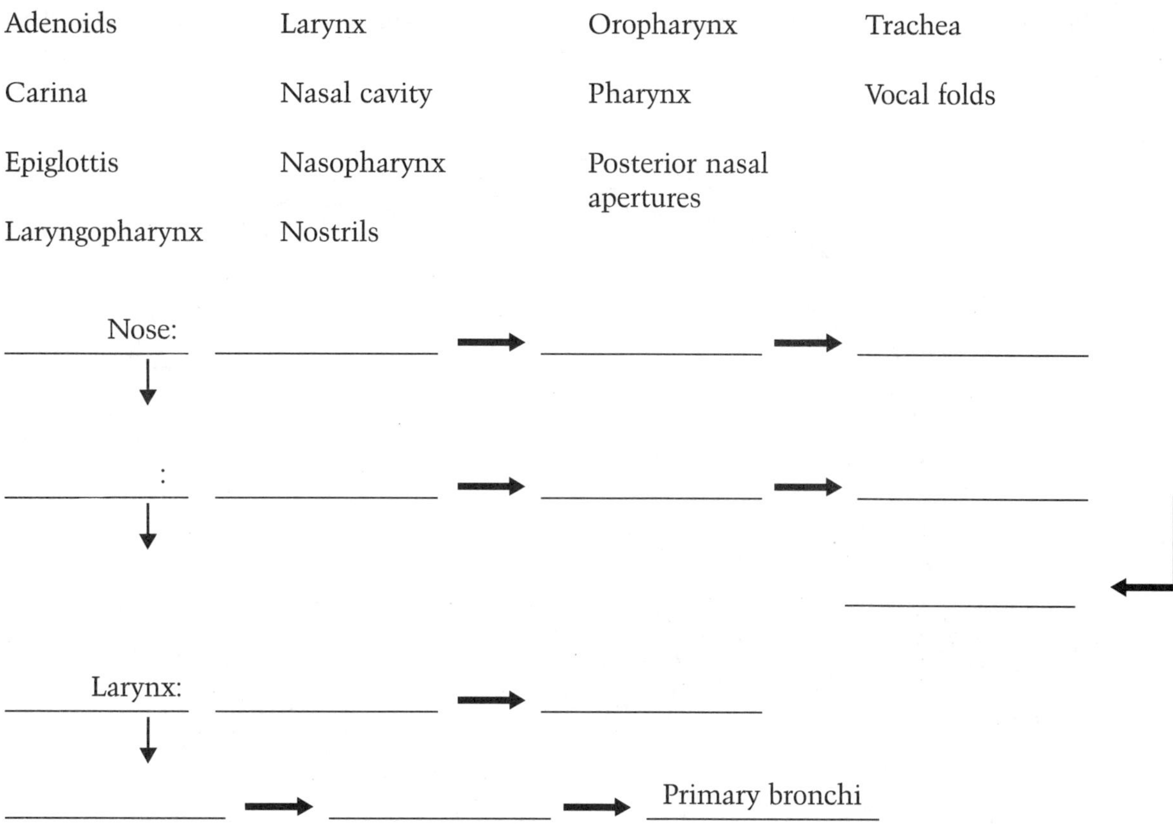

4. Figure 21.1 illustrates the skeletal framework of the nose. Identify the various bones and cartilages contributing to that framework by color coding and coloring them on the figure.

- ◯ Dense fibrous connective tissue
- ◯ Major alar cartilages
- ◯ Lateral process of septal cartilage
- ◯ Minor alar cartilages
- ◯ Maxillary bone (frontal process)
- ◯ Nasal bone
- ◯ Septal cartilage

Figure 21.1

5. Figure 21.2 is a sagittal view of the upper respiratory structures. First, correctly identify all structures with leader lines on the figure. Then, select different colors for the structures listed below and color the coding circles and the corresponding structures on the figure.

◯ Nasal cavity	◯ Larynx	◯ Tongue	◯ Thyroid cartilage
◯ Pharynx	◯ Paranasal sinuses	◯ Hard palate	◯ Cricoid cartilage
◯ Trachea			

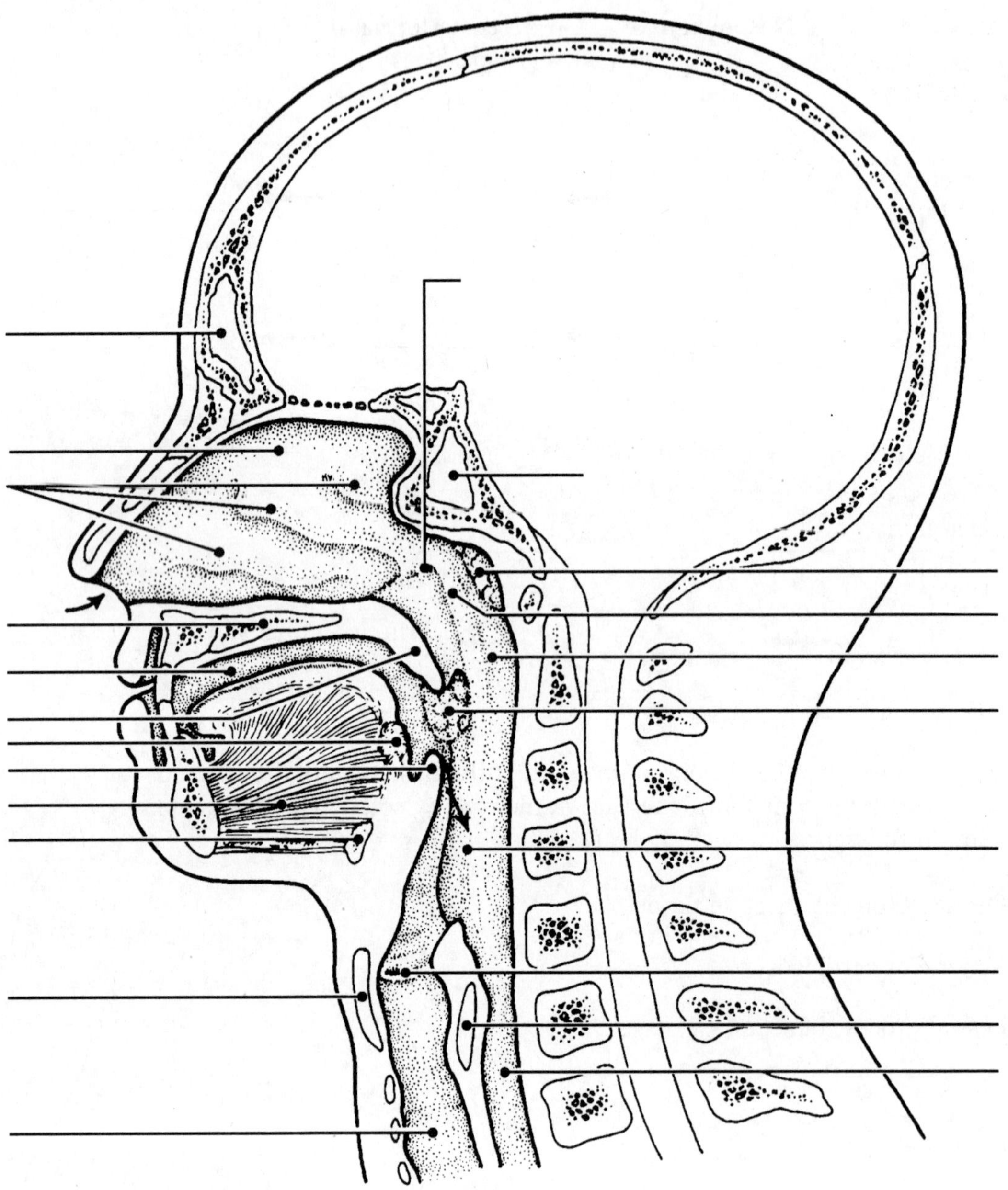

Figure 21.2

6. Complete the following statements concerning the nasal cavity and adjacent structures by inserting your answers in the answer blanks.

______________ 1.
______________ 2.
______________ 3.
______________ 4.
______________ 5.
______________ 6.
______________ 7.
______________ 8.
______________ 9.
______________ 10.

Air enters the nasal cavities of the respiratory system through the __(1)__. The nasal cavity is divided by the midline __(2)__. The nasal cavity mucosa has several functions. Its major functions are to __(3)__, __(4)__, and __(5)__ the incoming air. Mucous membrane-lined cavities called __(6)__ are found in several bones surrounding the nasal cavities. They make the skull less heavy and probably act as resonance chambers for __(7)__. The passageway common to the digestive and respiratory systems, the __(8)__, is often referred to as the throat; it connects the nasal cavities with the __(9)__ below. Clusters of lymphatic tissue, __(10)__, are part of the defensive system of the body.

7. Circle the term that does not belong in each of the following groupings.

1. Sphenoidal   Maxillary   Mandibular   Ethmoid   Frontal
2. Nasal cavity   Trachea   Alveolus   Bronchus   Dead air
3. Apex   Base   Hilus   Larynx   Pleura
4. Sinusitis   Peritonitis   Pleurisy   Tonsillitis   Laryngitis
5. External nose   Nasal bones   Major alar cartilages   Nasal septum
6. Conchae   Increase air turbulence   Turbinates   Choanae
7. Cuneiform cartilage   Corniculate cartilage   Tracheal cartilage   Arytenoid cartilage   Epiglottis
8. Laryngopharynx   Oropharynx   Transports air and food   Nasopharynx

8. Figure 21.3 is a diagram of the larynx and associated structures. Select a different color for each structure and color the coding circles and the corresponding structures on the figure. Identify, by adding a label and leader line, the laryngeal prominence. Then, answer the questions below the figure.

○ Hyoid bone   ○ Tracheal cartilages   ○ Thyrohyoid membrane

○ Cricoid cartilage   ○ Epiglottis   ○ Cricothyroid ligament

○ Thyroid cartilage   ○ Cricotracheal ligament

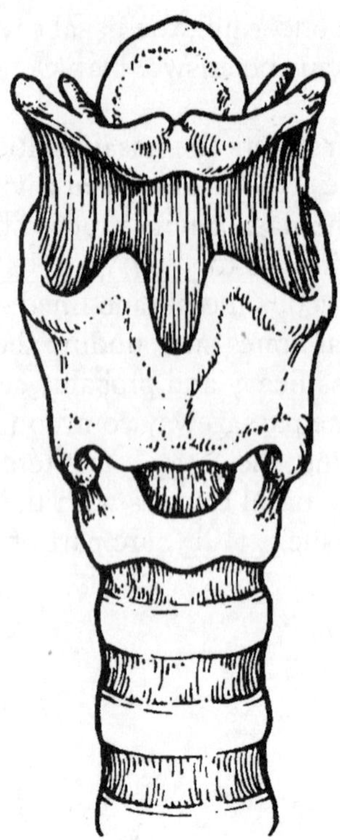

Figure 21.3

1. What are the three functions of the larynx? ____________________

____

____

____

2. What cartilages, not illustrated, anchor the vocal folds internally?

____

3. What type of cartilage forms the epiglottis? ____________________

4. What type of cartilage forms the other eight laryngeal cartilages? ____________________

5. Explain this difference. ____________________

____

____

6. What is the common name for the laryngeal prominence?

____

9. Use the key choices to match the following definitions with the correct terms.

Key Choices

A. Alveoli	E. Esophagus	I. Phrenic nerve	M. Visceral pleura
B. Bronchioles	F. Glottis	J. Primary bronchi	N. Uvula
C. Conchae	G. Palate	K. Trachea	O. Vocal folds
D. Epiglottis	H. Parietal pleura	L. Vagus nerve	

____________________ 1. Smallest respiratory passageways

____________________ 2. Separates the oral and nasal cavities

____________________ 3. Major nerve stimulating the diaphragm

____________________ 4. Food passageway posterior to the trachea

____________________ 5. Closes off the larynx during swallowing

____________________ 6. Windpipe

____________________ 7. Actual site of gas exchanges

____________________ 8. Pleural layer covering the thorax walls

____________________ 9. Autonomic nervous system nerve serving the thorax

____________________ 10. Lumen of the larynx

____________________ 11. Fleshy lobes in the nasal cavity that increase its surface area

____________________ 12. Close the glottis during the Valsalva maneuver

____________________ 13. Closes the nasopharynx during swallowing

____________________ 14. The cilia of its mucosa beat upward toward the larynx

10. Figure 21.4 shows a cross section through the trachea. First, label the layers indicated by the leader lines. Next, color the following: mucosa (including the cilia, epithelium, lamina propria)—light pink; area containing the submucosal seromucous glands—purple; hyaline cartilage ring—blue; trachealis muscle—orange; and adventitia—yellow. Then, respond to the questions following the figure.

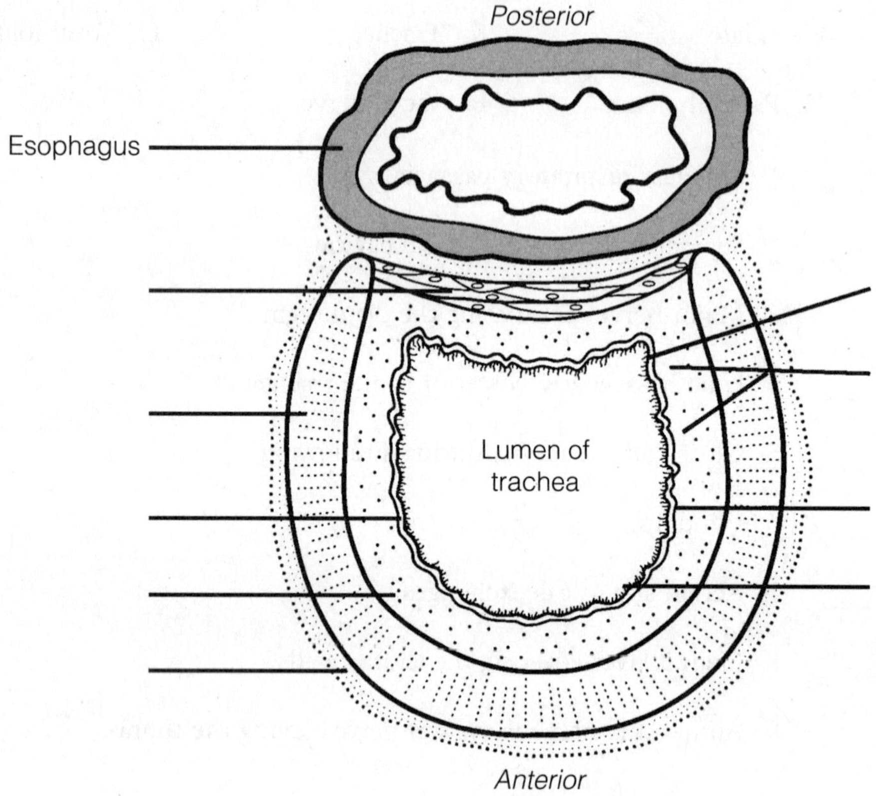

Figure 21.4

1. What important role is played by the cartilage rings that reinforce the trachea?

   ____

2. Of what importance is the fact that the cartilage rings are incomplete posteriorly?

   ____

3. What occurs when the trachealis muscle contracts and in what activities might this action be very helpful? ____

   ____

11. Complete the following statements by writing the missing terms in the answer blanks.

__________ 1.	Within the larynx are the __(1)__, which vibrate with exhaled air and allow an individual to __(2)__. Speech involves opening and closing the __(3)__. The __(4)__ muscles control the length of the true vocal cords by moving the __(5)__ cartilages. The tenser the vocal cords, the __(6)__ the pitch. To produce deep tones, the glottis is __(7)__. The greater the force of air rushing past the vocal cords, the __(8)__ the sound produced. Inflammation of the vocal cords is called __(9)__.
__________ 2.	
__________ 3.	
__________ 4.	
__________ 5.	
__________ 6.	
__________ 7.	
__________ 8.	
__________ 9.	

12. Using the key choices, match the proper type of lining epithelium with each respiratory structure listed below.

Key Choices

A. Stratified squamous

B. Pseudostratified ciliated columnar

C. Simple squamous

D. Stratified columnar

E. Simple cuboidal

_____ 1. Nasal cavity

_____ 2. Nasopharynx

_____ 3. Laryngopharynx

_____ 4. Trachea and primary bronchi

_____ 5. Bronchioles

_____ 6. Walls of alveoli (type I cells)

13. Match the bronchus or bronchiole type (at right) with the lung region supplied by that specific type of air tube (at left).

	Lung region	**Air tube**
_____	1. Bronchopulmonary segment	A. Main bronchus
_____	2. Lobule	B. Secondary bronchus
_____	3. Alveolar ducts and sacs	C. Segmental bronchus
_____	4. Entire lung	D. Large bronchiole
_____	5. Lung lobe	E. Respiratory bronchiole

14. Figure 21.5 illustrates the gross anatomy of lower respiratory structures. Intact structures are shown on the left; isolated respiratory passages are shown on the right. Select a different color for each of the respiratory system structures listed below and color the coding circles and the corresponding structures on the figure. Then complete the figure by labeling the areas and structures with leader lines. Be sure to include the pleural space, mediastinum, apex of right lung, diaphragm, clavicle, and the base of the right lung.

◯ Trachea	◯ Main (primary) bronchi	◯ Visceral pleura
◯ Larynx	◯ Lobar (secondary) bronchi	◯ Parietal pleura
◯ Intact lung	◯ Segmental (tertiary) bronchi	

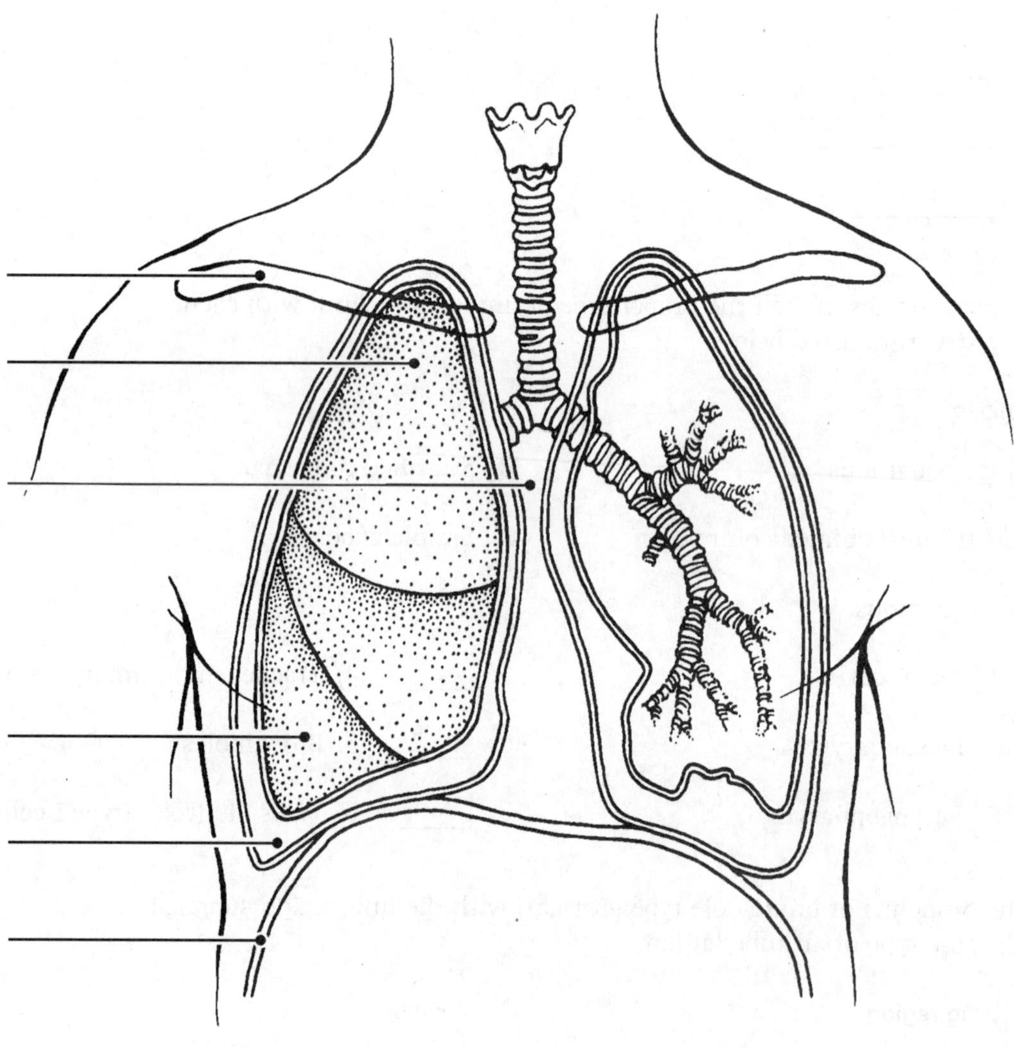

Figure 21.5

15. Figure 21.6 illustrates the microscopic structure of the respiratory unit of lung tissue. The external anatomy is shown in Figure 21.6A. Color the intact alveoli yellow, the pulmonary capillaries red, and the respiratory bronchiole green. Bracket and label an alveolar sac. Also label an alveolar duct.

    A cross section through an alveolus is shown in Figure 21.6B. Color the alveolar epithelium yellow, the capillary endothelium pink, and the red blood cells in the capillary red. Also, label the alveolar chamber and color it blue. Finally, add the symbols for oxygen gas (O_2) and carbon dioxide gas (CO_2) in the sites where they would be in higher concentration and add arrows showing their direction of movement through the respiratory membrane.

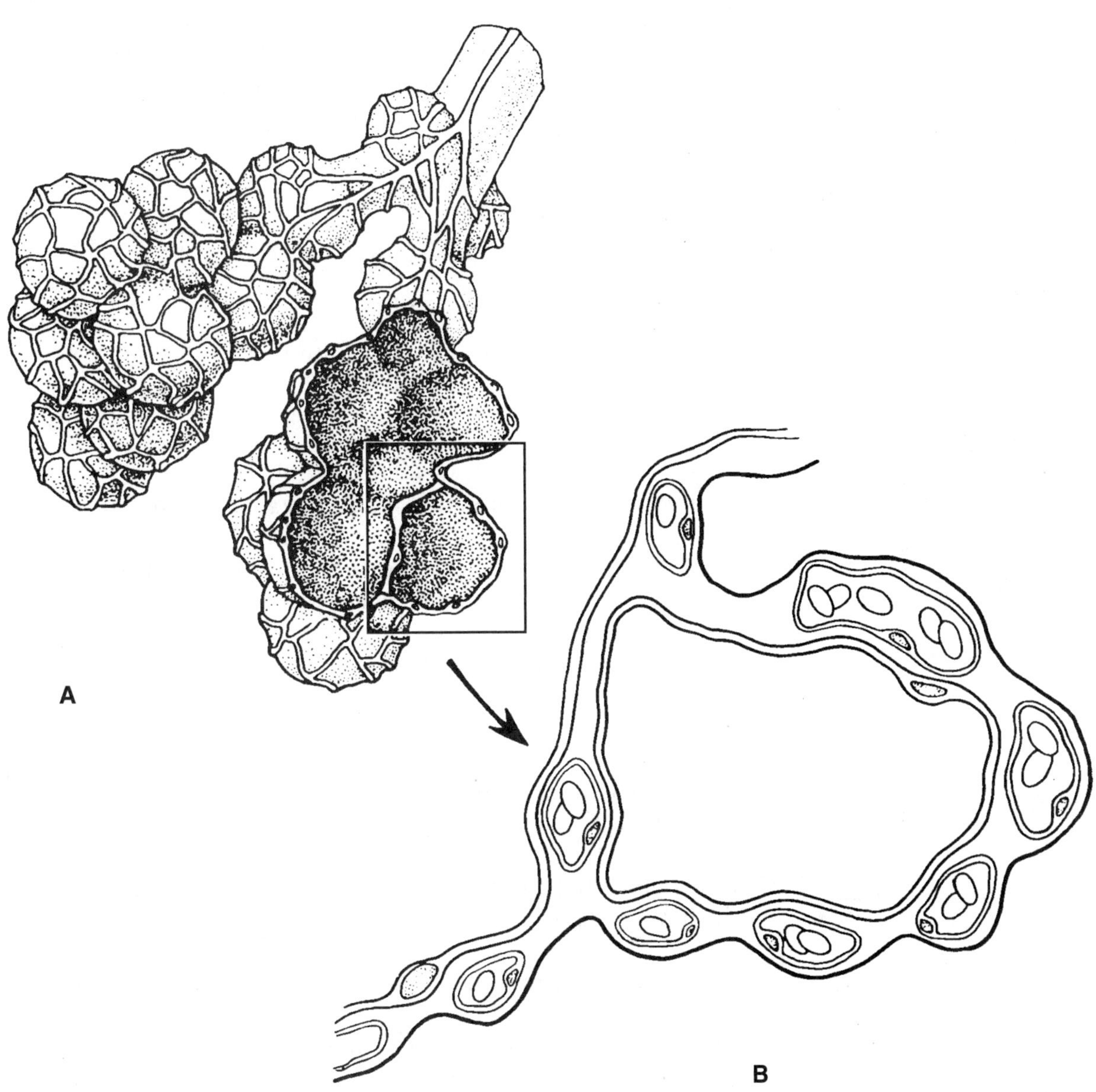

Figure 21.6

16. Complete the following paragraph on the alveolar cells and their roles by writing the missing terms in the answer blanks.

1. ____________________
2. ____________________
3. ____________________
4. ____________________
5. ____________________
6. ____________________
7. ____________________
8. ____________________

With the exception of the stroma of the lungs, which is __(1)__ tissue, the lungs are mostly air spaces, of which the alveoli comprise the greatest part. The bulk of the alveolar walls are made up of squamous epithelial cells called __(2)__ cells. Structurally, these cells are well suited for their __(3)__ function. The cuboidal cells of the alveoli, called __(4)__ cells, are much less numerous. These cells produce a fluid that coats the air-exposed surface of the alveolus and contains a lipid-based molecule called __(5)__ that functions to __(6)__ of the alveolar fluid. This fluid also contains __(7)__ that are important molecules in innate immunity. Although the pulmonary capillaries spiderweb over the alveolar surfaces, the nutritive blood supply of the lungs is provided by the __(8)__ arteries.

Mechanics of Breathing

1. Using the key choices, match the following facts about pressure with the correct terms.

Key Choices

A. Atmospheric pressure B. Intrapulmonary pressure C. Intrapleural pressure

____________________ 1. Barring pneumothorax, this pressure is always lower than atmospheric pressure (that is, is negative pressure)

____________________ 2. Pressure of air outside the body

____________________ 3. As it decreases, air flows into the passageways of the lungs

____________________ 4. As it increases over atmospheric pressure, air flows out of the lungs

____________________ 5. If this pressure becomes equal to the atmospheric pressure, the lungs collapse

____________________ 6. Rises well over atmospheric pressure during a forceful cough

____________________ 7. Also known as the intra-alveolar pressure

2. Many changes occur within the lungs as the diaphragm (and external intercostal muscles) contracts and then relaxes. These changes cause air to flow into and out of the lungs. The activity of the diaphragm is given in the left column of the following table. Several changes in internal thoracic conditions are listed in the column heads to the right. Complete the table by checking (✓) the appropriate column to correctly identify the change that would be occurring in each case relative to the stated diaphragm activity.

Activity of diaphragm (↑ = increased) (↓ = decreased)	Changes in							
	Internal volume of thorax		Internal pressure in thorax		Size of lungs		Direction of air flow	
	↑	↓	↑	↓	↑	↓	Into lung	Out of lung
Contracted, moves downward								
Relaxed, moves superiorly								

3. Various factors that influence pulmonary ventilation are listed in the key choices. Select the appropriate key choices to match the following descriptions about the lungs.

Key Choices

A. Respiratory passageway resistance

B. Alveolar surface tension forces

C. Lung elasticity

D. Lung compliance

________________ 1. The change in lung volume with a given change in transpulmonary pressure

________________ 2. Gas flow changes inversely with this factor

________________ 3. Essential for normal expiration

________________ 4. Leads to RDS (respiratory distress syndrome) when surfactant is absent

________________ 5. Diminished by age-related lung fibrosis and increasing rigidity of the thoracic cage

________________ 6. Its loss is the major pathology in emphysema

________________ 7. Reflects a state of tension at the surface of a liquid

________________ 8. Dramatically increased by asthma

________________ 9. Greatest in the medium-sized bronchi

4. During forced inspiration, accessory muscles are activated that raise the rib cage more vigorously than occurs during quiet inspiration.

1. Name three such muscles. ______________________________

2. Although normal quiet expiration is largely passive due to lung recoil, when expiration must be more forceful (or the lungs are diseased), muscles that increase the abdominal pressure or depress the rib cage are enlisted. Provide two examples of muscles that cause abdominal pressure to rise. ______________________________

3. Provide two examples of muscles that depress the rib cage. ______________________________

5. Figure 21.7 is a diagram showing respiratory volumes. Complete the figure by making the following additions.

1. Bracket the volume representing the vital capacity and color it yellow; label it VC.

2. Add green stripes to the area representing the inspiratory reserve volume and label it IRV.

3. Add red stripes to the area representing the expiratory reserve volume and label it ERV.

4. Identify and label the respiratory volume that is *now* yellow. Color the residual volume blue and label it appropriately on the figure.

5. Bracket and label the inspiratory capacity (IC).

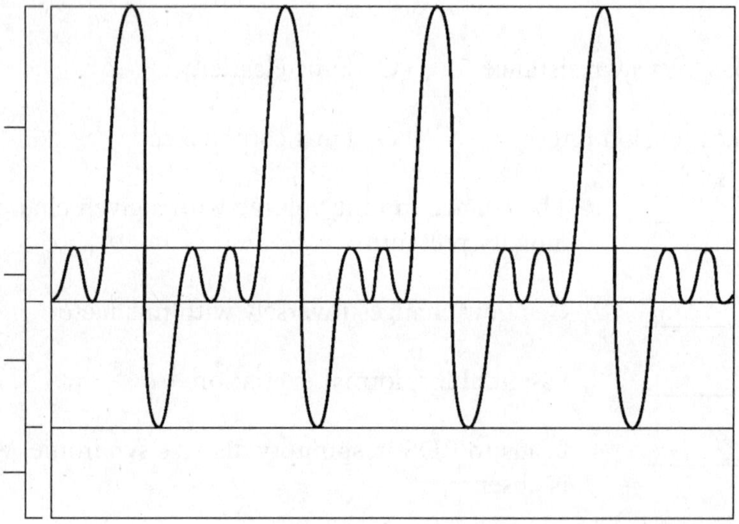

Figure 21.7

6. Check (✓) all factors that *decrease* the AVR.

_____ 1. Shallow breathing

_____ 2. Deep breathing

_____ 3. Rapid breathing

_____ 4. Slow breathing

_____ 5. Mucus in the respiratory passageways

7. This exercise concerns respiratory volume or capacity measurements. Using the key choices, select the terms that identify the respiratory volumes described by inserting the appropriate answers in the answer blanks.

Key Choices

A. Dead space volume	D. Residual volume (RV)	G. Vital capacity (VC)
B. Expiratory reserve volume (ERV)	E. Tidal volume (TV)	
C. Inspiratory reserve volume (IRV)	F. Total lung capacity (TLC)	

__________ 1. Respiratory volume inhaled or exhaled during normal breathing

__________ 2. Air in respiratory passages that does not contribute to gas exchange

__________ 3. Total amount of exchangeable air

__________ 4. Gas volume that allows gas exchange to go on continuously

__________ 5. Amount of air that can still be exhaled (forcibly) after a normal exhalation

__________ 6. Sum of all lung volumes

8. Complete the following statements by writing the missing terms in the answer blanks.

__________ 1.

__________ 2.

__________ 3.

__________ 4.

__________ 5.

Pulmonary function tests can be conducted using a(n) __(1)__, a device that measures respired air volumes. This type of test can distinguish between __(2)__, disorders that result from increased airway resistance, and __(3)__, which result from decreased lung capacity caused by changes such as tuberculosis. __(4)__ is decreased by obstructive pulmonary diseases, which reduce the rate of air flow. Restrictive diseases lower __(5)__, because the total inflation capacity is reduced.

9. Identify the three nonrespiratory movements described below.

__________ 1. Sudden inspiration, resulting from spasms of the diaphragm

__________ 2. A deep breath is taken, the glottis is closed, and air is forced out of the lungs against the glottis; clears the lower respiratory passageways

__________ 3. As just described, but clears the upper respiratory passageways

Gas Exchanges between the Blood, Lungs, and Tissues

1. This exercise describes some ideal gas laws that relate the physical properties of gases and their behavior in liquids. Match the key choices with the descriptions that follow.

Key Choices

A. Boyle's law B. Dalton's law C. Henry's law

__________ 1. Describes the fact that each gas in a mixture exerts pressure according to its percentage in the mixture

__________ 2. States that $P_1V_1 = P_2V_2$ when temperature is constant

__________ 3. States that when gases are in contact with liquids, each gas will dissolve in proportion to its partial pressure and solubility in the liquid

2. What therapy is used to treat cases of carbon monoxide poisoning, gas gangrene, and tetanus? This therapy represents a clinical application of which of the ideal gas laws?

____

____

3. What are the mechanism and consequences of oxygen toxicity? __________

____

____

4. Match the key choices with the following descriptions.

Key Choices

A. External respiration C. Inspiration E. Pulmonary ventilation (breathing)

B. Expiration D. Internal respiration

__________ 1. Period of breathing when air enters the lungs

__________ 2. Period of breathing when air leaves the lungs

__________ 3. Alternate flushing of air into and out of the lungs

__________ 4. Exchange of gases between alveolar air and pulmonary capillary blood

__________ 5. Exchange of gases between blood and tissue cells

5. Rank the following body regions by numbering them in the proper sequence in each column to indicate the relative partial pressure of the gases named in each column heading. Indicate the area of lowest gas partial pressure by the number 1.

PO_2	PCO_2	PN_2	PH_2O
_____ 1. Alveoli	_____ 6. Alveoli	_____ 11. Alveoli	_____ 13. Alveoli
_____ 2. External atmosphere	_____ 7. External atmosphere	_____ 12. External atmosphere	_____ 14. External atmosphere
_____ 3. Arterial blood	_____ 8. Arterial blood		
_____ 4. Venous blood	_____ 9. Venous blood		
_____ 5. Tissue cells	_____ 10. Tissue cells		

6. Use the key choices to complete the following statements about gas exchanges in the body.

Key Choices

A. Active transport

B. Air of alveoli to capillary blood

C. Carbon dioxide-poor and oxygen-rich

D. Capillary blood to alveolar air

E. Capillary blood to tissue cells

F. Diffusion

G. Higher concentration

H. Lower concentration

I. Oxygen-poor and carbon dioxide-rich

J. Tissue cells to capillary blood

__________________ 1.

__________________ 2.

__________________ 3.

__________________ 4.

__________________ 5.

__________________ 6.

__________________ 7.

__________________ 8.

__________________ 9.

All gas exchanges are made by __(1)__. When substances pass in this manner, they move from areas of their __(2)__ to areas of their __(3)__. Thus oxygen continually passes from the __(4)__ and then from the __(5)__. Conversely, carbon dioxide moves from the __(6)__ and from __(7)__. From there it passes out of the body during expiration. As a result of such exchanges, arterial blood tends to be relatively __(8)__, while venous blood is relatively __(9)__.

7. Circle the correct alternative in each of the following statements about alveolar air flow and blood flow coupling.

1. When alveolar ventilation is inadequate, the P_{O_2} is low/high.
2. Consequently, the pulmonary vessels dilate/constrict.
3. Additionally, when alveolar ventilation is inadequate, the P_{CO_2} is low/high.
4. Consequently, the bronchioles dilate/constrict.

Transport of Respiratory Gases by Blood

1. Complete the following paragraphs concerning gas transport in the blood by writing the missing terms in the answer blanks.

1. ____________
2. ____________
3. ____________
4. ____________
5. ____________
6. ____________
7. ____________
8. ____________
9. ____________
10. ____________
11. ____________
12. ____________

Most oxygen is transported bound to __(1)__ inside the red blood cells. A very small amount is carried simply dissolved in __(2)__. Most carbon dioxide is carried in the form of __(3)__ in the plasma. The reaction in which CO_2 is converted to this chemical occurs more rapidly inside red blood cells than in plasma because the cells contain the enzyme __(4)__. After the chemical is generated, it diffuses into the plasma, and __(5)__ diffuse into the red blood cells. This exchange is called the __(6)__. Smaller amounts of carbon dioxide are transported dissolved in plasma and bound to hemoglobin as __(7)__.

The amount of CO_2 carried in blood is greatly influenced by the degree of oxygenation of the blood, an effect called the __(8)__. In general, the lower the amount of oxygen in the blood, the __(9)__ CO_2 that can be transported. Conversely, as CO_2 enters the blood, it prompts oxygen to dissociate from __(10)__. This effect is called the __(11)__. Carbon monoxide poisoning is lethal because carbon monoxide competes with __(12)__ for binding sites.

2. List four factors that enhance O_2 loading onto hemoglobin.

____

3. What is the functional importance of the following facts?

1. Hemoglobin is over 90% saturated with oxygen well before the blood has completed its route through the pulmonary capillaries.

____

____

2. Venous blood still has a hemoglobin saturation of over 70%.

____

____

____

4. In the grid provided in Figure 21.8, plot the following points. Then draw a sigmoid curve to connect the points. Finally, answer the questions following the grid.

A. PO_2 = 100 mm Hg; saturation = 99%

B. PO_2 = 70 mm Hg; saturation = 98%

C. PO_2 = 40 mm Hg; saturation = 75%

D. PO_2 = 20 mm Hg; saturation = 35%

E. PO_2 = 10 mm Hg; saturation = 10%

F. PO_2 = 0 mm Hg; saturation = 0%

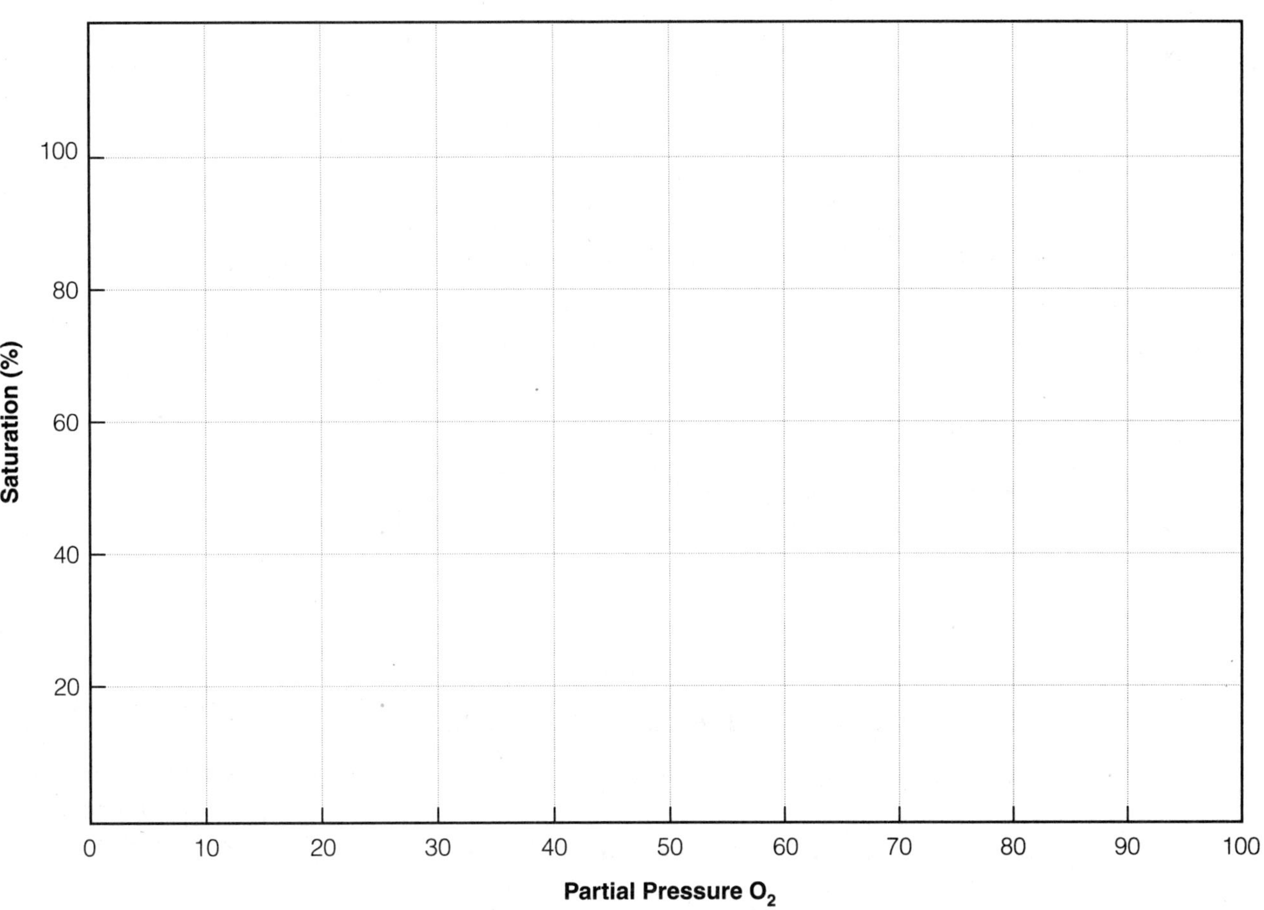

Figure 21.8

1. What is the curve just plotted called? ______________________________

____

2. What three factors will decrease the percent saturation of hemoglobin in the tissues?

____

____

5. Fetal hemoglobin (HbF) differs from that of an adult (HbA), as illustrated by the oxygen dissociation curve shown in Figure 21.9.

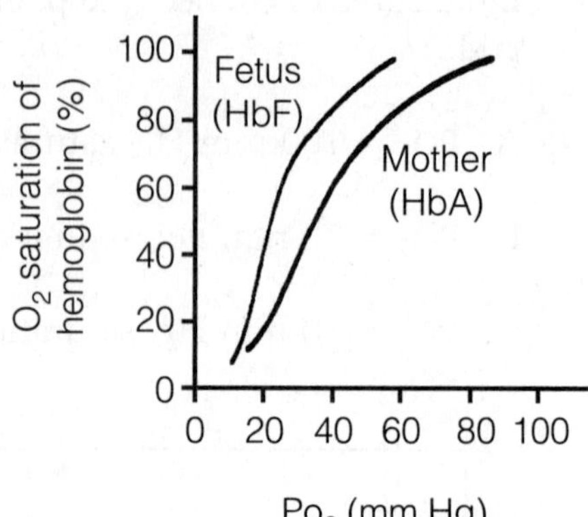

Figure 21.9

1. Explain what the differences in the graph mean or represent.

____

2. Explain the difference in the two hemoglobins that accounts for the variation.

____

Control of Respiration

1. There are several types of breathing controls. Match the structures in Column B with the descriptions in Column A.

	Column A	Column B
_____	1. Respiratory control center in the pons	A. Pontine respiratory group
_____, _____	2. Respiratory control centers in the medulla	B. Chemoreceptors in the aortic arch and carotid bodies
_____	3. Respond to overinflation of the lungs	C. VRG
_____	4. Respond to decreased oxygen levels in the blood	D. DRG
_____	5. Integrates input from peripheral stretch receptors and chemoreceptors	E. Stretch receptors in the lungs
_____	6. Medullary area that contains some neurons involved in inspiration and some that initiate expiration	
_____	7. The breathing "rhythm" center	
_____	8. Basically acts to fine-tune respiratory rhythms generated by the medulla	

2. Circle the term that does not belong in each of the following groupings.

1. Acidosis ↑ Carbonic acid ↓ pH ↑ pH
2. Acidosis Hyperventilation Hypoventilation CO_2 buildup
3. Apnea Cyanosis ↑ Oxygen ↓ Oxygen
4. ↓ Respiratory rate Exercise Anger ↓ CO_2 in blood
5. N narcosis The bends N toxicity Rapture of the deep
6. High altitude ↓ Po_2 ↑ Pco_2 ↓ Atmospheric pressure

Respiratory Adjustments

1. Answer the following questions concerning respiratory adjustments during exercise:

1. What are two differences between hyperpnea and hyperventilation?

____

____

2. What are three neural factors that trigger the abrupt increase in ventilation during exercise?

____

____

2. List three adaptive responses to high-altitude living.

1. ______________________________
2. ______________________________
3. ______________________________

Homeostatic Imbalances of the Respiratory System

1. Match the terms in Column B with the pathologic conditions described in Column A.

	Column A	Column B
________	1. Lack or cessation of breathing	A. Apnea
________	2. Normal breathing in terms of rate and depth	B. Asthma
		C. Chronic bronchitis
________	3. Labored breathing, or "air hunger"	D. Dyspnea
________	4. Chronic oxygen deficiency	E. Emphysema
________	5. Condition characterized by fibrosis of the lungs and an increase in size of the alveolar chambers	F. Eupnea
		G. Hypoxia
________	6. Condition characterized by increased mucus production, which clogs respiratory passageways and promotes coughing	H. Lung cancer
		I. Sleep apnea
________	7. Respiratory passageways narrowed by bronchiolar spasms	J. Tuberculosis
________	8. Together called COPD	
________	9. Incidence strongly associated with cigarette smoking; has increased dramatically in women recently	
________	10. Victims become barrel-chested because of air retention	
________	11. Infection spread by airborne bacteria; a recent alarming increase in cases among drug users and HIV-infected people	
________	12. Temporary cessation of breathing during sleep	

THE INCREDIBLE JOURNEY:

A Visualization Exercise for the Respiratory System

You carefully begin to pick your way down, using the cartilages as steps.

1. Complete the following narrative by writing the missing terms in the answer blanks.

1. ____________________
2. ____________________
3. ____________________
4. ____________________
5. ____________________
6. ____________________
7. ____________________
8. ____________________
9. ____________________
10. ____________________
11. ____________________

Your journey through the respiratory system is to be on foot. To begin, you simply will walk into your host's external nares. You are miniaturized, and your host is sedated lightly to prevent sneezing during your initial observations in the nasal cavity and subsequent descent.

You begin your exploration of the nasal cavity in the right nostril. One of the first things you notice is that the chamber is very warm and humid. High above, you see three large, rounded lobes, the __(1)__, which provide a large mucosal surface area for warming and moistening the entering air. As you walk toward the rear of this chamber, you see a large lumpy mass of lymphatic tissue, the __(2)__ in the __(3)__, or first portion of the pharynx. As you peer down the pharynx, you realize that it will be next to impossible to maintain your footing during the next part of your journey—it is nearly straight down, and the __(4)__ secretions are like grease. You sit down and dig in your heels to get started. After a quick slide, you land abruptly on one of a pair of flat sheetlike structures that begins to vibrate rapidly, bouncing you up and down helplessly. You are also conscious of a rhythmic hum during this jostling, and you realize that you have landed on a __(5)__. You pick yourself up and look over the superior edge of the __(6)__, down into the seemingly endless esophagus behind. You chastise yourself for not remembering that the __(7)__ and respiratory pathways separate at this point. Hanging directly over your head is a leaflike __(8)__ cartilage. Normally, you would not have been able to get this far because it would have closed off this portion of the respiratory tract. With your host sedated, however, this protective reflex did not work.

You carefully begin to pick your way down, using the cartilages as steps. When you reach the next respiratory organ, the __(9)__, your descent becomes much easier, because the structure's C-shaped cartilages form a ladderlike supporting structure. As you climb down the cartilages, your face is stroked rhythmically by soft cellular extensions, or __(10)__. You remember that their function is to move mucus laden with bacteria or dust and other debris toward the __(11)__.

12. ____________________
13. ____________________
14. ____________________
15. ____________________

You finally reach a point where the descending passageway splits into the two __(12)__, and since you want to control your progress (rather than slide downward), you choose the more horizontal __(13)__ branch. If you remain in the superior portion of the lungs, your return trip will be less difficult because the passageways will be more horizontal than steeply vertical. The passageways get smaller and smaller, slowing your progress.

16. ____________________
17. ____________________
18. ____________________
19. ____________________
20. ____________________

As you are squeezing into one of the smallest of the respiratory passageways, a __(14)__, you see a bright spherical chamber ahead. You scramble into this __(15)__, pick yourself up, and survey the area. Scattered here and there are lumps of a substance that looks suspiciously like coal, reminding you that your host is a smoker. As you stand there, a soft rustling wind seems to flow in and out of the chamber. You press your face against the transparent chamber wall and see disc-like cells, __(16)__, passing by in the capillaries on the other side. As you watch, they change from a somewhat bluish color to a bright __(17)__ color as they pick up __(18)__ and unload __(19)__.

You record your observations and then contact headquarters to let them know you are ready to begin your ascent. You begin your return trek, slipping and sliding as you travel. By the time you reach the inferior edge of the trachea, you are ready for a short break. As you rest on the mucosa, you begin to notice that the air is becoming close and very heavy. You pick yourself up quickly and begin to scramble up the trachea. Suddenly and without warning, you are hit by a huge wad of mucus and catapulted upward and out onto your host's freshly pressed handkerchief! Your host has assisted your exit with a __(20)__.

CHALLENGING YOURSELF

At the Clinic

1. Barbara is rushed to the emergency room after an auto accident. The eighth through tenth ribs on her left side have been fractured and have punctured the lung. What term is used to indicate lung collapse? Will both lungs collapse? Why or why not?
2. A young boy is diagnosed with cystic fibrosis. What effect will this have on his respiratory system?
3. Len LaBosco, a medical student, has recently moved to Mexico City for his internship. As he examines the lab report of his first patient, he notices that the RBC count is high. Why is this normal for this patient? (*Hint:* Mexico City is located on a high-altitude mesa.) What respiratory adjustments will Len have to make in his own breathing, minute ventilation, arterial P_{CO_2}, and hemoglobin saturation level?
4. Why must patients be forced to cough when recovering from chest surgery?
5. A mother, obviously in a panic, brings her infant who is feverish, hyperventilating, and cyanotic to the clinic. The infant is quickly diagnosed with pneumonia. What aspect of pneumonia has caused the cyanosis?
6. After a long bout of bronchitis, Mrs. Dupee complains of a stabbing pain in her side with each breath. What is her condition?
7. A patient with congestive heart failure has cyanosis coupled with liver and kidney failure. What type of hypoxia is this?

8. The Kozloski family is taking a long auto trip. Michael, who has been riding in the back of the station wagon, complains of a throbbing headache. Then, a little later, he seems confused and his face is flushed. What is your diagnosis of Michael's problem?

9. A new mother checks on her sleeping infant, only to find that it has stopped breathing and is turning blue. The mother quickly picks up the baby and pats its back until it starts to breathe. What tragedy has been averted?

10. Roger Proulx, a rugged four-pack-a-day smoker, comes to the clinic because of painful muscle spasms. Blood tests reveal hypocalcemia and elevated levels of calcitonin. What type of lung cancer is suspected?

11. A young man visiting his father in the hospital hears the clinical staff refer to the patient in the next bed as a "pink puffer." He notes the patient's barrel-shaped chest and wonders what the man's problem is. What is the patient's diagnosis, and why is he called a "pink puffer"?

12. Joanne Willis, a long-time smoker, is complaining that she has developed a persistent cough. What is your first guess as to her condition? What has happened to her bronchial cilia?

13. Mrs. Jackson, an elderly matron, has severe kyphosis (dowager's hump) of her spine. She is complaining of being breathless a lot and her vital capacity is below normal. What does her kyphosis have to do with her reduced vital capacity?

14. As a result of a stroke, the swallowing mechanism in an elderly woman is uncoordinated. What effects might this have on her respiratory system?

15. While diapering his 1-year-old boy (who puts almost everything in his mouth), Mr. Gregoire failed to find one of the small safety pins previously used. Two days later, his son developed a cough and became feverish. What probably had happened to the safety pin and where (anatomically) would you expect to find it?

16. Mr. and Ms. Rao took their sick 5-year-old daughter to the doctor. The girl was breathing entirely through her mouth, her voice sounded odd and whiny, and a puslike fluid was dripping from her nose. Which one of the four sets of tonsils was most likely infected in this child?

Stop and Think

1. When dogs pant, they inhale through the nose and exhale through the mouth. How does this provide a cooling mechanism?

2. A deep-sea expedition is being planned. Why is the location of the nearest hyperbaric chamber critically important?

3. Is the partial pressure of oxygen in expired air higher, lower, or the same as that of alveolar air? Explain your answer.

4. Consider the following: (a) Two girls in a high school cafeteria were giggling over lunch, and both accidentally sprayed milk out their nostrils at the same time. Explain in anatomical terms why swallowed fluids can sometimes come out the nose.

   (b) A boy in the same cafeteria then stood on his head and showed he could drink milk upside down without any of it entering his nasal cavity or nose. What prevented the milk from flowing downward into his nose?

5. A surgeon had to remove three adjacent bronchopulmonary segments from the left lung of Mr. Krigbaum, a patient with tuberculosis. Almost half of the lung was removed, yet there was no severe bleeding, and relatively few blood vessels had to be cauterized (closed off). Why was the surgery so easy to perform?

6. The cilia lining the upper respiratory passages (superior to the larynx) beat inferiorly while the cilia lining the lower respiratory passages (larynx and below) beat superiorly. What is the functional "reason" for this difference?

7. What is the function of the abundant elastin fibers that occur in the stroma of the lung and around all respiratory tubes from the trachea through the respiratory tree?

8. A man was choking on a piece of meat and no one was around to offer help. He saw a horizontal railing in his house that ran just above the level of his navel. He hurled himself against the railing in an attempt to perform the Heimlich maneuver on himself. Use logic to deduce whether this was a wise move that might save the man's life.

9. Why does an EMT administering a breathalyzer test for alcohol ask the person being tested to expel one deep breath instead of several shallow ones?

COVERING ALL YOUR BASES

Multiple Choice

Select the best answer or answers from the choices given.

1. Structures that are part of the respiratory zone include:
   A. terminal bronchioles
   B. respiratory bronchioles
   C. segmental bronchi
   D. alveolar ducts

2. Which of the following have a respiratory function?
   A. Olfactory mucosa C. Vibrissae
   B. Adenoids D. Auditory tubes

3. Which structures are associated with the production of speech?
   A. Cricoid cartilage C. Arytenoid cartilage
   B. Glottis D. Pharynx

4. The skeleton of the external nose consists of:
   A. cartilage and bone
   B. bone only
   C. hyaline cartilage only
   D. elastic cartilage only

5. Which of the following is not part of the conducting zone of the respiratory system?
   A. Pharynx D. Lobar bronchi
   B. Alveolar sac E. Larynx
   C. Trachea

6. The mucus sheets that cover the inner surfaces of the nasal cavity and bronchi are secreted by which cells?
   A. Serous cells in tubuloacinar glands
   B. Mucous cells in tubuloacinar glands and epithelial goblet cells
   C. Ciliated cells
   D. Alveolar type II cells

7. Select the single false statement about the true vocal cords:
   A. They are the same as the vocal folds.
   B. They attach to the arytenoid cartilages via the vocal ligaments.
   C. Exhaled air flowing through the glottis vibrates them to produce sound.
   D. They are also called the vestibular folds.

8. The function of alveolar type I cells is:
   A. to produce surfactant
   B. to propel mucous sheets
   C. phagocytosis of dust particles
   D. to allow rapid diffusion of respiratory gases

9. An examination of a lobe of the lung reveals many branches off the main passageway. These branches are:
   A. main bronchi C. tertiary bronchi
   B. lobar bronchi D. segmental bronchi

10. Microscopic examination of lung tissue shows a relatively small passageway with nonciliated simple cuboidal epithelium, abundant elastic fibers, absence of cartilage, and some smooth muscle. What is the classification of this structure?
    A. Lobar bronchus
    B. Segmental bronchus
    C. Respiratory bronchiole
    D. Terminal bronchiole

11. An alveolar sac:
    A. is an alveolus
    B. relates to an alveolus as a bunch of grapes relates to one grape
    C. is a huge, saclike alveolus in an emphysema patient
    D. is the same as an alveolar duct

12. The respiratory membrane (air-blood barrier) consists of:
   A. alveolar type I cell, basal membranes, endothelial cell
   B. air, connective tissue, lung
   C. type II cell, dust cell, type I cell
   D. pseudostratified epithelium, lamina propria, capillaries

13. Cells responsible for removing foreign particles from inspired air include:
   A. goblet cells   C. dust cells
   B. type II cells   D. ciliated cells

14. The root of the lung is at its:
   A. apex   C. hilum
   B. base   D. cardiac notch

15. Which of the following are characteristic of a bronchopulmonary segment?
   A. Removal causes collapse of adjacent segments
   B. Fed by a tertiary bronchus
   C. Supplied by its own branches of the pulmonary artery and vein
   D. Separated from other segments by its septum

16. During inspiration, intrapulmonary pressure is:
   A. greater than atmospheric pressure
   B. less than atmospheric pressure
   C. greater than intrapleural pressure
   D. less than intrapleural pressure

17. Lung collapse is prevented by:
   A. high surface tension of alveolar fluid
   B. high surface tension of pleural fluid
   C. high pressure in the pleural cavities
   D. high elasticity of lung tissue

18. Accessory muscles of inspiration include:
   A. external intercostals
   B. scalenes
   C. internal intercostals
   D. pectoralis major

19. The greatest resistance to gas flow occurs at the:
   A. terminal bronchioles
   B. alveolar ducts
   C. medium-sized bronchi
   D. respiratory bronchioles

20. Resistance is increased by:
   A. epinephrine
   B. parasympathetic stimulation
   C. inflammatory chemicals
   D. contraction of the trachealis muscle

21. Which of the following conditions will reduce lung compliance?
   A. Tuberculosis   C. IRDS
   B. Aging   D. Osteoporosis

22. Chemically, surfactant is classified as a:
   A. polar molecule
   B. hydrophobic molecule
   C. polysaccharide
   D. lipoprotein

23. Which of the following changes will accompany the loss of elasticity associated with aging?
   A. Increase in tidal volume
   B. Increase in inspiratory reserve volume
   C. Increase in residual volume
   D. Increase in vital capacity

24. What is the approximate alveolar ventilation rate in a 140-pound female with a tidal volume of 400 ml and a respiratory rate of 20 breaths per minute?
   A. 10,800 ml/min
   B. 8000 ml/min
   C. 5200 ml/min
   D. 2800 ml/mm

25. FEV is reduced but FVC is not affected in which disorder(s)?
   A. Asthma
   B. Tuberculosis
   C. Polio
   D. Cystic fibrosis

26. Disorders classified as COPDs include:
   A. pneumonia
   B. emphysema
   C. bronchitis
   D. sleep apnea

27. Considering an individual in good health, an increase in blood flow rate through the pulmonary capillaries so that blood is in the capillaries for about 0.5 sec instead of 0.75 sec:
   A. decreases the rate of oxygenation
   B. reduces oxygenation of the blood
   C. does not change the level of oxygen saturation of hemoglobin
   D. increases rate of diffusion of oxygen

28. Relatively high alveolar P_{O_2}, coupled with low P_{CO_2}, triggers:
   A. dilation of the supplying bronchiole
   B. constriction of the supplying bronchiole
   C. dilation of the supplying arteriole
   D. constriction of the supplying arteriole

29. In blood flowing through systemic veins of a person at rest:

A. hemoglobin is about 75% saturated
B. the blood is carrying about 20% vol oxygen
C. the Bohr effect has reduced hemoglobin's affinity for oxygen
D. on average, each hemoglobin molecule is carrying two oxygen molecules

30. Which of the following is (are) true concerning CO_2 transport in systemic venous blood in an exercising person?

A. The level of bicarbonate ion is higher than resting values.
B. Due to the Haldane effect, the amount of carbaminohemoglobin is increased.
C. The pH of the blood is higher.
D. The chloride shift is greater.

31. Stimulation of which of the following stimulates inspiration?

A. Dorsal respiratory group
B. Pontine respiratory group
C. Lung stretch receptors
D. Medullary chemoreceptors

32. The factor that has the greatest effect on the medullary chemoreceptors is:

A. acute hypercapnia
B. hypoxia
C. chronic hypercapnia
D. arterial pH

33. Which term refers to increase in depth, but not rate, of ventilation?

A. Eupnea
B. Dyspnea
C. Hyperpnea
D. Hyperventilation

34. Adjustment to high altitude involves:

A. increase in minute respiratory volume
B. hypersecretion of erythropoietin
C. dyspnea
D. increase in hemoglobin saturation

35. Compared to the interstitial fluid bathing active muscle fibers, blood in the arteries serving those fibers has a:

A. higher P_{O_2}
B. higher P_{CO_2}
C. higher HCO_3^- content
D. higher H^+ content

Word Dissection

For each of the following word roots, fill in the literal meaning and give an example, using a word found in this chapter.

Word root	Translation	Example
1. alveol		
2. bronch		
3. capn		
4. carin		
5. choan		
6. crico		
7. ectasis		
8. emphys		
9. flat		
10. nari		
11. nas		
12. pleur		
13. pne		
14. pneum		
15. pulmo		
16. respir		
17. spire		
18. trach		
19. ventus		
20. vestibul		
21. vibr		

22

The Digestive System

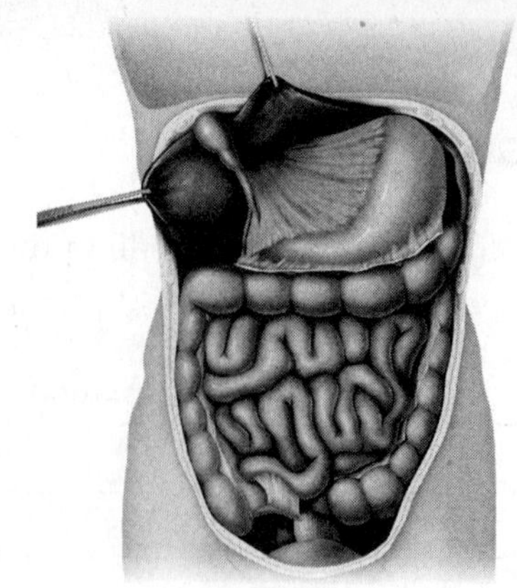

Student Objectives

When you have completed the exercises in this chapter, you will have accomplished the following objectives:

PART 1: OVERVIEW OF THE DIGESTIVE SYSTEM

1. Describe the function of the digestive system, and differentiate between organs of the alimentary canal and accessory digestive organs.
2. List and define the major processes occurring during digestive system activity.
3. Describe stimuli and controls of digestive activity.
4. Describe the location and function of the peritoneum.
5. Define retroperitoneal and name the retroperitoneal organs.
6. Define splanchic circulation.
7. Indicate the importance of the hepatic portal system.
8. Describe the tissue composition and the general function of each of the four layers of the alimentary canal.

PART 2: FUNCTIONAL ANATOMY OF THE DIGESTIVE SYSTEM

9. Describe the gross and microscopic anatomy and the basic functions of the mouth, pharynx, and esophagus.
10. Describe the composition and functions of saliva, and explain how salivation is regulated.
11. Explain the dental formula and differentiate clearly between deciduous and permanent teeth.
12. Describe the mechanisms of chewing and swallowing.
13. Identify structural modifications of the wall of the stomach that enhance the digestive process.
14. Name the cell types responsible for secreting the various components of gastric juice and indicate the importance of each component in stomach activity.
15. Describe stomach structure and indicate changes in the basic alimentary canal structure that aid its digestive function.
16. Explain how gastric secretion and stomach motility are regulated.
17. Define and account for the alkaline tide.
18. Identify and describe structural modifications of the wall of the small intestine that enhance the digestive process.
19. Differentiate between the roles of the various cell types of the intestinal mucosa.
20. Describe the function of intestinal hormones and paracrines.
21. Describe the histologic anatomy of the liver.
22. State the role of bile in digestion and describe how its entry into the small intestine is regulated.
23. Describe the role of the gallbladder.
24. State the role of pancreatic juice in digestion.
25. Describe how entry of pancreatic juice into the small intestine is regulated.
26. List the major functions of the large intestine.
27. Describe the regulation of defecation.

PART 3: PHYSIOLOGY OF CHEMICAL DIGESTION AND ABSORPTION

28. List the enzymes involved in chemical digestion; name the foodstuffs on which they act.
29. List the end products of protein, fat, carbohydrate, and nucleic acid digestion.
30. Describe the process of absorption of breakdown products of foodstuffs that occurs in the small intestine.

The digestive system processes food so that it can be absorbed and utilized by the body's cells. The digestive organs are responsible for food ingestion, digestion, absorption, and elimination of undigested remains from the body. In one sense, the digestive tract can be viewed as a disassembly line in which food is carried from one stage of its breakdown process to the next by muscular activity, and its nutrients are made available en route to the cells of the body. In addition, the digestive system provides for one of life's greatest pleasures—eating.

Chapter 22 reviews the anatomy of the alimentary canal and accessory digestive organs, the processes of mechanical and enzymatic breakdown, and absorption mechanisms. Cellular metabolism (utilization of foodstuffs by body cells) is considered in Chapter 23.

BUILDING THE FRAMEWORK

PART 1: OVERVIEW OF THE DIGESTIVE SYSTEM

1. The organs of the alimentary canal form a continuous tube from the mouth to the anus. List the organs of the alimentary canal, including the specific regions of the small and large intestine, from proximal to distal, in the blanks below.

   Mouth → (1) ____________________ → (2) ____________________ →

   (3) ____________________ → small intestine: (4) ____________________ →

   (5) ____________________ → (6) ____________________ → large intestine: cecum,

   appendix, (7) ____________________ → (8) ____________________ →

   (9) ____________________ → (10) ____________________ →

   (11) ____________________ → (12) ____________________ → anus

2. List the accessory structures associated with the mouth and the duodenum of the small intestine; mark each with an I if it is located inside the tract and an O if it is located outside the tract.

   Mouth: ____

   Duodenum: ____

3. Figure 22.1 is a frontal view of the digestive system. First, identify all structures with leader lines. Then, select a different color for each of the following organs and color the coding circles and the corresponding structures on the figure.

◯ Colon	◯ Pancreas	◯ Small intestine	◯ Tongue
◯ Esophagus	◯ Salivary glands	◯ Stomach	◯ Uvula
◯ Liver			

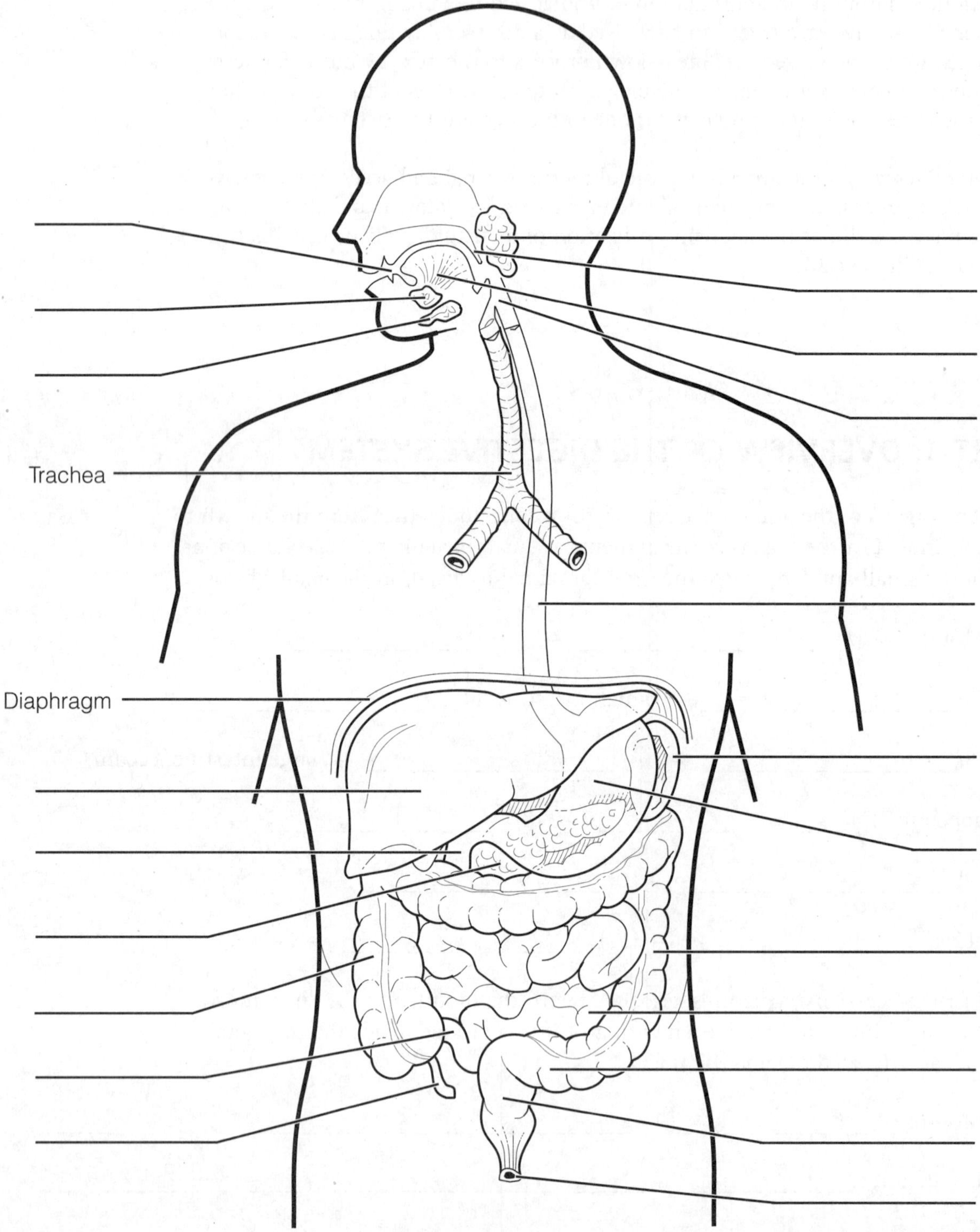

Figure 22.1

4. Match the descriptions in Column B with the appropriate terms referring to digestive processes in Column A.

	Column A	Column B
________	1. Ingestion	A. Transport of nutrients from lumen to blood
________	2. Propulsion	B. Enzymatic breakdown
________	3. Mechanical digestion	C. Elimination of feces
________	4. Chemical digestion	D. Eating
________	5. Absorption	E. Chewing
________	6. Defecation	F. Churning
		G. Includes swallowing
		H. Segmentation and peristalsis

5. Relative to neural controls of digestive activity, answer the following questions in the spaces provided:

1. What are three stimuli that elicit a response from sensors in the wall of the tract organs? ____________________

 ____

2. What are the two primary nerve plexuses regulating digestive function?

 ____

3. What are some differences between long and short reflexes?

 ____

 ____

 ____

6. Fill in the blanks below:

________ 1.
________ 2.
________ 3.
________ 4.

The body's most extensive serous membrane is the __(1)__. Connecting its visceral and parietal layers is a double serous sheet called __(2)__, which provides a route for blood vessels, lymphatics, and nerves. Organs behind the parietal layer are called __(3)__; those within this area are called __(4)__.

7. The walls of the alimentary canal have four typical layers, as illustrated in Figure 22.2. Identify each layer by placing its correct name in the answer blank before the appropriate description. Select different colors for each layer and color the coding circles and corresponding structures on the figure. Finally, assume the figure shows a cross-sectional view of the small intestine, and label the three structures with leader lines.

______________ ◯ 1. The secretory and absorptive layer

______________ ◯ 2. Layer composed of at least two muscle layers

______________ ◯ 3. Connective tissue layer, containing blood, lymph vessels, and nerves

______________ ◯ 4. Outermost layer of the wall; visceral peritoneum

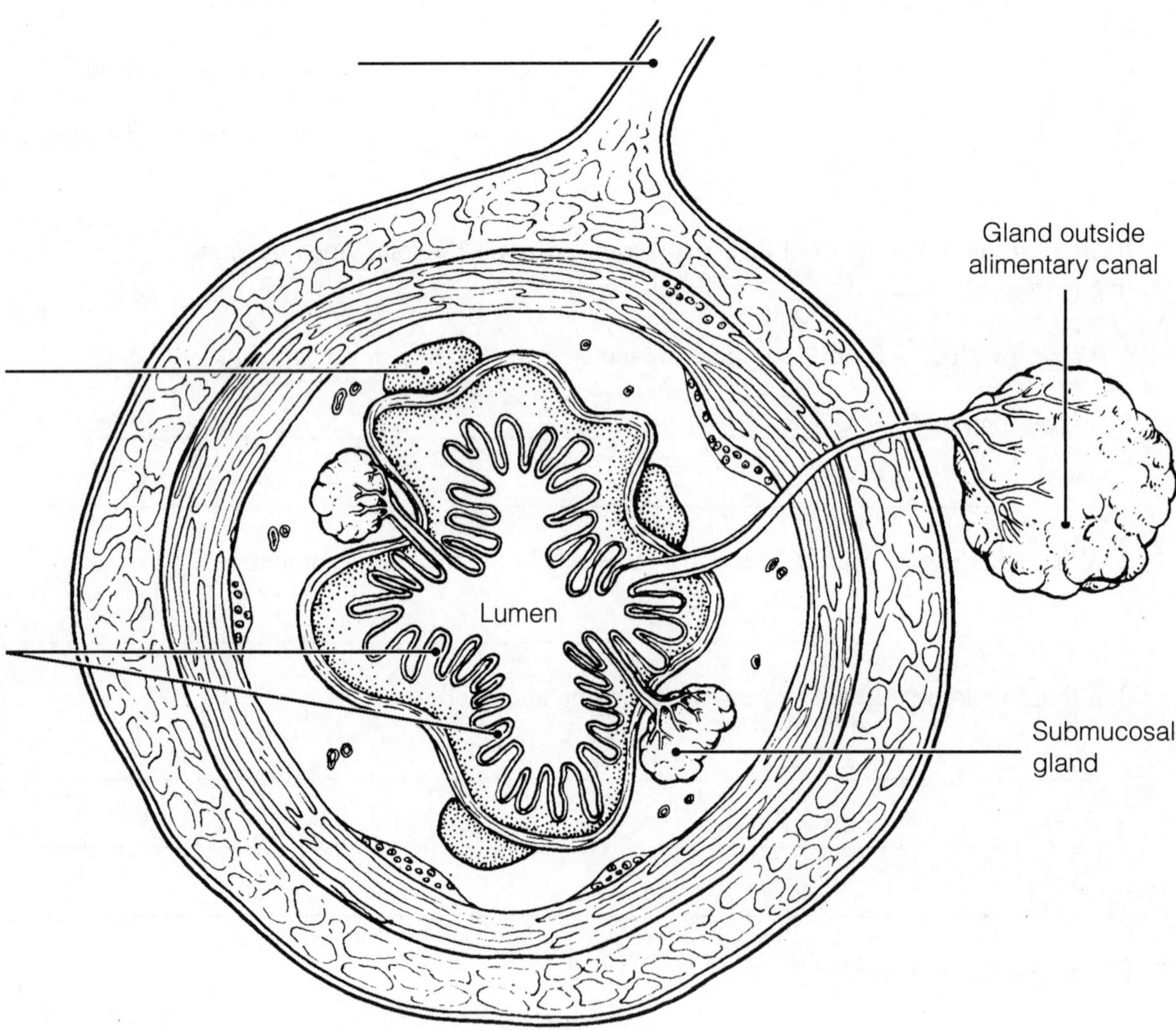

Figure 22.2

PART 2: FUNCTIONAL ANATOMY OF THE DIGESTIVE SYSTEM

1. Figure 22.3 illustrates oral cavity structures. First, identify all structures with leader lines. Then, color the structure that attaches the tongue to the floor of the mouth red, color the portions of the roof of the mouth unsupported by bone blue, color the structures that are essentially masses of lymphatic tissue yellow, and color the structure that contains the bulk of the taste buds pink.

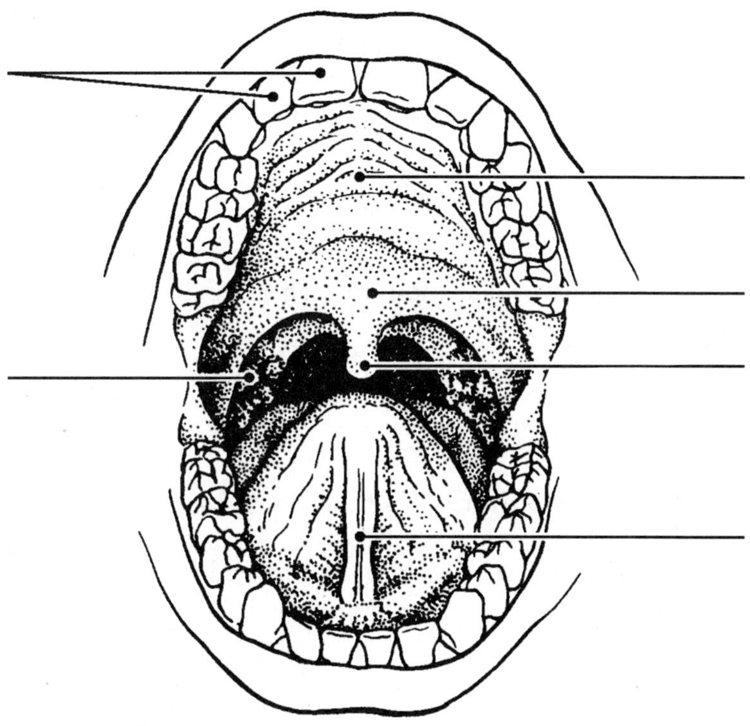

Figure 22.3

2. Answer the following questions concerning the salivary glands and saliva in the spaces provided:

1. What are four functions of saliva? ______________________________

____

____

2. What are the three main pairs of salivary glands, and in what part of the oral cavity do the ducts of each gland open? ______________________________

____

____

3. What are the two main cell types found in salivary glands, and what are their secretions? ______________________________

____

____

4. What two digestive enzymes are found in saliva?

______________________ and ______________________

5. What enzyme found in saliva inhibits bacterial growth?

____

3. Alternative choices separated by a slash (/) appear in each of the statements below concerning the regulation of salivation. In each case, circle the term that makes the statement correct.

1. Salivation is controlled primarily by the parasympathetic/sympathetic division of the autonomic nervous system.
2. Chemoreceptors are strongly activated by bitter/sour substances.
3. Salivation induced by the sight or thought of food is said to be a(n) conditioned/intrinsic reflex.

4. Complete the following statements referring to human dentition by writing the missing terms in the answer blanks.

The first set of teeth, called the __(1)__ teeth, begin to appear around the age of __(2)__ and usually have begun to be replaced by the age of __(3)__. The __(4)__ teeth are more numerous; that is, there are __(5)__ teeth in the second set as opposed to a total of __(6)__ teeth in the first set. If an adult has a full set of teeth, you can expect to find two __(7)__, one __(8)__, two __(9)__, and three __(10)__ in one side of each jaw. The most posterior molars in each jaw are commonly called __(11)__ teeth.

______________ 1.

______________ 2.

______________ 3.

______________ 4.

______________ 5.

______________ 6.

______________ 7.

______________ 8.

______________ 9.

______________ 10.

______________ 11.

5. Using the key choices, identify each tooth area described below and label the tooth in Figure 22.4. Then select different colors to represent the key structures with coding circles and color them on the figure. Finally, add labels to the figure to identify the crown, gingiva, and root of the tooth.

Key Choices

○ A. Cementum ○ C. Enamel ○ E. Pulp

○ B. Dentin ○ D. Periodontal membrane

______________________ 1. Material covering the tooth root

______________________ 2. Hardest substance in the body; covers tooth crown

______________________ 3. Attaches the tooth to bone and surrounding alveolar structures

______________________ 4. Forms the bulk of tooth structure; similar to bone

______________________ 5. A collection of blood vessels, nerves, and lymphatics

______________________ 6. Cells that produce this substance degenerate after tooth eruption

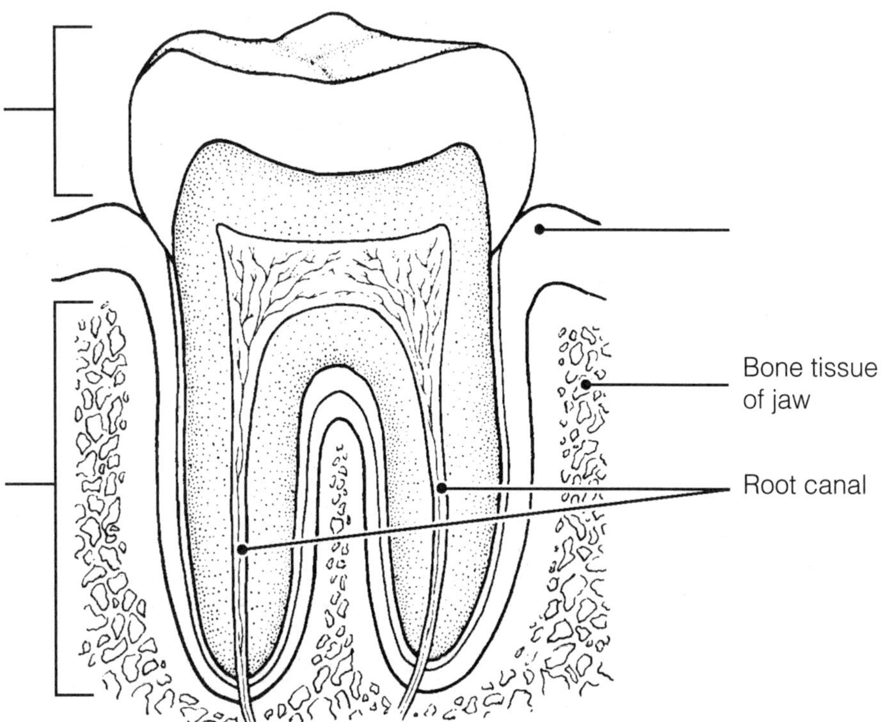

Figure 22.4

6. Using the key choices, match the terms with the descriptions of digestive system organs that follow by inserting the appropriate answers in the answer blanks.

Key Choices

A. Esophagus	D. Peyer's patches	G. Pyloric valve	J. Stomach
B. Microvilli	E. Pharynx	H. Rugae	K. Villi
C. Oral cavity	F. Plicae circulares	I. Small intestine	

__________________ 1. Projections of the intestinal mucosa that increase the surface area

__________________ 2. Large collections of lymph nodules found in the submucosa of the small intestine

__________________ 3. Folds of the small intestine wall

__________________ 4. Two anatomical regions involved in the mechanical breakdown of food

__________________ 5. A food chute; has no digestive or absorptive role

__________________ 6. Folds of the stomach mucosa and submucosa

__________________ 7. Common passage for food and air

7. Figure 22.5A is a longitudinal section of the stomach. Use the following terms to identify the regions with leader lines on the figure.

Body	Pyloric region	Greater curvature	Gastroesophageal sphincter
Fundus	Pyloric valve	Lesser curvature	

Select different colors for each of the following structures or areas and color the coding circles and corresponding structures or areas on the figure.

○ Oblique muscle layer　○ Longitudinal muscle layer　○ Serosa

○ Circular muscle layer　○ Area where rugae are visible

Figure 22.5B shows two types of secretory cells found in gastric glands. Identify the third type, called *chief cells,* by choosing a few cells deep in the glands and labeling them appropriately. Then, color the hydrochloric acid–secreting cells red, color the cells that produce mucus yellow, and color those that produce protein-digesting enzymes blue.

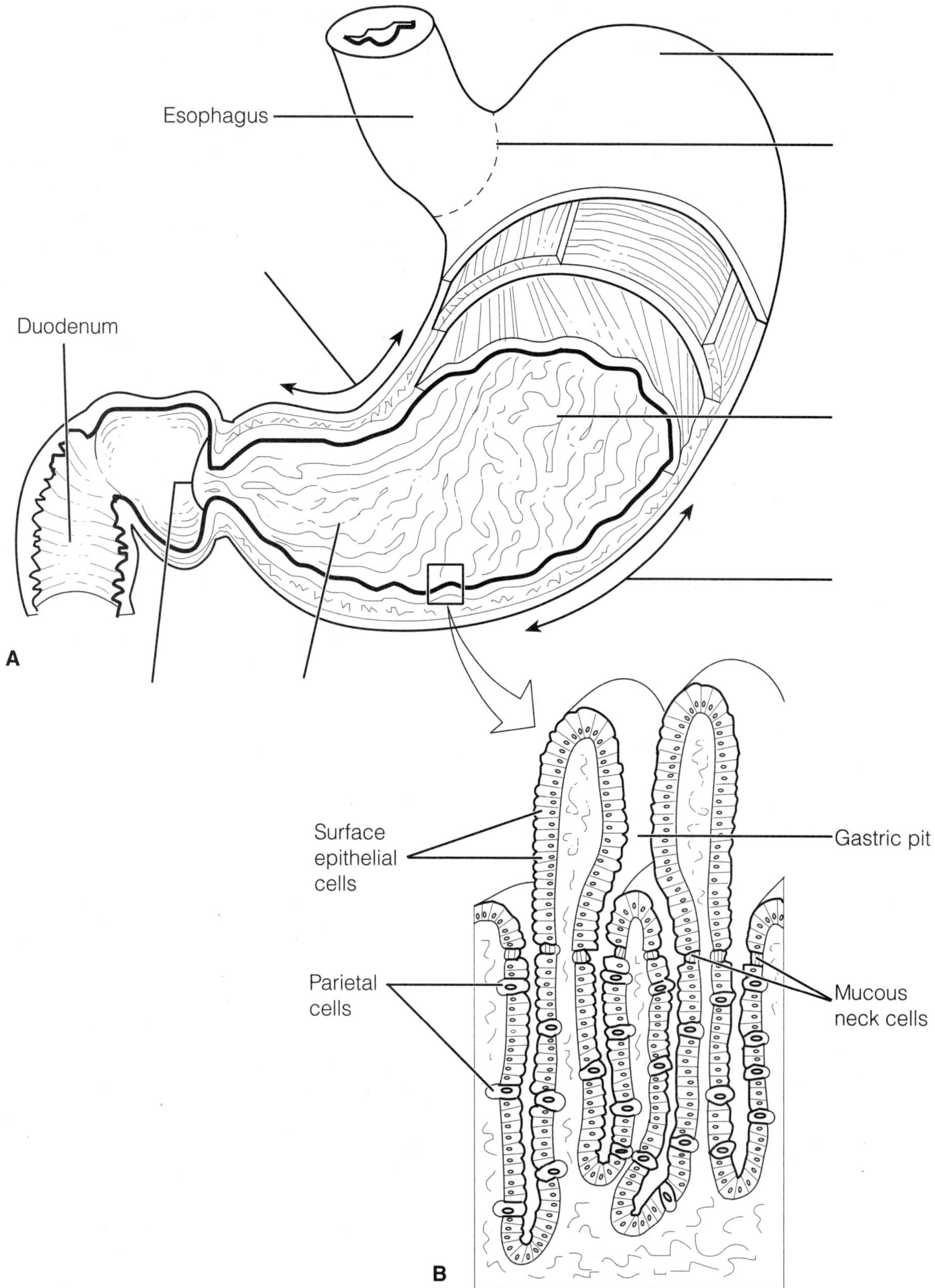
Esophagus
Duodenum
A
Surface epithelial cells
Gastric pit
Parietal cells
Mucous neck cells
B

Figure 22.5

8. Complete the following statements that describe mechanisms of food mixing and movement from the mouth to the stomach. Insert your responses in the answer blanks.

Common pathways for food and air are the __(1)__, which connects to the mouth, and __(2)__, which connects to the larynx. Epithelium in this region is __(3)__, which provides good protection from friction. Three bands of skeletal muscle, the __(4)__, propel food into the __(5)__. Entrance to the stomach is guarded by the __(6)__.

Swallowing, or __(7)__, occurs in two major phases, the __(8)__ and __(9)__. During the initial voluntary phase, the __(10)__ is used to push the food into the throat and the __(11)__ rises to close off the nasal passageways. As food is moved involuntarily through the pharynx, the __(12)__ rises to ensure that its passageway is covered by the __(13)__ so that ingested substances do not enter respiratory passages. It is possible to swallow water while standing on your head because the water is carried along the esophagus involuntarily by the process of __(14)__. The pressure exerted by food on the __(15)__ valve causes it to open so that food can enter the stomach.

1. ________________
2. ________________
3. ________________
4. ________________
5. ________________
6. ________________
7. ________________
8. ________________
9. ________________
10. ________________
11. ________________
12. ________________
13. ________________
14. ________________
15. ________________

9. Various types of glands secrete substances into the alimentary tube. Match the glands listed in Column B with the functions and locations described in Column A.

	Column A	Column B
________	1. Mucus-producing glands located in the submucosa of the small intestine	A. Duodenal glands
________	2. Secretory product contains amylase, a starch-digesting enzyme	B. Gastric glands
________	3. Sends a variety of enzymes in bicarbonate-rich fluid into the small intestine	C. Liver
________	4. Produces bile, which is transported to the duodenum via the bile duct	D. Pancreas
________	5. Produce hydrochloric acid and pepsinogen	E. Salivary glands

10. Figure 22.6 shows two views of the small intestine. First, label the villi in parts A and B and the circular folds in part A. Then select different colors to identify the following regions in part B and color them on the figure. Color the tunics in part A as desired.

◯ Simple columnar cells of surface epithelium

◯ Goblet cells of surface epithelium

◯ Lacteal

◯ Capillary network

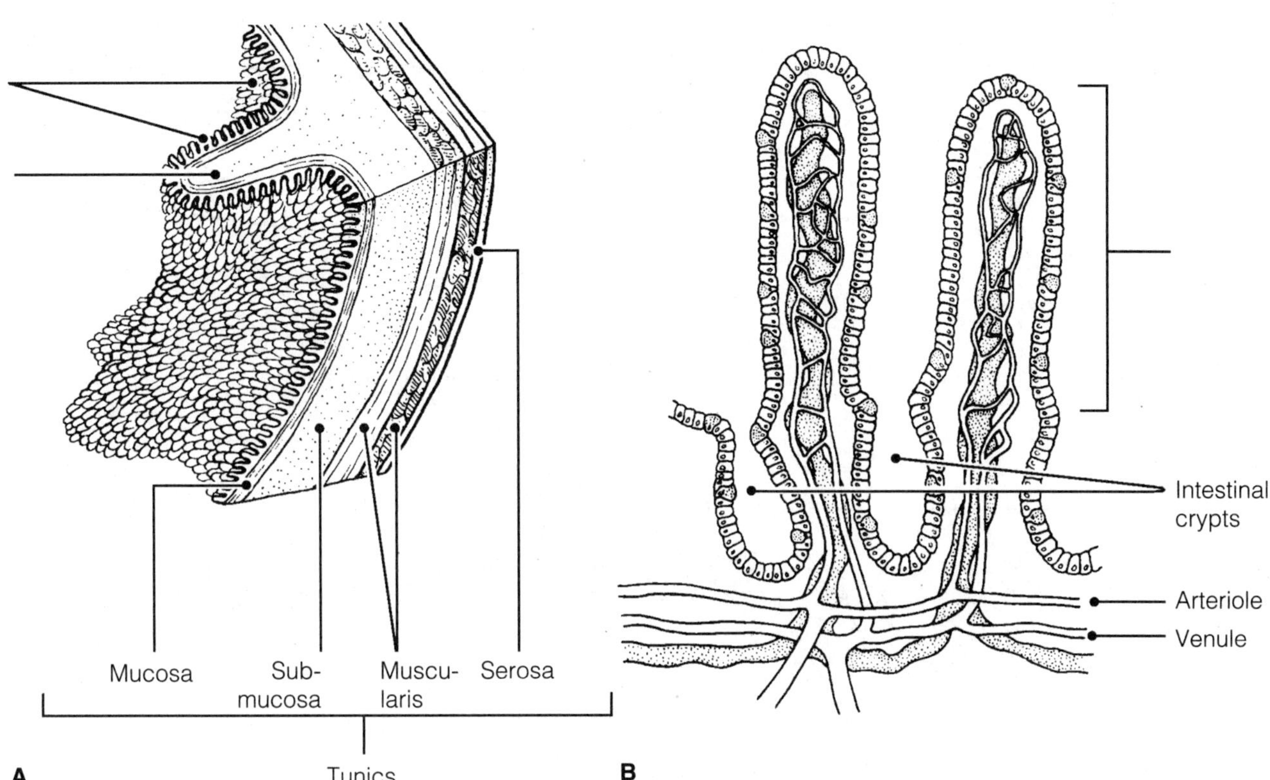

Figure 22.6

11. Circle the term that does not belong in each of the following groupings.

1. HCl secretion Acetylcholine Histamine Secretin Gastrin
2. Gastric emptying Slowed by fats Enterogastric reflex Solid foods
3. Cephalic phase Reflex phase Intestinal phase Psychological stimuli
4. Gastric juice Saliva Pancreatic juice Bicarbonate ions
5. Pacemaker cells Small intestine Esophagus Stomach
6. Stomach absorption Lipid-soluble drugs Salts Aspirin Alcohol

12. Three accessory organs are illustrated in Figure 22.7. Identify each of the three organs, their ducts, and the ligament with leader lines on the figure. Then select different colors for the following structures and color the coding circles and the corresponding structures on the figure.

- ○ Common hepatic duct
- ○ Bile duct
- ○ Cystic duct
- ○ Pancreatic duct

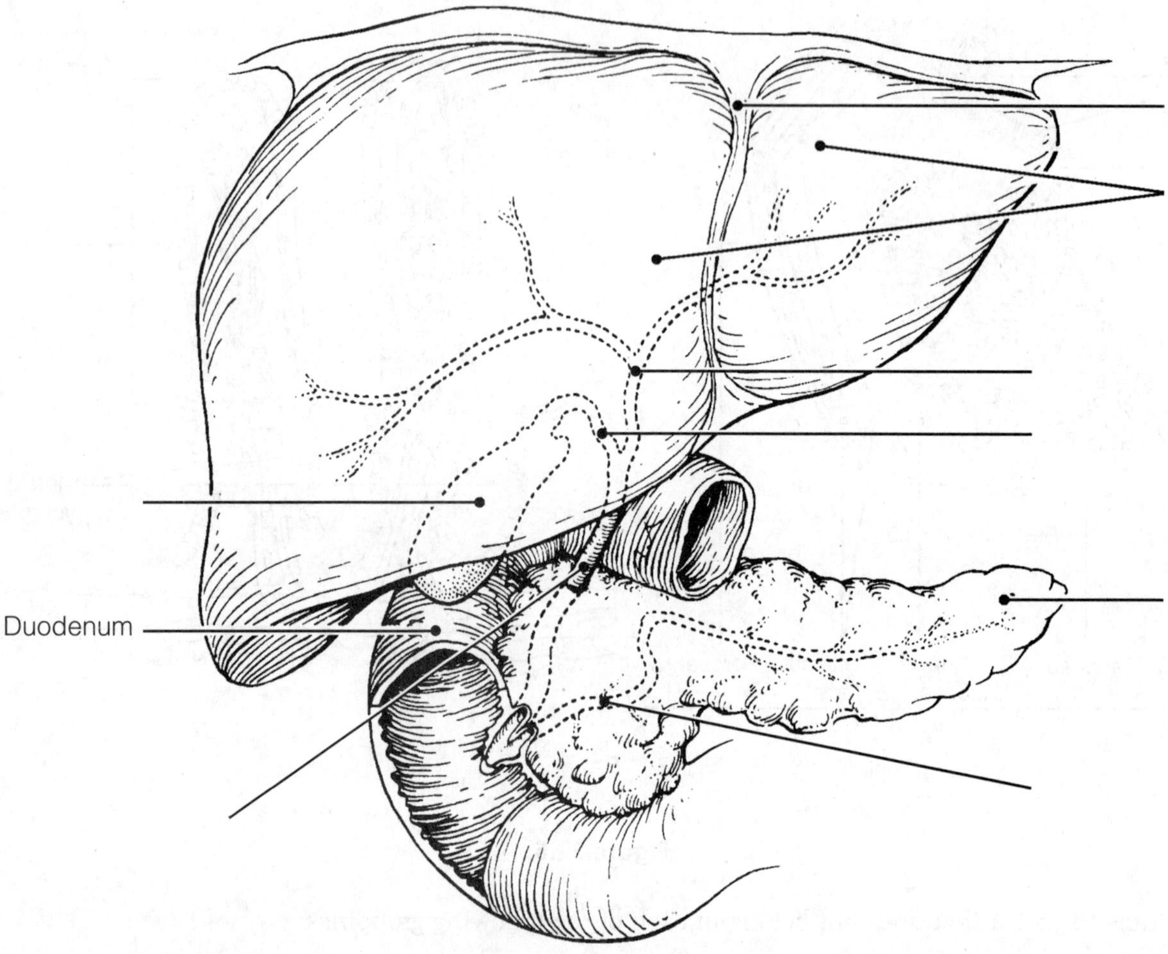

Figure 22.7

13. List the three major vessels of the portal triad: ____________, ____________, and ____________. Describe the location of the triads relative to the liver lobules.

____

____

14. Hormonal stimuli are important in digestive activities that occur in the stomach and small intestine. Using the key choices, identify the hormones that function as described in the following statements.

Key Choices

A. Cholecystokinin B. Gastrin C. Secretin D. Somatostatin

____________________ 1. These two hormones stimulate the pancreas to release its secretions.

____________________ 2. This hormone stimulates increased production of gastric juice.

____________________ 3. This hormone causes the gallbladder to release stored bile.

____________________ 4. This hormone causes the liver to increase its output of bile.

____________________ 5. These hormones inhibit gastric mobility and secretory activity.

15. This exercise concerns some aspects of food breakdown in the digestive tract. Using the key choices, select the appropriate terms to complete the following statements.

Key Choices

A. Bicarbonate-rich fluid D. HCl (hydrochloric acid) G. Mucus

B. Chewing E. Hormonal stimulus H. Psychological stimulus

C. Churning F. Mechanical stimulus

____________________ 1. The means of mechanical food breakdown in the mouth is __(1)__.

____________________ 2. The fact that the mere thought of a relished food can make your mouth water is an example of __(2)__.

____________________ 3. Many people chew gum to increase saliva formation when their mouth is dry. This type of stimulus is a __(3)__.

____________________ 4. Because living cells of the stomach (and everywhere) are largely protein, it is amazing that they arc not digested by the activity of stomach enzymes. The most important means of stomach protection is the __(4)__ it produces.

____________________ 5. The third layer of smooth muscle found in the stomach wall allows mixing and mechanical breakdown by __(5)__.

____________________ 6. The small intestine is protected from the corrosive action of HCl in chyme by __(6)__, which is ducted in from the pancreas.

16. Figure 22.8 shows a view of the large intestine. First, label a haustrum, the ileocecal valve, the external anal sphincter, the hepatic and splenic flexures, and the transverse mesocolon. Then, select different colors to identify the following parts in the figure:

◯ Anal canal	◯ Appendix	◯ Ascending colon
◯ Cecum	◯ Rectum	◯ Descending colon
◯ Sigmoid colon	◯ Teniae coli	◯ Transverse colon

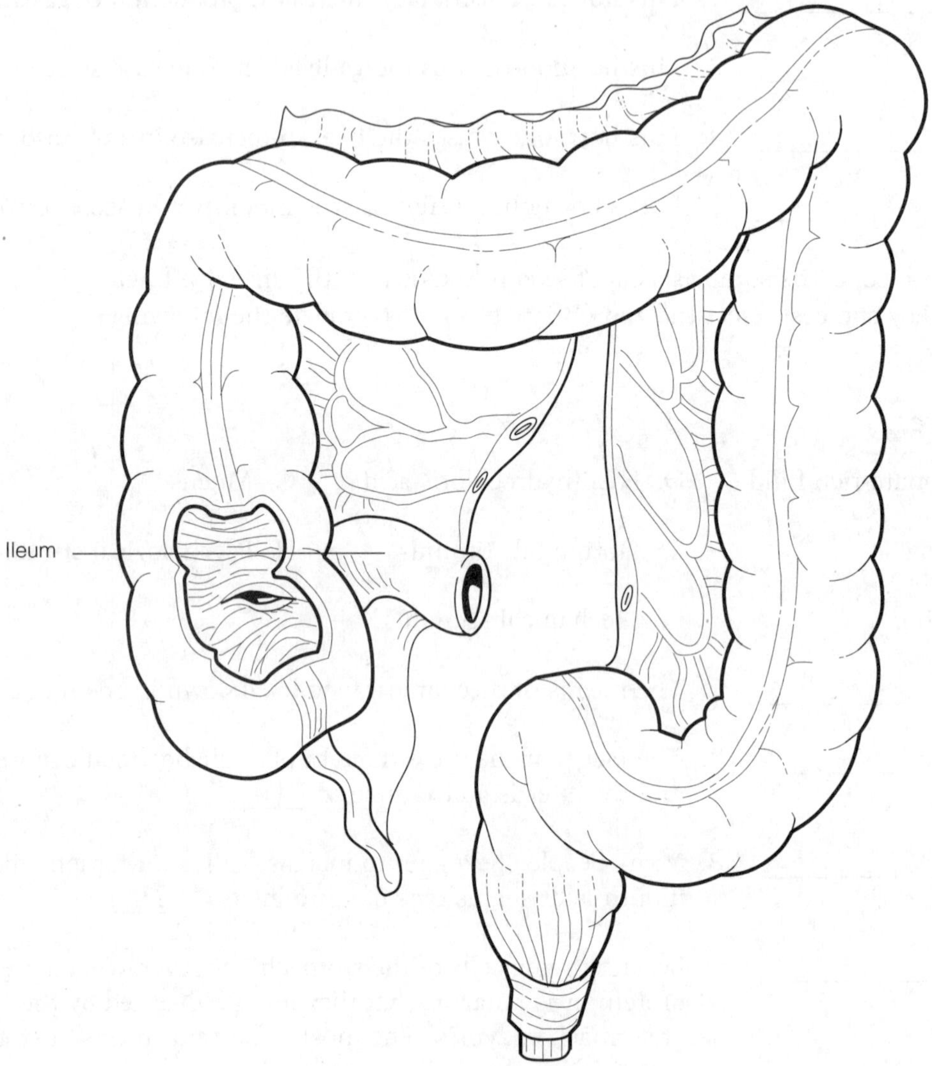

Ileum

Figure 22.8

17. Complete the paragraph relating to digestive system mobility by writing the missing terms in the answer blanks.

1. ____________________
2. ____________________
3. ____________________

The two major types of movements that occur in the small intestine are __(1)__ and __(2)__. One of these movements, __(3)__, acts to continually mix the food with digestive juices and, strangely, also plays the major role in propelling foods along the tract. Another type of movement seen only in the

____________ 4. large intestine, __(4)__, occurs infrequently and acts to move feces over relatively long distances toward the anus. The

____________ 5. presence of feces in the __(5)__ excites stretch receptors so that the __(6)__ reflex is initiated.

____________ 6.

18. Using the key choices, match the terms with the descriptions of digestive system organs that follow by inserting the appropriate answers in the answer blanks.

Key Choices

A. Anal canal	C. Colon	E. Haustrum	G. Plicae circulares
B. Appendix	D. Esophagus	F. Ileocecal valve	H. Small intestine

____________ 1. Saclike outpocketing of the large intestine wall

____________ 2. Prevents food from moving back into the small intestine once it has entered the large intestine

____________ 3. Blind-ended tube hanging from the cecum

____________ 4. Primarily involved in water absorption and feces formation

____________ 5. Region, containing two sphincters, through which feces are expelled from the body

19. Identify the following pathological conditions using the key choices.

Key Choices

A. Ankyloglossia	C. Gallstones	E. Hiatal hernia	G. Peritonitis
B. Appendicitis	D. Gingivitis	F. Jaundice	H. Ulcer

____________ 1. Inflammation of the abdominal serosa

____________ 2. Condition that may lead to reflux of acidic gastric juice into the esophagus, causing heartburn

____________ 3. Usually indicates liver problems or blockage of the biliary ducts

____________ 4. An erosion of the stomach or duodenal mucosa

____________ 5. Inflammation of the gums

____________ 6. Person is "tongue tied" because of a short lingual frenulum

PART 3: PHYSIOLOGY OF CHEMICAL DIGESTION AND ABSORPTION

1. Various types of foods are ingested in the diet and broken down to their building blocks. Use the key choices to complete the following statements. For some questions, more than one term is correct.

Key Choices

A. Amino acids	E. Fatty acids	H. Glucose	K. Meat/fish
B. Bread/pasta	F. Fructose	I. Lactose	L. Starch
C. Cheese/cream	G. Galactose	J. Maltose	M. Sucrose
D. Cellulose			

__________________ 1. Examples of dietary carbohydrates

__________________ 2. The three common simple sugars in our diet

__________________ 3. The most important of the simple sugars; referred to as "blood sugar"

__________________ 4. Disaccharides

__________________ 5. The only important *digestible* polysaccharide

__________________ 6. An indigestible polysaccharide that aids elimination because it adds fiber to the diet

__________________ 7. Protein-rich foods include these two food groups

__________________ 8. Protein foods must be broken down to these units before they can be absorbed

__________________ 9. Fats are broken down to these building blocks and glycerol

2. Dietary substances capable of being absorbed are listed below. If the substance is *most often* absorbed from the digestive tract by active transport processes, write A in the blank. If it is usually absorbed passively (by diffusion or osmosis), write P in the blank. In addition, circle the substance that is *most likely* to be absorbed into a lacteal rather than into the capillary bed of the villus.

_____ 1. Water _____ 3. Simple sugars _____ 5. Electrolytes

_____ 2. Amino acids _____ 4. Fatty acids

3. This exercise concerns chemical food breakdown and absorption in the digestive tract. Using the key choices, select the appropriate terms to complete the statements.

Key Choices

A. Amino acids	G. HCl	M. Nucleic acids
B. Bile	H. Intrinsic factor	N. Pepsin
C. Brush border enzymes	I. Lacteal	O. Proteases
D. Capillary	J. Lipases	P. Rennin
E. Fatty acids	K. Monosaccharides	Q. Salivary amylase
F. Glycerol	L. Monoglycerides	R. Vitamin D

__________________ 1. Starch digestion begins in the mouth when (1) is ducted in by the salivary glands.

__________________ 2. Protein foods are largely acted on in the stomach by (2) .

__________________ 3. A milk-coagulating enzyme found in children but not usually in adults is (3) .

__________________ 4. Intestinal enzymes are called (4) .

__________________ 5. For the stomach protein-digesting enzymes to become active, (5) is needed.

__________________ 6. A nonenzyme substance that emulsifies fats is (6) .

__________________ 7. Trypsin, chymotrypsin, aminopeptidase, and dipeptidase are all (7) .

__________________ 8. Chylomicrons are the products that ultimately enter the (8) when fats are being digested and absorbed.

__________________ 9. Unlike most other foodstuffs, (9) are broken down even further than their building blocks. The enzyme(s) involved in the final

__________________ 10. steps of that breakdown process is (are) (10) .

__________________ 11. Required for calcium absorption is (11) .

__________________ 12. __________________ 13. Transport for (12–14) is active and often coupled to the Na^+ pump.

__________________ 14.

4. Circle the term that does not belong in each of the following groupings.

1. Trypsin Dextrinase Maltase Aminopeptidase Intestinal enzymes
2. Micelles Accelerate lipid absorption Bile salts Chylomicrons
3. Ferritin Mucosal iron barrier Lost during hemorrhage Iron storage
4. Dextrinase Glucoamylase Maltase Nuclease Amylase
5. Hydrolysis Bond splitting Removal of water pH specific
6. Chymotrypsin Alkaline pH Pepsin Trypsin Dipeptidase
7. Vitamin A Vitamin C Passively absorbed Vitamin B_{12} Vitamin D

5. Fill in the table below with the specific information indicated in the headings:

Enzyme	Substrate	Product	Secreted by	Active in
Salivary amylase				
Pepsin				
Trypsin				
Dipeptidase				
Pancreatic amylase				
Lipase				
Maltase				
Carboxypeptidase				
Nuclease				

THE INCREDIBLE JOURNEY:

A Visualization Exercise for the Digestive System

. . . the passage beneath you opens, and you fall into a huge chamber with mountainous folds.

1. Complete the following narrative by writing the missing terms in the answer blanks.

1. ____________________
2. ____________________
3. ____________________
4. ____________________
5. ____________________
6. ____________________
7. ____________________
8. ____________________
9. ____________________
10. ____________________
11. ____________________
12. ____________________
13. ____________________
14. ____________________
15. ____________________
16. ____________________
17. ____________________

In this journey, you are to travel through the digestive tract as far as the appendix and then await further instructions. You are miniaturized as usual and provided with a wetsuit to protect you from being digested during your travels. You have a very easy entry into your host's open mouth. You look around and notice the glistening pink lining, or (1), and the perfectly cared-for teeth. Within a few seconds, the lips part and you find yourself surrounded by bread. You quickly retreat to the safety of the (2) between the teeth and the cheek to prevent getting chewed. From there you watch with fascination as a number of openings squirt fluid into the chamber, and the (3) heaves and rolls, mixing the bread with the fluid. As the bread begins to disappear, you decide that the fluid contains the enzyme (4).

You then walk toward the back of the oral cavity, and suddenly you find yourself being carried along by a squeezing motion of the walls around you. The name given to this propelling motion is (5). As you are carried helplessly downward, you see two openings, the (6) and the (7), below you. Just as you are about to straddle the solid area between them to stop your descent, the structure to your left moves quickly upward, and a trapdoorlike organ, the (8), flaps over the anterior opening. Down you go into the dark, yawning posterior opening, seeing nothing. Then the passage beneath you opens, and you fall into a huge chamber with mountainous folds. Obviously, you have reached the (9). The folds are very slippery, and you conclude that it must be the (10) coat that you read about earlier. As you survey your surroundings, juices begin to gurgle into the chamber from pits in the "floor," and your face begins to sting and smart. You cannot seem to escape this caustic fluid, and you conclude that it must be very dangerous to your skin because it contains (11) and (12). You reach down and scoop up some of the slippery substance from the folds and smear it on your face, confident that if it can protect this organ it can protect you as well! Relieved, you begin to slide toward the organ's far exit and squeeze through the tight (13) valve into the next organ. In the dim light, you see lumps of cellulose lying at your feet and large fat globules dancing lightly about. A few seconds later, your observations are interrupted by a wave of fluid pouring into the chamber from an opening high in the wall above you. The large fat globules begin to fall apart, and you decide that this enzyme flood has to contain (14) and the opening must be the duct from the (15). As you move quickly away to escape the deluge, you lose your footing and find yourself on a rollercoaster ride—twisting, coiling, turning, and diving through the lumen of this active organ. As you move, you are stroked by velvety, fingerlike projections of the wall, the (16). Abruptly your ride comes to a halt as you are catapulted through the (17) valve and fall into the appendix. Headquarters informs you that you are at the end of your journey. Your exit now depends on your own ingenuity and your sympathy for your host.

CHALLENGING YOURSELF

At the Clinic

1. A young boy is brought to the clinic wincing with pain whenever he opens his mouth. His parents believe he dislocated his jaw, which is obviously swollen on the left side. Luckily, the clinic has seen a number of similar cases in the past week, so the diagnosis is quick. What is the diagnosis?

2. Marv comments to his doctor at the clinic that he gets a "full," very uncomfortable feeling in his chest after every meal, as if the food were lodged there instead of traveling to his stomach. The doctor states that the description fits the condition called achalasia, in which the valve between the esophagus and stomach fails to open. What valve is involved?

3. Mr. Ashe, a man in his mid-60s, comes to the clinic complaining of heartburn. Questioning by the clinic staff reveals that the severity of his attacks increases when he lies down after eating a heavy meal. The man is about 50 pounds overweight. What is your diagnosis? Without treatment, what conditions might develop?

4. A young woman is put through an extensive battery of tests to determine the cause of her "stomach pains." She is diagnosed with gastric ulcers. An antihistamine drug is prescribed and she is sent home. What is the mechanism of her medication? What life-threatening problems can result from a poorly managed ulcer? Why did the clinic doctor warn the woman not to take aspirin?

5. Continuing from the previous question, the woman's ulcer got worse. She started complaining of back pain. The physician discovered that the back pain occurred because the pancreas was now damaged. Use logic to deduce how a perforating gastric ulcer could come to damage the pancreas.

6. An elderly man, found unconscious in the alley behind a local restaurant, is brought to the clinic in a patrol car. His abdomen is distended, his skin and scleras are yellow, and he reeks of alcohol. What do you surmise is the cause of his condition and what terms are used to describe the various aspects of his appearance?

7. Duncan, an inquisitive 8-year-old, saw his grandfather's dentures soaking overnight in a glass. He asked his grandfather how his teeth had fallen out. Reconstruct the kind of story the man is likely to have told.

8. A feverish 12-year-old girl named Kelly is brought to the clinic complaining of pain in the lower right abdominal quadrant. According to her parents, for the past week she has been eating poorly and often vomits when she does eat. What condition do you suspect? What treatment will she need? What complication will occur if treatment is given too late?

9. Eva, a woman in her 50s, complains of bloating, cramping, and diarrhea when she drinks milk. What is the cause of her complaint and what is a solution?

10. Mr. Erickson complains of diarrhea after eating and relates that it is most pronounced when he eats starches. Subsequent dietary variations reveal intolerance to wheat and other grains, but no problem with rice. What condition do you suspect?

11. A 21-year-old man with severe appendicitis did not seek treatment in time and died a week after his abdominal pain and fever began. Explain why appendicitis can quickly lead to death.

12. Pancreatitis often results from a gallstone. Another major cause of pancreatitis is chronic alcoholism, which can lead to precipitation of protein in the pancreatic duct. Why do you think such protein precipitation can result in problems?

Stop and Think

1. How do the following terms relate to the function of the smooth muscle of the digestive system? gap junction, stress-relaxation response, nonstriated
2. What happens to salivary amylase in the stomach? Pepsin in the duodenum?
3. How would the release of histamine by gastric enteroendocrine cells improve the *absorption* of nutrients?
4. Which would be more effective: a single "megadose" of calcium supplement once a day or a smaller supplement with each meal?
5. Trace the digestion and absorption of fats from their entrance into the duodenum to their arrival at the liver.
6. Clients are instructed not to eat before having blood tests run. How would a lab technician know if someone "cheated" and ate a fatty meal a few hours before having his blood drawn?
7. Why does liver impairment result in edema?
8. The location of the rapidly dividing, undifferentiated epithelial cells differs in the stomach and intestine. (a) Compare the locations of these cells in the two digestive organs. (b) What is the basic function of these dividing cells?
9. Bianca went on a trip to the Bahamas during spring vacation and did not study enough for her anatomy test that was scheduled early in the following week. On the test, she mixed up the following pairs of structures: (a) serous cells and serous membranes, (b) caries and canaliculi (bile canaliculi), (c) anal canal and anus, (d) diverticulosis and diverticulitis, (e) hepatic vein and hepatic portal vein. Can you help her by defining and differentiating all these sound-alike structures?
10. Name three organelles that are abundant in hepatocytes and explain how each of these organelles contributes to liver functions.
11. Which procedure would have the most detrimental effect on digestion—removal of the stomach, pancreas, or gallbladder? Explain your choice.

COVERING ALL YOUR BASES

Multiple Choice

Select the best answer or answers from the choices given.

1. Which of the following terms are synonyms?
   A. Gastrointestinal tract
   B. Digestive system
   C. Digestive tract
   D. Alimentary canal
2. A digestive organ that is not part of the alimentary canal is the:
   A. stomach
   B. liver
   C. small intestine
   D. large intestine
   E. pharynx
3. An organ that is not covered by a visceral peritoneum is the:
   A. stomach
   B. jejunum
   C. thoracic esophagus
   D. transverse colon
4. The GI tube layer responsible for the actions of segmentation and peristalsis is:
   A. serosa
   B. mucosa
   C. muscularis externa
   D. submucosa
5. The nerve plexus that stimulates secretion of digestive juices is the:
   A. submucosal plexus
   B. myenteric plexus
   C. subserous plexus
   D. local plexuses

6. Organs that are partly or wholly retroperitoneal include:
   A. stomach C. duodenum
   B. pancreas D. colon

7. Which of the following are part of the splanchnic circulation?
   A. Hepatic artery C. Hepatic portal vein
   B. Hepatic vein D. Splenic artery

8. Which alimentary canal tunic has the greatest abundance of lymph nodules?
   A. Mucosa C. Serosa
   B. Muscularis D. Submucosa

9. Proteins secreted in saliva include:
   A. mucin C. lysozyme
   B. amylase D. IgA

10. Which of these tissue types is most cellular?
    A. Enamel C. Dentin
    B. Pulp D. Cementum

11. The closure of which valve is assisted by the diaphragm?
    A. Ileocecal C. Gastroesophageal
    B. Pyloric D. Upper esophageal

12. Stratified squamous epithelium lines the:
    A. cardia of stomach C. anal canal
    B. esophagus D. rectum

13. Smooth muscle is found in the:
    A. tongue C. esophagus
    B. pharynx D. external anal sphincter

14. Rugae in the stomach encompass which tunics?
    A. Mucosa C. Serosa
    B. Muscularis D. Submucosa

15. Where in the stomach do the strongest peristaltic waves occur?
    A. Body C. Fundus
    B. Cardia D. Pylorus

16. The greater omentum:
    A. contains lymph nodes
    B. covers most of the anterior peritoneal cavity
    C. contains fat deposits
    D. attaches to the liver

17. Carbonic anhydrate is contained in:
    A. parietal cells of gastric glands
    B. chief cells of gastric glands
    C. pancreatic acini cells
    D. pancreatic duct cells

18. Sometimes the only treatment for chronic ileitis is surgical removal. What functions will be lost?
    A. Vitamin B_{12} absorption
    B. Absorption of calcium and iron
    C. Normal resistance to bacterial infection
    D. Enterohepatic circulation

19. Which statement is true about the peritoneal cavity?
    A. It is the same thing as the abdominopelvic cavity.
    B. This large cavity is filled with air.
    C. Like the pleural and pericardial cavities, it is a potential space containing serous fluid.
    D. It contains the pancreas and all the duodenum.

20. Which of these organs lies in the right hypochondriac region of the abdomen?
    A. Stomach C. Cecum
    B. Spleen D. Liver

21. Which phases of gastric secretion depend (at least in part) on the vagus nerve?
    A. Cephalic
    B. Gastric
    C. Intestinal (stimulatory)
    D. Intestinal (inhibitory)

22. Which of the following is (are) classified as enterogastrone(s)?
    A. Gastrin C. CCK
    B. GIP D. Secretin

23. After gastrectomy, it will be necessary to supplement:
    A. pepsin C. gastrin
    B. vitamin B_{12} D. intrinsic factor

24. Villi:
    A. are composed of both mucosa and submucosa
    B. contain lacteals
    C. can contract to increase circulation of lymph
    D. are largest in the duodenum

25. Brush border enzymes include:
    A. dipeptidase C. enteropeptidase
    B. lactase D. lipase

26. The portal triads include:
    A. branch of the hepatic artery
    B. central vein
    C. branch of the hepatic portal vein
    D. bile duct

27. Which of the following are in greater abundance in the hepatic vein than in the hepatic portal vein?
A. Urea
B. Aging RBCs
C. Plasma proteins
D. Fat-soluble vitamins

28. Jaundice may be a result of:
A. hepatitis
B. blockage of enterohepatic circulation
C. biliary calculi in the common bile duct
D. removal of the gallbladder

29. Release of CCK leads to:
A. contraction of smooth muscle in the duodenum
B. increased activity of hepatocytes
C. activity of the muscularis of the gallbladder
D. exocytosis of zymogenic granules in pancreatic acini

30. The pH of chyme entering the duodenum is adjusted by:
A. bile
B. intestinal juice
C. secretions from pancreatic acini
D. secretions from pancreatic ducts

31. Peristalsis in the small intestine begins:
A. with the cephalic phase of stomach activity
B. when chyme enters the duodenum
C. when CCK is released
D. when most nutrients have been absorbed

32. Relaxation of the ileocecal sphincter is triggered by:
A. gastrin
B. enterogastrones
C. presence of bile salts in the blood
D. gastroileal reflex

33. Relaxation of the teniae coli would result in loss of:
A. haustra
B. splenic and hepatic flexures
C. rectal valves
D. anal columns

34. Diverticulitis is associated with:
A. too much bulk in the diet
B. too little bulk in the diet
C. gluten intolerance
D. lactose intolerance

35. Which of the following are associated with the complete digestion of starch?
A. Maltose
B. Sucrose
C. Amylase
D. Oligosaccharides

36. Which of the following are tied to sodium transport?
A. Glucose
B. Fructose
C. Galactose
D. Amino acids

37. Endocytosis of proteins in newborns accounts for:
A. absorption of casein
B. absorption of antibodies
C. some food allergies
D. absorption of butterfat

38. Cell organelles in the intestinal epithelium that are directly involved in lipid digestion and absorption include:
A. smooth ER
B. Golgi apparatus
C. secretory vesicles
D. lysosomes

39. Excess iron is stored primarily in the:
A. liver
B. bone marrow
C. duodenal epithelium
D. blood

40. Which of the following correctly describes the flow of blood through the classical liver lobule and beyond? (More than one choice is correct.)
A. Portal vein branch to sinusoids to central vein to hepatic vein to inferior vena cava
B. Porta hepatis to hepatic vein to portal vein
C. Portal vein branch to central vein to hepatic vein to sinusoids
D. Hepatic artery branch to sinusoids to central vein to hepatic vein

41. Difficult swallowing is called:
A. ascites
B. dysphagia
C. ileus
D. stenosis

42. A digestive organ that has a head, neck, body, and tail is the:
A. pancreas
B. gallbladder
C. greater omentum
D. stomach

43. The hepatopancreatic ampulla lies in the wall of the:
A. liver
B. pancreas
C. duodenum
D. stomach

44. Two structures that produce alkaline secretions that neutralize the acidic stomach chyme as it enters the duodenum are:
A. intestinal flora and pancreatic acinar cells
B. gastric glands and gastric pits
C. lining epithelium of the intestine and Paneth cells
D. pancreatic ducts and duodenal glands

45. A 3-year-old girl was rewarded with a hug because she was now completely toilet trained. Which muscle had she learned to control?

A. Levator ani
B. Internal anal sphincter
C. Internal and external obliques
D. External anal sphincter

46. Which cell type fits this description? It occurs in the stomach mucosa, contains abundant mitochondria and many microvilli, and pumps hydrogen ions.

A. Absorptive cell
B. Parietal cell
C. Goblet cell
D. Mucous neck cell

47. The only feature in this list that is shared by both the large and small intestines is:

A. intestinal crypts
B. Peyer's patches
C. circular folds
D. teniae coli
E. haustra
F. intestinal villi

Word Dissection

For each of the following word roots, fill in the literal meaning and give an example, using a word found in this chapter.

Word root	Translation	Example
1. aliment		
2. cec		
3. chole		
4. chyme		
5. decid		
6. duoden		
7. enter		
8. epiplo		
9. eso		
10. falci		
11. fec		
12. fren		
13. gaster		
14. gest		
15. glut		
16. haustr		
17. hiat		
18. ile		
19. jejun		
20. micell		
21. oligo		
22. oment		
23. otid		
24. pep		
25. plic		
26. proct		
27. pylor		
28. ruga		
29. sorb		
30. splanch		
31. stalsis		
32. teni		

23

Nutrition, Metabolism, and Body Temperature Regulation

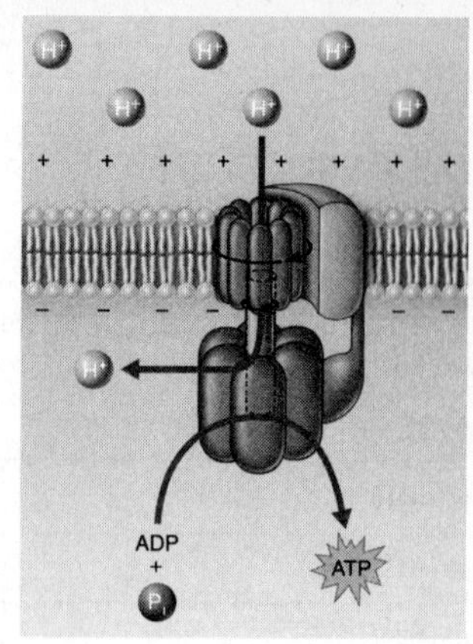

Student Objectives

When you have completed the exercises in this chapter, you will have accomplished the following objectives:

Diet and Nutrition

1. Define nutrient, essential nutrient, and calorie.
2. List the six major nutrient categories. Note important sources and main cellular uses.
3. Distinguish between simple and complex carbohydrate sources.
4. Indicate the major uses of carbohydrates in the body.
5. Indicate uses of fats in the body.
6. Distinguish between saturated, unsaturated, and trans fatty acid sources.
7. Distinguish between nutritionally complete and incomplete proteins.
8. Indicate uses of proteins in the body.
9. Define nitrogen balance and indicate possible causes of positive and negative nitrogen balance.
10. Distinguish between fat- and water-soluble vitamins, and list the vitamins in each group.
11. For each vitamin, list important sources, body functions, and important consequences of its deficit or excess.
12. List minerals essential for health; indicate important dietary sources and describe how each is used.

Overview of Metabolic Reactions

13. Define metabolism. Explain how catabolism and anabolism differ.
14. Define oxidation and reduction and indicate the importance of these reactions in metabolism.
15. Indicate the role of coenzymes used in cellular oxidation reactions.
16. Explain the difference between substrate-level phosphorylation and oxidative phosphorylation.

Metabolism of Major Nutrients

17. Summarize the important events and products of glycolysis, the Krebs cycle, and electron transport.
18. Define glycogenesis, glycogenolysis, and gluconeogenesis.
19. Describe the process by which fatty acids are oxidized for energy.
20. Define ketone bodies, and indicate the stimulus for their formation.
21. Describe how amino acids are metabolized for energy.
22. Describe the need for protein synthesis in body cells.

Metabolic States of the Body

23. Explain the concept of amino acid or carbohydrate–fat pools, and describe pathways by which substances in these pools can be interconverted.
24. List important events of the absorptive and postabsorptive states, and explain how these events are regulated.

Metabolic Role of the Liver

25. Describe several metabolic functions of the liver.

26. Differentiate between LDLs and HDLs relative to their structures and major roles in the body.

Energy Balance

27. Explain what is meant by body energy balance.

28. Describe some theories of food intake regulation.

29. Define basal metabolic rate and total metabolic rate. Name factors that influence each.

30. Distinguish between core and shell body temperature.

31. Describe how body temperature is regulated, and indicate the common mechanisms regulating heat production/retention and heat loss from the body.

In this chapter you will complete the studies of digestion and absorption presented in Chapter 22 by examining in detail the nature of nutrients and the roles they play in meeting the body's various needs. The breakdown products resulting from digestion of large nutrient molecules enter the cells. In cells, metabolic pathways may lead to the extraction of energy, the synthesis of new or larger molecules, or the interchange of parts of molecules. Unusable residues of molecules are eliminated as waste.

Subjects for study and review in Chapter 22 include nutrition, metabolism, control of the absorptive and postabsorptive states, the role of the liver, the regulation of body temperature, and important developmental aspects of metabolism.

BUILDING THE FRAMEWORK

Diet and Nutrition

1. Respond to the following questions by writing your answers in the answer blanks.

1. Name the six categories of nutrients. ______________________________

____

2. What is meant by the term *essential nutrients*? ______________________________

____

3. Define the unit used to measure the energy value of foods. ______________________________

____

4. Which carbohydrate is used by cells as the major ready source of energy? ______________

5. Which two types of cells rely almost entirely for their energy needs on the carbohydrate named in question 4? ______

6. What is meant by the term *empty calories*? ______

______

7. Name the major dietary source of cholesterol. ______

8. Which fatty acid(s) cannot be synthesized by the liver and must be ingested as a lipid?

______

9. Define *complete protein*. ______

______

______

10. Explain the metabolic importance of essential amino acids. ______

______

______

11. What is meant by the term *nitrogen balance*? ______

______

______

12. Name the two major uses of proteins synthesized in the body. ______

______

13. What is the function of most vitamins in the body? ______

______

14. Which organ chemically processes nearly every category of nutrient? ______

15. Which two minerals account for most of the weight of the body's minerals?

______

16. Which food category makes foods more tender or creamy and makes us feel full and satisfied? ______

2. Several descriptive statements concerning vitamins and the consequences of their excess and deficit appear below. Use the key choices to correctly identify each vitamin considered.

Key Choices

A. Vitamin A	E. Vitamin B_{12}	H. Vitamin E	K. Folic acid
B. Vitamin B_1	F. Vitamin C	I. Vitamin K	L. Niacin
C. Vitamin B_2	G. Vitamin D	J. Biotin	M. Pantothenic acid
D. Vitamin B_6			

______ 1. Most is synthesized in the colon by resident bacteria; essential for the liver's synthesis of clotting proteins.

______ 2. The most complex vitamin; contains cobalt; an essential coenzyme for DNA synthesis and production of choline.

______ 3. Also called pyridoxine; involved in amino acid metabolism; excess causes neurological defects (loss of sensation, etc.).

______ 4. Part of coenzyme cocarboxylase; deficit results in beriberi.

______ 5. Chemically related to sex hormones.

______ 6. Best extrinsic source is fortified milk; also made in the skin from cholesterol; deficit results in rickets in children and osteomalacia in adults.

______ 7. ______ 8. ______ 9. ______ 10. Fat-soluble vitamins.

______ 11. Functions in the body in the form of coenzyme A; essential for oxidation reactions and fat synthesis.

______ 12. Part of the coenzyme NAD required for many metabolic reactions; deficit results in pellagra.

______ 13. Found in large amounts in citrus fruits, strawberries, and tomatoes; essential for connective tissue matrix synthesis, blood clotting.

______ 14. The vitamin of vision; required for photopigment synthesis; deficit results in night blindness, clouding of cornea.

______ 15. Part of the coenzyme FAD; named for its similarity to ribose sugar; deficit causes cheilosis.

______ 16. Required for erythropoiesis; deficit causes macrocytic anemia.

______ 17. ______ 18. ______ 19. Act as antioxidants to disarm free radicals in the body.

______ 20. Its biologically active form in the body is tocopherol.

3. Match the minerals listed in the key choices to the appropriate descriptions. (*Note:* Not all of the trace minerals have been considered.)

Key Choices

A. Calcium	C. Iodine	E. Magnesium	G. Potassium	I. Sulfur
B. Chlorine	D. Iron	F. Phosphorus	H. Sodium	J. Zinc

_______________ 1. Required for normal growth, wound healing, taste, and smell

_______________ 2. Concentrates in the thyroid gland; essential for T_3 and T_4 synthesis

_______________ 3. Essential component of many proteins (e.g., insulin) as part of -S–S- bonds; found in virtually all of the denser connective tissues as part of the ground substance

_______________ 4. In the anion part of the calcium salts found in bone

_______________ 5. Principal intracellular cation; excesses result in muscle weakness and cardiac abnormalities

_______________ 6. The most abundant cation in extracellular fluid; a major factor in determining fluid shifts in the body; excesses lead to hypertension in some cases

_______________ 7. Its ion is the major extracellular anion

_______________ 8. An essential component of hemoglobin and some cytochromes; excesses cause liver damage

_______________ 9. Part of coenzymes that act in hydrolysis of ATP; excesses lead to diarrhea

4. Circle the term that does not belong in each of the following groupings.

1. Neutral fats   Triglycerides   Trisaccharides   Triacylglycerols
2. Meats   Eggs   Milk   Nuts   Fish
3. Mineral-rich foods   Vegetables   Fats   Milk   Legumes
4. Coenzymes B_2 Niacin   Vitamin C   Biotin
5. K^+ Osmotic pressure of blood Na^+ Cl^-
6. Stored for long-term use   Amino acids   Triacylglycerols   Glycogen
7. Vitamin D   Must be ingested   Ultraviolet radiation   Skin
8. Copper   Iron   Hemoglobin synthesis   Zinc
9. 0.8 g/kg of body weight   About 2 oz   About 16 oz   Minimum daily requirement of protein
10. Catabolic hormones   Anabolic hormones   Sex hormones   Pituitary growth hormones

5. Using the key choices, identify the foodstuffs used by cells in the following descriptions of cellular functions.

Key Choices

A. Carbohydrates B. Fats C. Proteins

__________ 1. The most-used substance for producing the energy-rich ATP

__________ 2. Important in building myelin sheaths and cell membranes

__________ 3. Tend to be conserved by cells

__________ 4. The second most important food source for making cellular energy

__________ 5. Form insulating deposits around body organs and beneath the skin

__________ 6. Used to make the bulk of structural and functional cell substances such as collagen, enzymes, and hemoglobin

__________ 7. Building blocks are amino acids

__________ 8. One of their building blocks, glycerol, is a sugar alcohol

Overview of Metabolic Reactions

1. Complete the following statements, which provide an overview of the metabolism of energy-containing nutrients, by writing the missing words in the answer blanks.

__________ 1.
__________ 2.
__________ 3.
__________ 4.
__________ 5.
__________ 6.
__________ 7.
__________ 8.
__________ 9.
__________ 10.

The general term for all chemical reactions necessary to maintain life is __(1)__. The breakdown of complex molecules to simpler ones is called __(2)__, and the reverse process, the synthesis of larger molecules from smaller ones, is called __(3)__. The term that means the extraction of energy from nutrient molecules is __(4)__. The chemical form of energy that cells generally use to drive their activities is __(5)__.

There are three states in the metabolism of energy-containing nutrients. Stage 1 occurs in the __(6)__, where large nutrient molecules are broken down to their absorbable forms (monomers). In Stage 2, the blood transports these monomers to tissue cells, where the monomers undergo catabolic or anabolic reactions in the cytoplasm of the cells.

Stage 3 occurs in the __(7)__ of cells, where generation of ATP requires the gas __(8)__ and where __(9)__ and __(10)__ gas are formed as the final waste products.

______________ 11.
______________ 12.
______________ 13.
______________ 14.
______________ 15.
______________ 16.

Reactions that generate ATP are called oxidation reactions. In most cases, the substrate molecules (the substance oxidized) lose __(11)__ and also __(12)__. Reactions in which substrate molecules gain energy by the addition of hydrogen atoms and electrons are called __(13)__ reactions. Because the enzymes regulating the oxidation-reduction (OR) reactions cannot accept (bind) hydrogen atoms removed during oxidation, molecules called __(14)__ are required. The abbreviations for two of the most important coenzymes in Stage 3 are __(15)__ and __(16)__.

Metabolism of Major Nutrients

1. Figure 23.1 is a simplified, highly schematic diagram of the major catabolic pathways taken by one glucose molecule entering a typical cell. Only the organelles involved with these reactions are included, and only *types* of molecules (*not quantities*) are shown. Dashed arrows indicate two or more steps in a series of reactions, and open arrows indicate the movement of molecules.

Using the key choices, complete the diagram by inserting the correct symbols or words in the empty boxes on the diagram. Also color code and color each box provided with a color coding circle. Then, referring to Figure 23.1, complete the statements that follow by inserting your answers in the answer blanks.

Key Choices

○ Acetyl CoA (Acetyl coenzyme A)	○ Krebs cycle
○ ADP (Adenosine diphosphate)	○ NADH + H^+ (Reduced NAD)
○ CO_2 (Carbon dioxide)	○ O_2 (Oxygen)
○ $FADH_2$ (Reduced FAD)	○ Pi (Inorganic phosphate)
○ Glucose-6-phosphate	○ Pyruvic acid
○ H_2O (Water)	

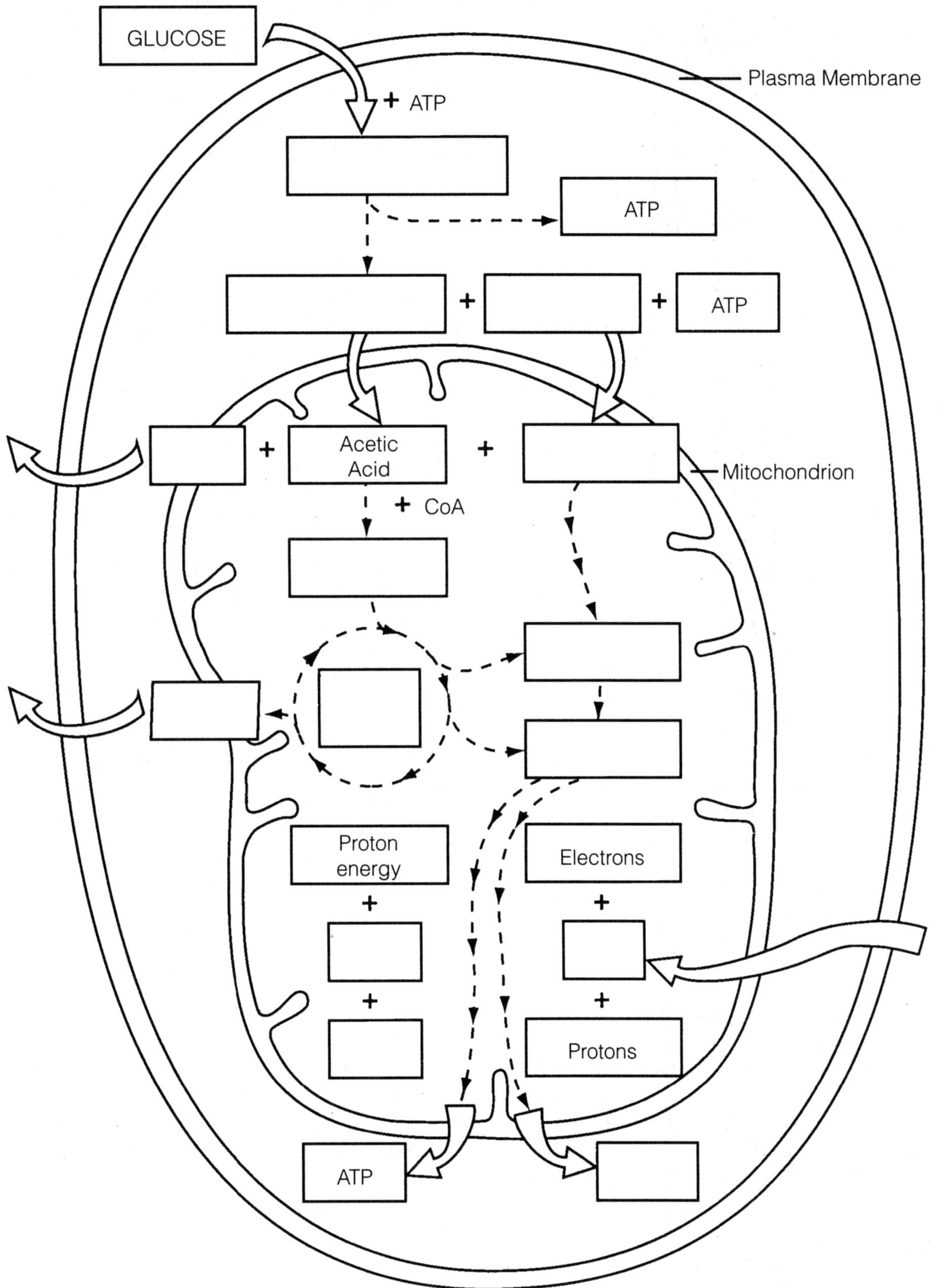
GLUCOSE
Plasma Membrane
+ ATP
ATP
+
+
ATP
+
Acetic Acid
+
Mitochondrion
+ CoA
Proton energy
Electrons
+
+
+
+
Protons
ATP

Figure 23.1

The catabolism of glucose into two pyruvic acid molecules is called __(1)__. This process occurs in the __(2)__ of the cell and is anaerobic, meaning that oxygen is __(3)__. Two molecules of reduced __(4)__ and a net gain of __(5)__ ATP molecules also result from this process. The first phase of glycolysis is __(6)__, in which glucose is converted to fructose-1,6-diphosphate, reactions energized by the hydrolysis of 2 ATP. The second phase, __(7)__, involves breaking the 6-C sugar into two 3-C compounds. The third phase is __(8)__, in which 2 pyruvic acid molecules are formed and 4 ATP (2 net ATP) are produced. The fate of the pyruvic acid depends on whether or not oxygen is available. If oxygen is absent, the two hydrogen atoms removed from pyruvic acid and transferred to NAD^+ combine again with __(9)__ to form __(10)__. Some cells, like __(11)__ cells, routinely undergo periods of anaerobic respiration and are not adversely affected, but __(12)__ cells are quickly damaged. Prolonged anaerobic respiration ultimately results in a decrease in blood __(13)__. __(14)__ cells undergo only glycolysis.

If oxygen is present, carbon is removed from pyruvic acid and leaves the cell as __(15)__, and pyruvic acid is oxidized by the removal of hydrogen atoms, which are picked up by __(16)__. The remaining two-carbon fragment (acetic acid) is carried by __(17)__ to combine with a four-carbon acid called __(18)__, forming the six-carbon acid __(19)__. This six-carbon acid is the first molecule of the __(20)__ cycle, which occurs in the __(21)__. The intermediate acids of this cycle are called __(22)__ acids. Each turn of the cycle generates __(23)__ molecules of carbon dioxide, __(24)__ molecules of reduced NAD^+, and __(25)__ molecules of reduced FAD. Krebs cycle reactions provide a metabolic pool, where two-carbon fragments from __(26)__ and __(27)__ as well as from glucose may be used to generate ATP.

The final reactions of glucose catabolism are those of the __(28)__ chain, which delivers electrons to molecular oxygen. Oxygen combines with __(29)__ to form __(30)__. Part of the energy released in the above reactions is used to form ATP by combining __(31)__ with inorganic phosphate (Pi). The combined overall reaction is called __(32)__. For each molecule of glucose that is completely oxidized, __(33)__ ATP molecules are generated.

1. ____________________
2. ____________________
3. ____________________
4. ____________________
5. ____________________
6. ____________________
7. ____________________
8. ____________________
9. ____________________
10. ____________________
11. ____________________
12. ____________________
13. ____________________
14. ____________________
15. ____________________
16. ____________________
17. ____________________
18. ____________________
19. ____________________
20. ____________________
21. ____________________
22. ____________________
23. ____________________
24. ____________________
25. ____________________
26. ____________________
27. ____________________
28. ____________________
29. ____________________
30. ____________________
31. ____________________
32. ____________________
33. ____________________

2. Figure 23.2 is a highly simplified diagram of a cross section of part of a mitochondrion. First, label the intermembrane space and the mitochondrial matrix on the appropriate lines on the diagram. (*Note:* The structures labeled EC represent enzyme complexes.) Next, select two colors and color the coding circles and the circles on the diagram that represent the electrically charged particles listed below. Remember that the *types* of particles are indicated, *not* the quantities of each. Then, referring to Figure 23.2, complete the statements that follow by inserting your answers in the answer blanks.

○ Electrons (Color the appropriate circles and insert e^- in each.)

○ Protons (Color the appropriate circles and insert H^+ in each.)

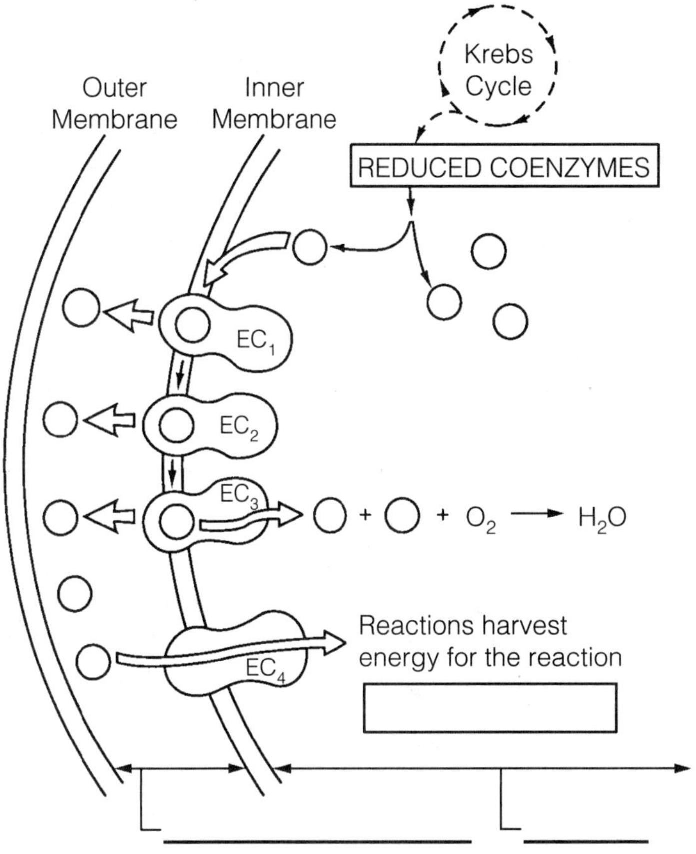

Krebs Cycle

Outer Membrane

Inner Membrane

REDUCED COENZYMES

EC_1

EC_2

EC_3

○ + ○ + O_2 → H_2O

EC_4

Reactions harvest energy for the reaction

Figure 23.2

________ 1.

________ 2.

________ 3.

________ 4.

________ 5.

________ 6.

________ 7.

________ 8.

________ 9.

________ 10.

________ 11.

________ 12.

________ 13.

Reduced coenzymes, generated by the Krebs cycle, release (1), which are split into (2) and (3). The electron acceptors are protein–(4) complexes located in the (5) mitochondrial membrane. The final electron acceptor is (6), which, upon combining with free (7) located in the matrix of the mitochondrion, forms water.

Some of the energy released from the electron transport chain is used by the enzyme complexes to pump protons from the (8) to the (9). As a consequence, the pH of the intermembrane space is (10) than that of the matrix. The electrochemical protein gradient across the inner membrane is a source of potential energy referred to as (11). The bottom enzyme complex on Figure 23.2 is named (12). It uses the energy of proton flow to drive the reaction (13). Also insert your answer to item 13 on the appropriate line in Figure 23.2.

3. Match the metabolic reactions in Column B to the terms in Column A.

	Column A	Column B
________________	1. Lipolysis	A. Glycerol + fatty acids → fats
________________	2. Ketogenesis	B. Fatty acids → acetyl CoA
________________	3. Beta oxidation	C. Fats → glycerol + fatty acids
________________	4. Lipogenesis	D. Fatty acids → ketone bodies

4. This exercise considers protein/amino acid metabolism. First, complete the text by writing the appropriate key choices in the answer blanks. Second, use terms from the key to complete the flowcharts shown in Figure 23.3. Third, after filling in the flowchart, respond to questions 8–12 following the completion section below.

Key Choices

A. Alpha ketoglutaric acid | D. Essential | G. Urea

B. Ammonia (NH_3) | E. Glutamic acid | H. Water (H_2O)

C. Oxidative deamination | F. Transamination

________________ 1.

________________ 2.

________________ 3.

________________ 4.

________________ 5.

________________ 6.

________________ 7.

Amino acids are actively accumulated by cells because proteins cannot be made unless all amino acid types are present. The nine amino acids that *must* be taken in the diet are called __(1)__ amino acids. The other amino acids may be synthesized by the process called __(2)__. When amino acids are oxidized to form cellular energy, their amino groups are removed and liberated as __(3)__. In the liver, this is combined with carbon dioxide to form __(4)__, which is removed from the body by the kidneys. The process by which an amine group is removed from an amino acid is called __(5)__. The keto acid that serves as the acceptor of the amine group is __(6)__. When this keto acid accepts the amine group, it becomes __(7)__.

8. Using a key choice, identify the pathway followed by the amino acid labeled I. ____________

9. Using a key choice, name the pathway followed by the amino acid labeled II. ____________

10. Why must ammonia (from the amine groups) be rapidly removed from the blood?

____

11. What is the general term for amino acids that lose their amine groups? ________________

12. Briefly explain why excess amino acids are oxidized for energy. ____________________

____

____

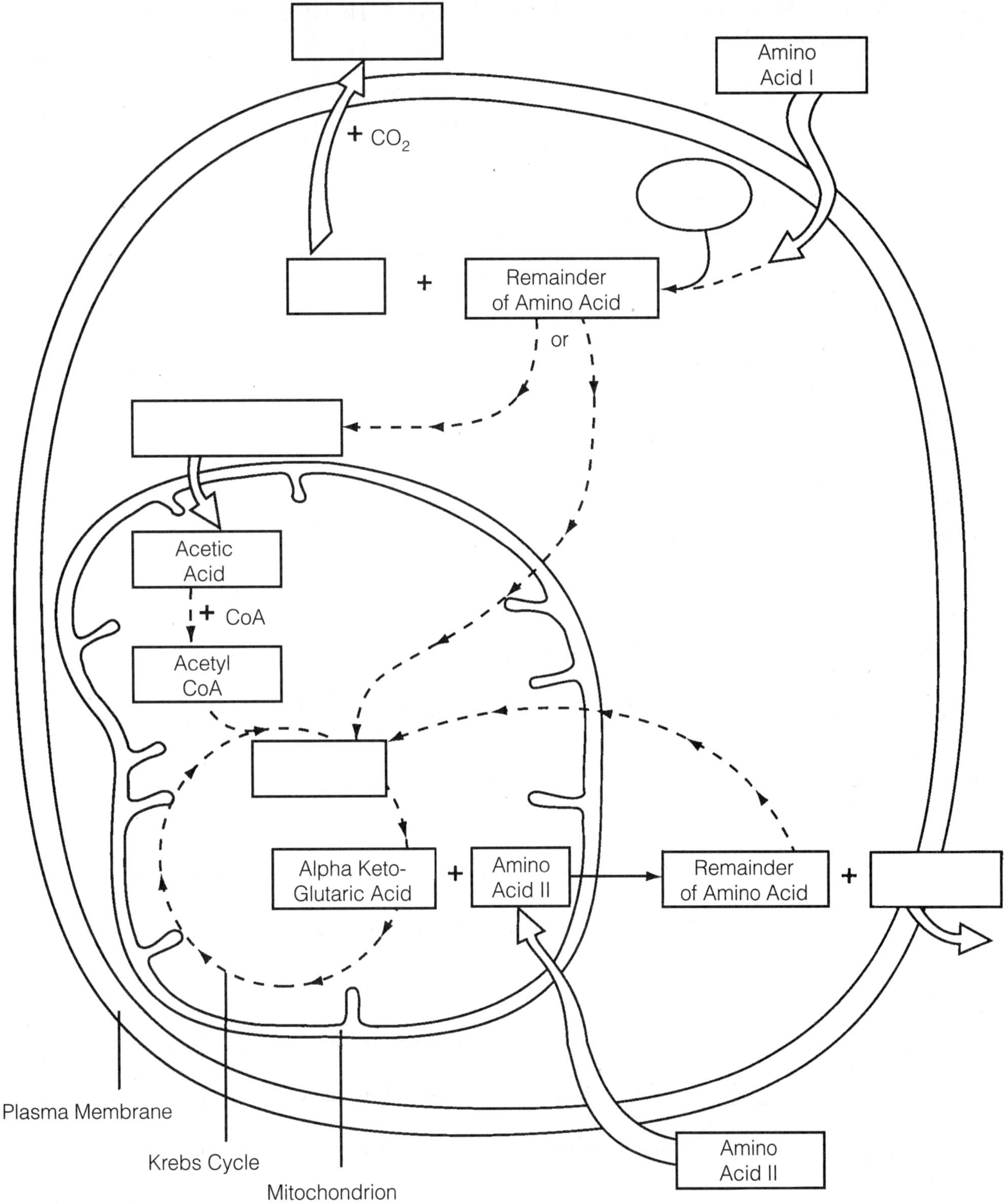
Amino Acid I
+ CO_2
+
Remainder of Amino Acid
or
Acetic Acid
+ CoA
Acetyl CoA
Alpha Keto-Glutaric Acid
+
Amino Acid II
Remainder of Amino Acid
+
Plasma Membrane
Krebs Cycle
Mitochondrion
Amino Acid II

Figure 23.3

5. Circle the term that does not belong in each of the following groupings pertaining to triglyceride (fat) metabolism.

1. 9 kcal/g   Glucose   Fat   Most concentrated form of energy
2. Chylomicrons   Transported in lymph   Hydrolyzed in plasma   Hydrolyzed in mitochondria
3. Cytoplasm   Beta oxidation   Two-carbon fragments   Mitochondria
4. Triglyceride synthesis   Lipogenesis   ATP deficit   Excess ATP
5. High dietary carbohydrates   Glucose conversion to fat   Low blood sugar level   Condensation of acetyl   CoA molecules

Metabolic States of the Body

1. Using the key choices, select the terms that match the following descriptions.

Key Choices

A. Absorptive state
B. Amino acid pool
C. Blood
D. Carbohydrate–fat pool
E. Glucose sparing
F. Molecular turnover (by degradation and synthesis)
G. Postabsorptive state

________________ 1. Results in the catabolic-anabolic steady state of the body

________________ 2. The nutrient-containing transport pool

________________ 3. Use by body cells of energy sources other than glucose

________________ 4. Molecules oxidized directly to produce cellular energy

________________ 5. Period when nutrients move from the GI tract into the bloodstream

________________ 6. Supply of building blocks available for protein synthesis

________________ 7. Period when body reserves are used for energy; GI tract is empty

________________ 8. Excess supply cannot be stored but is oxidized for energy or converted to fat or glycogen

2. If a statement is true, write the letter T in the answer blank. If a statement is false, change the underlined word(s) and write the correct word(s) in the answer blank.

_______________ 1. The efficiency of aerobic cellular respiration is almost 100%.

_______________ 2. Higher concentrations of ADP than ATP are maintained in the cytoplasm.

_______________ 3. If the cytoplasmic concentration of calcium ions (Ca^{2+}) is too high, the gradient energy for ATP production is used instead to pump Ca^{2+} ions out of the mitochondria.

_______________ 4. The storage of glucose molecules in long glycogen chains results from the process called glycolysis.

_______________ 5. The cells of the pancreas and heart muscle are most active in glycogen synthesis.

3. If a statement about the *absorptive* state is true, write the letter T in the answer blank. If a statement is false, change the underlined word(s) and write the correct word(s) in the answer blank.

_______________ 1. Major metabolic events are catabolic.

_______________ 2. The major energy fuel is glucose.

_______________ 3. Excess glucose is stored in muscle cells as glucagon or in adipose cells as fat.

_______________ 4. The major energy source of skeletal muscle cells is glucose.

_______________ 5. In a balanced diet, most amino acids pass through the liver to other cells, where they are used for the generation of ATP.

_______________ 6. The hormone most in control of the events of the absorptive state is insulin.

_______________ 7. The sympathetic nervous system signals the pancreatic beta cells to secrete insulin.

_______________ 8. Insulin stimulates the active transport of glucose into tissue cells.

_______________ 9. Insulin inhibits lipolysis and glycogenesis.

_______________ 10. Most amino acids in the hepatic portal blood remain in the blood for uptake by hepatocytes.

_______________ 11. Amino acids move into tissue cells by active transport.

4. Respond to the following questions about the *postabsorptive* state by writing your answers in the answer blanks.

1. What is the homeostatic range of blood glucose levels? Give units.

____

2. Why must blood glucose be maintained at least at the lower levels?

____

3. Name four sources of glucose in order from the most to the least available when blood glucose levels are falling. Include the tissues of origin and the name of the metabolic process involved.

____

____

____

____

4. Briefly explain the indirect metabolic pathway by which skeletal muscle cells help maintain blood glucose levels.

____

____

____

5. Trace the metabolic pathway by which lipolysis produces blood glucose.

____

____

____

____

6. How are tissue proteins used as a source of blood glucose?

____

____

7. During the period of glucose sparing, which energy source is available to organs other than the brain? ______________________________

8. After four or five days of fasting, accompanied by low blood glucose levels, the brain begins to use a different energy source. What is this energy source?

____

9. Name the two major hormones controlling events of the postabsorptive state.

____

10. Name the origin and targets of glucagon.

____

11. Name two humoral factors that stimulate the release of glucagon.

____

____

12. Why is the release of glucagon important when blood levels of amino acids rise (as after a high-protein, low-carbohydrate meal)?

____

____

13. Briefly describe the role of the sympathetic nervous system. ____________________

____

____

5. Complete the table below, which indicates the effects of several hormones on the characteristics indicated at the top of each column. Use an up arrow (↑) to indicate an increase in a particular value or a down arrow (↓) to indicate a decrease. No response is required in the boxes with XXX.

Hormone	Blood glucose	Blood amino acids	Glycogenolysis	Lipogenesis	Protein synthesis
Insulin					
Glucagon		XXX			XXX
Epinephrine		XXX			XXX
Growth hormone					
Thyroxine		XXX			

Metabolic Role of the Liver

1. The liver has many functions in addition to its digestive function. Complete the following statements that elaborate on the liver's function by inserting the correct terms in the answer blanks.

Answer blanks
1. ______________
2. ______________
3. ______________
4. ______________
5. ______________
6. ______________
7. ______________
8. ______________
9. ______________
10. ______________
11. ______________
12. ______________
13. ______________
14. ______________
15. ______________
16. ______________
17. ______________
18. ______________
19. ______________
20. ______________
21. ______________
22. ______________
23. ______________
24. ______________
25. ______________
26. ______________
27. ______________

The liver is the most important metabolic organ in the body. In its metabolic role, the liver uses amino acids from the nutrient-rich hepatic portal blood to make many blood proteins such as (1), which helps to hold water in the bloodstream, and (2), which prevent blood loss when blood vessels are damaged. The liver also makes a steroid substance that is released to the blood. This steroid, (3), has been implicated in high blood pressure and heart disease. Additionally, the liver acts to maintain homeostatic blood glucose levels. It removes glucose from the blood when blood levels are high, a condition called (4), and stores it as (5). Then, when blood glucose levels are low, a condition called (6), liver cells break down the stored carbohydrate and release glucose to the blood once again. This latter process is termed (7). When the liver makes glucose from noncarbohydrate substances such as fats or proteins, the process is termed (8). In addition to its processing of amino acids and sugars, the liver plays an important role in the processing of fats. Other functions of the liver include the (9) of drugs and alcohol, and its (10) cells protect the body by ingesting bacteria and other debris.

The liver forms small complexes called (11), which are needed to transport fatty acids, fats, and cholesterol in the blood because lipids are (12) in a watery medium. The higher the lipid content of the lipoprotein, the (13) is its density. Thus, low-density lipoproteins (LDLs) have a (14) lipid content. (15) are lipoproteins that transport dietary lipids from the GI tract. The liver produces VLDLs (very low density lipoproteins) to transport (16) to adipose tissue.

After releasing their cargo, VLDL residues convert to LDLs, which are rich in (17). The function of LDLs is transport of cholesterol to peripheral tissues, where cells use it to construct their plasma (18) or to synthesize (19). The function of HDLs (high-density lipoproteins) is transport of cholesterol to the (20), where it is degraded and secreted as (21), which are eventually excreted.

High levels of cholesterol in the plasma are of concern because of the risk of (22). The type of dietary fatty acid affects cholesterol levels. (23) fatty acids promote excretion of cholesterol. When cholesterol levels in plasma are measured, high levels of (24) DLs indicate that the fate of the transported cholesterol is catabolism to bile salts.

Two other important functions of the liver are the storage of vitamins (such as vitamins (25)) and of the metal (26) (as ferritin) and the processing of bilirubin, which results from the breakdown of (27) cells.

Energy Balance

1. Circle the term that does not belong in each of the following groupings.

1. BMR TMR Rest Postabsorptive state
2. Thyroxine Iodine ↓ Metabolic rate ↑ Metabolic rate
3. Obese person ↓ Metabolic rate Women Child
4. 4 kcal/gram Fats Carbohydrates Proteins
5. ↑ TMR Fasting Muscle activity Ingestion of protein
6. Body's core Lowest temperature Heat loss surface Body's shell
7. Radiation Evaporation Vasoconstriction Conduction

2. Using the key choices, select the factor regulating food intake that best fits the following conditions. Also, if appropriate, state whether hunger is depressed (use ↓), or stimulated (use ↑), or both (use ↑ ↓) by each condition described. Insert your letter answers and the arrows in the answer blanks.

Key Choices

A. Body temperature C. Nutrient signals related to total energy stores

B. Hormones D. Psychological factors E. ARC nucleus of hypothalamus

__________ 1. Site of major appetite and food-getting centers

__________ 2. Release of leptin and insulin

__________ 3. Secretions of insulin or cholecystokinin

__________ 4. Low plasma levels of amino acids

__________ 5. The sight, taste, smell, or thought of food

__________ 6. Secretions of glucagon or epinephrine

__________ 7. Increasing amounts of fatty acids and glycerol in the blood

3. Respond to the following by writing your answers in the answer blanks.

1. Briefly explain cell metabolism in terms of the first law of thermodynamics (energy can neither be created nor destroyed).

____

____

____

2. Nearly all the energy derived from food is eventually converted to ________________.

3. How does the balance between energy input and energy output affect body weight?

____

____

4. Using the key choices, select the terms that match the following descriptions pertaining to body temperature regulation. Insert the appropriate answers in the answer blanks.

Key Choices

A. Blood	D. Heat	G. Hypothermia	J. Pyrogens
B. Vasoconstriction in skin	E. Hyperthermia	H. Perspiration	K. Shivering
C. Frostbite	F. Hypothalamus	I. Radiation	

________________ 1. By-product of cell metabolism

________________ 2. Means of conserving or increasing body heat

________________ 3. Medium that distributes heat to all body tissues

________________ 4. Site of the body's thermostat

________________ 5. Chemicals released by injured tissue cells and macrophages that cause resetting of the thermostat

________________ 6. Death of cells deprived of oxygen and nutrients, resulting from withdrawal of blood from the skin circulation

________________ 7. Means of liberating excess body heat

________________ 8. Extremely low body temperature

________________ 9. Fever

________________ 10. Cause release of hypothalamic prostaglandins

5. If a statement about the regulation of body temperature is true, write the letter T in the answer blank. If a statement is false, change the underlined word(s) and write the correct word(s) in the answer blank.

________________ 1. One of the most important ways the body regulates temperature is by changing <u>visceral organ</u> activity.

________________ 2. The homeostatic range of body temperature is from 35.8°C (96°F) to <u>43°C (110°F)</u>.

________________ 3. With each 1°C rise in body temperature, the rate of enzymatic catalysis increases by <u>1%</u>.

__________________ 4. When the body temperature exceeds the homeostatic range, carbohydrates begin to break down.

__________________ 5. Convulsions occur when the body temperature reaches 41°C (106°F).

__________________ 6. The cessation of all thermoregulatory mechanisms may lead to heat exhaustion.

CHALLENGING YOURSELF

At the Clinic

1. Mary Maroon comes to the clinic to get information on a vegetarian diet. What problems may arise when people make uninformed decisions on what to eat for a vegetarian diet? What combination of vegetable foods will provide Mary with all the essential amino acids?

2. A group of new parents is meeting to learn about good nutrition for their children. The topic is "empty calories." What does the term mean and what are some foods with empty calories?

3. Mrs. Piercecci, a woman in her mid-40s, complains of extreme fatigue and faintness about four hours after each meal. What hormone might be deficient and what condition is indicated?

4. Zena, a teenager, has gone to the sports clinic for the past two years to have her fat content checked. This year, her percent body fat is up and tissue protein has not increased. Questioning reveals that Zena has been on crash diets four times since the last checkup, only to regain the weight (and more) each time. She also admits sheepishly that she "detests" exercise. How does cyclic dieting, accompanied by lack of exercise, cause an increase in fat and a decrease in protein?

5. Benny, a 6-year-old child who is allergic to milk, has extremely bowed legs. What condition do you suspect and what is the connection to not drinking milk?

6. Katy, an anorexic girl, shows high levels of acetone in her blood. What is this condition called and what has caused it?

7. There has been a record heat wave lately, and many elderly people are coming to the clinic complaining that they "feel poorly." In most cases, their skin is cool and clammy, and their blood pressure is low. What is their problem? What can be done to alleviate it?

8. During the same period, Bert Winchester, a construction worker, is rushed in unconscious. His skin is hot and dry, and his coworkers say that he suddenly keeled over on the job. What is Bert's condition and how should it be handled?

9. The mother of a 1-year-old confides in the pediatrician (somewhat proudly) that she has started her child on skim milk to head off any future problems with atherosclerosis. Why is this not a good practice?

10. A blood test shows cholesterol levels at about 280 mg/100 ml. What is this condition called? What cholesterol level is safe? What are some complications of high cholesterol?

11. Mrs. Rodriguez has a bleeding ulcer and has lost her appetite. She appears pale and lethargic when she comes in for a physical. She proves to be anemic and her RBCs are large and pale. What mineral supplements should be ordered?

12. Mr. Hodges, pretty much a recluse, unexpectedly appears at the clinic seeking help with a change he has experienced. He has been losing weight, his skin is full of bruises, and his gums are bleeding. However, what really "drove" him to the clinic is a sore that won't heal. What is his problem called? What vitamin deficiency causes it? What dietary changes would you suggest?

13. Oliver is complaining about the fact that he is obese but it is "not his fault" because he eats very few simple carbohydrates and fats. Explain how this situation might occur.

Stop and Think

1. Amino acids must be metabolized; they cannot be stored. What would be the osmotic and acid-base effects of accumulating amino acids?
2. What would be the diffusion and osmotic effects of failing to turn glucose into glycogen and adipose tissue for storage?
3. What is the *specific* mechanism by which oxygen starvation causes death?
4. Does urea production increase or decrease when essential amino acids are lacking in the diet? Explain your answer.
5. What is the difference in the effect on blood glucose of a meal of simple sugars versus a meal of an equal amount of complex carbohydrates?
6. Does loss of mental function accompany the deterioration associated with diabetes mellitus? Why or why not?
7. What would happen to blood clotting if excess omega-3 fatty acids were ingested?
8. Why is the factor for calculating BMR higher for males than it is for females?
9. Why might PKU sufferers have poor vision?
10. What supplements would you suggest for someone taking diuretics?
11. What is the metabolic link between diabetes mellitus and arteriosclerosis?
12. Does hypothyroidism result in hyperthermia or hypothermia? Explain.
13. Why do we feel hot when we exercise vigorously?
14. In the mid-1980s, a calorie-free fat substitute (olestra) that is neither digested nor absorbed hit the market shelves in the United States. At that time, there was concern that vitamin deficiencies might ensue from the use of olestra. What type of vitamins were of concern and why?
15. As indicated in Figure 23.4 concerning the Krebs cycle, succinic acid is converted to fumaric acid. This conversion process involves the liberation of H^+, a detail not shown in the figure. During a laboratory exercise dealing with this reaction, you are working with a suspension of bean-cell mitochondria and a blue dye that loses its color as it absorbs hydrogen ions. You know from previous enzyme experiments that the higher the concentration of the substrate (succinic acid in this case), the more rapidly the reaction proceeds (the faster the dye will decolor). Your experimental tubes contain three different succinic acid concentrations—0.1 mg/L, 0.2 mg/L, and 0.3 mg/L. Which of the graphs shown—A, B, or C—represents the expected results and why?

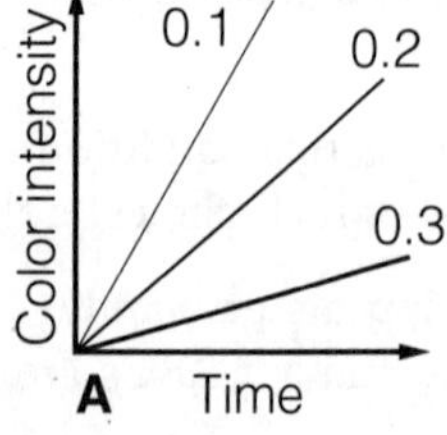

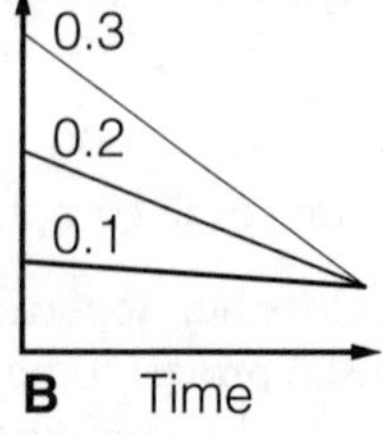

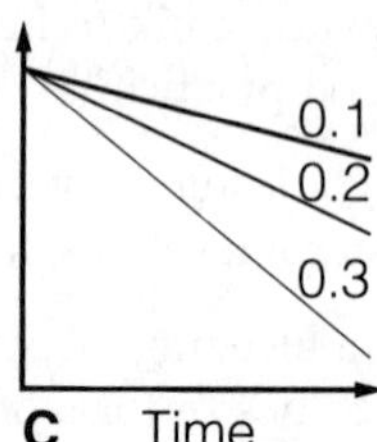

Figure 23.4

16. The body easily converts excess carbohydrates into fats. However, it needs input from dietary proteins to convert carbohydrates or fats alone into proteins. What do proteins in the diet contribute?

COVERING ALL YOUR BASES

Multiple Choice

Select the best answer or answers from the choices given.

1. Which of the following are "essential" nutrients?
A. Glucose
B. Linoleic acid
C. Cholesterol
D. Leucine

2. Which factors will work to promote (and allow) protein synthesis?
A. Presence of all 20 amino acids
B. Sufficiency of caloric intake
C. Negative nitrogen balance
D. Secretion of anabolic hormones

3. Which of the listed vitamins is not found in plant foods?
A. Folic acid
B. Biotin
C. Cyanocobalamin (B_{12})
D. Pantothenic acid

4. Deficiency of which of these vitamins results in anemia?
A. Thiamin
B. Riboflavin
C. Biotin
D. Folic acid

5. Vitamins that act as coenzymes in the Krebs cycle include:
A. riboflavin
B. niacin
C. biotin
D. pantothenic acid

6. Absorption of which of the following minerals parallels absorption of calcium?
A. Zinc
B. Magnesium
C. Iron
D. Copper

7. Components of the electron transport system include:
A. cobalt
B. copper
C. iron
D. zinc

8. Substrate-level phosphorylation occurs during:
A. glycolysis
B. beta-oxidation
C. Krebs cycle
D. electron transport

9. Which processes occur in the mitochondria?
A. Glycogenesis
B. Gluconeogenesis
C. Beta oxidation
D. Formation of acetyl CoA

10. The chemiosmotic process involves:
A. buildup of hydrogen ion concentration
B. electron transport
C. oxidation and reduction
D. ATP synthase

11. Which of the following molecules has the highest energy content?
A. Glucose
B. Acetyl CoA
C. Dihydroxyacetone phosphate
D. ATP

12. Krebs cycle intermediates include:
A. glutamic acid
B. pantothenic acid
C. succinic acid
D. fumaric acid

13. Chemicals that can be used for gluconeogenesis include:
A. amino acids
B. glycerol
C. fatty acids
D. alpha ketoglutaric acid

14. Ingestion of saturated fatty acids (as opposed to equal quantities of polyunsaturated fatty acids) is followed by:
A. increased cholesterol formation
B. increased cholesterol excretion
C. increased production of chylomicrons
D. increased formation of bile salts

15. Which events occur during the absorptive state?
A. Use of amino acids as a major source of energy
B. Lipogenesis
C. Beta oxidation
D. Increased uptake of glucose by skeletal muscles

16. Hormones that act to decrease blood glucose level include:
A. insulin
B. glucagon
C. epinephrine
D. growth hormone

17. During the postabsorptive state:
A. glycogenesis occurs in the liver
B. fatty acids are used for fuel
C. amino acids are converted to glucose
D. lipolysis occurs in adipose tissue

18. Which processes contribute to gluconeogenesis?
A. Conversion of pyruvic acid to glucose by skeletal muscle
B. Release of lactic acid by skeletal muscle
C. Formation of glucose from triglycerides in adipose tissue
D. Release of glycerol from adipose tissue

19. Only the liver functions to:
A. store iron C. produce plasma proteins
B. form urea D. form ketone bodies

20. Quadruple the RDA daily of which of the following may be lethal?
A. Vitamin C C. Selenium
B. Vitamin A D. Vitamin E

21. Which transport particles carry cholesterol destined for excretion from the body?
A. HDL C. LDL
B. Chylomicron D. VLDL

22. Transport of triglycerides from the liver to adipose tissue is the function of:
A. LDL C. VLDL
B. HDL D. all lipoproteins

23. Glucose (or its metabolites) can be converted to:
A. glycogen C. nonessential amino acids
B. triglycerides D. starch

24. Basal metabolic rate:
A. is the lowest metabolic rate of the body
B. is the metabolic rate during sleep
C. is measured as kcal per square meter of skin per hour
D. increases with age

25. Body temperature is higher:
A. in children than in adults
B. in cases of hyperthyroidism
C. after eating
D. when sleeping

26. Which of the following types of heat transfer involves heat loss in the form of infrared waves?
A. Conduction C. Evaporation
B. Convection D. Radiation

27. Body temperature increases:
A. during chemical thermogenesis
B. when pyrogens are released
C. when sweating and flushing occur
D. with heat exhaustion

28. PKU is the result of inability to metabolize:
A. tyrosine C. ketone bodies
B. melanin D. phenylalanine

29. A wasting disease due to deficiency of both protein and carbohydrates is:
A. marasmus C. pica
B. kwashiorkor D. cystic fibrosis

Word Dissection

For each of the following word roots, fill in the literal meaning and give an example, using a word found in this chapter.

Word root	Translation	Example
1. acet		
2. calor		
3. flav		
4. gluco		
5. kilo		
6. lecith		
7. linol		
8. nutri		
9. pyro		

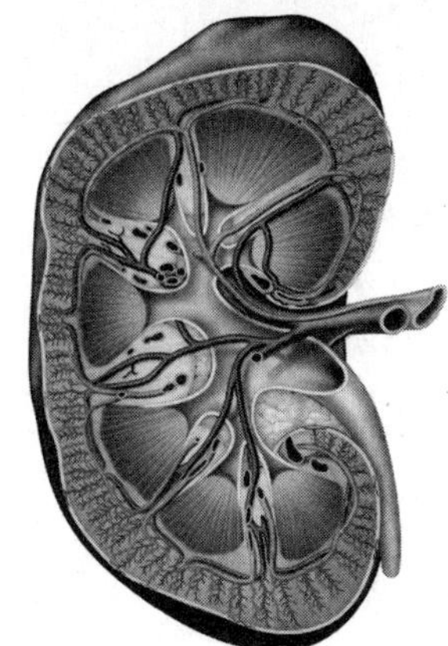

24

The Urinary System

Student Objectives

When you have completed the exercises in this chapter, you will have accomplished the following objectives:

Kidney Anatomy

1. Describe the gross anatomy of the kidney and its coverings.
2. Trace the blood supply through the kidney.
3. Describe the anatomy of a nephron.

Kidney Physiology: Mechanisms of Urine Formation

4. Describe the forces (pressures) that promote or counteract glomerular filtration.
5. Compare the intrinsic and extrinsic controls of the glomerular filtration rate.
6. Describe the mechanisms underlying water and solute reabsorption from the renal tubules into the peritubular capillaries.
7. Describe how sodium and water reabsorption is regulated in the distal tubule and collecting duct.
8. Describe the importance of tubular secretion and list several substances that are secreted.
9. Describe the mechanisms responsible for the medullary osmotic gradient.
10. Explain formation of dilute versus concentrated urine.
11. Define renal clearance and explain how this value summarizes the way a substance is handled by the kidney.

Urine

12. Describe the normal physical and chemical properties of urine.
13. List several abnormal urine components, and name the condition characterized by the presence of detectable amounts of each.

Ureters, Urinary Bladder, and Urethra

14. Describe the general location, structure, and function of the ureters.
15. Describe the general location, structure, and function of the urinary bladder.
16. Describe the general location, structure, and function of the urethra.
17. Compare the course, length, and functions of the male urethra with those of the female.

Micturition

18. Define micturition and describe its neural control.

Metabolism of nutrients by body cells produces various wastes, such as carbon dioxide and nitrogenous wastes (creatine, urea, and ammonia), as well as imbalances of water and essential ions. The metabolic wastes and excesses must be eliminated from the body. Essential substances are retained to ensure internal homeostasis and proper body functioning.

Although several organ systems are involved in excretory processes, the urinary system is primarily responsible for removal of nitrogenous wastes from the blood. In addition to this purely excretory function, the kidneys maintain the electrolyte, acid-base, and fluid balances of the blood. Kidneys are thus major homeostatic organs of the body. Malfunction of the urinary system, particularly the kidneys, leads to a failure of homeostasis, resulting (unless corrected) in death.

Student activities in Chapter 24 are concerned with identification of urinary system structures, the composition of urine, and the physiological processes involved in urine formation.

BUILDING THE FRAMEWORK

1. Name three metabolic activities of the kidneys unrelated to urine production. ____________

____

____

____

2. Define and give the cause of:

ptosis: ____

____

hydronephrosis: ____

____

3. Figure 24.1 is a frontal view of the entire urinary system. Excluding red and blue, select different colors to color the coding circles and the corresponding organs on the figure. Color the aorta and its branches red; color the inferior vena cava and its tributaries blue.

◯ Kidney ◯ Bladder ◯ Ureters ◯ Urethra

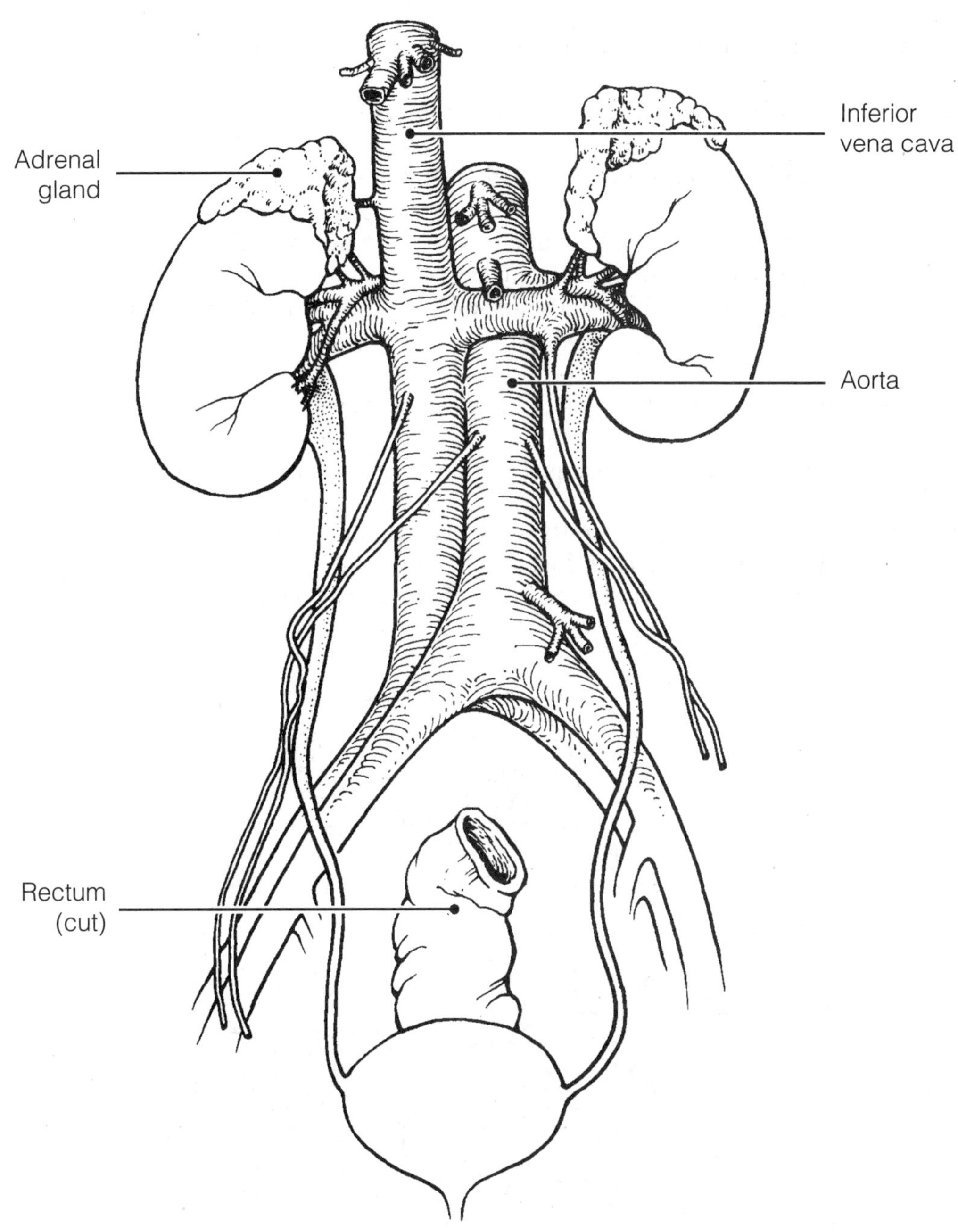

Figure 24.1

Kidney Anatomy

1. Figure 24.2 is a longitudinal section of a kidney. First, using the correct anatomical terminology, identify and label the following regions and structures indicated by leader lines on the figure.

1. Fibrous membrane immediately surrounding the kidney
2. Basinlike area of the kidney that is continuous with the ureter
3. Cuplike extension of the pelvis that drains the apex of a pyramid
4. Area of cortexlike tissue running through the medulla

Then, excluding the color red, select different colors to identify the following areas and structures. Color the coding circles and the corresponding areas and structures on the figure; label these regions using the correct anatomical terms.

○ Area of the kidney that contains the greatest proportion of nephron structures

○ Striped-appearing structures formed primarily of collecting ducts

Finally, beginning with the renal artery, draw part of the vascular supply to the cortex on the figure. Include and label the segmental artery, interlobar artery, arcuate artery, and cortical radiate artery. Color the vessels bright red.

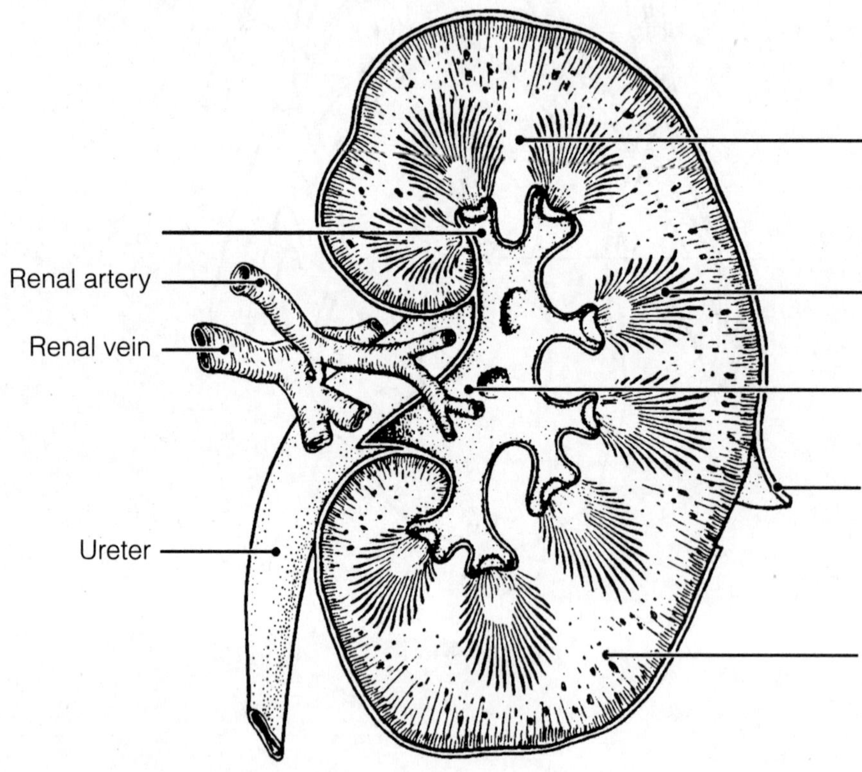

Figure 24.2

2. Figure 24.3A shows a renal corpuscle (glomerulus plus glomerular capsule) on the left and a segment of a glomerular capillary and the associated visceral part of Bowman's capsule on the right. Label each structure that is indicated by a leader line. Color blood vessels red, the parietal layer of the renal corpuscle and parts of the renal tubule yellow, and the podocytes orange. When you have completed part A of the figure, identify specifically the locations of the cross sections of the renal tubules shown in part B, using one of the choices provided below. Then, on the lines that follow, explain the rationale or reasoning behind your choices.

Proximal convoluted tubules (PCT)	Collecting tubule
Distal convoluted tubules (DCT)	Ascending loop of Henle (thin segment)

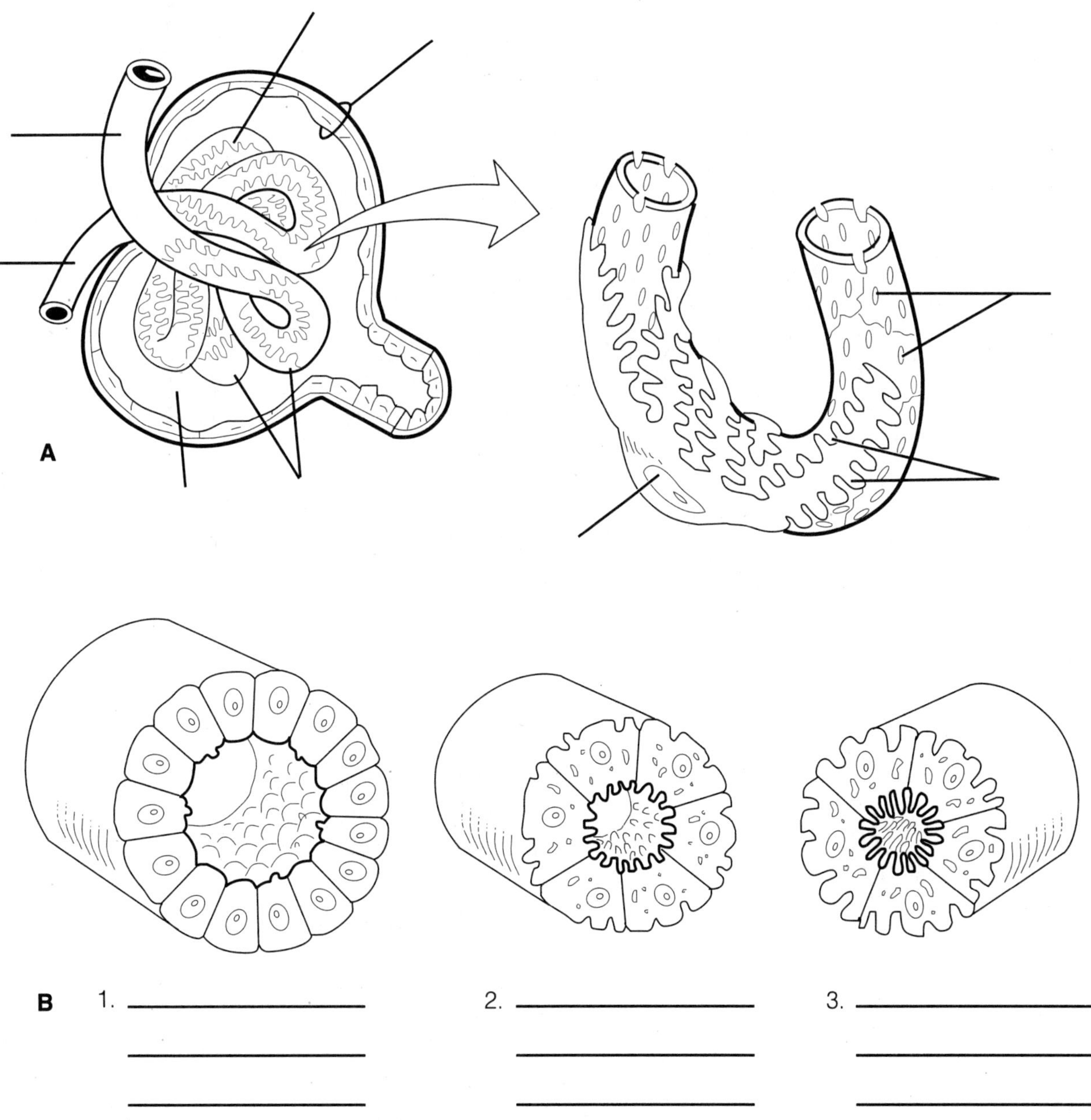

Figure 24.3

3. Figure 24.4 is a diagram of the nephron and associated blood supply. First, match each of the numbered structures on the figure to one of the terms following the figure. Write the terms in the numbered answer blanks. Then, color the structure on the figure that contains podocytes green, the filtering apparatus red, the capillary bed that directly receives the reabsorbed substances from the tubule cells blue, the structure into which the nephron empties its urine product yellow, and the tubule area that is the primary site of tubular reabsorption orange.

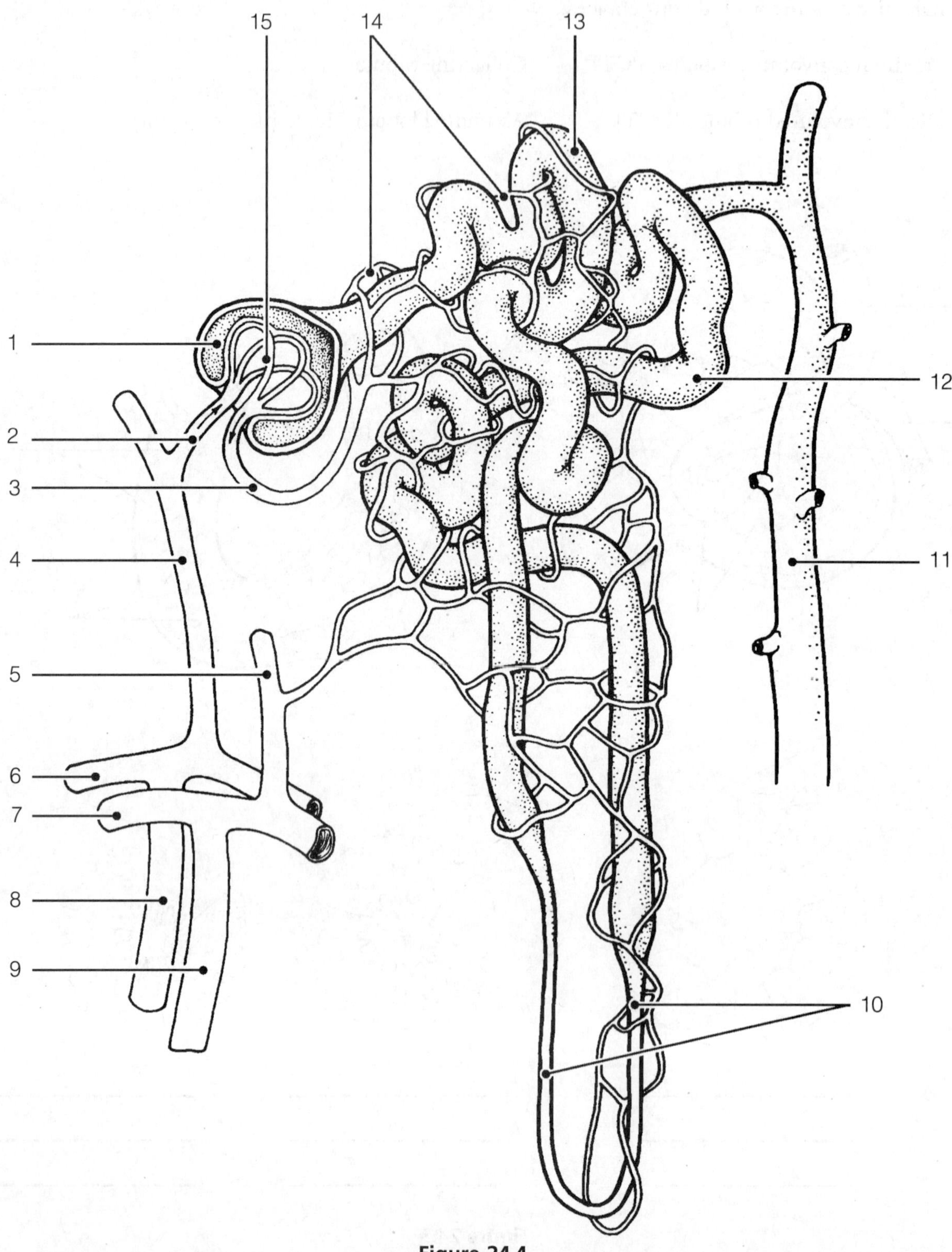

Figure 24.4

Afferent arteriole	Efferent arteriole	Cortical radiate artery
Arcuate artery	Glomerular capsule	Cortical radiate vein
Arcuate vein	Glomerulus	Loop of Henle
Collecting duct	Interlobar artery	Proximal convoluted tubule (PCT)
Distal convoluted tubule (DCT)	Interlobar vein	Peritubular capillaries

________ 1.	________ 6.	________ 11.
________ 2.	________ 7.	________ 12.
________ 3.	________ 8.	________ 13.
________ 4.	________ 9.	________ 14.
________ 5.	________ 10.	________ 15.

4. Circle the term that does not belong in each of the following groupings.

1. Renal pelvis   Renal sinus   Renal pyramid   Minor calyx   Major calyx
2. Perirenal fat   Fibrous capsule   Renal corpuscle   Renal fascia
3. Juxtaglomerular apparatus   Granular cells   Glomerulus   Macula densa cells
4. Glomerulus   Peritubular capillaries   Vasa recta   Collecting duct
5. Cortical nephrons   Vasa recta   Juxtamedullary nephrons   Long loops of Henle
6. Nephron   Proximal convoluted tubule   Distal convoluted tubule   Collecting duct
7. Medullary pyramids   Glomeruli   Renal pyramids   Collecting ducts
8. Glomerular capsule   Podocytes   Loop of Henle   Glomerulus
9. Ureter   Renal artery   Renal pyramids   Renal hilum

5. Assign each of these urine-carrying ducts a number from 1 to 5, according to the path taken by the urine (the urine first passes through #1, then #2, etc.).

_____ 1. renal pelvis   _____ 3. papillary collecting duct   _____ 5. minor calyx

_____ 2. ureter   _____ 4. major calyx

Kidney Physiology: Mechanisms of Urine Formation

1. Complete the following statements by inserting your answers in the answer blanks.

The glomerulus is a unique high-pressure capillary bed because the __(1)__ arteriole feeding it is larger in diameter than the __(2)__ arteriole draining the bed. Glomerular filtrate is very similar to __(3)__, but it has fewer proteins. Mechanisms of tubular reabsorption include __(4)__ and __(5)__. As an aid for the reabsorption process, the cells of the proximal convoluted tubule have dense __(6)__ on their luminal surface, which increase the surface area dramatically. Other than reabsorption, an important tubule function is __(7)__, which is important for ridding the body of substances not already in the filtrate. Blood composition depends on __(8)__, __(9)__, and __(10)__. In a day's time, 180 liters of blood plasma are filtered into the kidney tubules, but less than __(11)__ liters of urine are usually produced. __(12)__ is responsible for the normal yellow color of urine. The three major nitrogenous wastes found in the blood, which must be disposed of, are __(13)__, __(14)__, and __(15)__. The kidneys determine how much water is to be lost from the body. When water loss via vaporization from the __(16)__ or __(17)__ of perspiration from the skin is excessive, urine output __(18)__. If the kidneys become nonfunctional, __(19)__ is used to cleanse the blood of impurities.

____________ 1.
____________ 2.
____________ 3.
____________ 4.
____________ 5.
____________ 6.
____________ 7.
____________ 8.
____________ 9.
____________ 10.
____________ 11.
____________ 12. ____________ 16.
____________ 13. ____________ 17.
____________ 14. ____________ 18.
____________ 15. ____________ 19.

2. Use the three components of the filtration membrane in the key choices to match the descriptions that follow.

Key Choices

A. Capillary endothelium B. Basement membrane C. Podocyte (visceral) membrane of glomerular capsule

____________ 1. Repels anions and prevents their filtration

____________ 2. Its pores prevent filtration of blood cells

____________ 3. Its filtration slits appear to play little or no role in restricting passage of substances into the tubule

____________ 4. Prevents passage of all but the smallest proteins

3. Figure 24.5 is a diagram of a nephron. Add colored arrows on the figure to show the location and direction of the following processes.

1. Black arrows at the site of filtrate formation
2. Red arrows at the major site of amino acid and glucose reabsorption
3. Green arrows at the sites most responsive to action of ADH (show direction of water movement)
4. Yellow arrows at the sites most responsive to the action of aldosterone (show direction of Na^+ movement)
5. Blue arrows at the major sites of tubular secretion

Then, label the proximal convoluted tubule (PCT), distal convoluted tubule (DCT), loop of Henle, glomerular capsule, and glomerulus on the figure. Also label the collecting duct (not part of the nephron).

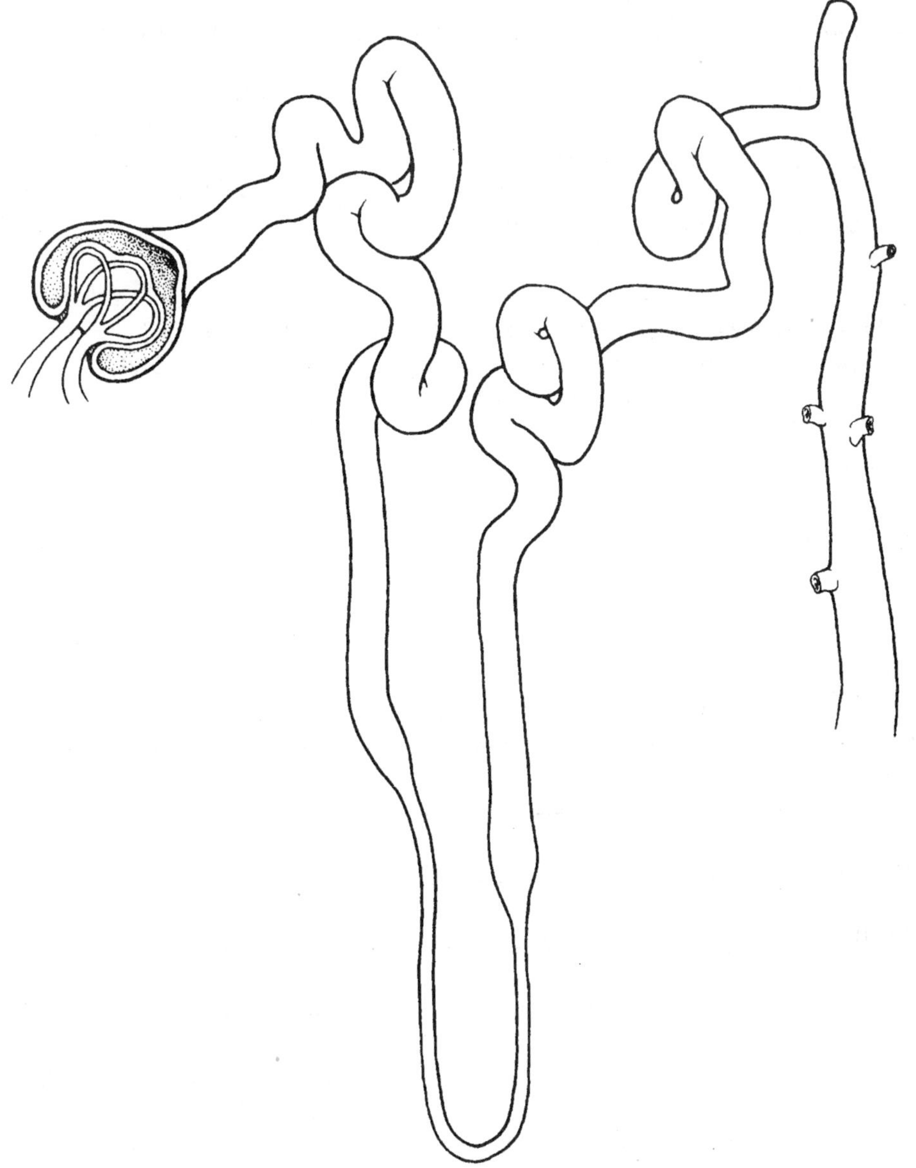

Figure 24.5

4. Complete the following equation referring to net filtration pressure (NFP).

1. NFP = Glomerular hydrostatic pressure – (______________________________).

2. NFP = ____________________ mm Hg

3. Which of these pressures is equivalent to blood pressure? ____________________

4. Which of these pressures is a consequence of proteins in blood? ____________________

5. List below the two intrinsic mechanisms of renal autoregulation, along with the structure(s) responsible for each:

____

____

6. Renin release by the granular cells of nephrons plays an important role in increasing systemic blood pressure. Check (✓) all factors that promote renin release.

_____ 1. Drop in systemic blood pressure

_____ 2. Slowly flowing renal filtrate

_____ 3. Increased stretch of the afferent arterioles

_____ 4. Sympathetic nervous system stimulation of granular cells

_____ 5. Low filtrate Na^+ concentration

7. For each of the substances listed, write whether it is actively reabsorbed, passively reabsorbed in response to an electrochemical gradient, or passively reabsorbed in response to an osmotic gradient.

1. Amino acids ______________________________
2. Bicarbonate ions ______________________________
3. Chloride ions ______________________________
4. Fatty acids ______________________________
5. Glucose ______________________________
6. Lactate ______________________________
7. Potassium ions ______________________________
8. Sodium ions ______________________________
9. Urea ______________________________
10. Vitamins ______________________________
11. Water ______________________________

8. A glucose tolerance test is given to a woman suspected of having diabetes mellitus. Urine and blood samples are taken periodically (typically before testing and at 30, 60, 120, and 180 minutes) after the patient drinks a liquid containing a known amount of glucose. Graph the data given in the table below, using red for blood glucose and yellow for urine glucose. Then answer the questions that follow.

Sample number	Blood glucose, mg/100 ml	Urine glucose, mg/min
1	100	0
2	225	0
3	290	0
4	375	0
5	485	90

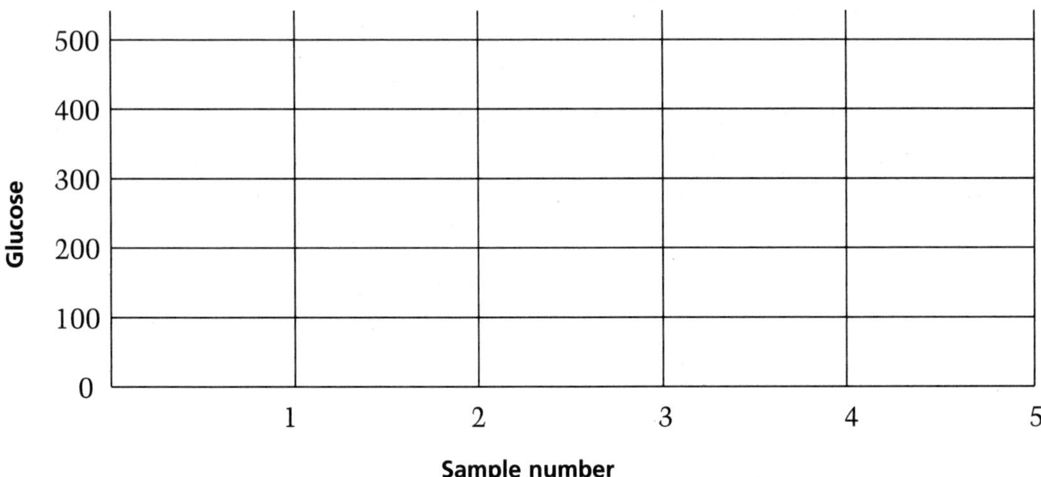

1. According to the graphed information, what is the transport maximum for glucose? ________

2. Is the woman diabetic? ________

9. Name two cations and two nitrogenous wastes that are secreted by the tubule cells into the filtrate.

____

10. Of what importance is the medullary osmotic gradient?

____

____

____

____

11. Complete the statements by filling in the blanks below:

1. ____________
2. ____________
3. ____________
4. ____________
5. ____________
6. ____________
7. ____________
8. ____________

The __(1)__ limb of the loop of Henle is relatively impermeable to solutes but freely permeable to water. The __(2)__ limb of the loop is permeable to solutes but not to water; its __(3)__ segment, in particular, has an active pump that moves salt (NaCl) out of the filtrate into the medullary interstitial fluid. The medullary regions of the collecting ducts are permeable to __(4)__, a small solute that enhances the osmolarity of the medullary interstitial fluid. The __(5)__ acts as a countercurrent exchanger to maintain the osmotic gradient while delivering a nutrient blood supply. The result of this mechanism is to maintain a high solute content in the __(6)__ of the medullary pyramids of about __(7)__ mOsm/L. The permeability of the distal portion of the DCT and the collecting ducts to water is controlled by the hormone __(8)__.

12. Figure 24.6 depicts some of the regulatory events that control processes occurring in the collecting ducts, particularly concerning water balance in the body. Use the terms provided in the list just below to fill in the empty boxes in the diagram. Also, identify by coloring and by adding appropriate leader lines and labels: the hypothalamus, the collecting duct, and the posterior pituitary. (*Note:* ⊕ indicates a stimulatory effect and ⊖ indicates an inhibitory effect.)

Rising blood osmolality	Water reabsorption	Rising blood volume	ADH
Falling blood osmolality	Increases porosity	Falling blood volume	

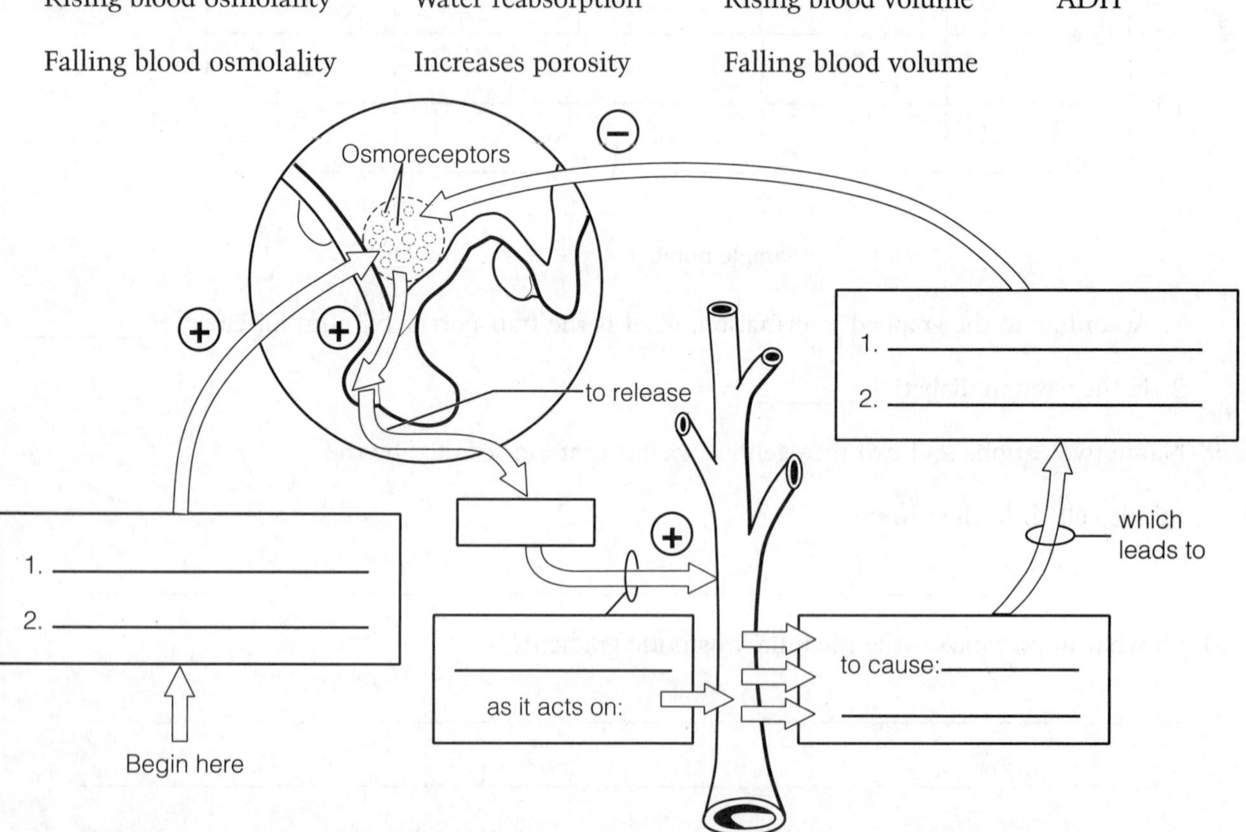

Figure 24.6

13. Circle the term that does not belong in each of the following groupings. (*Note:* BP = blood pressure.)

1. Hypothalamus ADH Aldosterone Osmoreceptors
2. Glomerulus Secretion Filtration Podocytes
3. Aldosterone ↑ Na^+ reabsorption ↑ K^+ reabsorption ↑ BP
4. ADH ↓ BP ↑ Blood volume ↑ Water reabsorption
5. ↓ Aldosterone Edema ↓ Blood volume ↓ K^+ retention

14. Respond to the following by inserting your answer into the blank following each question:

1. What is the glomerular filtration rate of an individual who was injected with inulin at a plasma concentration of 1 mg/ml, and who has a flow rate of urine formation of 2 ml/min, and a urine concentration of inulin of 90 mg/ml? __________
2. Is this a normal, low, or high GFR? __________

Assuming a normal GFR, state whether each of the following substances would be partially reabsorbed or secreted into the urine:

3. creatinine, RC = 145 ml/min ______________________________
4. uric acid, RC = 15 ml/min ______________________________
5. urea, RC = 75 ml/min ______________________________
6. penicillin, RC = 230 ml/min ______________________________

Urine

1. Decide whether the following conditions would result in an increase or a decrease in the specific gravity of urine. Write I in the answer blank to indicate an increase and D to indicate a decrease.

_____ 1. Drinking excessive fluids

_____ 2. Chronic renal failure

_____ 3. Pyelonephritis

_____ 4. Using diuretics

_____ 5. Limited fluid intake

_____ 6. Fever

2. Decide whether the following conditions would cause urine to become more acidic or more basic. If more acidic, write A in the blank; if more basic, write B in the blank.

_____ 1. Protein-rich diet

_____ 2. Bacterial infection

_____ 3. Starvation

_____ 4. Diabetes mellitus

_____ 5. Vegetarian diet

3. Assuming normal conditions, note whether each of the following substances would be in greater concentration in the urine than in the glomerular filtrate (G), in lesser concentration in the urine than in the glomerular filtrate (L), or absent in both urine and glomerular filtrate (A). Place the correct letter in each answer blank.

_____ 1. Water

_____ 2. Urea

_____ 3. Uric acid

_____ 4. Pus (white blood cells)

_____ 5. Glucose

_____ 6. Albumin

_____ 7. Creatinine

_____ 8. Hydrogen ions

_____ 9. Potassium ions

_____ 10. Red blood cells

_____ 11. Sodium ions

_____ 12. Amino acids

4. Several specific terms are used to indicate the presence of abnormal urine constituents. Identify each of the following abnormalities with the term that names the condition. Then, for each condition, provide one possible cause of the condition.

1. Presence of red blood cells ____________ Cause ____________
2. Presence of ketones ____________ Cause ____________
3. Presence of albumin ____________ Cause ____________
4. Presence of pus ____________ Cause ____________
5. Presence of bile pigments ____________ Cause ____________
6. Presence of "sand" ____________ Cause ____________
7. Presence of glucose ____________ Cause ____________

5. Glucose and albumin are both normally absent from urine, but the reasons for their absence differ.

1. Explain the reason for the absence of glucose in urine. ____________

   ____

2. Explain the reason for the absence of albumin in urine. ____________

   ____

6. What pigment is responsible for the yellow color of urine? ____________

____

Ureters, Urinary Bladder, and Urethra

1. Complete the following statements by inserting your answers in the answer blanks.

1. ____________________
2. ____________________
3. ____________________
4. ____________________
5. ____________________
6. ____________________
7. ____________________
8. ____________________

Urine continuously formed by the __(1)__ is routed down the __(2)__ by the mechanism of __(3)__ to a storage organ called the __(4)__. Eventually the urine is conducted to the body exterior by the __(5)__. In the male, this tubelike structure is approximately __(6)__ inches long and transports both urine and __(7)__. In the female, this conduit is approximately __(8)__ inches long and transports only urine.

2. Circle the term that does not belong in each of the following groupings.

1. Bladder   Kidney   Transitional epithelium   Detrusor muscle   Urine storage
2. Trigone   Ureteric orifices   Urethral opening   Bladder   Forms urine
3. Surrounded by prostate gland   Drains bladder   Continuous with renal pelvis   Urethra   Contains internal and external sphincters
4. Renal calculi   Urinary retention   Lithotripsy   Calcium salts   Blocks urethra
5. Prostatic   Male   Female   Membranous   Spongy

Micturition

1. Complete the following statements by writing the missing terms in the answer blanks.

1. ____________________
2. ____________________
3. ____________________
4. ____________________
5. ____________________
6. ____________________

Micturation is another term that means __(1)__. As urine accumulates in the bladder, __(2)__ are activated. This results in a reflex that causes the detrusor muscle of the bladder to __(3)__, and urine flows through the open __(4)__ sphincter. The __(5)__ sphincter is controlled __(6)__; thus an individual can temporarily postpone emptying the bladder until it has accumulated more urine.

(continues on next page)

______________________ 7.
______________________ 8.
______________________ 9.
______________________ 10.
______________________ 11.
______________________ 12.

__(7)__ is a condition in which voiding cannot be voluntarily controlled. It is normal in __(8)__ because nervous control of the voluntary sphincter has not yet been achieved. Other causative conditions include __(9)__ and __(10)__. __(11)__, a condition in which one is unable to void, is a common problem in elderly men due to __(12)__ enlargement.

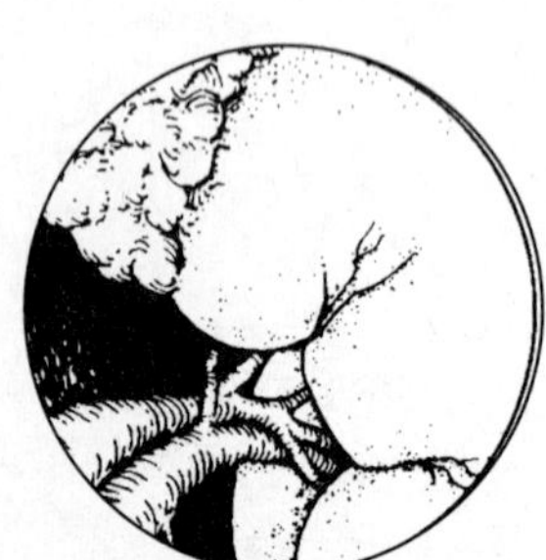

THE INCREDIBLE JOURNEY:

A Visualization Exercise for the Urinary System

Soon you see the kidney looming brownish red through the artery wall.

1. Complete the following narrative by writing the missing terms in the answer blanks.

______________________ 1.
______________________ 2.
______________________ 3.
______________________ 4.
______________________ 5.
______________________ 6.

For your journey through the urinary system, you must be made small enough to filter through the filtration membrane from the bloodstream into a renal __(1)__. You will be injected into the subclavian vein and must pass through the heart before entering the arterial circulation. As you travel through the systematic circulation, you have at least 2 minutes to relax before reaching the __(2)__ artery supplying a kidney. Soon you see the kidney looming brownish red through the artery wall. Once you have entered the kidney, the blood vessel conduits become increasingly smaller until you finally reach the __(3)__ arteriole feeding into the filtering device, or __(4)__.

______________________ 7.
______________________ 8.
______________________ 9.

Once in the filter, you maneuver yourself so that you are directly in front of a pore. Within a fraction of a second, you are swept across the filtration membrane into the __(5)__ part of the renal tubule. Drifting along, you lower the specimen cup to gather your first filtrate sample for testing. You study the readout from the sample and note that it is very similar in composition to __(6)__, with one exception—there are essentially no __(7)__. Your next sample doesn't have to be collected until you reach the "hairpin," or, using the proper terminology, the __(8)__ part of the tubule. As you continue your journey, you notice that the tubule cells have dense fingerlike projections extending from their surfaces into the lumen of the tubule. These are __(9)__, which increase the

9. ____________________
10. ____________________
11. ____________________
12. ____________________
13. ____________________
14. ____________________
15. ____________________
16. ____________________
17. ____________________
18. ____________________
19. ____________________
20. ____________________
21. ____________________
22. ____________________
23. ____________________
24. ____________________

surface area of tubule cells, because this portion of the tubule is very active in the process of (10). Soon you collect your second sample, and later, in the distal convoluted tubule, your third sample. When you read the computer's summary of the third sample, you make the following notes in your register:

1. Virtually no nutrients such as (11) and (12) are left in the filtrate.
2. The pH is acid, 6.0. This is quite a change from the pH of (13) recorded for the newly formed filtrate.
3. There is a much higher concentration of (14) wastes here.
4. There are fewer (15) ions but more (16) ions than there were earlier.
5. The color of the filtrate is yellow, indicating a high relative concentration of the pigment (17).

Gradually you become aware that you are moving along much more quickly. You see that the water level has dropped dramatically and that the stream is turbulent and rushing. As you notice this, you realize that the hormone (18) must have been released recently to cause this water level drop. You take an abrupt right turn and then drop straight downward. You realize that you must be in a(n) (19). Within a few seconds, you land with a splash in what appears to be a large tranquil sea with an ebbing tide toward a dark area at the far shore. You drift toward the dark area, confident that you are in the kidney (20). As you reach and enter the dark tubelike structure, your progress becomes rhythmic—something like being squeezed through a sausage skin. Then you realize that your progress is being regulated by the process of (21). Suddenly, you free-fall and land in the previously stored (22) in the bladder, where the air is very close. Soon the walls of the bladder begin to gyrate, and you realize you are witnessing a(n) (23) reflex. In a moment, you are propelled out of the bladder and through the (24) to exit your host.

CHALLENGING YOURSELF

At the Clinic

1. Mimi is brought to the clinic by her parents. Though she is 23 years old and 5' 5" tall, she weighs only 90 pounds. She insists there is nothing wrong, except for some lower back pains. The clinic staff quickly realizes that the young woman is anorexic. What kidney problems might result from extremely low body weight?
2. A man was admitted to the hospital after being trampled by his horse. He received crushing blows to his lower back on both sides, and is in considerable pain. His chart shows a urine output of 70 ml in the last 24 hours. What is this specific symptom called? What will be required if the effects of his trauma persist?

3. A young woman has come to the clinic with dysuria and frequent urination. What is the most likely diagnosis?

4. The parents of 3-year-old Mitchell bring him to the clinic because his "wee-wee" (urine) is tinged with blood. His face and hands are swollen, and a check of his medical record shows he was diagnosed with strep throat two days before. What is the probable cause of Mitchell's current kidney problem?

5. What single test will reveal the cause of polyuria to be insulin-related? What other hormone-related disorder results in polyuria? What will be the specific gravity and nature of the urine in this second disorder?

6. Four-year-old Eddie is a chronic bed-wetter. The clinical name for his disorder, noted on his chart, is ______________________. What might explain his problem?

7. An elderly diabetic woman with several vascular problems is admitted to the hospital; uremia indicates involvement of the kidneys. What is a likely problem in this case and what is a common cause?

8. Emmanuel is terribly malnourished and in negative nitrogen balance. How does this condition affect his GFR?

9. Felicia, a medical student, arrived late to surgery where she was to observe removal of an extensive abscess from the external surface of a patient's kidney. Felicia was startled to find the surgical team working to reinflate the patient's lung and remove air from the pleural cavity. How could renal surgery lead to such an event?

Stop and Think

1. A typical capillary has a high-pressure (arterial) and a low-pressure (venous) end. How does a glomerular capillary compare to this? What provides the low-pressure end for the glomerulus and how does this arrangement increase the ability of the kidney to filter the blood?

2. What will happen to GFR and urine output if plasma protein production decreases (due to hepatitis, cirrhosis, or dietary protein deficit)?

3. Assume you could peer into cells of the proximal convoluted tubule to examine their organelles. What clues could you use to help determine whether movement of solutes through the apical and basolateral cell membranes is passive or active?

4. What happens to the medullary osmotic gradient when ADH is secreted? How is the gradient reestablished?

5. Why does excess alcohol intake result in "cotton mouth" the next morning?

6. Does caffeine cause production of a more dilute urine, as in diabetes insipidus, or loss of excess solute?

7. What happens to the rate of RBC production in a patient on dialysis with total renal failure? What could be given to the patient to counteract such a problem?

8. What is the effect on the micturition reflex of spinal cord transection above the pelvic splanchnic nerves?

9. In hemodialysis, blood diverted from the body is filtered and then allowed to flow on one side of a semipermeable membrane. A fluid called the dialysate flows along the other side of the membrane. What components must the initial dialysate have or not have to assure that the reabsorption and secretion functions of a normal kidney are mimicked during hemodialysis?

COVERING ALL YOUR BASES

Multiple Choice

Select the best answer or answers from the choices given.

1. A radiologist is examining an X ray of the lumbar region of a patient. Which of the following is (are) indicative of normal positioning of the right kidney?
   A. Slightly higher than the left kidney
   B. More medial than the left kidney
   C. Closer to the inferior vena cava than the left kidney
   D. Anterior to the 12th rib

2. Which of the following encloses both kidney and adrenal gland?
   A. Renal fascia
   B. Perirenal fat capsule
   C. Fibrous capsule
   D. Visceral peritoneum

3. Which structures are inflamed in pyelitis?
   A. Renal pelvis C. Fibrous capsule
   B. Minor calyces D. Ureter

4. Microscopic examination of a section of the kidney shows a thick-walled vessel with renal corpuscles scattered in the tissue on one side of the vessel but not on the other side. What vessel is this?
   A. Interlobar artery C. Cortical radiate vein
   B. Cortical radiate artery D. Arcuate artery

5. Structures that are at least partly composed of simple squamous epithelium include:
   A. collecting tubules C. glomerular capsule
   B. glomerulus D. loop of Henle

6. Which structures are freely permeable to water?
   A. Distal convoluted tubule
   B. Thick segment of ascending limb of the loop of Henle
   C. Descending limb of the loop of Henle
   D. Proximal convoluted tubule

7. Cells with large numbers of mitochondria are found in the:
   A. thick segment of the loop of Henle
   B. proximal convoluted tubule
   C. distal convoluted tubule
   D. collecting tubule

8. Low-pressure capillary beds include:
   A. glomerulus C. peritubular capillaries
   B. vasa recta D. renal plexus

9. Which statement(s) is (are) true concerning the JG apparatus?
   A. The macula densa secretes renin.
   B. The granular cells are part of the renal corpuscle.
   C. The macula densa monitors filtrate concentration.
   D. The granular cells are mechanoreceptors.

10. The barriers in the filtration membrane consist of:
    A. fenestrations, which exclude blood cells
    B. anionic basement membrane, which repels plasma proteins
    C. filtration slits, which are smaller than plasma proteins
    D. a "molecular sieve," which excludes nutrients

11. The GFR would be increased by:
    A. glomerular hydrostatic pressure of 70 mm Hg
    B. glomerular osmotic pressure of 35 mm Hg
    C. sympathetic stimulation of the afferent arteriole
    D. secretion of aldosterone

12. Which stimulates the contractile myogenic mechanism of renal autoregulation?
    A. Increased systemic blood pressure
    B. Low filtrate osmolarity
    C. Secretion of renin
    D. Sympathetic stimulation of afferent arterioles

13. Consequences of hypovolemic shock include:
    A. dysuria C. anuria
    B. polyuria D. pyelitis

14. Sodium deficiency hampers reabsorption of:
    A. glucose C. creatinine
    B. albumin D. water

15. Which substance has a high transport maximum?
    A. Urea C. Glucose
    B. Inulin D. Lactate

16. Facultative water reabsorption mainly occurs in the:
    A. glomerular capsule
    B. distal convoluted tubule
    C. collecting duct
    D. proximal convoluted tubule

17. Effects of aldosterone include:
    A. increase in sodium ion excretion
    B. increase in water retention
    C. increase in potassium ion concentration in the urine
    D. higher blood pressure

18. Which of the following is dependent on tubular secretion?
    A. Clearing penicillin from the blood
    B. Removal of nitrogenous wastes that have been reabsorbed
    C. Removal of excess potassium ions from the blood
    D. Control of blood pH

19. Which is (are) involved in establishing the medullary osmotic gradient?
    A. Movement of water out of the descending limb of the loop of Henle
    B. Movement of water into the ascending limb of the loop of Henle
    C. Active transport of urea into the interstitial space
    D. Impermeability of the collecting tubule to water

20. To calculate renal clearance, one must know:
    A. the concentration of the substance in plasma
    B. the concentration of the substance in urine
    C. the glomerular filtration rate
    D. the rate of urine formation

21. Normal urine constituents include:
    A. uric acid
    B. urochrome
    C. acetone
    D. potassium ions

22. Which of the following are normal values?
    A. Urine output of 1.5 L/day
    B. Specific gravity of 1.5
    C. pH of 6
    D. GFR of 125 ml/hour

23. The ureter:
    A. is continuous with the renal pelvis
    B. is lined by the renal capsule
    C. exhibits peristalsis
    D. is much longer in the male than in the female

24. The urinary bladder:
    A. is lined with transitional epithelium
    B. has a thick, muscular wall
    C. receives the ureteral orifices at its superior aspect
    D. is innervated by the renal plexus

25. Which of the following are controlled voluntarily?
    A. Detrusor muscle
    B. Internal urethral sphincter
    C. External urethral sphincter
    D. Levator ani muscle

26. The main function of transitional epithelium in the ureter is:
    A. protection against kidney stones
    B. secretion of mucus
    C. reabsorption
    D. stretching

27. Jim was standing at a urinal in a crowded public restroom and a long line was forming behind him. He became anxious (sympathetic response) and found he could not micturate no matter how hard he tried. Use logic to deduce Jim's problem.
    A. His internal urethral sphincter was constricted and would not relax.
    B. His external urethral sphincter was constricted and would not relax.
    C. His detrusor muscle was contracting too hard.
    D. He almost certainly had a burst bladder.

28. A major function of the collecting ducts is:
    A. secretion
    B. filtration
    C. concentrating urine
    D. lubrication with mucus

29. What is the glomerulus?
    A. The same as the renal corpuscle
    B. The same as the uriniferous tubule
    C. The same as the nephron
    D. Capillaries

30. Urine passes through the ureters by which mechanism?
    A. Ciliary action
    B. Peristalsis
    C. Gravity alone
    D. Suction

Word Dissection

For each of the following word roots, fill in the literal meaning and give an example, using a word found in this chapter.

Word root	Translation	Example
1. azot		
2. calyx		
3. diure		
4. glom		
5. gon		
6. mictur		
7. nephr		
8. pyel		
9. ren		
10. spado		
11. stroph		
12. trus		

25

Fluid, Electrolyte, and Acid-Base Balance

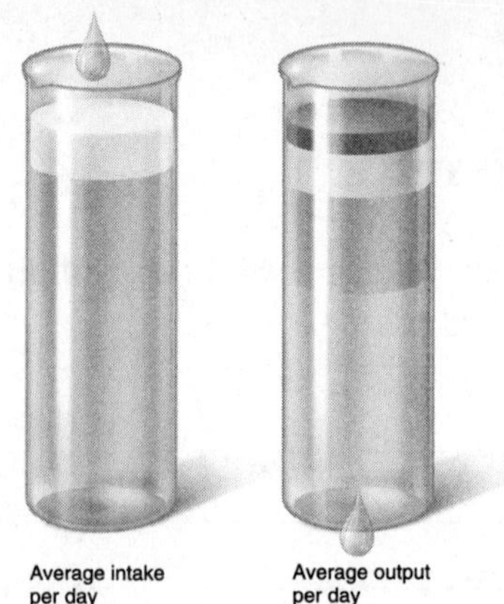

Student Objectives

When you have completed the exercises in this chapter, you will have accomplished the following objectives:

Body Fluids

1. List the factors that determine body water content and describe the effect of each factor.
2. Indicate the relative fluid volume and solute composition of the fluid compartments of the body.
3. Contrast the overall osmotic effects of electrolytes and nonelectrolytes.
4. Describe factors that determine fluid shifts in the body.

Water Balance and ECF Osmolality

5. List the routes by which water enters and leaves the body.
6. Describe feedback mechanisms that regulate water intake and hormonal controls of water output in urine.
7. Explain the importance of obligatory water losses.
8. Describe possible causes and consequences of dehydration, hypotonic hydration, and edema.

Electrolyte Balance

9. Indicate routes of electrolyte entry and loss from the body.
10. Describe the importance of ionic sodium in fluid and electrolyte balance of the body, and indicate its relationship to normal cardiovascular system functioning.
11. Describe mechanisms involved in regulating sodium balance, blood volume, and blood pressure.
12. Explain how potassium, calcium, and anion balances in plasma are regulated.

Acid-Base Balance

13. List important sources of acids in the body.
14. Name the three major chemical buffer systems of the body and describe how they resist pH changes.
15. Describe the influence of the respiratory system on acid-base balance.
16. Describe how the kidneys regulate hydrogen and bicarbonate ion concentrations in the blood.
17. Distinguish between acidosis and alkalosis resulting from respiratory and metabolic factors. Describe the importance of respiratory and renal compensations to acid-base balance.

Scientists believe that life began in a salty sea and that the migration of living organisms from the sea to the land was accomplished by taking the chemistry of the sea into the body. Our trillions of cells are continually bathed in this "sea of life." (How salty is this sea of life? Taste your tears!) The functioning of each cell depends on the chemical composition of its immediate environment, which must remain roughly constant, for only then can each cell maintain its own composition and fluid volume. When the body's chemistry is in balance, the whole body exists in a steady, stable state, which we call health. This secret of life was discovered by the great physiologist Claude Bernard, who applied the term *homeostasis* to describe the constant state of the body's internal environment.

This chapter focuses on the composition of the internal environment and the organs and mechanisms involved in maintaining its homeostasis. Subjects include the composition of the major fluid compartments of the body, the water/electrolyte/acid-base balance of these fluid compartments and their interrelationships, and the related developmental changes occurring throughout the life span.

BUILDING THE FRAMEWORK

Body Fluids

1. Circle the term that does not belong in each of the following groupings. (*Note:* ECF = extracellular fluid compartment; ICF = intracellular fluid compartment.)

1. Female adult   Male adult   About 50% water   Less skeletal muscle

2. Obese adult   Lean adult   Less body water   More adipose tissue

3. Claude Bernard's "internal environment"   ICF   About two-thirds of total body water   All cells

4. ECF   Interstitial fluid   ICF   Plasma   CSF

5. Electric charge   Nonelectrolyte   Covalent bonding   No electric charge

2. Determine if the following descriptions refer to electrolytes (E) or to nonelectrolytes (N).

_____ 1. Lipids, monosaccharides, and neutral fats

_____ 2. Have greater osmotic power at equal concentrations

_____ 3. The most numerous solutes in the body's fluid compartments

_____ 4. Salts, acids, and bases

_____ 5. Most of the *mass* of dissolved solutes in the body's fluid compartments

_____ 6. Each molecule dissociates into two or more ions

3. Using the data in the table below, calculate how many milliequivalents per liter (mEq/L) of each ion are normally present in plasma. Insert your answers in the spaces in the table. Then, referring to the table, answer the questions that follow.

	Bicarbonate HCO_3^-	Calcium Ca^{2+}	Chloride Cl^-	Magnesium Mg^{2+}	Hydrogen phosphate HPO_4^{2-}	Potassium K^+	Sodium Na^+
Average concentration (mg/L) in plasma	1647	90	3798	42	39	164	3266
Atomic weight (daltons)	61	40	35	24	96	39	23
mEq/L in plasma							

1. Electrolyte concentrations in the two extracellular compartments are similar, except for the high concentration of which *organic* molecules in plasma?

2. Name the chief cation in the extracellular compartments.

3. Name the two major anions in the extracellular compartments.

4. Indicate which electrolytes are in higher concentrations (use ↑) and which are in lower concentrations (use ↓) in intracellular fluids as compared with extracellular fluids.

____ HCO_3^- ____ Cl^- ____ HPO_4^{2-} ____ Na^+ ____ Ca^{2+} ____ Mg^{2+} ____ K^+

4. Figure 25.1 illustrates the three major fluid compartments of the body. Arrows indicate direction of fluid flow. First, select three different colors and color the coding circles and the fluid compartments on the figure. Then, referring to Figure 25.1, respond to the statements that follow. If a statement is true, write T in the answer blank. If a statement is false, change the underlined word(s) and write the correct word(s) in the answer blank.

○ Interstitial fluid ○ Intracellular fluid ○ Plasma

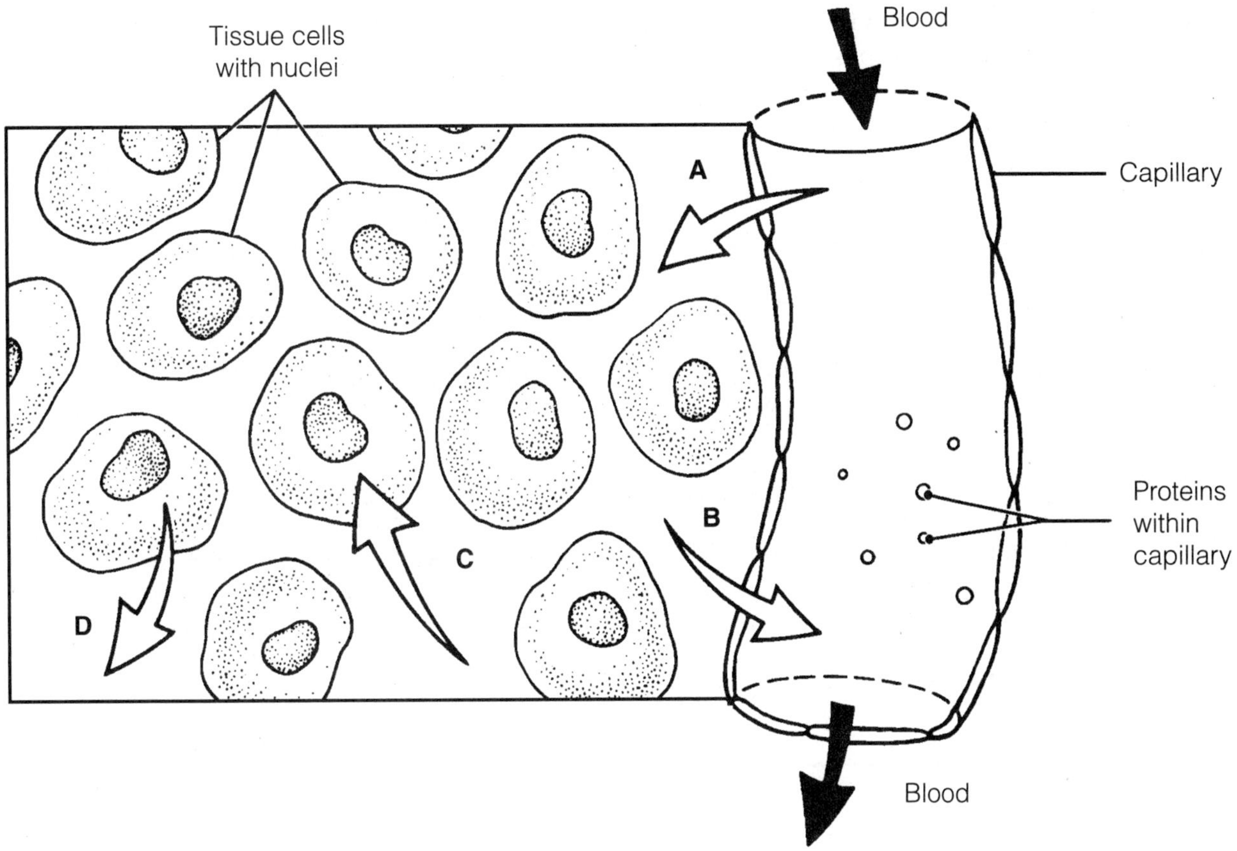

Tissue cells with nuclei
Blood
Capillary
A
B
C
D
Proteins within capillary
Blood

Figure 25.1

____________ 1. Exchanges between plasma and interstitial fluid compartments take place across the capillary membranes.

____________ 2. The fluid flow indicated by arrow A is driven by active transport.

____________ 3. The fluid flow indicated by arrow B is driven by osmotic pressure.

____________ 4. The excess of fluid flow at arrow A over that at arrow B normally enters the tissue cells.

____________ 5. Exchanges between the interstitial and intracellular fluid compartments occur across capillary membranes.

__________ 6. Interstitial fluid serves as the link between the body's external and internal environments.

__________ 7. The osmolarities of all body fluids are nearly always unequal.

__________ 8. If the osmolarity of the extracellular fluid is increased, the fluid flow at arrow C will occur.

Water Balance and ECF Osmolality

1. Respond to the following questions about body water by writing your answers in the answer blanks.

1. Name three sources of body water and specify which source accounts for the bulk of body water.

____

____

2. Name four routes by which water is lost from the body and specify which route accounts for the greatest water loss.

____

____

____

3. Explain the term *insensible water loss* and give two examples.

____

____

2. Complete the following statements about the regulation of water intake and output by writing the missing terms in the answer blanks.

A dry mouth excites the thirst center, located in the __(1)__. __(2)__ in the volume and/or __(3)__ in the osmolality of plasma, as well as __(4)__ and rising blood levels of __(5)__, stimulate the __(6)__ of the thirst center. After water intake, moistening of the mouth and throat mucosae and activation of __(7)__ receptors in the __(8)__ send signals that inhibit the thirst center (negative feedback). This mechanism rapidly satisfies thirst and prevents excessive __(9)__ of body fluids.

1. __________
2. __________
3. __________
4. __________
5. __________
6. __________
7. __________
8. __________
9. __________

__________ 10.
__________ 11.
__________ 12.
__________ 13.
__________ 14.
__________ 15.

After thirst is quenched, a(n) (10) in plasma osmolality causes the kidneys to begin eliminating excess water. Inhibition of the release of the hormone (11) leads to excretion of large volumes of (12) urine. Regulation of water balance by kidney activity is closely related to (13) ion concentration. Uncontrollable water losses in urine are called (14) water losses. To maintain blood homeostasis, the kidneys must excrete a minimum of (15) ml of water per day to get rid of about 1000 mOsm of water-soluble solutes.

3. Using the key choices, select the disorder of water balance that applies to each of the following descriptions and write the letter of your answer in the answer blank.

Key Choices

A. Dehydration B. Edema C. Hypotonic hydration

__________ 1. Abnormal increase of interstitial fluid

__________ 2. Abnormal increase of intracellular fluid

__________ 3. May lead to decrease of blood volume and falling blood pressure

__________ 4. Signs are thirst, sticky oral mucosa, dry skin, and decreased urine volume

__________ 5. If not remedied, leads to cerebral edema, coma, and death

__________ 6. Usually follows severe burns or prolonged vomiting

__________ 7. Extracellular fluid compartment loses water

__________ 8. Markedly decreased electrolyte concentrations in the ECF, especially Na^+

__________ 9. May be caused by renal insufficiency

__________ 10. Causes may be increased capillary hydrostatic pressure and hypoproteinemia

Electrolyte Balance

1. Complete the following statements concerning electrolyte balance by writing the missing terms in the answer blanks.

Electrolyte balance usually refers to the balance of __(1)__, which enter the body in __(2)__ and leave the body in three ways: __(3)__, __(4)__, and __(5)__. Regulating the electrolyte balance of the body is one of the most important functions of the __(6)__.

The most important and most abundant cation in the ECF is __(7)__. It is the only cation in the ECF that exerts significant __(8)__ pressure. As a consequence, it controls ECF __(9)__ and the distribution of __(10)__ in the body.

1. ____________________
2. ____________________
3. ____________________
4. ____________________
5. ____________________
6. ____________________
7. ____________________
8. ____________________
9. ____________________
10. ____________________

2. Figure 25.2 illustrates one mechanism that controls the release of aldosterone. Complete the flowchart by responding to the questions in the boxes or filling in the missing terms. Write your answers in the answer blanks in each box. Then complete the statements that follow.

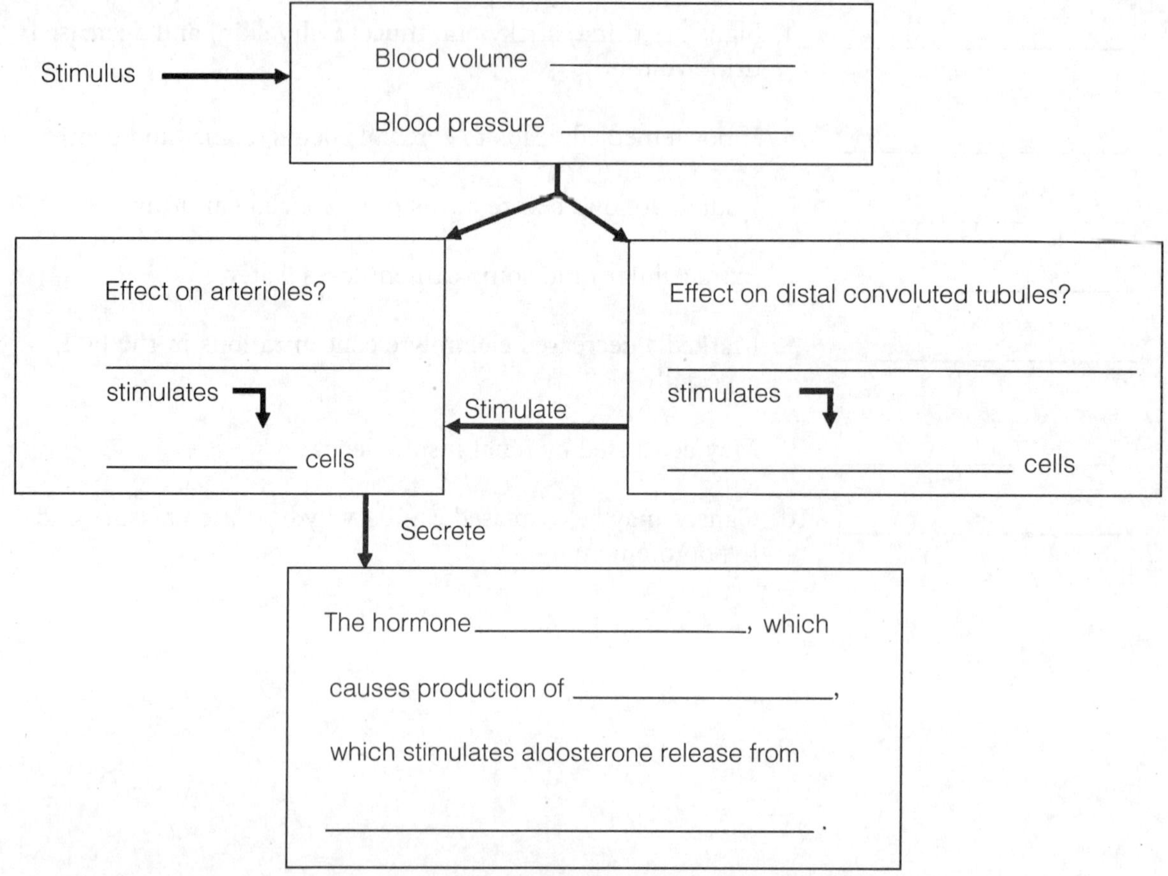

Figure 25.2

______ 1.
______ 2.
______ 3.
______ 4.
______ 5.
______ 6.
______ 7.
______ 8.
______ 9.
______ 10.
______ 11.
______ 12.
______ 13.
______ 14.

The juxtaglomerular apparatus is made up of the __(1)__ and __(2)__ cells. Aldosterone promotes more __(3)__ of sodium and __(4)__. Consequently, blood volume and blood pressure __(5)__. However, aldosterone acts slowly, and the length of time for the effects of aldosterone to occur is measured in __(6)__.

A faster reversal of low ECF volume and low blood pressure is effected by the __(7)__ system. __(8)__, located in the heart, carotid arteries, and aorta, sense the low blood pressure and alert the __(9)__, which causes sympathetic impulses to be sent to the __(10)__ of the body. These vessels __(11)__, increasing the peripheral resistance and, thus, the blood pressure. Sympathetic fibers also stimulate the __(12)__ of the kidneys to release __(13)__, which promotes an increase in blood pressure by promoting a(n) __(14)__ in blood volume.

3. From the following electrolytes, Ca^{2+}, Cl^-, K^+, HCO_3^-, and Na^+, select the ions that apply to each of the following descriptions. Insert the symbols of the ions in the answer blanks.

______ 1. Ion excretion that is accelerated by aldosterone

______ 2. Balance controlled primarily by parathyroid hormone

______ 3. During acidosis, is excreted in place of bicarbonate ion

______ 4. Renal effect on ion is chiefly excretory; little retention ability

______ 5. The major ionic determinant of body water flows

______ 6. Ion that accompanies Na^+ reabsorption under normal circumstances

______ 7. For cation balance, this ion moves across cell membrane in opposition to H^+

______ 8. Intestinal absorption of ion enhanced by active form of vitamin D

______ 9. Relative ECF/ICF concentrations of ion directly affect resting membrane potential

______ 10. Renal tubules conserve this ion while excreting phosphate ion

4. Circle the term that does not belong in each of the following groupings.

1. ↓ H^+ secretion ↓ Blood pH ↑ H^+ secretion ↓ K^+ secretion
2. ↑ Water output ↓ Plasma volume ↑ ADH ↓ ADH
3. Atrial natriuretic peptide Vasodilation ↑ Na^+ excretion ↑ Na^+ retention
4. Aldosterone ↑ Na^+ reabsorption ↑ K^+ reabsorption ↑ BP
5. ADH ↓ BP ↑ Blood volume ↑ Water reabsorption
6. ↑ ADH Dilute urine Aquaporins inserted in principal cells Dehydration

Acid-Base Balance

1. Match the pH values in Column B with the conditions described in Column A.

	Column A	Column B
______	1. Normal pH of arterial blood	A. pH < 7.00
______	2. Normal pH of intracellular fluid	B. pH = 7.00
______	3. Normal pH of venous blood and interstitial fluid	C. pH < 7.35
______	4. Physiological alkalosis (arterial blood)	D. pH = 7.35
______	5. Physiological acidosis (arterial blood)	E. pH = 7.40
______	6. Chemical neutrality; neither acidic nor basic	F. pH > 7.45
______	7. Chemical acidity	

2. Indicate the chief sources of the following acids found in the body.

1. Lactic acid ______________________________
2. Carbonic acid ______________________________
3. Hydrochloric acid ______________________________
4. Fatty acids and ketones ______________________________
5. Phosphoric acid ______________________________

____

6. Citric acid and acetic acid ______________________________

3. By what three methods is H^+ concentration in body fluids regulated? Give the approximate time for each method to respond to pH changes.

1. ______________________________

2. ______________________________

3. ______________________________

4. Using the key choices, characterize the activity of the following chemical species when they are dissolved in water. Write the appropriate answers in the answer blanks.

Key Choices

A. Neutral   C. Strong base   E. Weak base

B. Strong acid   D. Weak acid

______________	1. H_2CO_3	______________	6. NaCl
______________	2. NH_3	______________	7. HCl
______________	3. H_2SO_4	______________	8. HAc (Acetic)
______________	4. Na_2HPO_4	______________	9. NaOH
______________	5. NaH_2PO_4	______________	10. $NaHCO_3$

5. Describe the following substances with respect to their activity related to H^+ ions.

1. Weak acid ______________________________

2. Strong acid ______________________________

3. Weak base ______________________________

4. Strong base ______________________________

5. Chemical buffer system ______________________________

____

6. Figure 25.3 is a cutaway drawing of a kidney tubule and peritubular capillary. This highly simplified drawing illustrates renal regulation of H^+ secretion, Na^+ and HCO_3^- conservation, and control of ECF pH. Select three colors and color the coding circles. Then, complete the reactions in the drawing by writing in the answer blanks the appropriate chemical symbols in color. Color the rest of the figure as you wish. Next, referring to Figure 25.3, complete the statements that follow by writing your responses in the answer blanks.

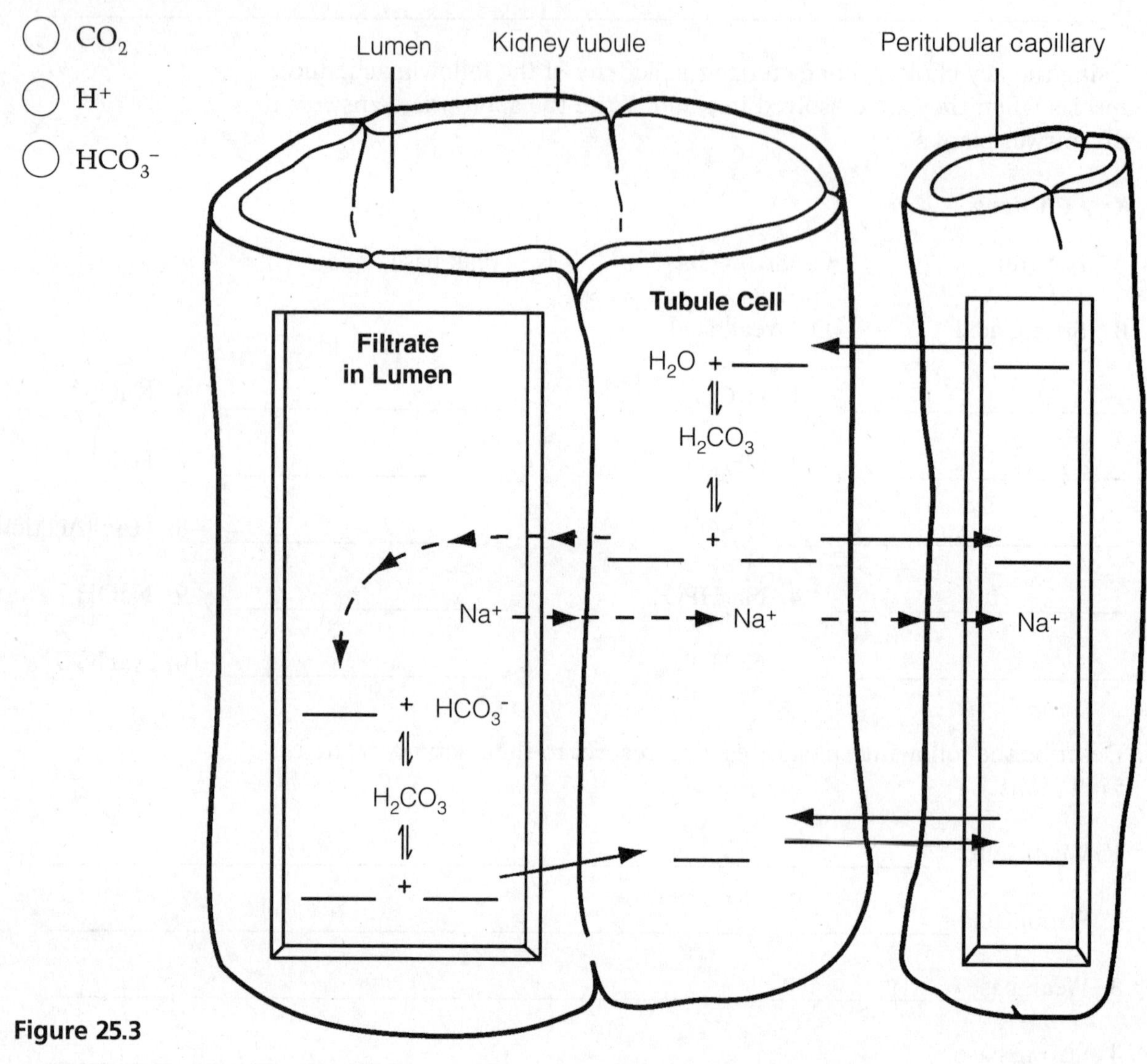

Figure 25.3

______________ 1.
______________ 2.
______________ 3.
______________ 4.
______________ 5.

When the pH of the ECF falls, the concentration of CO_2 in the peritubular capillary (1). CO_2 diffuses into the tubule cell, shifting the equilibrium to form (2), which quickly dissociates, liberating H^+ and HCO_3^-. Tubule cells cannot reabsorb HCO_3^- from the (3), but they can move HCO_3^- formed (4) into the (5), thus conserving HCO_3^-.

6. ____________
7. ____________
8. ____________
9. ____________

At the same time, the H^+ liberated in the tubule cell is secreted into the __(6)__, where it combines with __(7)__ in the filtrate to form __(8)__, generating both __(9)__. This excess H^+ is therefore tied up in the form of __(10)__. For every H^+ secreted into the lumen of the kidney tubule by the kidney cell, __(11)__ Na^+ is reabsorbed and moved back into the __(12)__. The tubule cells secrete H^+ and reabsorb Na^+ by __(13)__ transport.

10. ____________
11. ____________
12. ____________

When all of the filtered bicarbonate has been reclaimed, additional secreted H^+ begins to be __(14)__, but first they must be buffered. The two most important urine buffers are the __(15)__ and the __(16)__ buffer systems.

13. ____________ 15. ____________

14. ____________ 16. ____________

7. Circle the term that does not belong in each of the following groupings. (*Note:* NH_3 = ammonia; NH_4^+ = ammonium ion.)

1. Large amounts of H^+ excreted H^+ in H_2O Phosphate buffer system   Glutamine-NH_4^+ buffer system
2. Na_2HPO_4 H^+ excretion Na_2CO_3 NaH_2PO_4
3. ↑ H^+ in filtrate   ↓ H^+ in filtrate   ↑ $NaHCO_3$ reabsorbed   ↑ NH_4^+ secretion
4. ↓ Urine pH   ↑ H^+ in urine   ↑ HCO_3^- in urine   ↑ Ketones
5. Carbonic anhydrase   Plasma   Red blood cells   Rapid CO_2 loading
6. ↓ Plasma H^+ Respiratory centers in medulla   Deep, rapid breathing   ↑ CO_2 retention
7. Acidosis   ↑ Carbonic acid   ↓ pH   ↑ pH
8. Glutamine oxidation NH^+ in urine HCO_3^- in urine HCO_3^- reabsorbed

8. Respond to the following statements concerning buffer systems of the body. If a statement is true, write T in the answer blank. If a statement is false, change the underlined word(s) and write the correct word(s) in the answer blank.

________________ 1. Weak acids are important in chemical buffer systems because they dissociate <u>completely</u>.

________________ 2. The bicarbonate, phosphate, and protein buffer systems act <u>independently</u>.

________________ 3. If NaOH is added to the bicarbonate buffer system, the equilibrium will shift to the <u>right</u>.

________________ 4. When NaOH is added to the bicarbonate buffer system, the chemical species H_2O and <u>Na_2CO_3</u> will be formed.

________________ 5. Addition of HCl to the phosphate buffer system causes an equilibrium shift so that the weak base <u>HPO_4^{2-}</u> is produced.

________________ 6. The most abundant and powerful buffers in the ICF and plasma are the <u>phosphates</u>.

________________ 7. The group of atoms on a protein molecule that acts as a base is <u>R-COOH</u>.

________________ 8. The single most important blood buffer is the <u>phosphate</u> buffer system.

________________ 9. The ultimate acid-base regulatory organs are the <u>lungs</u>.

9. Match the key choices with the acid-base balance abnormalities described below. (Note: P_{CO_2} = partial pressure of CO_2.)

Key Choices

A. Metabolic acidosis

B. Metabolic alkalosis

C. Normal range

D. Respiratory acidosis

E. Respiratory alkalosis

________________ 1. Indicated by plasma HCO_3^- levels above the normal range

________________ 2. P_{CO_2} = 35–45 mm Hg; pH = 7.35–7.45

________________ 3. A common result of hyperventilation

________________ 4. Indicated by plasma P_{CO_2} levels above the normal range

________________ 5. A common cause is excess HCO_3^- loss resulting from prolonged diarrhea

________________ 6. Imbalance where HCO_3^- levels are high; compensated by slow, shallow breathing (therefore high P_{CO_2})

CHALLENGING YOURSELF

At the Clinic

1. Jack Tyler exhibits excessive thirst and polyuria, with hypernatremia. Will you check his pituitary or adrenal glands? Do you suspect hyposecretion or hypersecretion, and of what hormone? What is the name of his condition?
2. A patient is diagnosed with bronchogenic carcinoma complicated by ectopic secretion of ADH. Is he more likely to suffer from dehydration or hypotonic hydration? Will he be hyponatremic or hypernatremic?
3. A woman with Cushing's disease has regular checkups at the clinic. Because of the associated hyperaldosteronism, will she more likely suffer from dehydration, hypotonic hydration, or edema?
4. If a tumor of the glucocorticoid-secreting cells of the adrenal cortex crowds out the cells that produce aldosterone, what is the likely effect on urine composition and volume?
5. Ben Ironfield, an elderly patient, suffers from chronic renal hypertension. What are its effects on his urine output?
6. Ms. Schmid comes to the clinic suffering from extreme premenstrual syndrome; her estrogen level is high. Will a diuretic relieve some of her symptoms?
7. A hypertensive patient named George Handel is given potassium replacement to counteract the K^+ depletion caused by his diuretic. He now exhibits hyperventilation. His blood pH is 7.33, his P_{CO_2} is 33 mm Hg, and his blood bicarbonate level is 20 mEq/L. Clinically, what name is given to his state of acid-base balance? What system is compensating? What is the connection with potassium and how does it relate to his blood pH?
8. A young boy has been diagnosed with a hypersecretory tumor of a parathyroid gland. What condition will result?
9. A particularly virulent intestinal virus has triggered repeated vomiting and severe diarrhea in Christine. After 12 hours of feeling helpless, her parents bring her to the clinic. Along with rehydration, what electrolytes must be replaced? What pH adjustment will be necessary if Christine begins to hyperventilate? Shortly after Christine has been diagnosed and treated, her infant cousin starts to exhibit the same symptoms. Three hours later, the baby's anterior fontanel and eyes appear sunken. Should treatment be sought, or is it safe to wait and see if symptoms subside without treatment?

Stop and Think

1. Saline solution is used to reverse hypotonic hydration. Are body cell membranes permeable to saline? Explain your response.
2. What are the cardiac effects of dehydration?
3. High levels of stress stimulate secretion of glucocorticoids. What effect will this have on blood volume?
4. When protein synthesis is increased (such as during postsurgical healing), is hypokalemia or hyperkalemia more common?
5. Red blood cells break down in stored blood. What ion imbalance will be seen following rapid transfusion of stored blood?

6. Much of the calcium carried in the blood is bound to albumin, and calcium and hydrogen ions exchange places when albumin buffers pH changes. That is, as H^+ is bound, Ca^{2+} is liberated. Will alkalosis be correlated with hypocalcemia or hypercalcemia? Will hypoalbuminemia be correlated with hypo- or hypercalcemia?

7. Calcium phosphate is not freely ionizable at high concentrations. How does this fact relate to the increase in phosphate ion excretion when calcium ion is absorbed at the kidneys?

8. Crush injuries cause the release of intracellular K^+ into the ECF. What are some possible consequences of this sequence?

COVERING ALL YOUR BASES

Multiple Choice

Select the best answer or answers from the choices given.

1. Which is a normal value for percent of body weight that is water for a middle-aged man?
   A. 73%
   B. 50%
   C. 45%
   D. 60%

2. The smallest fluid compartment is the:
   A. ICF
   B. ECF
   C. plasma
   D. IF

3. Which of the following are electrolytes?
   A. Glucose
   B. Lactic acid
   C. Urea
   D. Bicarbonate

4. One mEq/L of calcium chloride equals:
   A. 1 mEq/L of sodium chloride
   B. 1 mEq/L of magnesium chloride
   C. 1 mOsm
   D. 0.5 mOsm

5. The chloride ion concentration in plasma:
   A. equals that in ICF
   B. equals that in IF
   C. equals the sodium ion concentration in plasma
   D. is less than that in IF

6. In movement between IF and ICF:
   A. water flow is bidirectional
   B. nutrient flow is unidirectional
   C. ion flow is selectively permitted
   D. ion fluxes are not permitted

7. Insensible water loss includes:
   A. metabolic water
   B. vaporization from lungs
   C. water of oxidation
   D. evaporation from skin

8. The thirst mechanism involves:
   A. crenation of osmoreceptors
   B. hypoactivity of salivary glands
   C. damping of the mechanism when osmotic balance is reestablished
   D. stretch receptors in the stomach

9. Which of the following will result in hyponatremia?
   A. SIADH
   B. Hyperaldosteronism
   C. Renal compensation for alkalosis
   D. Excessive water ingestion

10. Disorders that result in edema include:
    A. kwashiorkor
    B. congestive heart failure
    C. systemic inflammatory response
    D. acute glomerulonephritis

11. Sodium retention is "uncoupled" from water retention by the action of which hormone?
    A. Aldosterone
    B. Atrial natriuretic peptide
    C. Antidiuretic hormone
    D. Estrogen

12. People with Cushing's disease:
    A. retain sodium
    B. hypersecrete aldosterone
    C. suffer from potassium deficiency
    D. are hypertensive

13. Inhibition of ADH:
    A. is caused by alcohol consumption
    B. results in edema
    C. is caused by diarrhea
    D. is triggered by hyponatremia

14. Hyperkalemia:
    A. triggers secretion of aldosterone
    B. may result from severe alcoholism
    C. disturbs acid-base balance
    D. results from widespread tissue injury

15. Renal tubular secretion of potassium is:
    A. obligatory
    B. increased by aldosterone
    C. balanced by tubular reabsorption
    D. increased in alkalosis

16. Resorption of calcium from the skeleton:
    A. is increased by calcitonin
    B. is paralleled by resorption of phosphate
    C. results in transiently increased buffering capacity of the blood
    D. is enhanced by vitamin D

17. Chloride ion reabsorption:
    A. exactly parallels sodium ion reabsorption
    B. fluctuates according to blood pH
    C. increases during acidosis
    D. is controlled directly by aldosterone

18. Of the named fluid compartments, which has the lowest pH?
    A. Arterial blood C. Intracellular fluid
    B. Venous blood D. Interstitial fluid

19. Which buffer system(s) is (are) not important urine buffers?
    A. Phosphate C. Protein
    B. Ammonium D. Bicarbonate

20. Which of the following is a volatile acid?
    A. Carbonic C. Lactic
    B. Hydrochloric D. Phosphoric

21. Respiratory acidosis occurs in:
    A. asthma C. barbiturate overdose
    B. emphysema D. cystic fibrosis

22. Only the kidneys can remove excesses of:
    A. phosphoric acid C. keto acids
    B. carbonic acid D. uric acid

23. Regulation of acid-base balance by the kidneys involves:
    A. activity of carbonic anhydrase
    B. buffering by deamination
    C. secretion of hydrogen ions into the urine
    D. pH-dependent selective reabsorption of chloride or bicarbonate from the filtrate

24. Given the blood values pH = 7.31, P_{CO_2} = 63 mm Hg, and HCO_3^- = 30 mEq/L, it would be accurate to conclude that:
    A. the patient exhibits respiratory acidosis
    B. the patient exhibits metabolic acidosis
    C. renal compensation is occurring
    D. respiratory compensation is occurring

25. Bicarbonate ion:
    A. is usually conserved
    B. is usually secreted
    C. is unable to penetrate the luminal border of the tubule cells
    D. is reabsorbed into the blood as H^+ is secreted

26. In the carbonic acid-bicarbonate buffer system, strong acids are buffered by:
    A. carbonic acid
    B. water
    C. bicarbonate ion
    D. the salt of the strong acid

27. Relative to generation of new bicarbonate ions that can be absorbed into the blood, in one mechanism:
    A. NH_4^+ is excreted in urine
    B. two HCO_3^- are generated per glutamine metabolized in the PCT cells
    C. type B intercalated cells are involved
    D. type A intercalated cells are involved

Word Dissection

For each of the following word roots, fill in the literal meaning and give an example, using a word found in this chapter.

Word root	Translation	Example
1. kal		
2. natri		
3. osmol		

26

The Reproductive System

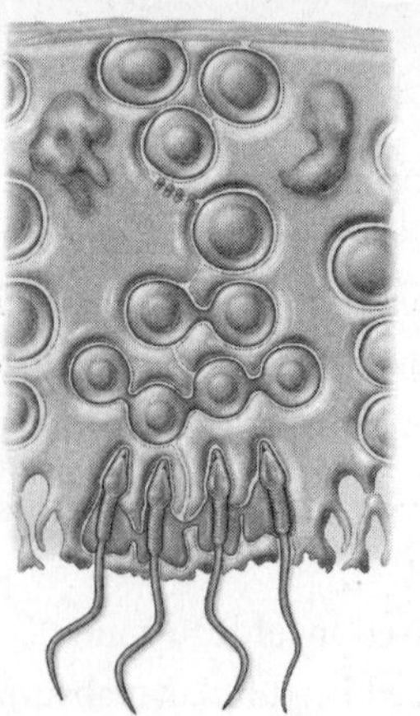

Student Objectives

When you have completed the exercises in this chapter, you will have accomplished the following objectives:

Anatomy of the Male Reproductive System

1. Describe the structure and function of the testes, and explain the importance of their location in the scrotum.
2. Describe the structure of the penis and indicate its role in the reproductive process.
3. Describe the location, structure, and function of the accessory reproductive organs of the male.
4. Discuss the sources and functions of semen.

Physiology of the Male Reproductive System

5. Describe the phases of the male sexual response.
6. Define meiosis. Compare and contrast it to mitosis.
7. Outline events of spermatogenesis.
8. Discuss hormonal regulation of testicular function and the physiological effects of testosterone on male reproductive anatomy.

Anatomy of the Female Reproductive System

9. Describe the location, structure, and function of the ovaries.
10. Describe the location, structure, and function of each of the organs of the female reproductive duct system.
11. Describe the anatomy of the female external genitalia.
12. Discuss the structure and function of the mammary glands.

Physiology of the Female Reproductive System

13. Describe the process of oogenesis and compare it to spermatogenesis.
14. Describe the ovarian cycle phases, and relate them to events of oogenesis.
15. Describe the regulation of the ovarian and uterine cycles.
16. Discuss the physiological effects of estrogens and progesterone.
17. Describe the phases of the female sexual response.

Sexually Transmitted Infections

18. Describe the infectious agents and modes of transmission of gonorrhea, syphilis, chlamydia, trichomoniasis, genital warts, and genital herpes.

The biological function of the reproductive system is to provide the means for producing offspring. The essential organs are those producing the germ cells (testes in males and ovaries in females). The male manufactures sperm and delivers them to the female's reproductive tract; the female, in turn, produces eggs. If the time is suitable, the fusion of egg and sperm produces a fertilized egg, which is the first cell of the new individual. Once fertilization has occurred, the female uterus protects and nurtures the developing embryo.

In this chapter, student activities test understanding of the structures of the male and female reproductive systems, germ cell formation, and the menstrual cycle.

BUILDING THE FRAMEWORK

Anatomy of the Male Reproductive System

1. The location of the testes in the scrotum is essential to provide the ideal temperature for sperm formation, which is approximately 3°C lower than that in the abdominal cavity.

1. How do the scrotal muscles (dartos and cremaster) help maintain temperature homeostasis of the testes?

____

____

____

2. What is the location and presumed function of the myoid cells of the testes?

Location: _______________

Function: _______________

3. How does the pampiniform venous plexus surrounding the testicular artery help maintain temperature homeostasis of the testes?

____

____

2. Figure 26.1 is a sagittal view of the male reproductive structures. First, identify the following organs in the figure by writing each term at the appropriate leader line.

Bulbourethral gland	Glans penis	Prepuce	Spongy urethra
Corpus cavernosum	Ejaculatory duct	Prostate	Testis
Corpus spongiosum	Epididymis	Scrotum	
Ductus deferens	Bulb of penis	Seminal vesicle	

Next, select different colors for the structures that correspond to the following descriptions and color the coding circles and the structures on the figure.

○ Spongy tissue that is engorged with blood during erection

○ Portion of the duct system that also serves the urinary system

○ Structure that provides the ideal temperature conditions for sperm formation

○ Structure cut or cauterized during a vasectomy

○ Structure removed in circumcision

○ Gland whose secretion contains sugar to nourish sperm

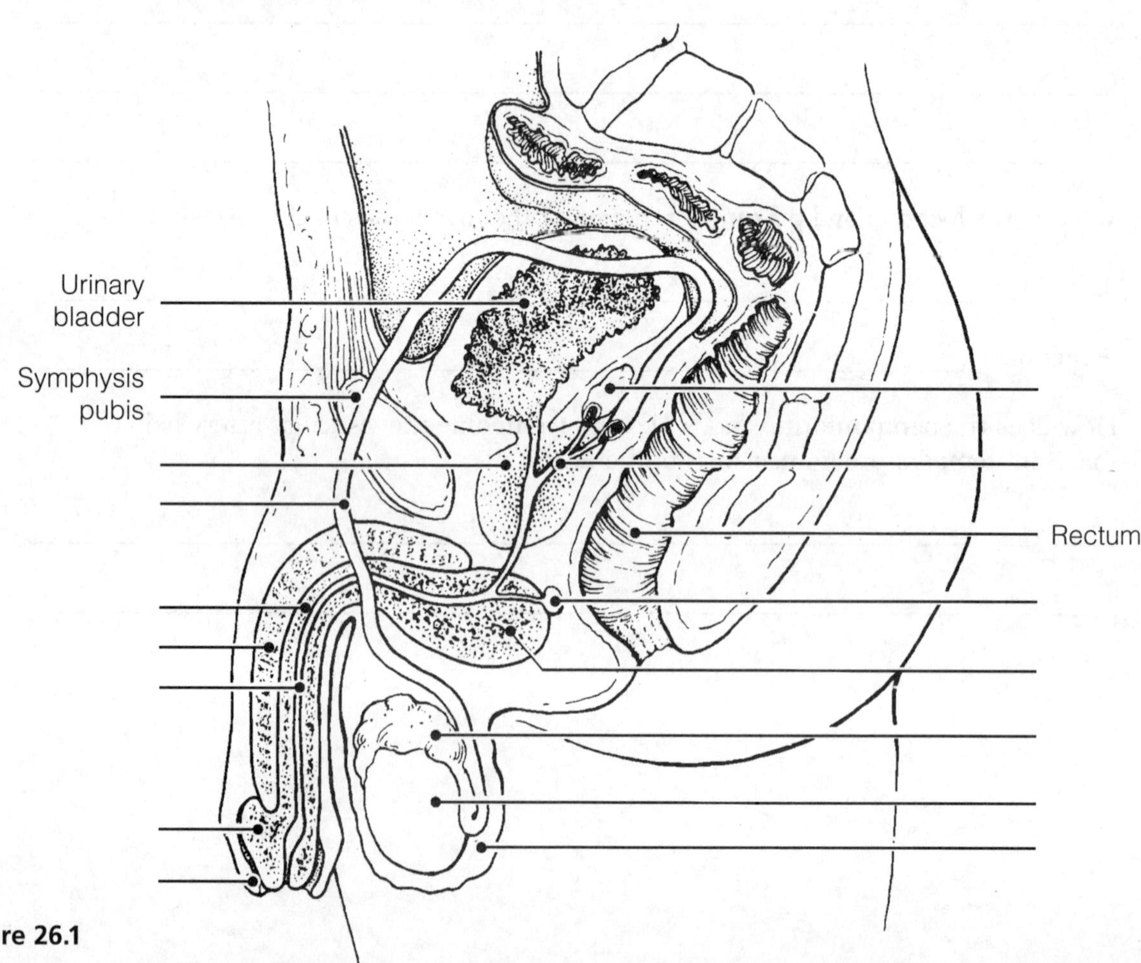

Figure 26.1

3. Using the key choices, match each term with the appropriate anatomical descriptions. Insert the correct answers in the answer blanks.

Key Choices

A. Bulbourethral glands	E. Penis	I. Scrotum
B. Epididymis	F. Prepuce	J. Spermatic cord
C. Ductus deferens	G. Prostate	K. Testes
D. Glans penis	H. Seminal vesicles	L. Urethra

________________ 1. Organ that delivers semen to the female reproductive tract

________________ 2. Site of sperm and testosterone production

________________ 3. Passageway for conveying sperm from the epididymis to the ejaculatory duct

________________ 4. Conveys both sperm and urine down the length of the penis

________________ ________________ 5. Organs that contribute to the formation of semen

________________ ________________

________________ 6. External skin sac that houses the testes

________________ 7. Tubular storage site for sperm; hugs the lateral aspect of the testes

________________ 8. Cuff of skin encircling the glans penis

________________ 9. Surrounds the urethra at the base of the bladder; produces a milky, slightly acid fluid

________________ 10. Produces over half of the seminal fluid

________________ 11. Empties a lubricating mucus into the urethra

________________ 12. Connective tissue sheath enclosing the ductus deferens, blood vessels, and nerves

4. Using the following terms in their proper order, trace the pathway of sperm from the testis to the urethra: rete testis, epididymis, seminiferous tubule, ductus deferens.

____________ → ____________ → ____________ → ____________

5. Figure 26.2 is a longitudinal section of a testis. First, select different colors for the structures that correspond to the following descriptions. Then, color the coding circles and color and label the structures on the figure that have leader lines. Complete the labeling of the figure by adding the following terms: lobule, rete testis, and septum.

   ◯ Site(s) of spermatogenesis

   ◯ Tubular structure in which sperm mature and become motile

   ◯ Fibrous coat protecting the testis

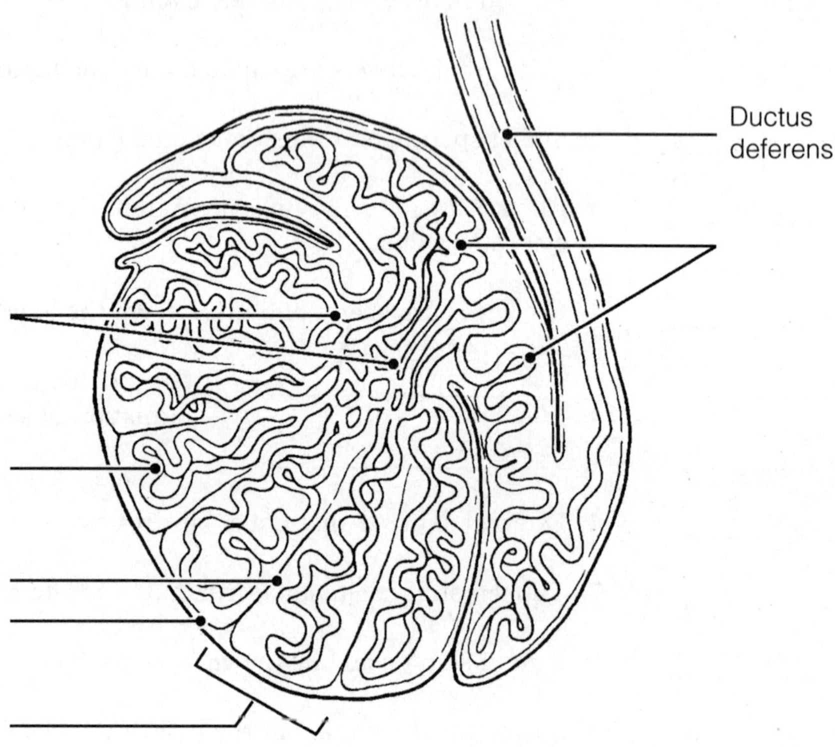

Figure 26.2

Physiology of the Male Reproductive System

1. The following statements refer to events that occur during cellular division. Using the key choices, indicate in which type of nuclear division each event occurs by writing the correct answers in the answer blanks.

Key Choices

A. Mitosis B. Meiosis C. Both mitosis and meiosis

________________ 1. Final product is two daughter cells, each with 46 chromosomes

________________ 2. Final product is four daughter cells, each with 23 chromosomes

________________ 3. Involves the phases prophase, metaphase, anaphase, and telophase

_______________ 4. Occurs in all body tissues

_______________ 5. Occurs only in the gonads

_______________ 6. Increases the cell number for growth and repair

_______________ 7. Daughter cells have the same number and types of chromosomes as the mother cell

_______________ 8. Daughter cells are different from the mother cell in their chromosomal makeup

_______________ 9. Chromosomes are replicated before the division process begins

_______________ 10. Provides cells for the reproduction of offspring

_______________ 11. Consists of two consecutive divisions of the nucleus; chromosomes are not replicated before the second division

_______________ 12. Involves synapsis and crossing over of the homologous chromosomes

2. Figure 26.3 illustrates a single sperm. On the figure, bracket and label the head and the midpiece, and circle and label the tail. Then, select different colors for the structures described below. Color the coding circles and color and label the structures, using correct terminology, on the figure.

○ The DNA-containing area

○ The enzyme-containing sac that aids sperm penetration of the egg

○ Metabolically active organelles that provide ATP to energize sperm movement

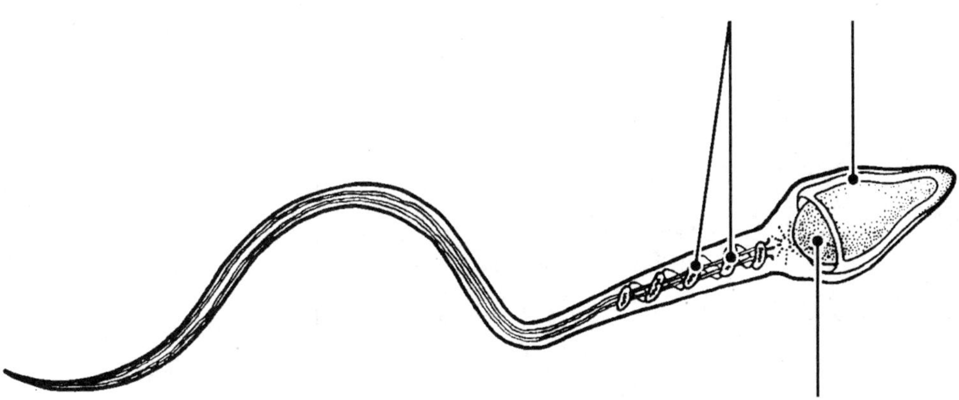

Figure 26.3

3. This exercise considers the process of sperm production in the testis. Figure 26.4 represents a portion of a cross-sectional view of a seminiferous tubule in which spermatogenesis is occurring. First, using the key choices, select the terms that match the following descriptions.

Key Choices

A. Follicle-stimulating hormone

○ B. Primary spermatocyte

○ C. Secondary spermatocyte

○ D. Spermatogonium

○ E. Sperm

○ F. Spermatid

G. Testosterone

_____ 1. Primitive stem cell

_____ 2. Contains 23 chromosomes

_____ 3. Product of meiosis I

_____ 4. Product of meiosis II

_____ 5. Functional motile gamete

_____ 6. Hormones necessary for sperm production

Then, select different colors for the cell types with coding circles listed in the key choices and color them on the figure. Label all the structures with leader lines on the figure. In addition, label and color the cells that produce testosterone.

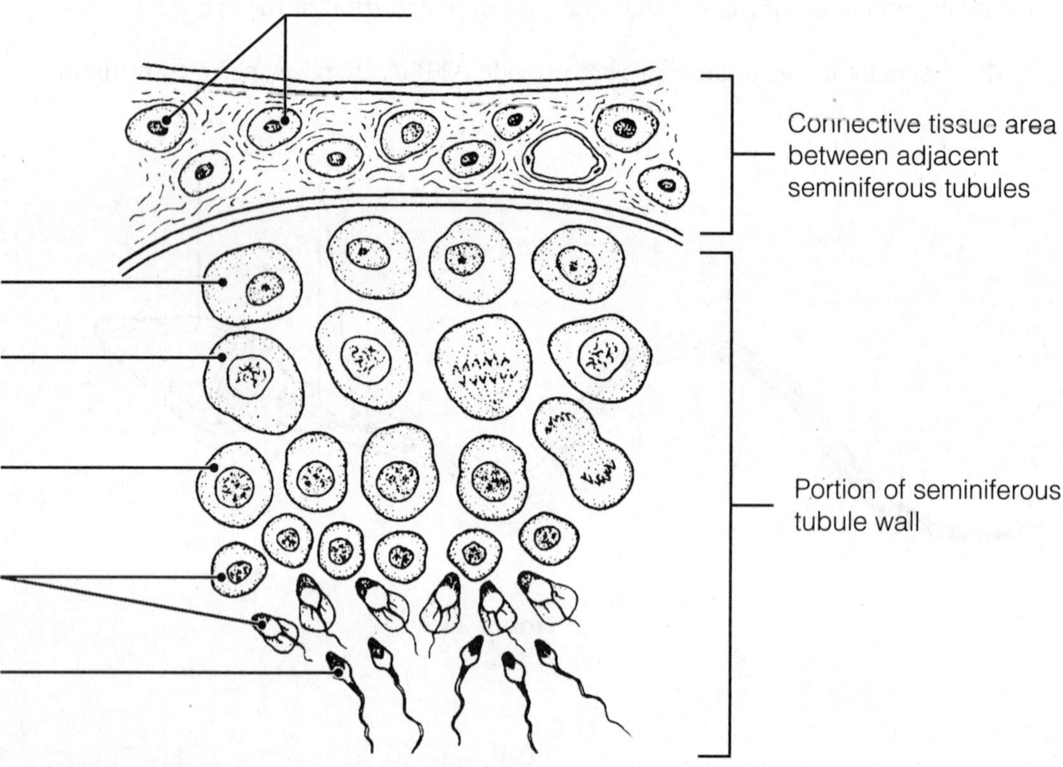

Figure 26.4

4. Match the key terms with the following statements relating to the hormonal regulation of male reproductive function.

Key Choices

A. Androgen-binding protein (ABP) D. Inhibin

B. Follicle-stimulating hormone (FSH) E. Luteinizing hormone (LH)

C. Gonadotropin-releasing hormone (GnRH) F. Testosterone

________________ 1. Substance released by sustentacular cells that enhances spermatogenesis by keeping testosterone concentration in the area high

________________ 2. Hormone that prompts the interstitial cells to produce testosterone

________________ 3. Hormone released by the hypothalamus that stimulates the anterior pituitary to release FSH and LH

________________ 4. Hormones that feed back to inhibit hypothalamic and/or anterior pituitary secretory activity

5. The process by which spermatids are converted to mature sperm is called spermiogenesis. Answer the following questions about this process.

1. What exactly does spermiogenesis involve? ________________________________

____

____

2. What cells aid in this process by disposing of excess cytoplasm and providing nutrition to the maturing sperm?

____

3. What is the blood-testis barrier and why is it important?

____

____

____

____

____

6. Name four male secondary sex characteristics.

1. ______________________________ 3. ______________________________

2. ______________________________ 4. ______________________________

7. Circle the term that does not belong in each of the following groupings.

1. Semen Acidic Relaxin Fibrinolysin
2. Spermatogonium Type B cell Type A cell Germ cell
3. Sertoli cells Sustentacular cells Blood-testis barrier Spermatogenic
4. Parasympathetic nervous system Ejaculation Climax Rhythmic contractions of bulbospongiosus muscle
5. Testosterone Inhibin Cholesterol nucleus Dihydrotestosterone
6. Nitric oxide Vasoconstriction Parasympathetic reflex Erection

Anatomy of the Female Reproductive System

1. Figure 26.5 is an inferior view of the female external genitalia. Label the anus, clitoris, labia minora, urethral orifice, hymen, mons pubis, and vaginal orifice on the figure. These structures are indicated with leader lines. Then color the homologue of the male penis blue, color the membrane that partially obstructs the vagina yellow, and color the vestibule red. Mark the extent of the perineum with a dashed line.

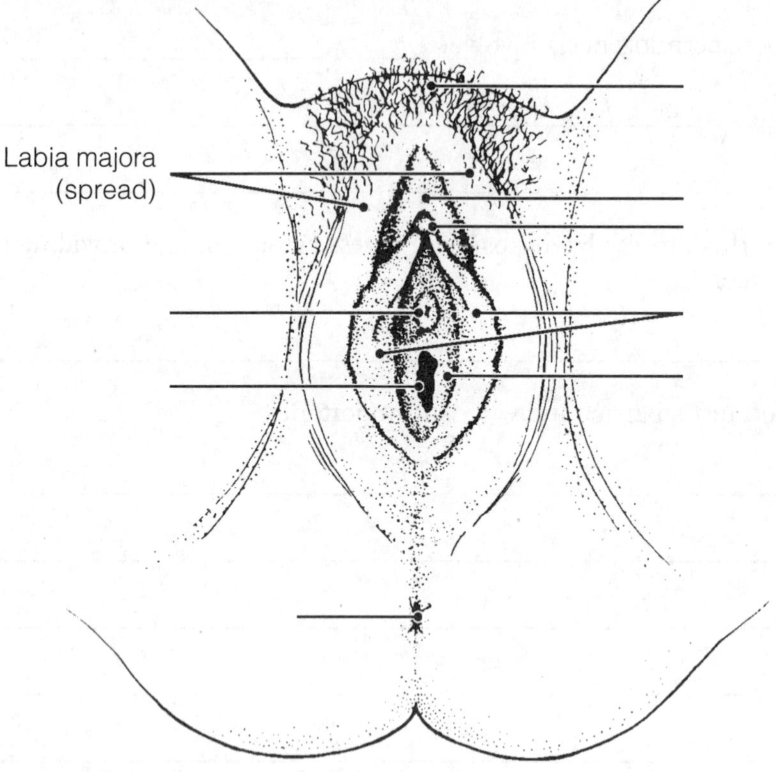

Figure 26.5

2. Figure 26.6 is a sagittal view of the female reproductive organs. First, label all structures on the figure with leader lines. Then, select different colors for the following structures and color the coding circles and corresponding structures on the figure.

- ◯ Lining of the uterus, endometrium
- ◯ Muscular layer of the uterus, myometrium
- ◯ Pathway along which an egg travels from the time of its release to its implantation
- ◯ Ligaments suspending the reproductive organs
- ◯ Structure forming female hormones and gametes
- ◯ Homologue of the male scrotum

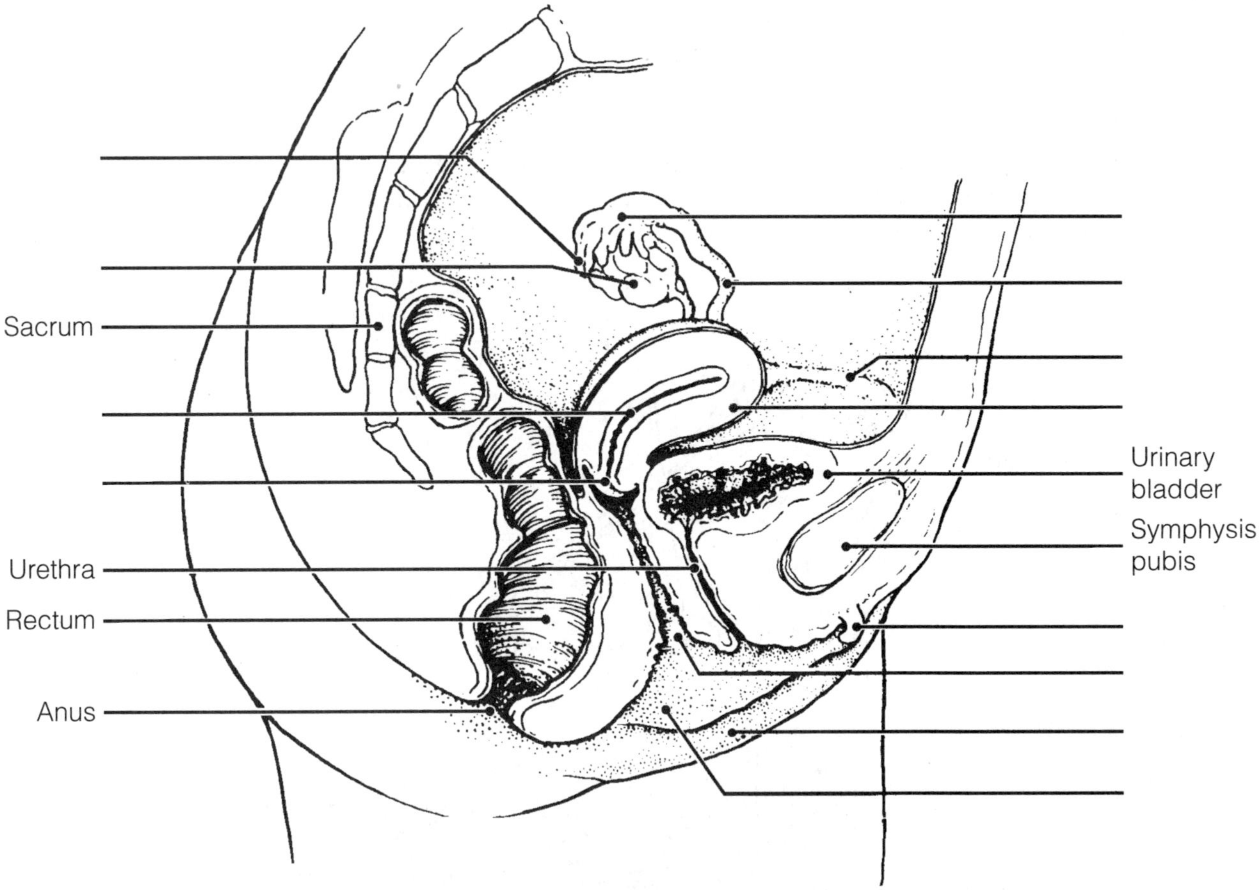

Figure 26.6

3. What is the significance of the fact that the uterine tubes are not structurally continuous with the ovaries? Address this question from both reproductive and health aspects.

____

____

____

4. Identify the following female anatomical structures.

_______________ 1. Chamber that houses the developing fetus

_______________ 2. Canal that receives the penis during sexual intercourse

_______________ 3. Usual site of fertilization

_______________ 4. Becomes erect during sexual stimulation

_______________ 5. Duct through which the ovum travels to reach the uterus

_______________ 6. Membrane that partially closes the vaginal canal

_______________ 7. Primary female reproductive organ

_______________ 8. Move to create fluid currents to draw the ovulated "egg" into the uterine tube

5. Figure 26.7 is a sagittal section of a breast. First, use the following terms to correctly label all structures with leader lines on the figure: alveolar glands, areola, lactiferous duct and sinus, and nipple. Then color the structures that produce milk blue and the fatty tissue of the breast yellow.

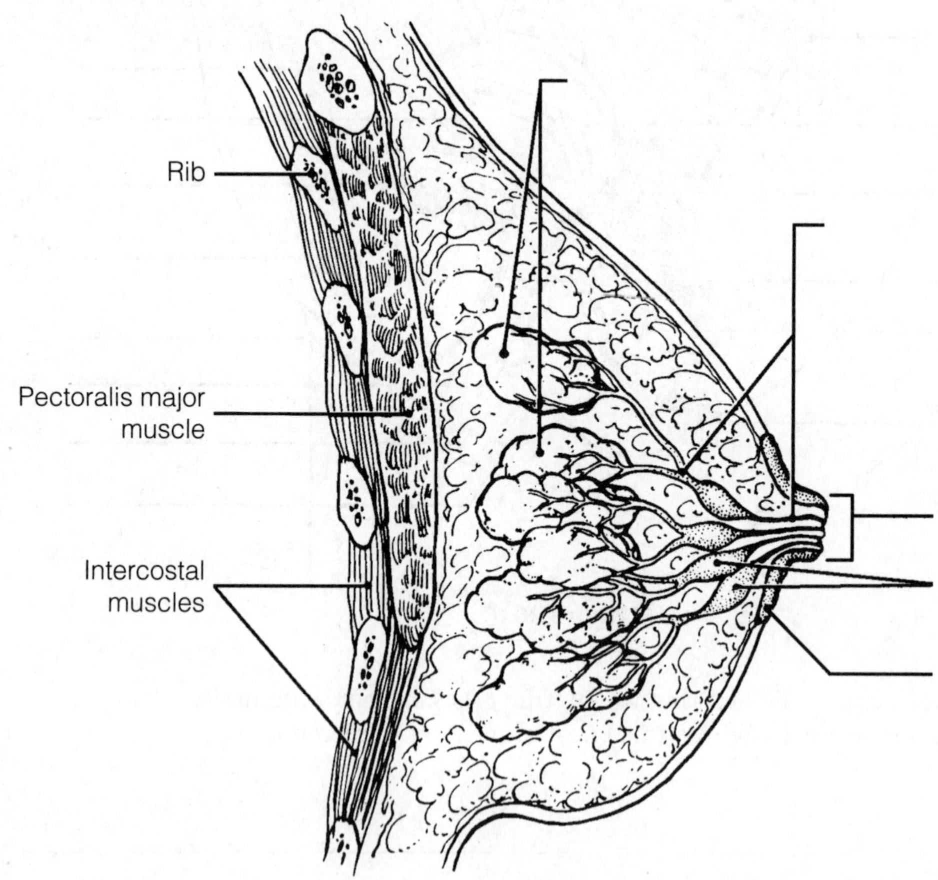

Figure 26.7

Physiology of the Female Reproductive System

1. The following statements deal with anterior pituitary and ovarian hormonal interrelationships. Name the hormone(s) described in each statement. Place your answers in the answer blanks.

______________ 1. Promotes growth of ovarian follicles and production of estrogen

______________ 2. Triggers ovulation

______________ and ______________ 3. Inhibit follicle-stimulating hormone (FSH) release by the anterior pituitary

______________ 4. Stimulates luteinizing hormone (LH) release by the anterior pituitary

______________ 5. Converts the ruptured follicle into a corpus luteum and causes it to produce progesterone and estrogen

______________ 6. Maintains the hormonal production of the corpus luteum

2. Figure 26.8 is a sectional view of the ovary. First, identify all structures with leader lines on the figure. Second, select different colors for the following structures and color the coding circles and corresponding structures on the figure. Then, answer the questions that follow on the next page.

◯ Cells that produce estrogen ◯ All oocytes ◯ Structure that produces progesterone

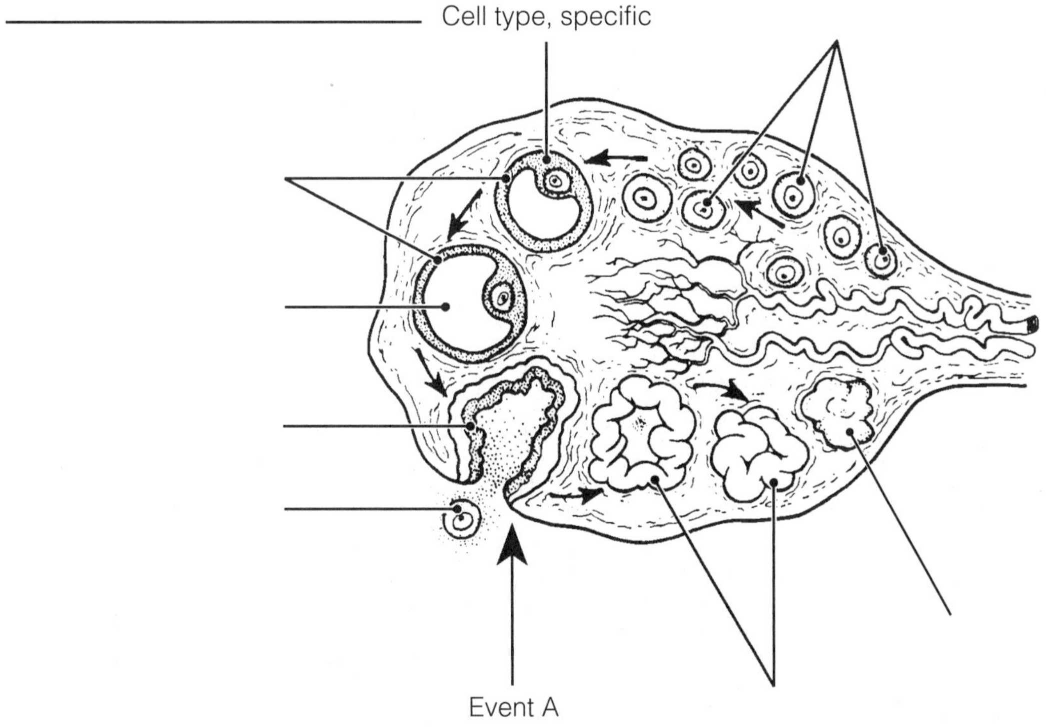

Figure 26.8

1. Name the event depicted as Event A on the figure. ______________________

2. Into what area is the ovulated cell released? ______________________

3. When is a mature ovum (egg) produced in humans? ______________________

4. What structure in the ovary becomes a corpus luteum? ______________________

5. Name the final cell types and indicate the number of each cell type produced by oogenesis in the female. ______________________

6. What happens to the tiny cells nearly devoid of cytoplasm ultimately produced during oogenesis and why does it happen?

____

3. Using the key choices, identify the cell type you would expect to find in the following structures.

Key Choices

A. Oogonium B. Ovum C. Primary oocyte D. Secondary oocyte

______________ 1. Forming part of the primary follicle in the ovary

______________ 2. In the uterine tube before fertilization

______________ 3. In the mature, or graafian, follicle of the ovary

______________ 4. In the uterine tube shortly after sperm penetration

4. The following exercise refers to Figure 26.9.

In Figure 26.9A, the blood levels of two gonadotropic hormones (FSH and LH) of the anterior pituitary are indicated. Identify each hormone by labeling the blood level lines on the figure. Then select different colors for each of the blood level lines and color them on the figure.

In Figure 26.9B, identify the blood level lines for the ovarian hormones estrogen and progesterone. Then select different colors for each blood level line, and color them on the figure.

In Figure 26.9C, color the following structures, using different colors for each.

○ Primordial follicle ○ Secondary follicle ○ Ovulating follicle

○ Vesicular follicle ○ Corpus luteum

In Figure 26.9D, identify the endometrial changes occurring during the menstrual cycle by coloring the areas depicting the following phases. Use different colors to represent each phase.

○ Secretory phase ○ Menses ○ Proliferative phase

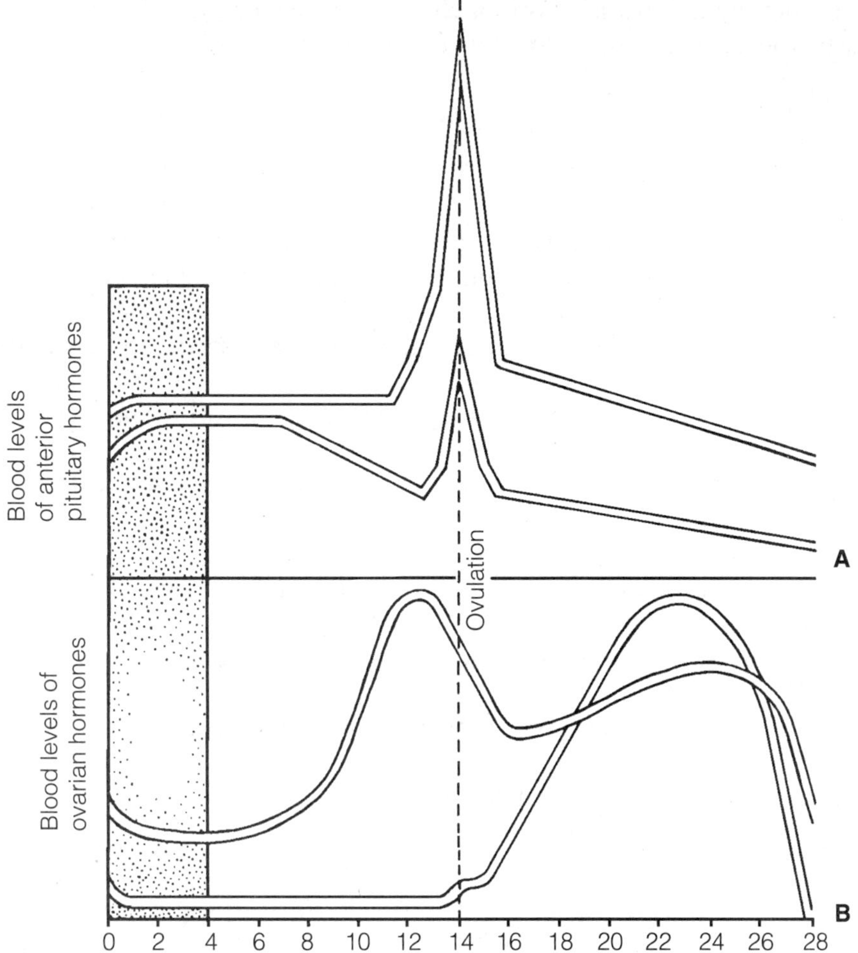

Blood levels of anterior pituitary hormones
Blood levels of ovarian hormones
Ovulation
A
B
0 2 4 6 8 10 12 14 16 18 20 22 24 26 28

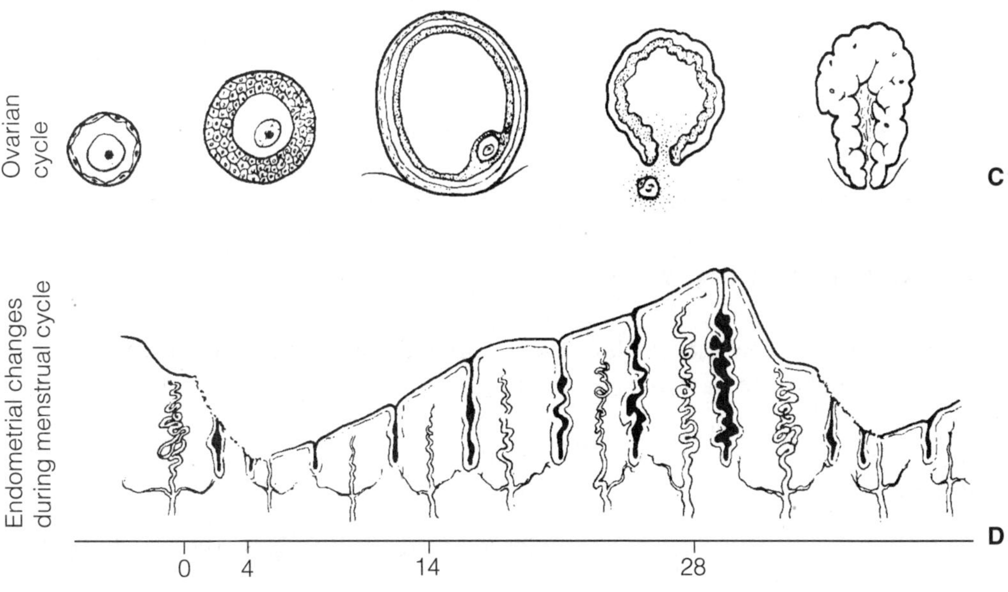

Ovarian cycle
C
Endometrial changes during menstrual cycle
D
0 4 14 28

Figure 26.9

5. Use the key choices to identify the ovarian hormone(s) responsible for the following events. Insert the correct answer(s) in the answer blanks.

Key Choices

A. Estrogen B. Progesterone

__________ 1. Absence causes the spiral arteries of the endometrium to go into spasms and kink and the endometrium to slough off (menses)

__________ 2. Causes the endometrial glands to begin secreting nutrients

__________ 3. The endometrium is repaired and grows thick and velvety

__________ 4. Maintains the myometrium in an inactive state if implantation of an embryo has occurred

__________ 5. Stimulates gland formation in the endometrium

__________ 6. Responsible for the secondary sex characteristics of females

__________ 7. Causes the cervical mucus to become viscous

6. Name four secondary sex characteristics of females.

1. __________ 3. __________

2. __________ 4. __________

7. Compare and contrast the two phases of sexual response in females and males.

____

____

____

____

____

Sexually Transmitted Infections

1. Sexually transmitted infections (STIs) are infectious diseases spread through sexual contact. Match the names of these diseases in the key choices with the appropriate descriptions.

Key Choices

A. Chlamydia B. Genital herpes C. Genital warts D. Gonorrhea E. Syphilis

__________ 1. Marked by exacerbations and remissions.

__________ 2. Its destructive tertiary lesions are called gummas.

_______________ 3. Most common symptom in males is urethritis and penile drip; may be asymptomatic in females.

_______________ 4. Not recognized as a health problem until the 1970s; accounts for 25% to 50% of all pelvic inflammatory disease.

_______________ 5. Congenital forms can cause severe fetal malformations.

_______________ 6. Newborns of infected mothers may have conjunctivitis and respiratory tract infections.

_______________ 7. Like AIDS, a viral rather than a bacterial disease.

_______________ 8. Only diagnosed by cell culture techniques; treated with tetracycline.

_______________ 9. Disease routinely treated with penicillin.

_______________ 10. Typically caused by the human papilloma virus; different strains linked to invasive cervical cancer.

THE INCREDIBLE JOURNEY:

A Visualization Exercise for the Reproductive System

...you hear a piercing sound coming from the almond-shaped organ as its wall ruptures.

1. Complete the following narrative by writing the missing terms in the answer blanks.

_______________ 1.

_______________ 2.

_______________ 3.

This is your final journey. You are introduced to a hostess who has agreed to have her cycles speeded up by megahormone therapy so that all of your observations can be completed in less than a day. Your instructions are to observe and document as many events of the two female cycles as possible.

You are miniaturized to enter your hostess through a tiny incision in her abdominal wall (this procedure is called a laparotomy, or, more commonly, "belly button surgery"), and you end up in her peritoneal cavity. You land on a large, pear-shaped organ in the abdominal cavity midline, the __(1)__. You survey the surroundings and begin to make organ identifications and notes of your observations. Laterally and way above you and on each side is an almond-shaped __(2)__. Each of these almond-shaped organs is suspended by a ligament and almost touched by featherduster-like projections of a tube snaking across the abdominal cavity toward it. The projections appear to be almost still, which is puzzling because you thought that they were the __(3)__, or fingerlike projections, of the uterine tubes, which are supposed to be in motion. You walk toward the end of one of the uterine tubes to take a better look. As you study the end of the tube more closely, you discover that the featherlike projections are now moving more rapidly, as if they are trying to

______________________ 4.

______________________ 5.

______________________ 6.

______________________ 7.

______________________ 8.

______________________ 9.

______________________ 10.

______________________ 11.

______________________ 12.

______________________ 13.

______________________ 14.

______________________ 15.

______________________ 16.

______________________ 17.

______________________ 18.

coax something into the uterine tube. Then you spot a reddened area on the almond-shaped organ, which seems to be enlarging even as you watch. As you continue to observe the area, you waft gently up and down in the peritoneal fluid. Suddenly you feel a gentle but insistent sucking current, drawing you slowly toward the uterine tube. You look upward and see that the reddened area now looks like an angry boil, and the fingerlike projections are gyrating and waving frantically. You realize that you are about to witness __(4)__. You try to get still closer to the opening of the uterine tube, when you hear a piercing sound coming from the almond-shaped organ as its wall ruptures. Then you see a ball-like structure, with a "halo" of tiny cells enclosing it, being drawn into the uterine tube. You have just seen the __(5)__ surrounded by its capsule of __(6)__ cells entering the uterine tube. You hurry into the uterine tube behind it and hold onto one of the tiny cells to follow it to the uterus. The cell mass that you have attached to has no way of propelling itself, yet you are being squeezed along toward the uterus by a process called __(7)__. You also notice that there are __(8)__, or tiny hairlike projections of the tubule cells, that are all waving in the same direction as you are moving.

For a while you are carried tranquilly along, but then you are startled by a deafening noise. Suddenly there are thousands of tadpole-like __(9)__ swarming all around you and the sphere of cells. Their heads seem to explode as their __(10)__ break and liberate digestive enzymes. The cell mass now has hundreds of openings in it, and some of the small cells are beginning to fall away. As you peer through the rather transparent cell "halo," you see that one of the tadpole-like structures has penetrated the large central cell. The penetrated cell then begins to divide. You have just witnessed the second __(11)__ division. The products of this division are one large cell, the __(12)__, and one very tiny cell, the __(13)__, which is now being ejected. This cell will soon be __(14)__ because it has essentially no cytoplasm or food reserves. As you continue to watch, the sperm nucleus and that of the large central cell fuse, an event called __(15)__. You note that the new cell just formed by this fusion is called a(n) __(16)__, the first cell of the embryonic body.

As you continue to move along the uterine tube, the central cell divides so fast that virtually no cell growth occurs between the divisions. Thus the number of cells forming the embryonic body increases, but the cells become smaller and smaller. Finally, the uterine chamber looms before you. As you drift into its cavity, you scrutinize its lining, the __(17)__. You notice that it is thick and velvety in appearance and that the fluids you are drifting in are slightly sweet. The embryo makes its first contact with the lining, detaches, and then makes a second contact at a slightly more inferior location. This time it sticks, and as you watch, the lining of the organ begins to erode away. The embryo is obviously beginning to burrow into the rich cushiony lining.

You now leave the embryo and propel yourself well away from it. As you float in the cavity fluids, you watch the embryo disappear from sight beneath the uterine lining. Then you continue to travel downward through your hostess' reproductive tract, exiting her body at the external opening of the __(18)__.

CHALLENGING YOURSELF

At the Clinic

1. A young man with a painless, hard lump on his left testicle has been admitted to the hospital and scheduled for surgery. What do you suspect?
2. Unable to afford a professional mover, Gary enlisted the help of his friends to move into a new apartment. The apartment was on the third floor, so the move was a difficult, laborious one, but it was complete by the end of the day. Exhausted, Gary fell into bed but awoke the next morning with pain in his groin. What do you think happened?
3. Gunter, a 70-year-old man, has the problem of urine retention, and a digital rectal examination reveals an enlargement inferior to his urinary bladder. What is this condition and what are some likely causes?
4. Mrs. Ginko's Pap smear result shows some abnormal cells. What possibility should be investigated?
5. Marylou, a sexually active young woman, comes to the clinic after several weeks of more-or-less continuous pelvic pain and heavy bleeding. What is the doctor's diagnosis? What does he need to know before prescribing medication?
6. A 53-year-old woman has come to the clinic complaining of abdominal discomfort, increased flatulence, and bloating. Palpation during a pelvic exam reveals a massive left ovary. What do you think is wrong?
7. After bearing six children in eight years, a woman complains of perineal discomfort whenever she increases the abdominal pressure (sneezing, bowel movement). She says it feels like "her insides are coming out." What is her problem?
8. Julio is diagnosed as infected with gonorrhea and chlamydia. What clinical name is given to the general class of reproductive system infections and why is it crucial to inform his partners of his infection?

Stop and Think

1. In what ways does the reproductive system contribute to the body's homeostasis?
2. Why might testicular mumps in an adult, but not in a prepubertal, male result in sterility?
3. What would have to happen to trigger circulating antibodies to sperm antigens?
4. Based on your knowledge of male reproductive anatomy, explain whether or not a vasectomy greatly diminishes the volume of the ejaculate.
5. In medieval times, young boys with beautiful soprano voices were sometimes castrated so that their singing voice would not change. Explain the cause and effect relationship here.

COVERING ALL YOUR BASES

Multiple Choice

Select the best answer or answers from the choices given.

1. Which of the following are accessory sex structures in the male?
   A. Gonads
   B. Gametes
   C. Broad shoulders
   D. Seminal vesicles
2. Within the spermatic cord are the:
   A. ductus deferens
   B. dartos muscle
   C. testicular artery
   D. gubernaculum
3. Which is (are) associated with varicocele?
   A. Low sperm count
   B. Malfunctional cremaster muscle
   C. Low testosterone levels
   D. Pampiniform plexus
4. Stereocilia:
   A. are found in the ducts of the epididymis
   B. are elongated cilia
   C. supply nutrients to stored sperm
   D. supply activating enzymes during ejaculation
5. Which of the following structures have a region called the ampulla?
   A. Ductus deferens
   B. Uterine tube
   C. Ejaculatory duct
   D. Lactiferous duct
6. Seminal vesicle secretions have:
   A. a low pH
   B. fructose
   C. prostaglandins
   D. sperm-activating enzymes
7. The structure(s) superior to the urogenital diaphragm is (are):
   A. bulbourethral glands
   B. bulb of the penis
   C. prostate
   D. membranous urethra
8. Seminalplasmin:
   A. activates sperm
   B. digests the intercellular glue in the corona radiata
   C. has bacteriostatic properties
   D. neutralizes vaginal acidity
9. A chromosome count in a spermatogenic cell within the adluminal compartment of the seminiferous tubule shows it to be diploid. The cell could be a:
   A. spermatogonium
   B. spermatid
   C. sustentacular cell
   D. primary spermatocyte
10. As a result of crossover:
    A. maternal genes can end up on a paternal chromosome
    B. synapsis occurs
    C. a tetrad is formed
    D. no two spermatids have exactly the same genetic makeup
11. Prominent structures in a cell undergoing spermiogenesis include:
    A. chromosomes
    B. Golgi apparatus
    C. rough ER
    D. mitochondria
12. Treatment for which disorders will reduce fertility?
    A. Chlamydia
    B. Testicular cancer
    C. Cryptorchidism
    D. Gonorrhea
13. Components of the brain-testicular axis include:
    A. parasympathetic impulses of the erection reflex triggered by pleasurable thoughts
    B. stimulation of gonadotropin release
    C. negative feedback from the sustentacular cells
    D. maintenance of normal sperm count
14. Which of the following attach to the ovary?
    A. Fimbriae
    B. Mesosalpinx
    C. Suspensory ligaments
    D. Broad ligament
15. Which structures are involved in estrogen secretion?
    A. Primordial follicle
    B. Theca
    C. Granulosa cells
    D. Corpus luteum
16. Each month, only one:
    A. primordial follicle is stimulated
    B. follicle secretes estrogen
    C. vesicular follicle undergoes ovulation
    D. ovary is stimulated
17. On day 17 of a woman's monthly cycle:
    A. FSH levels are rising
    B. progesterone is being secreted
    C. the ovary is in the ovulatory phase
    D. the uterus is in the proliferative phase

18. Cells that are functional on day 8 of a woman's monthly cycle include:
    A. granulosa cells C. luteal cells
    B. theca cells D. secondary oocyte

19. A sudden decline in estrogen and progesterone levels:
    A. causes spasms of the spiral arteries
    B. triggers ovulation
    C. ends inhibition of FSH release
    D. causes fluid retention

20. Dimensions of which of the following are most important when a diaphragm or cervical cap is being fitted?
    A. Hymen C. Fornix
    B. Vestibule D. Vaginal diameter

21. Which of the female structures listed below have a homologue in the adult male?
    A. Labia majora C. Uterine tubes
    B. Clitoris D. Labia minora

22. An STI that is more easily detected in males than females, is treatable with penicillin, and can cause lesions in the nervous and cardiovascular systems is:
    A. gonorrhea C. syphilis
    B. chlamydia D. herpes

23. If the uterine tube is a trumpet ("salpinx"), what part of it represents the wide, open end of the trumpet?
    A. Isthmus C. Infundibulum
    B. Ampulla D. Flagellum

24. The myometrium is the muscular layer of the uterus, and the endometrium is the ________________ layer.
    A. serosa C. submucosa
    B. adventitia D. mucosa

25. All of the following are true of the gonadotropins except that they are:
    A. secreted by the pituitary gland
    B. LH and FSH
    C. hormones with important functions in both males and females
    D. the sex hormones secreted by the gonads

26. The approximate area between the anus and clitoris in the female is the:
    A. peritoneum C. vulva
    B. perineum D. labia

27. A test to detect cancerous changes in cells of the uterus and cervix is:
    A. pyelogram C. D&C
    B. Pap smear D. laparoscopy

28. In humans, separation of the cells at the two-cell stage following fertilization may lead to the production of twins, which in this case would be:
    A. of different sexes C. fraternal
    B. identical D. dizygotic

29. Human ova and sperm are similar in that:
    A. about the same number of each is produced per month
    B. they have the same degree of motility
    C. they are about the same size
    D. they have the same number of chromosomes

30. A person who has had this type of mastectomy will have great difficulty adducting the arms, as when doing push-ups.
    A. Radical mastectomy
    B. Simple mastectomy
    C. Lumpectomy

31. Which of these is mismatched?
    A. Vagina—penis
    B. Testis—ovary
    C. Labia majora—scrotum
    D. Oviduct—ductus deferens

32. After ovulation, the ruptured follicle:
    A. degenerates
    B. becomes a corpus luteum
    C. sloughs off as waste material
    D. mends and produces another oocyte

33. Select the false statement about the uterine cervix.
    A. It is the superiormost part of the uterus.
    B. Its entire mass projects into and lies within the vagina.
    C. Its cervical glands secrete mucus.
    D. It contains the cervical canal.

34. Some cells in both the ovary and the testis respond to follicle-stimulating hormone. These cells, which are said to correspond to one another, are:
    A. granulosa (follicle) cells and sustentacular cells
    B. oogonia and spermatogonia
    C. theca and interstitial cells
    D. germinal epithelium and tunica vaginalis

Word Dissection

For each of the following word roots, fill in the literal meaning and give an example, using a word found in this chapter.

Word root	Translation	Example
1. acro		
2. alb		
3. apsi		
4. arche		
5. cremaster		
6. didym		
7. estro		
8. forn		
9. gamet		
10. gen		
11. gest		
12. gono		
13. gubern		
14. gyneco		
15. herp		
16. hymen		
17. hyster		
18. inguin		
19. lut		
20. mamma		
21. mei		
22. men		
23. menstru		
24. metr		
25. oo		
26. orchi		
27. ov		
28. pampin		
29. penis		
30. pub		
31. pudend		
32. salpin		
33. scrot		
34. semen		
35. uter		
36. vagin		
37. vener		
38. vulv		

Epilogue: A Day in the Life

Well, we've spent nearly a year taking the human body apart (in the academic sense, of course). You have been asked to critically examine each organ system, often in minute detail, to learn about its contributions to body function and homeostasis. The problem with this traditional approach is that once the body has been "taken apart" for such an examination, it is never put back together again. Hence, the goal of this epilogue is to illustrate how seemingly simple everyday activities—like eating, eliminating urine, and walking—are brought about by scores of physiological events operating both independently and cooperatively.

A description of John's day and activities, from the time he wakes up one morning until just before he wakes up the next morning, is followed on the left-hand column both in pictures and in a simple narrative. You are sure to find nearly all of his activities very familiar (and part of your own daily routine). On the adjacent column is a replication of a computer printout listing many of the physiological activities that are going on in John's body at that particular moment. Obviously this computer printout is not a complete listing. As long as you are alive, you are breathing, your heart is pumping, your mind is active, and your hormone levels are waxing and waning. Many of these activities are mentioned only when they need be enhanced, as during strenuous exercise. For the most part, our goal is to focus on specific activities and to indicate what other body systems are important in helping to effect or moderate those activities.

After completing the Epilogue, I realized it could be used as another learning tool. For example:

1. It provides a study guide by spelling out which physiological events are occurring during everyday activities, like riding a bicycle. It is certainly clear that riding a bike involves the muscular system (skeletal muscles), but the Epilogue also points out how the nervous system, cardiovascular system, respiratory system, and others are indispensable to muscle activity and maintaining homeostasis.
2. You can expand your understanding of the events described in the Epilogue by figuring out what would happen if one or more of the variables in certain events were changed. For example, what would happen to John's equilibrium if he had more than the one beer? What would happen to his immune system and skin cells if he stayed out all day on the beach instead of just part of the day?
3. Use the type of approach in the Epilogue to examine the impact of what you are doing on your own physiology. For instance, work out the pathways of physiological events and the various system connections involved when you are performing with your town's theater group, swimming, or just relaxing. Or ask yourself questions. "What if I did 75 crunches instead of 20? What would change, and how? What if I ate five jelly doughnuts instead of cereal and fruit for breakfast? What effect would that have on my endocrine system, metabolism, and weight?"

You will probably be a bit overwhelmed with the number and kinds of processes that do (and must) occur to keep the body functioning smoothly and effectively as we go about our daily tasks. Even better, you will be amazed at how much you have learned as you worked your way through the course in A&P. Enjoy—it's time to see the forest instead of the trees!

A DAY in THE LIFE
COMPLETE
WITH COMPUTER PRINTOUT

7:00 AM AS JOHN'S EYES OPEN, HE CHECKS THE TIME, THEN CRAWLS DEEPER UNDER THE COVERS.

7:10 AM AFTER A FEW MINUTES OF TOSSING, HE REALIZES HE HAS TO URINATE.

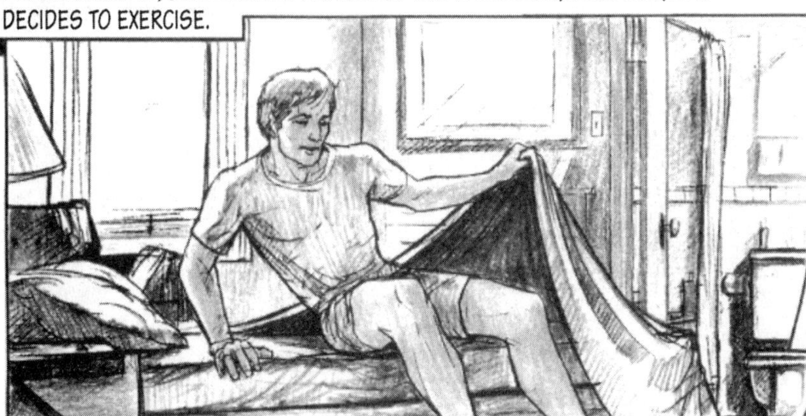

7:12 AM ARISING, JOHN MAKES HIS WAY TO THE BATHROOM, URINATES, AND DECIDES TO EXERCISE.

Awakening occurs as:

1. Core body temperature rises.
2. EEG changes from stage 4 to stage 1 pattern.
3. Norepinephrine levels rise and serotonin (sleep neurotransmitter) levels decline in the brain; cortical activity rises. Neurons of dorsal raphe nuclei of midbrain reticular formation are firing at their maximal rate.

Full arousal and recognition of urge to urinate occurs as:

1. Accumulating urine distends the urinary bladder wall, initiating the micturition reflex. Detrusor muscle of bladder contracts and internal urethral sphincter relaxes, allowing urine to enter the proximal (prostatic) urethra.
2. Afferent impulses sent to brain cause recognition of urge to urinate.

Blood pressure homeostasis maintained while standing up:

1. Blood pressure drops as erect posture is achieved.
2. Baroreceptors are activated, causing reflex activation of the sympathetic nervous system vasomotor center in the medulla. Sympathetic vasomotor fibers release norepinephrine, causing most arterioles of the body to constrict.
3. Blood pressure rises, restoring homeostasis.

Mobilization of voluntary muscles to walk to the bathroom:

1. Voluntary activity is initiated by precommand regions of the brain (cerebellum and basal nuclei).
2. Sitting, standing, and walking are monitored by the cerebellum, which receives continuous feedback on body balance and tension of muscles and tendons from visual, proprioceptor, and vestibular receptors. The cerebellum dispatches nerve impulses to the motor and premotor cortical regions to maintain balance and gait.
3. Locomotion and rhythmic activities of the limbs are regulated by control neurons in the brain stem, which turn on the central pattern generators in the spinal cord.*

Blood supply to skeletal muscles increases:

1. Sympathetic nervous system activity dilates arterioles in active skeletal muscles.
2. Precapillary sphincters open, allowing blood to enter the true capillaries to serve the muscle cells as their metabolic activity increases.

*All voluntary skeletal muscle activities involve this interplay of physiologic events. From this point on they will be indicated by the letters VSMA (voluntary skeletal muscle activity).

7:30 AM BECAUSE HE HAS A BUSY DAY PLANNED, JOHN DECIDES TO GET TO HIS CYCLING EARLY. HE PULLS ON HIS CYCLING CLOTHES, STRETCHES, AND BEGINS SLOWLY TO GIVE HIS MUSCLES A CHANCE TO WARM UP.

7:40 AM JOHN, TRYING TO INCREASE HIS SPEED, ENDURANCE, AND STRENGTH, ALTERNATES BETWEEN CYCLING FLAT OUT (AS FAST AS HE POSSIBLY CAN), CRUISING AT A MODERATE PACE, AND PEDALING UPHILL IN A HIGH GEAR.

Warmup:

1. Muscle fibers begin to contract and generate heat, increasing the efficiency of the muscle enzyme systems.
2. VSMA occurs.
3. The sympathetic nervous system promotes vasodilation in the working muscles; precapillary sphincters begin to open. Catecholamines are released and enhance the sensitivity of the stretch reflexes.
4. Skeletal muscle and respiratory pumps enhance venous return (hence, cardiac output and blood pressure).

Effects of training on muscle:

1. Sustained aerobic exercise produces muscle modifications that ultimately enhance endurance: more red nonfatiguable fibers, more capillaries, mitochondria, and myoglobin in the skeletal muscles.
2. High intensity (speed or resistance) exercise produces faster-acting, stronger muscles (by increasing muscle size and the numbers of white fibers).

Effects of training on the other organ systems (occur gradually):

1. Cardiovascular efficiency increases as the heart hypertrophies, venous return and stroke volume increase, and enhanced blood levels of LDLs clear fatty substances from the blood vessel walls.
2. Respiratory efficiency increases as VO_{max} increases.
3. The skeleton becomes stronger due to mechanical stresses of muscles pulling.
4. Neuromuscular coordination improves.
5. Lymphatic drainage improves due to the skeletal muscle pump.
6. Tissue cells become more responsive to insulin, helping to prevent diabetes mellitus.

Respiration increases to enhance gas exchange in the lungs:

1. Respiratory rate rises as exercise begins, stabilizes, then remains high until oxygen debt is repaid after exercise.

Muscle pH homeostasis maintained:

1. During periods of intense activity, aerobic mechanisms are unable to keep pace and anaerobic glycolysis becomes the means of generating ATP.
2. Lactic acid is produced, moves from the muscles to the blood, and is sequestered by the liver. This prevents muscle and blood pH from falling too low during athletic activity.

Temperature homeostasis maintained:

1. Heat generated by muscles warms the blood. Central temperature sensors in hypothalamus respond by activating sympathetic-mediated heat loss mechanisms: dermal blood vessels dilate to allow heat to radiate from skin surface, and sweat glands are activated to cool the body via evaporation of sweat.

9:35 AM AFTER COOLING DOWN AT A SLOWER PACE, THEN STRETCHING TO MAINTAIN FLEXIBILITY AND PREVENT CRAMPING, JOHN HEADS FOR THE SHOWER.

9:45 AM AFTER DRYING OFF, JOHN AMBLES ON TO THE KITCHEN WHERE HE POURS HIMSELF A GLASS OF OJ AND A BOWL OF CEREAL AND BEGINS TO EAT.

Body temperature homeostasis maintained during showering:

1. Hot shower washes away "skin-dwelling" bacteria and warms the skin surface.
2. Peripheral and central thermoreceptors are activated: sympathetic fibers dilate dermal blood vessels.
3. After the shower, radiation from the skin surface causes rapid heat loss.
4. Thermoreceptors initiate heat promoting mechanisms by hypothalamus. Sympathetic fibers stimulate vasoconstriction of cutaneous blood vessels and increase metabolic rate via released norepinephrine. Brain centers controlling muscle tone are activated. Shivering begins, generating large amounts of heat.
5. Temperature homeostasis is restored and shivering stops.

Ambulation and manipulation of food:

1. VSMA occurs.

Food intake activities

1. Salivation and secretion of stomach juices are initiated by sight of a liked food.
2. After voluntary activation of chewing muscles, mastication continues reflexively in response to pressure stimuli in the mouth.
3. Salivary amylase in saliva digests starch (in cereal) to simpler sugars.
4. Swallowing initiated voluntarily by tongue. Beyond the oropharynx, swallowing occurs reflexively. Food propelled to the stomach by peristaltic contractions.

Digestion in stomach and intestine:

1. Rhythm of peristaltic contractions in the stomach initiated by local pacemaker cells.
2. Food in stomach activates pressure and chemical receptors that increase gastric mobilization and secretory activity.
3. HCl-activated pepsin digests cereal proteins: ingested food is converted to chyme.
4. Gastric gland secretion of mucus protects the gastric mucosa from self-digestion.
5. Chyme enters the small intestine and activates pressure and chemical receptors that reflexively cause secretion of intestinal juice, CCK, and secretin. Bile and pancreatic juice enter the duodenum.
6. HCO_3^- rich pancreatic juice provides the proper alkaline pH for digestion in small intestine. Pancreatic amylase, lipases, proteases, nucleases, and intestinal brush border enzymes complete digestion.
7. Digested foodstuffs, minerals, vitamins, and water are absorbed as the chyme is mixed and moved along the length of the small intestine by segmentation.
8. Digestion is completed by end of ileum; peristaltic waves sweep undigested debris into the large intestine.
9. Foodstuff entering the large intestine is dehydrated. B and K vitamins made by enteric bacteria are absorbed.

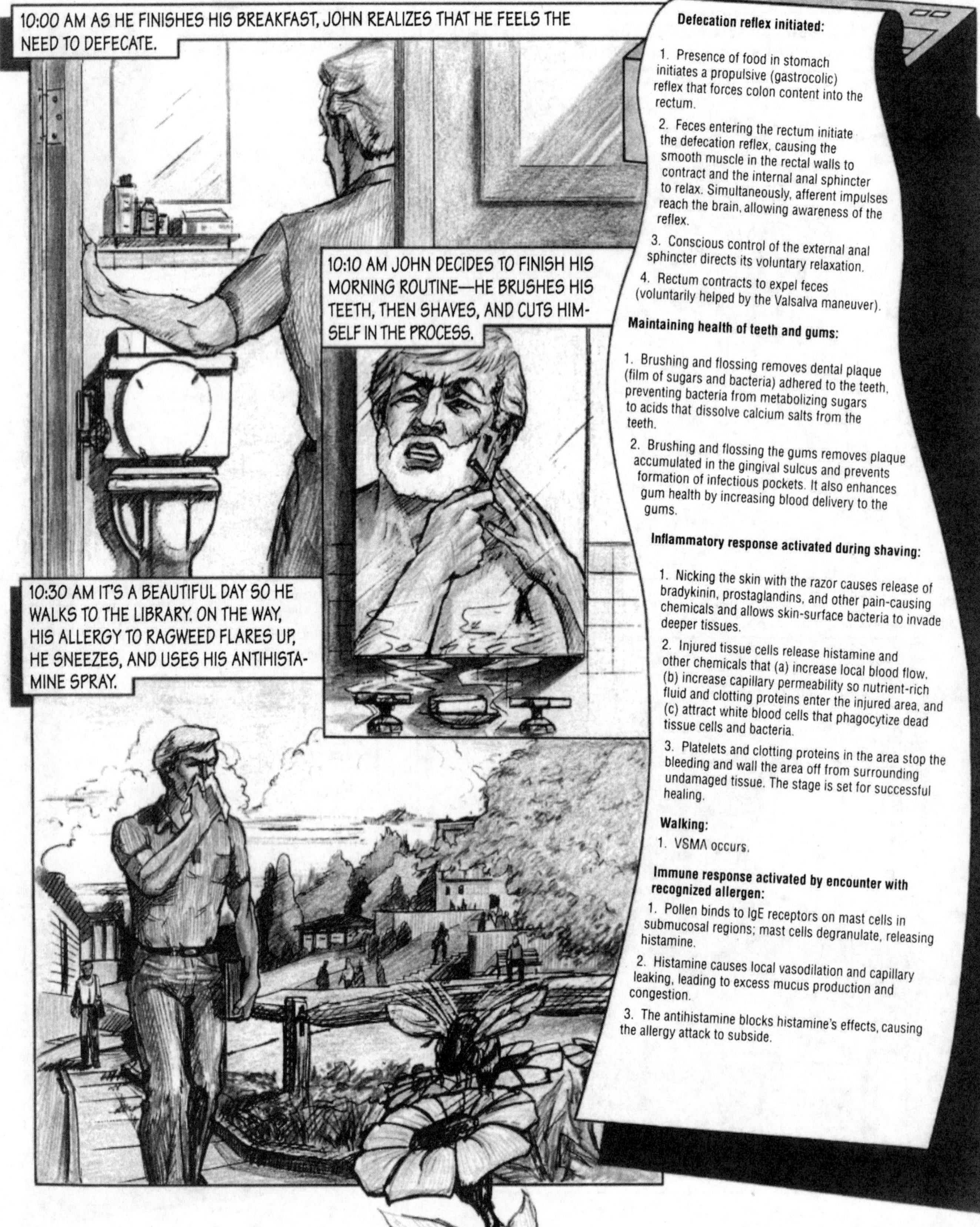
10:00 AM AS HE FINISHES HIS BREAKFAST, JOHN REALIZES THAT HE FEELS THE NEED TO DEFECATE.
10:10 AM JOHN DECIDES TO FINISH HIS MORNING ROUTINE—HE BRUSHES HIS TEETH, THEN SHAVES, AND CUTS HIMSELF IN THE PROCESS.
10:30 AM IT'S A BEAUTIFUL DAY SO HE WALKS TO THE LIBRARY. ON THE WAY, HIS ALLERGY TO RAGWEED FLARES UP, HE SNEEZES, AND USES HIS ANTIHISTAMINE SPRAY.
Defecation reflex initiated:
1. Presence of food in stomach initiates a propulsive (gastrocolic) reflex that forces colon content into the rectum.
2. Feces entering the rectum initiate the defecation reflex, causing the smooth muscle in the rectal walls to contract and the internal anal sphincter to relax. Simultaneously, afferent impulses reach the brain, allowing awareness of the reflex.
3. Conscious control of the external anal sphincter directs its voluntary relaxation.
4. Rectum contracts to expel feces (voluntarily helped by the Valsalva maneuver).
Maintaining health of teeth and gums:
1. Brushing and flossing removes dental plaque (film of sugars and bacteria) adhered to the teeth, preventing bacteria from metabolizing sugars to acids that dissolve calcium salts from the teeth.
2. Brushing and flossing the gums removes plaque accumulated in the gingival sulcus and prevents formation of infectious pockets. It also enhances gum health by increasing blood delivery to the gums.
Inflammatory response activated during shaving:
1. Nicking the skin with the razor causes release of bradykinin, prostaglandins, and other pain-causing chemicals and allows skin-surface bacteria to invade deeper tissues.
2. Injured tissue cells release histamine and other chemicals that (a) increase local blood flow, (b) increase capillary permeability so nutrient-rich fluid and clotting proteins enter the injured area, and (c) attract white blood cells that phagocytize dead tissue cells and bacteria.
3. Platelets and clotting proteins in the area stop the bleeding and wall the area off from surrounding undamaged tissue. The stage is set for successful healing.
Walking:
1. VSMA occurs.
Immune response activated by encounter with recognized allergen:
1. Pollen binds to IgE receptors on mast cells in submucosal regions; mast cells degranulate, releasing histamine.
2. Histamine causes local vasodilation and capillary leaking, leading to excess mucus production and congestion.
3. The antihistamine blocks histamine's effects, causing the allergy attack to subside.

10:45 AM JOHN READS THE ASSIGNED CHAPTER, STUDIES MAKING CONNECTIONS, AND COMPLETES THE END-OF-CHAPTER QUESTIONS.

12:30 PM JOHN MEETS HIS BUDDIES AT THE CAMPUS CENTER FOR LUNCH OUTSIDE AT A SUNNY TABLE.

Reading the textbook:

1. The pupils of the eyes are constricted by oculomotor nerves (III) to prevent entry of the most divergent light rays and increase clarity of vision.
2. The lens is bulged for close vision and the eyes are converged; the real image is focused on the fovea centralis of the retina.
3. Eyes move across lines of print (superior colliculi and extrinsic eye muscles involved).
4. Photoreceptors are activated; nerve impulses pass to the thalamus where synapse occurs; visual pathway continues to the visual cortex in the occipital lobes where vision occurs.

Comprehending and consolidating the material:

1. Impulses are sent from the visual cortex to language areas of the brain for comprehension of the written words.
2. Comprehension requires retrieval of information previously learned and stored in long term memory depots in the cerebral cortex to provide a guideline or comparison point for current understanding.
3. Transfer of new information to long-term memory and consolidation activate the hippocampus, amygdala, and frontal forebrain. These processes occur best when new material can be related to existing information.

Food intake and digestion occurring.

pH, water, and electrolyte balance of the blood is maintained:

1. As nutrients enter the blood, blood buffers resist pH changes.
2. Renal mechanisms reabsorb or fail to reabsorb bicarbonate ions as necessary to maintain blood pH, and maintain water and electrolyte balance in response to hormones (particularly ADH, aldosterone, and ANP).

Body temperature maintained within physiological limits.

Keratinocytes protected from UV radiation:

1. UV rays of sunlight stimulate melanocytes to produce more melanin; keratinocytes engulf the melanin to protect their nuclei from UV damage.
2. UV exposure temporarily disables the immune response by inhibiting the antigen-presenting Langerhans cells.

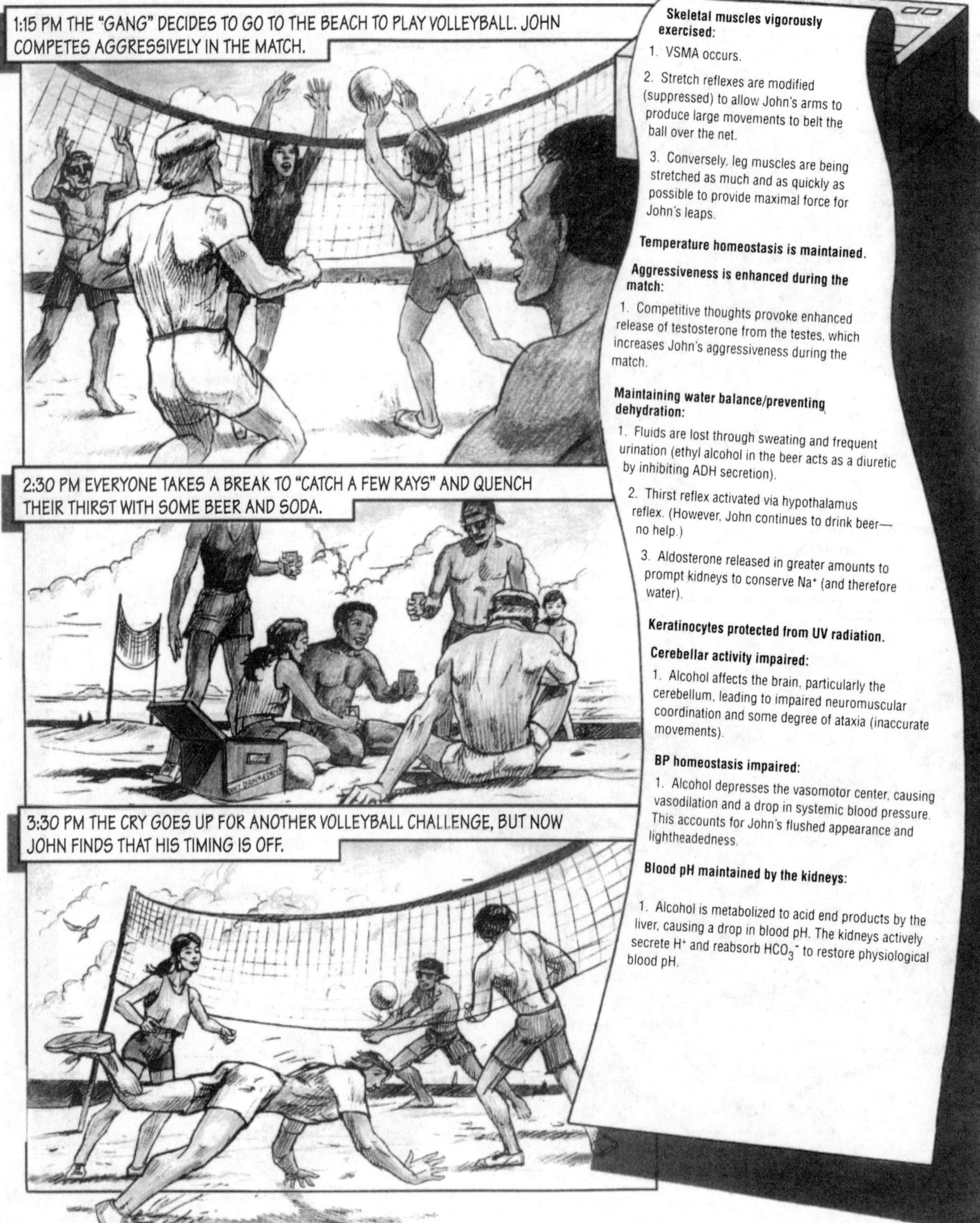
1:15 PM THE "GANG" DECIDES TO GO TO THE BEACH TO PLAY VOLLEYBALL. JOHN COMPETES AGGRESSIVELY IN THE MATCH.
2:30 PM EVERYONE TAKES A BREAK TO "CATCH A FEW RAYS" AND QUENCH THEIR THIRST WITH SOME BEER AND SODA.
3:30 PM THE CRY GOES UP FOR ANOTHER VOLLEYBALL CHALLENGE, BUT NOW JOHN FINDS THAT HIS TIMING IS OFF.
Skeletal muscles vigorously exercised:
1. VSMA occurs.
2. Stretch reflexes are modified (suppressed) to allow John's arms to produce large movements to belt the ball over the net.
3. Conversely, leg muscles are being stretched as much and as quickly as possible to provide maximal force for John's leaps.
Temperature homeostasis is maintained.
Aggressiveness is enhanced during the match:
1. Competitive thoughts provoke enhanced release of testosterone from the testes, which increases John's aggressiveness during the match.
Maintaining water balance/preventing dehydration:
1. Fluids are lost through sweating and frequent urination (ethyl alcohol in the beer acts as a diuretic by inhibiting ADH secretion).
2. Thirst reflex activated via hypothalamus reflex. (However, John continues to drink beer—no help.)
3. Aldosterone released in greater amounts to prompt kidneys to conserve Na^+ (and therefore water).
Keratinocytes protected from UV radiation.
Cerebellar activity impaired:
1. Alcohol affects the brain, particularly the cerebellum, leading to impaired neuromuscular coordination and some degree of ataxia (inaccurate movements).
BP homeostasis impaired:
1. Alcohol depresses the vasomotor center, causing vasodilation and a drop in systemic blood pressure. This accounts for John's flushed appearance and lightheadedness.
Blood pH maintained by the kidneys:
1. Alcohol is metabolized to acid end products by the liver, causing a drop in blood pH. The kidneys actively secrete H^+ and reabsorb HCO_3^- to restore physiological blood pH.

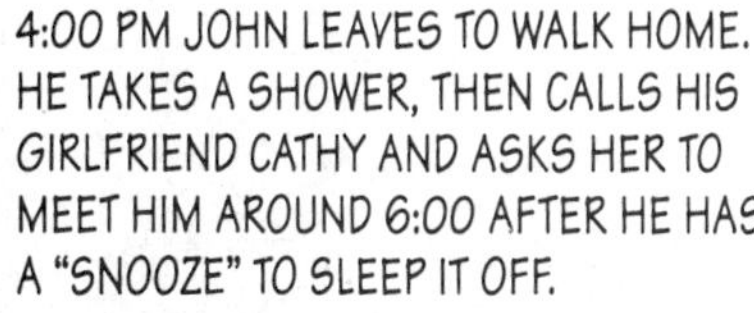

4:00 PM JOHN LEAVES TO WALK HOME. HE TAKES A SHOWER, THEN CALLS HIS GIRLFRIEND CATHY AND ASKS HER TO MEET HIM AROUND 6:00 AFTER HE HAS A "SNOOZE" TO SLEEP IT OFF.

6:00 PM WHEN THEY MEET, JOHN COMPLAINS THAT HE'S SO HUNGRY HIS "BACKBONE" IS TOUCHING HIS BELLYBUTTON." THEY GRAB A QUICK TACO BEFORE CATCHING A MOVIE.

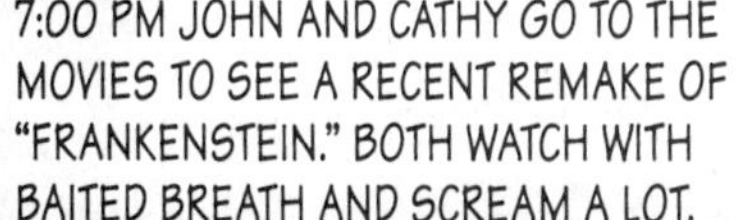

7:00 PM JOHN AND CATHY GO TO THE MOVIES TO SEE A RECENT REMAKE OF "FRANKENSTEIN." BOTH WATCH WITH BAITED BREATH AND SCREAM A LOT.

Events of postabsorptive state—Maintaining plasma glucose homeostasis:

1. Blood glucose levels dropping. Brain continues to use glucose as its energy fuel; other body organs using fatty acids (glucose sparing).

2. Glucagon released; promotes glycogenolysis and gluconeogenesis which cause glucose (and fatty acids) to be released to the blood.

Absorptive period—Processing of nutrient-laden blood:

1. Rising glucose and amino acid levels in plasma cause pancreatic beta cells to release insulin.

2. Blood draining from the digestive viscera is carried to the liver via the hepatic portal vein; liver cells respond to insulin by converting glucose to glycogen, liver sequesters amino acids to make plasma proteins.

3. Insulin accelerates glucose entry into tissue cells and speeds uptake of amino acids by tissue cells.

Vision and comprehension of viewed scenes.

Hearing:

1. The sound reaches both ears at the same time. The eardrums vibrate, transmit the vibration to the ossicles, which in turn transmit the vibratory motion to the fluids of the cochlea. Hair cells (hearing receptors) are activated according to their position on the basilar membrane and the sound frequencies.

2. Impulses transmitted to the auditory cortex in the temporal lobes where interpretation of the tones as words, screams, and other sounds occurs.

Sensory inputs evoke emotional responses:

1. Nearly all sensory pathways make synapses with limbic system structures where the sensory perception is coupled to emotional content on the basis of past experience. Hence, scenes recognized as threatening evoke screams or produce modifications of respiration (breath holding).

10:30 PM AFTER THE MOVIE, JOHN WALKS CATHY HOME. JOHN FINDS CATHY BEAUTIFUL AND EXCITING, AND AFTER A FEW KISSES, THEY DECIDE THAT THE BETTER PART OF WISDOM IS TO CALL IT A NIGHT.

11:30 PM AFTER HIS LONG AND BUSY DAY, JOHN IS REALLY TIRED AND DECIDES TO TURN IN. HE FALLS ASLEEP ALMOST AS SOON AS HIS HEAD HITS THE PILLOW.

Walking.

1. VSMA occurs.

Sexual arousal:

1. Cathy's fragrance, and the feel of her body against his as they nuzzle and kiss, arouse John sexually. Parasympathetic fibers cause vasodilation of the penile blood vessels. As the corpora cavernosa begins to fill with blood, erection begins to occur.
2. As John leaves, interruption of sexually pleasurable stimuli results in vasoconstriction and resumption of the flaccid penile state.

Sleep:

1. Initial drowsiness and floating feeling (indicated by alpha waves) gives way to sleep spindles, and finally delta wave sleep about 30 minutes later. During this progression of NREM sleep stages, vital signs (blood pressure, temperature, respiration rate, pulse) decline to lowest levels of the day. Gastric mobility increases during delta wave sleep; norepinephrine levels decline and serotonin levels rise in the brain.
2. About 90 minutes after sleep begins, REM occurs. The EEG backtracks to alpha waves, vital signs increase, and gastric mobility declines. Skeletal muscles, except the ocular muscles (which move rapidly under the lids), are temporarily paralyzed. Norepinephrine and ACh levels rise in the brain. Dreaming occurs. (Sweet dreams, John.)
3. NREM and REM sleep alternate throughout the night, with REM periods becoming closer together and body temperature rising as time of arousal nears.

Appendix: Answers

Chapter 1 The Human Body: An Orientation

BUILDING THE FRAMEWORK

An Overview of Anatomy and Physiology

1. 1. Regional anatomy: All structures (bones, muscles, blood vessels, etc.) in a particular body *region* studied at the same time. Systemic anatomy: All organs of an organ system studied simultaneously; for example, if one is studying the skeletal system, all the bones of the body are studied. 2. Gross anatomy: Studies easily visible structures (bones, muscles, etc.). Microscopic anatomy: Studies structures that cannot be viewed without a microscope—for example, cells and tissues. 3. Developmental anatomy: Studies the changes in body structure that occur throughout life. Embryology: Considers only those changes that occur from conception to birth. 4. Histology: Study of tissues, collections of cells with similar structure and function. Cytology: Study of parts of a cell and the functions of those parts.
2. Physiological study: C, D, E, F, G, H, J   Anatomical study: A, B, I, K, L, M
3. complementarity

Levels of Structural Organization

1. cells, tissues, organs, organ systems
2. 1. atom   2. epithelium   3. heart   4. digestive system
3.

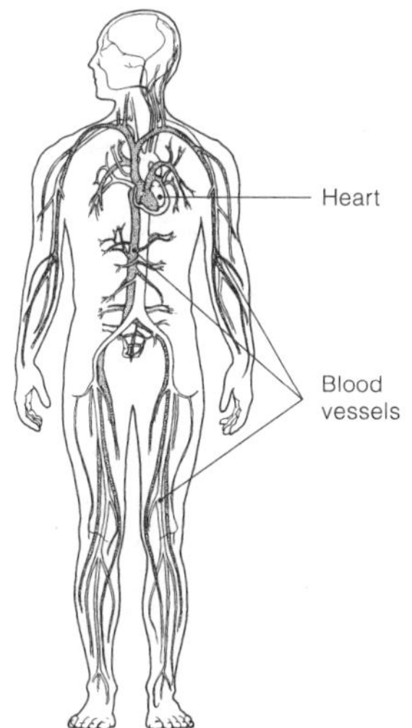

Figure 1.1
Cardiovascular system

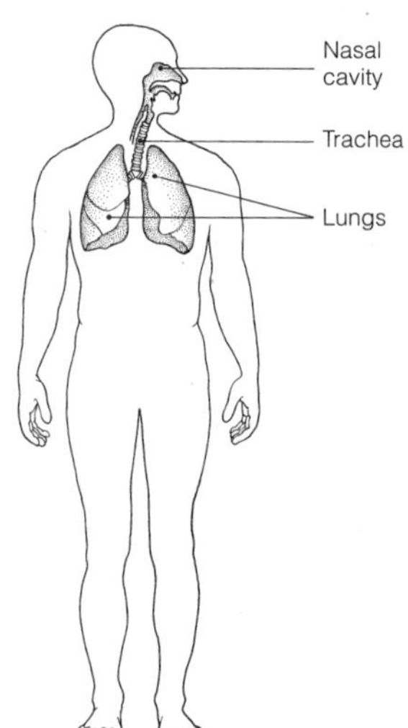

Figure 1.2
Respiratory system

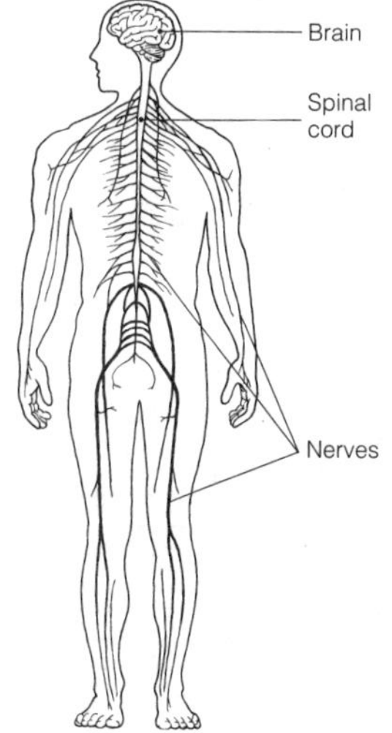

Figure 1.3
Nervous system

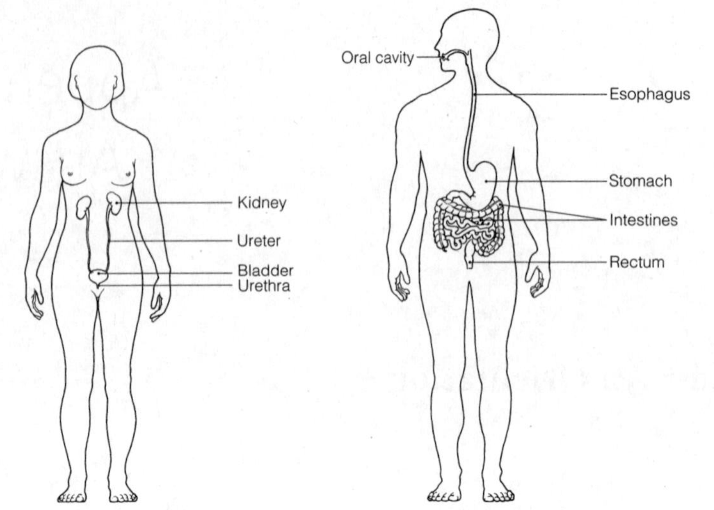

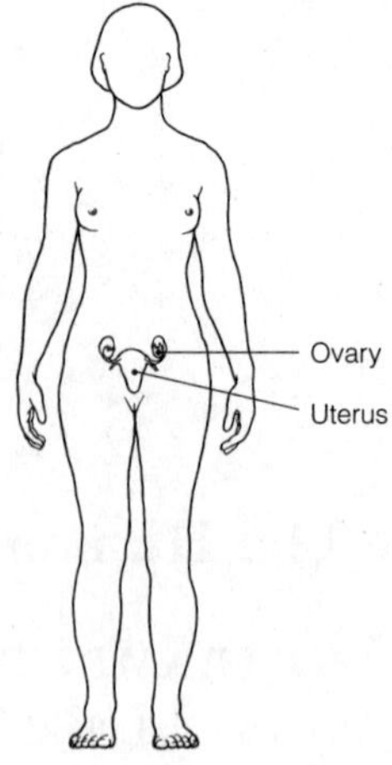

Figure 1.4
Urinary system

Figure 1.5
Digestive system

Figure 1.6
Reproductive system

4. 1. K 2. C 3. J 4. A 5. D 6. E 7. B 8. I 9. A 10. F 11. K 12. C, H 13. C 14. D

5. 1. A 2. C 3. K 4. H 5. B 6. J 7. G

Maintaining Life

1. 1. D 2. H 3. C 4. A 5. B 6. G 7. F 8. E 9. D

2. 1. C 2. B 3. E 4. D 5. E 6. A

Homeostasis

1. A dynamic state of equilibrium in which internal conditions vary within narrow limits.

2.

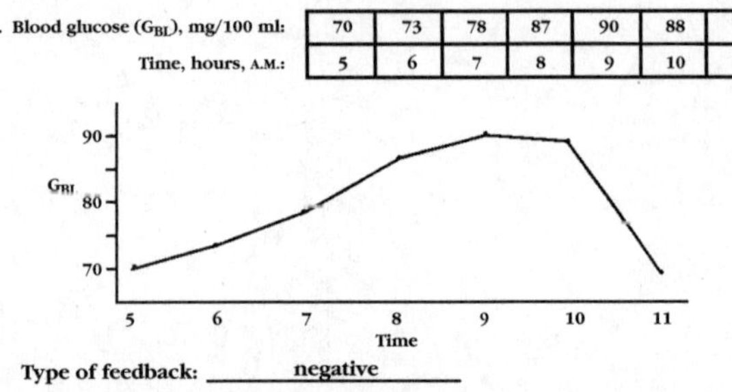

1. Blood glucose (G_{BL}), mg/100 ml:	70	73	78	87	90	88	76
Time, hours, A.M.:	5	6	7	8	9	10	11

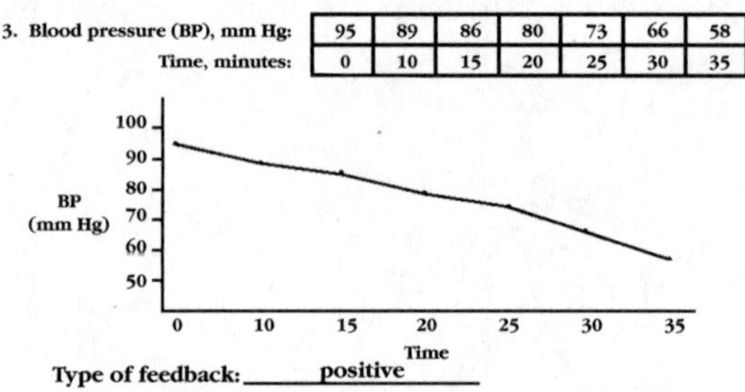

3. Blood pressure (BP), mm Hg:	95	89	86	80	73	66	58
Time, minutes:	0	10	15	20	25	30	35

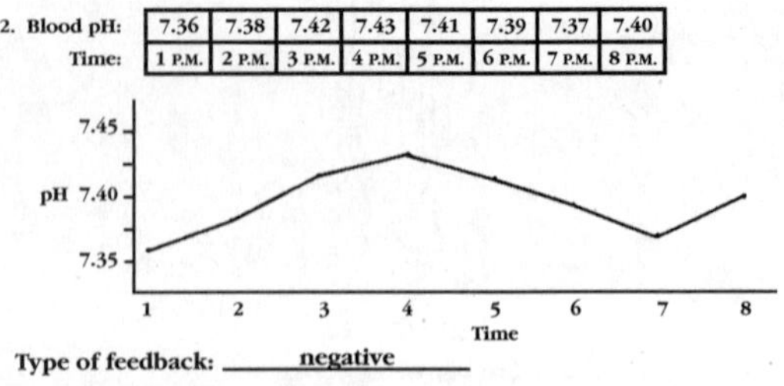

2. Blood pH:	7.36	7.38	7.42	7.43	7.41	7.39	7.37	7.40
Time:	1 P.M.	2 P.M.	3 P.M.	4 P.M.	5 P.M.	6 P.M.	7 P.M.	8 P.M.

The Language of Anatomy

1. 1. A 2. G 3. D 4. D 5. F 6. A 7. F 8. H 9. B 10. A 11. D 12. I 13. K 14. C 15. K 16. J 17. A

2. 1. elbow 2. shoulder 3. forehead 4. muscles 5. knee 6. small intestine

3. 1. distal 2. antecubital 3. brachial 4. left upper quadrant 5. ventral cavity

4. 1. B 2. H 3. I 4. F 5. P 6. G 7. L 8. J 9. K 10. D

5. **Section A:** midsagittal **Section B:** transverse The hands are incorrectly depicted; the palms should face anteriorly.

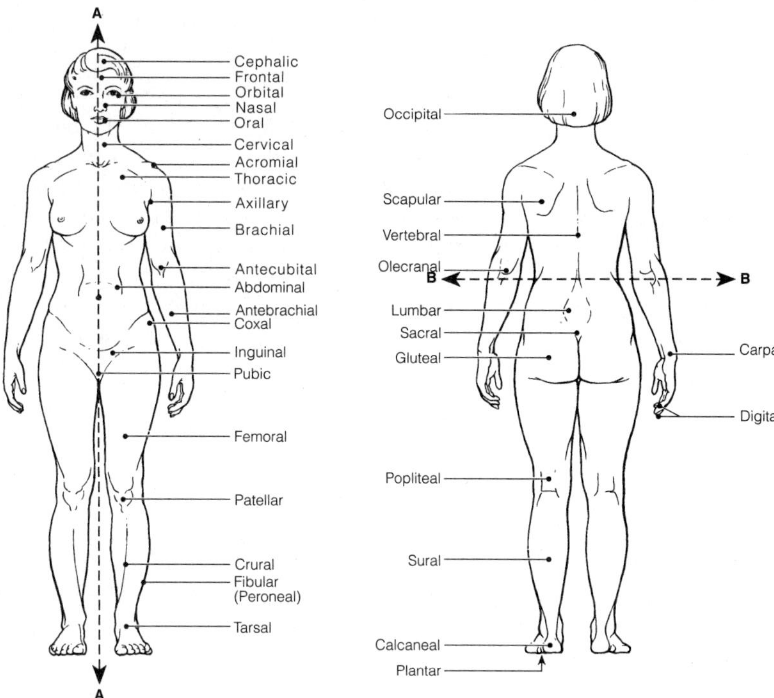

Figure 1.7

6. 1. ventral 2. dorsal 3. dorsal

7. 1. pleura 2. pericardium 3. peritoneum

8. 1. A 2. A 3. A 4. A 5. A 6. C 7. A 8. D 9. D 10. B 11. A 12. A

9. 1. 2, 3, 7, 11, 12 2. 2, 3 3. 2 4. 1, 2, 3, 5 5. 2, 3

10.

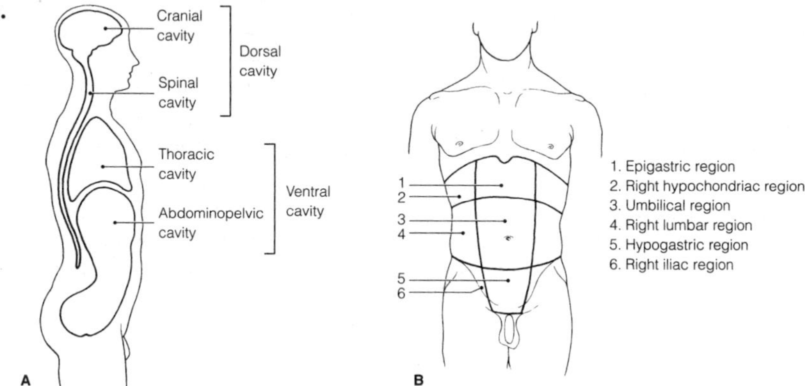

Figure 1.8

11.

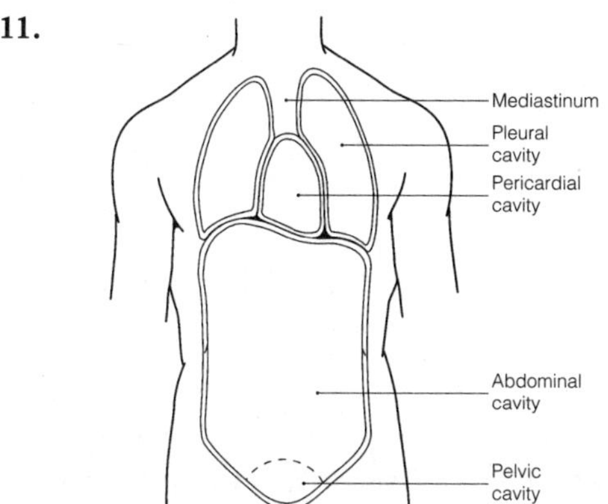

Figure 1.9

12. 1. F, D 2. F, E 3. C, B 4. F, A 5. F, A

CHALLENGING YOURSELF

At the Clinic

1. skeletal, muscular, cardiovascular, integumentary, nervous
2. the need for nutrients and water
3. The high blood pressure increases the workload on the heart. Circulation of blood decreases, and the heart itself begins to receive an inadequate blood supply. As the heart weakens further, the backup in the veins worsens and the blood pressure rises even higher leading to severe edema. Without intervention, circulation becomes so sluggish that organ failure sets in. A heart-strengthening medication will increase the force of the heartbeat so that more blood is pumped out with each beat. The heart can then pump more blood, reducing the backup and increasing circulation. The blood supply to the heart musculature also improves and the heart becomes stronger.
4. Fever, the elevation of body temperature, helps the immune system ward off infection, apparently by inhibiting the reproduction of infectious organisms. The stimulus (presence of infectious agents) triggers a response (fever) that reduces the stimulus (presence of foreign invaders) by calling the immune system into play. Thus, the temporary increase in body temperature returns the body to homeostasis.
5. right side, below the rib cage
6. 1 (c), 2 (a), 3 (b), 4 (d)
7. The physician will probably use ultrasound because it will provide the information desired (approximation of head size) with the least danger to the fetus.
8. Perhaps the best choice—CT scans, which can localize structures of the brain very precisely—is not one of the options. Of the options given, MRI is the best choice because it visualizes gray and white matter of the brain very clearly.
9. The anterior and lateral aspects of the abdomen have no bony (skeletal) protection.
10. The secretion of the hypothalamic releasing factor, ICSH, and testosterone are all reduced.

Stop and Think

1. 1. protection 2. absorb light 3. absorption 4. thick, hollow, muscular organ 5. thin, elastic tissue
2. integumentary—boils, acne, athlete's foot; digestive—gingivitis, colitis, gastritis; respiratory—cold and flu; urinary—cystitis and kidney infections; reproductive—sexually transmitted infections
3. maintenance of boundaries, movement, responsiveness, digestion, metabolism, excretion, and growth
4. nutrients, oxygen, and water; nutrients can be stored
5. Negative means opposite: a parameter that moves away from its optimum value will trigger changes that move it in the opposite direction, back toward its optimum.
6. A serous membrane forms as a hollow sphere that wraps around an organ and then continues on to cover the walls of the cavity that house the organ. Since the organ does not puncture through its covering serous membrane to gain entry to the space between the two serous membrane layers, the result is a double membrane with fluid between the layers. The layers slide over each other with very little friction.
7. Anatomists describe positions of body parts with extreme precision; the smaller, more numerous abdominopelvic regions localize the parts very accurately. Medical personnel are often rushing to locate an injury and often rely on the patient's description of the site of pain. Quadrants give sufficient information and take less time to describe.
8. frontal; sagittal; transverse

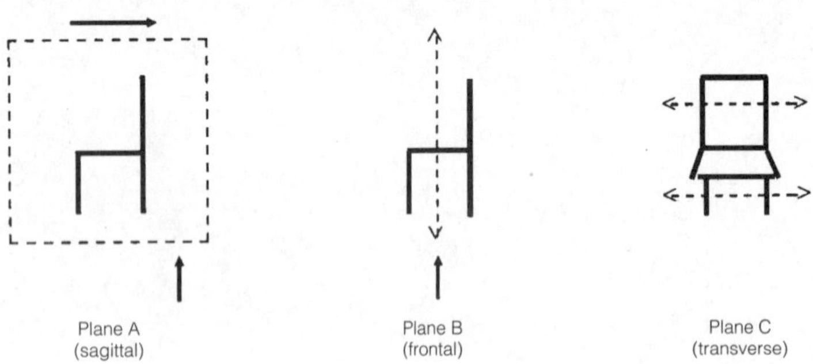

9. Dominic's hernia is in his groin; his lower back just above the buttocks is painful due to the infected kidney; the perineal area encompasses anal and genital structures between the legs.

COVERING ALL YOUR BASES

Multiple Choice

1. D **2.** E **3.** B **4.** C **5.** D **6.** A, B, C, D, E **7.** C **8.** C **9.** A, B, C, D **10.** A, D **11.** B **12.** A **13.** B **14.** B **15.** B **16.** C **17.** C **18.** A **19.** D **20.** B **21.** C **22.** D **23.** A, C, E **24.** B, C, D, E **25.** A, B, D **26.** A, B, D **27.** D **28.** B

Word Dissection

	Word root	Translation	Example
1.	ana	apart	anatomy
2.	chondro	cartilage	hypochondriac
3.	corona	crown	coronal
4.	cyto	cell	cytology
5.	epi	upon	epigastric
6.	gastr	stomach, belly	hypogastric
7.	histo	tissue	histology
8.	homeo	same	homeostasis
9.	hypo	below, under	hypogastric
10.	lumbus	loin	lumbar
11.	meta	change	metabolism
12.	ology	study of	physiology
13.	org	living	organism
14.	para	near	parasagittal
15.	parie	wall	parietal
16.	pathy	disease	pathology
17.	peri	around	pericardium
18.	stasis	standing still	homeostasis
19.	tomy	to cut	anatomy
20.	venter	belly	ventral
21.	viscus	organ	visceral

Chapter 2 Chemistry Comes Alive

BUILDING THE FRAMEWORK

PART 1: BASIC CHEMISTRY

Definition of Concepts: Matter and Energy

1. 1. B, D 2. A, B, C, D 3. A, B
2. 1. C 2. B, D 3. C 4. A 5. D

Composition of Matter: Atoms and Elements

1.

Particle	Location	Electrical charge	Mass
Proton	Nucleus	+1	1 amu
Neutron	Nucleus	0	1 amu
Electron	Orbitals	−1	0 amu

2. 1. O 2. C 3. K 4. I 5. H 6. N 7. Ca 8. Na 9. P 10. Mg 11. Cl 12. Fe
3. 1. E 2. F 3. C 4. B 5. B 6. D 7. A 8. G 9. I 10. J 11. H 12. I
4. 1. Ca, P 2. C, O, H, N 3. Fe 4. Ca, K, Na 5. I 6. P 7. Cl

How Matter Is Combined: Molecules and Mixtures

1. 1. Molarity 2. Mixture 3. Solution 4. Suspension 5. Colloid 6. Colloid 7. Suspension
2. 1. H_2O_2 is a molecule of a compound. $2OH^-$ are two negative ions. 2. $2O^{2-}$ are two negative ions. O_2 is a molecule of an element. 3. $2H^+$ are two positive ions. H_2 is a molecule of an element.
3. 1. water = 18 2. ammonia = 17 3. carbonic acid = 62
4. 1. Perspiration 2. $Ca_3(PO_4)_2$ 3. Scatters light 4. Water 5. One-molar NaCl 6. Salt water

Chemical Bonds

1. 1. 6 2. 12 amu 3. carbon 4. 4 5. ±4 6. active 7. isotope 8. radioisotope 9. true

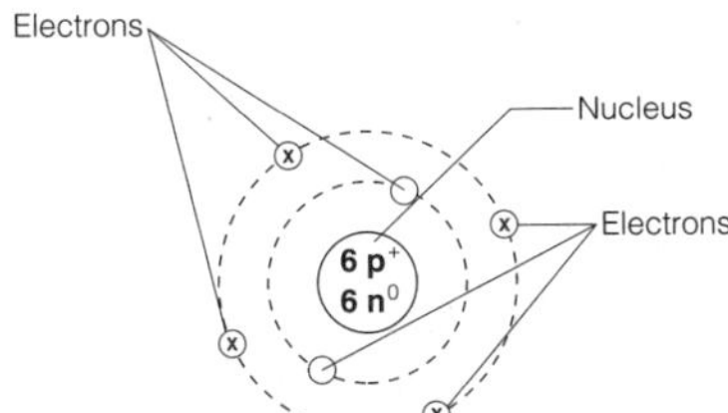

Figure 2.1

2. A. Bond = ionic; name of compound = LiF (lithium fluoride)
   B. Bond = covalent; name of compound = HF (hydrogen fluoride)

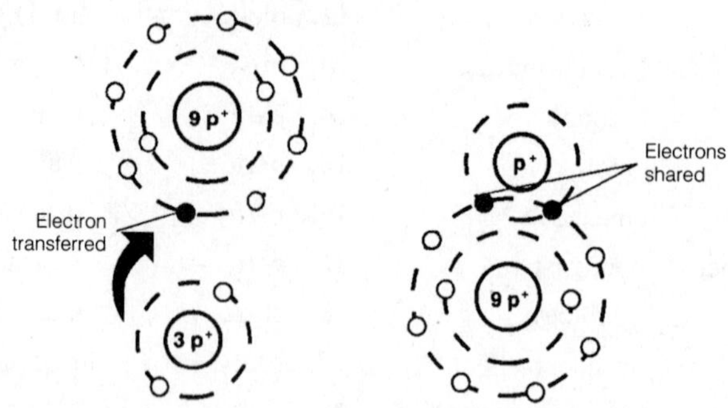

Figure 2.2

3.

Chemical symbol	Atomic number	Atomic mass	Electron distribution
H	1	1	1, 0, 0
C	6	12	2, 4, 0
N	7	14	2, 5, 0
O	8	16	2, 6, 0
Na	11	23	2, 8, 1
Mg	12	24	2, 8, 2
P	15	31	2, 8, 5
S	16	32	2, 8, 6
Cl	17	35	2, 8, 7
K	19	39	2, 8, 8, 1
Ca	20	40	2, 8, 8, 2

4. (Table completion: Atoms forming ionic bonds)

Chemical symbol	Loss/gain of electrons	Electrical charge
H	Loses 1	+1
Na	Loses 1	+1
Mg	Loses 2	+2
Cl	Gains 1	−1
K	Loses 1	+1
Ca	Loses 2	+2

5. (Table completion: atoms that form covalent bonds)

Chemical Symbol	Number of electrons shared	Number of bonds made
H	1	1
C	4	4
N	3	3
O	2	2

6.

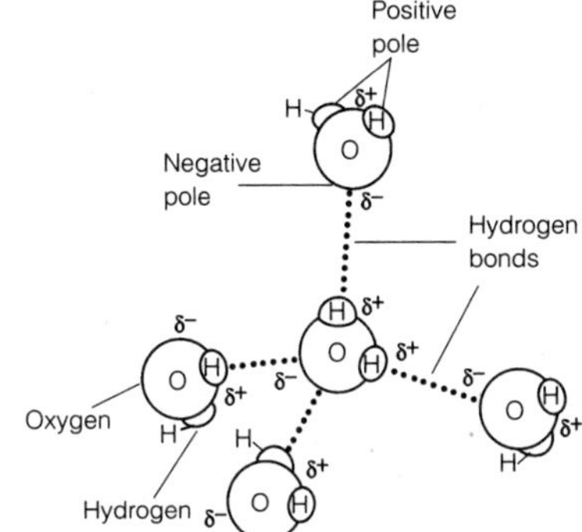

Figure 2.3

7. Circle B, C, E

Chemical Reactions

1. 1. A 2. C 3. B 4. C and D

2. 1. low reactant concentration 2. large particles 3. reactant 4. electron acceptor 5. simple → complex

PART 2: BIOCHEMISTRY

1. Organic: 3, 5, 6, 8

Inorganic Compounds

1. 1. heat capacity 2. water (aqueous solution) 3. 70% (60% to 80%) 4. hydrogen bonds 5. heat of vaporization 6. hydrolysis (digestion) 7. synthesis (dehydration)

2. 1. Milk of magnesia 2. Urine (pH 6.2) 3. $NaHCO_3$ 4. Organic 5. pH 4 6. pH 2

3. 1. A 2. B 3. D 4. B 5. A 6. D 7. D 8. A 9. C 10. B

4. a measure of acidity or H^+ ion concentration

5. Weak acid: B, C, E; strong acid: A, D, E, F, G

6. 1. solvent 2. solutes 3. acidic

Organic Compounds

1. 1. G 2. D, E 3. A 4. F 5. H 6. B 7. C 8. G 9. C 10. A 11. F 12. F 13. C 14. H 15. B 16. C 17. H 18. C

2. 1. T 2. triglycerides 3. T 4. polar 5. T 6. ATP 7. T 8. peptide 9. glucose 10. T 11. phosphate

3. 1. starch or glycogen 2. A, E 3. tertiary 4. C, D 5. E

Figure 2.4: A. monosaccharide B. globular protein C. polysaccharide D. fat E. nucleotide

4. 1. T 2. substrates 3. -ase 4. T 5. decrease 6. T 7. secondary and possibly primary

5. 1.C 2. C 3. B 4. A

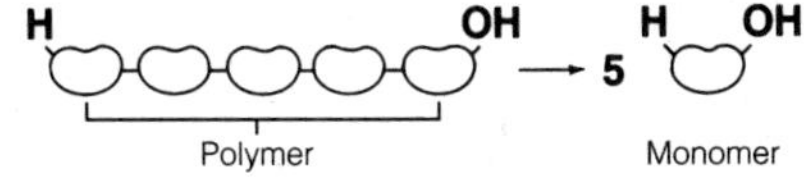

Figure 2.5

6. X all but #3.

7.

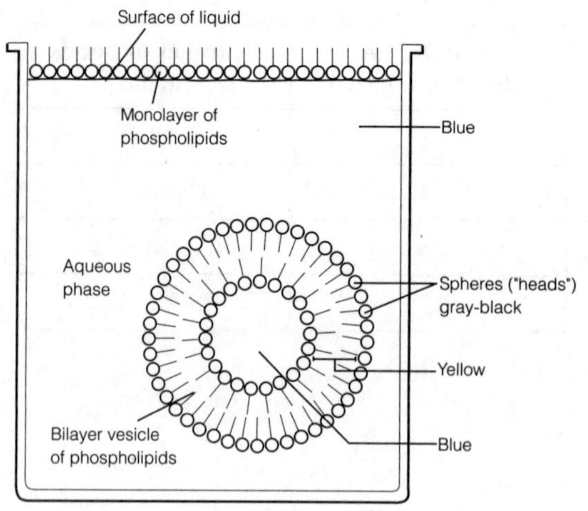

Figure 2.6

8. Unnamed nitrogen bases: thymine (T) and guanine (G)

1. hydrogen bonds 2. double helix 3. 12 4. complementarity

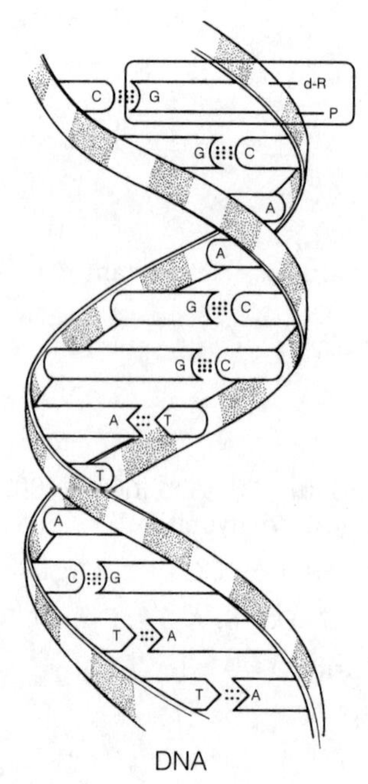

Figure 2.7

9. 1. Glucose 2. Ribose 3. Glycogen 4. Glycerol 5. Glucose 6. Hydrolysis 7. Carbon 8. Denaturation

10. A. cyclic AMP B. ATP

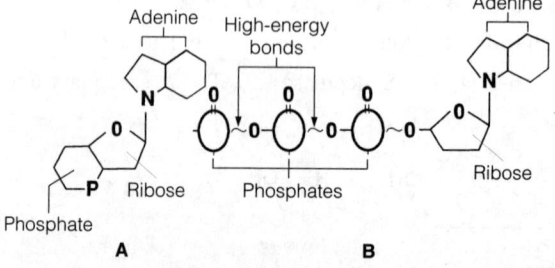

Figure 2.8

11.

Name of protein?	Secondary or tertiary structure?	Molecular shape: fibrous or globular?	Water-soluble or insoluble?	Function of protein?	Location in body?
Collagen	Secondary	Fibrous	Insoluble	Structure	Cartilage and bone
Proteinase	Tertiary	Globular	Soluble	Enzyme-hydrolysis	Stomach
Hemoglobin	Quaternary	Globular	Soluble	Carries O_2	Red blood cells
Keratin	Secondary	Fibrous	Insoluble	Structure; also prevents water loss	Skin and hair

The Incredible Journey

1. 1. negative 2. positive 3. hydrogen bonds 4. red blood cells 5. protein 6. amino acids 7. peptide 8. H^+ and OH^- 9. hydrolysis or digestion 10. enzyme 11. glucose 12. glycogen 13. dehydration or synthesis 14. H_2O 15. collagen 16. hydration (water) 17. increase 18. evaporate 19. thermal 20. vaporization

CHALLENGING YOURSELF

At the Clinic

1. Acidosis means blood pH is below the normal range. The patient should be treated with something to raise the pH.
2. Lack of sufficient ventilation would cause CO_2 to accumulate in the blood, thus increasing the amount of carbonic acid formed. This addition of acid would decrease the blood pH, causing acidosis.
3. As hydrogen ion concentration builds, causing acidosis, Hugo will hyperventilate. This will lower the amount of CO_2 in the blood and reduce the amount of carbonic acid formed. The decrease in carbonic acid will help offset the increase in acid retention by his diseased kidneys.
4. An inability to absorb fat will reduce vitamin D absorption from the diet. Because cholesterol absorption also will be reduced, the amount of vitamin D formed in the body may also decrease. Without sufficient levels of vitamin D (which aids calcium uptake by intestinal cells), calcium absorption will also decrease. Consequently, the bones will lose calcium, resulting in soft bones.
5. Each of the 20 amino acids has a different chemical group called the R group. The R group on each amino acid determines how it will fit in the folded, three-dimensional, tertiary structure of the protein and the bonds it may form. If the wrong amino acid is inserted, its R group might not fit into the tertiary structure properly, or required bonds might not be made; hence the entire structure might be altered. Because function depends on structure, this means the protein will not function properly.
6. Chris should have added a catalyst to reduce the amount of activation energy needed.
7. Heat increases the kinetic energy of molecules. Vital biological molecules, like proteins and nucleic acids, are denatured (rendered nonfunctional) by excessive heat because intramolecular bonds essential to their functional structure are broken. Because all enzymes are proteins, their destruction is lethal.
8. Stomach discomfort is frequently caused by excess stomach acidity ("acid indigestion"). An antacid contains a weak base that will neutralize the excess H^+.
9. Some enzymes require a mineral or a vitamin as a cofactor. In such cases, if the particular vitamin or mineral is not available, the enzymes will not function properly.

Stop and Think

1. Breaking the ATP down to ADP and P_i releases the amount of energy stored in the bonds. Only part of that potential energy is actually used for cellular use; the rest is lost as heat. Nonetheless, the total amount of energy released (plus activation energy) must be absorbed to remake the bonds of ATP.
2. Not all chemical bonds are easily explained by the planetary models. The bonds in ozone and carbon monoxide do not fit into the explanation provided by the octet rule, as there is no way to satisfy the proper bond number for each atom in those molecules.
3. After helium (atomic number 2) would come atomic number 10 (2 + 8), then atomic number 18 (2 + 8 + 8).
4. The negative surface charges on the proteins make them repel each other. If the proteins settled out, they would come in closer contact than their electrical charges would allow. Thus, the repulsion prevents settling. Additionally, water molecules orient to the charged proteins, keeping them separated by hydration layers. Hence, cellular fluid is a colloid.
5. Sodium chloride dissociates into two ions when it is dissolved in water; glucose does not dissociate. Because a one-molar solution of any substance has the same number of molecules, the original salt added was comparable to the amount of sugar. Once the salt dissociates, however, there are 2 moles of particles in solution: 1 mole of

sodium ions, and 1 mole of chloride ions. Because calcium chloride consists of three ions (one calcium ion and two chloride ions), a one-molar solution will have 3 moles of ions, giving it three times as many particles in solution as a one-molar glucose solution.

6. Calcium carbonate, $CaCO_3$.
7.

```
                                          H
H       H      H-O                        |    O
 \     /          \                       |  //
  N - N            C=O              H - C - C
 /     \          /                       |    \
H       H      H-O                        |     O-H
                                          H
```

8. If water molecules were closer together in ice than in liquid water, ice would be denser than water and would sink instead of floating. As a lake surface froze, the ice would sink. This would expose additional lake water to freezing conditions, and the additional ice formed would also sink. The process would continue until the lake froze from the bottom up. Complete thawing would be extremely slow because the warm air and sunlight would not reach the lake bottom easily. Because ice floats, a layer of ice insulates the water beneath it, slowing the freezing process and usually preserving some liquid water for aquatic life to survive the winter.
9. Because ammonia is a polar molecule, the large nitrogen atom is slightly electronegative and can attract hydrogen ions.
10. Raising the temperature increases the kinetic energy of molecules, so water dissociates more at 38°C than at room temperature. The increased dissociation results in a greater number of free hydrogen ions, which, because pH is a measure of free hydrogen ions, will be reflected in a lower pH. Thus, although water is neutral by definition (hydrogen ions always equal hydroxyl ions), the numerical standard of pH 7 is not always precisely neutral.
11. They are the most energetic electrons of an atom and the only ones to participate in bonding behavior.
12. The fact that fats float indicates that triglycerides (nonpolar compounds) and water (polar molecules) do not mix; hence fats will not go into aqueous solution (dissolve in water). Thus, they can be stored in an "inactive" state that does not interfere with solute concentrations needed for other chemical reactions.
13. "a" = without; "tom" = cut; atom means "indivisible"
14. Chlorine can make covalent bonds. If an atom of chlorine combines with an atom with similar electronegativity (such as another chlorine atom), one covalent bond will be formed, as it needs only one electron to fill its valence shell.
15. Because the relative concentrations of reactants and products govern the direction of a reversible reaction, a cell can control direction by controlling concentration. Cells can manipulate concentrations of either reactants or products.
16. Water is an acid because it releases a hydrogen ion when it dissociates. Water is a base because the negative pole at the oxygen atom can attract a hydrogen ion, in the same way as ammonia.
17. The pH is 6; this solution is 10 times as acidic as a solution of pH 7.
18.

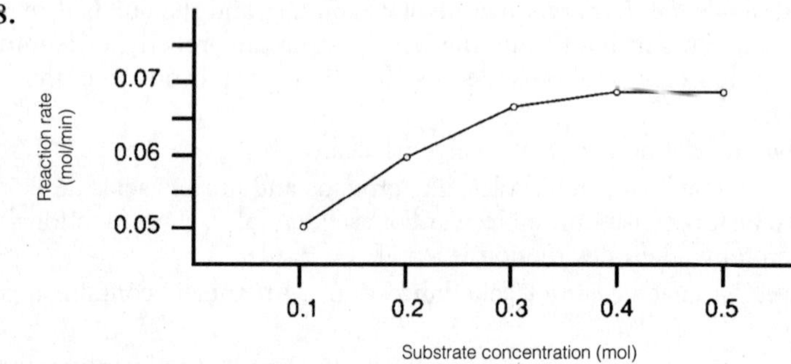

    1. The plateau indicates saturation of enzyme.
    2. Interference with active site would decrease reaction rate.

19. Energy does not occupy space and has no mass. Energy is described in terms of its effect on matter, i.e., its ability to do work or put matter into motion.
20. Unsaturated fats are oils; they tend to have shorter fatty acid chains containing one or more carbons that have double or triple bonds. Saturated fats are solid (like butter or lard); their fatty acids are longer and their contained carbon atoms are fully saturated with hydrogen (all single bonds).
21. The reactants are X and YC; the products are XC and Y. All are 1 mole quantities.
22. The system will never reach equilibrium because the reaction continues to go to the right.
23. Radioisotopes are used to destroy cancerous tissue, in scans for diagnostic purposes, for tracing out metabolic pathways, and for carbon dating.
24. Evelyn has less "insulation" (subcutaneous fat) and a larger relative surface area, so she loses heat to the environment much more readily than Barbara.

COVERING ALL YOUR BASES

Multiple Choice

1. B, E **2.** C **3.** B, C **4.** A, C, D **5.** D **6.** B **7.** C, D, E **8.** A **9.** A, B, E **10.** A **11.** E **12.** A, B, C, D, E **13.** E **14.** D **15.** A **16.** B **17.** C, E **18.** C, D **19.** C **20.** D **21.** A, E **22.** A **23.** A, B, C, D, E **24.** A, D **25.** B, C **26.** C **27.** D **28.** B, C

Word Dissection

	Word root	Translation	Example		Word root	Translation	Example
1.	ana	up	anion, anabolic	**8.**	iso	equal	isotope
2.	cata	down	cation, catabolic	**9.**	kin	to move	kinetic
3.	di	two	dipole, disaccharide	**10.**	lysis	break	hydrolysis
4.	en	in	endergonic	**11.**	mono	one	monosaccharide
5.	ex	out	exergonic	**12.**	poly	many	polysaccharide
6.	glyco	sweet	glycogen	**13.**	syn	together	synthesis
7.	hydr	water	dehydration	**14.**	tri	three	triglyceride

Chapter 3 Cells: The Living Units

BUILDING THE FRAMEWORK

Overview of the Cellular Basis of Life

1. 1. A cell is the basic structural and functional unit; an organism's activity is dependent on cellular activity; biochemical activity determines and is determined by subcellular structure; continuity of life is based on cell reproduction. 2. cubelike, tilelike, disk-shaped, spherical, branching, cylindrical 3. plasma membrane, cytoplasm, nucleus 4. A model that describes a cell in terms of common features/functions that all cells share.

The Plasma Membrane: Structure and Functions

1.

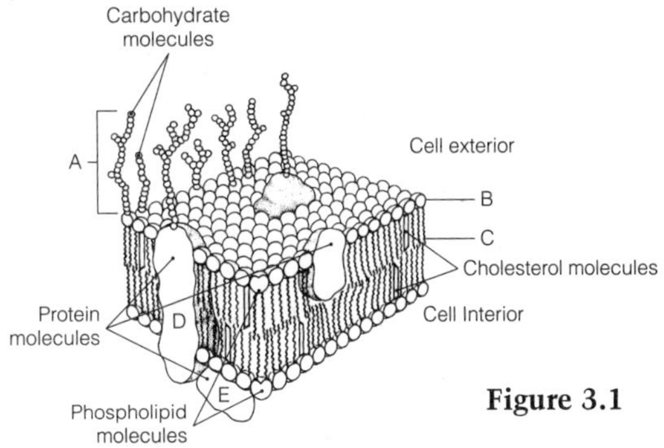

Figure 3.1

1. fluid mosaic model 2. to stabilize the plasma membrane by wedging between the phospholipid "tails" 3. glycocalyx 4. C 5. hydrophobic 6. D: integral; E: peripheral

2.

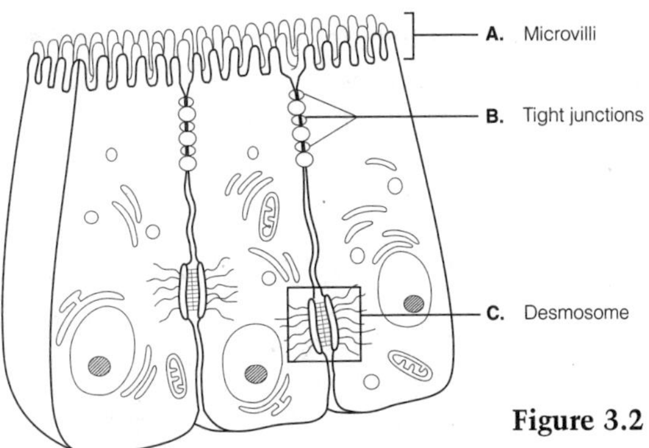

Figure 3.2

1. Microvilli increase the surface area of the plasma membrane. 2. Microvilli are found on cells involved in secretion and/or absorption. 3. actin 4. the glycocalyx and tongue-in-groove folding of adjacent plasma membranes 5. tight junction 6. desmosome 7. desmosome 8. gap junction 9. connexons

3.

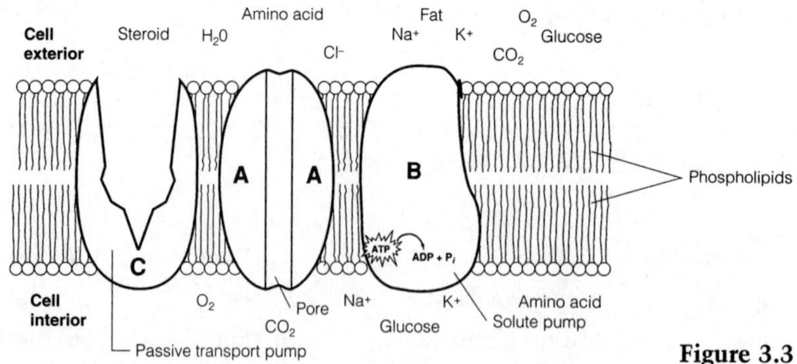

Figure 3.3

Arrow for Na^+ should be red and shown leaving the cell; those for glucose, Cl^-, O_2, fat, and steroids (except cholesterol, which enters by receptor-mediated endocytosis) should be blue and entering the cell. CO_2 (blue arrow) should be leaving the cell and moving into the extracellular fluid. Amino acids and K^+ (red arrows) should be entering the cell. Water (H_2O) moves passively (blue arrows) through the membrane (in or out) depending on local osmotic conditions.
1. fat, O_2, CO_2, some H_2O 2. glucose 3. H_2O, (probably) Cl^- 4. Na^+, K^+, amino acids

4. 1. D, E, G, H 2. B, C 3. A 4. B, C 5. H 6. E, F, G 7. H 8. D, E, F, G 9. E 10. D

5. 1. A. hypertonic B. isotonic C. hypotonic 2. A. crenated B. normal (discoid) C. spherical, some hemolyzed 3. same solute concentration inside and outside the cell 4. Water is moving by osmosis into the cell from the site of higher water concentration (cell exterior) to the site of lower water concentration (cell interior). 5. Tonicity deals with the effect of nonpenetrating solutes on the movement of water into or out of cells; osmolarity is a measure of the total solute concentration, both penetrating and nonpenetrating solutes.

6.

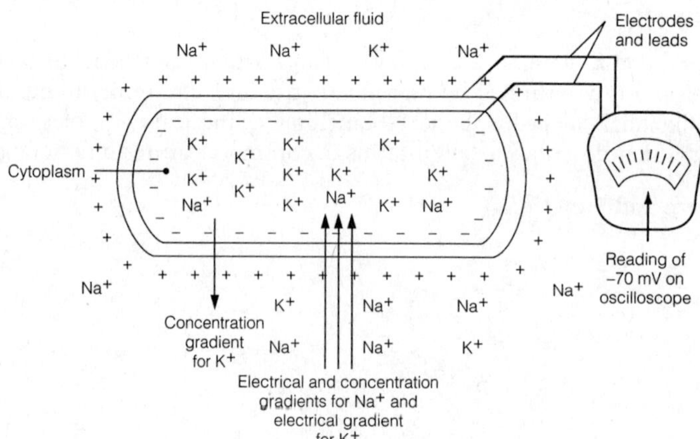

Figure 3.5

Potassium (K^+) is most important in determining the resting membrane potential.

7. 1. Communication between adjacent cells 2. Impermeable junction 3. Impermeable intercellular space 4. High extracellular K^+ concentration 5. Protein anions move out of cell 6. More K^+ pumped out than Na^+ carried in 7. Carbohydrate chains on cytoplasmic side of membrane 8. Nonselective 9. Exocytosis 10. CAMs 11. NO 12. Phospholipids

The Cytoplasm

1. The unstructured gel-like part of the cytoplasm that contains biological molecules in solution and organelles.
2. Organelles: the highly structured functional structures within the cell. Inclusions: stored nutrients, crystals of various types, secretory granules, and waste products in the cell.
3.

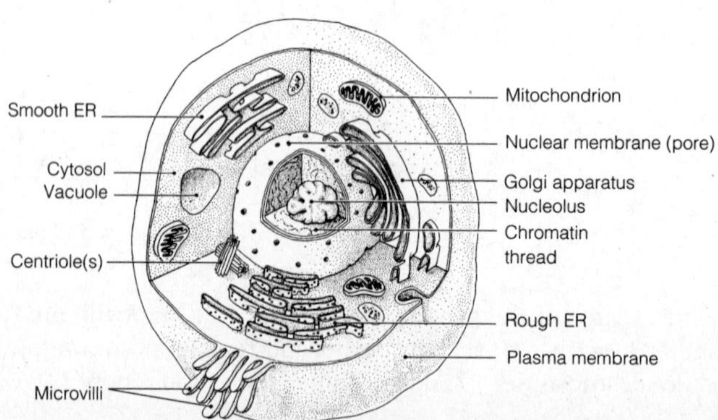

Figure 3.6

4.

Cell structure	**Location**	**Function**
Plasma membrane	External boundary of the cell	Confines cell contents; regulates entry and exit of materials
Lysosome	Scattered in cytoplasm	Digests ingested materials and worn-out organelles
Mitochondria	Scattered throughout the cell	Controls release of energy from foods; forms ATP
Microvilli	Projections of the plasma membrane	Increase the membrane surface area
Golgi apparatus	Near the nucleus (in the cytoplasm)	Packages proteins to be exported from the cell; packages lysosomal enzymes
Centrioles	Two rod-shaped bodies near the nucleus	"Spin" the mitotic spindle
Smooth ER	In the cytoplasm	Site of steroid synthesis and lipid metabolism
Rough ER	In the cytoplasm	Transports proteins (made on its ribosomes) to other sites in the cell; site of membrane lipid synthesis
Ribosomes	Attached to ER membranes or scattered in the cytoplasm	Synthesize proteins
Cilia	Extensions of cell to exterior	Act collectively to move substances across cell surface in one direction
Microtubules	Internal structure of centrioles; part of the cytoskeleton	Important in cell shape; suspend organelles
Peroxisomes	Throughout cytoplasm	Detoxify alcohol and free radicals accumulating from normal metabolism
Microfilaments	Throughout cytoplasm; part of cytoskeleton	Contractile protein (actin); moves cell or cell parts; core of microvilli
Intermediate filaments	Part of cytoskeleton	Act as internal "guy wires"; help form desmosomes
Inclusions	Dispersed in the cytoplasm	Provide nutrients; represent cell waste products, etc.

5. 1. Centrioles 2. Cilia 3. Smooth ER 4. Vitamin A storage 5. Mitochondria 6. Ribosomes 7. Lysosomes

6. 1. microtubules 2. intermediate filaments 3. microtubules 4. microfilaments 5. intermediate filaments 6. microtubules

7. 1. B 2. F 3. D 4. E 5. C, H 6. G 7. A

8. The nuclear and plasma membranes and essentially all organelles except mitochondria and cytoskeletal elements make up the endomembrane system. The components of this system act together to synthesize, store, and export cell products and to degrade or detoxify ingested or harmful substances.

The Nucleus

1.

Nuclear structure	General location/appearance	Function
Nucleus	Usually in center of the cell; oval or spherical	Storehouse of genetic information; directs cellular activities
Nucleolus	Dark spherical body in the nucleus	Storehouse and assembly site for ribosomes
Chromatin	Dispersed in the nucleus; threadlike	Contains genetic material (DNA); coils during mitosis
Nuclear membrane	Encloses nuclear contents; double membrane penetrated by pores	Regulates entry/exit of substances to and from the nucleus

2. 1. Cluster of histone proteins 2. The components are believed to play a role in regulating DNA function by exposing or not exposing certain DNA fragments.

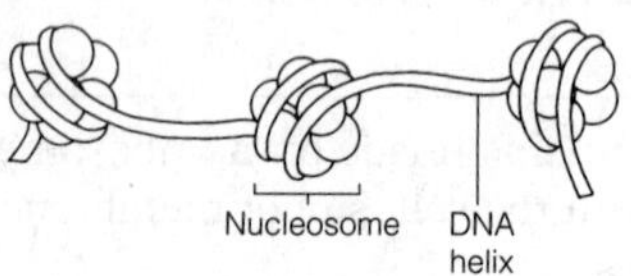

Figure 3.7

Cell Growth and Reproduction

1. The cell life cycle consists of interphase and the mitotic phase, during which the cell divides.

Phase of interphase	Important events
G_1	Cell grows rapidly and is active in its normal metabolic activities. Centrioles begin replicating.
S	Cell growth continues. DNA is replicated, new histone proteins are made, and chromatin is assembled.
G_2	Brief phase when remaining enzymes (or other proteins) needed for cell division are synthesized; centriole replication completed.

2.

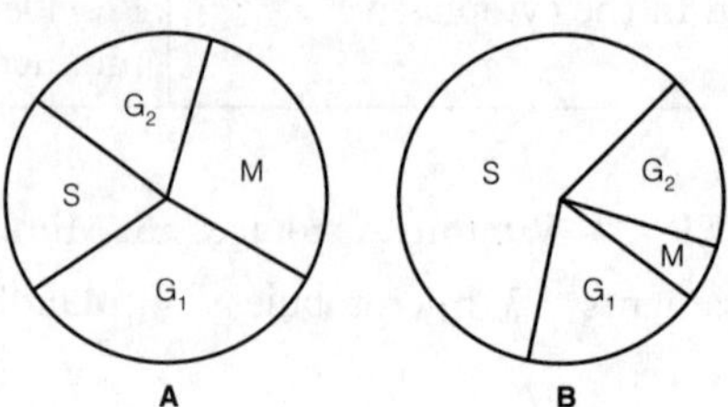

Figure 3.8

3. 1. MPF (M-phase promoting factor) 2. Cdk (a kinase enzyme) 3. Cyclin

4. 1. C 2. A 3. D 4. D 5. B 6. C 7. C 8. E 9. B, C 10. C 11. E 12. A, B 13. E 14. A

5. Figure 3.9: A. Prophase B. Anaphase C. Telophase D. Metaphase

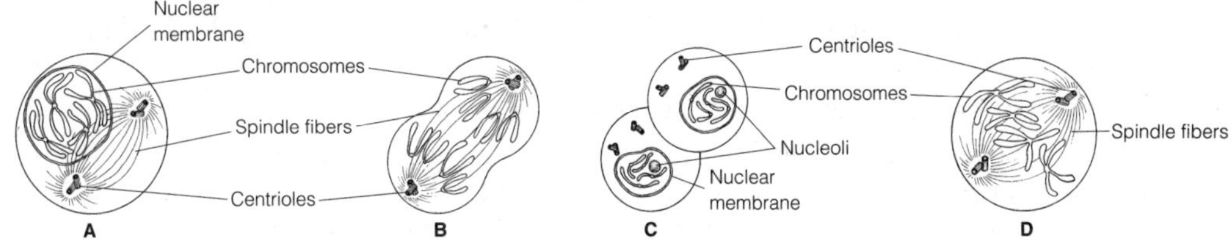

Figure 3.9

Figure 3.10: A. Prophase B. Metaphase C. Anaphase (early) D. Telophase E. Anaphase (late)

6. 1. P 2. K 3. O 4. T 5. C 6. B 7. E 8. F 9. V 10. S 11. Q 12. M 13. U 14. L 15. N 16. H 17. I 18. R

7. 1. transcription 2. translation 3. DNA 4. anticodon; triplet 5. One or more segments of a DNA strand that programs one polypeptide chain (or one tRNA or rRNA)

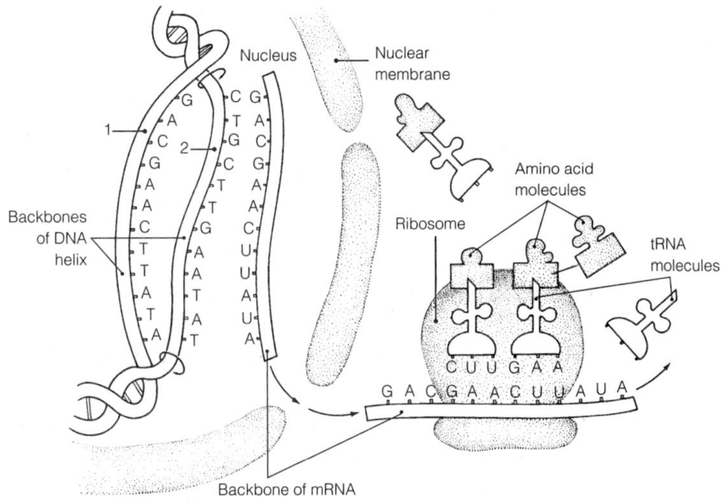

Figure 3.11

8. 1. nucleus 2. cytoplasm 3. coiled 4. centromeres 5. binucleate cell 6. spindle 7. interphase

Extracellular Materials

1. 1. Body fluids: e.g., blood plasma, interstitial fluid, cerebrospinal fluid, eye humors. Important in transport in body and as solvents.

2. Cellular secretions: e.g., saliva, gastric juice, mucus, tears. Important as lubricants and some aid the digestive process.

3. Extracellular matrix: e.g., glycoprotein intercellular "glue," and the matrix (ground substance and fibers) secreted by connective tissue cells. Important in binding cells together and providing strong structures (bones, cartilages, etc.) that can support or protect other body organs.

The Incredible Journey

1. 1. cytosol 2. nucleus 3. mitochondrion 4. ATP 5. ribosomes 6. rough endoplasmic reticulum 7. nuclear pores 8. chromatin 9. DNA 10. nucleoli 11. Golgi apparatus 12. lysosome (or peroxisome)

CHALLENGING YOURSELF

At the Clinic

1. Normally, lactose (a disaccharide) is digested to monosaccharides by the enzyme lactase in the intestine. Lactase deficiency prevents this digestion. Consequently, lactose remains in the intestinal lumen and acts as an osmotic agent to attract water and prevent its absorption. Hence, diarrhea (or watery feces) occurs. Adding lactase to milk hydrolyzes lactose to glucose and galactose, which can move into intestinal cells by facilitated diffusion. Water follows.

2. Increased capillary permeability causes more blood plasma to filter into the tissue spaces from the bloodstream. On a small scale, this causes the localized swelling (edema) associated with inflammation. In anaphylaxis, the systemic reaction causes so *much* fluid to leave the bloodstream that there is insufficient blood volume (fluid) to maintain circulation to the vital organs.
3. LDL is removed from the circulation by receptor-mediated endocytosis.
4. Glucose is reclaimed from the forming urine by carrier molecules in the plasma membranes of cells forming the tubules of the kidneys. If too much glucose is present, the carrier molecules are *saturated.* Hence, some glucose will not be reabsorbed and returned to the blood, but instead will be lost to the body in urine.
5. In blood stasis, the filtration pressure at the capillaries increases. Albumin (being a small protein) can be forced into the interstitial fluid under such conditions, increasing its osmotic pressure. If the osmotic pressure of the interstitial fluid exceeds that of the blood plasma, more of the water filtering out of the capillary will *remain* in the tissue spaces. This increase in interstitial fluid volume leads to edema.
6. Lysosomal destruction releases acid hydrolases into the cytoplasm, killing the cell. When the cell lyses, inflammation is triggered. Hydrocortisone is an anti-inflammatory steroid that stabilizes lysosomal membranes. Because hydrocortisone is a steroid, it is soluble in an oil base rather than a water base. Steroids can diffuse through the lipid bilayer of cell membranes, so they can be administered topically.
7. Streptomycin inhibits bacterial protein synthesis. If the cells are unable to synthesize new proteins (many of which would be essential enzymes), they will die.
8. Phagocytes ingest debris—living (bacteria) or nonliving (carbon particles)—both types of which would be present in large amounts in a smoker's lungs.

Stop and Think

1. Because the phospholipids orient themselves so that polar-to-polar and nonpolar-to-nonpolar regions are aligned, any gaps in the membrane are quickly sealed.
2. The molecular weights are ammonium hydroxide, 35; sulfuric acid, 98. Because sulfuric acid is about three times the weight of ammonium hydroxide, it diffuses at about 1/3 the rate. The precipitate forms about 1/4 of a meter from the sulfuric acid end.
3. Skin cells lose their desmosomes as they age. By the time they reach the free surface, they are no longer tightly bound to each other.
4. Conjugation (attachment) of the lipid product to a non–lipid-soluble substance, such as protein, will trap the product.
5. The plasma membrane folds (microvilli); the internal membranes are folded (ER and Golgi apparatus); the inner membranes of mitochondria are folded (cristae).
6. Because mitochondria contain DNA and RNA, mitochondrial ribosomes could be expected, too, and do indeed, exist. Mitochondrial nucleic acids and ribosomes are extremely similar to those of bacteria.
7. Avascular tissues rely on the diffusion of nutrients from surrounding fluids. As this is not as efficient or effective as a vascular supply, these tissues do not get very thick. An exception is the epidermis of the skin, which is quite thick, but cells in the skin layers farthest from the underlying blood vessels are dead.
8. A: This cell secretes a protein product, as evidenced by large amounts of rough ER, Golgi apparatus, and secretory vesicles near the cell apex. B: The cell secretes a lipid or steroid product, as evidenced by the lipid droplets surrounded by the abundant smooth ER.
9. 1. Fructose will diffuse into the cell. 2. Glucose will diffuse out of the cell. 3. Water enters the cell by diffusing along its concentration gradient. 4. The cell swells due to water entry.
10. 1.

		2. After deletion:	
DNA:	TAC GCA TCA CIT TTG ATC	DNA:	TAG ATC
mRNA:	AUG CGU AGU GAA AAC UAG	mRNA:	AUC UAG
amino acids:	Met Arg Ser Glu Asn Stop	amino acids:	none; a nonsense message

COVERING ALL YOUR BASES

Multiple Choice

1. E **2.** C **3.** B **4.** A **5.** C, D **6.** C **7.** B **8.** B, D **9.** B **10.** C **11.** A **12.** C **13.** A, D **14.** A, B, C, D **15.** C **16.** E **17.** D, E **18.** C **19.** A **20.** E **21.** B **22.** D **23.** C **24.** D **25.** C **26.** A **27.** A **28.** A, B, C **29.** A **30.** A **31.** B

Word Dissection

	Word root	Translation	Example
1.	chondri	lump	mitochondria
2.	chrom	color	chromosome
3.	crist	crest	cristae
4.	cyto	cell	cytology
5.	desm	bond	desmosome
6.	dia	through	dialysis
7.	dys	bad	dysplasia
8.	flagell	whip	flagellum
9.	meta	between	metaphase
10.	mito	thread	mitosis, mitochondria
11.	nucle	little nut	nucleus
12.	onco	a mass	oncogene
13.	osmo	pushing	osmosis
14.	permea	pass through	permeable
15.	phag	eat	phagocyte
16.	philo	love	hydrophilic
17.	phobo	fear	hydrophobic
18.	pin	drink	pinocytosis
19.	plasm	to be molded	cytoplasm
20.	telo	end	telophase
21.	tono	tension	isotonic
22.	troph	nourish	atrophy
23.	villus	hair	microvilli

Chapter 4 Tissue: The Living Fabric

BUILDING THE FRAMEWORK

Overview of Body Tissues

1. 1. Areolar 2. Cell 3. Elastic fibers 4. Bones 5. Nervous 6. Blood 7. Vascular 8. Keratin 9. Vascular 10. Fluid matrix

2. A. simple squamous epithelium B. simple cuboidal epithelium C. cardiac muscle D. dense regular connective tissue E. bone F. skeletal muscle G. nervous tissue H. hyaline cartilage I. smooth muscle tissue J. adipose (fat) tissue K. stratified squamous epithelium L. areolar connective tissue

The noncellular areas of D, E, H, J, and L are matrix.

3. 1. B 2. C 3. D 4. A 5. B 6. D 7. C 8. B 9. A 10. A 11. C 12. A 13. D

Epithelial Tissue

1. protection, absorption, filtration, excretion, secretion, and sensory reception

2. cellularity, specialized contacts (junctions), polarity, avascularity, regeneration, supported by connective tissue

3. 1. B 2. F 3. A 4. D 5. G 6. K 7. H 8. J 9.I 10. L

4. A=1; C=2; E=3; D=4; B=5

5. B=1; D=2; C=3; A=4

6.

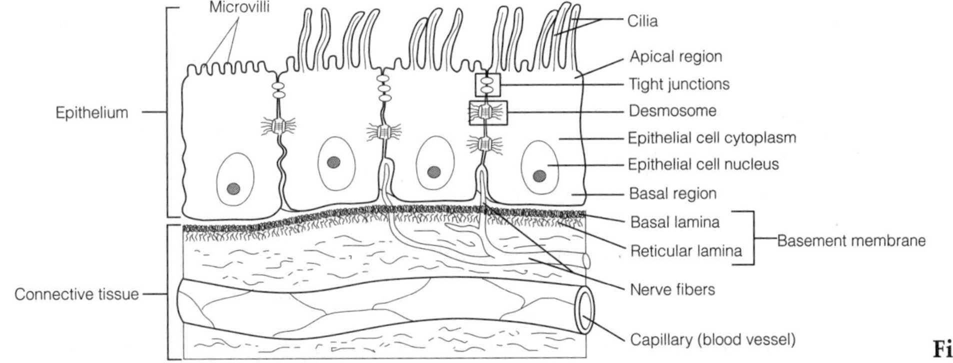

Figure 4.2

7. 1. simple alveolar gland 2. simple tubular gland 3. tubuloalveolar gland 4. compound gland

8. 1. T 2. glandular cells secrete their product 3. merocrine 4. merocrine 5. T 6. repair their damage 7. merocrine 8. T

9. 1. B 2. A 3. A 4. B 5. C 6. B

10. 1. epithelial 2. goblet cells 3. mucin 4.–6. duct (from epithelium), secretory unit, and supportive connective tissue 7. branching 8. tubular

Connective Tissue

1. 1. C 2. A 3. D 4. J 5. B 6. H 7. A 8. H 9. J 10. G 11. K 12. I 13. E 14. L 15. F
2.

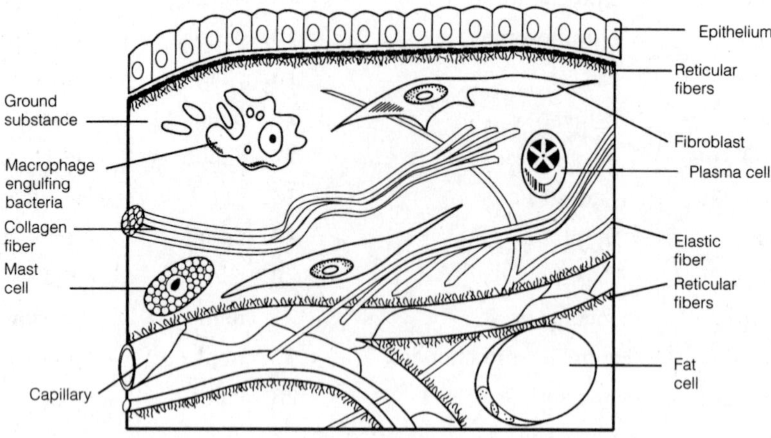

Figure 4.3

3. B=1; C=2; A=3
4. 1. H 2. E 3. C 4. F 5. L 6. G 7. E 8. B 9. D 10. E 11. A 12. I 13. K

Nervous Tissue

1. The neuron has long cytoplasmic extensions that promote its ability to transmit impulses long distances within the body.
2. Conduct

Muscle Tissue

1. Skeletal: 1, 3, 5, 6, 7, 11, 13 Cardiac: 2, 3, 4 (typically), 10, 12, 14, 15 Smooth: 2, 4, 7, 8, 9, 14
2. Step 1: The cells are not striated; identifies smooth muscle.
   Step 2: The striated cells exhibit intercalated discs; identifies cardiac muscle

Covering and Lining Membranes

1. In each case, the visceral layer of the serosa covers the external surface of the organ, and the parietal layer lines the body cavity walls.

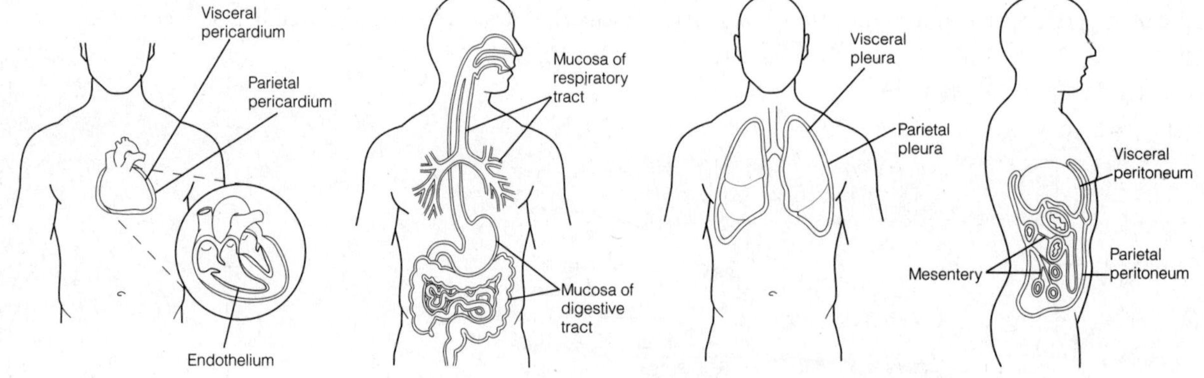

Figure 4.4

2.

Membrane	Tissue type (epithelial/connective)	Common locations	Functions
Mucous	Epithelial sheet with underlying connective tissue lamina propria	Lines respiratory, digestive, and reproductive tracts	Protection, lubrication, secretion, absorption
Serous	Epithelial sheet (mesothelium), scant areolar connective tissue	Lines internal ventral cavities and covers their organs	Lubrication for decreased friction during organ movements

2. *(continued)*

Membrane	Tissue type (epithelial/connective)	Common locations	Functions
Cutaneous	Epithelial (stratified squamous, keratinized) underlain by a dense fibrous connective tissue dermis	Covers the body exterior; is the skin	Protection from external insults; protection from water loss

Tissue Repair

1. 1. inflammation 2. clotting proteins 3. granulation 4. T 5. regeneration 6. T 7. collagen 8. T 9. T 10. bacteria-inhibiting 11. fibrosis

CHALLENGING YOURSELF

At the Clinic

1. Individual adipocytes easily lose and gain cell volume by losing or accumulating more fat.
2. Granulation tissue secretes bacteria-inhibiting substances.
3. Recovery will be long and painful because collagenous structures like tendons are so poorly vascularized.
4. Cartilage is largely avascular, and tendons are poorly vascularized. The more ample the blood supply, the quicker tissue heals.
5. Hypersecretion of serous fluid and its accumulation in the pleural space displaces lung volume. Because the lungs are not able to expand fully, respiration is more difficult. Additionally, when there is more fluid in the tissue spaces, diffusion of nutrients to the tissue cells (from the bloodstream) takes much longer and can impair tissue health and viability.
6. The peritoneum will be inflamed and infected. Because the peritoneum encloses so many richly vascularized organs, a spreading peritoneal infection can be life threatening.

Stop and Think

1. Simple epithelium is thinner than the diameter of the smallest vessel and must overlie all other tissues. Stratified epithelium forms the protective covering of deeper tissues. If blood vessels were within the epithelial layer, protection of the vessels would be reduced. Cartilage undergoes severe compression, which could damage vessels. Tendons are designed to withstand a strong pull; vessels would weaken the structure of tendons and would be at risk for rupture during extreme tension.
2. Mitochondria are important sites of ATP production, and muscles use large amounts of ATP during contraction. After macrophages ingest bacteria or other debris, it must be broken down. Lysosomes are the site of that breakdown.
3. Microvilli are found on virtually all simple cuboidal and simple columnar epithelia. The increase in surface area contributed by the microvilli offsets the decrease in transport rate by the thicker layers. Ciliated tissue can include all three types of simple epithelium but is found predominantly as simple columnar and pseudostratified.
4. Endocrine glands, like exocrine glands, are derived from the surface epithelium. In embryonic stages, these glands grow down from the surface, but eventually they lose their surface connection.
5. A basal lamina is only part of the basement membrane, which consists of epithelial cell secretions (the basal lamina) plus connective tissue fibers (the reticular lamina). A mucous membrane is an epithelial membrane composed of an epithelial sheet and underlying connective tissue. A sheet of mucus is a layer of secreted mucus (a cell product).
6. Large arteries must be able to expand in more than one direction; hence, they have irregularly arranged elastic connective tissue.
7. A bone shaft, like a skeletal cartilage, is surrounded by a sheet of dense irregular connective tissue.
8. Skeletal muscle cells are quite large compared to other muscle types. Typically, the multinucleate condition indicates that the demands for protein synthesis could not be handled by a single nucleus.
9. The nervous system develops by invagination from the ectoderm, much as glands develop internally from superficial epithelium.

COVERING ALL YOUR BASES

Multiple Choice

1. B **2.** D **3.** A **4.** B **5.** C **6.** B **7.** D, E **8.** C **9.** A **10.** D **11.** B **12.** A, C, D **13.** C **14.** D **15.** B **16.** D **17.** A **18.** B **19.** C **20.** B **21.** E **22.** B, C, D **23.** C **24.** C **25.** B, E **26.** A, B, D, E **27.** A, B, C, D, E **28.** A **29.** A, B

Word Dissection

	Word root	Translation	Example
1.	ap	tip	apical, apocrine
2.	areola	space	areolar connective
3.	basal	foundation	basal lamina
4.	blast	forming	fibroblast
5.	chyme	juice	mesenchyme
6.	crine	separate	endocrine
7.	endo	within	endoderm, endothelium
8.	epi	upon, over	epithelium
9.	glia	glue	neuroglia, microglia
10.	holo	whole	holocrine
11.	hormon	excite	hormone
12.	hyal	glass	hyaline, hyaluronic
13.	lamina	thin plate	reticular lamina
14.	mero	part	merocrine
15.	meso	middle	mesoderm, mesothelium
16.	retic	network	reticular tissue
17.	sero	watery fluid	serous, serosa
18.	squam	a scale	squamous
19.	strat	layer	stratified

Chapter 5 The Integumentary System

BUILDING THE FRAMEWORK

The Skin

1. 1. stratified squamous epithelium 2. mostly dense irregular connective tissue
2. 1. They are increasingly farther from the blood supply in the dermis. 2. They are increasingly laden with (water-resistant) keratin and surrounded by glycolipids, which hinder their ability to receive nutrients by diffusion.
3. 1. lines of cleavage 2. striae 3. flexure lines 4. wrinkles
4.

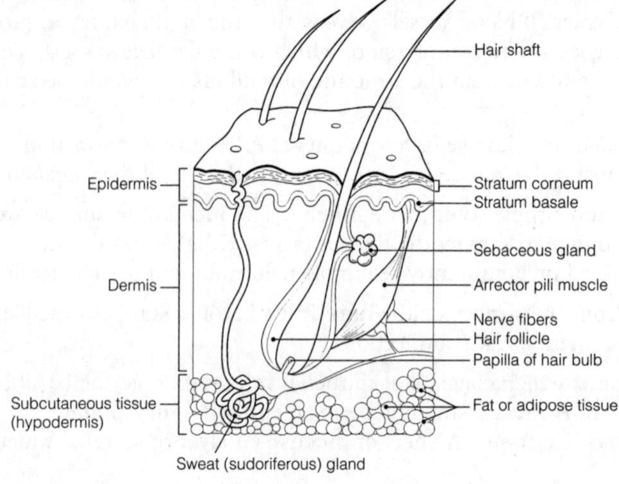

Figure 5.1

5. 1. D 2. B, D 3. F 4. I 5. A 6. B 7. I 8. A 9. I 10. J 11. A 12. A 13. C 14. B 15. I 16. B
6. 1. Keratin 2. Wart 3. Stratum basale 4. Arrector pili 5. Elastin 6. Melanocytes 7. Lamellated granules 8. Fibroblast
7. 1. C 2. A 3. C 4. B 5. C 6. A 7. B
8. 1. A 2. E 3. D 4. C 5. B

Appendages of the Skin

1. The papilla contains blood vessels that nourish the growth zone of the hair. The sebaceous gland secretes sebum into the follicle. The arrector pili muscle pulls the follicle into an upright position during cold or fright. The bulb is the actively growing region of the hair. See diagram B for localization of the dermal and epidermal root sheaths.

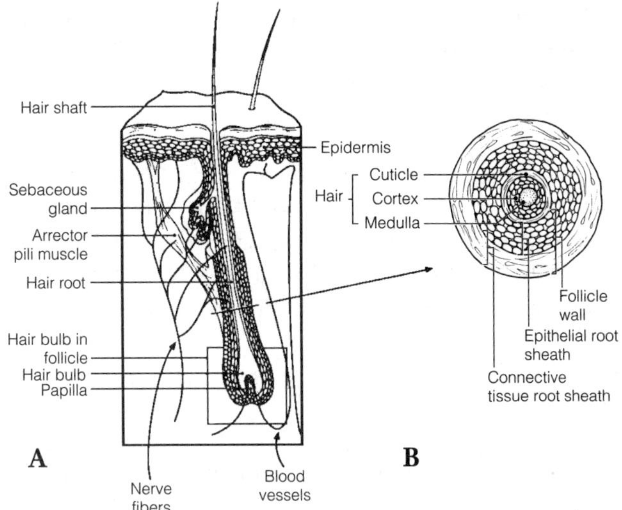

Figure 5.2

2. 1. Poor nutrition 2. Keratin 3. Stratum corneum 4. Eccrine glands 5. Vellus hair 6. Desquamation 7. Growth phase
3. Alopecia
4. Genetic factors; emotional trauma; some drugs (chemotherapy agents); ringworm infection
5. 1. cuticle 2. The stratum germinativum is thicker here, preventing the rosy cast of blood from flushing through

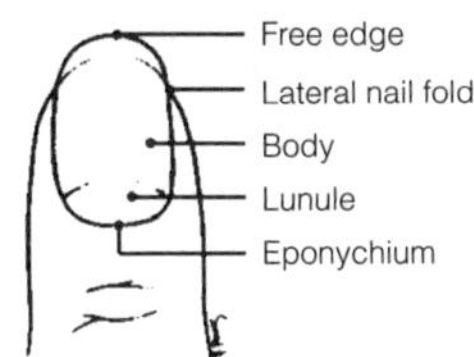

6. 1. A 2. B, C 3. B 4. C (B) 5. A 6. A 7. A, B 8. A 9. B 10. C

Functions of the Integumentary System

1. 1. B 2. M 3. C 4. M (C) 5. C 6. C
2. All *but* water-soluble substances
3. It depresses immune response by inhibiting the activity of macrophages that aid in the immune response.
4. Sweat glands are regulated by the sympathetic division of the nervous system. When it is hot, blood vessels in the skin become flushed with warm blood, allowing heat to radiate from the skin surface, and perspiration output increases. Evaporation of sweat from the body surface removes large amounts of body heat.
5. 1. nervous 2.–5. cold, heat, deep pressure, and light pressure 6. cholesterol 7. dermal blood vessels 8. ultraviolet 9. calcium

Homeostatic Imbalances of Skin

1. Loss of body fluids containing proteins and essential electrolytes, which can lead to dehydration, shock, and renal shutdown
2. 1. C 2. B 3. A 4. B 5. C 6. C
3. It allows the estimation of the extent of burns so that fluid volume replacement can be correctly calculated.

4. 1. squamous cell carcinoma 2. basal cell carcinoma 3. malignant melanoma
5. Pigmented areas that are Asymmetric, have irregular Borders, exhibit several Colors, and have a Diameter greater than 6 mm are likely to be cancerous.

The Incredible Journey

1. 1. collagen 2. elastin 3. dermis 4. phagocyte (macrophage) 5. connective tissue root 6. epidermis 7. stratum basale 8. melanin 9. keratin 10. squamous (stratum corneum) cells

CHALLENGING YOURSELF

At the Clinic

1. The sun's radiation can cause damage to DNA, in spite of the protection afforded by melanin. Without repair mechanisms, the DNA damage is permanent, and the effects are cumulative.
2. The baby has seborrhea, or cradle cap, a condition of overactive sebaceous glands. It is not serious; the oily deposit is easily removed with attentive washing and soon stops forming.
3. A Mongolian spot is a pigmented area that is the result of escaped melanin being scavenged by dermal macrophages. It is not a sign of child abuse.
4. Systemic hives would indeed be worrisome because it might indicate a possibly life-threatening plasma loss and correlated drop in blood volume.
5. Probably not. Organic solvents are lipid-soluble and thus can pass through the skin by dissolving in the lipids of the cells' plasma membranes. Hence, a shower (which would wash away substances soluble in water) will not remove the danger of poisoning. Kidney failure and brain damage can result from poisoning by organic solvents.
6. Fluid/electrolyte replacement and the prevention of massive infection.
7. The most likely diagnosis is basal cell carcinoma, because it causes ulcers and is slow to metastasize. Years of exposure to sunlight probably caused his cancer.
8. Subcutaneous tissue is mostly fat, which is a good insulator. Hence, infants and the elderly are more sensitive to cold.
9. Because there are fewer connections to underlying tissues, wounds to the face tend to gape more.
10. Carotene can accumulate in the stratum corneum and the hypodermis, giving the skin an orange cast. Awareness of diet is particularly important in very young children who tend to go on "food jags," refusing all but one food, to determine whether "liked" foods are high in carotene.
11. The drugs used to treat cancer exert their major cell-killing effects on rapidly dividing cells. Like cancer cells, the hair follicles are targets of these drugs, which accounts for the children's baldness.
12. Healing is much cleaner and faster when the incision is made along or parallel to a cleavage line (which represents a separation between adjacent muscle bundles).

Stop and Think

1. The term "membrane" refers to the sheetlike *structure* of the skin; all epithelial membranes are at least simple organs. The term "organ" means there are at least *two types of tissue* in a structure's composition, and at least simple functions can be performed.
2. The undulating folds (formed by the dermal papillae) at the surface between the epidermis and the dermis increase the surface area at the site of their union. Increased surface area provides more space for capillary networks (increasing the area for diffusion of nutrients to and wastes from the avascular epidermis) and is more secure, making it harder for the epidermis to tear away from the dermis.
3. There is very little hypodermal tissue underlying the skin of the shins. Consequently, the skin is quite securely and tightly fastened to the underlying structure (bone) there.
4. The stratum corneum is so far removed from the underlying blood supply in the dermis that diffusion of nutrients is insufficient to keep the cells alive. Also, the accumulation of glycolipids in the intercellular spaces retards diffusion of water-soluble nutrients. The accumulated keratin within the cells inhibits life functions. In terms of benefit, bacterial or other infections cannot become established easily in the dead, water-insoluble layer. Dead cells provide a much more effective mechanical barrier than living cells.
5. The nerve endings located closest to the surface are the most sensitive to light touch. Tactile discs, in the epidermis, are very responsive to light touch; hair follicle receptors, although not as close, are stimulated by the movement of hairs on the skin, which project from the surface.
6. When skin peels as a result of sunburn damage, the deeper layers of the stratum corneum, which still maintain their desmosomal connections, detach. The desmosomes hold the cells together within the layers.
7. Unless there is a protein deficiency, additional dietary protein will not increase protein utilization by hair- and nail-forming cells.

8. Tissue overgrowth (hyperplasia) occurred in response to the continual irritation. Although there are some atypical cells (dysplasia), there was no evidence of neoplasia (tumor formation). No, he does not have cancer of the mouth.
9. The apocrine sweat glands, associated with axillary and pubic hair follicles, appear to secrete chemicals that act as sexually important signals.
10. Because of the waterproofing substance in the epidermis (both in and between the cells), the skin is relatively impermeable to water entry.
11. Connective tissue ground substance is gel-like and embedded within it are various types of cells (raisins and almonds) and fibers (shredded cabbage).

COVERING ALL YOUR BASES

Multiple Choice

1. B, D **2.** D **3.** B **4.** A **5.** D **6.** D **7.** A **8.** D **9.** C **10.** D, E **11.** E **12.** A, B, D **13.** A, B **14.** A, B **15.** B **16.** A, C, D **17.** D **18.** C **19.** A, B, D **20.** B, C, D **21.** B **22.** A, D, E **23.** B, C, D **24.** C **25.** C **26.** B **27.** A, C, D **28.** B, C, D, E

Word Dissection

	Word root	Translation	Example
1.	arrect	upright	arrector pili
2.	carot	carrot	carotene
3.	case	cheese	vernix caseosa
4.	cere	wax	cerumen
5.	corn	horn, horny	stratum corneum
6.	cort	bark, shell	cortex
7.	cutic	skin	cutaneous, cuticle
8.	cyan	blue	cyanosis
9.	derm	skin	dermis, epidermis
10.	folli	a bag	follicle
11.	hemato	blood	hematoma
12.	hirsut	hairy	hirsutism
13.	jaune	yellow	jaundice
14.	kera	horn	keratin
15.	lanu	wool, down	lanugo
16.	lunul	crescent	lunule
17.	medull	marrow	medulla
18.	melan	black	melanin
19.	pall	pale	pallor
20.	papilla	nipple	dermal papillae
21.	pili	hair	arrector pili
22.	plex	network	plexus
23.	rhea	flow	seborrhea
24.	seb	grease	sebaceous gland
25.	spin	spine, thorn	stratum spinosum
26.	sudor	sweat	sudoriferous
27.	tegm	cover	integument
28.	vell	fleece, wool	vellus

Chapter 6 Bones and Skeletal Tissues

BUILDING THE FRAMEWORK

Skeletal Cartilages

1. 1. C 2. A 3. C 4. B 5. A 6. C 7. C 8. B
2. 1. T 2. F 3. T 4. F 5. F 6. T 7. F
3. Although collagen fibers form cartilage's supporting framework, its proteoglycans and hyaluronic acid attract and organize huge amounts of water, which becomes the main component of cartilage. This water "saves space" for bone development and is responsible for cartilage's resilience throughout life.

Classification of Bones

1. 1. S 2. F 3. L 4. L 5. F 6. L 7. L 8. F 9. I

Functions of Bones

1. 1. They support the body by providing a rigid skeletal framework. 2. They protect the brain, spinal cord, lungs, and other internal organs. 3. They act as levers and provide attachments for muscles in movement. 4. They serve as a reservoir for fat and minerals. 5. They are the sites of formation of blood cells.

Bone Structure

1.

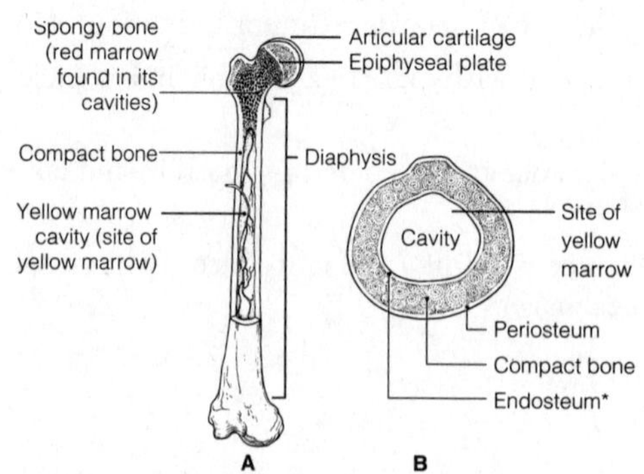

Figure 6.1

2. 1. C 2. A 3. C, D 4. A 5. E 6. B 7. B 8. F

3. 1. internal layer of spongy bone 2. endosteum

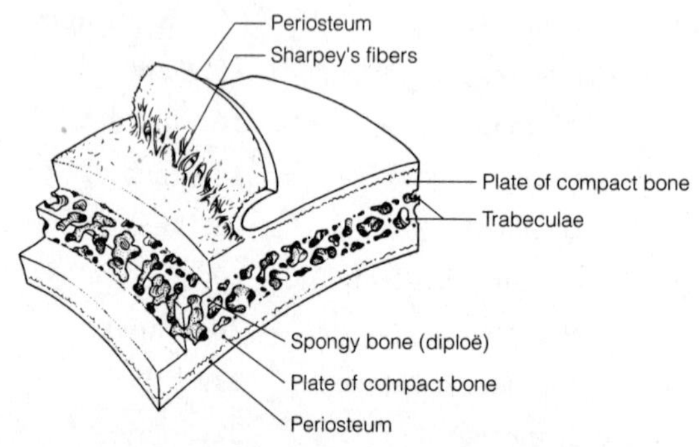

Figure 6. 2

4. 1. B 2. C 3. A 4. E 5. D Matrix is the nonliving part of bone shown as unlabeled white space on the figure.

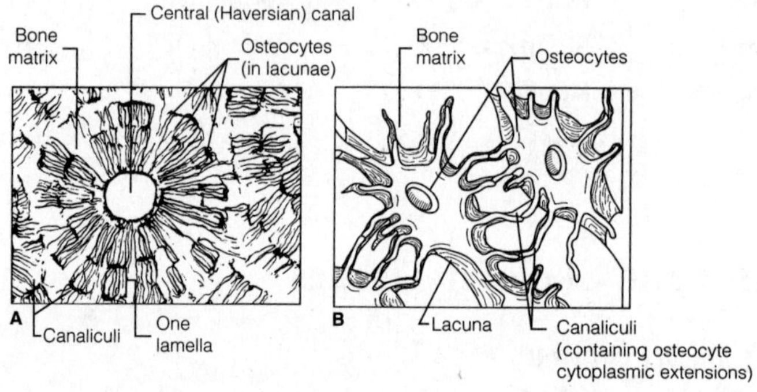

Figure 6.3

5. 1. P 2. P 3. O 4. O 5. P 6. O 7. P 8. P 9. P 10. D 11. D 12. O

6. 1. Collagen 2. Yellow marrow 3. Osteoclasts 4. Marrow cavity 5. Periosteum 6. Perichondrium 7. Lamellar

Bone Development

1. If arranged in the correct sequence, the listed elements would be numbered: 3, 2, 4, 1, 5, 6 (note: events 2 and 3 may occur simultaneously).

2. 1. intramembranous 2. cartilage 3. T 4. calcium salts 5. T 6. mesenchymal cells or osteoblasts 7. T 8. secondary 9. hyaline cartilage 10. endosteal 11. T

3. 1. mesenchymal cells 2. ossification center 3. osteoblasts 4. osteoid 5. mineralization 6. bone matrix 7. woven 8. mesenchyme 9. periosteum 10. diploë

4. 1. chondroblasts 2. enlarging cartilage cells (chondrocytes) 3. C 4. C 5. osteoblasts 6. bone

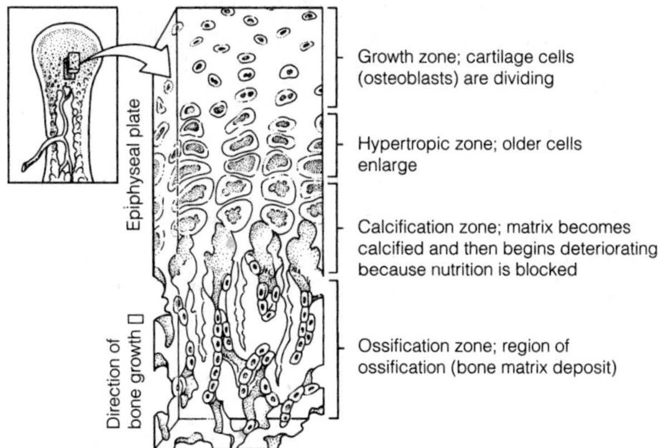

Figure 6. 4

Bone Homeostasis: Remodeling and Repair

1. 1. G 2. F 3. A 4. H 5. D 6. B 7. E 8. C

2. 1. Bone resorption 2. Calcium salts 3. Hypocalcemia 4. Blood calcium levels 5. Growth in length

3. Because the points of maximal compression and tension are on opposing sides of the shaft, they cancel each other out, so there is essentially no stress in the middle (center) of the shaft. Consequently, the shaft can contain spongy (rather than compact) bone in its center. The impact of striking the ball stresses the bones of his serving arm, causing them to thicken in accord with Wolff's law.

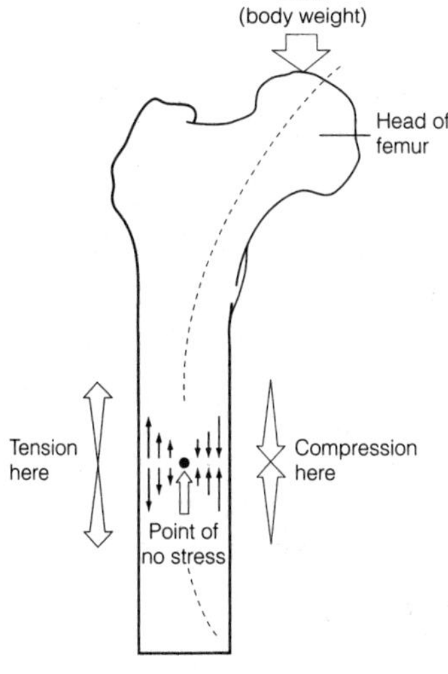

Figure 6.5

4. 1. D 2. A 3. E

5. 1. H 2. A 3. I 4. F 5. C 6. D 7. G 8. E 9. H 10. I 11. B

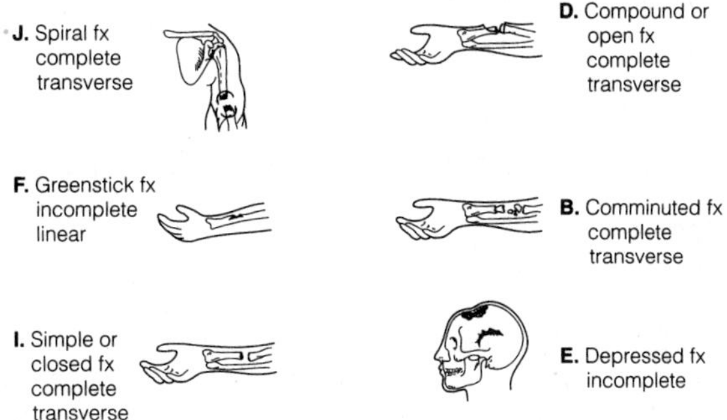

Figure 6.6

6. 1. T 2. T 3. phagocytes (macrophages) 4. fibroblasts 5. periosteum 6. T 7. spongy

Homeostatic Imbalances of Bone

1. 1. Osteoporosis 2. Osteomalacia 3. Paget's disease 4. Porous bones 5. Osteomalacia 6. Calcified epiphyseal discs

The Incredible Journey

1. 1. femur 2. spongy 3. stress 4. red blood cells 5. red marrow 6. nerve 7. central (Haversian) 8. compact 9. canaliculi 10. lacunae (osteocytes) 11. matrix 12. osteoclast

CHALLENGING YOURSELF

At the Clinic

1. The woman's vertebral compression fractures and low bone density indicate osteoporosis. This condition in postmenopausal women is at least partially due to lack of estrogen production. It can be treated with supplements of calcium and vitamin D, proper nutrition, and possibly an estrogen-progestin protocol or fluoride and etidronate.
2. Infection can spread to the bone. The doctor will examine the X ray for evidence of osteomyelitis, which is notoriously difficult to manage.
3. The curving of the line indicates a spiral fracture.
4. Abnormal collagen production coupled with short stature indicates achondroplasia.
5. Osteoporosis is the deterioration and breakdown of bone matrix. Osteoclasts are the cells that break bone matrix down.
6. The woman is suffering from hypercalcemia, which is causing metastatic calcification (i.e., calcium salt deposits in soft tissues of the body).
7. The doctor will check for hypersecretion of growth hormone.
8. Paget's disease can cause abnormal thickening of bone such as that experienced by Egil.
9. The epiphyseal plate remains the same thickness during the growth period because as cartilage on the distal epiphyseal plate face is formed, that on the proximal face is replaced by bone.

Stop and Think

1. The blood vessels determine the pattern for the development of the lamellae, so blood vessels in the osteon region appear first.
2. Bone tissue is too rigid for interstitial growth; there is only appositional growth after ossification. Before ossification, growth can occur in membrane or in cartilage templates.
3. Long bones tend to be arranged and to articulate end to end. Even if two long bones are side by side, they still articulate most prominently at their ends. Short bones are arranged and articulate side to side as well as end to end.
4. The medullary cavity's diameter must increase to keep pace with the appositional growth of the surrounding bone collar. Osteoclasts will remove the bone on the interior of the collar to widen the cavity. Likewise, osteoclasts are active on the internal surfaces of cranial bones to broaden their curvatures to accommodate the expanding brain.
5. Bone tissue is highly vascular. Fibrous tissue is poorly vascularized, and cartilage is mostly avascular. In fact, the vascularization of the template is part of the process that triggers ossification.
6. The bone will heal faster because bone is vascularized and has an excellent nutrient delivery system (canaliculi). Cartilage lacks both of these characteristics.
7. Yes, he has cause to worry. The boy is about to enter his puberty-adolescence growth spurt, and interference with the growth plate (epiphyseal plate) is probable.
8. According to Wolff's law, growth of a bone is a response to the amount of compression and tension placed on it, and a bone will grow in the direction necessary to best respond to the stresses applied.
9. Signs of healing of the bone at the surgical incision (usually a hole in the skull) indicate that the patient lived at least for a while.
10. The high water content helps explain both points. (a) Cartilage is largely water plus resilient ground substance and fibers (no calcium salts). (b) Water "saves room" for bones to develop in.
11. There is no gravity in space and that is one of the factors that stresses bones, causing them to grow stronger.

COVERING ALL YOUR BASES

Multiple Choice

1. A, B, C **2.** A, B, D **3.** A, D **4.** A, B, D **5.** A, D **6.** A, C, D **7.** D **8.** C **9.** A, B, C, E **10.** A, B **11.** A, B, C, D **12.** A **13.** B, D **14.** A, C, D, E **15.** D **16.** C **17.** A **18.** A, C **19.** A, D **20.** B, E **21.** A **22.** A, B, C, D, E **23.** C **24.** C **25.** D **26.** B, C **27.** C **28.** B

Word Dissection

	Word root	Translation	Example		Word root	Translation	Example
1.	call	hardened	callus	7.	myel	marrow	osteomyelitis
2.	cancel	latticework	cancellous bone	8.	physis	growth	diaphysis
3.	clast	break	osteoclast	9.	poie	make	hematopoiesis
4.	fract	break	fracture	10.	soma	body	somatomedin
5.	lamell	small plate	lamellar bone	11.	trab	beam	trabeculae
6.	malac	soft	osteomalacia				

Chapter 7 The Skeleton

BUILDING THE FRAMEWORK

1. Bones of the skull, vertebral column, and thoracic cage are parts of the axial skeleton. All others belong to the appendicular skeleton.

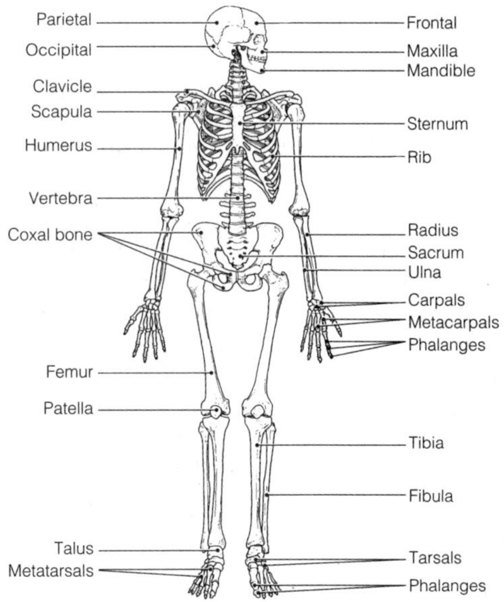

Figure 7.1

2. 1.B, H 2. A 3. G 4. E 5. F 6. D 7. C

PART 1: THE AXIAL SKELETON

The Skull

1. A, B, I, K, L, and M should be circled.

2.

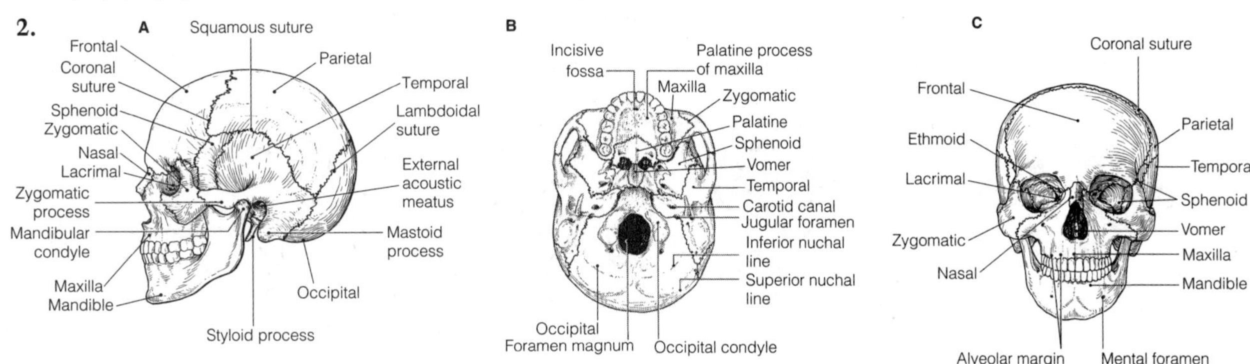

Figure 7.2

3. 1. B, K 2. I, K 3. K, M 4. K 5. O 6. K 7. A 8. J 9. I 10. A 11. I 12. L 13. M 14. B 15. F 16. F 17. M 18. F 19. C 20.–23. A, B, G, L 24. A 25. L 26. E 27. N 28. M

4. 1. Mucosa-lined air-filled cavities in bone. 2. They lighten the skull and serve as resonance chambers for speech. 3. Their mucosa is continuous with that of the nasal passages, into which they drain. 4. To moisten and warm incoming air and trap particles and bacteria in sticky mucus.

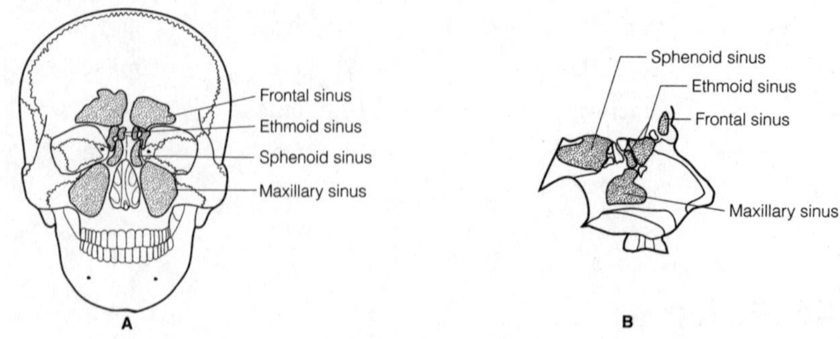

Figure 7.3

5.

Bone marking(s)	Skull bone	Function of bone marking
Coronoid process	Mandible	Attachment for temporalis muscle
Carotid canal	Temporal	Passes internal carotid artery
Sella turcica	Sphenoid	Supports, partly encloses pituitary gland
Jugular foramen	Temporal	Passes jugular vein and cranial nerves IX, X, and XI
Palatine processes	Maxillae	Form anterior two-thirds of hard palate
Styloid process	Temporal	Attachment site for several neck muscles
Optic canals	Sphenoid	Pass optic nerves and ophthalmic arteries
Cribriform plates	Ethmoid	Pass olfactory nerve fibers
External occipital crest	Occipital	Attachment for ligamentum nuchae and neck muscles
Mandibular condyles	Mandibles	Form freely movable joints with temporal bones
Mandibular fossa	Temporal	Receives condyle of mandible
Stylomastoid foramen	Temporal	Where facial nerve leaves skull

The Vertebral Column

1. 1. F 2. A 3. C, E 4. A, E 5. B

2. 1. Cervical, C_1 to C_7 2. Thoracic, T_1 to T_{12} 3. Lumbar, L_1 to L_5 4. Sacrum, fused 5. Coccyx, fused 6. Atlas, C_1 7. Axis, C_2

3. 1. A, B, C 2. G 3. C 4. A 5. F 6. E 7. B 8. D 9. E 10. G 11. A, B

4. 1. J 2. F 3. G 4. E 5. H A. Cervical, atlas B. Cervical C. Thoracic D. Lumbar

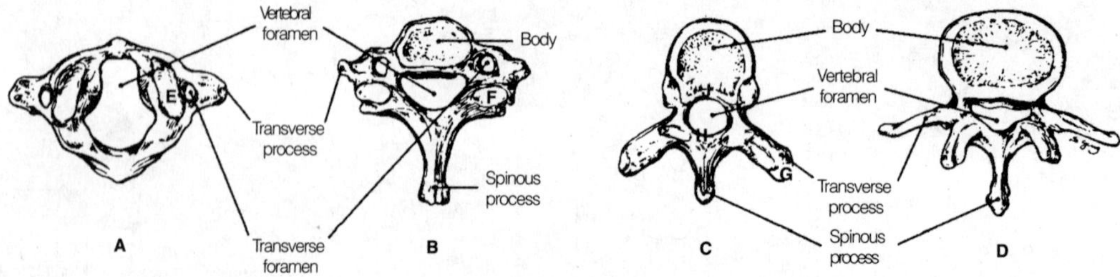

Figure 7.5

5.

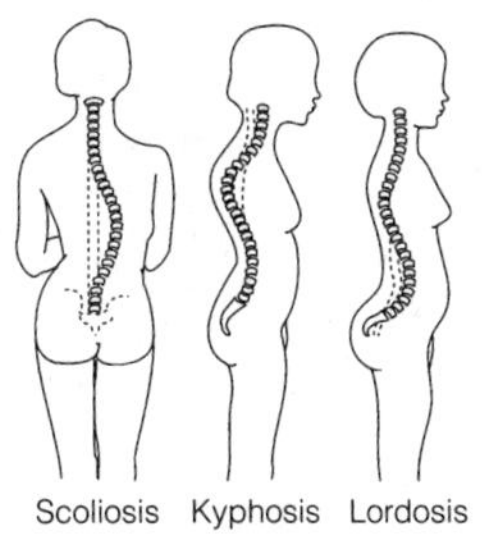

Figure 7.6

The Thoracic Cage

1. 1. lungs 2. heart 3. true 4. false 5. floating 6. thoracic vertebrae 7. sternum 8. an inverted cone

2. Ribs 1–7 on each side are vertebrosternal (true) ribs; ribs 8–10 on each side are vertebrochondral ribs; and ribs 11–12 on each side are vertebral (floating) ribs. (Ribs 8–12 collectively are also called false ribs.)

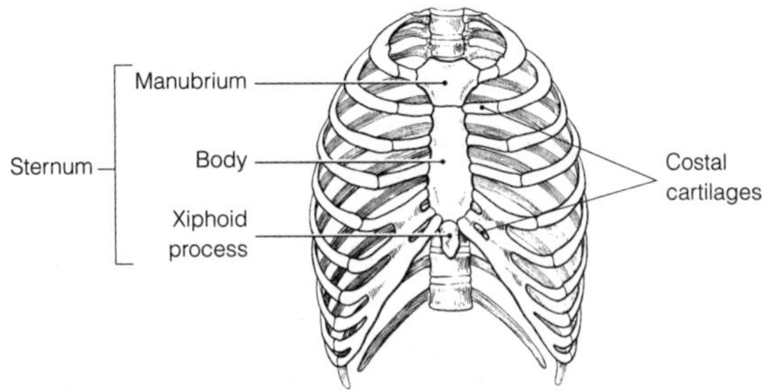

Figure 7.7

PART 2: THE APPENDICULAR SKELETON

The Pectoral (Shoulder) Girdle

1. 1. pelvic girdle 2. pectoral girdle 3. support 4. to protect organs 5. to allow manipulation movements 6. locomotion

2.

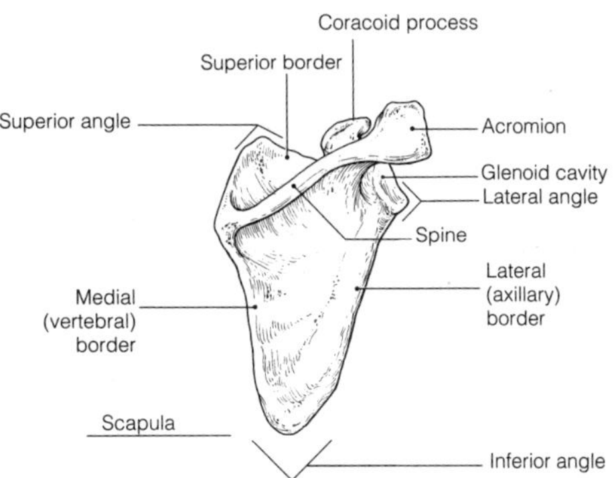

Figure 7.8

The Upper Limb

1.

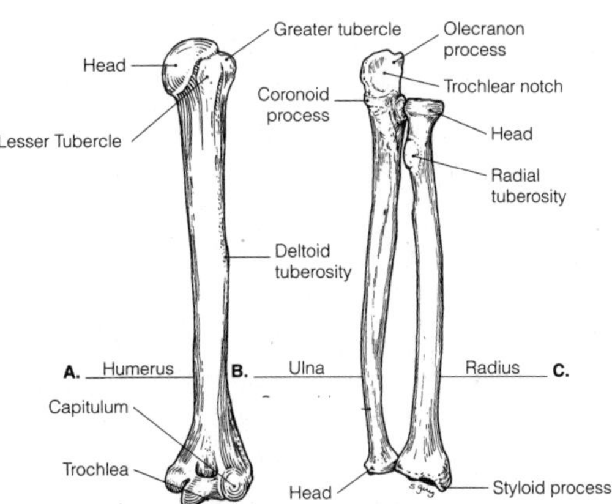

Figure 7.9

2.

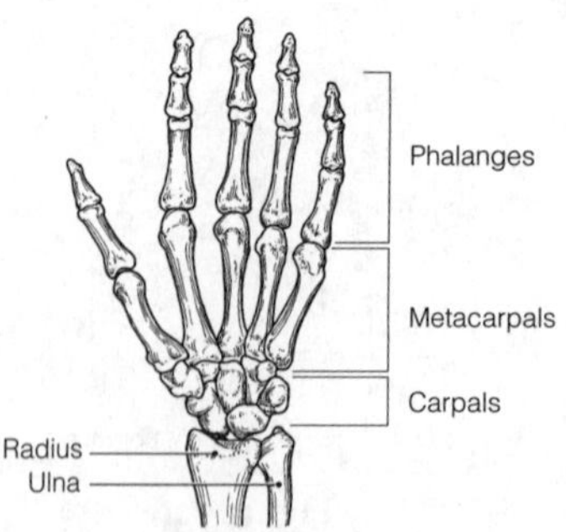

Figure 7.10

3. 1. A, C, D 2. B, E, F

4. 1. G 2. I 3. D 4. P 5. O 6. T 7. A 8. P 9. D 10. H 11. E 12. D 13. S 14. T 15. B 16. F 17. T 18. P 19. Q 20. C 21. M 22. J

The Pelvic (Hip) Girdle

1. 1. Female inlet is larger and more circular. 2. Female sacrum is less curved; pubic arch is rounder. 3. Female ischial spines are shorter and less sharp; pelvis is lighter and shallower.

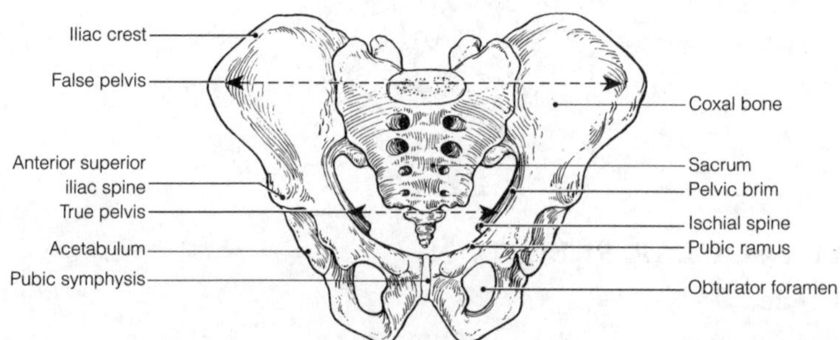

Figure 7.11

The Lower Limb

1.

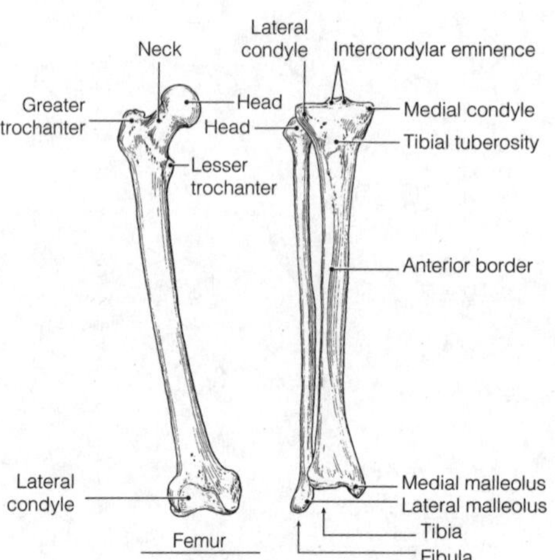

Figure 7.12

2. 1. I, K, S 2. J 3. R 4. H 5. A 6. T 7. C 8. D 9. W 10. C, Q, W 11. X 12. Q 13. W 14. N 15. L 16. B 17. V 18. O 19. P 20. E, G 21. U 22. S 23. C

CHALLENGING YOURSELF

At the Clinic

1. The cribriform plates of the ethmoid bone, which surround the olfactory nerves. These plates are quite fragile and are often crushed by a blow to the front of the skull. This severs the nerves, which cannot grow back.
2. No; the palatine bones are posterior to the palatine processes of the maxillae. If the palatine processes do not fuse, then the palatine bones remain unfused as well.
3. The weakening of thoracic vertebrae and corresponding muscles on *one side* of the body favors the development of scoliosis.
4. The interosseous membrane, which connects the radial and ulnar shafts, was torn.
5. Truck drivers frequently suffer from herniated discs, because they spend so many hours in a seated position.
6. The condition is called lordosis and it is an accentuated lumbar curvature. The added weight of Mr. Ogally's pot-belly changed his center of gravity, forcing the discs of the weight-bearing lumbar vertebrae to adapt to that change.
7. The man is of an age at which the xiphoid process finally ossifies. In its cartilaginous state, the xiphoid process is not as easily felt.
8. Seven bones contribute to the orbit: frontal, sphenoid, zygomatic, maxilla, palatine, lacrimal, and ethmoid bones.
9. Malcolm's shoulder has "fallen" anteromedially. Because of its curvatures, the clavicle typically fractures anteriorly, thus sparing the subclavian artery, which lies posteriorly.
10. The grandmother's bones have lost bone mass and are more porous and brittle than those of the child, placing her weakened ribs at greater risk for fracture when stressed by physical trauma.
11. Most likely the paranasal sinuses on the right side of the face.

Stop and Think

1. The foramen magnum allows the spinal cord to connect with the brain stem within the skull. It has nothing to do with eating or food.
2. The turbinates increase the surface area of the mucous membrane of the nasal cavity, which lies over them. This increases the contact between air and membrane for filtering, warming, and humidifying the air.
3. External acoustic meatus, petrous region, middle ear cavity, inner ear cavity, and internal acoustic meatus all play a role in hearing.
4. Some snakes have jaws that are "designed" to dislocate; others have double articulations at the jaw.
5. No. Nodding "yes" involves the joint between the occipital condyles and the superior articular surfaces of the atlas. Shaking "no" involves the joint between the atlas and the dens of the axis.
6. The largest ribs are easiest to fracture. They protrude more, and don't have the protection of the clavicle that the first pairs have. The floating ribs do not fracture easily; lacking an anterior bony attachment, they can move (or be displaced) by forceful trauma.
7. The thoracic cage must maintain flexibility for breathing; for that, it must sacrifice the strong but rigid protection afforded by solid bone casing such as that of the cranium.
8. The acetabulum must provide greater strength and resist stress that might dislocate its limb bone. The glenoid cavity must provide for more flexibility.
9. Scaphoid, lunate, triquetral, pisiform, trapezium, trapezoid, capitate, hamate. These are the carpals of the hand, and the first four bones named are more proximal.
10. Anterior cranial fossa: frontal, ethmoid, and sphenoid bones. Middle cranial fossa: sphenoid and temporal bones. Posterior cranial fossa: temporal and occipital bones.
11. If the person has attained his/her final height, the epiphyseal plate will no longer be present (replaced by the epiphyseal line).

COVERING ALL YOUR BASES

Multiple Choice

1. C **2.** B, C, E **3.** A, B, C, D **4.** A, C, D **5.** D **6.** A, B, C, D, E **7.** A **8.** A, C, D **9.** A, C, D, E **10.** B, D, E **11.** B, C, D **12.** A, C, D, E **13.** B, C **14.** A, C, D **15.** A, B, D **16.** B **17.** C **18.** B, C **19.** D **20.** B, E **21.** B, D, E **22.** A, C, D **23.** A, C, D **24.** B, C, E **25.** B **26.** B **27.** A **28.** A **29.** B **30.** A **31.** E **32.** B **33.** B, C

Word Dissection

	Word root	Translation	Example		Word root	Translation	Example
1.	acetabul	wine cup	acetabulum	**13.**	pal	shovel	palate
2.	alve	socket	alveoli, alveolar	**14.**	pect	the breast	pectoral girdle
3.	append	hang to	appendicular	**15.**	pelv	a basin	pelvic girdle
4.	calv	bald	calvaria	**16.**	pis	pea	pisiform
5.	clavicul	a key	clavicle	**17.**	pter	wing	pterygoid process
6.	crib	a sieve	cribriform plate	**18.**	scaph	a boat	scaphoid
7.	den	tooth	dens	**19.**	skeleto	dried body	skeleton
8.	ethm	a sieve	ethmoid	**20.**	sphen	a wedge	sphenoid
9.	glab	smooth	glabella	**21.**	styl	pointed instrument	styloid process
10.	ham	hooked	hamate	**22.**	sutur	a seam	suture, sutural bone
11.	ment	chin	mental foramen	**23.**	vert	turn, joint	vertebra
12.	odon	tooth	odontoid process	**24.**	xiph	a sword	xiphoid process

Chapter 8 Joints

BUILDING THE FRAMEWORK

Classification of Joints

1. 1. To bind bones together and permit movement 2. The type of material binding the bones; whether a joint cavity is present; degree of movement permitted 3. Fibrous: synarthroses, immovable; cartilaginous: amphiarthroses, slightly movable; synovial: diarthroses, freely movable

Fibrous, Cartilaginous, and Synovial Joints

1. 1. A; 5 2. C 3. B; 1, 2, 4, and 6 4. B; 2 and 6 5. C 6. C 7. A; 3 8. A; 3 9. C 10. C 11. B; 6 12. C 13. A; 5 14. C 15. B; 1 16. A; 5 17. B; 6 18. B; 4

2. Synovial. Synovial joints are ideal where mobility and flexibility are the goals. Most joints of the axial skeleton provide immobility (or at most slight mobility) because the bones of the axial skeleton typically *protect* or *support* internal (often fragile) organs.

3. 

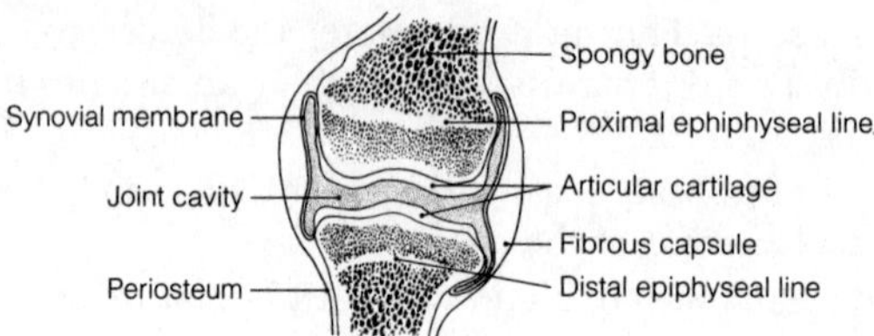

Figure 8.1

4. 1. A 2. C, D 3. B 4. B 5. C, D

5. 1. I 2. A 3. P 4. J 5. C 6. F 7. Q 8. E 9. K 10. N 11. A, B, C, E, H, I, L

6. 1. A, C, E 2. G 3. D, I 4. N 5. J 6. M 7. F 8. K 9. B, L 10. H

7. 1. B 2. E 3. E 4. E 5. A 6. B 7. E 8. F 9. C 10. D 11. C, D 12. B, F 13. A 14. E

8. 1. Multiaxial joint 2. Saddle joint 3. Tibiofibular joints 4. Multiaxial joint 5. Elbow joint 6. Amphiarthrotic 7. Bursae 8. Condyloid joint 9. Biaxial joint

9. 1. D 2. C 3. A 4. B

10. A. hip joint B. knee joint

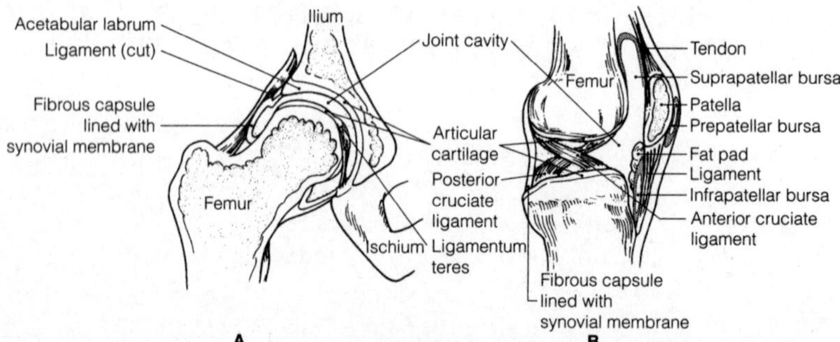

Figure 8.3

Homeostatic Imbalances of Joints

1. 1. dislocation 2. T 3. osteoarthritis 4. bursae 5. acute 6. vascularized 7. rheumatoid 8. gouty 9. T 10. synovial membrane 11. rheumatoid 12. menisci 13. reduce friction 14. osteoarthritis 15. synovial
2. Ligaments and tendons (injured during sprains) are poorly vascularized and cartilages are avascular. Hence, these structures heal slowly and often poorly.

The Incredible Journey

1. 1. meniscus 2. femur 3. tibia 4. joint cavity 5. synovial fluid 6. lubricate 7. condyles 8. anterior cruciate 9. patella 10. bursa 11. lateral (outer) 12. tibia 13. ligament 14. blood supply 15. meniscus

CHALLENGING YOURSELF

At the Clinic

1. Jenny's elbow has been dislocated.
2. The temporomandibular joint is the source of his problem.
3. Because the clavicles brace the scapulae, without clavicles the scapulae are extremely movable and the shoulders appear droopy. Although the condition is not serious, it reduces the shoulders' resistance to stress.
4. The rotator cuff consists of ligaments that are the main reinforcers of the shallow shoulder joint. Its rupture reduces outward movements, such as abduction.
5. The ligamentum teres contains a blood vessel that supplies the head of the femur. The femoral head ordinarily also receives nourishment from vessels reaching it from the diaphysis, but a fracture would cut off that blood supply.
6. Bertha's clavicle fractured.
7. The fibula is usually used for bone grafts of this sort.
8. Typically a "torn cartilage" is an articular disc, which is fibrocartilage.
9. Rheumatoid arthritis, fairly common in middle-aged women, causes this type of deformity.

Stop and Think

1. Diarthrotic joints in the axial skeleton: temporomandibular joint, atlanto-occipital joint, atlantoaxial joint, intervertebral joints (between articular processes), vertebrocostal joints, sternoclavicular joints (which include one axial bone, the sternum, and one appendicular bone, the clavicle), and sacroiliac joint (which also includes an appendicular bone).
2. Pronation and supination involve not only rotation at the proximal radioulnar joint (a diarthrotic joint) but also crossing of the shafts of the radius and the ulna (an amphiarthrotic syndesmosis). The wrist bones are not involved.
3. The coxal bone develops from the ossification centers in the ilium, ischium, and pubis within a fused hyaline cartilage template. Thus the joint is classified as a synchondrosis. After ossification of the entire coxal bone, the joint classification is synostosis.
4. Interdigitation and interlocking increase the articulating surface area between adjoining bones. Such an increased area of attachment means more ligament between the bones, increasing the resistance to movement at the joints. Further, the interlocking of bony regions prevents separation of the bones.
5. The periosteum is a dense irregular connective tissue, and ligaments are dense regular connective tissue. Microscopic examination would reveal a transition from collagenous fibers, which lie in different orientations to those arranged in parallel bundles.
6. The synovial membrane, which lines the joint cavity, is composed of loose connective tissue only. Hence, it is not an epithelial membrane.
7. The purpose of menisci is to more securely seat the bones within the joint and to absorb shock. Although such activities are uncommon, the jaw (temporomandibular) joint can withstand great pressure as when you bite the cap off a glass bottle or hang by your teeth in a circus act. The sternoclavicular joint absorbs shock from your entire arm when you use your hand to break a fall.
8. Chronic slumping forward stretches the posterior intervertebral ligaments; once stretched, ligaments do not resume their original length. Eventually, the overextended ligaments are unable to hold the vertebral column in an erect position.
9. The knee joint flexes posteriorly.
10. The head is able to move in a manner similar to circumduction, but the term more accurately applies to movement at a single joint. Circular movement of the head involves the atlanto-occipital and atlantoaxial joints, as well as amphiarthrotic movements of the cervical vertebrae.
11. A typical hinge joint has a single articulation with a convex condyle fitting into a concave fossa. The knee joint has a double condylar articulation, with tibial (concave) condyles replacing the fossa. The doubling of the condyloid joints prevents movement in the frontal plane (abduction and adduction).

COVERING ALL YOUR BASES

Multiple Choice

1. E and possibly D **2.** B, D **3.** A **4.** A, B, C, D **5.** A, B, D **6.** A, B **7.** B, D **8.** A, B, D **9.** D **10.** A, B, C, D **11.** C **12.** D **13.** A, B **14.** A, B **15.** C **16.** D **17.** A **18.** B, D **19.** C **20.** B, C, D **21.** A, B, C, D **22.** D **23.** B **24.** B **25.** A, D **26.** B **27.** D **28.** C, D

Word Dissection

	Word root	Translation	Example
1.	ab	away	abduction
2.	ad	toward	adduction
3.	amphi	between	amphiarthrosis
4.	ankyl	crooked	ankylosis
5.	arthro	joint	arthritis
6.	artic	joint	articulation
7.	burs	purse	bursa
8.	cruci	cross	cruciate ligament
9.	duct	lead, draw	abduction
10.	gompho	nail	gomphosis
11.	labr	lip	glenoid labrum
12.	luxa	dislocate	luxation
13.	menisc	crescent	meniscus
14.	ovi	egg	synovial
15.	pron	bent forward	pronation
16.	rheum	flux	rheumatism
17.	spondyl	vertebra	spondylitis
18.	supine	lying on the back	supination

Chapter 9 Muscles and Muscle Tissue

BUILDING THE FRAMEWORK

Overview of Muscle Tissues

1. 1. A, B 2. A, C 3. B 4. C 5. A, B 6. A 7. C. 8. C 9. C 10. B

2. A. smooth muscle B. cardiac muscle C. skeletal muscle

3. 1. Bones 2. Promotes labor during birth 3. Contractility 4. Contractility 5. Stretchability 6. Promotes growth

Skeletal Muscle Anatomy

1. 1.G 2. B 3. I 4. D 5. A 6. H 7. F 8. E 9. K 10. C

The endomysium is the fine connective tissue surrounding each muscle cell (fiber) in the figure.

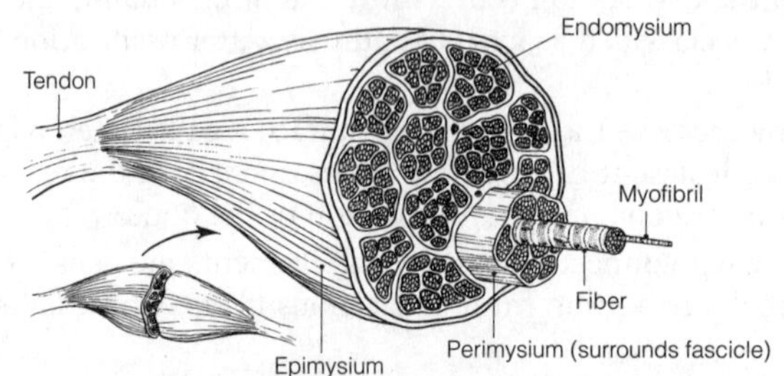

Figure 9.2

2. Matching: B1 = C2 B2 = C3 B3 = C1

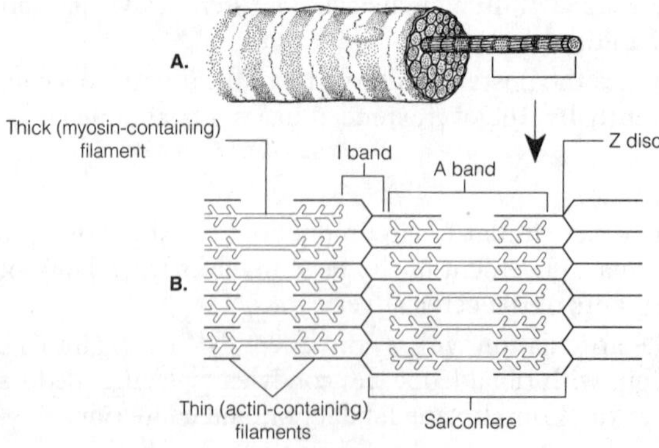

Figure 9.3

3. Myosin heads attach to actin myofilaments, moving them toward the center of the sarcomere.

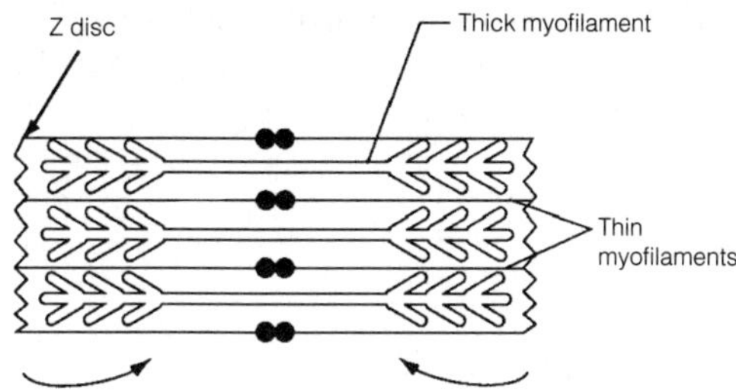

4. 1. C 2. B 3. A
5.

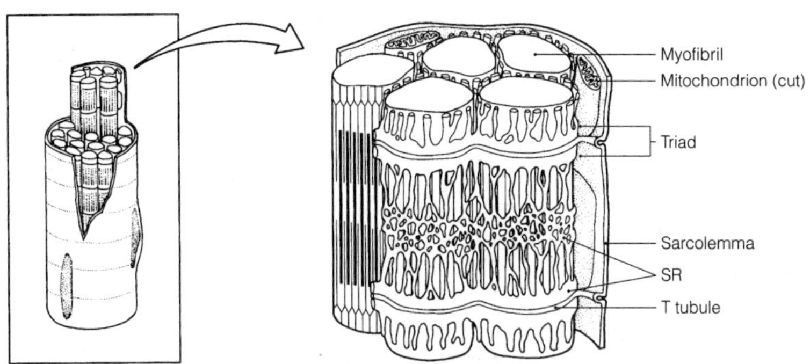

Figure 9.4

6. 1. foot proteins 2. calcium channels 3. voltage sensors 4. action potential
7. Check 1 and 3.

Physiology of Skeletal Muscle Fibers and Skeletal Muscles

1.

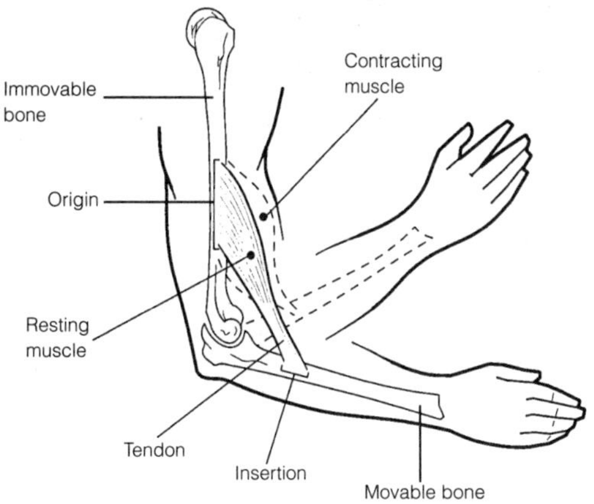

Figure 9.5

2. 1. 4 2. 9 3. 10 4. 1 5. 11 6. 8 7. 2 8. 6 9. 5 10. 7 11. 3
3. The ACh receptors are located in the junctional folds of the sarcolemma.

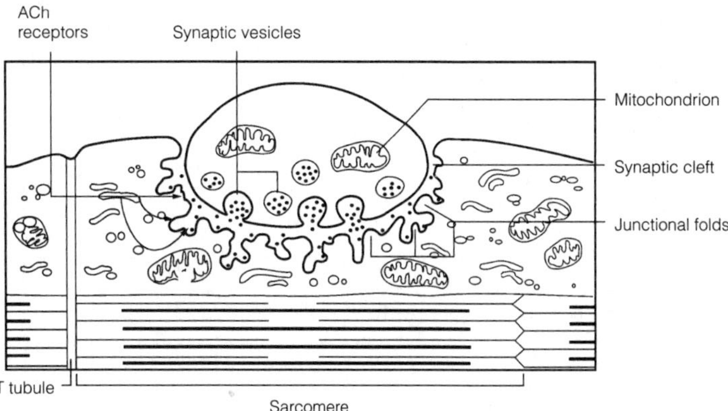

Figure 9.6

4. Tropomyosin is located in the groove between the F–actin strands. The troponin complexes are attached at specific sites to tropomyosin.

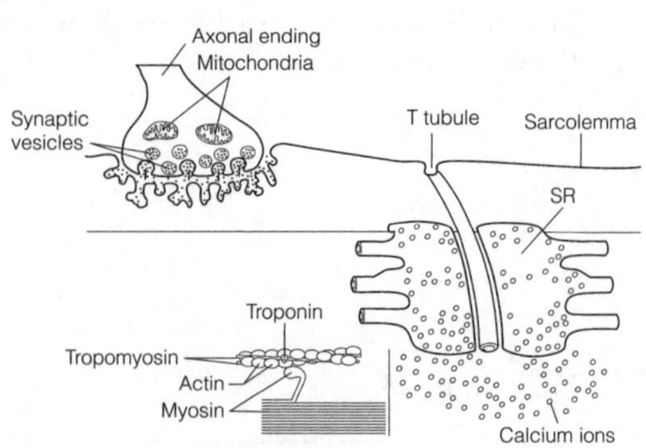

Figure 9.7

5. 1. 1 2. 3 3. 4 4. 7 5. 5 6. 2 7. 6
6. One motor neuron and all the muscle cells it innervates.
7. 1. Troponin 2. Relies on aerobic respiration 3. End product is lactic acid 4. Fuels extended periods of strenuous activity 5. Psychological factors 6. Twitch contraction 7. Muscle hypertrophy 8. Increased endurance 9. Increased muscle strength
8.

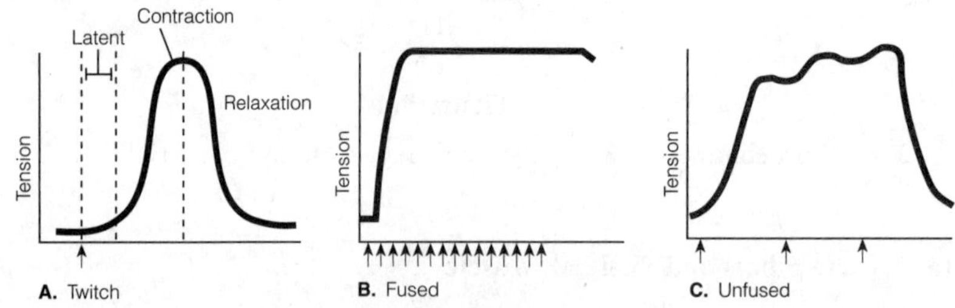

Figure 9.8

9. 1. E 2. B 3. G 4. F 5. A 6. D 7. H 8. B
10. 1. C, D, E, F 2. A, B
11. 1. fatty acids 2. mitochondria 3. carbon dioxide 4. creatine 5. glycogen 6. glucose 7. pyruvic acid 8. lactic acid
12.

Characteristics	Slow oxidative fibers	Fast glycolytic fibers	Fast oxidative fibers
Rapid twitch rate		✓	✓
Fast myosin ATPases		✓	✓
Use mostly aerobic metabolism	✓		✓
Large myoglobin stores	✓		✓
Large glycogen stores		✓	Intermediate
Fatigue slowly	✓		Intermediate
Fibers are white		✓	
Fibers are small	✓		
Fibers contain many capillaries and mitochondria	✓		✓

13. 1. Myofilaments do not overlap; cross bridges cannot attach to generate sliding. 2. Myofilaments touch and interfere with one another; sliding is not possible. 3. The muscle is fatigued.

14. Correct choices: 1, 3, 4, 7.

15. You breathe more deeply and possibly more rapidly until the deficit has been repaid.

Smooth Muscle

1. Answers checked: 5, 8, 10, 12, 13, 15. Answers circled: 1, 2, 3, 4, 14.

2. 1. endomysium 2. smooth muscle 3. single-unit (visceral) 4. two 5. longitudinally 6. circularly 7. peristalsis 8. gap junctions 9. action potentials 10. pacemaker 11. varicosities 12. diffuse junctions 13. extracellular space 14. calmodulin 15. myosin (light chain) kinase 16. phosphate

3. 1. B 2. A 3. B 4. B 5. A 6. A 7. B 8. B 9. A, B 10. B 11. B

4. Smooth muscle is in the walls of hollow organs, many of which store substances temporarily (for example, urine, a baby). When smooth muscle is stretched, it responds (by contraction) only briefly and then relaxes to accommodate filling.

5. 1. Both are innervated independently by their "own" axonal endings and exhibit summation. 2. Multiunit muscle is regulated by the autonomic division (rather than the somatic division) of the nervous system and also responds to hormonal stimuli.

The Incredible Journey

1. 1. endomysium 2. motor unit 3. neuromuscular (myoneural) 4. acetylcholine 5. sodium 6. action potential 7. calcium 8. actin 9. myosin 10. calcium

CHALLENGING YOURSELF

At the Clinic

1. Curare binds to ACh receptors in the neuromuscular junctions and prevents ACh attachment. This blocks transmission of the excitatory stimulus to the muscle fibers, preventing contraction.

2. Acetylcholinesterase is the enzyme that breaks down acetylcholine. Thus it turns off the signal telling the muscle cells to contract and allows them to relax. When organophosphates interfere with the action of the enzyme, the muscle cells are continually stimulated to contract and cannot return to their resting state.

3. The symptoms indicate myasthenia gravis. If a blood test shows antibodies to ACh receptors, the disease is confirmed.

4. When a body part increases in bulk very quickly, the dermis of the skin may tear, resulting in striae (stretch marks). The young man needs to reduce the emphasis on his thighs and add exercises to work on his calves (jogging or resistance exercises) and his arms.

5. Considering Charlie's age, low-impact aerobic exercise would be better for his joints; his articular cartilage probably is not very resilient, and he could suffer from arthritis if he put too much stress on his joints. His bones would benefit from weight-bearing exercise such as walking. He needs to progress from a slow pace to a faster one, giving his cardiovascular and muscular systems time to accommodate to the increased demands.

6. RICE: rest, ice, compression, and elevation (of her leg).

7. Tragically, the chances are good that the boy is afflicted with muscular dystrophy, probably Duchenne muscular dystrophy. There is currently no confirmed treatment for this condition, although transplants of myoblasts are being used experimentally. The condition is fatal when it progresses to the muscles of respiration.

8. Charlie is probably suffering from intermittent claudication, caused when his poorly vascularized out-of-shape muscles contract and squeeze the blood vessels to the point of collapse. This condition is less of a problem in people with a healthy, vigorous cardiovascular system. Charlie should continue his workout to improve his heart function, but he might have to proceed more slowly than initially planned.

9. Ca^{2+} is the stimulus for cross bridge activity, and contraction will occur as long as sufficient amounts of Ca^{2+} are in the cell and ATP is present so that myosin heads can be released from the actin sites.

Stop and Think

1. All three types of muscle produce movement of some sort. Skeletal muscles move bones or skin; cardiac muscle propels blood out of the heart and into blood vessels; smooth muscle raises hairs on cold skin, pushes food along the digestive tract, and dilates the pupil of the eye, just to name a few of its functions. However, only skeletal muscle maintains posture, stabilizes joints, and is important in body heat production.

2. Extensibility is the ability to lengthen. If a muscle on one side of a bone didn't lengthen, the muscle on the other side wouldn't be able to shorten. Each muscle has to stretch when its opposing muscle contracts, or else the bone will not move. For stabilizing joints, muscles on both sides of the bone contract to make sure the bone does not move.

3. Ducks are probably slower fliers but they can fly for a much longer time.

4. A cross section through a muscle fiber at the T tubules would reveal that this tubular system branches to encircle each myofibril.

5. Cross bridge attachment is asynchronous; some cross bridges attach and pull while others detach and recock. The holding cross bridges detach just as the others start to pull. Thus the thin myofilaments are always held in place before being pulled closer in, just as a person pulling a rope grabs hold with one hand before letting go with the other. In a precision rowing team, all oars lift in unison. If all cross bridges detached in unison, nothing would hold the thin myofilaments in place. They would slide back apart after each power stroke, and no gain would be made.
6. Calcium is moved into the central, tubular network of the SR by active transport involving calsequestrin. The resulting decrease in calcium concentration in the sarcoplasm increases the rate of detachment of calcium from troponin, leading to relaxation.
7. The elasticity of the muscle fiber, coupled with the internal tension built up when the fiber shortened, causes the myofilaments to slide back apart. Each sarcomere returns to its resting length. The H zone reappears and the I band lengthens.
8. Synaptic vesicles fusing with the axonal membrane create a high surface area for release of ACh, increasing its rate of diffusion through the synaptic cleft to its receptors. The junctional folds of the sarcolemma at the synapse increase the membrane surface for ACh receptors, increasing the speed and strength of response of the muscle cell to ACh stimulation. The elaborate T system increases the communication between the cell's interior and the sarcolemma, relative to the electrical events of the sarcolemma. In the SR, the tubular arrangement between the terminal cisternae increases the surface area for active transport of calcium, speeding the onset of relaxation. Thus, high surface area increases the number of membrane transport opportunities for ACh, sodium, potassium, and calcium and increases the sensitivity of the cell to ACh and electrical activity.
9. The best reason is (c). Smooth muscle cells contract too slowly to carry out many of the tasks expected of skeletal muscle. Re (a): Smooth muscle is very strong, as strong as skeletal muscle, size for size. Re (b): This answer is incorrect. Smooth muscle cells do contract by sliding filament mechanism. Re (d): Compared to skeletal muscle, smooth muscle is far more fatigue resistant, and continues to contract for much longer periods of time without giving up the ghost.
10. If the muscle cell remained refractory, it could not be restimulated for additional contractions at short intervals. Hence, tetanic contraction would not be possible.
11. Because the motor units contract in turn, each has time to recover before its next turn. If the same motor unit was stimulated continually, its muscle fibers would fatigue. Even keeping your eyelids open all day would be tiring. Muscle tone is accomplished by asynchronous motor unit summation.
12. As a muscle contracts, the blood vessels within it are squeezed and can collapse. In moderate activity, collapse is unlikely or of short duration. As activity becomes more vigorous, contraction is stronger and more prolonged and blood supply can actually decrease. Active muscles rely on oxygen stored in myoglobin. Training encourages the growth of more capillaries and increases myoglobin stores. Activity also increases blood pressure, making vessel collapse less likely and resulting in the delivery of more blood with each heartbeat.
13. Muscles are extensible, but the connective tissue that replaces atrophied muscle is not. Unless the joints are moved through their full range routinely and/or braced in an extended position, the fibrous tissue will be shorter than the original muscle and the joint will be contorted.
14. Calcium binds to troponin in skeletal muscle, triggering the attachment of myosin cross bridges and the power stroke. In smooth muscle, which does not have troponin, calcium binds to calmodulin, which activates a kinase enzyme. This triggers cross bridge attachment.
15. Without an elaborate SR, calcium flux into the cell is relatively slow and requires that Ca^{2+} enter from the cell exterior. This slows and prolongs both the contractile period and the relaxation period, as expulsion of calcium is also a membrane-limited phenomenon.
16. The heart has its own "in-house" (intrinsic) pacemaker that will continue to set the pace of the heartbeat. However, without neural controls, the heart rate could not be increased substantially during vigorous activity. There will be some increase in rate, however, due to released epinephrine and NE by the adrenal glands.

COVERING ALL YOUR BASES

Multiple Choice

1. C, D **2.** A, B, C **3.** B **4.** B, C **5.** A, B, C, D, E **6.** C, E **7.** C **8.** D **9.** A, C, D **10.** A, C **11.** A, D **12.** C **13.** C **14.** A, C, D **15.** A, C **16.** C **17.** B **18.** C **19.** A, D **20.** A **21.** B **22.** A, B, D **23.** B **24.** B, D **25.** B, D **26.** A, B, D **27.** A, B, C **28.** A, C, D **29.** A, C **30.** C, D **31.** A, B, C

Word Dissection

	Word root	Translation	Example
1.	fasci	bundle	fascicle
2.	lemma	sheath	sarcolemma
3.	metric	measure	isometric
4.	myo	mouse	myofibril
5.	penna	feather, wing	unipennate
6.	raph	seam	raphe
7.	sarco	flesh	sarcoplasm
8.	stalsis	constriction	peristalsis
9.	stria	streak	striation
10.	synap	union	synapse
11.	tetan	rigid, tense	tetanus

Chapter 10 The Muscular System

BUILDING THE FRAMEWORK

Interactions of Skeletal Muscles in the Body

1. 1. B 2. C 3. A 4. D

Naming Skeletal Muscles

1. 1. A, F 2. D 3. G 4. A, F 5. B, C 6. A 7. B, C 8. C, F 9. C, E 10. D

Muscle Mechanics: Importance of Fascicle Arrangement and Leverage

1. 1. muscles 2. bones 3. the muscle inserts 4. joint

2. Circle: fast, moves a small load over a large distance, requires minimal shortening, and muscle force greater than the load.

3. The brachialis, which is inserted closer to the fulcrum (elbow) than the load, is structured for mechanical disadvantage; the brachioradialis, which is inserted far from the elbow, works by mechanical advantage but is a weak elbow flexor at best.

4.

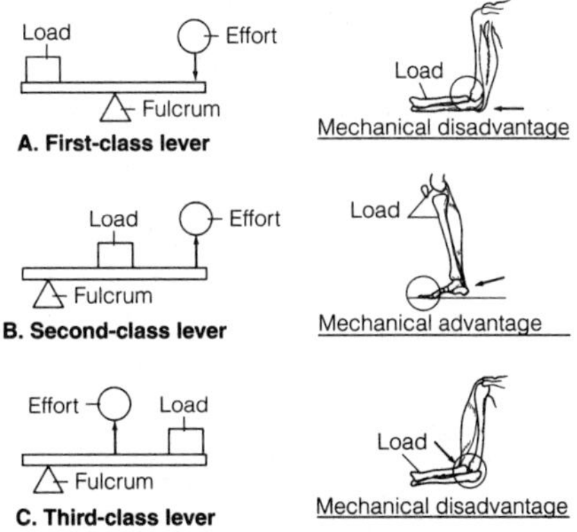

Figure 10.1

5. 1. Bipennate; C 2. Circular; D 3. Parallel; A 4. Convergent; B. Examples of A include the sartorius muscle of the thigh, nearly all forearm muscles, and the rectus abdominis. Examples of B are the pectoralis major and minor. Examples of C include the rectus femoris and the flexor hallucis longus. Examples of D are the orbicularis oris and oculi muscles. Consult muscle tables for more examples.

Major Skeletal Muscles of the Body

1. 1. E 2. G 3. A 4. F 5. C 6. B 7. H

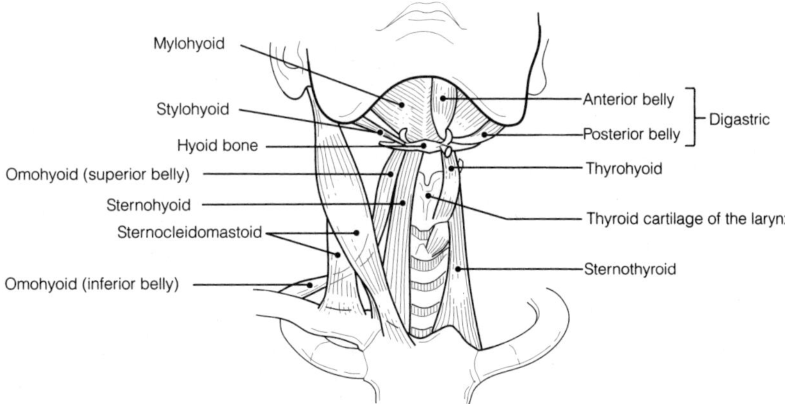

Figure 10.3

2. 1. E 2. C 3. H 4. F 5. B 6. D 7. G

3. 1. zygomatic 2. buccinator 3. orbicularis oculi 4. frontal belly of the epicranius 5. orbicularis oris 6. masseter 7. temporalis 8. splenius 9. sternocleidomastoid 10. pterygoid muscle

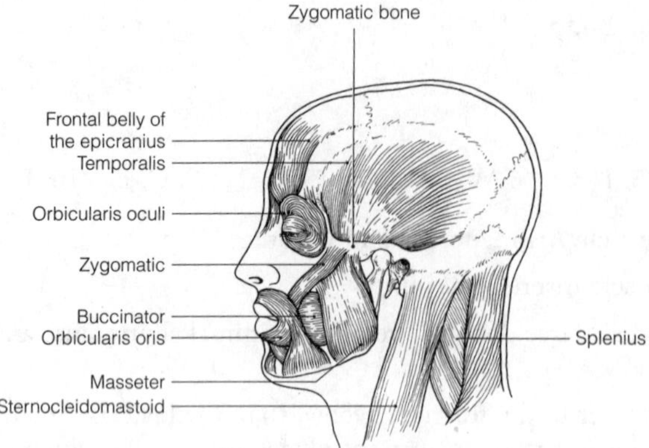

Figure 10.4

4. They work as a group during swallowing to propel a food bolus to the esophagus.

5. 1. rectus abdominis 2. pectoralis major 3. deltoid 4. external oblique 5. sternocleidomastoid 6. serratus anterior 7. rectus abdominis, external oblique, internal oblique, transversus abdominis 8. linea alba 9. transversus abdominis 10. external intercostals 11. diaphragm 12. pectoralis minor

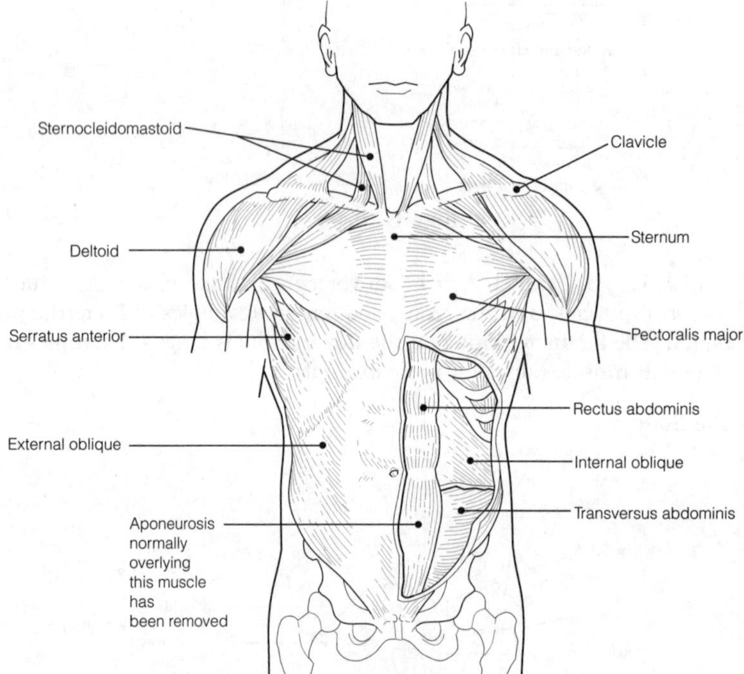

Figure 10.5

6. Muscles in the ventral leg either dorsiflex the ankle or extend the toes, movements that do not work against gravity. By contrast, posterior calf muscles plantar-flex the ankle against the pull of gravity.

7. 1. trapezius 2. latissimus dorsi 3. deltoid 4. erector spinae 5. quadratus lumborum 6. rhomboids 7. levator scapulae 8. teres major 9. supraspinatus 10. infraspinatus 11. teres minor

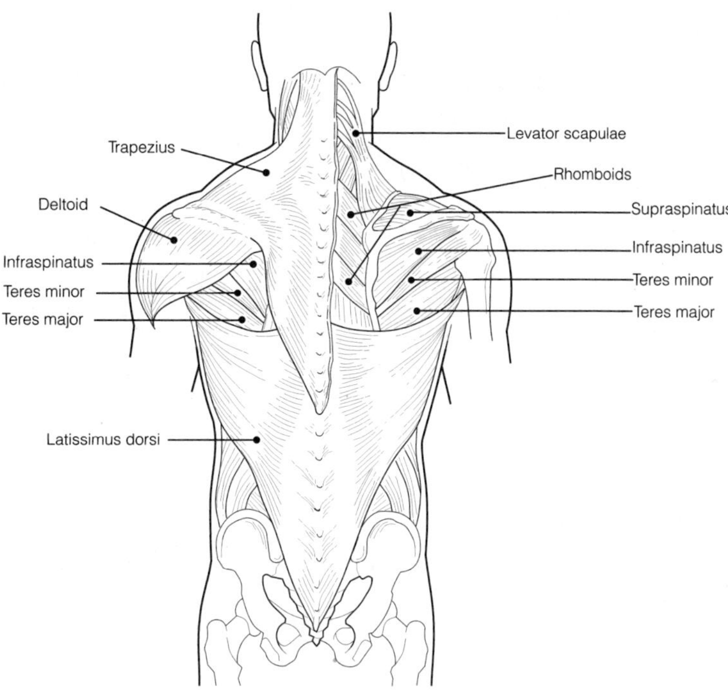

Figure 10.6

8. 1. C 2. D 3. B 4. A

9. 1. B, E 2. C, F 3. A, D, G 4. F 5. A 6. E 7. D

10. 1. C 2. A 3. B 4. D 5. F 6. E

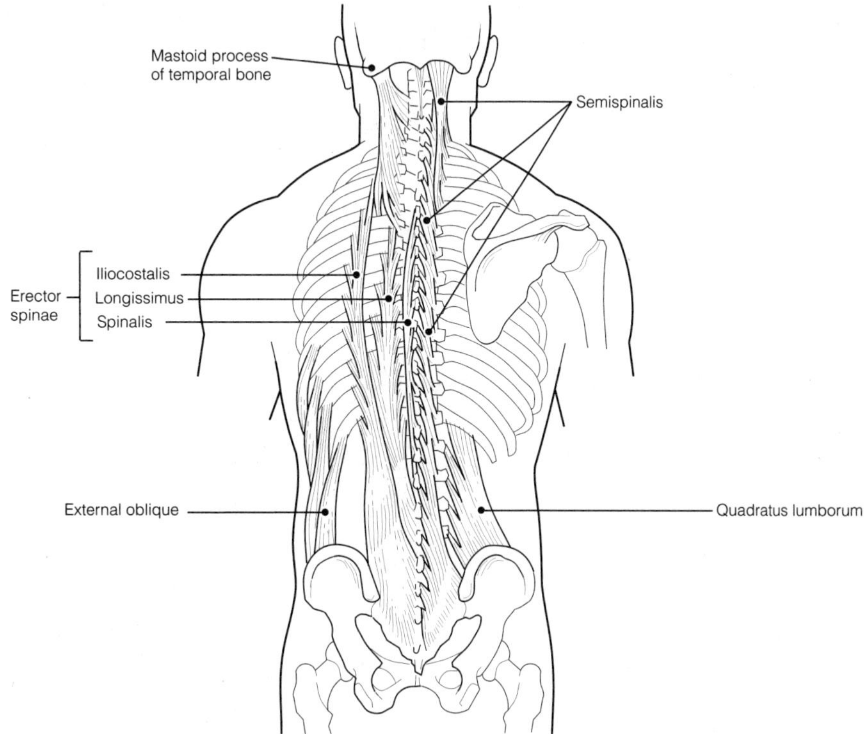

Figure 10.7

11. 1.H 2. K 3. G 4. E, F 5. I 6. A 7. C 8. D 9. L 10. J

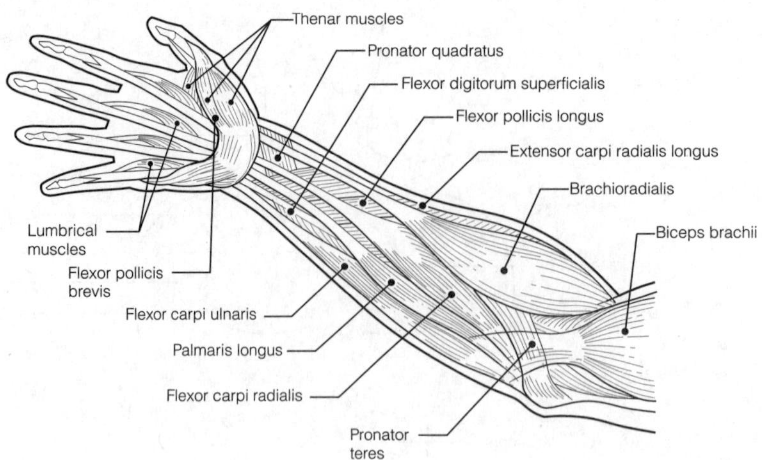

Figure 10.8

12. 1. flexor carpi ulnaris 2. extensor digitorum 3. biceps brachii 4. brachioradialis 5. triceps brachii 6. anconeus 7. brachialis 8. extensor carpi radialis brevis 9. extensor carpi radialis longus 10. abductor pollicis longus 11. extensor pollicis brevis 12. deltoid

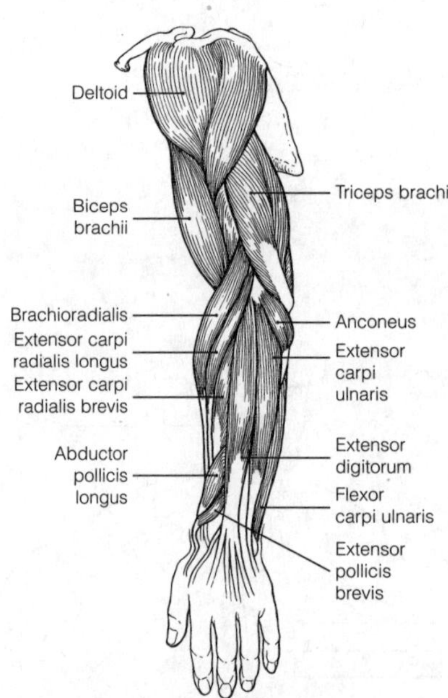

Figure 10.9

13. 1. iliopsoas 2. gluteus maximus 3. triceps surae 4. tibialis anterior 5. adductors 6. quadriceps 7. hamstrings 8. gluteus medius 9. gracilis 10. tensor fasciae latae 11. extensor digitorum longus 12. fibularis longus 13. tibialis posterior

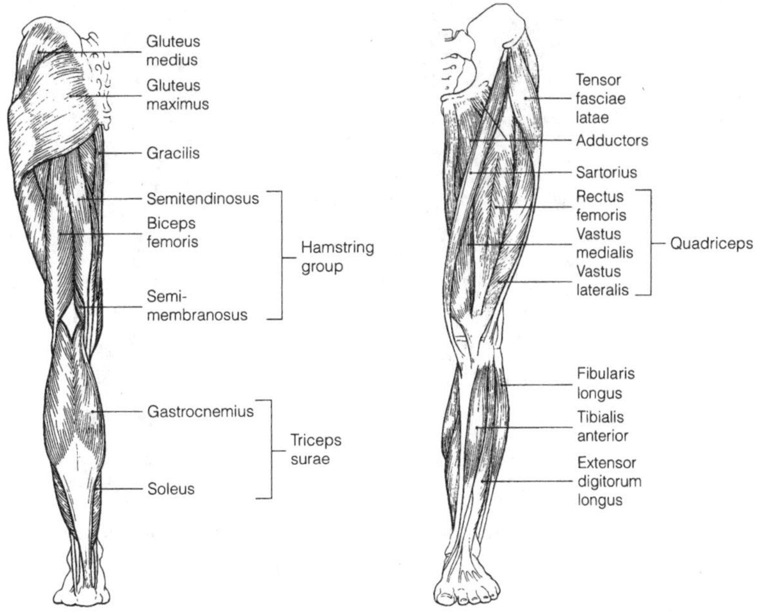

Figure 10.10

14. 1. deltoid 2. gluteus maximus or medius 3. vasti 4. quadriceps 5. calcaneal or Achilles 6. proximal 7. forearm 8. anterior 9. posteriorly 10. knee 11. flex

15. 1. iliopsoas, rectus femoris, (adductors longus and brevis) 2. quadriceps 3. tibialis anterior (extensor digitorum longus, fibularis tertius, extensor hallucis longus)

16. 1. D 2. F 3. C 4. A 5. H 6. G 7. B

17. 1. E 2. D 3. F 4. A 5. G 6. C 7. B

18. 1. Biceps femoris 2. Antagonists 3. Gluteus minimus 4. Vastus medialis 5. Supraspinatus 6. Fibularis longus 7. Teres major 8. Gastrocnemius/soleus

19. 1. 4 2. 5 3. 17 4. 16 5. 7 6. 6 7. 20 8. 14 9. 19 10. 12 11. 11 12. 10 13. 22 14. 1 15. 2 16. 3 17. 15 18. 21 19. 13 20. 9 21. 18 22. 8

20. 1.1 2. 2 3. 5 4. 9 5. 7 6. 4 7. 6 8. 3 9. 8 10. 10 11. 11

21.

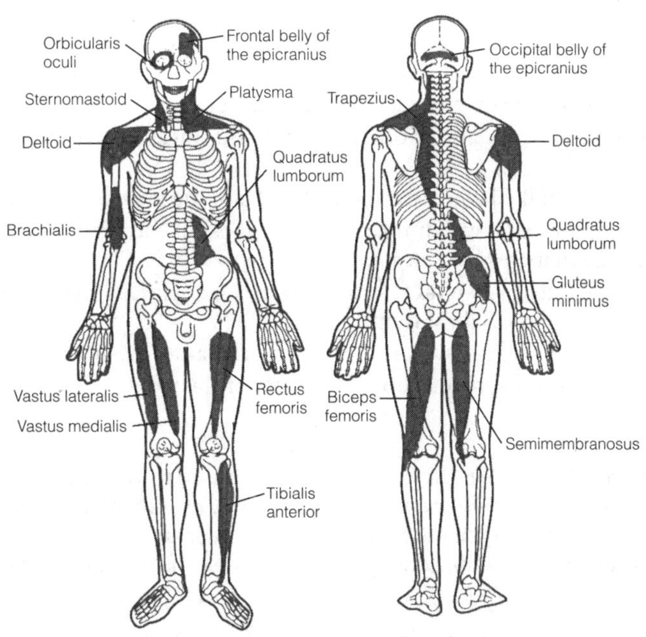

Figure 10.13

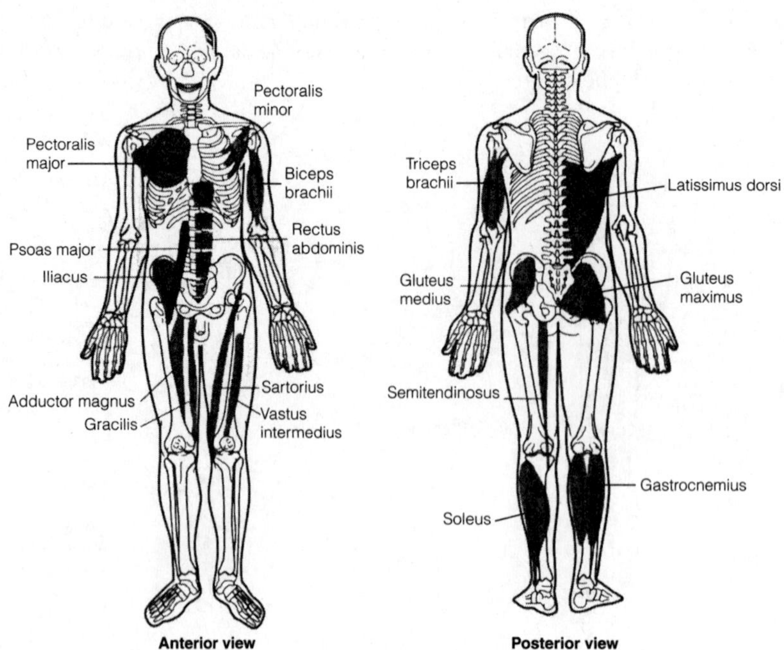

Figure 10.14

CHALLENGING YOURSELF

At the Clinic

1. The facial nerve, which innervates the muscles of the face.
2. When we are in the fully bent-over (hip flexed) position, the erector spinae are relaxed. When that movement is reversed, they are totally inactive, leaving the gluteus maximus and hamstring muscles to initiate the action. Thus sudden or improper lifting techniques are likely to injure both back ligaments and the erector spinae, causing the latter to go into painful spasms.
3. The phrenic nerve innervates the diaphragm; the intercostal nerves supply the intercostal muscles. Paralysis of these essential breathing muscles is fatal.
4. The unilateral weakening of the abdominal wall will result in scoliosis because the stronger contraction of the muscles of the opposite side will pull the spine out of alignment.
5. The rectus abdominis is a narrow, medially placed muscle that does not extend completely across the iliac regions. No, if the incision was made as described, the rectus abdominis was not cut.
6. The muscle of the urogenital diaphragm, which surrounds the vaginal opening, through which uterine prolapse would occur, is the deep transverse perineus. Lateral to the vagina is the pubococcygeus of the levator ani (part of the pelvic diaphragm).
7. The deltoid and the pectoralis major have broad origins on the inferior border of the clavicle. The sternocleidomastoid inserts on the medial superior border of the clavicle; the trapezius inserts on the lateral third of the clavicle. An arm sling inhibits use of these clavicle-moving muscles.
8. Tendons attaching at the anterior wrist are involved in wrist and finger flexion. Malcolm will lose his ability to make a fist and grasp a baseball.
9. The hamstrings can be strained (pulled) when the hip is flexed and the knee is vigorously extended at the same time.
10. The latissimus dorsi and the trapezius, which together cover most of the superficial surface of the back, are receiving most of the massage therapist's attention.
11. Gluteus maximus—action prevented is lateral rotation of the thigh; gluteus medius and minimus—actions prevented are thigh abduction and medial rotation.
12. Steroid use might end the adolescent growth spurt by causing premature closure of the epiphyseal discs.
13. The muscles of the external urethral sphincter must be kept contracted.

Stop and Think

1. Very few muscles don't attach to bone at least at one end. The muscles in the tongue (intrinsic tongue muscles), which change the shape of the tongue, are among those few. Sphincter muscles, such as the orbicularis muscles of the face, the sphincter ani, and the external urethral sphincter, cause constriction of an opening. Muscles of facial expression mostly originate on bone but insert on fascia or skin; consequently, they alter the position of the soft tissues of the face.

2. With the help of fixators, an insertion can be the immovable end and the origin can be moved in selected cases. For example, the trapezius originates on the occipital bone and inserts on the scapula, but fixing the scapula will allow the trapezius to assist in extension of the head. Chin-ups give another example: the forearm is fixed when grasping a chin-up bar, so the bones of origin (the shoulder girdle)—and the body along with them—are moved.
3. Muscles pull bones toward their origin as they contract: muscles originating on the front of the body pull bones anteriorly (usually flexion); muscles on the back pull bones posteriorly (usually extension); laterally placed muscles abduct (or, with the foot, evert); medially located muscles adduct (or invert the foot). Muscles *usually* move the bone beyond (distal to) their bellies: the muscles on the thorax move the arm; the muscles of the arm move the forearm; the muscles of the forearm move the wrist and fingers. Likewise, the muscles of the hip move the thigh, the muscles of the thigh move the leg, and so on.
4. The latissimus dorsi inserts on the anterior surface of the humerus, getting there via its medial side. In rotation, the latissimus pulls the anterior surface of the humerus toward the body—this is medial rotation.
5. The tendon of the tibialis anterior is extensive, starting at the distal third of the tibia and continuing around the medial side of the ankle, finally inserting on the plantar surface of the foot. Because it inserts on the bottom of the foot, it can pull the sole of the foot medially (inversion). Likewise, on the lateral side, the fibularis longus inserts on the plantar surface, so it can pull the sole laterally (eversion).

COVERING ALL YOUR BASES

Multiple Choice

1. A **2.** B **3.** B, D **4.** A, B, C, D **5.** A, B, C **6.** B, C, D **7.** A, D **8.** B, C **9.** A, C, D **10.** A, C, D **11.** A, B **12.** A, B **13.** C **14.** D **15.** A, B, C, E **16.** A, B, C, D **17.** B, D **18.** B **19.** A, B, D **20.** C **21.** B, C **22.** B, D **23.** A, B **24.** B, C, D **25.** A, B, C, D **26.** D **27.** D **28.** D **29.** B **30.** A, B, C, D

Word Dissection

	Word root	Translation	Example		Word root	Translation	Example
1.	agon	contest	agonist, antagonist	6.	glossus	tongue	genioglossus
2.	brevis	short	peroneus brevis	7.	pectus	chest, breast	pectoralis major
3.	ceps	head, origin	biceps brachii	8.	perone	fibula	peroneus longus
4.	cleido	clavicle	sternocleidomastoid	9.	rectus	straight	rectus abdominis
5.	gaster	belly	gastrocnemius				

Chapter 11 Fundamentals of the Nervous System and Nervous Tissue

BUILDING THE FRAMEWORK

Functions and Divisions of the Nervous System

1. 1. It monitors all information about changes occurring both inside and outside the body. 2. It processes and interprets the information received and integrates it in order to make decisions. 3. It commands responses by activating muscles, glands, and other parts of the nervous system.

2.

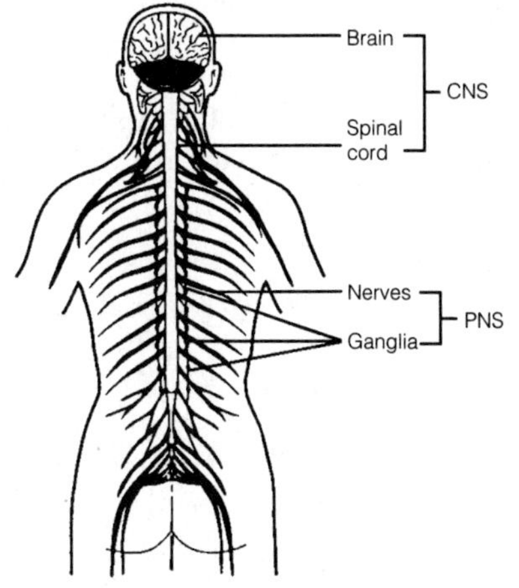

Figure 11.1

3. 1. B 2. D 3. C 4. A 5. B 6. C

Histology of Nervous Tissue

1. 1. T 2. neuroglia 3. PNS 4. neurons 5. neuroglia 6. microglia 7. ependymal cells 8. oligodendrocytes 9. astrocytes

2. 1. B 2. D 3. E 4. C 5. A 6. C 7. D 8. A 9. E 10. F 11. E 12. C

3. 1. Centrioles 2. Mitochondria 3. Glycogen 4. Output 5. Axon collateral

4.

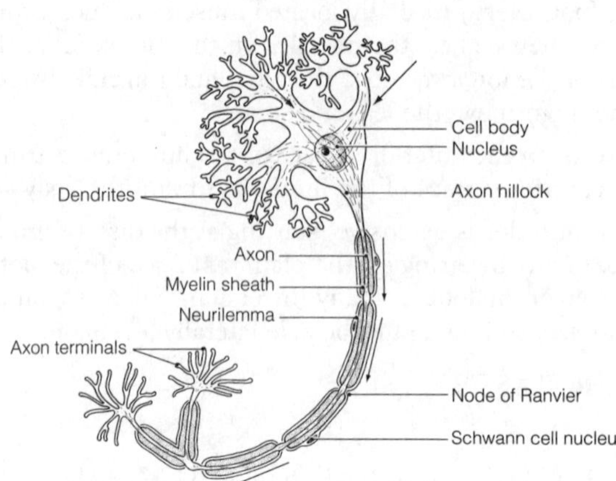

Figure 11.2

5. 1. D 2. B 3. A 4. C

6. Neurons *have extreme longevity,* they are *amitotic,* and they have a *high metabolic rate.*

7. 1. G 2. C 3. E 4. I 5. A 6. F 7. H 8. D

8. 1. Cell body 2. Centrioles 3. Clusters of glial cells 4. Dendrites 5. Gray matter

9. A. Bipolar, sensory B. Unipolar, sensory C. Multipolar, motor or association

10. 1. motor 2. multipolar 3. association 4. multipolar 5. bipolar 6. multipolar 7. sensory 8. unipolar

11. A. Unmyelinated B. Myelinated

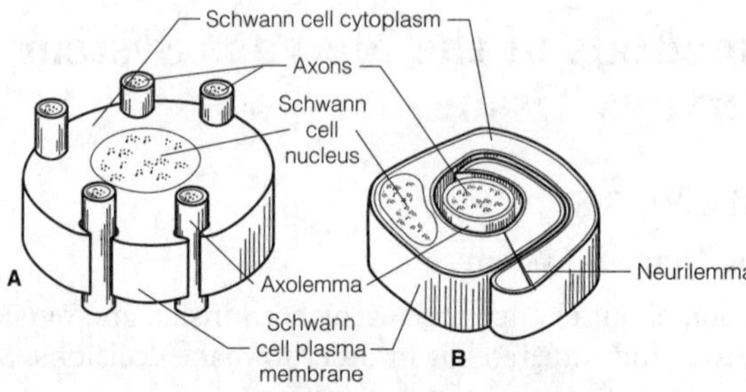

Figure 11.4

Membrane Potentials

1. 1. Ohm's 2. I = V/R 3. ions 4. voltage 5. plasma membrane 6. proteins 7. passive or leakage 8. gated 9. chemical 10. electrical 11. mechanically gated channels

2. Check 1, 4, 5, 6, 9

3.

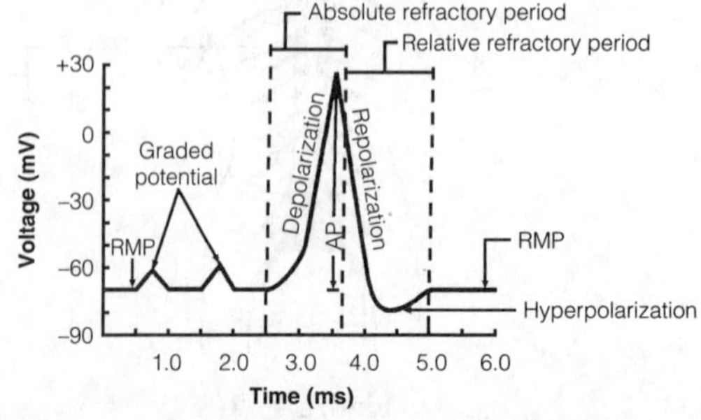

Figure 11.5

4. 1. H 2. C 3. G 4. I 5. B 6. A 7. J 8. L 9. K 10. B 11. D 12. L 13. F 14. E 15. B 16. C 17. B 18. E 19. B

5. 1. High Na^+ 2. High Cl^- 3. $3Na^+$ in / $2K^+$ out 4. Nerve impulse 5. Hyperpolarization 6. Cell bodies

6. 1. faster 2. sodium 3. T 4. sodium 5. presynaptic 6. a neurotransmitter binds 7. a potential difference 8. T 9. metabotropic

Synapses and Neurotransmitters and Their Receptors

1. 1. calcium 2. presynaptic axonal membrane 3. neurotransmitter 4. synaptic cleft 5. receptors 6. membrane potential 7. excitatory 8. sodium and potassium 9. positive 10. action potentials 11. action potential 12. hillock 13. inhibitory 14. negative 15. hyperpolarization 16. summation 17. summate 18. axoaxonic 19. neuromodulators

2. 1. A 2. B 3. B 4. A 5. B 6. A 7. B 8. B 9. A

3. 1. B, F, G 2. D, F, H, J, K 3. A, E, G 4. C, E, F, H, I, L

4. 1. Slow response 2. Short-lived effect 3. Weak stimuli 4. Pain receptors 5. ACh

5.

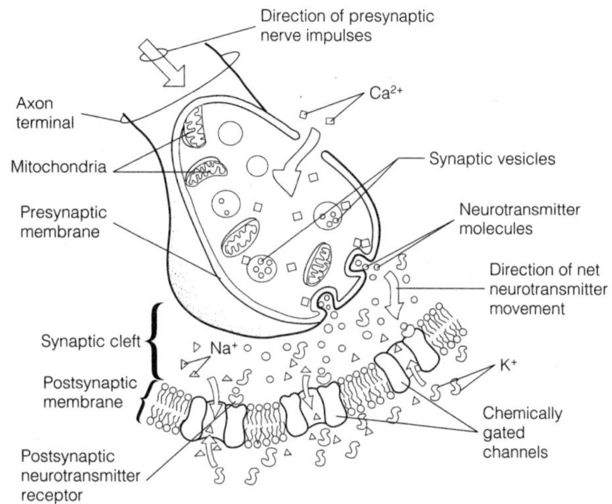

Figure 11.6

6. 1. A 2. E 3. D 4. B 5. E 6. C 7. A 8. C 9. B

7.

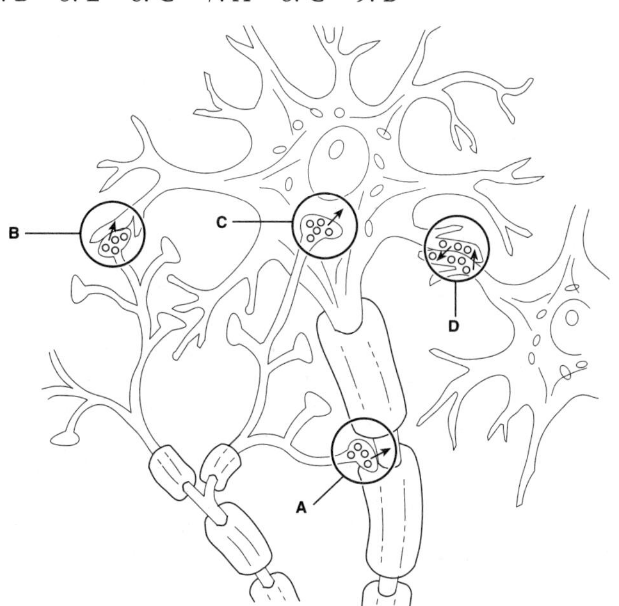

Figure 11.7

Basic Concepts of Neural Integration

1. 1. PNS 2. All-or-nothing response 3. Variety of stimuli
2. 1. pin-prick pain 2. skeletal muscle 3. two (third with muscle)

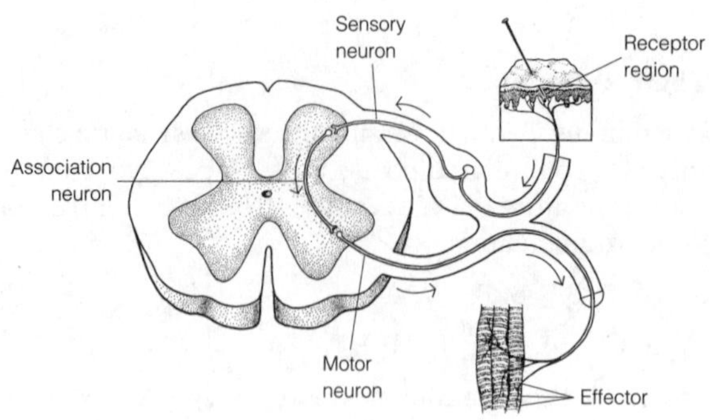

Figure 11.8

3. 1. B 2. A 3. B 4. D 5. A 6. D 7. C 8. C

CHALLENGING YOURSELF

At the Clinic

1. Diuretics often cause a potassium deficiency, which would lead to hyperpolarization of neurons' receptive areas. This would inhibit impulse generation and reduce neuronal output.
2. Many viruses, including some herpesviruses, travel up axons by a retrograde type of transport used for recycling of cellular materials to the cell body.
3. Because nerve cells are amitotic, the tumor must be a glioma, developing from one of the types of neuroglia.
4. Myelination is essential for saltatory conduction. Multiple sclerosis is characterized by loss of myelination, interrupting the rapid node-to-node propagation of action potentials.
5. GABA is an inhibitory neurotransmitter that counterbalances the excitation of motor and integrative function. Seizures result from hyperactivity of neurons that have escaped normal GABA-induced inhibition.
6. Caffeine is a stimulant; it lowers the threshold of neurons and makes them more irritable, decreasing the amount of stimulation required to trigger an action potential.
7. The continuous pressure of the bannister on his axillary blood vessels impaired the delivery of O_2 and other nutrients to the nerve fibers in his right upper limb, leading to their inability to conduct nerve impulses.

Stop and Think

1. Microglia are involved in phagocytosis of infected and damaged cells. Hence, trauma (due to stroke or a physical blow) and infection can induce an increase in the concentration of microglia.
2. A high degree of branching increases the surface area of a cellular process, as is seen in dendritic spines and in telodendria. The increase in surface area increases the space available for synapses.
3. Increased axon diameter at the nodes of Ranvier increases the surface area of the axonal membrane there. This provides more room for sodium channels and increases the speed of impulse propagation and recovery.
4. Motor proteins (kinesin, dynein, myosin) and microtubules ("tracks").
5. Neurons without axons can come into close contact with dendrites of other neurons and pass along their excitation by direct contact.
6. Excitatory stimulation (EPSPs) opens channels that allow simultaneous passage of sodium and potassium. Inhibition (IPSPs) opens potassium and/or chloride gates. The net sodium flux (occurring at excitatory synapses) generates an electrical charge opposite that of the inhibitory potassium and chloride fluxes. If the ion fluxes exactly balance, the membrane voltage will not differ from the resting membrane potential and will graph as a flat line.
7. The pathway with longer axons will have fewer synapses. Because synaptic delay significantly slows impulse transmission, the pathway with longer fibers will be faster. In a parallel after-discharge circuit, the parallel pathways have different numbers of synapses. This arrangement ensures that the final postsynaptic cell will be stimulated sequentially instead of simultaneously by the presynaptic neurons.

8. If chloride gates open (due to inhibitory stimulation), more negative chloride ions will diffuse into the cell from the extracellular space (following their concentration gradient). Hence, the chloride flux hyperpolarizes the cell (inside becomes more negative).
9. The effect of neurotransmitter is to create a *graded* potential, not an *action* potential. Depolarization and refractory period are characteristics of the action potential. Persistence of neurotransmitter can increase the frequency of action potentials but will not affect the character of the action potential.
10. False. Most neurons make and release more than one neurotransmitter, simultaneously or at different times. It is quite likely that these neurotransmitters have very different effects. For example, release of an amino acid will most likely open ion channels and exert its effect only briefly, whereas a peptide neurotransmitter is more likely to produce a more prolonged response (mediated by a G protein second-messenger system).
11. Formation of new synapses can increase the influence of a particular presynaptic neuron over a postsynaptic neuron. Neurons in the facilitated zone have fewer synapses and require more impulses to move the postsynaptic cell to threshold. Increasing the number of synaptic connections can change such a neuron's zone designation to "discharge zone." Fewer impulses may then be sufficient to push the postsynaptic cell to threshold.
12. Yes, a neuron can be the postsynaptic cell of a converging circuit and the presynaptic cell of a diverging circuit, just to name one possibility. Formation of axon collaterals increases the likelihood of a single neuron being part of multiple circuits.
13. It's an axon because (1) dendrites do not generate and conduct action potentials (peripheral processes do generate and conduct APs) and (2) dendrites are never myelinated (many, if not most, peripheral processes of sensory neurons—particularly somatic sensory neurons—are myelinated). It's a dendrite because the electrical current is moving *toward* the cell body. (This is the older definition of dendrite/axon, i.e., dendrites conduct electrical currents toward the cell body; axons conduct an impulse away from the cell body.) Function follows structure; if the peripheral process looks and acts like an axon functionally, then it must be an axon.
14. After birth, the number of neurons actually decreases. In the case of a lost ability, the increased dependence on other senses can slow the loss of neurons involved in the "enhanced sense" pathways and facilitate those pathways. It is neuron retention and increased efficiency of neurotransmission, not an increase in neurons, that causes the enhancement.
15. The self-propagating change in membrane potential that travels along the membrane from the point of stimulation.
16. The somatic division is involved in stretching, doing situps, walking, and brushing her teeth. The autonomic nervous system is involved in the gurgling of her stomach.

COVERING ALL YOUR BASES

Multiple Choice

1. D **2.** D **3.** C **4.** B **5.** B, D **6.** A **7.** B, D **8.** B, C **9.** D **10.** B **11.** C **12.** A, B **13.** A, C, D **14.** B, C **15.** A **16.** B **17.** B, D **18.** A **19.** D **20.** B, D **21.** A, B **22.** C, D **23.** C **24.** D **25.** A **26.** A, B, C **27.** A, B **28.** B, D **29.** C **30.** C **31.** D **32.** B

Word Dissection

	Word root	Translation	Example
1.	dendr	branch	dendrite
2.	ependy	tunic	ependymal
3.	gangli	knot, swelling	ganglion
4.	glia	glue	neuroglia
5.	neur	cord	neuron
6.	nom	law	autonomic
7.	oligo	few	oligodendrocyte
8.	salta	leap, dance	saltatory conduction
9.	syn	clasp, join	synapse

Chapter 12 The Central Nervous System

BUILDING THE FRAMEWORK

The Brain

1.

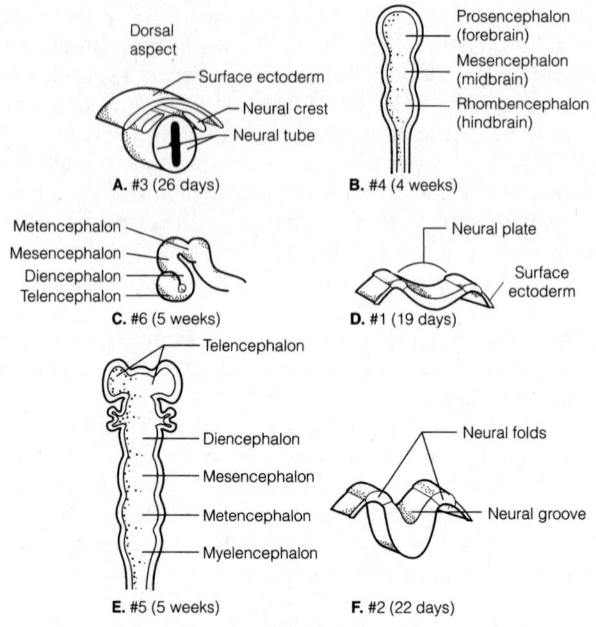

Figure 12.1

2.

Secondary brain vesicle	Adult brain structures	Neural canal regions
Telencephalon	Cerebral hemispheres (cortex, white matter, basal nuclei)	Lateral ventricles; superior portion of third ventricle
Diencephalon	Diencephalon (thalamus, hypothalamus, epithalamus)	Most of third ventricle
Mesencephalon	Brain stem: midbrain	Cerebral aqueduct
Metencephalon	Brain stem: pons; cerebellum	Fourth ventricle
Myelencephalon	Brain stem: medulla oblongata	Fourth ventricle

3. 1. D 2. L 3. F 4. C 5. K 6. B 7. E 8. A 9. I 10. H 11. J 12. G Areas B and C should be striped.

4. Color B, D, F, and H blue

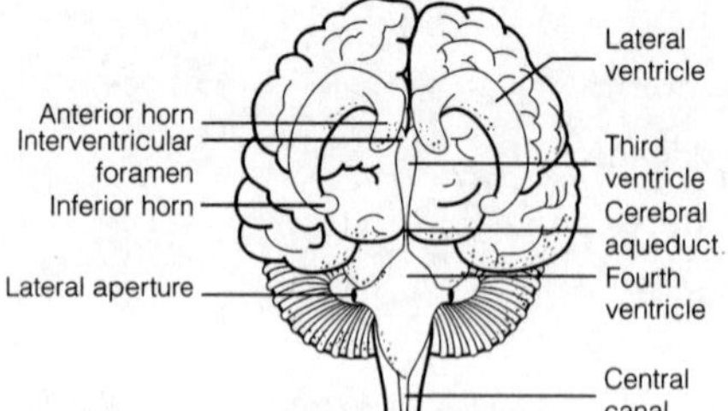

Figure 12.3

5.

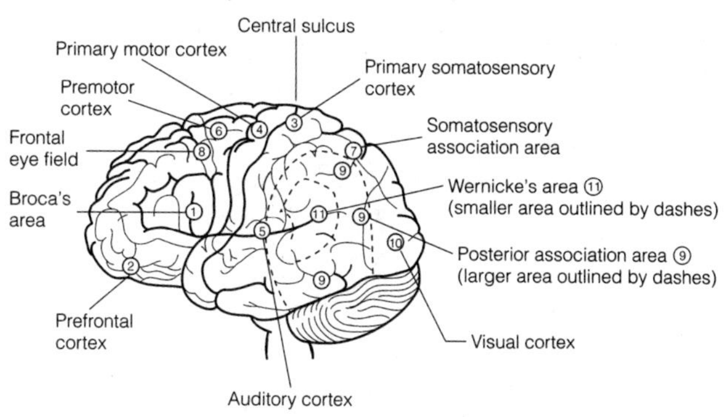

Figure 12.4

6. 1. G 2. W 3. W 4. W 5. W 6. G 7. G 8. W

7. 1. postcentral 2. temporal 3. frontal 4. Broca's 5. left 6. T 7. precentral 8. premotor 9. fingertips 10. somatosensory association cortex 11. occipital 12. T 13. prefrontal 14. posterior association 15. affective 16. language 17. right

8. 1. hypothalamus 2. medulla oblongata 3. cerebellum 4. thalamus 5. medulla oblongata 6. cerebral peduncle 7. pineal body 8. pons 9. corpora quadrigemina

9. 1. L 2. N 3. Q 4. O 5. B 6. F 7. A 8. D 9. M 10. I 11. K 12. G 13. P 14. H 15. J 16. C 17. E Figure 12.5: Brain stem areas O, J, and K should be colored blue. The gray dotted areas contain cerebrospinal fluid and should be colored yellow.

10.

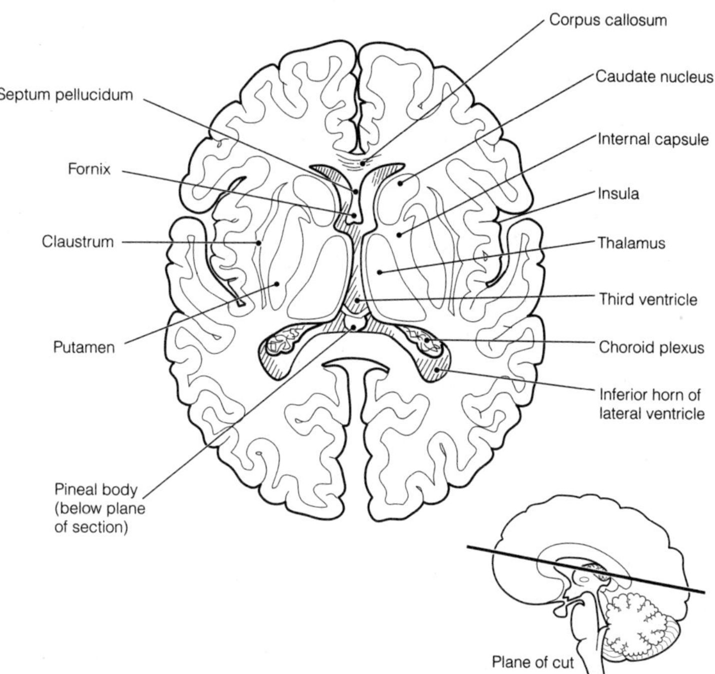

Figure 12.6

11. 1. B, C, D, F 2. H 3. A 4. I 5. B 6. G 7. E 8. D 9. F

12.

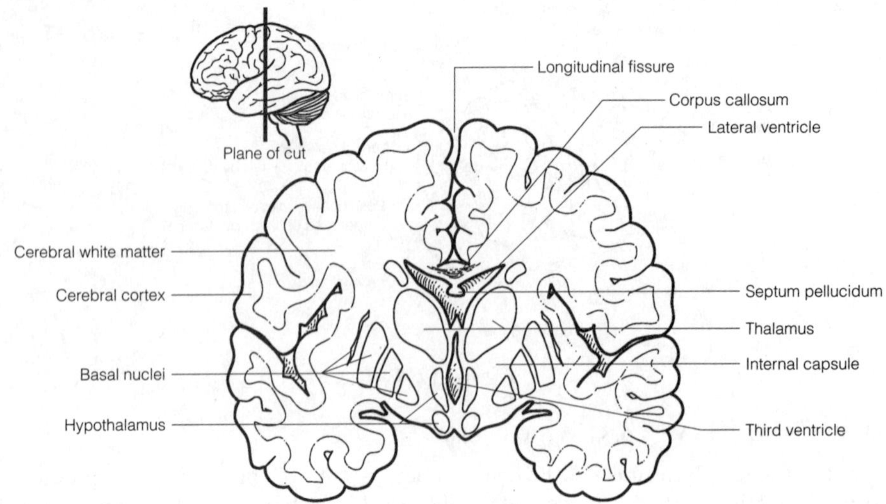

Figure 12.7

13. The cranial nerves include all named nerves (except those designated as spinal nerves). The midbrain is the area including C, I, and K. The diencephalon is the region from which the mammillary bodies and the infundibulum project. The pons is the enlarged region between K and A; the medulla is the region from A to just below D where the spinal cord begins.

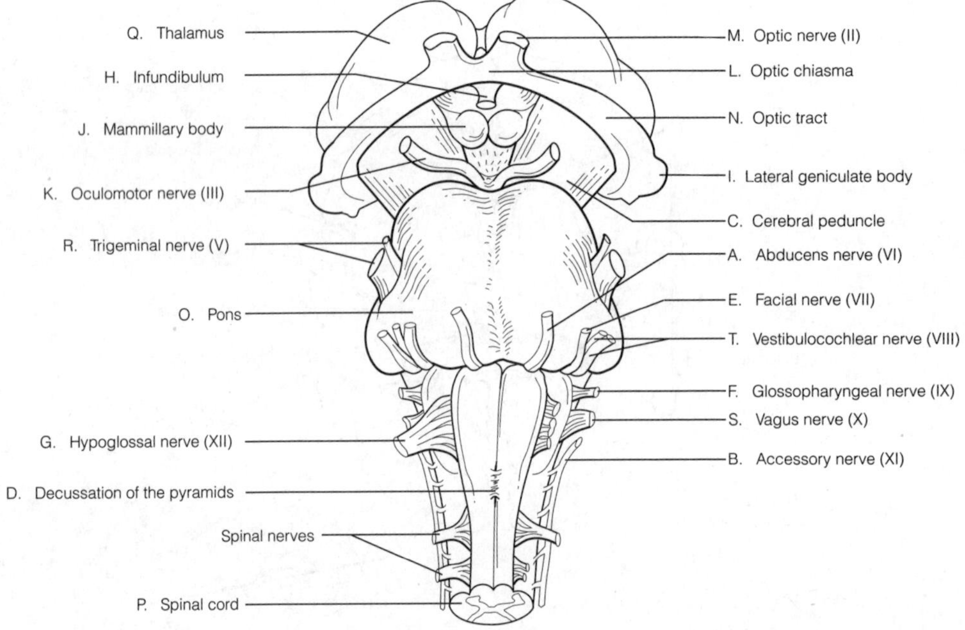

Figure 12.8

14. 1. Cerebral peduncles 2. Pineal gland 3. Pons 4. Paralysis 5. Speech disorder 6. Motor impairment 7. Association fibers 8. Internal capsule 9. Caudate nucleus 10. Parietal lobe

15. 1. 5 2. 2 3. 5 4. 1 5. 3 6. 4

16. 1. R 2. R 3. L 4. L 5. L 6. R 7. L 8. R 9. R 10. L 11. R 12. L 13. L

Higher Mental Functions

1. 1. A 2. D 3. A 4. C 5. C 6. B
2. 1. electroencephalogram (EEC) 2. frequencies 3. Hz (hertz or cycles per second) 4. amplitude 5. sleep 6. cerebral cortical function 7. epilepsy 8. tonic-clonic (grand mal) 9. brain death
3. 1. 24 2. coma 3. brain stem 4. hypothalamus 5. increase 6. T 7. NREM 8. acetylcholine 9. REM 10. T 11. narcolepsy
4. Consciousness encompasses perception of sensations, initiation of voluntary movement, and higher mental processing. It is evidence of holistic information processing by the brain. Consciousness involves large areas of

simultaneous cerebral cortical activity, is superimposed on other types of neural activity, and is totally interconnected throughout the cerebrum.

5. alertness, drowsiness, stupor, coma
6. 1. S 2. S 3. L 4. L 5. S 6. L
7. 1. An alert, aroused state of consciousness. 2. Repetition or rehearsal of facts or skills. 3. Association of new information with stored information. 4. Time for chemical or structural changes to occur.
8. 1. Declarative memory: the learning of precise, detailed information; related to conscious thoughts when consolidation into long-term memory is associated with already learned facts. 2. Nondeclarative memory: the learning of specific motor activities; not usually related to conscious thoughts; best acquired by practice and remembered in repeated performance. Types of nondeclarative memory include procedural memory (playing a violin), motor memory (riding a bike), and emotional memory (becoming scared when confronted by a large dog, for example).
9. 1. subcortical structures 2. prefrontal cortex 3. a more lasting memory 4. anterograde amnesia 5. yes 6. ACh 7. basal nuclei, thalamus, and premotor cortex 8. dopamine
10. It analyzes incoming word sounds, and produces outgoing word sounds and grammatical structures.

Protection of the Brain

1.

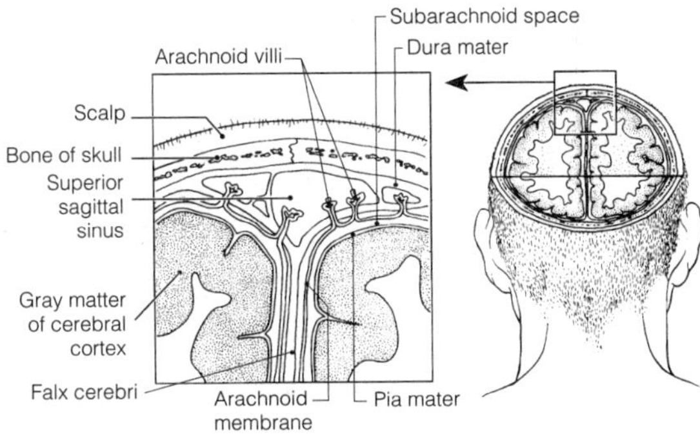

Figure 12.9

2. 1. pia mater 2. arachnoid villi 3. dura mater 4. subarachnoid space 5. subarachnoid space 6. superior sagittal sinus 7. falx cerebri
3. 1. blood plasma 2. hydrogen ions (H^+) 3. choroid plexuses 4. ventricle 5. ependymal 6. cerebral aqueduct 7. central canal 8. subarachnoid space 9. fourth ventricle 10. arachnoid villi 11. hydrocephalus 12. brain damage
4. 1. tight junctions 2. oxygen 3. potassium 4. hypothalamus 5. lipid-soluble 6. T
5. 1. E 2. F 3. D 4. G 5. B 6. C 7. I 8. A 9. H

The Spinal Cord

1. 1. C 2. A 3. F 4. D 5. E 6. B
2.

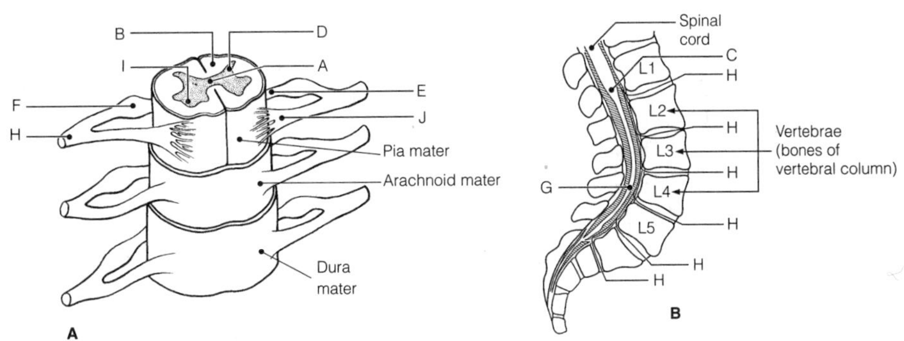

Figure 12.10

3. 1. foramen magnum 2. lumbar 3. lumbar tap or puncture 4. 31 5. 8 6. 12 7. 5 8. 5 9. cauda equina
4. 1. meningeal 2. epidural 3. cerebrospinal fluid 4. anterior median fissure 5. white 6. gray commissure 7. ventral 8. lateral 9. sensory 10. motor 11. neural crest
5. 1. SS 2. VS 3. VM 4. SM 5. VS 6. SS 7. VM

6. The pathway could be accurately identified by the process of elimination. It continues all the way to the somatosensory cortex, so it is not the spinocerebellar pathway. It ascends in the *lateral* white column, so it is not either the fasciculus cuneatus or gracilis, which are dorsal column pathways (and constitute the *specific* system). Circle the junctions (synapses) between the first- and second-order neurons and between the second- and third-order neurons.

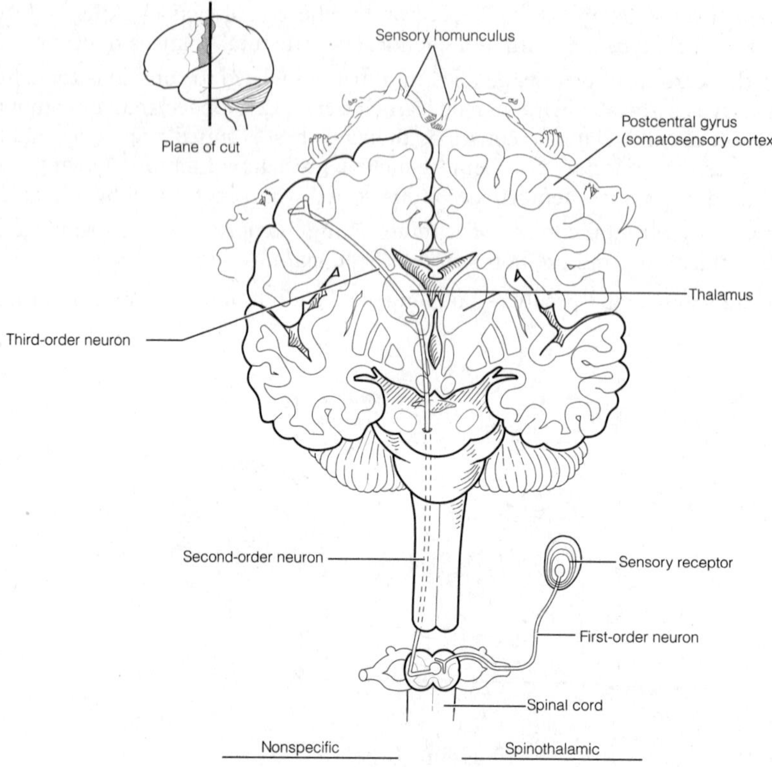

Figure 12.11

7.

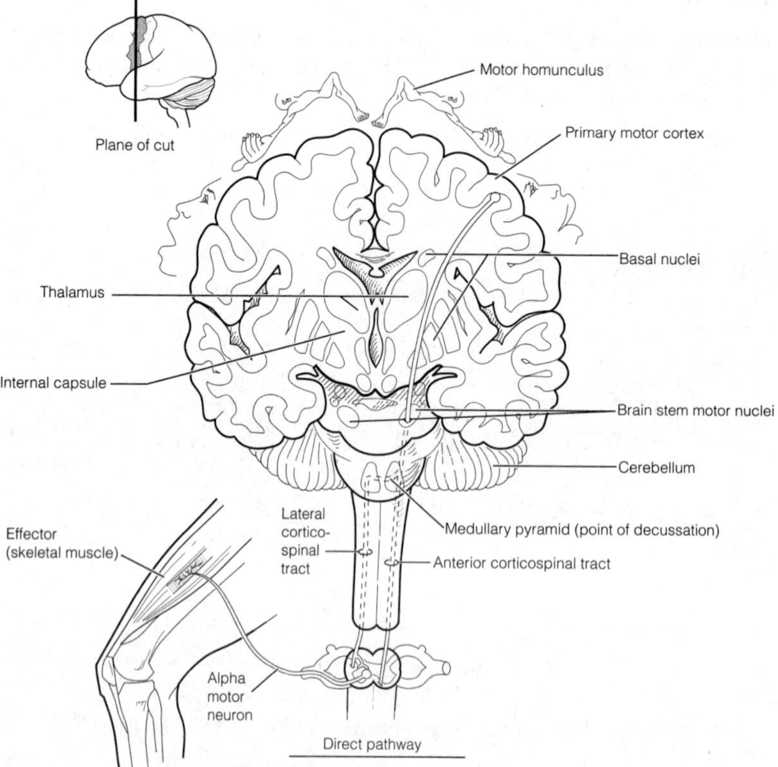

Figure 12.12

8. 1. E 2. G 3. C 4. I 5. D 6. H 7. F 8. B

Diagnostic Procedures for Assessing CNS Dysfunction

1. 1. T 2. cerebral angiogram 3. PET scan 4. above

The Incredible Journey

1. 1. cerebellum 2. medulla oblongata 3. hypothalamus 4. memories 5. temporal 6. a motor speech area 7. reasoning 8. frontal 9. vagus 10. dura mater 11. subarachnoid space 12. fourth

CHALLENGING YOURSELF

At the Clinic

1. The cerebellum.
2. Meningitis; by examining a sample of CSF for the presence of microbes.
3. The RAS of the reticular formation of the brain stem.
4. Somatosensory cortex.
5. Broca's area and adjacent regions of the motor cortex.
6. Left motor cortex.
7. Parkinson's disease.
8. Decreases.
9. Epilepsy (absence seizures); the condition will probably resolve itself without treatment.
10. Tonic-clonic (grand mal) epilepsy.
11. Intracranial hemorrhage.
12. Lower motor neuron damage. Upper motor neuron damage will leave spinal reflexes intact (spastic paralysis), which will maintain the muscle mass.
13. Parkinson's disease; levodopa or perhaps a transplant of dopamine-secreting tissue.
14. Four hours of sleep is not abnormal for an elderly person. A sleeping aid will reduce slow-wave (restorative) sleep, which compounds the problem.
15. Retrograde amnesia; this does not affect skill memory.

Stop and Think

1. Humans rely primarily on vision for assessment of the external environment; consequently, much of the cerebral cortex is devoted to this sense.
2. Short-term memory is affected by ECT. Long-term memory does not rely solely on the continuation of electrical output; rather, it is based on physical or chemical changes to the cells in the memory pathway.
3. Yes. All tracts from the cerebrum pass as projection fibers alongside (or into) the thalamus, and from there to the midbrain, pons, and medulla—all the brain stem regions.
4. The primary cortex receives input directly from the eye; damage here will cause blindness. Damage to the association cortex will cause inability to process visual images and make sense of them, but visual sensations will still occur.
5. Yes. If blockage of CSF flow is internal, the ventricular spaces above the blockage will enlarge, making the brain larger and more hollow and putting pressure on neural tissue from the inside. If the blockage is in the arachnoid villi, CSF will accumulate in the subarachnoid space, putting external pressure on the brain. The skull will enlarge, but the brain will not.
6. Because mannitol causes crenation and fluid loss from cells, it must be increasing the osmotic pressure of the solution bathing (outside) the cells; that is, it increases blood's solute osmotic concentration in comparison to that of the cells of the capillary wall. Osmosis hinges on differential permeability. If the cells' plasma membranes were permeable to mannitol, it would diffuse into the cells until it reached equilibrium, and osmosis would not occur. Thus, the cells must be impermeable to mannitol.
7. Peripheral nerves of the somatic nervous system carry both motor and sensory fibers so you might expect the cut to affect both types of function.
8. The correct choice is (c); the injury was at spinal cord level L_3. Recall that the spinal cord levels are several segments superior to vertebral levels.
9. The spinal cord is so narrow that any damage severe enough to cause paralysis will most likely affect both sides of the cord. In the brain, motor control of the two sides of the body is sufficiently separated so that unilateral damage and paralysis are more likely.
10. The dorsal side of the spinal cord carries predominantly ascending sensory tracts, so damage here would likely cause paresthesia.
11. The primary somatosensory cortex is being stimulated, so (c) is the most appropriate response.

COVERING ALL YOUR BASES

Multiple Choice

1. B **2.** A **3.** B **4.** C, D **5.** B **6.** A **7.** A, B, C, D **8.** C **9.** B, C **10.** B, C, D **11.** C **12.** B, C **13.** C **14.** A, C **15.** A, C **16.** A, C, D **17.** A, B, C **18.** A, B, C, D **19.** D **20.** A, B, D **21.** B, D **22.** A, B, C **23.** A, C **24.** C **25.** B **26.** D **27.** A, B, D **28.** D **29.** B **30.** B **31.** D **32.** A, B, C, D **33.** A, C, D **34.** D **35.** D **36.** A, B, C, D **37.** B, C, D **38.** A, B, C, D **39.** A, B, C, D **40.** A

Word Dissection

	Word root	Translation	Example
1.	campo	sea animal	hippocampus
2.	collicul	little hill	superior colliculi
3.	commis	united	commissure
4.	cope	cut	syncope
5.	enceph	within the head	mesencephalon
6.	epilep	laying hold of	epilepsy
7.	falx	sickle	falx cerebri
8.	forn	arch	fornix
9.	gyro	turn, twist	gyrus
10.	hippo	horse	hippocampus
11.	infundib	funnel	infundibulum
12.	isch	suppress	ischemia
13.	lemnisc	ribbon	medial lemniscus
14.	nigr	black	substantia nigra
15.	rhin	nose	rhinencephalon
16.	rostr	snout	rostral
17.	uncul	little	homunculus
18.	uncus	hook	uncus (olfactory)

Chapter 13 The Peripheral Nervous System and Reflex Activity

BUILDING THE FRAMEWORK

1. 1. sensory 2. afferent 3. motor 4. efferent 5. somatic 6. autonomic 7. somatic motor 8. autonomic motor

2. 1. receptor → 2. sensory nerve fiber → 3. spinal nerve → 4. dorsal root of spinal nerve → 5. spinal cord → 6. motor nerve fiber → 7. ventral root → 8. spinal nerve → 9. ventral ramus of spinal nerve → 10. effector

PART 1: SENSORY RECEPTORS AND SENSATION

1. 1. sensation 2. perception 3. electrical energy 4. permeability 5. graded 6. threshold 7. action 8. spinocerebellar 9. thalamus

2. 1. primary sensory cortex 2. frequency 3. T 4. thalamus 5. dorsal 6. first-order 7. T

3. 1. C, 2 2. B, 1 3. A, 5 (3) 4. A, 4 5. A, 2 6. A, 3 7. B, 2 8. B, 2 9. A, 1

4. 1. Tendon stretch 2. Numerous in muscle 3. Light touch 4. Thermoreceptor 5. Muscle spindle 6. Nociceptors 7. Tactile discs 8. Free dendritic endings 9. Tonic receptors

5.

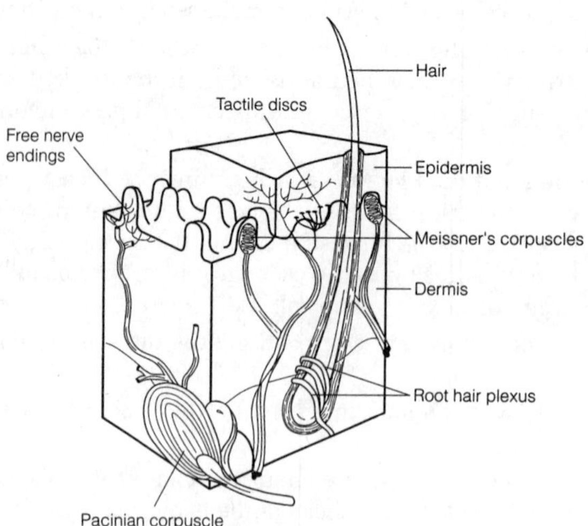

Figure 13.1

6. 1. A, F 2. A, F 3. B, E, F, G 4. B, E, G 5. C, D, H 6. A 7. C, D

The Eye and Vision

7. 1. extrinsic 2. eyelids 3. tarsal or Meibomian 4. chalazion

8.

Accessory eye structures	Secretory product
Conjunctiva	Mucus
Tarsal glands	Oily secretion
Lacrimal glands	Saline solution with mucus, lysozyme, antibodies (IgA), and enkephalins

9. 1. 2 2. 4 3. 3 4. 1

10. 1. Superior rectus turns eye superiorly. 2. Inferior rectus turns eye inferiorly. 3. Superior oblique turns eye inferiorly and laterally. 4. Lateral rectus turns eye laterally. 5. Medial rectus turns eye medially. 6. Inferior oblique turns eye superiorly and laterally.

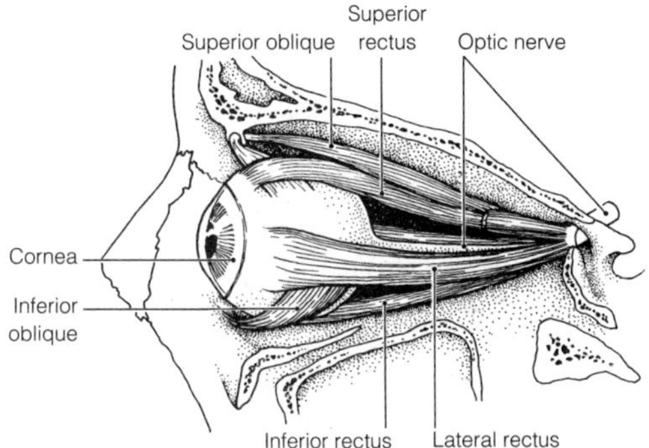

Figure 13.2

11. 1. E 2. G 3. B 4. A 5. D 6. F

12. 1. Vitreous humor 2. Iris 3. Far vision 4. Mechanoreceptors 5. Iris 6. Pigmented retina 7. Pigmented layer 8. Macula lutea 9. Richly vascular 10. Inner segment 11. Tarsal plate

13.

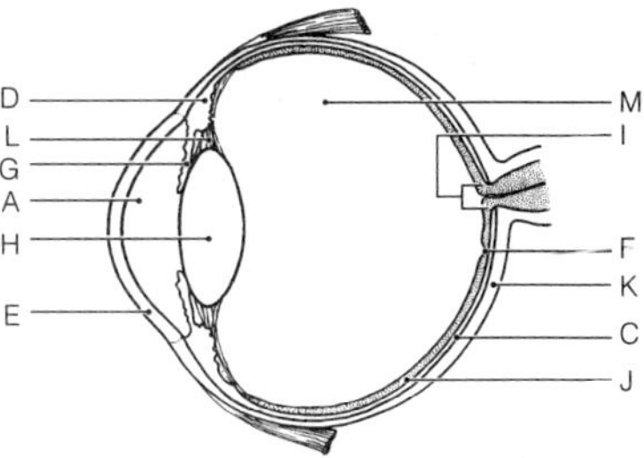

Figure 13.3

14. 1. L 2. A 3. K 4. I 5. D 6. C 7. B 8. J 9. M 10. A 11. D 12. G 13. F 14.–17. E, H, M, A 18. E 19. G

15. 1. E 2. C 3. A 4. D 5. B 6. F

16. 1. In distant vision the ciliary muscle is relaxed, the lens convexity is decreased, and the degree of light refraction is decreased. 2. In close vision the ciliary muscle is contracted, the lens convexity is increased, and the degree of light refraction is increased.

17. During reading, all intrinsic muscles and the extrinsic medial recti are contracted. Distant vision does not require accommodation and relaxes these muscles.

18. 1. E 2. F 3. D 4. A 5. C

19. 1. L 2. A 3. F 4. H 5. K 6. D 7. I 8. C 9. G 10. E 11. B 12. J

20. 1. 3 2. blue 3. green 4. red 5. at the same time 6. total color blindness 7. males 8. rods 9. fovea centralis 10. retinal periphery

21. 1. opsin; rhodopsin 2. all-*trans*; bleaching of the pigment 3. vitamin A; liver 4. Na^+; leaks into outer segments, causing depolarization and local currents that lead to neurotransmitter release. 5. Membrane becomes impermeable to Na^+; hyperpolarization occurs; neurotransmitter release stops.

22. Retina → optic nerve → optic chiasma → optic tract → synapse in thalamus → optic radiation → optic cortex.

23. Check 1, 2, 3, 4, 11, 12

24. The visual fields partially overlap, and inputs from both eyes reach each visual cortex, allowing for depth perception.

25. 1. ganglion cells 2. bipolar 3. photoreceptors 4. on-center 5. off-center 6. horizontal 7. amacrine 8. contrasts 9. lateral geniculate nuclei of thalamus 10. striate 11. P 12. cones 13. M 14. "what" 15. "where"

26.

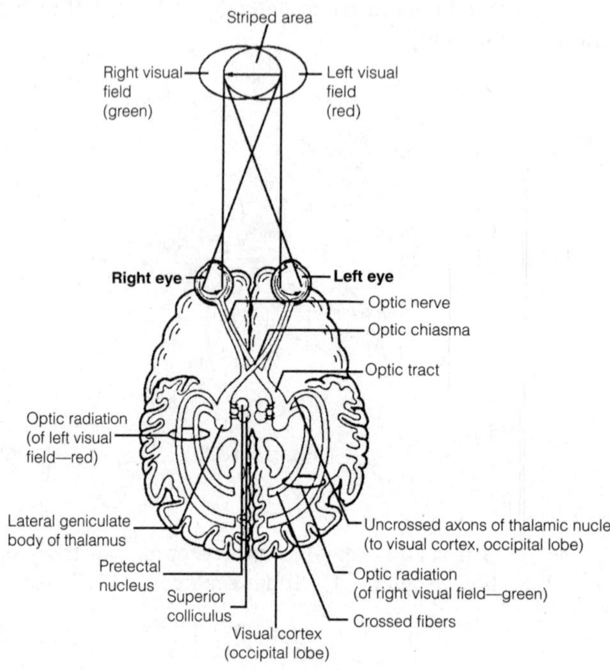

Figure 13.4

The Chemical Senses: Taste and Smell

27. 1. chemoreception 2. bitter 3. olfactory 4. roof 5. limbic 6. thalamus

28.

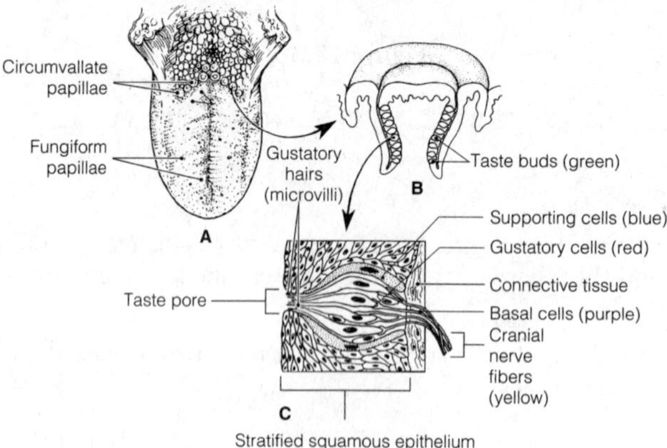

Figure 13.5

29. 1. Mucus 2. glomeruli 3. bipolar 4. amygdala, hypothalamus, and other limbic system structures

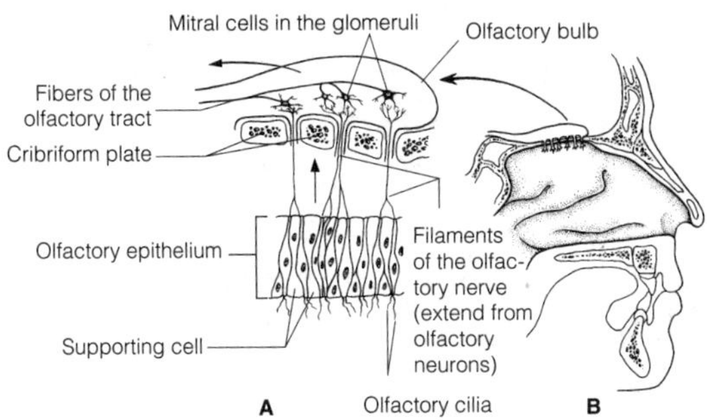

Figure 13.6

30. 1. Musky 2. Epithelial cell 3. Yellow-tinged epithelium 4. Olfactory nerve 5. Four receptor types 6. Metal ions 7. H^+ 8. Anosmia

31. 1. Receptor 2. Cyclic AMP 3.–4. Na^+, Ca^{2+} 5. Impulse generation

The Ear: Hearing and Balance

32. 1.–3. E, I, M 4.–6. C, K, N 7.–9. A, F, L 10. K, N 11. B 12. M 13. C 14. B 15.–16. K, N 17. G 18. D 19. H 20. L

33.

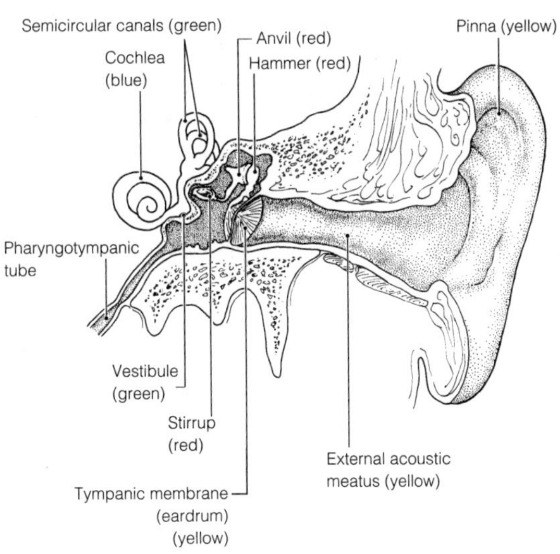

Figure 13.7

34. Eardrum → hammer → anvil → stirrup → oval window → perilymph → membrane → endolymph → hair cells.

35. 1. C 2. F 3. E 4. D 5. B 6. G 7. I 8. J 9. A

36. Stapedius attached to stapes; tensor tympani attached to malleus. When loud sounds assault the ears, these muscles contract and limit the movements of the stapes and malleus against their respective membranes.

37.

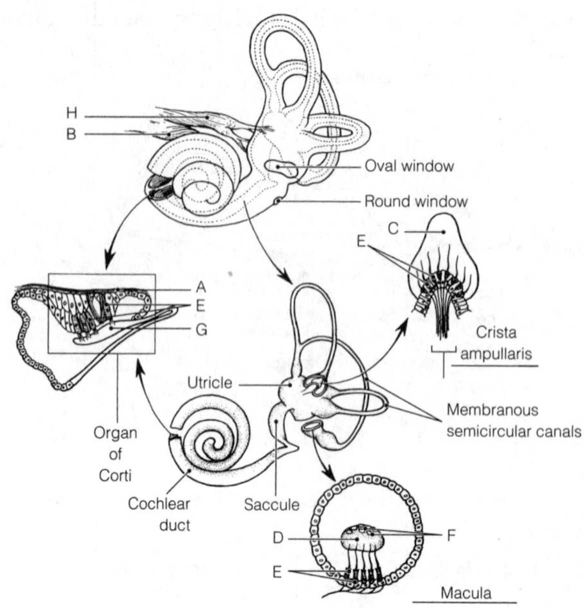

Figure 13.8

38. The ossicles deliver the total force impinging upon the eardrum to the oval window, which is about one-twentieth the size of the eardrum, in effect amplifying the sound.

39. 1. vibrating object 2. elastic 3–4. compression and rarefaction 5. wavelength 6. short 7. long 8. pitch 9. loudness 10. decibels

40. Short basilar membrane fibers resonate with higher frequency sounds, while larger fibers resonate with lower frequency sounds. At points of maximal basilar membrane vibration, the hair cells are excited.

41. 1. C 2. S 3. C 4. S 5.C, S 6. C 7. S

42. 1. C 2. I 3. A 4. D 5. B 6. J 7. H 8. K 9. E 10. G 11. L

43. 1. nausea 2. dizziness 3. balance problems

44. 1. Pinna 2. Tectorial membrane 3. Sound waves 4. Pharyngotympanic tube 5. Optic nerve 6. Retina 7. Inner hair cells

PART 2: TRANSMISSION LINES: NERVES AND THEIR STRUCTURE AND REPAIR

1. somatic afferent, somatic efferent, visceral afferent, visceral efferent

2.

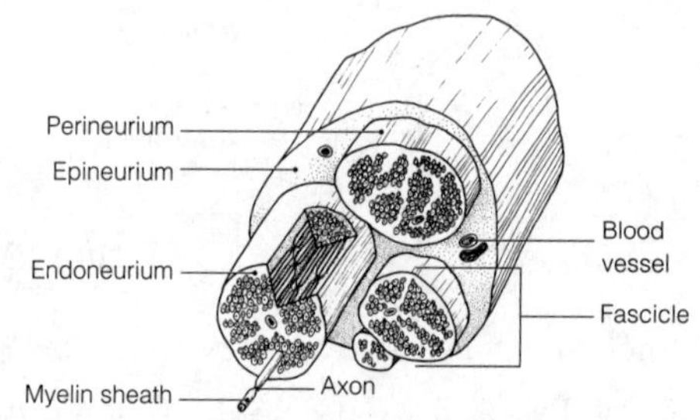

Figure 13.9

3. 1. divide 2. peripheral 3. cell body 4. degenerate 5. phagocytes (macrophages) 6. cords 7. Schwann cells 8. central 9. astrocytes and microglia 10. macrophages 11. oligodendrocytes 12. bear inhibitory proteins

4. 1. H 2. K 3. I 4. L 5. B 6. C 7. G, H, L 8. D 9. A 10. F 11. J 12. D, E 13. G 14. C

5. 1. Accessory (XI) 2. Olfactory (I) 3. Oculomotor (III) 4. Vagus (X) 5. Facial (VII) 6. Trigeminal (V) 7. Vestibulocochlear (VIII) 8. Glossopharyngeal (IX) 9. III, IV, VI 10. Trigeminal (V) 11. Optic (II) 12. Vestibulocochlear (VIII) 13. Hypoglossal (XII) 14. Facial (VII) 15. Olfactory (I)

6. 1. A 2. F 3. D 4. C 5. E 6. B

7.

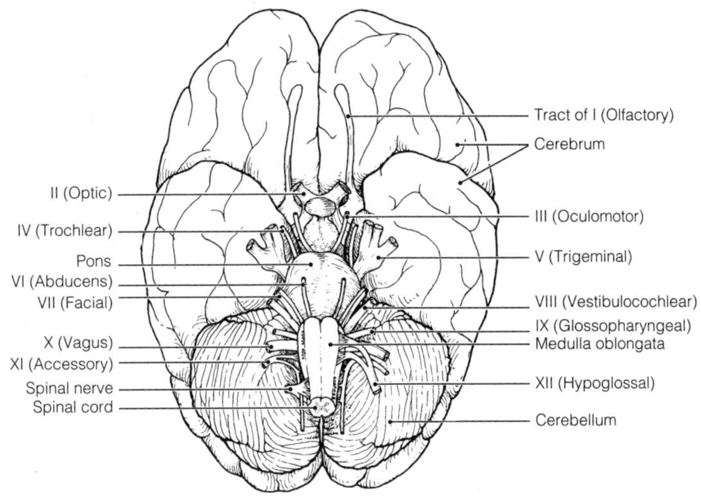

Figure 13.10

8. Temporal, zygomatic, buccal, mandibular, and cervical.

9. 1. dorsal 2. ventral roots 3. rami 4. plexuses 5. limbs 6. thorax 7. posterior trunk

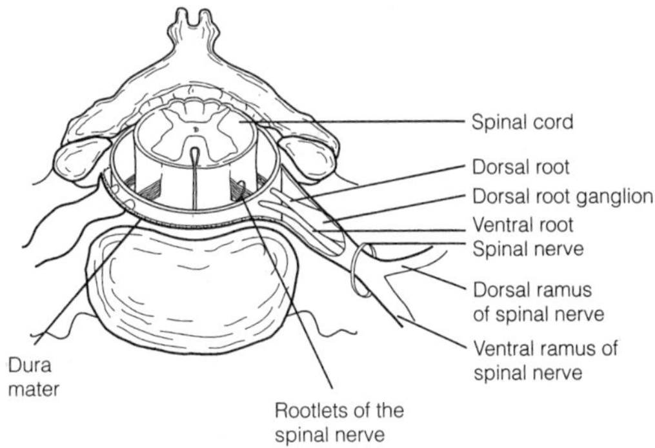

Figure 13.11

10. 1. Two roots 2. Thoracic roots 3. Afferent fibers 4. Ventral root 5. Two spinal roots

11. 1. cervical plexus 2. phrenic nerve 3. sciatic nerve (tibial division) 4. fibular, tibial 5. median 6. musculocutaneous 7. lumbar plexus 8. femoral 9. ulnar 10. brachial plexus 11. sciatic nerve 12. sacral plexus 13. radial nerve 14. femoral nerve 15. obturator nerve 16. inferior gluteal

12.

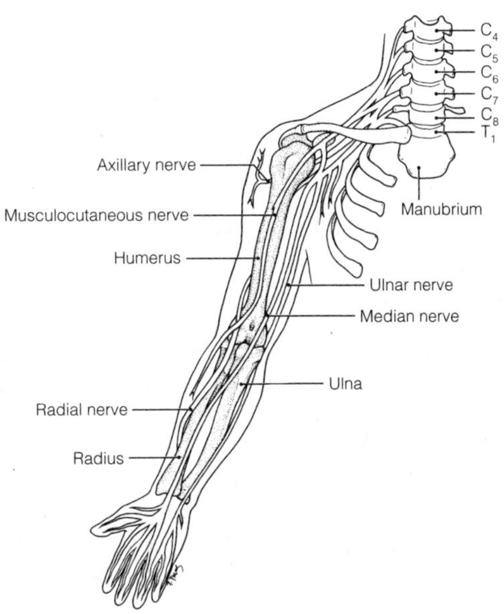

Figure 13.12

13.

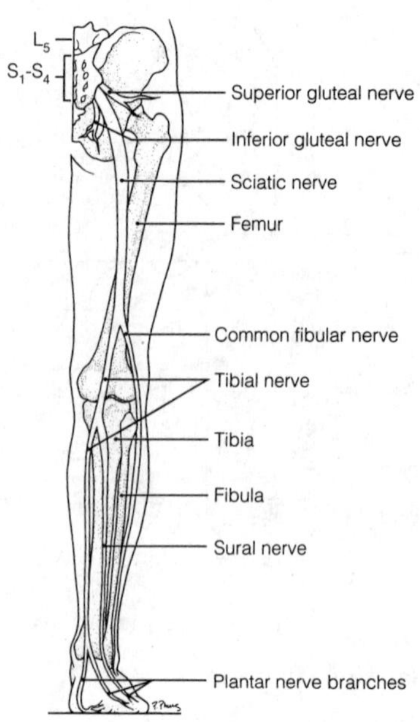

Figure 13.13

14. 1. radial 2. T 3. median 4. ulnar 5. T

PART 3: MOTOR ENDINGS AND MOTOR ACTIVITY

1. Check 1, 3, 4, 8

2. 1. C 2. B, G 3. E 4. D 5. H 6. C 7. A 8. A, D

3.

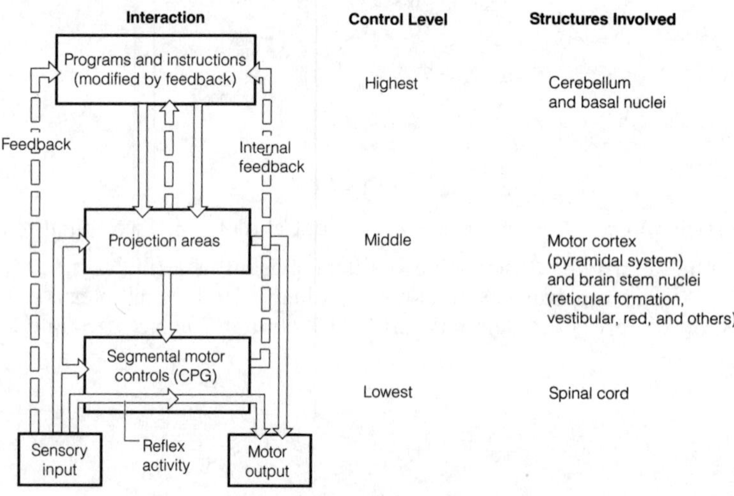

Figure 13.14

4. 1. Consciously considered actions 2. Command neurons 3. Cerebellum 4. Precommand areas 5. Basal nuclei 6. Multineuronal tracts

PART 4: REFLEX ACTIVITY

1. 1. A rapid, involuntary, predictable response to a stimulus. 2. receptor, sensory neuron, integration center, motor neuron, effector 3. somatic reflexes and autonomic reflexes 4. spinal reflex

2. 1. A 2. B 3. A 4. B 5. A 6. B 7. B 8. A

3. 1. muscle spindles 2. intrafusal fibers 3. primary sensory 4. type Ia fibers 5. secondary sensory 6. type II fibers 7. gamma efferent fibers 8. extrafusal fibers 9. alpha efferent fibers 10. alpha motor neurons 11. reciprocal inhibition

4. 1. primary sensory endings 2. alpha motor neuron 3. synapse 4. spinal cord 5. muscle spindle 6. intrafusal fiber 7. capsule of the receptor 8. ascending fibers 9. skeletal muscle (extrafusal fibers) 10. motor ending to extrafusal fiber

5. Stretch: 1, 3, 5, 6, 8, 9 Golgi tendon: 2, 3, 4, 7, 9, 10

6. 1. E 2. D 3. B 4. A, B, C, D, F 5. C, E 6. E 7. B 8. A, F 9. A, C, D, F 10. A, F 11. D 12. E 13. F 14. A

The Incredible Journey

1. 1. bony labyrinth 2. perilymph 3. saccule 4. utricle 5. (gel) otolithic membrane 6. otoliths 7. macula 8. static 9. cochlear duct 10. organ of Corti 11. hearing 12. cochlear division of cranial nerve VIII 13. semicircular canals 14. cupula 15. crista ampullaris 16. dynamic

CHALLENGING YOURSELF

At the Clinic

1. Abducens.
2. Uncinate fit; irritation of the olfactory pathway sometimes follows brain surgery.
3. Pinkeye (infectious conjunctivitis); yes, because infectious conjunctivitis is caused by a bacterial infection.
4. Patching the strong eye to force the weaker eye muscles to become stronger.
5. Trigeminal nerve; trigeminal neuralgia (tic douloureux).
6. Damage to the facial nerve. The symptoms are facial paralysis, loss of taste, drooping of lower eyelid; the mouth sags; the eye tears and can't completely close.
7. Glossopharyngeal.
8. Vagus; complete vagal paralysis results in death.
9. Trapezius, sternocleidomastoid.
10. Brachial plexus.
11. Median nerve.
12. In suturing the nerve back together, there is no guide to ensure that each nerve fiber continues across the transection into the same neurilemma in which it started. Nerve fibers can grow into pathways different from their original ones and establish new synapses. The brain cannot keep track of which nerve fibers have grown into different pathways and projects sensations back to the original point of origin.
13. The olfactory nerve (I) fibers, which are located in the roof of the nasal cavity.
14. The right hypoglossal nerve (cranial XII).
15. Spinal cord transection at L_5 (c). The posterior femoral cutaneous nerve, which arises from ventral rami S_1–S_3, provides the sensory supply in the areas described. The motor nerves for those areas also receive fibers from the ventral rami of L_4 and L_5, explaining his lack of motor problems.
16. Check the gag and swallowing reflexes (IX and X) and for weakness in the sternocleidomastoid and trapezius muscles (XI).
17. The femoral nerve has been damaged during Clancy's fall.
18. Nyctalopia; vitamin A; photoreceptors of retina.
19. Vestibular apparatus (including the semicircular canals).
20. Glaucoma; inadequate drainage of aqueous humor; blindness due to compression of the retina and optic disc.

Stop and Think

1. Nissl bodies are composed of rough ER and ribosomes. Because most of the protein synthesis will be devoted to synthesizing the internal proteins needed for repair and replacement of the damaged axon, the ribosomes will predominantly be free, rather than attached. The ribosomes of the Nissl bodies will separate from the ER membrane, and the characteristic appearance of the Nissl body will be lost until repair is nearly complete. The nucleolus functions in ribosome assembly, so as the demand for ribosomes increases, nucleolar function and appearance will become more significant. The accumulation of raw materials for protein synthesis may increase the osmotic pressure of the cell, causing it to attract water and take on a swollen appearance.
2. Pricking the finger excites the nonspecific pathway for pain, beginning with the dendritic endings of a first-order neuron, a nociceptor, in the finger. This neuron synapses with the second-order neuron in the spinal cord, and the axon of the second-order neuron (after crossing over in the spinal cord) ascends in the anterolateral pathway through the lateral spinothalamic tract. Synapse with third-order neurons occurs in thalamic nuclei, but there are many synapses with nuclei of the reticular formation as well. Pain is perceived, and the cortex is aroused as processing at the perceptual level occurs. The somatosensory cortex pinpoints the sensation; the sensory association area perceives the sensation as undesirable. Integrative input to the prefrontal area sends signals to the motor speech area, which directs the primary motor cortex to activate the muscles of speech to say "ouch."
3. Exteroceptors that are not cutaneous receptors include the chemoreceptors of the tongue and nasal mucosa, the photoreceptors of the eyes, and the mechanoreceptors of the inner ear. These all monitor changes in the external environment, so they are classified as exteroceptors.
4. An inability to feel the parts of your body is the result of accommodation by the proprioceptors. Although these adapt to only a minimal extent, they can slowly accommodate if you don't change your body position for a very long time.
5. Starting from twelve o'clock, for the right eye: lateral movement (abducens), inferior movement, medial movement (oculomotor), superior movement (oculomotor, trochlear). For the left eye: oculomotor for medial and inferior movements, abducens for lateral movement, oculomotor and trochlear for superior movement.

6. The precommand areas *ready* the primary motor cortex for action, but the actual command (nerve impulse) to the muscles is issued by the motor cortex.
7. No. The bundle of "nerves" in the cauda equina really consists of dorsal and ventral roots, which don't merge into the spinal nerve until after exiting the vertebral column.
8. The origin of the phrenic nerve so high up the spinal cord means that the distance between brain and phrenic is short. This decreases the likelihood of spinal cord damage above the phrenic nerve, as there isn't much spinal cord above that nerve's origin.
9. Initially, as muscle spindles are stretched, the reflex sends impulses back to contract the muscle. With prolonged stretching, accommodation decreases the vigor of the stretch reflex somewhat, and the muscle can relax and stretch more, reducing the risk of tearing muscle tissue during exercise.
10. The nerves serving the ankle joint are those serving the muscles that cause movement at the ankle joint—the posterior femoral cutaneous, common fibular, and tibial.
11. The nurse has forgotten the phenomenon of adaptation, a condition in which sensory receptor transmission declines or stops when the stimulus is unchanging. She has had her hand in the water while filling the tub, and the thermoreceptors in *her* skin have accommodated to the hot water.
12. Use word roots to help you remember. Glossopharyngeal means tongue-pharynx, and that nerve serves those regions. Hypoglossal means "below the tongue," and this nerve serves the small muscles that move the tongue.
13. Abdul was more correct. The facial nerve is almost entirely a motor nerve.
14. Smell receptors are chemoreceptors that respond to chemicals in solution and adapt quickly when the stimulus is unchanging. Motion receptors of the inner ear are mechanoreceptors involved in the sense of balance. Loss of balance can have dire consequences here. These receptors, like all proprioceptors, do not adapt.
15. Most somatic reflexes involve fast A fibers, while pain sensations travel over slow C fibers. Pain perception that occurs *after* the somatic reflex is triggered helps reinforce the proper behavioral response (learning not to touch hot objects).

COVERING ALL YOUR BASES

Multiple Choice

1. C **2.** A, C, D **3.** D **4.** C, D **5.** C, D **6.** A, C, D **7.** B **8.** B **9.** A, B, C **10.** D **11.** A, B, C, D **12.** A, C **13.** A, C, D **14.** B, D **15.** C **16.** C **17.** A, B **18.** B **19.** A **20.** D **21.** A **22.** C **23.** D **24.** A **25.** D **26.** D **27.** B, C **28.** B, D **29.** B, C, D **30.** A, B **31.** C **32.** A **33.** B **34.** C **35.** A, D **36.** A **37.** C **38.** A, B, C **39.** D **40.** B **41.** B, D **42.** B **43.** A, B, D **44.** A, C **45.** A, C **46.** B **47.** B, C, D **48.** A, B **49.** A, C, D **50.** A, C, D **51.** A **52.** A, C **53.** A, B, D **54.** B **55.** A, D **56.** C **57.** C **58.** A, C, D **59.** B, C, D **60.** A, B, C, D **61.** A, C, D **62.** B, C **63.** C **64.** C **65.** A, B, C, D **66.** C **67.** A, B, D **68.** C **69.** A, C **70.** A, B, C **71.** A, C, D **72.** B **73.** A, B

Word Dissection

	Word root	Translation	Example
1.	ampulla	flask	crista ampullaris
2.	branchi	gill	branchial groove
3.	caruncl	bit of flesh	caruncle of eye
4.	cer	wax	ceruminous gland
5.	cochlea	snail shell	cochlear duct
6.	esthesi	sensation	kinesthetic
7.	fove	small pit	fovea centralis
8.	glauc	gray	glaucoma
9.	glosso	tongue	glossopharyngeal
10.	gust	taste	gustatory hair
11.	kines	movement	kinesthetic
12.	lut	yellow	macula lutea
13.	macula	spot	macula lutea
14.	metr	measure	emmetropic
15.	modiol	water wheel bucket	modiolus of cochlea
16.	noci	harmful	nociceptors
17.	olfact	smell	olfactory cell
18.	palpebra	eyelid	superior palpebra
19.	papill	nipple	fungiform papilla
20.	pinn	wing	pinna of ear
21.	presby	old	presbycusis
22.	propri	one's own	proprioceptors
23.	puden	shameful	pudendal nerve
24.	sacc	sack	saccule of inner ear
25.	scler	hard	sclera of eye
26.	tars	flat	tarsal plate
27.	trema	hole	helicotrema
28.	tympan	drum	tympanic membrane
29.	vagus	wanderer	vagus nerve

Chapter 14 The Autonomic Nervous System

BUILDING THE FRAMEWORK

Introduction

1. 1. S, A 2. S 3. A 4. S, A 5. S 6. S 7. A 8. A 9. S, A 10. A 11. A 12. S 13. S 14. S 15. A 16. S, A

2.

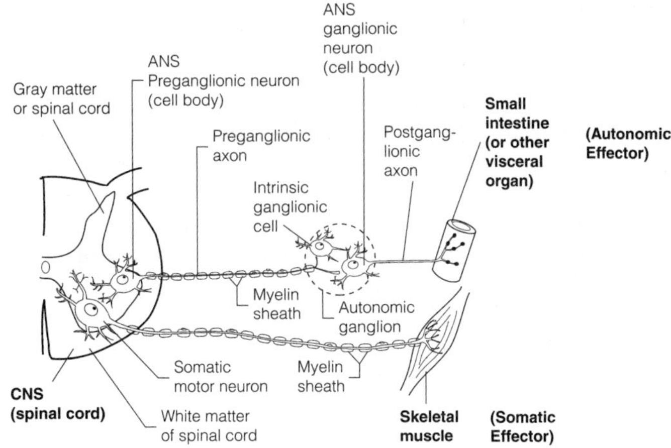

Figure 14.1

ANS Anatomy

1. Sweat glands, the adrenal medulla, and most blood vessel walls are provided with sympathetic fibers only.

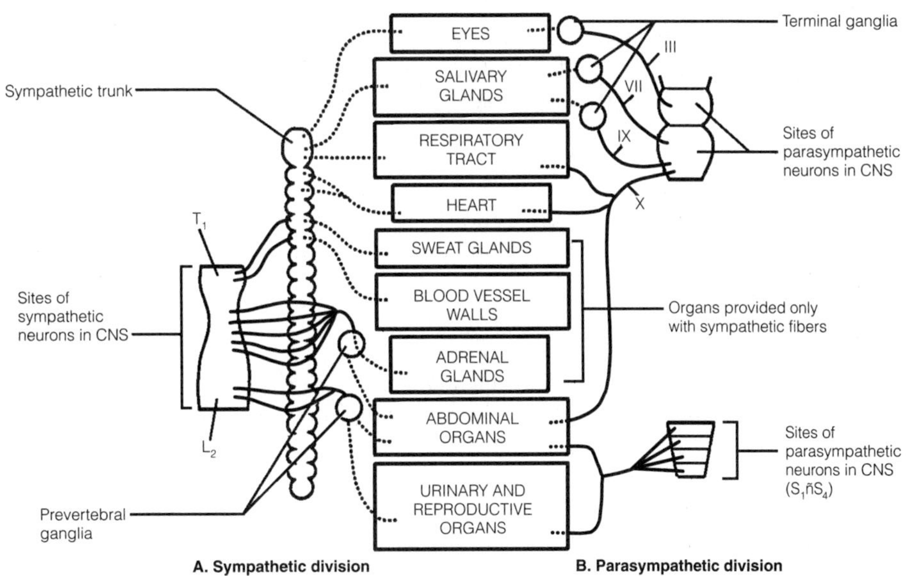

Figure 14.2

2. 1. lateral 2. T_1–L_2 3. spinal cord 4. white (myelinated) 5. paravertebral ganglia 6. sympathetic 7. postganglionic 8. rami communicantes 9. unmyelinated 10. smooth muscle 11. cervical 12. pupil (iris) 13. salivary 14. skin 15. prevertebral (collateral) 16. thoracic splanchnic 17. celiac and superior mesenteric 18. abdominal 19. adrenal 20. lumbar 21. urinary 22. brain stem 23. sacral 24. terminal 25. oculomotor (III) 26. VII (facial) 27. IX (glossopharyngeal) 28. vagus (X) 29. plexuses 30. thorax 31. abdomen 32. sacral region
3. 1.S 2. P 3. S 4. P 5. P 6. P 7. S 8. P 9. P 10. S 11. S 12. S, P

ANS Physiology

1. Sympathetic: 2, 4, 7, 9 Parasympathetic: 1, 3, 5, 6, 8
2. 1. Fibers that release acetylcholine 2. Nicotinic and muscarinic receptors
3. 1. Fibers that release norepinephrine 2. Alpha and beta adrenergic receptors
4. All ganglionic neurons and the hormone-producing cells of the adrenal medulla
5. 1.C 2. A 3. C 4. D 5. B 6. A 7. B 8. B
6. Sympathetic: 2, 3, 4, 6 Parasympathetic: 1, 5, 7
7. 1. NE 2. Nicotinic receptor 3. Heart rate slows 4. Parasympathetic tone 5. Rapid heart rate 6. Sympathetic tone

Homeostatic Imbalances of the ANS

1. 1. smooth muscle 2. T 3. sympathectomy 4. adrenergic

CHALLENGING YOURSELF

At the Clinic

1. Parasympathetic.
2. The sympathetic division causes vasoconstriction of vessels supplying the digestive organs. Stress-induced overstimulation of sympathetic nerves can cause an almost total shutdown of blood supply to the stomach. Lack of blood leads to tissue death and necrosis, which causes the ulcers.
3. Achalasia.
4. The condition is the result of poorly developed parasympathetic nerve supply to the lower colon. As a result, the muscle of the colon fails to contract, and fecal matter cannot be expelled. Because the nerve deficiency cannot be corrected, surgical removal of the distended area is the only course of treatment. After removal of the underdeveloped region, fecal material will be moved along healthy portions of the colon normally.
5. A sympathectomy will be performed. It will result in dilation of the distal blood vessels as their vasoconstrictor fibers are cut.
6. Sweating is stimulated by sympathetic nerves; after they are cut, anhidrosis will result.
7. The condition is autonomic hyperreflexia. The staff will watch for hypertension-induced cerebrovascular accident.
8. Bertha probably has a urinary tract infection that has spread to her kidneys. Kidney inflammation causes pain to be referred to the lower back.
9. Antiadrenergic drugs (such as phentolamine) that interfere with the activity of the sympathetic vasomotor fibers.
10. Epigastric region.
11. Back to the laboratory for our mad scientist. His victim would suffer from vision problems (unable to control iris and ciliary muscle regulating the lens), have dry eyes, dry mouth, and dry nasal mucosa. Hardly fatal.
12. The pulse would rise because parasympathetic innervation of the heart antagonizes sympathetic innervation, which raises the pulse.

Stop and Think

1. No. Without sensory feedback to inform the CNS of the physical and chemical status of the viscera, the motor output of the autonomic nerves can be inappropriate and even life threatening.
2. Parasympathetic.
3. Spinal reflex, autonomic reflex.
4. Pelvic splanchnic nerves to the inferior hypogastric plexus.
5. Cerebral input to the preganglionic neuron in the sacral spinal cord is inhibitory; cerebral stimulation of motor neurons controlling the more inferior skeletal muscle (voluntary) sphincter can shut off urine flow.
6. Transection above the level of the reflex would cut off the possibility for cerebral inhibition. Voiding would become entirely involuntary and totally reflexive.
7. Shunting of blood to the vessels of the hands reduces blood flow to the brain, thus reducing the vessel distension that causes the pain of a migraine headache.
8. The cholinergic preganglionic axon runs from a lateral horn of the gray matter in the upper thoracic spinal cord to the ventral root of the spinal nerve, through the white ramus communicans to the paravertebral ganglion of the thoracic region of the sympathetic chain, and up the chain (without synapsing) to the cervical region to synapse in the superior cervical ganglion. There it releases ACh to excite the postganglionic neuron. The adrenergic postganglionic axon runs from the ganglion to join with fibers of the oculomotor nerve to reach the iris of the eye. Release of NE at the neuroeffector junctions stimulates the radial smooth muscle layer of the iris, which results in dilation of the pupil.
9. From the cardiac center in the medulla to the nucleus of the vagus nerve, where the cholinergic preganglionic axon leaves in the vagus nerve and runs to the cardiac plexuses. From there it continues on to the heart itself, ending at an intramural ganglion, where ACh is released. ACh excites the postganglionic neurons, whose cholinergic postganglionic axons travel to the target cells, where they release ACh to inhibit cardiac function.

COVERING ALL YOUR BASES

Multiple Choice

1. D **2.** D **3.** B **4.** D **5.** B, C **6.** A, B, C **7.** A **8.** B **9.** A **10.** A **11.** C **12.** D **13.** C **14.** C
15. B **16.** A **17.** B **18.** B **19.** C **20.** C **21.** A **22.** A **23.** A **24.** A **25.** A **26.** B

Word Dissection

	Word root	Translation	Example		Word root	Translation	Example
1.	adren	toward the kidney	adrenaline	5.	ortho	straight	orthostatic
2.	chales	relaxed	achalasia	6.	para	beside	parasympathetic
3.	epinephr	upon the kidney	epinephrine	7.	pathos	feeling	sympathetic
4.	mural	wall	intramural ganglia	8.	splanchn	organ	splanchnic nerve

Chapter 15 The Endocrine System

BUILDING THE FRAMEWORK

The Endocrine System: An Overview

1. 1. K 2. H 3. D 4. G 5. A 6.–8. B, I, L 9. E 10. J

2. Figure 16.1: A. Pineal B. Anterior pituitary C. Posterior pituitary D. Thyroid E. Thymus F. Adrenal gland G. Pancreas H. Ovary I. Testis J. Parathyroids K. Placenta L. Hypothalamus

Hormones

1. 1. I 2. N 3. A 4. L 5. K 6. G 7. C 8. D 9. F 10. B 11. J 12. E 13. H

2. A circulating hormone enters the bloodstream or lymph and often affects distant body targets. A local hormone (autocrine or paracrine), like a circulating hormone, is released into the interstitial fluid but it exerts its effects nearby on cells in the local area.

3. 1. A 2. B 3. B 4. A 5. C 6. C 7. A 8. B

4. 1. first 2. a G protein 3. adenylate cyclase 4. ATP 5. second 6. protein kinases 7. phosphate 8–10. target cell type, the kinases the target cells contain, substrates available for phosphorylation 11.–13. diacylglycerol, inositol triphosphate, cyclic GMP 14. Ca^{2+}

5. Hormones are inactivated by enzymes in their target cells as well as by liver and kidney enzymes.

6. 1.–3. A, G, I 4. C 5.–8. B, E, H, K 9. L 10. D 11. F

7. 1.B 2. C 3. A

8. 1.D 2. B 3. A 4. C

The Pituitary Gland and Hypothalamus

1.

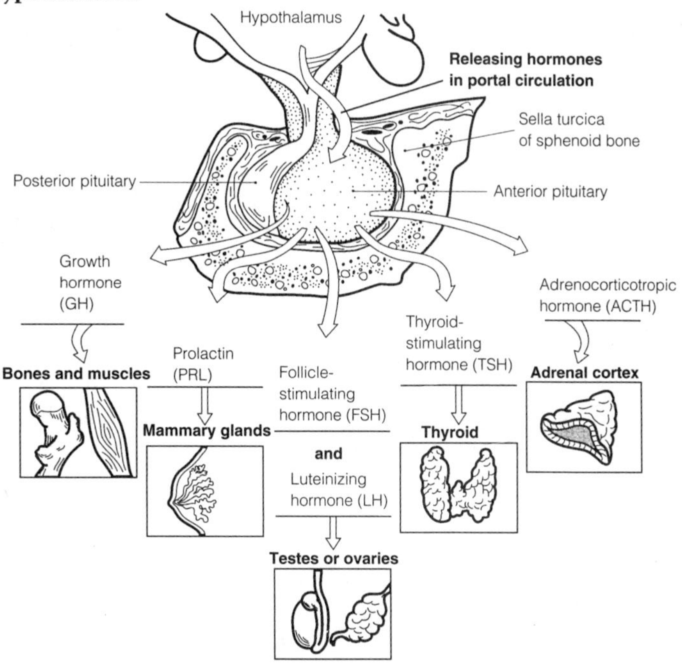

Figure 15.2

2. 1. Anterior lobe 2. Posterior lobe 3. Oxytocin 4. Hypophyseal portal system

3. In high amounts, it promotes blood vessel constriction, increasing blood pressure (it has a pressor effect).

4. Check 1, 4, 6, 7, 10, 11. Circle 2, 3, 5, 8, 9.

Major Endocrine Organs

1. 1. C 2. D 3. A 4. L 5. A 6. B 7. E, I (A) 8. C 9. F 10. C 11. F 12. C 13. H 14. C 15. D 16. E, I (A) 17. C 18. G 19. J (A) 20. K 21. L 22. C

2. 1. thyroxine/T_3 2. thymosin 3. PTH 4. cortisone (glucocorticoids) 5. epinephrine 6. insulin 7.–10. TSH, FSH, ACTH, LH 11. glucagon 12. ADH 13. FSH 14. LH 15. estrogens 16. progesterone 17. aldosterone 18. prolactin 19. oxytocin

3. 1. Increases blood Ca^{2+} levels 2. Depresses glucose uptake 3. Thyroxine 4. Cortisol

4. 1. Kidneys: Causes retention of Ca^{2+} and enhances excretion of PO_4^{3-}. Promotes activation of vitamin D. 2. Intestine: Indirect effect. Via its effect on the kidney's activation of vitamin D, PTH causes increased absorption of Ca^{2+} from foodstuffs. 3. Bones: Causes enhanced release of Ca^{2+} from bone matrix.

5. 1. 2 2. 7 3. 8 4. 5 5. 9 6. 1 7. 4 8. 3 9. 6

6. 1. Polyuria—high sugar content in kidney filtrate (acts as an osmotic diuretic and) causes large amounts of water to be lost in the urine. 2. Polydipsia—thirst due to large volumes of urine excreted. 3. Polyphagia—hunger because blood sugar cannot be used as a body fuel even though its levels are high.

7. 1. B 2. A 3. A 4. B 5. B 6. A 7. A 8. B 9. B 10. B 11. B 12. A

8. 1. A, E 2. B 3. C 4. E 5. A 6. E 7. A 8. B, E 9. A, E 10. D

9. 1. B 2. A 3. B 4. A 5. A 6. A

10. 1. B 2. D 3. C 4. A

11. 1. estrogens/testosterone 2. PTH 3. ADH 4. thyroxine 5. thyroxine 6. insulin 7. growth hormone 8. estrogens/progesterone 9. thyroxine

12. 1. growth hormone 2. thyroxine 3. PTH 4. glucocorticoids 5. growth hormone 6. androgens

Hormone Secretion by Other Organs

1.

Hormone	Chemical makeup	Source	Effects
Gastrin	Peptide	Stomach	Stimulates stomach glands to secrete HCl
Secretin	Peptide	Duodenum	Stimulates the pancreas to secrete HCO_3^--rich juice and the liver to release more bile; inhibits stomach glands
Cholecystokinin	Peptide	Duodenum	Stimulates the pancreas to secrete enzyme-rich juice and the gallbladder to contract; relaxes sphincter of Oddi
Erythropoietin	Glycoprotein	Kidney in response to hypoxia	Stimulates production of red blood cells by bone marrow
Cholecalciferol (vitamin D_3)	Steroid	Skin; activated by kidneys	Enhances intestinal absorption of calcium
Atrial natriuretic peptide (ANP)	Peptide	Heart atrial cells	Inhibits Na^+ reabsorption by kidneys; inhibits renin and aldosterone release
Leptin	Peptide	Adipose tissue	Suppresses appetite; increases energy expenditure

The Incredible Journey

1. 1. insulin 2. pancreas 3. hypothalamus 4. ADH 5. parathyroid 6. calcium 7. adrenal medulla 8. epinephrine 9. T_3/T_4

CHALLENGING YOURSELF

At the Clinic

1. Pituitary dwarfs (deficient in GH) are short in stature but have fairly normal proportions; cretins (deficient in thyroxine) retain childlike body proportions and have thick necks and protruding tongues.
2. Hypothyroidism; iodine deficiency (treated by dietary iodine supplement) or thyroid cell burnout (treated by hormonal supplement).
3. Hypersecretion of ADH causes diabetes insipidus, and insufficiency of insulin (or lack of response to insulin) causes diabetes mellitus. Both involve polyuria and consequent polydipsia. Glucose in the urine will indicate diabetes mellitus. (More complicated tests are used to diagnose diabetes insipidus.)
4. Adrenal cortex.
5. Hyperparathyroidism (resulting in hypercalcemia and undesirable calcium salt deposit), probably from a parathyroid tumor.

6. Cushing's syndrome, most likely caused by tumor; anterior pituitary (hypersecretion of ACTH) or adrenal cortex (hypersecretion of cortisol).
7. Hyperthyroidism; anterior pituitary tumor (hypersecretion of TSH) or thyroid tumor (hypersecretion of thyroxine).
8. Epinephrine and norepinephrine; pheochromocytoma, a tumor of the chromaffin cells of the adrenal medulla; exophthalmos and goiter would not be present.
9. Hypoglycemia; overdose of insulin.
10. Prolactin; tumor of the adenohypophysis.
11. Lying between the pinealocytes in the gland are grains of calcium salts. This "pineal sand," like the calcium salts of bone, is radiopaque.
12. Her PTH level is high because her fetus is calling for calcium to build its bones. Maryanne's poor diet has elevated her PTH level and her bones are losing calcium.

Stop and Think

1.

Protein hormones	*Steroid hormones*
(a) rough ER, Golgi apparatus, secretory vesicles	smooth ER
(b) storage in secretory vesicles	no storage
(c) manufactured constantly	manufactured only as needed
(d) secretion by exocytosis	leaves cell by diffusion through lipid bilayer
(e) most (but not all) transported dissolved in plasma	transport may require a protein transport molecule
(f) receptors on plasma membrane	receptors in cytoplasm or nucleus
(g) second messenger used	no second messenger
(h) immediate activation	lag time for protein synthesis to occur
(i) effects end as soon as hormone is metabolized	effects prolonged after hormone is metabolized

2. (a) Growth hormone-stimulating hormone, growth hormone-inhibiting hormone (somatostatin), prolactin-inhibiting hormone (dopamine), TSH, ACTH, FSH, LH.

   (b) Calcium, glucose, sodium, and potassium blood levels.
3. Hormone-producing cells located in organs of the digestive tract. They are called paraneurons because many of their hormones are identical to certain neurotransmitters released by neurons.
4. Down-regulation.
5. These hormones are lipid soluble; they must be "tied down" to prevent their escape through the plasma membrane.
6. Corticotropin-releasing hormone secreted by hypothalamus → capillaries of the hypophyseal portal system → anterior pituitary gland → release of ACTH by anterior pituitary → general circulation to zona fasciculata of the adrenal cortex → glucocorticoids (cortisol, etc.) released.
7. The thyroid gland contains thyroid follicles surrounded by soft connective tissue. The follicular cells of the follicles produce the thyroglobulin colloid from which T_3 and T_4 are split. At the external aspects of the follicles are *parafollicular* (C) cells that produce calcitonin, a completely different hormone. Usually structurally associated with the thyroid gland are the tiny parathyroid glands, which look completely different histologically. The parathyroid cells produce PTH, a calcitonin antagonist.
8. For the giant, GH is being secreted in excess by the anterior pituitary, resulting in extraordinary height. For the dwarf, GH is deficient, resulting in very small stature but normal body proportions. For the fat man, T_3 and T_4 are not being adequately produced, resulting in depressed metabolism and leading to obesity (myxedema). The bearded lady has a tumor of her adrenal cortex (androgen-secreting area), leading to hirsutism.
9. Phosophodiesterase degrades cAMP. Hence, hyperactivity of the hormone stimulus would occur.

COVERING ALL YOUR BASES

Multiple Choice

1. B, C **2.** A, C **3.** C **4.** A, C, D **5.** C **6.** A, B, D **7.** C **8.** C **9.** B **10.** D **11.** B **12.** A, B, C, D **13.** A, B, C, D **14.** A, B, C, D **15.** D **16.** A, B, C **17.** A, B, C, D **18.** D **19.** A, C **20.** C **21.** A, B, C, D **22.** A, B, C **23.** B **24.** B **25.** B **26.** D **27.** A **28.** A

Word Dissection

	Word root	Translation	Example		Word root	Translation	Example
1.	adeno	gland	adenohypophysis	6.	hormon	excite	hormonal
2.	crine	separate	endocrine	7.	humor	fluid	humoral control
3.	dips	thirst, dry	polydipsia	8.	mell	honey	diabetes mellitus
4.	diure	urinate	diuretic	9.	toci	birth	oxytocin
5.	gon	seed, offspring	gonad	10.	trop	turn, change	tropic hormone

Chapter 16 Blood

BUILDING THE FRAMEWORK

Overview: Blood Composition and Functions

1. 1. connective tissue 2. formed elements 3. plasma 4. blood clotting 5. red blood cells 6. hematocrit 7. plasma 8. white blood cells 9. platelets 10. 1 11. oxygen

2. 1. Distributes (1) O_2 from lungs to tissue cells and CO_2 from tissues to lungs; (2) metabolic wastes to elimination sites, such as kidneys; (3) hormones; (4) heat throughout the body. 2. Regulatory: (1) blood buffers maintain normal blood pH; (2) blood solutes maintain blood volume for normal circulation. 3. Protection: (1) clotting proteins and platelets prevent blood loss; (2) certain plasma proteins and white blood cells defend the body against foreign substances that have gained entry.

Blood Plasma

1. 1. nutrients 2. electrolytes 3. plasma proteins

2. Proteins found in plasma:

Constituent	Description/importance
Albumin	60% of plasma proteins; important for osmotic balance
Fibrinogen	4% of plasma proteins; important in blood clotting
Globulins • alpha and beta • gamma	36% of plasma proteins: • transport proteins • antibodies
Nonprotein nitrogenous substances	Lactic acid, urea, uric acid, creatinine, ammonium salts
Nutrients	Organic chemicals absorbed from the digestive tract
Respiratory gases	O_2, CO_2

Formed Elements

1. 1. 6 2. 1 3. 4 4. 5 5. 7 6. 3 7. 2 Circle reticulocyte, underline normoblast.

2. 1. 100–120 days 2. iron, vitamin B_{12}, and folic acid 3. They are engulfed by macrophages of the liver, spleen, or bone marrow. 4. Iron is salvaged and stored, and amino acids from the breakdown of globin are conserved. The balance of hemoglobin is degraded to bilirubin, which is secreted in bile by the liver, and leaves the body in feces.

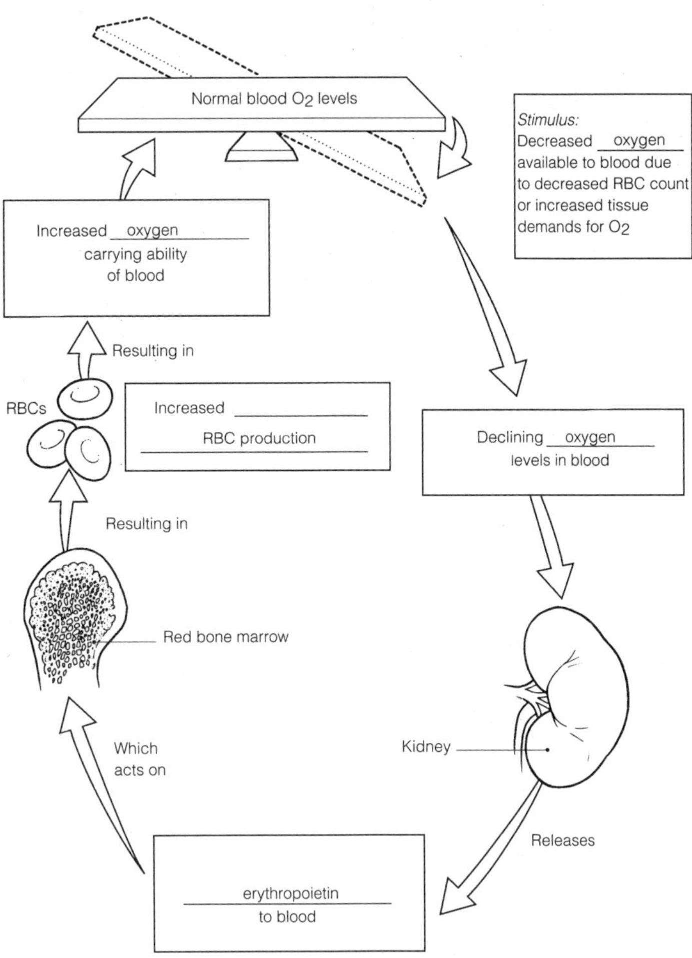

Figure 16.1

3. Check 1, 2, 3
4. 1. biconcave disk 2. anucleate 3. virtually none 4. hemoglobin molecules 5. spectrin, a protein attached to the cytoplasmic face of the plasma membrane 6. anaerobic 7. no mitochondria 8. globin protein (2 alpha and 2 beta chains) and 4 heme groups 9. heme (iron) 10. globin (amino acids) 11. lungs 12. tissues 13. oxyhemoglobin 14. reduced hemoglobin or deoxyhemoglobin
5. 1.F 2. B 3. A 4. C 5. B 6. D 7. E
6. 1. diapedesis 2. differential 3. kidneys 4. globin parts 5. 5.5 6. T 7. T 8. hematocrit 9. less 10. lymphocytes 11. T 12. transferrin
7. 1. 2 2. 5 3. 1 4. 4 5. 3
8. 1. G 2.–4. A, B, G 5. I 6. F 7. G 8. D 9. F 10. E 11. C 12. B 13. A 14. D 15. I 16. H, I 17. F 18. H 19. I 20–24. A, B, D, F, G 25. G 26. D 27. F 28. G
9. Figure 17.2: A. neutrophil B. monocyte C. eosinophil D. lymphocyte
10. 1. Erythrocytes 2. Monocytes 3. Antibodies 4. Lymphocyte 5. Platelets 6. Aneurysm 7. Increased hemoglobin 8. Hemoglobin 9. Lymphocyte
11. They are chemicals that promote the proliferation of white blood cells. Macrophages, T lymphocytes.
12. 1. granulocyte colony-stimulating factor 2. interleukin 3 3. erythropoietin (EPO)

Hemostasis

1. 1. A 2. F 3. H 4. J 5. E 6. G 7. I 8. D 9. C 10. B
2. 1. I 2. I 3. B 4. B 5. E 6. B 7. E 8. I 9. E 10. B 11. B
3. 1. A 2. C 3. H 4. E 5. G 6. I 7. D 8. B 9. J 10. L 11. K 12. F

Transfusion and Blood Replacement

1.

Blood type	Agglutinogens or antigens	Agglutinins or antibodies in plasma	Can donate blood to type	Can receive blood from type
Type A	A	Anti-B	A, AB	A, O
Type B	B	Anti-A	B, AB	B, O
Type AB	A, B	None	AB	A, B, AB, O
Type O	None	Anti-A, anti-B	A, B, AB, O	O

2. No; A+
3. A situation in which plasma antibodies attach to and lyse red blood cells different from your own.

Diagnostic Blood Tests

1.

Characteristic	Normal value or normal range
% of body weight	8%
Blood volume	4–5 L (female), 5–6 L (male)
Arterial blood pH	7.4 (7. 35–7. 45)
Blood temperature	38°C
RBC count	4.3–5.2 million cells/mm^3 μl; (f) 5.1–5.8 million cells/mm^3
Hematocrit	45%; 37± 5% (f); 42 ± 5% (m)
Hemoglobin	12–16 g/100 ml (f); 13–18 (m)
WBC count	4,800–10,800/μl
Differential WBC count #5 neutrophils	50–70%
#2 eosinophils	1–4%
#1 basophils	<1%
#4 lymphocytes	25–40%
#3 monocytes	3–8%
Platelet count	150,000–400,000/μl

2. Tests for certain nutrients
   1. Glucose level (glucose tolerance test)
   2. Fat content

   Tests for clotting
   1. Prothrombin time
   2. Platelet count

   Tests for mineral levels
   1. Calcium assay
   2. Potassium assay
   3. Sodium (NaCl) assay

   Tests for anemia
   1. Hematocrit
   2. Total RBC count
   3. Hb determination

The Incredible Journey

1. 1. hematopoiesis 2. hemostasis 3. hemocytoblasts 4. neutrophil 5. phagocyte 6. erythropoietin 7. red blood cells 8. hemoglobin 9. oxygen 10. lymphocytes 11. antibodies 12.–15. basophils, eosinophils, monocytes, platelets 16. endothelium 17. platelets 18. serotonin 19. fibrin 20. clot 21. prothrombin activator 22. prothrombin 23. thrombin 24. fibrinogen 25. embolus

CHALLENGING YOURSELF

At the Clinic

1. Erythropoietin.
2. Aplastic anemia; short-term: transfusion; long-term: bone marrow transplant; packed red cells.
3. Hemorrhagic anemia.
4. Polycythemia vera.
5. Acute leukemia.
6. Vitamin K.
7. Other antigens; for long-term transfusions, typing and cross-matching of blood for these other antigens is also necessary.
8. 1. Hemolytic disease of the newborn (erythroblastosis fetalis). 2. Red blood cells have been destroyed by the mother's antibodies, so the baby's blood is carrying insufficient oxygen. 3. She must have received mismatched Rh^+ blood in a previous transfusion. 4. Give the mother RhoGAM to prevent her from becoming sensitized to the Rh^+ antigen. 5. Fetal progress will be followed in expectation of hemolytic disease of the newborn; intrauterine transfusions will be given if necessary, as well as complete blood transfusion to the newborn.
9. Although red marrow is found in the cavities of most bones of young children, red marrow and active hematopoietic sites are found only in flat bones of adults. The bones most often chosen for obtaining a marrow sample in adults are those flat bones that are most easily accessible, the sternum and ilium.
10. It seems that Rooter has "gifted" the Jones family members (at least the daughters) with pinworms. Eosinophils are the body's most effective defense against infestations by fairly large parasites (such as pinworms, tapeworms, and the like), which explains the girls' elevated eosinophil counts.
11. The stomach cells are the source of gastric intrinsic factor, which is essential as part of the carrier system for the absorption of vitamin B_{12} by the intestinal cells. Apparently, insufficient intrinsic factor–producing cells remained after the surgery, requiring that this patient receive vitamin B_{12} injections. There would be no point in giving the vitamin orally because it is not absorbable in the absence of intrinsic factor. If he refuses the shots, he will develop pernicious anemia.
12. Infectious mononucleosis: total WBC count; differential WBC count. Bleeding problems: prothrombin time, clotting time, platelet count (tests for certain procoagulants, vitamin K levels, etc.).
13. Tests to determine whether fat absorption from the intestine is normal will indicate if vitamin K is being absorbed normally, and a history of a recent course of antibiotics may reveal that enteric bacteria, which synthesize vitamin K, have been destroyed. A low platelet count is diagnostic for thrombocytopenia.
14. The diagnosis is thalassemia. It can be treated with a blood transfusion.

Stop and Think

1. Crushing of the artery elicits the release of more chemical clotting mediators and a stronger vascular spasm, which is more effective in reducing blood loss.
2. Only hemocytoblasts form blood elements. If most hemocytoblasts are diverted to WBC production, fewer are available to form RBCs and platelets.
3. poly = many; cyt = cells; emia = blood (many cells in the blood). dia = through; ped = foot; sis = act of (putting a foot through).
4. The extrinsic "shortcut" provides for faster clotting; the multiple steps of the intrinsic pathway result in an amplifying cascade.
5. Sodium EDTA is an anticoagulant; removal of calcium ions *prevents* clotting because ionic calcium is required for nearly every step of hemostasis.
6. Fetal hemoglobin F must receive oxygen from maternal hemoglobin A; the higher affinity of hemoglobin F ensures that it can load oxygen from hemoglobin A.
7. When an athlete is exercising vigorously and body temperature rises, the warmed blood increases in volume. A blood test done at this time to determine iron content would indicate iron-deficiency anemia. This "illusion" of anemia is called athlete's anemia.
8. RBC antigens that occur only in specific families; that is, they are highly restricted in the general population.
9. It will rise. Because CO competes with oxygen for binding sites on heme, the oxygen-carrying capacity of John's blood will decline, thereby stimulating erythropoiesis.

COVERING ALL YOUR BASES

Multiple Choice

1. A, D **2.** B **3.** B, D **4.** A, D **5.** A, B, D **6.** D **7.** A, B, D **8.** A, B, C **9.** C **10.** B **11.** D **12.** B, C **13.** D **14.** C **15.** B **16.** C **17.** B, C, D **18.** A, B, C **19.** C **20.** C **21.** B, C **22.** C **23.** B, C **24.** C **25.** B, C, D **26.** A **27.** B **28.** A, B, C, D **29.** A, D **30.** D **31.** B **32.** A **33.** B, C **34.** A

Word Dissection

	Word root	Translation	Example
1.	agglutin	glued together	agglutination
2.	album	egg white	albumin
3.	bili	bile	bilirubin
4.	embol	wedge	embolus
5.	emia	blood	anemia
6.	erythro	red	erythrocyte
7.	ferr	iron	ferritin
8.	hem	blood	hemoglobin
9.	karyo	nucleus	megakaryocyte
10.	leuko	white	leukocyte
11.	lymph	water	lymphatic system
12.	phil	love	neutrophil
13.	poiesis	make	hemopoiesis
14.	rhage	break out	hemorrhage
15.	thromb	clot	thrombocyte

Chapter 17 The Cardiovascular System: The Heart

BUILDING THE FRAMEWORK

Heart Anatomy

1. 1. mediastinum 2. diaphragm 3. second 4. midsternal line 5. fibrous 6. visceral 7. epicardium 8. friction 9. myocardium 10. cardiac muscle 11. fibrous skeleton 12. endocardium 13. endothelial 14. 4 15. atria 16. ventricles 17. papillary muscles 18. trabeculae carneae

2.

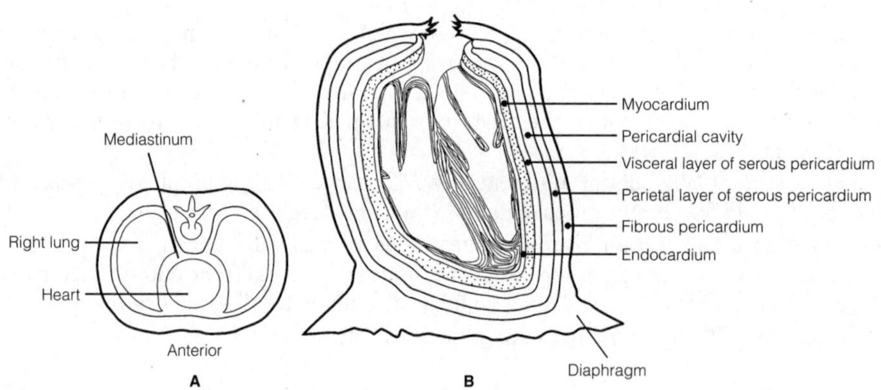

Figure 17.1

3. 1. right ventricle 2. pulmonary valve 3. pulmonary arteries 4. lungs 5. right and left pulmonary veins 6. left atrium 7. mitral (bicuspid) 8. left ventricle 9. aortic 10. aorta 11. capillary beds 12. superior vena cava 13. inferior vena cava

In answer for Figure 17.2, the white areas represent regions transporting O_2-rich blood. The gray vessels transport O_2-poor blood.

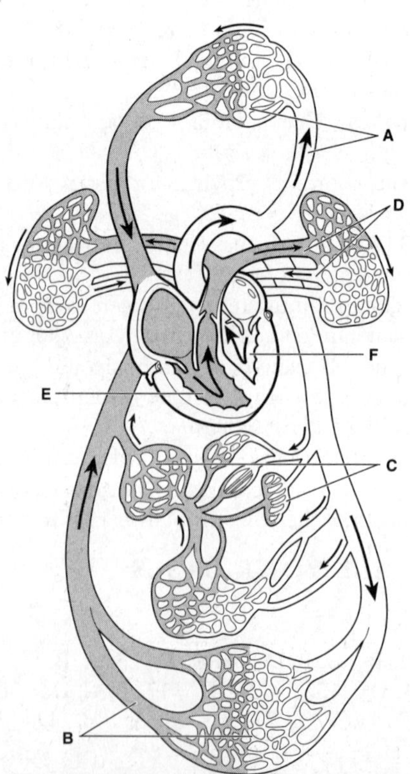

Figure 17.2

4. Figure 17.3: 1. right atrium 2. left atrium 3. right ventricle 4. left ventricle 5. superior vena cava 6. inferior vena cava 7. aorta 8. pulmonary trunk 9. left pulmonary artery 10. right pulmonary artery 11. right pulmonary veins 12. left pulmonary veins 13. vessels of coronary circulation 14. apex of heart 15. ligamentum arteriosum

5.

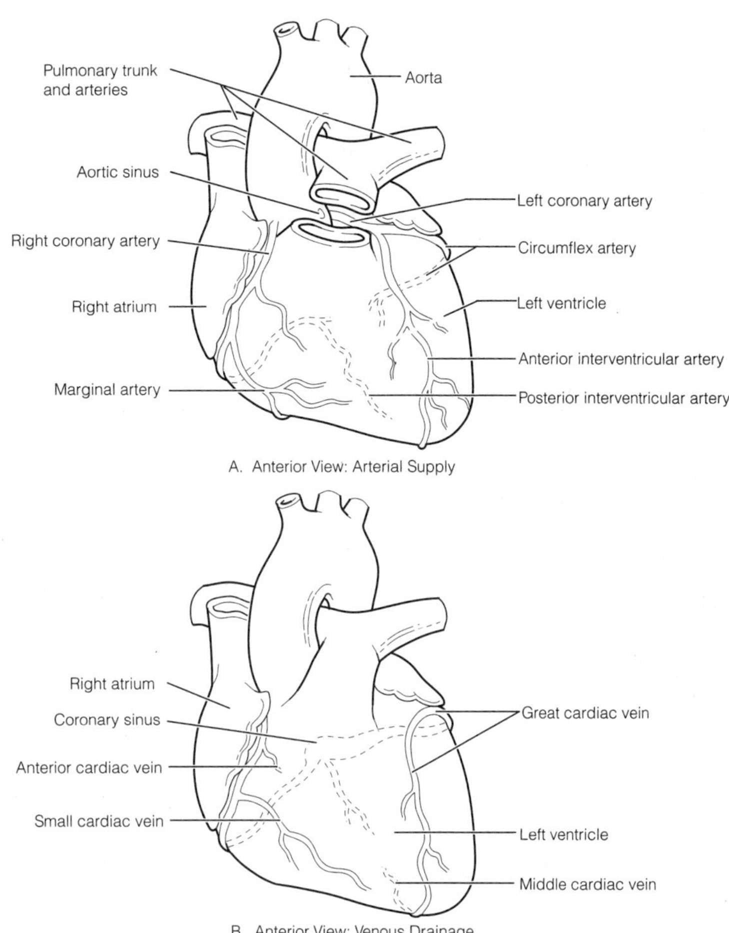

Figure 17.4

Cardiac Muscle Fibers

1. 1. C 2. C 3. C 4. C 5. S 6. S 7. C 8. S 9. C 10. S 11. C 12. C

2.

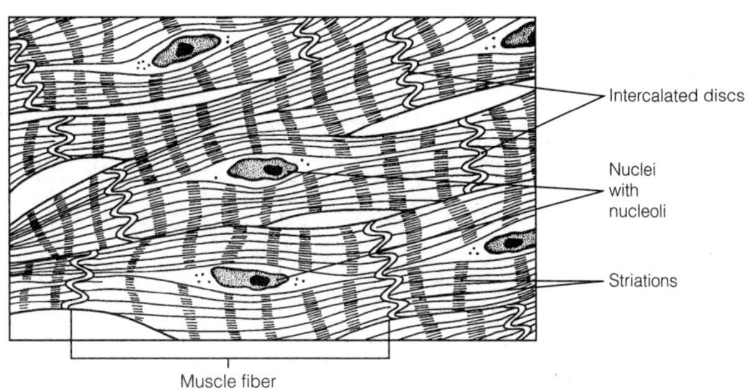

Figure 17.5

1. endomysium 2. prevent separation of adjacent cells 3. allow impulse (ions) to pass from cell to cell 4. functional syncytium 5. gap junctions

3.

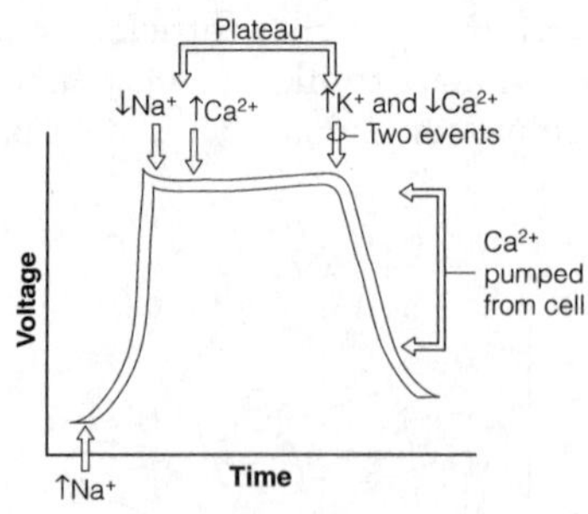

Figure 17.6

Heart Physiology

1. 1. spontaneously 2. resting membrane potential 3. action potential generation 4. pacemaker potentials 5. K^+ 6. Na^+ 7. into 8. Ca^{2+} channels

2. A. 6 B. 7 C. 8 D. 9 E. 9 F. 8 G. 1 H. 2

1. SA node 2. AV node 3. AV bundle or bundle of His 4. bundle branches 5. Purkinje fibers 6. pulmonary valve 7. aortic valve 8. mitral valve 9. tricuspid valve

Figure 17.7: Red arrows should be drawn from the left atrium to the left ventricle and out the aorta. Blue arrows should be drawn from the superior and inferior venae cavae into the right atrium, then into the right ventricle and out the pulmonary trunk. Green arrows should be drawn from 1 to 5 in numerical order. The cords, called chordae tendineae, should run from the edges of the flaps of the AV valves to the inferior ventricular walls.

3. 1. sinus rhythm 2. SA node 75 beats/min; AV node 50 beats/min; AV bundle 30 beats/min; Purkinje fibers 30 beats/min 3. 0.3–0.5 m/s 4. 0.22

4.

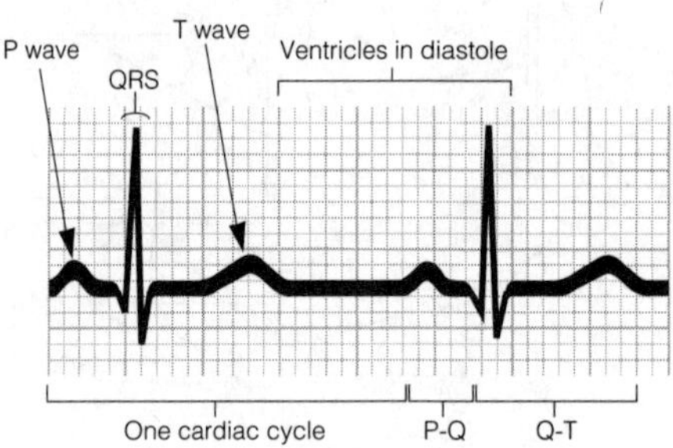

Figure 17.8

5. Figure 17.9: 1. B has extra P waves 2. C shows tachycardia 3. A has an abnormal QRS complex

6. 1. systole 2. diastole 3. lub-dup 4. atrioventricular 5. semilunar 6. ventricles 7. atria 8. atria 9. ventricles 10. murmurs

7.

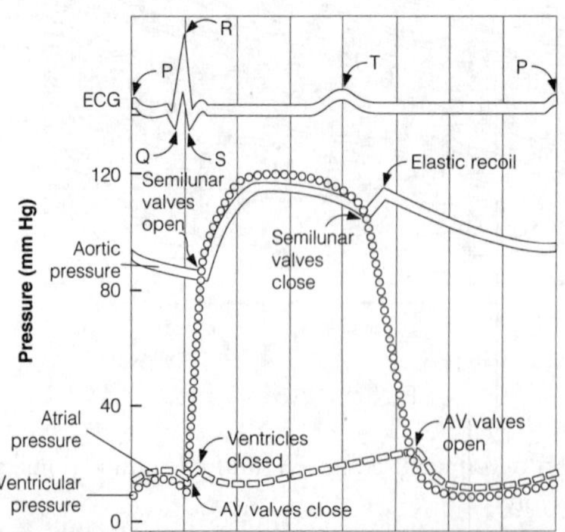

Figure 17.10

8. 1. AV valves open 2. SA node is pacemaker 3. Ventricular systole 4. Semilunar valves open 5. Heart sound after valve closes 6. AV valve open

9. 1. cardiac output 2. heart rate 3. stroke volume 4. 75 beats per minute 5. 70 ml per beat 6. 5250 ml per minute 7. minute 8. 120 ml 9. 50 ml 10. cardiac reserve 11. running, climbing, or any aerobic activity 12. end diastolic volume (EDV) 13. end systolic volume (ESV) 14. preload 15. cross–bridge 16. force 17. venous blood 18. ventricles

10. 1. Increased force of heartbeat that is independent of EDV. 2. Sympathetic nervous system activation; glucagon; thyroxine; epinephrine. 3. Exercise enhances the effectiveness of the respiratory and skeletal muscle "pumps" and activates the sympathetic nervous system.

11. Check 1, 2, 4, 5, 6, 8, 9, 10, 13, 15, 19

12. 1. sympathetic 2. T 3. increases 4. high 5. fetal 6. resting heart rate 7. systolic 8. left 9. T 10. T 11. T

13. 1.H 2. D 3. G 4. C 5. B 6. E 7. J 8. F 9. A 10. I

14. 1. C 2. A 3. B 4. D

CHALLENGING YOURSELF

At the Clinic

1. Cardiac tamponade; compression of the heart by excess pericardial fluid reduces the space for ventricular activity and impairs ventricular filling.
2. Cor pulmonale; pulmonary embolism.
3. Mitral valve prolapse; valve replacement.
4. Myocarditis caused by the strep infection.
5. Valvular stenosis; heart murmur will be high-pitched during systole.
6. Angina pectoris.
7. Complete heart block; conduction pathway between SA node and AV node is damaged.
8. An MI is an area of dead cardiac muscle that may have been replaced by scar tissue. A clot lodged in a coronary vessel is the usual cause. Because part of the conduction pathway is obliterated by an MI, it takes longer to depolarize the heart and disturbs its rhythm.
9. Zero; myocardial infarction. The posterior interventricular artery supplies much of the left ventricle, the systemic pump.
10. Bradycardia, which results from excessive vagal stimulation of the heart, can be determined by taking pulse.
11. Peripheral congestion due to right heart failure.
12. BP measurement: A high blood pressure may hint that the patient has hypertensive heart disease; when BP is chronically elevated, the heart has to work harder. Blood lipid and cholesterol levels: High levels of triglycerides and cholesterol are risk factors for coronary heart disease. Electrocardiogram: Will indicate if the pacing and electrical events of the heart are abnormal. Chest X ray: Will reveal if the heart is enlarged or abnormally located in the thorax.
13. A defective valve would be detected during auscultation. Because the valve flaps are not electrically excitable, an ECG does not reveal a valvular problem.

Stop and Think

1. They will not be compressed when the ventricles contract, as are the vessels within the myocardium.
2. Contraction from the top down is like squeezing a tube of toothpaste from the bottom up. The blood is moved in the direction of the outflow at the AV valves. Ventricles contract from the bottom up, due to the arrangement of the conduction system, from apex up the lateral walls. This also moves blood toward the valves but, in this case, against the pull of gravity.
3. Referred pain from damage to the heart is felt in the left chest and radiates down the medial side of the left arm to the fifth digit in severe cases.
4. Prolonged contraction of the myocardium allows sufficient tension to build to overcome the inertia of the blood and actually move the fluid out of the chambers.
5. It takes about one minute, the time it takes for the entire blood volume to circulate through the heart/body.
6. With exercise, blood flow to the myocardium is maintained and expanded, and the heart becomes stronger and more efficient. With CHP, myocardial blood flow is diminished, and the heart weakens. A healthy heart attains sufficient ventricular pressure to provide sufficient stroke volume with a reduced heart rate. The weakened, congested heart is thin-walled and flabby and cannot exert sufficient force to eject much blood during ventricular systole, hence heart rate often increases.
7. Hypothyroidism leads to reduced heart rate.
8. The flap over the foramen ovale acts as a valve, preventing backflow from the left atrium to the right atrium. The groove channels blood from the inferior vena cava to the left atrium (via the foramen ovale), providing the systemic circulation with most of the freshly oxygenated blood.

9. Exercise increases heart rate only during the period of increased activity; during rest (which is most of the time), the heart rate decreases as a result of regular exercise. If the resting heart rate drops from 80 to 60 bpm, and one hour of exercise a day causes heart rate to be elevated to 180 bpm, the net change in number of heart beats per day can be calculated as follows:
Sedentary: 80 beats/min x 60 min/h x 24 h/day = 115, 200 beats/day
Exerciser: 60 beats/min x 60 min/h x 23 h/day +
180 beats/min x 60 min/h x 1 h/day = 93, 600 beats/day
As you can see, an "exerciser" gains, not loses, time.

COVERING ALL YOUR BASES

Multiple Choice

1. D **2.** A, D **3.** A, D **4.** A, B, C, D **5.** A, B, D **6.** C **7.** A, D **8.** A, C **9.** A **10.** C, D **11.** A, C, D **12.** A, D **13.** C **14.** C **15.** C **16.** D **17.** C **18.** B, D **19.** C **20.** A, C **21.** D **22.** B **23.** B **24.** A, D **25.** C **26.** A, B, C **27.** A, B, C, D **28.** B **29.** B **30.** D **31.** B **32.** B

Word Dissection

	Word root	Translation	Example		Word root	Translation	Example
1.	angina	choked	angina pectoris	8.	ectop	displaced	ectopic focus
2.	baro	pressure	baroreceptor	9.	intercal	insert	intercalated disc
3.	brady	slow	bradycardia	10.	pectin	comb	pectinate muscle
4.	carneo	flesh	trabeculae carneae	11.	sino	hollow	sinoatrial node
5.	cusp	pointed	tricuspid valve	12.	stenos	narrow	mitral stenosis
6.	diastol	stand apart	diastole	13.	systol	contract	systole
7.	dicro	forked	dicrotic notch	14.	tachy	fast	tachycardia

Chapter 18 The Cardiovascular System: Blood Vessels

BUILDING THE FRAMEWORK

PART 1: OVERVIEW OF BLOOD VESSEL STRUCTURE AND FUNCTION

1. 1. femoral artery 2. brachial artery 3. popliteal artery 4. facial artery 5. radial artery 6. posterior tibial artery 7. temporal artery 8. dorsalis pedis
2. 1. A 2. B 3. A 4. A 5. C 6. B 7. C 8. A
A. artery; relatively thick media; small round lumen B. vein; (relatively) thin media; large lumen; valves
C. capillary; single layer of endothelium

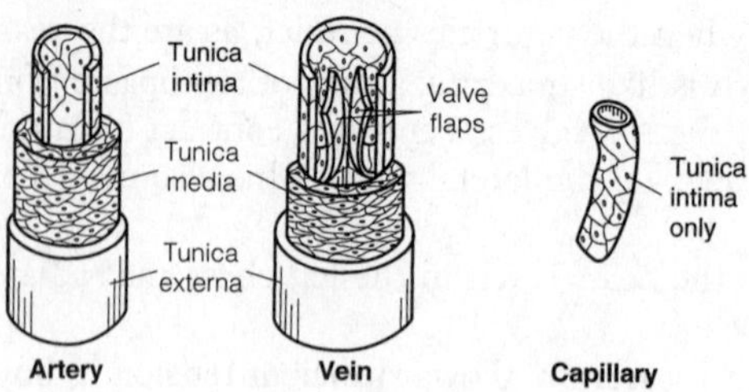

Figure 18.1

3. 1. A 2. E 3. G 4. B 5. C 6. D 7. F 8. H 9. K 10. I 11. J 12. E 13. G
4. Arterial anastomoses provide alternate pathways for blood to reach a given organ. If one branch is blocked, an alternate branch can still supply the organ.

5.

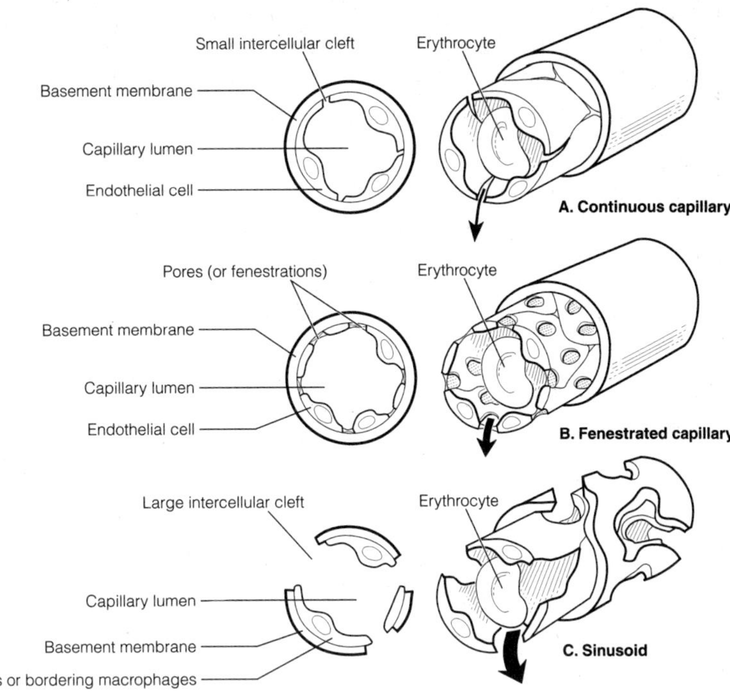

Figure 18.2

6. 1. Elastic 2. Pressure points 3. Heart 4. Gap junctions 5. Kidney 6. High pressure 7. End arteries 8. Thick media 9. True capillaries

7. The venous valves prevent backflow of blood.

8. Veins have large lumens and thin walls and can hold large volumes of blood. At any time, up to 65% of total blood volume can be contained in veins. These reservoirs are most abundant in the skin and visceral organs (particularly the digestive viscera).

PART 2: PHYSIOLOGY OF CIRCULATION

1. 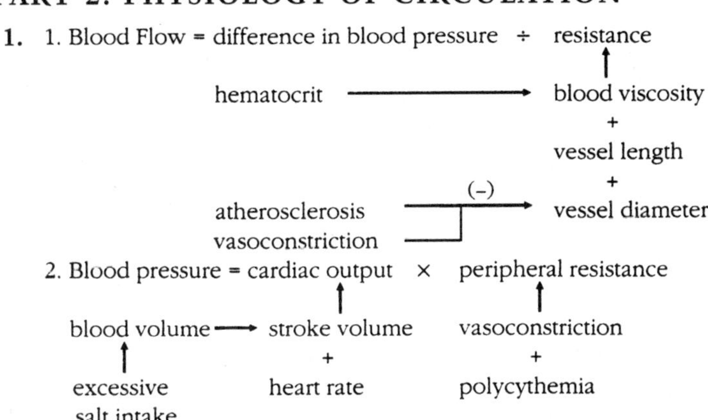

2. When the heart contracts, blood is forced into the large arteries, stretching the elastic tissue in their walls. During diastole, the artery walls recoil against the blood, maintaining continuous pressure and blood flow.

3. 1. H 2. B 3. C 4. D 5. D 6. J 7. E 8. A 9. G (A) 10. I

4. 1. D 2. I 3. I 4. I 5. I 6. D 7. D 8. I 9. D 10. D 11. D 12. I 13. I 14. D 15. I 16. I 17. D 18. I 19. I 20. D 21. I 22. I

5. 1. Cardiac output 2. Blood flow 3. Low viscosity 4. Blood pressure 5. High blood pressure 6. Vasodilation 7. Vein 8. 120 mm Hg 9. Cardiac cycle 10. Inactivity 11. Sympathetic activity

6. 1. increase 2. orthostatic 3. brain 4. stethoscope 5. low 6. vasoconstricting 7. hypertension 8. arterioles 9. medulla/brain 10. T 11. reduction 12. vasoconstriction 13. blood vessel length 14. capillaries 15. arterial system 16. autoregulation 17. T

7. 1. interstitial fluid 2. concentration gradient (via diffusion) 3. fat-soluble substances like fats and gases 4. water and water-soluble substances like sugars and amino acids 5. through the metarteriole-thoroughfare channels 6. A 7. capillary blood 8. capillary hydrostatic (blood) pressure (Hp_c) 9. blood pressure 10. capillary colloid osmotic pressure (Op_c) 11. albumin 12. at the arterial end 13. It is picked up by lymphatic vessels for return to the bloodstream.

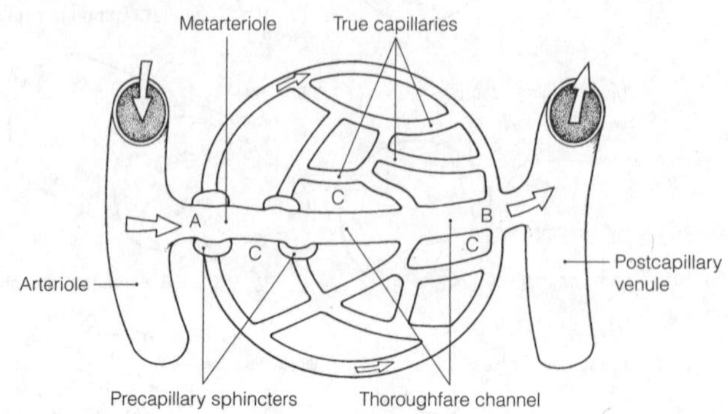

Figure 18.3

8. Hypovolemic shock is a decrease in total blood volume, as from acute hemorrhage. Increased heart rate and vasoconstriction result. In vascular shock, blood volume is normal, but pronounced vasodilation expands the vascular bed, resistance decreases, and blood pressure falls. A common cause of vascular shock is bacterial infection.

9. 1. capillaries 2. arteries 3. capillaries 4. arteries 5. veins

10. 1. F 2. A 3. D 4. B 5. A 6. D 7. C 8. E 9. B 10. D 11. C 12. A 13. A 14. A 15. F 16. E

PART 3: CIRCULATORY PATHWAYS: BLOOD VESSELS OF THE BODY

1. The right atrium and ventricle and all vessels with "pulmonary" in their name should be colored blue; the left atrium and ventricle and the aortic arch and lobar arteries should be colored red.

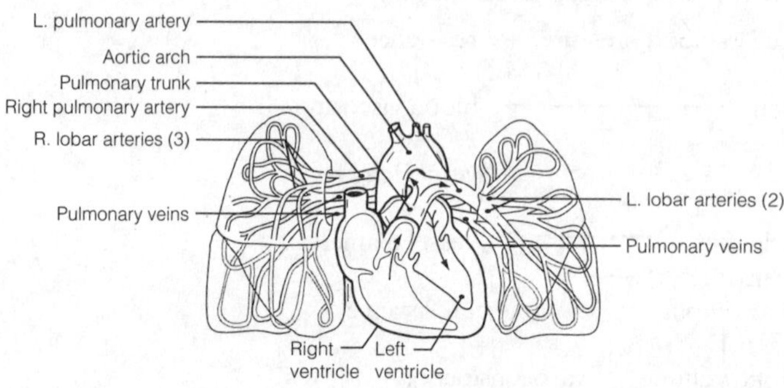

Figure 18.4

2.

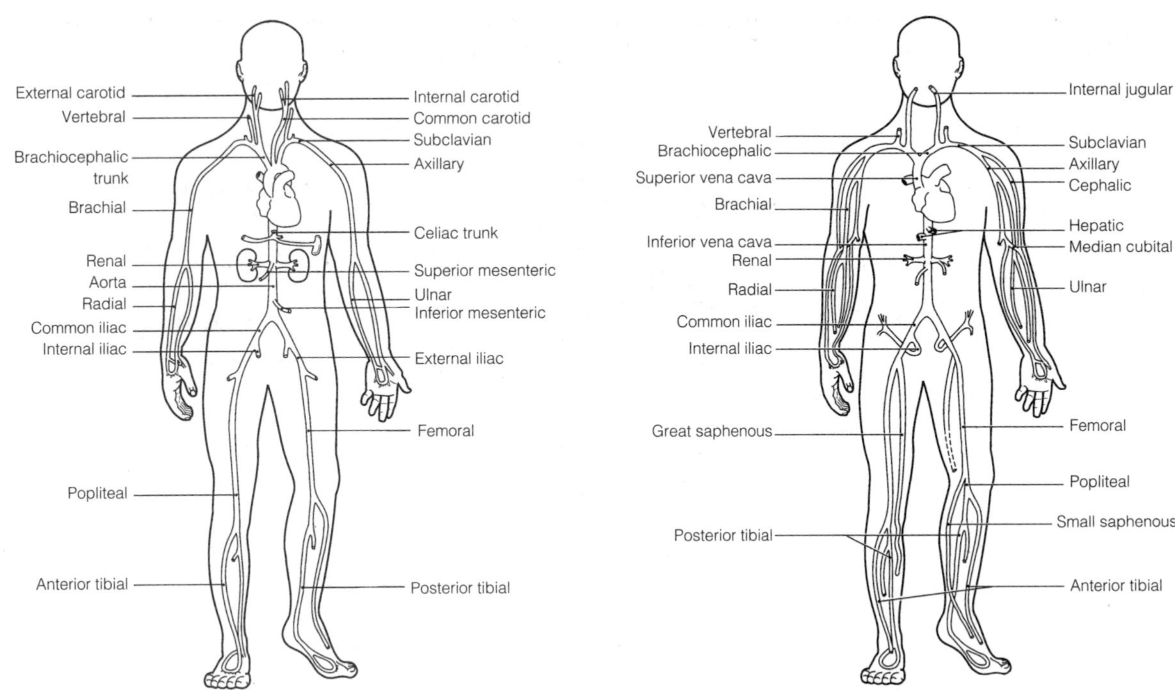

Figure 18.5 Arteries

Figure 18.6 Veins

3.

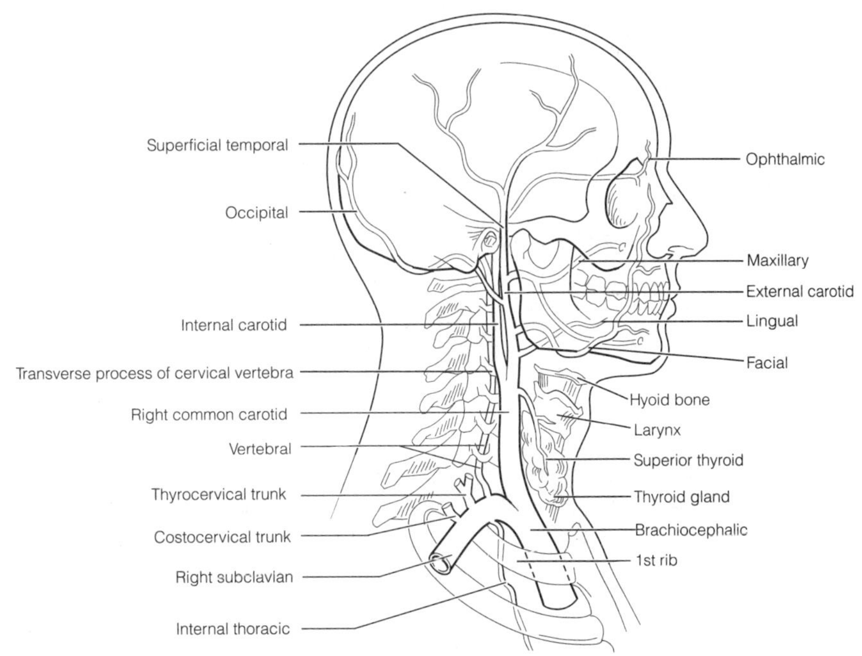

Figure 18.7 Arteries of the head and neck

4. 1. F 2. C, D 3. A The cerebral arterial circle consists of the communicating arteries and those parts of the cerebral arteries that complete the arterial anastomosis around the pituitary.

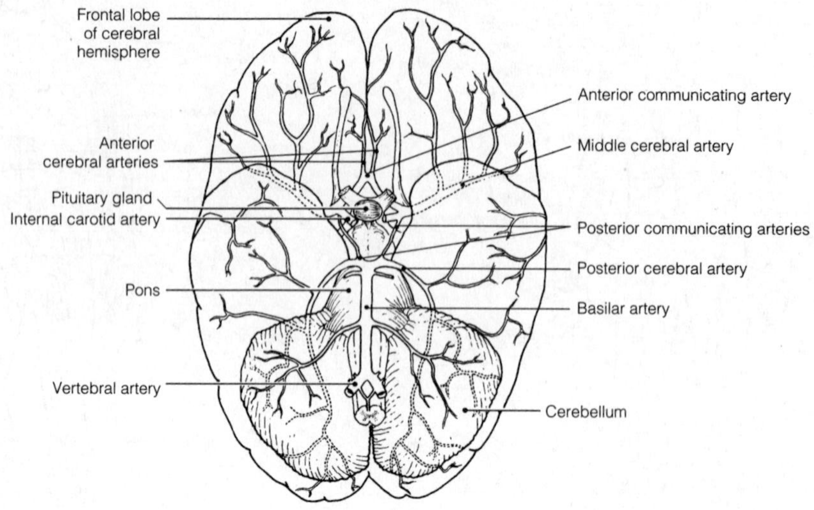

Figure 18.8

5. 1. F 2. W 3. H 4. P 5. Y 6. B 7. J 8. I 9. S 10. C 11. C 12. N 13. Q 14. L 15. C 16. X 17. G 18. E 19. K 20.–22. A, R, T 23. U

6.

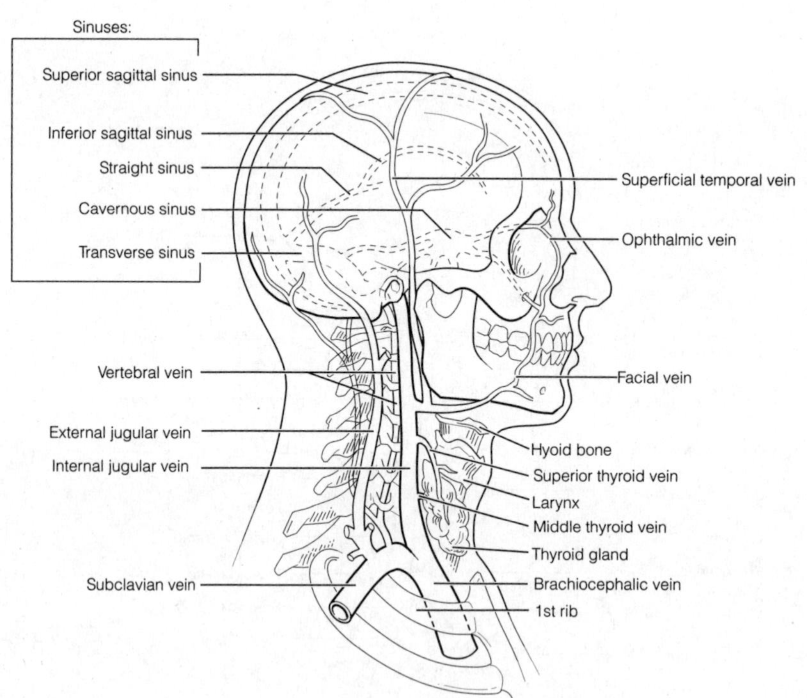

Figure 18.9 Veins of the head and neck

7.

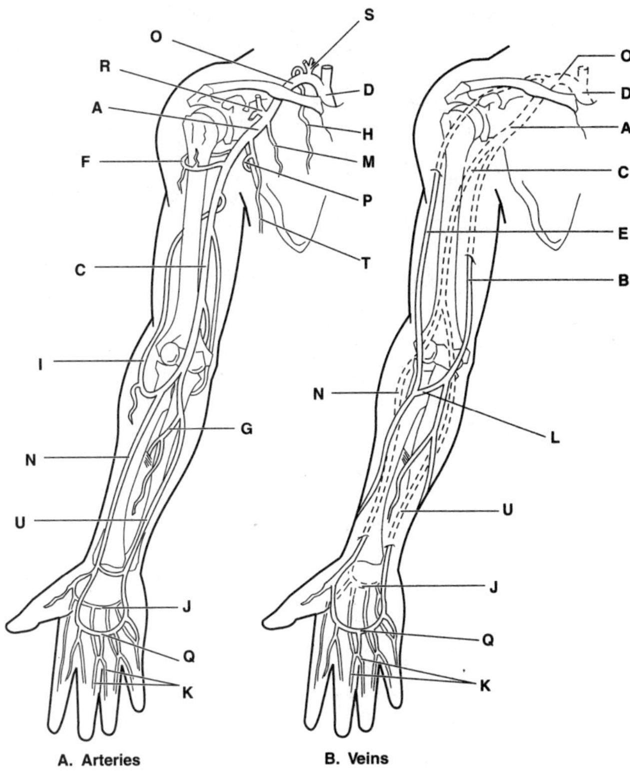

Figure 18.10

8.

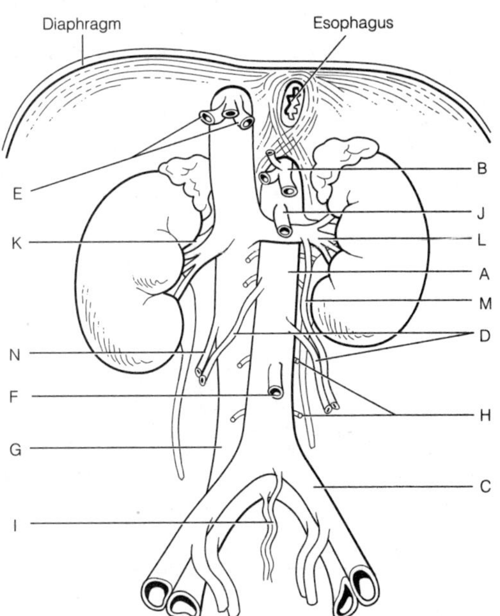

Figure 18.11

9.

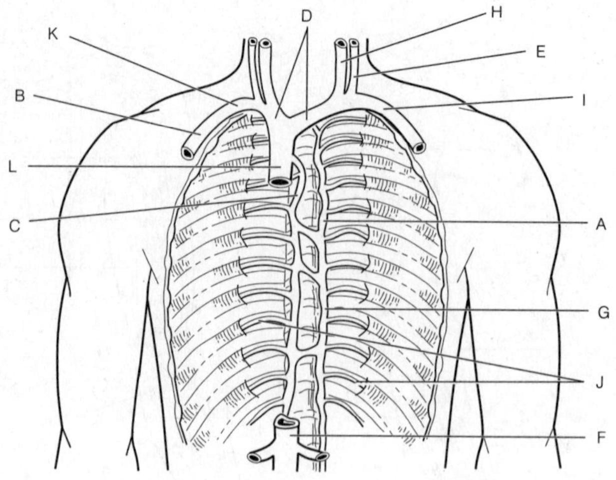

Figure 18.12

10.

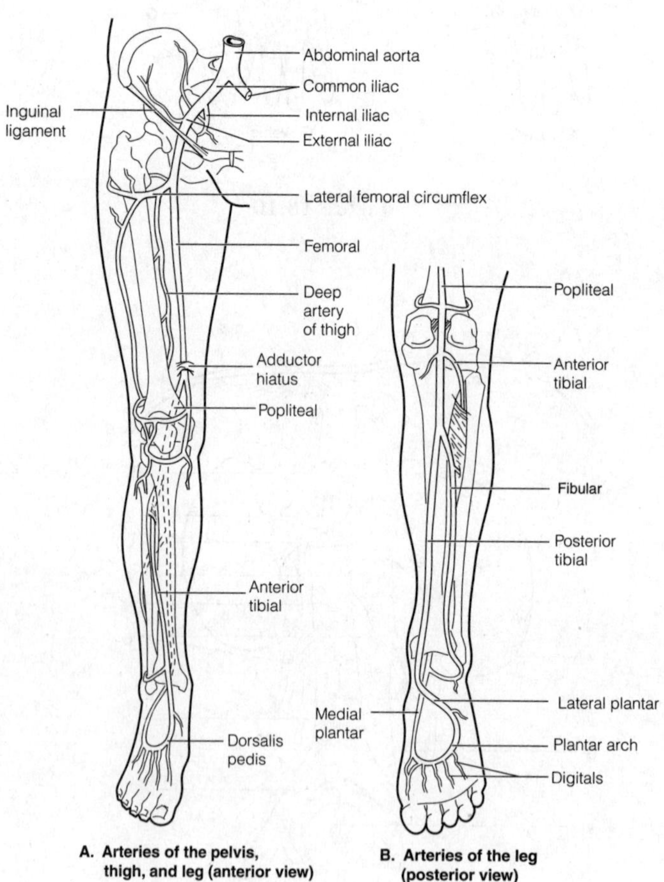

Figure 18.13

11.

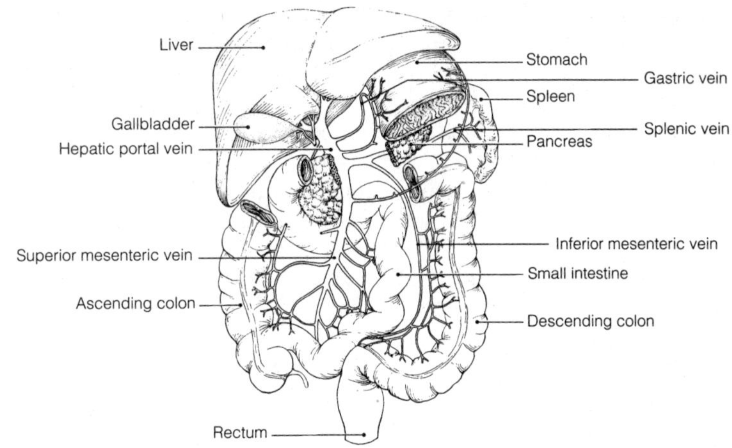

Figure 18.14

12. 1. S 2. X 3. U 4. E 5. T 6. Q 7. D 8. A 9. R 10. M 11. F 12. J 13. B 14. O 15. L 16.–18. I, N, V 19. K 20. G 21. H

The Incredible Journey

1. 1. left atrium 2. left ventricle 3. mitral (bicuspid) 4. chordae tendineae 5. diastole 6. systole/contraction 7. aortic semilunar 8. aorta 9. superior mesenteric 10. endothelial 11. superior mesenteric vein 12. splenic 13. nutrients 14. phagocytic (van Kupffer) 15. hepatic 16. inferior vena cava 17. right atrium 18. pulmonary 19. lungs 20. capillaries 21. gas (O_2 and CO_2) 22. subclavian

CHALLENGING YOURSELF

At the Clinic

1. The veins, particularly the superficial saphenous veins, become very prominent and tortuous. Typically it is due to failure of the venous valves. The problem here is the restriction of blood flow by her enlarged uterus during her many pregnancies. Elevate the legs whenever possible; avoid standing still.
2. Thrombosis, atherosclerosis; arterial anastomosis (circle of Willis).
3. High; polycythemia increases blood viscosity (thus peripheral resistance), which increases blood pressure.
4. Hypovolemic shock due to blood loss.
5. Loss of vasomotor tone due to damage to the vasomotor center in the medulla could cause extreme vasodilation, resulting in vascular shock.
6. Chronically elevated; chronic hypersecretion of ADH will trigger vasoconstriction (due to the vasopressin effect) and excessive water retention, both of which increase peripheral resistance.
7. Occlusion of the renal blood supply will reduce blood pressure in the kidneys, triggering the release of renin, which in turn precipitates activation of angiotensin, a powerful vasoconstrictor.
8. Nicotine is a vasoconstrictor; a high dose of nicotine in a body not accustomed to it can reduce blood flow to the brain and cause dizziness.
9. Balloon (or another type of) angioplasty.
10. Septicemia; bloodborne bacterial toxins trigger widespread vasodilation, which causes blood pressure to plummet (vascular shock).
11. Cardiogenic shock.
12. The dorsalis pedis pulse is the most distal palpable pulse. Chances are if his popliteal and pedal pulses are strong, his right limb is getting adequate circulation. Also, the skin will be warm and *non*cyanotic if the blood supply is adequate.
13. Inflammation of the veins; thrombophlebitis, which can lead to pulmonary embolism.
14. Pulse pressure = 20 mm Hg; low; loss of elasticity.
15. Beta blockers prevent the effect of epinephrine, which constricts the arterioles and raises the blood pressure. Diuretics increase urine output, which reduces blood volume. Hence, the treatment will result in lower blood pressure.
16. His fears are unfounded because there is no artery in the exact middle of the anterior forearm. Most likely the median vein of the forearm has been cut, but it is still a fairly small superficial vein and venous bleeding is much slower than arterial bleeding.

Stop and Think

1. A thin layer is sufficient to withstand the frictional forces of the blood. A thicker layer would reduce the size of the lumen and require more nutrients to maintain it.
2. Decrease; vasoconstriction increases resistance, which decreases blood flow.

3. Open; all true capillaries will be flushed with blood. In the face exposed to cold, the capillary networks are closed off. The cells deplete oxygen and nutrients, and waste products build up. Precapillary sphincters relax, but no blood flows to the capillaries because the supplying arteriole is closed off. In warm air, the arterioles vasodilate, and blood flows into all the capillaries.
4. In a warm room, the skin capillaries are flushed with blood, which helps lose heat from the body. Rerouting blood to the head takes longer in this circumstance than in a cold room, in which many skin capillary beds are being bypassed by the metarteriole-thoroughfare channel shunts.
5. Pressure of the fetus on the inferior vena cava will reduce the return of blood from the lower torso and lower limbs.
6. Pericardial sac.
7. Organs supplied by end arteries have no collateral circulation. If the supplying end artery is blocked, no other circulatory route is available.
8. Formation of new tissues (adipose, muscle).
9. Vertebral arteries to basilar artery to posterior cerebral artery to posterior communicating artery to R. internal carotid artery beyond the blockage (hopefully).
10. Reduced plasma proteins (particularly albumin) will reduce the osmotic pressure of the blood. Consequently, less fluid will be drawn back at the venous end of capillaries, and edema will result.
11. Decreased cardiac output would reduce blood pressure to the brain. If compensatory mechanisms are inadequate to maintain blood supply to the vasomotor center, vasomotor tone will be reduced. The resulting vasodilation will cause vascular shock, which will reduce venous return to the heart. This will lead to even lower cardiac output, which will further diminish blood delivery to the vasomotor center.
12. An aneurysm is a ballooned-out and weakened area in a blood vessel. The primary problem in all cases is that it might rupture and cause a fatal stroke. The second problem in the patient discussed here is that the enlarged vessel is pressing on brain tissue and nerves. Because neural tissue is very fragile, it is susceptible to irreversible damage from physical trauma as well as from deprivation of a blood supply (another possible consequence of the compression of nervous tissue). Surgery was done to replace the weakened region of the vessel with inert plastic tubing.
13. There is no "great choice" to make here. Try to compress the subclavian artery that runs just deep to the clavicle by forcing your fingers inferiorly just posterior to the clavicular midline and lateral to the sternocleidomastoid muscle. Blockage of the subclavian artery would prevent the blood from reaching the axilla.
14. Erythrocytes are 7–8 μm in diameter, thus five of them side by side would measure a lumen diameter of 35–40 μm (the approximate size of the observed vessel). As the average capillary is 8–10 μm, one erythrocyte would just about fill a capillary's entire lumen. Hence, the vessel is most likely a postcapillary venule.

COVERING ALL YOUR BASES

Multiple Choice

1. A, B, C **2.** A, B, C, D **3.** A, C **4.** A, B **5.** C **6.** B **7.** A, B, C, D **8.** B, C **9.** B, D **10.** C **11.** A, B, C **12.** B, C, D **13.** C, D **14.** A, B, D **15.** A, B, C, D **16.** D **17.** B, C, D **18.** B **19.** A, B, D **20.** A, C **21.** A, B, C, D **22.** C **23.** A, C, D **24.** B **25.** A **26.** A, B, C, D **27.** A, B, C **28.** A, B **29.** C **30.** B **31.** D **32.** A **33.** C **34.** B **35.** D **36.** D

Word Dissection

	Word root	Translation	Example
1.	anastomos	coming together	anastomoses
2.	angio	vessel	angiogram
3.	aort	great artery	aorta
4.	athera	gruel	atherosclerosis
5.	auscult	listen	auscultation
6.	azyg	unpaired	azygos vein
7.	capill	hair	capillary
8.	carot	stupor	carotid artery
9.	celia	abdominal	celiac artery
10.	entero	intestine	mesenteric arteries
11.	epiplo	membrane	epiploic artery
12.	fenestr	window	fenestrated capillary
13.	jugul	throat	jugular vein
14.	ortho	straight	orthostatic hypotension
15.	phleb	vein	phlebitis
16.	saphen	clear, apparent	great saphenous vein
17.	septi	rotten	septicemia
18.	tunic	covering	tunica externa
19.	vaso	vessel	vasodilation
20.	viscos	sticky	viscosity

Chapter 19 The Lymphatic System and Lymphoid Organs and Tissues

BUILDING THE FRAMEWORK

Lymphatic Vessels

1. 1. pump 2. arteries 3. veins 4. valves 5. lymph 6. 3 liters
2. Lymphatic vessels pick up fluid leaked from the cardiovascular system. The overlapping edges of the endothelial cells form minivalves that allow fluid to enter, but not leave, the lymph capillaries. When the interstitial pressure rises, gaps are exposed between the endothelial flaps, and fluid enters. Collapse of the vessel is prevented by fine filaments that anchor the lymph capillaries to surrounding tissues.
3.

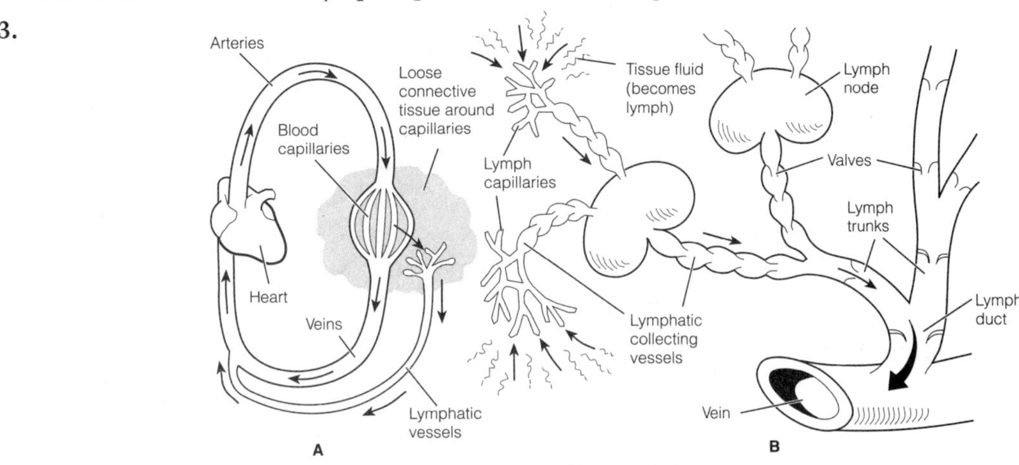

Figure 19.1

4. 1. Fat-free lymph 2. Blood capillary 3. Abundant supply of lymphatics 4. High-pressure gradient 5. Impermeable
5.

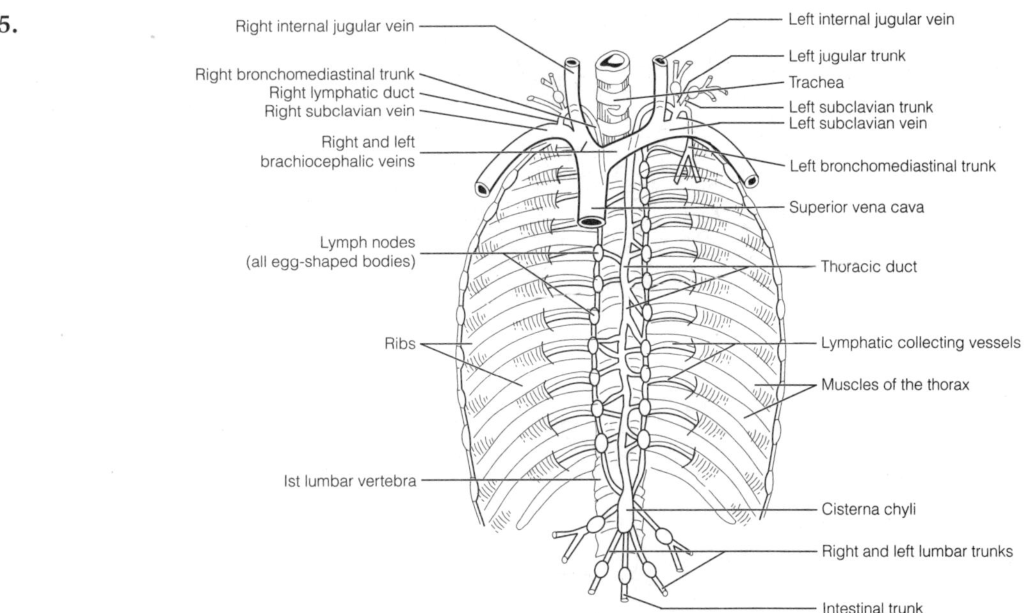

Figure 19.2

Lymphoid Cells and Tissues

1. 1. lymphocytes 2. antigens 3. T cells 4. B cells 5. macrophages (and dendritic cells) 6. phagocytizing 7. reticular cells 8. reticular 9. diffuse 10. nodular or follicular

2. The right side of head and upper torso, and the right upper limb should be colored green.

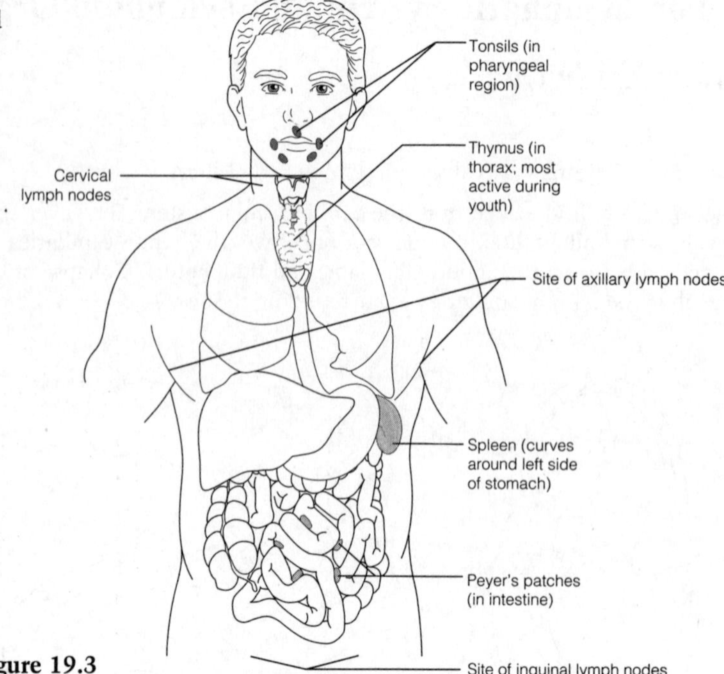

Figure 19.3

Lymph Nodes

1. 1. B lymphocytes 2. They produce and release antibodies. 3. T lymphocytes 4. macrophages 5. phagocytes 6. reticular 7. This slows the flow of lymph through the node, allowing time for immune cells and macrophages to respond to foreign substances present in the lymph. 8. valves in the afferent and efferent lymphatics 9. cervical, axillary, inguinal 10. They act to protect the body by removing bacteria or other debris from the lymphatic stream.

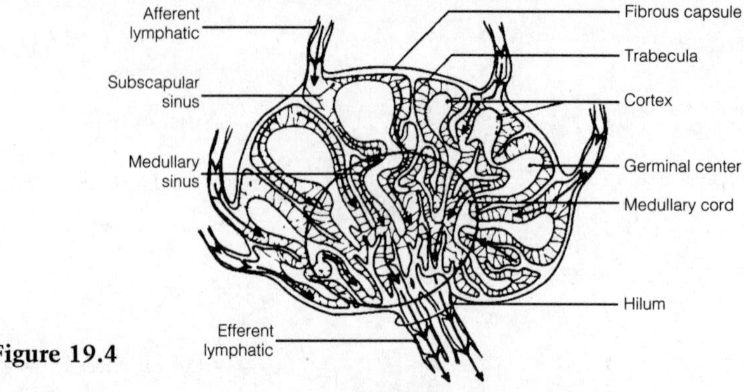

Figure 19.4

Other Lymphoid Organs

1. 1. C 2. A 3. D 4. B, E 5. C 6. C 7. D 8. E 9. B 10. E

2. 1. iron 2. platelets 3. RBCs 4. erythropoiesis 5. red 6. sinuses 7. RBCs (erythrocytes) 8. white 9. immune

3.

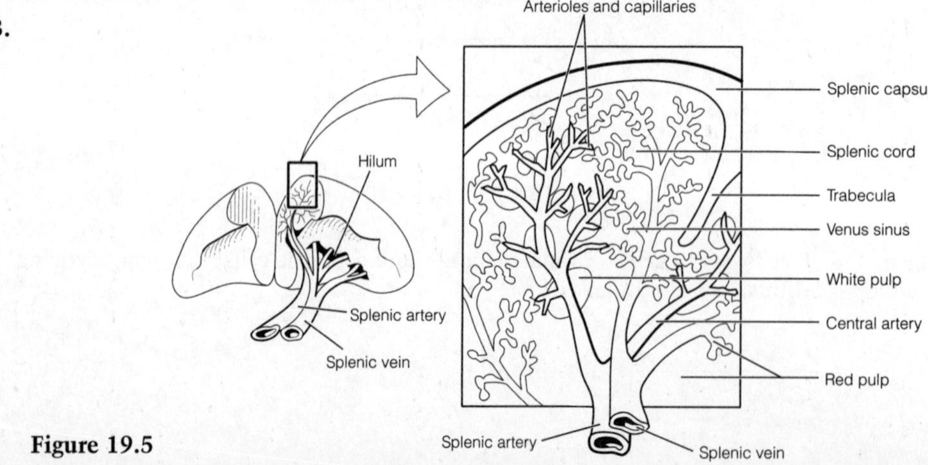

Figure 19.5

4. MALT, means mucosa-associated lymphatic tissue. Located just deep to the mucosa, MALT acts as a sentinel to protect the upper respiratory and digestive tracts from entry of foreign matter.
5. palatine tonsils, pharyngeal tonsils, and lingual tonsils

CHALLENGING YOURSELF

At the Clinic

1. Lymphangitis (inflammation of the lymphatic vessels) due to infection.
2. Lymphedema; no, the lymphatic vessels will be replaced by venous budding.
3. Bubo; bubonic plague.
4. Her lymph nodes are enlarged but firm, and not tender to the touch.
5. Hemorrhage; the spleen is a blood reservoir, and the circulatory pattern in the spleen makes it impossible to close off blood vessels and prevent bleeding. No; the liver, bone marrow, and other tissues can take over the spleen's functions.
6. Pharyngeal tonsils, which lie close to the orifices of the auditory tubes in the pharynx.
7. Angelica's enlarged spleen is due to septicemia. Remember, the spleen has a blood-cleansing function. As it removes disease-causing microorganisms and they are killed by phagocytes, the resulting inflammation causes swelling.

Stop and Think

1. Only lymph nodes filter lymph.
2. Cancer cells that have entered lymphatic vessels can get trapped in the lymph node sinuses.
3. Macrophages are found in the endothelial lining of organs like the liver and are associated with reticular connective tissue in lymph nodes.
4. If proteins accumulated in the interstitial fluid, the osmotic gradient of the capillaries would be lost. Fluid would be retained in the tissue spaces, resulting in edema.
5. Charlie is "right on." Lymph is excess tissue fluid that has leaked from the blood in the vascular system, and lymphatic vessels act to return that excess fluid to the blood.
6. White pulp consists of tight clusters of lymphocytes associated with reticular fibers, which is, in fact, the definition of a lymph nodule or follicle.

COVERING ALL YOUR BASES

Multiple Choice

1. C, D **2.** A, B, C, D **3.** A, B **4.** C **5.** D **6.** A **7.** B **8.** A, B **9.** A, B, C **10.** A, B, C, D **11.** A, B, D **12.** A, B, D **13.** B, D **14.** A, B, C, D **15.** D **16.** A, C, D **17.** A, B, C, D **18.** A, C **19.** A, B, C, D **20.** C

Word Dissection

	Word root	Translation	Example
1.	adeno	a gland	lymphadenopathy
2.	angi	vessel	lymphangiography
3.	chyle	juice	cisterna chyli
4.	lact	milk	lacteals
5.	lymph	water; clear water	lymphatic vessels

Chapter 20 The Immune System: Innate and Adaptive Body Defenses

BUILDING THE FRAMEWORK

PART 1: INNATE DEFENSES

1. 1. surface membrane barriers, mucosae 2. natural killer cells 3. chemicals

Surface Barriers: Skin and Mucosae

1. 1. tears; saliva 2. stomach; vagina 3. sebaceous glands; skin 4. goblet cells; digestive

2. 1. A, B, E, F 2. C, G 3. A, B, D, E, F 4. D 5. A–G

3. They propel mucus laden with trapped debris superiorly away from the lungs to the throat, where it can be spat out.

Internal Defenses: Cells and Chemicals

1. Phagocytosis is ingestion and destruction of particulate material. To occur, the phagocyte must adhere to the particle; this is difficult if the particle is smooth. The rougher the particle, the more easily it is ingested.

2. Neutrophils release potent oxidizing chemicals to the interstitial fluid, which, besides killing pathogens, also damages them.

3. 1.F 2. D 3. E 4. A 5. C 6. G 7. B 8. I 9. H

4. 1. Nausea 2. Natural killer cells 3. Neutrophils 4. Interferon 5. Antimicrobial 6. Macrophages 7. Defensins

5.

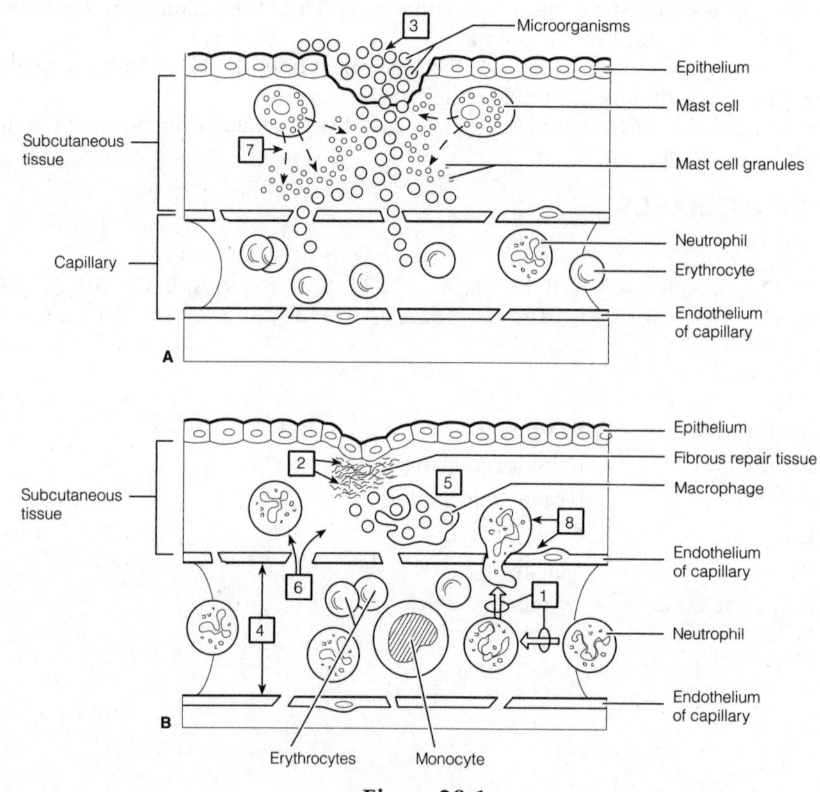

Figure 20.1

6. In response to inflammatory chemicals in the local area, the capillary endothelial cells sprout CAMs that bind specifically with certain other CAMs on the surfaces of neutrophils. This binding directs the neutrophils to the specific area where their phagocytic services are needed.

7. 1. Dilutes harmful chemicals in the area. 2. Brings in oxygen and nutrients needed for repair. 3. Allows entry of clotting proteins. 4. Also forms a structural basis for permanent repair.

8. 1. 20 2. classical 3. antigen-antibody 4. alternative 5. P 6. polysaccharide 7. C3b 8. membrane attack complex (MAC) 9. lysis 10. opsonization 11. C3a 12. C-reactive protein 13. classical 14. opsonization

9. Interferon is synthesized in response to viral infection of a cell. The cell produces and releases interferon proteins, which diffuse to nearby cells and attach to their surface receptors. This binding event prevents viruses from multiplying within those cells.

10. Check 1, 3, 4.

PART 2: ADAPTIVE DEFENSES

Antigens

1. 1. immune system 2. immunogenicity 3. reactivity 4. proteins 5. haptens 6. size 7. antigenic determinants 8. simple 9. rejected
2. 1. MHC proteins 2. major histocompatibility complex 3. immune system

Cells of the Adaptive Immune System: An Overview

1. antigen-specific; systemic; has memory
2. The ability to recognize foreign substances in the body by binding to them.
3. 1. The appearance of antigen-specific receptors on the membrane of the lymphocyte. 2. fetal stage 3. its genes 4. binding to "its" antigen 5. self

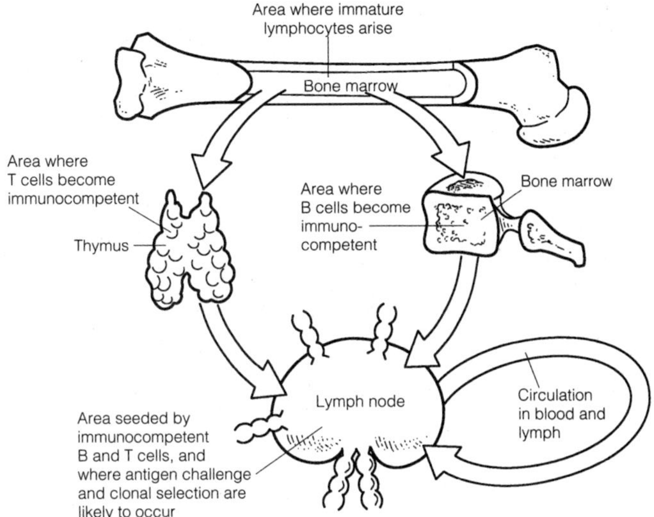

Figure 20.2

4. 1. A 2. F 3. D 4. B 5. J 6. I 7. C 8. G 9. H
5. T cell: 1, 3, 4, 5, 7, 8, 9 B cell: 1, 2, 4, 6, 7, 8

Humoral Immune Response

1. 1. The V part 2. The C part

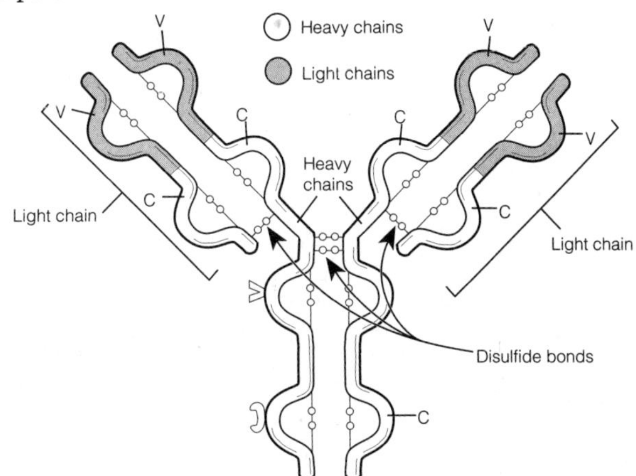

Figure 20.3

2. 1. B, E 2. D 3. E 4. D, E 5. D 6. C 7. A 8. E
3. 1. A 2. P 3. P 4. P 5. A 6. A
4. 1. P 2. P 3. S 4. P 5. S
5. 1. antigen 2. complement fixation and activation 3. neutralization 4. agglutination 5. M 6. precipitation 7. phagocytes

Cell-Mediated Immune Response

1. 1. B 2. B 3. D 4. A 5. C 6. B 7. A

2. 1. G 2. C 3. B 4. J 5. F 6. E 7. D 8. E 9. A 10. I

3. 1. Cytokines 2. Liver 3. T cell activation 4. Xenograft 5. Natural killer cells 6. Na^+-induced polymerization 7. Class I MHC proteins 8. B cell activation

4.

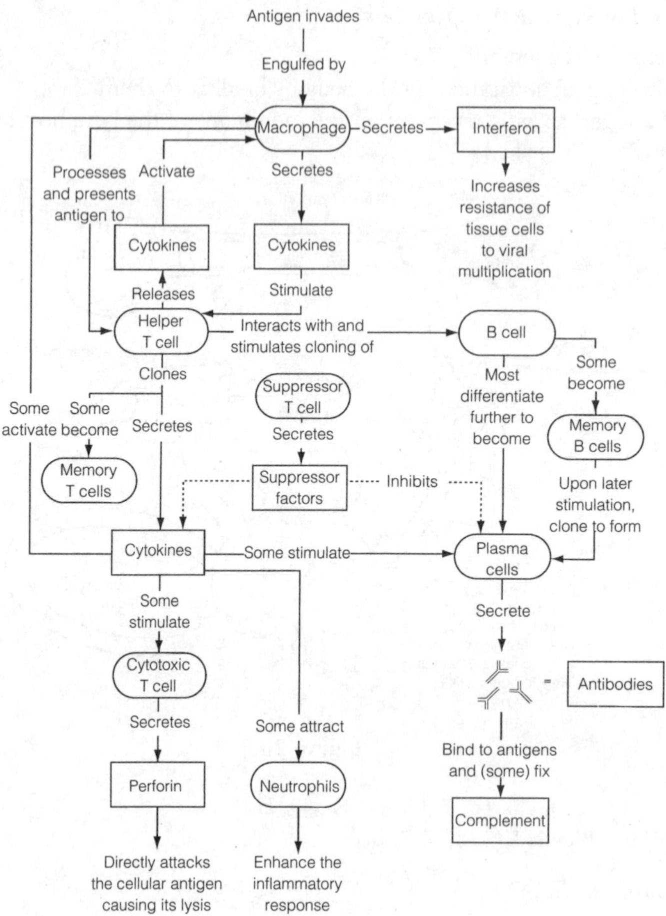

Figure 20.4

5. 1. cell-surface receptors 2. chemicals 3. anergy 4. destruction

6. 1. allografts 2. an unrelated person 3. cytotoxic T cells and macrophages 4. To prevent rejection, the recipient's immune system most be suppressed. 5. The patient is unprotected from foreign antigens, and bacterial or viral infection is a common cause of death.

Homeostatic Imbalances of Immunity

1. 1. C 2. A 3. A 4. C 5. B 6. C 7. B 8. C 9. B

2. 1. A, B, C, E 2. A, B 3. C 4. E 5. D 6. A 7. A, B 8. A, B 9. D 10. E 11. A, B 12. D 13. D 14. C, E

The Incredible Journey

1. 1. protein 2. lymph node 3. B cells 4. plasma cell 5. antibodies 6. macrophage 7. antigens 8. antigen presenters 9. T 10. clone 11. immunological memory

CHALLENGING YOURSELF

At the Clinic

1. Active artificial immunity.
2. Mother: Rh negative; father: Rh positive; IgG antibodies.
3. Valve replacement surgery with pig valves.
4. They are encountering different viral strains and their immune systems must recognize new antigenic determinants.
5. Anaphylactic shock; epinephrine (adrenaline).
6. Contact dermatitis (delayed hypersensitivity), probably caused by a reaction to the chemicals in the detergent used to launder the diapers.
7. Thrombocytopenia; autoimmune disorder.
8. Hashimoto's thyroiditis (an autoimmune response to the formerly "hidden" antigens in the thyroid hormone colloid).
9. Digestive complaints often indicate a food allergy.
10. Sterility associated with mumps occurs only after production of sperm antigens begins, after the onset of puberty.
11. The plasma cells are the antibody-producing "machinery," so David's antibody count is increasing.

Stop and Think

1. The acidity of the vaginal tract inhibits bacterial growth. Bacterial establishment would trigger an inflammatory response, resulting in vaginitis.
2. No; salmonella and other pathogens such as intestinal viruses survive passage through the stomach. Parasitic worms' eggs or cysts provide protection.
3. Swelling of the nasal and pharyngeal mucosa or glands inhibits drainage from the sinuses and middle ear cavity. Without a drainage route, pathogens in the surface secretions accumulate instead of being flushed out.
4. Chewing anything stimulates saliva secretion. Saliva washes over the gums and teeth, helping to flush out pathogens; lysozyme in saliva can kill bacteria.
5. Histamine: hist, tissue; amine, amino acid; diapedesis: dia, through; ped, foot; interleukin: inter, between; leuk, white (blood cells).
6. Olfactory receptors.
7. Lipid-soluble; it is absorbed through the skin, which is characteristic of lipid-soluble but not water-soluble molecules.
8. The constant recycling of lymphocytes through blood, lymph nodes, and lymph. Its role is to provide for the broadest and quickest sampling of the body for antigen recognition and interception.

COVERING ALL YOUR BASES

Multiple Choice

1. A, B, C **2.** B, D **3.** A, B, C, D **4.** D **5.** A **6.** C **7.** B **8.** A, C **9.** A, B, C, D **10.** A, C **11.** B, C, D **12.** C **13.** B **14.** A, B, C, D **15.** B, D **16.** B, C, D **17.** A, C **18.** A, B, C, D **19.** B **20.** C, D **21.** B, C, D **22.** A, C, D **23.** D **24.** A, B, C, D **25.** A **26.** A, B **27.** A, B, C, D **28.** B, C **29.** A **30.** A, B, D

Word Dissection

	Word root	Translation	Example
1.	hapt	fastened	hapten
2.	humor	fluid	humoral immunity
3.	macro	big, large	macrophage
4.	opso	delicacy	opsonization
5.	penta	five	pentamer
6.	phylax	preserve, guard	anaphylaxis
7.	phago	eat	phagocyte
8.	pyro	fire	pyrogen
9.	vacc	cow*	vaccine

* The first vaccine was produced using the cowpox virus.

Chapter 21 The Respiratory System

BUILDING THE FRAMEWORK

Functional Anatomy of the Respiratory System

1. pulmonary ventilation, external respiration, transport of respiratory gases, and internal respiration
2. 1. respiratory bronchioles, alveolar ducts, and alveoli 2. gas exchange 3. nasal cavity, pharynx, larynx, trachea, bronchi, and all of their branches except those of the respiratory zone
3. Nose: nostrils → nasal cavity → posterior nasal aperture
   ↓
   Pharynx: nasopharynx → adenoids → oropharynx → laryngopharynx
   ↓
   Larynx: epiglottis → vocal folds
   ↓
   Trachea: → carina → primary bronchi
4.

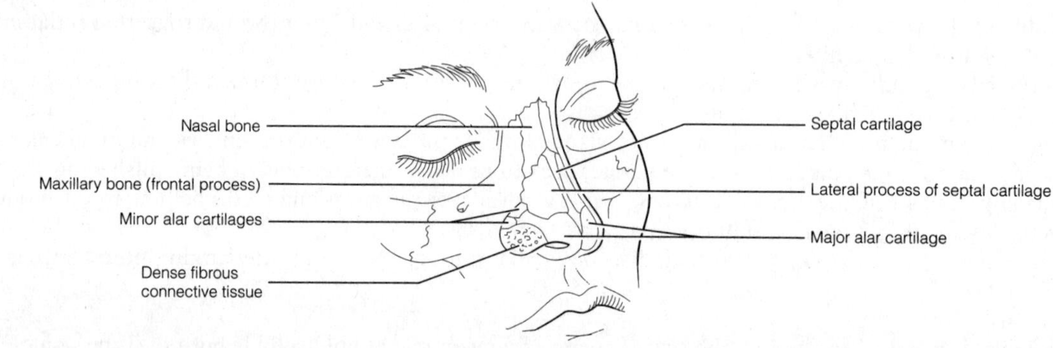

Figure 21.1

5. Note that the frontal and sphenoidal sinuses should be colored with the same color. Likewise, the subdivisions of the pharynx (the nasopharynx, oropharynx, and laryngopharynx) should be identified visually with a single color.

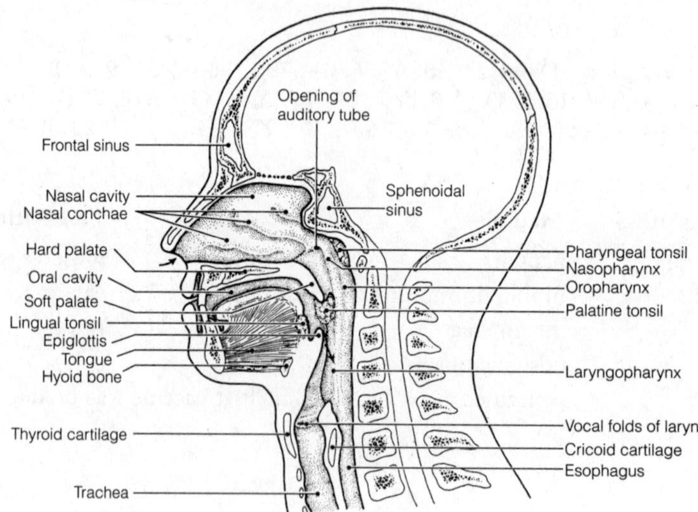

Figure 21.2

6. 1. Nostrils or nares 2. nasal septum 3.–5. (in any order): warm; moisten; trap debris in 6. paranasal sinuses 7. speech 8. pharynx 9. larynx 10. tonsils

7. 1. Mandibular 2. Alveolus 3. Larynx 4. Peritonitis 5. Nasal septum 6. Choanae 7. Tracheal cartilage 8. Nasopharynx

8. 1. Provides a patent airway; serves as a switching mechanism to route food into the posterior esophagus; voice production location (contains vocal cords). 2. arytenoid cartilages 3. elastic 4. hyaline 5. The epiglottis has to be flexible to be able to flap over the glottis during swallowing. The more rigid hyaline cartilages support the walls of the larynx. 6. Adam's apple

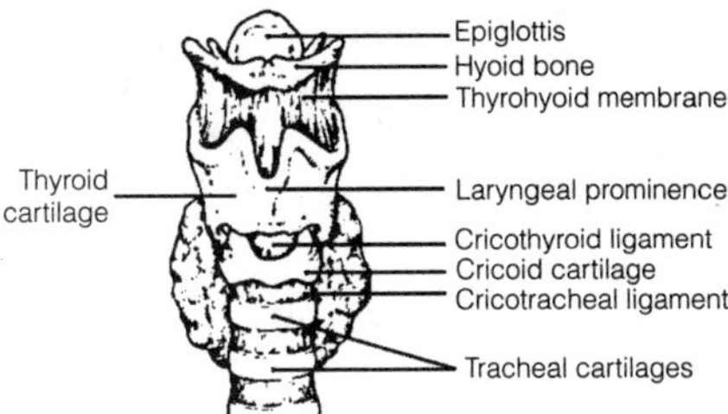

Figure 21.3

9. 1.B 2. G 3. I 4. E 5. D 6. K 7. A 8. H 9. L 10. F 11. C 12. O 13. N 14. K

10.

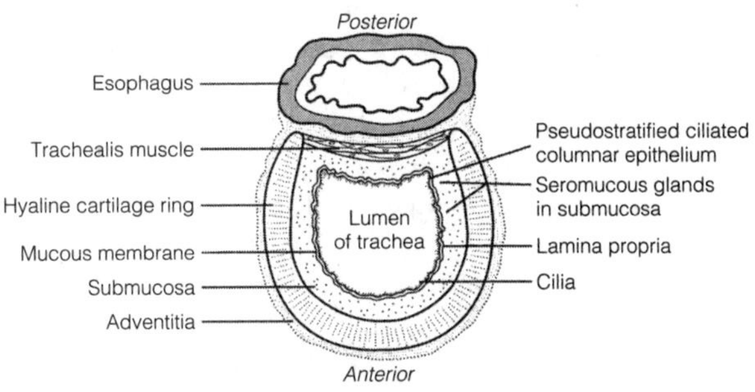

Figure 21.4

1. The C-shaped cartilage rings prevent the trachea from collapsing during the pressure changes occurring during breathing. 2. Allows the esophagus to expand anteriorly when we swallow food or fluids. 3. It constricts the tracheal lumen, causing air to be expelled with more force—coughing, yelling.

11. 1. vocal folds (true vocal cords) 2. speak 3. glottis 4. intrinsic laryngeal 5. arytenoid 6. higher 7. wide 8. louder 9. laryngitis

12. 1.B 2. B 3. A 4. B 5. E 6. C

13. 1.C 2. D 3. E 4. A 5. B

14.

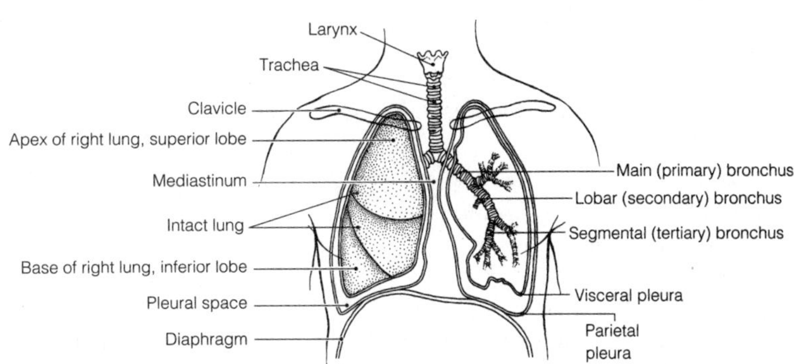

Figure 21.5

15.

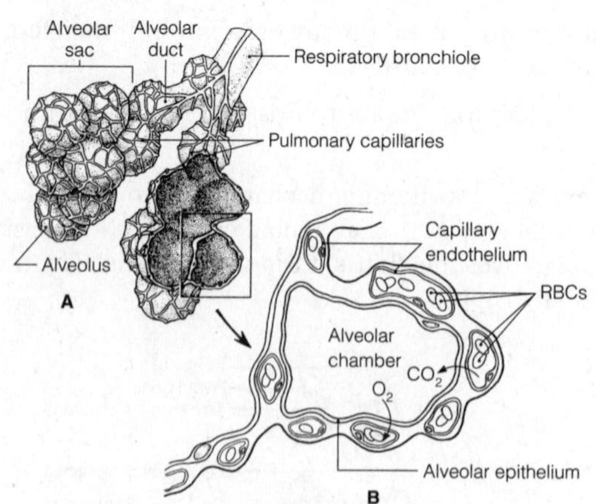

Figure 21.6

16. 1. elastic connective 2. type I 3. gas exchange 4. type II 5. surfactant 6. decrease the surface tension 7. antimicrobial proteins 8. bronchial

Mechanics of Breathing

1. 1. C 2. A 3. B 4. B 5. C 6. B 7. B

2. When the diaphragm is contracted, the internal volume of the thorax increases, the internal pressure in the thorax decreases, the size of the lungs increases, and the direction of air flow is into the lungs. When the diaphragm is relaxed, the internal volume of the thorax decreases, the internal pressure in the thorax increases, the size of the lungs decreases, and the direction of air flow is out of the lungs.

3. 1. D 2. A 3. C 4. B 5. D 6. C 7. B 8. A 9. A

4. 1. scalenes, sternocleidomastoids, pectorals 2. obliques and transversus muscles 3. internal intercostals and latissimus dorsi

5.

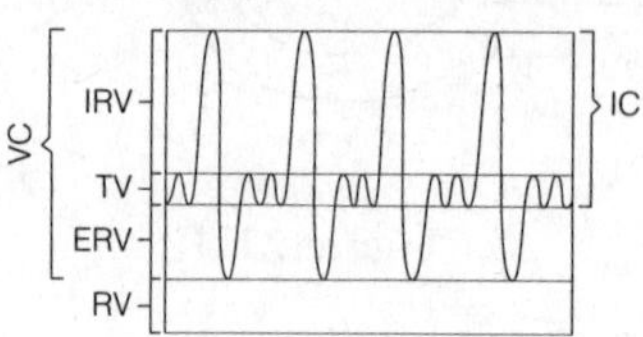

Figure 21.7

6. Check 1, 3, 5.

7. 1. E 2. A 3. G 4. D 5. B 6. F

8. 1. spirometer 2. obstructive pulmonary disease 3. restrictive disorders 4. FEV or forced expiratory volume 5. FVC or forced vital capacity

9. 1. hiccup 2. cough 3. sneeze

Gas Exchanges between the Blood, Lungs, and Tissues

1. 1. B 2. A 3. C

2. Hyperbaric oxygen; Henry's law.

3. Mechanism: generation of free radicals; consequences: CNS disturbances, coma, death.

4. 1. C 2. B 3. E 4. A 5. D

5. 1. 4 2. 5 3. 3 4. 2 5. 1 6. 2 7. 1 8. 3 9. 4 10. 5 11. 1 12. 2 13. 2 14. 1

6. 1. F 2. G 3. H 4. B 5. E 6. J 7. D 8. C 9. I

7. 1. low 2. constrict 3. high 4. dilate

Transport of Respiratory Gases by Blood

1. 1. hemoglobin 2. plasma 3. bicarbonate ion 4. carbonic anhydrase 5. chloride ions 6. chloride shift 7. carbamino Hb 8. Haldane effect 9. more 10. hemoglobin 11. Bohr effect 12. oxygen

2. high Po_2 of blood; low temperature; alkalosis; low levels of 2,3-BPG

3. 1. Hb can still be nearly completely saturated at lower atmospheric Po_2 or in those with respiratory disease. 2. Much more oxygen can be unloaded to the tissues without requiring cardiovascular or respiratory system adjustments to meet increased tissue demands.

4. 1. oxygen-hemoglobin dissociation curve 2. lower pH; higher temperature, increase in BPG

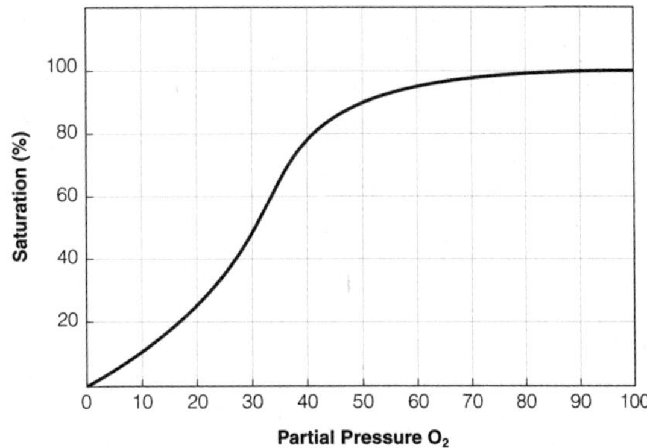

Figure 21.8

5. 1. HbF saturation is greater at a lower Po_2. 2. Po_2 levels are always lower in umbilical (and fetal) blood, and HbF has a greater affinity for O_2 than does HbA.

Control of Respiration

1. 1. A 2. C, D 3. E 4. B 5. D 6. C 7. C 8. A
2. 1. ↑ pH 2. Hyperventilation 3. ↑ Oxygen 4. ↓ CO_2 in blood 5. N toxicity 6. ↑ Pco_2

Respiratory Adjustments

1. 1. Hyperpnea increases depth of ventilation but not necessarily the rate of ventilation; hyperventilation is deep, and typically rapid, ventilation. Hyperpnea does not lead to significant changes in blood levels of O_2 and CO_2, whereas hyperventilation may result in hypocapnia. 2. Psychological stimuli, cortical motor activation, and proprioceptors.
2. 1. Minute ventilation increases. 2. Hemoglobin's affinity for oxygen declines so that more oxygen is unloaded to the tissues. 3. Enhanced erythropoiesis.

Homeostatic Imbalances of the Respiratory System

1. 1. A 2. F 3. D 4. G 5. E 6. C 7. B 8. C, E 9. H 10. E 11. J 12. I

The Incredible Journey

1. 1. nasal conchae 2. pharyngeal tonsil 3. nasopharynx 4. mucus 5. vocal fold 6. larynx 7. digestive 8. epiglottis 9. trachea 10. cilia 11. oral cavity 12. main bronchi 13. left 14. bronchiole 15. alveolus 16. red blood cells 17. red 18. oxygen 19. carbon dioxide 20. cough

CHALLENGING YOURSELF

At the Clinic

1. Atelectasis; the lungs are contained in separate pleural sacs, so only the left lung will collapse.
2. The mucus secreted in the conducting zone will be abnormally thick and difficult to clear. Consequently, the respiratory passageways tend to become blocked and infection is more likely.
3. The lower oxygen pressure of high altitudes prompts renal secretion of erythropoietin, leading to accelerated RBC production. Len will notice that he will begin to hyperventilate. His minute ventilation will increase by about 2–3 L/min. His arterial Pco_2 will be lower than the normal 40 mm Hg, and his hemoglobin saturation will be only about 67%.
4. Chest surgery causes painful breathing, and many patients try not to cough because the pain can be intense. However, coughing is necessary to clear mucus; if allowed to accumulate, mucus can cause blockage and increase the risk of infection.
5. Inflammation in the alveoli causes fluid to accumulate in the air spaces, which increases the apparent thickness of the respiratory membrane and reduces the lungs' ability to oxygenate the blood.
6. Pleurisy.
7. Stagnant hypoxia.
8. Michael most likely is suffering from carbon monoxide poisoning.
9. Sudden infant death syndrome.
10. Small cell (oat cell) carcinoma of the lungs.
11. Emphysema; although ventilation is difficult, oxygenation is sufficient, so the skin is a pink (normal) color early on. Cyanosis does not occur until late in the disease progress.
12. Chronic bronchitis; smoking inhibits ciliary action.
13. Kyphosis is an exaggerated thoracic curvature that reduces the ability to inflate the lungs fully.

14. Failure of the epiglottis and soft palate to close the respiratory channels completely during swallowing. The former will place the patient at risk for aspiration pneumonia.
15. The baby most likely swallowed the safety pin (babies put everything in their mouths). It is probably in the R. primary bronchus, which is larger in diameter and runs more vertically.
16. Pharyngeal tonsils.

Stop and Think

1. When air is inhaled through the nose, it is warmed by the nasal mucosa. Exhaling through the mouth expels the warmed air, resulting in heat loss.
2. Deep-sea divers risk nitrogen narcosis, the bends, and gas emboli. Effective treatment requires a hyperbaric chamber.
3. Expired air has a higher oxygen partial pressure than does alveolar air. Expired air is a mixture of (fresh) air from the dead air space and oxygen-depleted air from the alveoli.
4. a. Normally, during swallowing, the soft palate reflects superiorly to seal the nasopharynx and prevent food or drink from entering the nasal cavity. During giggling, however, this sealing mechanism sometimes fails to operate (because giggling demands that air be forced out of the nostrils), and swallowed fluids may enter the nasal cavity, then exit through the nostrils.
   b. Even though standing on his head, the boy made certain that he swallowed carefully, so that his soft palate correctly sealed the entrance to his nasal cavity. Then, his swallowing muscles directed the milk, against gravity, through his esophagus to the stomach.
5. Because each bronchopulmonary segment is isolated by connective tissue septa and has its own vascular supply.
6. Those in the upper passageways move mucus toward the *esophagus* to be swallowed; those in the lower passageways do *likewise* but in the opposite direction, which prevents mucus from "pooling" in the lungs.
7. The elastin is responsible for the natural elasticity of the lungs, which allows them to recoil passively during expiration.
8. It could be if he was hitting the right spot (just inferior to the rib cage) with a vigorous upward thrust, which would rapidly propel the air out of his lungs upward through the respiratory passageways. (Of course, he might also break a few ribs or rupture his spleen or liver in the process.)
9. Shallow breaths flush air out of dead space areas where the air does not participate in gas exchange. A deeper breath is more likely to include air containing alcohol that is vaporizing from the blood into the alveoli.

COVERING ALL YOUR BASES

Multiple Choice

1. B, D **2.** C **3.** B, C, D **4.** A **5.** B **6.** B **7.** D **8.** D **9.** D **10.** B **11.** B **12.** A **13.** A, C, D **14.** C **15.** B, C, D **16.** B, C **17.** B **18.** B, D **19.** C **20.** B, C, D **21.** A, B, C **22.** D **23.** C **24.** C **25.** A, D **26.** B, C **27.** C **28.** B, C **29.** A, C **30.** A, B, D **31.** A, D **32.** A **33.** C **34.** A, B **35.** A

Word Dissection

	Word root	Translation	Example		Word root	Translation	Example
1.	alveol	cavity	alveolus	12.	pleur	side; rib	pleura
2.	bronch	windpipe	bronchus	13.	pne	breath	eupnea, apnea
3.	capn	smoke	hypercapnia	14.	pneum	air, lungs	pneumothorax
4.	carin	keel	carina	15.	pulmo	lung	pulmonary
5.	choan	funnel	choanae	16.	respir	breathe	respiration
6.	crico	ring	cricoid cartilage	17.	spire	breathe	inspiration
7.	ectasis	dilation	atelectasis	18.	trach	rough	trachea
8.	emphys	inflate	emphysema	19.	ventus	wind	ventilation
9.	flat	blow, blown	inflation	20.	vestibul	porch	vestibule
10.	nari	nostril	nares	21.	vibr	shake, vibrate	vibrissae
11.	nas	nose	nasal				

Chapter 22 The Digestive System

BUILDING THE FRAMEWORK

PART 1: OVERVIEW OF THE DIGESTIVE SYSTEM

1. 1. pharynx 2. esophagus 3. stomach 4. duodenum 5. jejunum 6. ileum 7. ascending colon 8. transverse colon 9. descending colon 10. sigmoid colon 11. rectum 12. anal canal

2. Mouth: (I) teeth, tongue; (O) salivary glands and ducts.
Duodenum: (O) liver, gallbladder, bile ducts, pancreas.

3. The sublingual, submandibular, and parotid glands are all salivary glands.

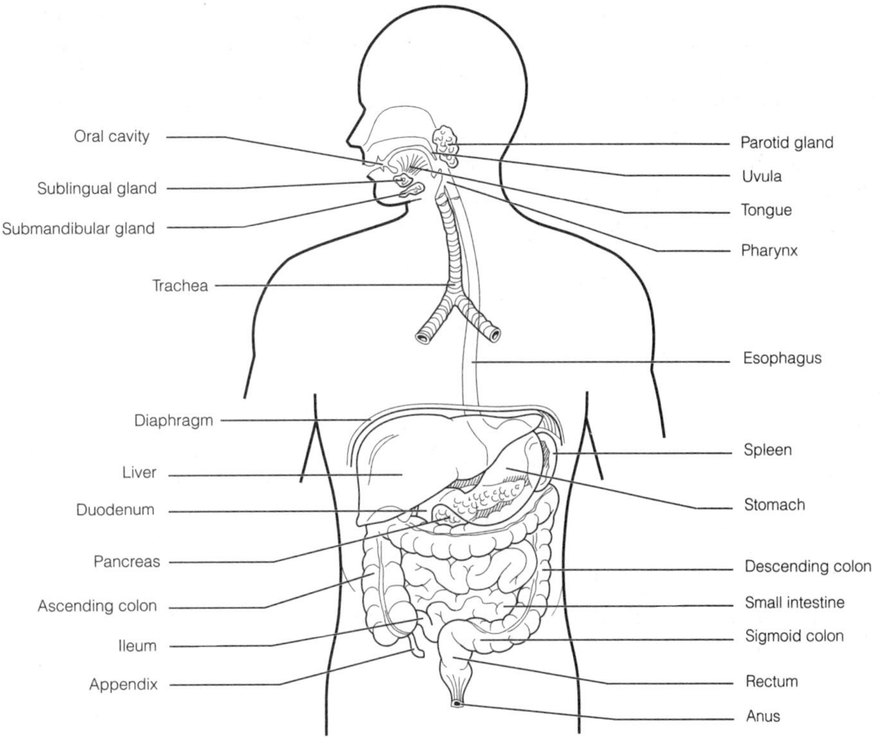

Figure 22.1

4. 1. D 2. G, H 3. E, F, H 4. B 5. A 6. C

5. 1. stretch, osmolarity/pH, and presence of substrates and end products of digestion 2. submucosal plexus and myenteric plexus 3. long reflexes involve CNS centers and extrinsic autonomic nerves and can involve long regions of the tract; short reflexes involve only local (enteric) plexuses, and are very limited in their range

6. 1. peritoneum 2. mesentery 3. retroperitoneal organs 4. intraperitoneal or peritoneal organs

7. 1. mucosa 2. muscularis externa 3. submucosa 4. serosa

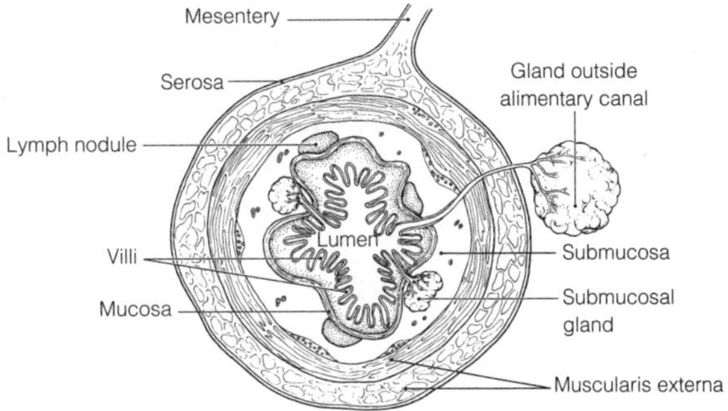

Figure 22.2

PART 2: FUNCTIONAL ANATOMY OF THE DIGESTIVE SYSTEM

1.

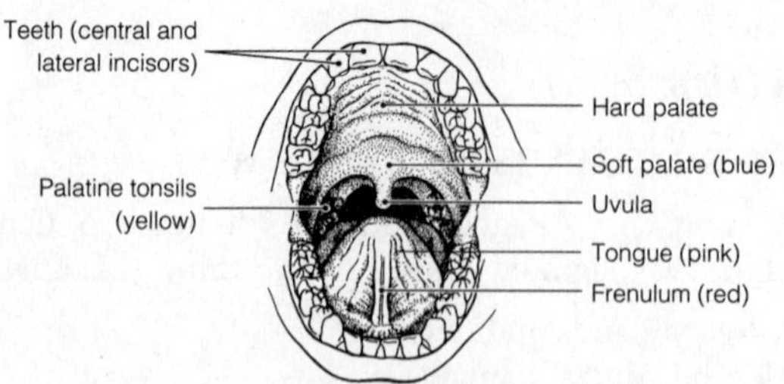

Figure 22.3

2. 1. Cleans the mouth, dissolves chemicals for tasting, moistens and compacts food, contains starch-digesting enzymes that begin the chemical digestion of starch. 2. Parotid glands, open next to second upper molar; submandibular glands, open at base of lingual frenulum; sublingual glands, open along floor of mouth 3. Serous cells secrete watery, enzyme-containing fluid: mucous cells secrete mucus 4. Salivary amylase and lingual lipase 5. Lysozyme

3. Circle: 1. parasympathetic 2. sour 3. conditioned

4. 1. deciduous 2. 6 months 3. 6 years 4. permanent 5. 32 6. 20 7. incisors 8. canine (eyetooth) 9. premolars (bicuspids) 10. molars 11. wisdom

5. 1. A 2. C 3. D 4. B 5. E 6. C

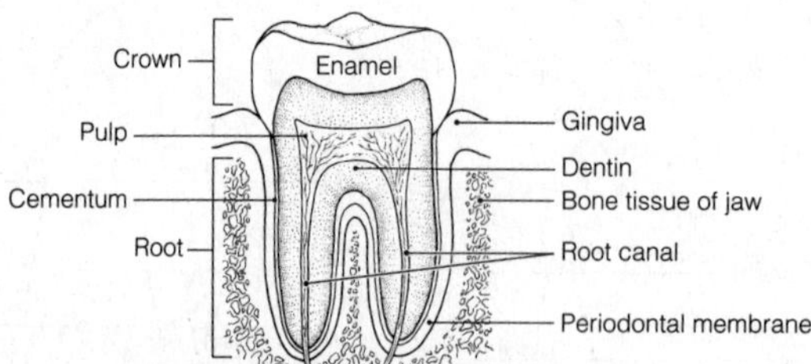

Figure 22.4

6. 1. K 2. D 3. F 4. C, J 5. A 6. H 7. E

7. On part B, the HCl-secreting parietal cells should be colored red, the mucous neck cells yellow, and the cells identified as chief cells (which produce protein-digesting enzymes) blue.

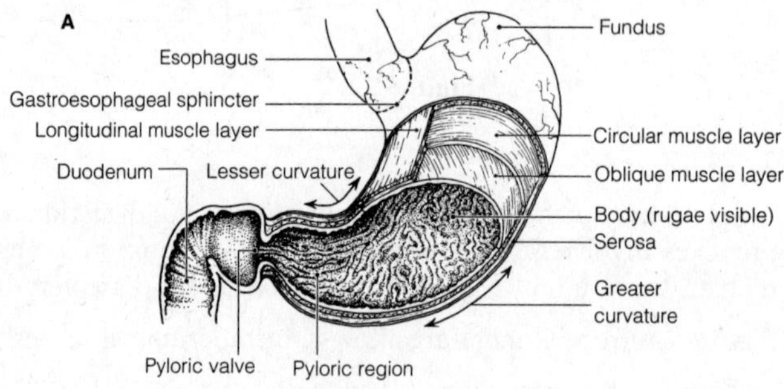

Figure 22.5

8. 1. oropharynx 2. laryngopharynx 3. stratified squamous 4. pharyngeal constrictor muscles 5. esophagus 6. gastroesophageal sphincter 7. deglutition 8. buccal 9. pharyngeal-esophageal 10. tongue 11. uvula 12. larynx 13. epiglottis 14. peristalsis 15. gastroesophageal

9. 1. A 2. D, E 3. D 4. C 5. B

10.

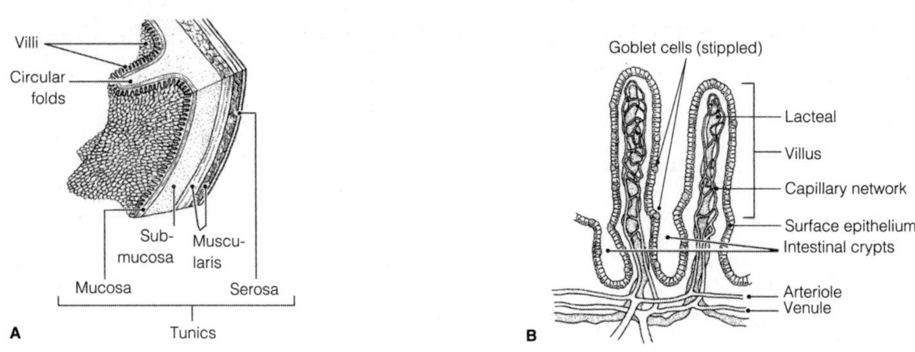

Figure 22.6

11. 1. Secretin 2. Gastric emptying 3. Intestinal phase 4. Gastric juice 5. Esophagus 6. Salts

12.

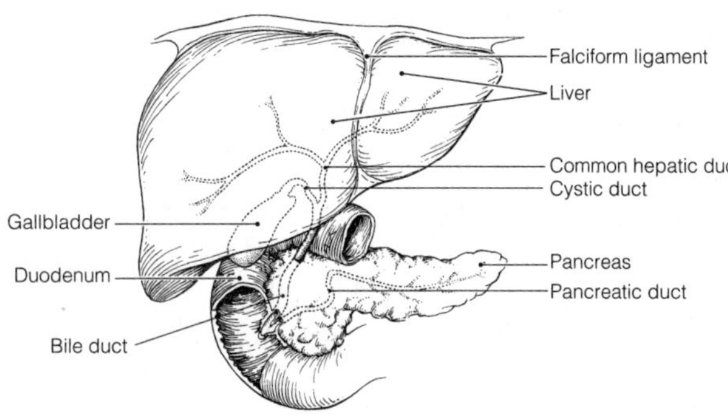

Figure 22.7

13. Branches of hepatic artery, hepatic vein, and bile duct. Located at the lobule periphery in the surrounding connective tissue.

14. 1. A, C 2. B 3. A 4. C 5. A, C, D

15. 1. B 2. H 3. F 4. G 5. C 6. A

16.

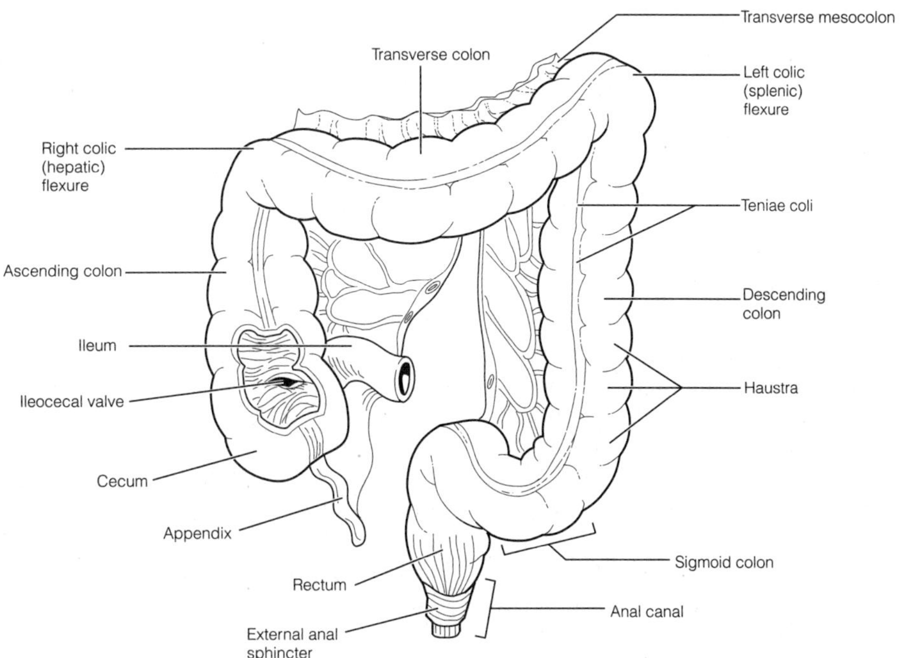

Figure 22.8

17. 1. peristalsis 2. segmentation 3. segmentation 4. mass movements 5. rectum 6. defecation 7. emetic 8. vomiting

18. 1. E 2. F 3. B 4. C 5. A

19. 1. G 2. E 3. F 4. H 5. D 6. A

PART 3: PHYSIOLOGY OF CHEMICAL DIGESTION AND ABSORPTION

1. 1. B, D, F, G, H, I, J, L, M 2. F, G, H 3. H 4. I, J, M 5. L 6. D 7. C, K 8. A 9. E
2. 1. P 2. A 3. A 4. P 5. A. Circle 4, Fatty acids.
3. 1. Q 2. N 3. P 4. C 5. G 6. B 7. O 8. I 9. M 10. C 11. R 12.–14. A, F, K
4. 1. Trypsin 2. Chylomicrons 3. Lost during hemorrhage 4. Nuclease 5. Removal of water 6. Pepsin 7. Vitamin B_{12}
5.

Enzyme	Substrate	Product	Secreted by	Active in
Salivary amylase	Starch	Maltose and oligosaccharides	Salivary glands	Mouth
Pepsin	Protein	Polypeptides	Gastric glands	Stomach
Trypsin	Protein, peptides	Dipeptides, small peptides	Pancreas	Small intestine
Dipeptidase	Dipeptides	Amino acids	Not secreted; brush border enzyme	Small intestine
Pancreatic amylase	Starch, oligosaccharides	Maltose	Pancreas; salivary glands (small amounts)	Small intestine
Lipase	Triglycerides	Monoglycerides, fatty acids, glycerol	Pancreas	Small intestine Stomach
Maltase	Maltose	Glucose	Not secreted; brush border enzyme	Small intestine
Carboxy-peptidase	Polypeptide	Amino acids	Not secreted; brush border enzyme	Small intestine
Nuclease	Nucleic acids	Pentose sugars, nitrogenous bases, phosphate ions	Pancreas (and brush border enzymes)	Small intestine

The Incredible Journey

1. 1. mucosa 2. vestibule 3. tongue 4. salivary amylase 5. peristalsis 6. esophagus 7. larynx 8. epiglottis 9. stomach 10. mucus 11. pepsin 12. hydrochloric acid 13. pyloric 14. lipase 15. pancreas 16. villi 17. ileocecal

CHALLENGING YOURSELF

At the Clinic

1. Mumps.
2. Gastroesophageal sphincter.
3. Heartburn due to a hiatal hernia; esophagitis and esophageal ulcers.
4. Histamine is one of the chemical stimuli for HCl secretion; thus an antihistamine drug will inhibit HCl secretion; perforation, peritonitis, and massive hemorrhage. She was told not to take aspirin because it can cause stomach bleeding.
5. Leakage of HCl and pepsin from perforating gastric ulcer will literally erode and digest away any other tissues with which these chemicals come into contact.
6. Probably alcoholic cirrhosis; ascites (distended abdomen); jaundice (yellow skin).
7. Gramps probably told Duncan that germs "ate away" at the roots of his teeth so the teeth fell out. He concluded with the warning "So brush 3 times a day!"
8. Appendicitis; surgical removal; burst appendix with life-threatening peritonitis.
9. Lack of lactase (lactose intolerance); addition of lactase to milk.
10. Gluten enteropathy (adult celiac disease).
11. Appendicitis is caused by bacterial infection. If untreated, bacterial overgrowth may cause the appendix to rupture, resulting in fecal contamination of the peritoneal cavity.
12. The precipitated proteins may block the pancreatic ducts. Some of the proenzymes will be activated in the ducts over time, and these in turn will cause the pancreas to produce more digestive enzymes, causing digestion of the ducts.

Stop and Think

1. *Gap junctions* provide for a continuous contraction along the digestive tract (peristalsis) because each muscle cell stimulates the next. The *stress-relaxation reflex* is important when mass peristalsis delivers fecal material to the rectum at an inconvenient moment. After stretching, the rectal smooth muscle will relax to accommodate to the increased mass. Because smooth muscle is *nonstriated,* it can stretch considerably without losing its ability to contract.
2. Salivary amylase is denatured by the acidity of the stomach; because it is a protein, it is digested along with dietary proteins. In the duodenum, pepsin is inactivated by the relatively high pH there and will be digested along with other proteins in the intestinal lumen.
3. Because histamine increases blood flow and capillary permeability, it might enhance the blood's ability to pick up nutrients from the digestive epithelium.
4. Because virtually all calcium is absorbed in the duodenum, a massive load of calcium can saturate the calcium transport mechanism. Smaller, more frequent doses will decrease likelihood of saturation and result in better absorption of the entire daily dose.
5. Fats are *emulsified* by bile salts. Digestion of triglycerides typically involves hydrolysis of two of the three fatty acids by *pancreatic lipase,* resulting in monoglycerides and free fatty acids. Bile salts form *micelles* that transport fat end products to the intestinal epithelium. *Absorption* occurs passively through the phospholipid bilayer. Triglycerides are re-formed in the epithelial cells and packaged with other lipids into protein-coated *chylomicrons,* which enter the *lacteals* to circulate from the lymphatic system to the general circulation. Entrance to the liver is via the hepatic artery of the systemic arterial system.
6. Examination of the plasma would quickly reveal the presence of chylomicrons, which give the plasma a milky-white appearance.
7. The liver manufactures albumin, which is essential in maintaining the osmotic balance of the blood. An albumin deficiency reduces the blood's osmotic pressure, and fluid is retained in the tissue spaces.
8. (a) In the stomach, these cells are found at the junctions of the gastric pits with the gastric glands. In the small intestine, they are at the base of the intestinal glands (crypts). (b) Their common function is to replace the epithelial cells (exposed to the harsh conditions of the digestive tract) as they die and slough off.
9. (a) Serous (gland) cells produce a watery enzyme-rich secretion; serous membranes produce a lubricating fluid within the ventral body cavity. (b) Caries are cavities (decayed areas) in teeth; (bile) canaliculi are tiny canals in the liver that carry bile. (c) Anal canal is the last portion of the tubular alimentary canal; runs from the rectum to the anus (external opening). (d) Diverticulosis is a condition in which the (weakened) walls of the large intestine pouch out; diverticulitis is inflammation of these diverticuli. (e) Hepatic vein drains venous blood from the liver; hepatic portal vein brings nutrient-rich venous blood to the liver from the digestive viscera.
10. Rough ER and Golgi apparatus: Liver makes tremendous amounts of proteins for export. Smooth ER: Liver is an important site of fat metabolism and cholesterol synthesis and breakdown. Peroxisomes: Liver is an important detoxifying organ.
11. Removal of the pancreas because it provides enzymes that digest *all* foods. The stomach digests only proteins for the most part. If the gallbladder is removed, the cystic duct will expand to store bile.

COVERING ALL YOUR BASES

Multiple Choice

1. A, C, D **2.** B **3.** C **4.** C **5.** A **6.** B, C, D **7.** A, B, C, D **8.** D **9.** A, B, C, D **10.** B **11.** C **12.** B, C **13.** C **14.** A, D **15.** D **16.** A, B, C **17.** A, D **18.** A, C, D **19.** A, C **20.** D **21.** A, B **22.** B, C, D **23.** B **24.** B, C, D **25.** A, B, C **26.** A, C, D **27.** A, C **28.** A, C **29.** C, D **30.** A, B, D **31.** D **32.** A, D **33.** A **34.** B **35.** A, C, D **36.** A, C, D **37.** B, C **38.** A, B, C **39.** C **40.** A, D **41.** B **42.** A **43.** C **44.** D **45.** D **46.** B **47.** A

Word Dissection

	Word root	Translation	Example
1.	aliment	nourish	alimentary canal
2.	cec	blind	cecum
3.	chole	bile	cholecystokinin
4.	chyme	juice	chyme
5.	decid	falling off	deciduous teeth
6.	duoden	twelve each	duodenum
7.	enter	intestine	mesentery
8.	epiplo	thin membrane	epiploic appendages
9.	eso	within, inward	esophagus
10.	falci	sickle	falciform ligament
11.	fec	dregs	feces
12.	fren	bridle	lingual frenulum
13.	gaster	stomach	gastric juice
14.	gest	carry	digestion
15.	glut	swallow	deglutition
16.	haustr	draw up	haustra
17.	hiat	gap	hiatal hernia
18.	ile	intestine	ileum
19.	jejun	hungry	jejunum
20.	micell	a little crumb	micelle
21.	oligo	few	oligosaccharides
22.	oment	fat skin	greater omentum
23.	otid	ear	parotid gland
24.	pep	digest	pepsin
25.	plic	fold	plicae circulares
26.	proct	anus, rectum	proctodeum
27.	pylor	gatekeeper	pylorus
28.	ruga	wrinkle	rugae
29.	sorb	suck in	absorb
30.	splanch	the viscera	splanchnic circ.
31.	stalsis	constriction	peristalsis
32.	teni	ribbon	teniae coli

Chapter 23 Nutrition, Metabolism, and Body Temperature Regulation

BUILDING THE FRAMEWORK

Diet and Nutrition

1. 1. carbohydrates, lipids, proteins, vitamins, minerals, water 2. Molecules the body cannot make and must ingest. 3. 1 kcal = the heat necessary to raise the temperature of 1 kg of water by 1° centigrade. 4. glucose 5. brain cells and red blood cells 6. Carbohydrate foods that provide energy sources but no other nutrients. 7. egg yolk 8. linoleic acid (ingested as lecithin) and (possibly) linolenic acid 9. A protein containing all the different kinds of amino acids required by the body. 10. The essential amino acids must be available for protein synthesis to occur, and the body cannot make them rapidly enough to meet its needs. If not, amino acids will be "burned" for energy. 11. The amount of nitrogen ingested in proteins equals the amount of nitrogen excreted in urine or feces (rate of protein synthesis equals rate of protein degradation and loss). 12. structural and functional uses 13. Most vitamins function as coenzymes. 14. liver 15. calcium and phosphorus 16. Fats (lipids)

2. 1. I 2. E 3. D 4. B 5. H 6. G 7.–10. A, G, H, I 11. M 12. L 13. F 14. A 15. C 16. K 17.–19. A, F, H 20. H

3. 1. J 2. C 3. 1 4. F 5. G 6. H 7. B 8. D 9. E

4. 1. Trisaccharides 2. Nuts 3. Fats 4. Vitamin C 5. K^+ 6. Amino acids 7. Must be ingested 8. Zinc 9. About 16 oz 10. Catabolic hormones

5. 1. A 2. B 3. C 4. B 5. B 6. C 7. C 8. B

Overview of Metabolic Reactions

1. 1. metabolism 2. catabolism 3. anabolism 4. cellular respiration 5. ATP 6. GI tract 7. mitochondria 8. oxygen 9. water 10. carbon dioxide 11. hydrogen atoms 12. electrons 13. reduction 14. coenzymes 15. NAD^+ 16. FAD

Metabolism of Major Nutrients

1. 2. glycolysis 2. cytosol 3. not used 4. NAD^+ 5. two 6. sugar activation 7. sugar cleavage 8. sugar oxidation 9. pyruvic acid 10. lactic acid 11. skeletal muscle 12. brain 13. pH 14. red blood 15. carbon dioxide 16. NAD^+ 17. coenzyme A 18. oxaloacetic acid 19. citric acid 20. Krebs (or citric acid) 21. mitochondrion 22. keto 23. two 24. three 25. one 26. fatty acids 27. amino acids 28. electron transport 29. protons and electrons 30. water 31. ADP 32. oxidative phosphorylation 33. 32–36

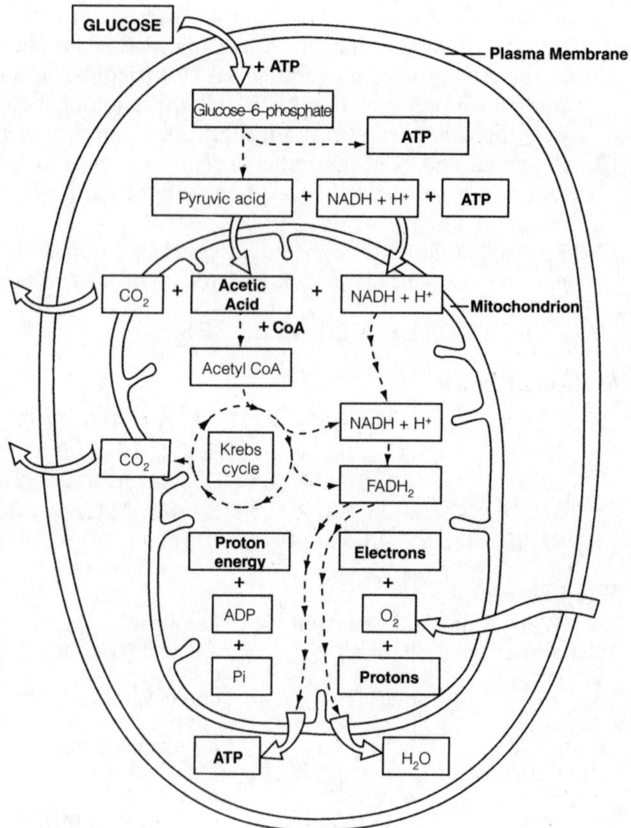

Figure 23.1

2. 1. hydrogen atoms 2. electrons 3. protons 4. metal 5. inner 6. oxygen 7. protons 8. mitochondrial matrix 9. intermembrane space 10. lower 11. proton motive force (or proton gradient) 12. ATP synthase 13. ADP + $P_i \rightarrow$ ATP

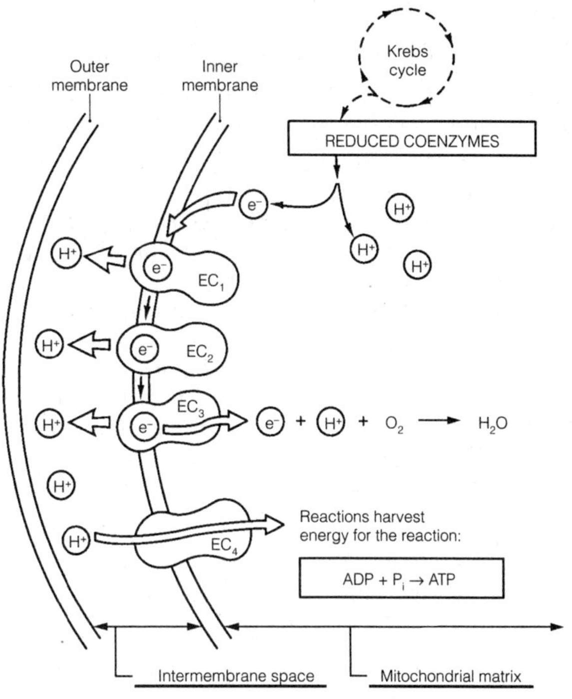

Figure 23.2

3. 1. C 2. D 3. B 4. A

4. 1. D 2. F 3. B 4. G 5. C 6. A 7. E 8. C 9. F 10. Ammonia is toxic to body cells; also blood pH may rise because it acts as a base. 11. Keto acids 12. There are no storage depots for amino acids in the body. Proteins continually degrade and newly ingested amino acids are required. If an essential amino acid is not consumed, the remaining ones cannot be used for protein synthesis and are oxidized for energy.

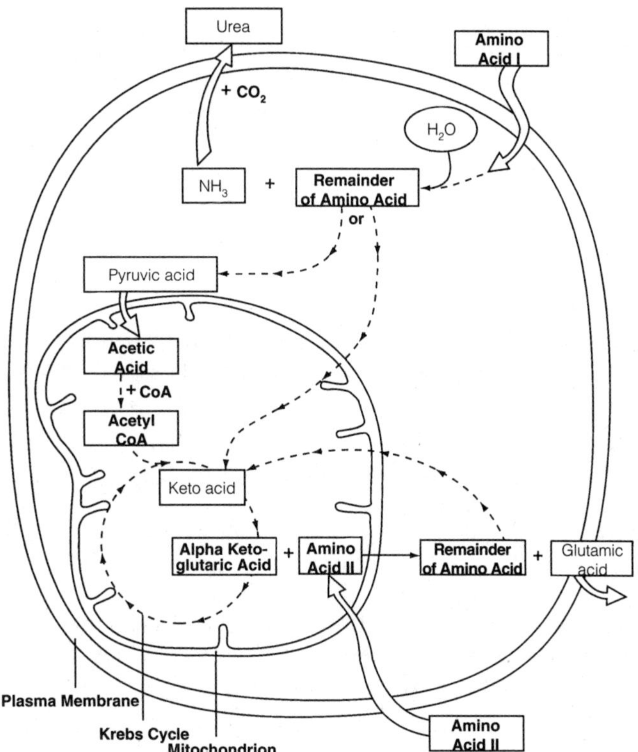

Figure 23.3

5. 1. Glucose 2. Hydrolyzed in mitochondria 3. Cytoplasm 4. ATP deficit 5. Low blood sugar level

Metabolic States of the Body

1. 1. F 2. C 3. E 4. D 5. A 6. B 7. G 8. B

2. 1. about 38% 2. ATP than ADP 3. into 4. glycogenesis 5. liver and skeletal muscle

3. 1. anabolic 2. T 3. glycogen 4. triglycerides 5. protein synthesis 6. T 7. rising blood glucose levels 8. facilitated diffusion 9. glycogenolysis and gluconeogenesis 10. other body cells 11. T

4. 1. 80–100 mg glucose per 100 ml blood 2. The brain prefers glucose as its energy source. 3. The liver, by glycogenolysis; skeletal muscle, by glycogenolysis; adipose tissue and liver, by lipolysis; tissue cells, by protein catabolism. 4. Skeletal muscle cells cannot dephosphorylate glucose, but they can produce pyruvic acid or lactic acid. The liver reconverts these to glucose, which is released to the blood. 5. Lipolysis produces fatty acids and glycerol. Only the glycerol can be converted directly to glucose by the liver. (Acetyl CoA, from fatty acid oxidation, is produced beyond the reversible steps of glycolysis and cannot be converted to glucose.) 6. The liver deaminates amino acids and converts the residues to glucose, which is released to the blood. 7. Fats 8. Ketone bodies from the oxidation of fats by the liver. 9. Epinephrine and glucagon 10. Alpha cells of pancreatic islets produce glucagon, which targets the liver and adipose tissue. 11. Falling glucose levels and rising amino acid levels. 12. High amino acid levels in blood stimulate release of insulin. Glucagon counteracts effects of insulin and prevents abrupt hypoglycemia. 13. Hypothalamic receptors sense declines in blood glucose levels. Stimulation of the adrenal medulla causes release of epinephrine, which produces the same effects as glucagon.

5.

Hormone	Blood glucose	Blood amino acids	Glycogenolysis	Lipogenesis	Protein synthesis
Insulin	↓	↓	↓	↑	↓
Glucagon	↑	XXX	↑	↓	XXX
Epinephrine	↑	XXX	↑	↓	XXX
Growth hormone	↑ (over the long term)	↓	↑	↓	↑
Thyroxine	↓	XXX	↑	↓	↑

Metabolic Role of the Liver

1. 1. albumin 2. clotting proteins 3. cholesterol 4. hyperglycemia 5. glycogen 6. hypoglycemia 7. glycogenolysis 8. gluconeogenesis 9. detoxification 10. phagocytic 11. lipoproteins 12. insoluble 13. lower 14. high 15. chylomicrons 16. triglycerides (fats) 17. cholesterol 18. membranes 19. steroid hormones 20. liver 21. bile salts 22. atherosclerosis 23. unsaturated 24. H (high) 25. A, D, and B_{12} 26. iron 27. red blood

Energy Balance

1. 1. TMR 2. ↓ Metabolic rate 3. Child 4. Fats 5. Fasting 6. Body's core 7. Vasoconstriction

2. 1. E 2. B ↓ 3. B ↓ 4. C ↑ 5. D ↓↑ 6. B ↑ 7. C ↓

3. 1. The intake of energy by the catabolism of food equals the total energy output of heat plus work plus energy storage. 2. heat 3. Body weight remains stable when energy intake equals energy outflow. Weight change accompanies energy inequality.

4. 1. D 2. B, K 3. A 4. F 5. J 6. C 7. H, I 8. G 9. E 10. J

5. 1. skeletal muscle 2. 38.2°C (100°F) 3. 10% 4. proteins 5. T 6. heat stroke

CHALLENGING YOURSELF

At the Clinic

1. Many vegetables contain incomplete proteins. Unless complete proteins are ingested the value of the dietary protein for anabolism is lost, because the amino acids will be oxidized for energy. Beans and grains.

2. "Empty calories" means simple carbohydrates with no other nutrients, such as candy, cookies, and soft drinks.

3. Glucagon; hypoglycemia.

4. Dieting triggers catabolism of proteins as well as fat, resulting in weight loss due to loss of muscle and adipose tissue. Subsequent weight gain without exercise usually means regaining only fat, not muscle. Continual cycling of muscle/adipose loss followed by adipose gain can significantly alter body fat composition and "rev up" metabolic systems that increase the efficiency of fat storage and retention.
5. Rickets; deficiency of vitamin D and calcium.
6. Ketosis; her self-starvation has resulted in deficiency of carbohydrate fuels (and oxaloacetic acid). This deficiency promotes conversion of acetyl CoA to ketone bodies such as acetone.
7. Heat exhaustion; they should drink a "sports drink" containing electrolytes or lemonade to replace lost fluids.
8. Bert has heat stroke. Heavy work in an environment that restricts heat loss results in a spiraling upward of body temperature and cessation of thermoregulation. Bert should be immersed in cool water immediately to bring his temperature down and avert brain damage.
9. Children have a greater requirement for fat than adults, particularly up to age 2 or 3 when myelination of the nervous system is still a major consideration.
10. Hypercholesterolemia; <200 mg/100 ml blood; atherosclerosis, strokes, and heart attacks.
11. Iron. She has hemorrhagic anemia compounded by iron loss.
12. Mr. Hodges has scurvy due to vitamin C deficiency. Recommend citrus fruits and plenty of tomatoes.
13. Over the long term, the body converts excess calories to fat regardless of the source of those calories.

Stop and Think

1. Osmotic effect would be retention of water in cells maintaining a pool of amino acids; pH would decrease.
2. A higher concentration of glucose-phosphate in a cell would attract water, and the cell would swell.
3. Oxygen is required as the final electron acceptor in the electron transport chain. Lack of oxygen brings electron transport to a halt and backs up the Krebs cycle as well, because NAD^+ and FAD cannot be recycled. ATP production grinds to a halt in cells that cannot rely on anaerobic respiration (such as brain cells), and cell death occurs.
4. Urea production increases. Protein synthesis will not proceed unless all necessary amino acids are available. Hence, any amino acids that are not utilized are deaminated and oxidized for energy or converted to fat and stored. Excess amino groups will be converted to urea.
5. Absorption of simple sugars is quite rapid compared to absorption of complex carbohydrates, which must be digested first. The result is a much more rapid rise in blood glucose level when simple sugars are ingested.
6. No; the ability of brain cells to take in glucose is not regulated by insulin.
7. Because omega-3 fatty acids reduce the stickiness of platelets, excessive bleeding might result from ingesting an excess of these lipids.
8. Males tend to have a higher ratio of muscle to fat than females. Because muscle tissue has a higher BMR than fat, a higher proportion of muscle requires a higher calculation factor.
9. Because PKU sufferers cannot manufacture melanin, pigment in the interior of the eye could be deficient, resulting in poor vision. (Albinos likewise have poor vision.)
10. Depending on the diuretic, different electrolytes have an increased rate of excretion. A common type of diuretic promotes potassium excretion, requiring potassium supplementation.
11. The excess of glucose can form abnormal cross-bridges between protein fibers in the vessel walls, resulting in hardening of the arteries.
12. Hypothyroidism would result in hypothermia, because thyroid hormones are thermogenic (heat generating).
13. When muscles are "at work" they are generating and using large amounts of ATP to power the sliding of their myofilaments. Because some heat is "lost" in every chemical reaction, a large amount of heat is generated at the same time.
14. Deficiency of fat-soluble vitamins (A, D, K), which are absorbed along with the fat molecules.
15. Graph C, because the sample with the highest concentration of succinic acid is decolorizing fastest, going from a high-intensity, blue dye to a nearly colorless dye. A is wrong because the graph shows the color intensity of all the samples increasing with time. B is incorrect because the color intensities of the tubes are different to begin with and no differences in dye content of the samples were mentioned.
16. Nitrogen via the amine group.

COVERING ALL YOUR BASES

Multiple Choice

1. B, D **2.** A, B, D **3.** C **4.** D **5.** A, B, C, D **6.** B **7.** B, C **8.** A, C **9.** C, D **10.** A, B, C, D **11.** A **12.** C, D **13.** A, B **14.** A **15.** B, D **16.** A **17.** B, C, D **18.** B, D **19.** B, D **20.** B **21.** A **22.** C **23.** A, B, C **24.** C **25.** A, B, C **26.** D **27.** A, B, D **28.** D **29.** A

Word Dissection

	Word root	Translation	Example
1.	acet	vinegar	acetyl CoA
2.	calor	heat	calorie
3.	flav	yellow	riboflavin
4.	gluco	sweet	glucose
5.	kilo	thousand	kilocalorie
6.	lecith	egg yolk	lecithin
7.	linol	flax oil	linoleic acid
8.	nutri	feed, nourish	nutrient
9.	pyro	fire	pyrogen

Chapter 24 The Urinary System

BUILDING THE FRAMEWORK

1. Formation of renin (involved in blood pressure regulation) and erythropoietin (which stimulates erythropoiesis); activation of vitamin D; carries out gluconeogenesis during prolonged fasting
2. Ptosis is the dropping of the kidneys to a more inferior position, usually caused by loss of support by the adipose capsule. Hydronephrosis is a backing up of urine in the kidney, leading to pressure on the renal tissue. Its usual cause is kinking of a ureter due to ptosis.
3. 

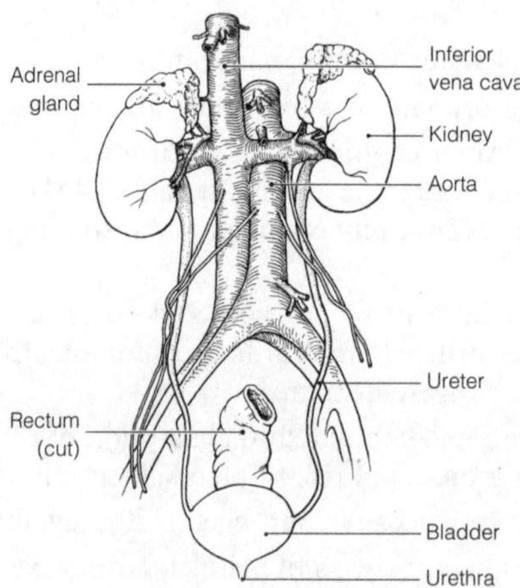

Figure 24.1

Kidney Anatomy

1. The fibrous membrane surrounding the kidney is the fibrous capsule; the basinlike area is the pelvis; a cuplike extension of the pelvis is a calyx; the cortexlike tissue running through the medulla is the renal column. The cortex contains the bulk of the nephron structures; the medullary pyramids are primarily formed by collecting ducts.

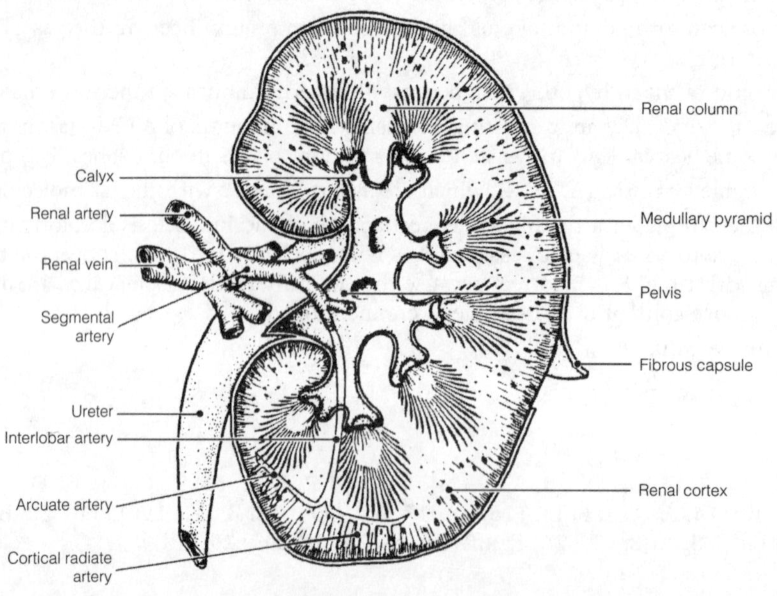

Figure 24.2

2.

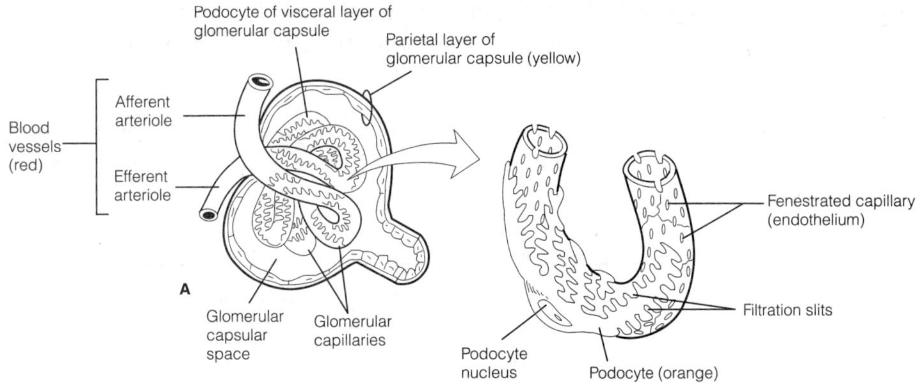

Figure 24.3

B1 is a collecting duct; few microvilli, large diameter. B2 is a DCT; has microvilli but less abundant than those in the PCTs. B3 is a PCT; very dense and long microvilli.

3. 1. glomerular capsule 2. afferent arteriole 3. efferent arteriole 4. cortical radiate artery 5. cortical radiate vein 6. arcuate artery 7. arcuate vein 8. interlobar artery 9. interlobar vein 10. loop of Henle 11. collecting duct 12. DCT 13. PCT 14. peritubular capillaries 15. glomerulus Number 1 is green, 15 is red, 14 is blue, 11 is yellow, 13 is orange.

4. 1. Renal pyramid 2. Renal corpuscle 3. Glomerulus 4. Collecting duct 5. Cortical nephrons 6. Collecting duct 7. Glomeruli 8. Loop of Henle 9. Renal pyramids

5. 1. 4 2. 5 3. 1 4. 3 5. 2

Kidney Physiology: Mechanisms of Urine Formation

1. 1. afferent 2. efferent 3. blood plasma 4. diffusion 5. active transport 6. microvilli 7. secretion 8.–10. diet, cellular metabolism, urine output 11. 1.5 12. urochrome 13.–15. urea, uric acid, creatinine 16. lungs 17. evaporation 18. decreases 19. dialysis (artificial kidney)

2. 1. B 2. A 3. C 4. B

3. Black arrows are at the site of filtrate formation, the glomerulus. Arrows leave the glomerulus and enter the glomerular capsule. Red arrows are at the site of amino acid and glucose reabsorption. They go from the PCT interior and pass through the PCT walls to the capillary bed surrounding the PCT. Nutrients leave the filtrate. Green arrows are at the site of ADH action. Arrows indicating water movement leave the interior of the collecting duct. Water passes through its walls to enter the peritubular capillary bed. Water leaves the filtrate. Yellow arrows are at the site of aldosterone action. Arrows indicating Na^+ movement leave the collecting duct (and DCT) and pass through their walls into the surrounding capillary bed. Na^+ leaves the filtrate. Blue arrows at the site of tubular secretion enter the PCT, the late DCT, and the cortical part of the collecting duct to enter the filtrate.

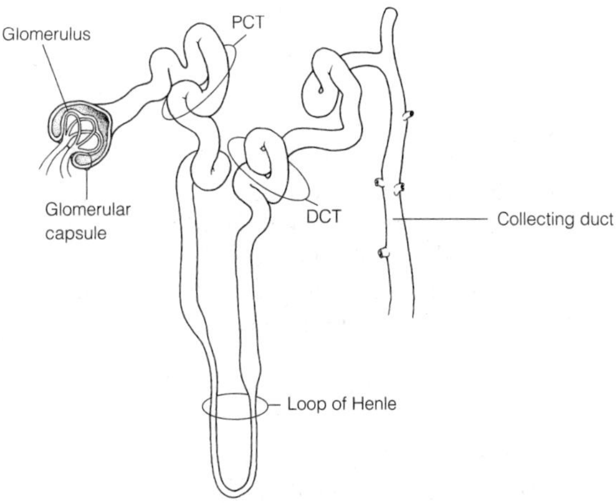

Figure 24.5

4. 1. Glomerular osmotic pressure (OP_g) + Capsular hydrostatic pressure (HP_c) 2. 10 3. Glomerular hydrostatic pressure (HP_g) 4. Glomerular osmotic pressure (OP_g)

5. Myogenic mechanism: smooth muscle cells in afferent arterioles respond to degree of stretch. Tubuloglomerular feedback mechanism: macula densa of the juxtaglomerular apparatus in the distal part of the ascending loop of Henle.
6. Check 1, 2, 4, 5.
7. 1. active 2. and 3. passive, electrochemical gradient; active in some sites 4. passive, electrochemical gradient 5. active 6. active 7. active 8. active 9. passive, electrochemical gradient 10. active 11. passive, osmotic gradient
8. 1. 375 mg/100 ml or slightly higher 2. yes

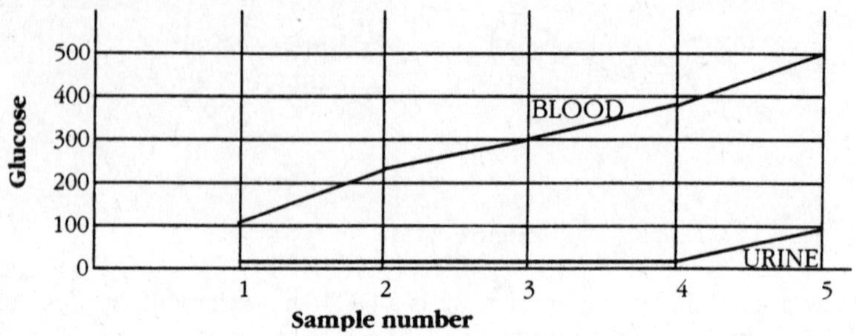

9. H^+, K^+, creatinine, ammonia (as ammonium ion)
10. In the absence of ADH, it permits very dilute urine to be excreted, thus decreasing blood volume. In the presence of ADH, it allows small amounts of very concentrated urine to be excreted, thus increasing blood volume.
11. 1. descending 2. ascending 3. thick 4. urea 5. vasa recta 6. papilla (apecies) 7. 1200 8. ADH
12.

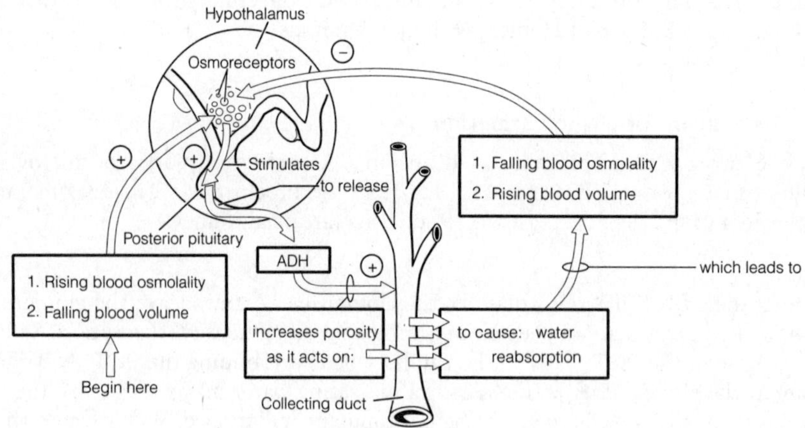

Figure 24.6

13. 1. Aldosterone 2. Secretion 3. ↑ K^+ reabsorption 4. ↓ BP 5. ↓ K^+ retention
14. 1. 180 ml/min 2. high 3. secreted 4., 5. reabsorbed 6. secreted

Urine

1. 1. D 2. D 3. I 4. D 5. I 6. I
2. 1. A 2. B 3. A 4. A 5. B
3. 1. L 2. G 3. G 4. A 5. L 6. A 7. G 8. G 9. G 10. A 11. G 12. L
4. 1. Hematuria is caused by bleeding in the urinary tract. 2. Ketonuria is caused by diabetes mellitus or starvation. 3. Albuminuria is caused by glomerulonephritis or pregnancy. 4. Pyuria is caused by urinary tract infection. 5. Bilirubinuria is caused by liver disease. 6. The presence of "sand" has no official terminology and is caused by kidney stones. 7. Glycosuria is caused by diabetes mellitus.
5. 1. Glucose is usually totally reabsorbed by tubule cells. 2. Albumin usually does not pass through the glomerular filter.
6. Urochrome is responsible for the yellow color of urine.

Ureters, Urinary Bladder, and Urethra

1. 1. kidneys 2. ureters 3. peristalsis 4. urinary bladder 5. urethra 6. 8 7. sperm/semen 8. 1.5
2. 1. Kidney 2. Forms urine 3. Continuous with renal pelvis 4. Blocks urethra 5. Female

Micturition

1. 1. voiding 2. stretch receptors 3. contract 4. internal urethral sphincter 5. external urethral sphincter 6. voluntarily 7. incontinence 8. infants, toddlers 9. emotional or neural problems 10. pressure (pregnancy) 11. urinary retention 12. prostate

The Incredible Journey

1. 1. tubule 2. renal 3. afferent 4. glomerulus 5. Glomerular capsule 6. blood plasma 7. proteins 8. loop of Henle 9. microvilli 10. reabsorption 11. glucose 12. amino acids 13. 7.35–7.45 14. nitrogenous 15. sodium 16. potassium 17. urochrome 18. antidiuretic hormone 19. collecting duct 20. pelvis 21. peristalsis 22. urine 23. micturition 24. urethra

CHALLENGING YOURSELF

At the Clinic

1. Renal ptosis and subsequent hydronephrosis.

2. Anuria; renal dialysis.

3. Urinary tract infection (urethritis and cystitis), probably by *E. coli.*

4. Acute glomerulonephritis due to a reaction to *Streptococcus* bacteria.

5. Test for glucose in the urine; diabetes insipidus; 1.001 mg with no sugar in the urine.

6. Nocturnal enuresis; perhaps Eddie is a very heavy sleeper and is, thus, unresponsive to the "urge" to urinate.

7. Renal infarct; blockage of one or more interlobar arteries.

8. It would increase it because Emmanuel's colloid osmotic pressure, which opposes filtration, would be lower (due to fewer blood proteins).

9. The kidneys lie retroperitoneal in the superior lumbar region and receive some protection from the rib cage. If the parietal pleura (adjacent) is penetrated during surgery, the lung will collapse.

Stop and Think

1. The glomerular capillaries are a high-pressure bed, considerably higher than typical capillaries. There is no low-pressure end on the glomerulus; rather, the peritubular capillaries provide the low-pressure bed for reabsorption of filtrate. Compared to a typical capillary bed, the combination of the glomerulus and the peritubular capillaries provides an enormous surface area for filtration and reabsorption.

2. A decrease in plasma proteins will reduce the glomerular colloid osmotic pressure. The change in the net filtration pressure will increase GFR, decrease blood volume (and therefore flow rate) in the peritubular capillaries, and increase flow rate in the renal tubule. Also, the colloid osmotic pressure in the peritubular capillaries will be low, so reabsorption of water into the peritubular capillaries will be reduced. The high flow rate in the proximal convoluted tubule will decrease the reabsorption of solutes and thus reduce the obligatory reabsorption of water. Urine output will increase.

3. The mitochondria congregate along the basal side of the cells, indicating that active transport is occurring on that side.

4. Secretion of ADH increases reabsorption of water from the collecting ducts, thereby briefly diluting the interstitial fluid of the medullary pyramids. But equilibration with the blood in the vasa recta then dilutes the blood, carrying off the excess water and reestablishing the osmotic gradient.

5. Alcohol inhibits ADH secretion, resulting in excess water loss by the formation of dilute urine. The resulting dehydration leads to decreased secretion by body glands, including the salivary glands, making the mouth dry.

6. Caffeine increases GFR by causing vasodilation of the glomerular capillaries. The increase in flow rate through the renal tubule reduces the time available for reabsorption of solutes, resulting in dilute urine.

7. A dialysis patient has a deficiency of erythropoietin, resulting in a low RBC count. Erythropoietin supplementation will counteract the problem.

8. Transection above the pelvic splanchnic nerves does not alter the reflex, but it prevents conscious inhibition of micturition. Urinary incontinence results as voiding becomes totally reflexive.

9. Because the blood contains Na^+ and other ions essential to the body, appropriate ion concentrations must be in the dialysate to prevent their loss from the body. Urea, creatinine, and other wastes should be absent from the dialysate to promote their removal from blood.

COVERING ALL YOUR BASES

Multiple Choice

1. C, D **2.** A **3.** A, B **4.** D **5.** B, C, D **6.** C, D **7.** B, C **8.** B, C **9.** C, D **10.** A, B **11.** A, D **12.** A, C **13.** C **14.** A, D **15.** C, D **16.** C **17.** B, C, D **18.** A, B, C, D **19.** A **20.** A, B, D **21.** A, B, D **22.** A, C, D **23.** A, C **24.** A, B, D **25.** C, D **26.** D **27.** A **28.** C **29.** D **30.** B

Word Dissection

	Word root	Translation	Example
1.	azot	nitrogen	azotemia
2.	calyx	cup-shaped	minor calyx
3.	diure	urinate	diuretic
4.	glom	ball	glomerulus
5.	gon	angle	trigone
6.	mictur	urinate	micturition
7.	nephr	kidney	nephron
8.	pyel	pelvis	pyelitis
9.	ren	kidney	renal
10.	spado	castrated	hypospadias
11.	stroph	twist, turn	exstrophy
12.	trus	thrust, push	detrusor

Chapter 25 Fluid, Electrolyte, and Acid-Base Balance

BUILDING THE FRAMEWORK

Body Fluids

1. 1. Male adult 2. Lean adult 3. Claude Bernard's "internal environment" 4. ICF 5. Electric charge

2. 1. N 2. E 3. E 4. E 5. N 6. E

3. mEq/L in plasma: bicarbonate, 27; calcium, 4.5; chloride, 108; magnesium, 3.5; hydrogen phosphate, 0.8; potassium, 4.2; sodium, 142. 1. proteins 2. Na^+ 3. Cl^- and HCO_3^- 4. ↓ are HCO_3^-, Ca^{2+}, Cl^-, and Na^+; the others are ↑.

4. 1. T 2. hydrostatic pressure 3. T 4. lymphatic vessels 5. tissue cell 6. plasma 7. equal 8. D

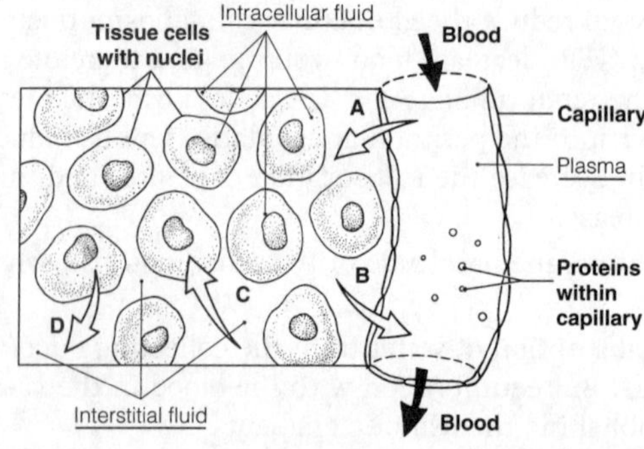

Figure 25.1

Water Balance and ECF Osmolality

1. 1. Most water (60%) comes from ingested fluids. Other sources are moist foods and cellular metabolism. 2. The greatest water loss (60%) is from excretion of urine. Other routes are as water vapor in air expired from lungs, through the skin in perspiration, and in feces. 3. Insensible water loss is water loss of which we are unaware. This type continually occurs via evaporation from skin and in water vapor that is expired from the lungs. It is uncontrollable.

2. 1. hypothalamus 2. decrease 3. increase 4. increase in baroreceptor input 5. angiotensin II 6. osmoreceptors 7. stretch 8. stomach and intestines 9. dilution 10. decrease 11. ADH 12. dilute 13. Na^+ 14. obligatory 15. 500

3. 1. B 2. C 3. A, B 4. A 5. C 6. A 7. A 8. C 9. C 10. B

Electrolyte Balance

1. 1. salts 2. foods and fluids 3.–5. in feces, in perspiration, in urine 6. kidneys 7. Na^+ 8. osmotic 9. volume 10. water

2. 1. granular 2. macula densa 3. retention/reabsorption 4. water 5. increase 6. hours to days 7. sympathetic nervous 8. baroreceptors 9. brain stem 10. arterioles 11. constrict 12. granular cells 13. renin 14. rise

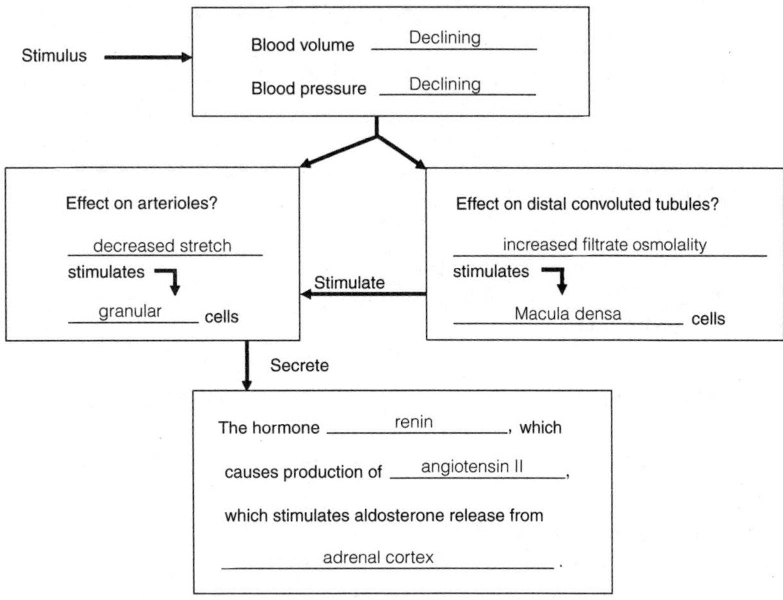

Figure 25.2

3. 1. K^+ 2. Ca^{2+} 3. Cl^- 4. K^+ 5. Na^+ 6. HCO_3^- 7. K^+ 8. Ca^{2+} 9. K^+ 10. Ca^{2+}

4. 1. ↓ H^+ secretion 2. ↑ ADH 3. ↑ Na^+ retention 4. ↑ K^+ reabsorption 5. ↓ BP 6. Dilute urine

Acid-Base Balance

1. 1. E 2. B 3. D 4. F 5. C 6. B 7. A

2. 1. anaerobic respiration 2. combination of carbon dioxide (from cellular respiration) and water 3. stomach secretions 4. fat metabolism 5. oxidation of phosphorus-containing substances, such as proteins and nucleic acids 6. events of the Krebs cycle

3. 1. Chemical buffering; response in less than 1 second. 2. Adjustment in respiratory rate and depth; response in minutes. 3. Regulation by kidneys, response in hours or days.

4. 1. D 2. E 3. B 4. E 5. D 6. A 7. B 8. D 9. C 10. E

5. 1. Weak acids only partially dissociate (free some H^+) in solution, but can be forced to dissociate more by an alkaline environment. 2. Strong acids dissociate completely and free all H^+ regardless of pH. 3. Weak bases partially dissociate and bind relatively few H^+. 4. Strong bases dissociate completely and bind all available H^+. 5. A chemical buffer system includes molecular species that quickly resist large changes in pH by releasing or binding H^+.

6.

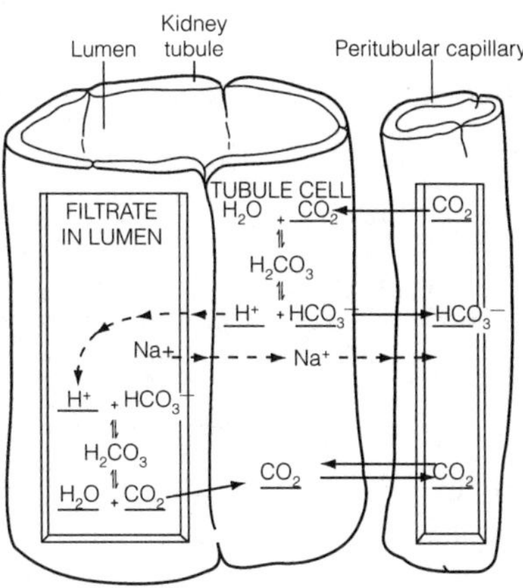

Figure 25.3

1. rises 2. H_2CO_3 3. filtrate 4. within them 5. peritubular capillary blood 6. tubule lumen (filtrate) 7. HCO_3^- 8. H_2CO_3 9. H_2O and CO_2 10. H_2O 11. one 12. peritubular capillary blood 13. active 14. excreted 15. phosphate 16. glutamine-ammonium ion

7. 1. H^+ in H_2O 2. Na_2CO_3 3. ↑ H^+ in filtrate 4. ↑ HCO_3^- in urine 5. plasma 6. ↓ plasma H^+ 7. ↑ pH 8. HCO_3^- in urine

8. 1. incompletely 2. together 3. T 4. $NaHCO_3$ 5. weak acid, $H_2PO_4^-$ 6. proteins 7. R-NH_2 8. bicarbonate 9. kidneys

9. 1. B 2. C 3. E 4. D 5. A 6.B

CHALLENGING YOURSELF

At the Clinic

1. Pituitary; hyposecretion of ADH; diabetes insipidus.
2. Hypotonic hydration; hyponatremia.
3. Edema.
4. High sodium content and copious urine volume (although the glucocorticoids can partially take over the role of aldosterone).
5. Increased due to suppression of the renin-angiotensin-aldosterone mechanism.
6. Yes; estrogen has aldosteronelike effects.
7. Metabolic acidosis with partial respiratory compensation; the high potassium level has triggered increased renal secretion of potassium and a resulting retention of hydrogen ions.
8. Hypercalcemia.
9. Sodium, potassium, chloride, and bicarbonate; hyperventilation indicates acidosis, which can be corrected with bicarbonate. Because infants are both more susceptible to and more at risk from dehydration and acidosis, treatment should be sought immediately. Sunken fontanelles in an infant are an obvious sign of severe dehydration.

Stop and Think

1. Cells are poorly permeable to saline. It remains largely in the capillaries, where it can draw the excess water into the bloodstream for removal by the kidneys.
2. Decreased blood volume leads to a lower stroke volume. Consequently, homeostatic regulatory mechanisms will increase the heart rate in an attempt to maintain normal cardiac output.
3. Because high levels of glucocorticoids have aldosteronelike effects, blood volume will increase.
4. Protein synthesis increases cellular uptake of potassium, reducing its concentration in the ECF and causing hypokalemia.
5. Release of ICF will increase the potassium concentration in the plasma; rapid or massive transfusion can cause transient hyperkalemia.
6. Compensation for alkalosis by plasma protein buffer components will mean release of hydrogen ions that have been bound to the proteins. This will simultaneously increase binding of calcium to the proteins, causing hypocalcemia. Hypoalbuminemia means fewer plasma proteins on which calcium can attach; this will result in hypocalcemia.
7. A high ratio of phosphate to calcium would increase the formation of insoluble calcium phosphate. The kidneys maintain the required high level of dissociated calcium ions by simultaneously reducing phosphate levels.
8. Undesirable depolarization of neurons and muscle fibers throughout the body due to a reduced membrane potential. This is often followed by inability to activate these excitable cells.

COVERING ALL YOUR BASES

Multiple Choice

1. D **2.** C **3.** B, D **4.** B, D **5.** D **6.** A, B, C **7.** B, D **8.** A, B, D **9.** A, D **10.** A, B, C, D **11.** C **12.** A, B, C, D **13.** A, D **14.** A, B, C, D **15.** A, B, D **16.** B, D **17.** B **18.** C **19.** C, D **20.** A **21.** A, B, C, D **22.** A, C, D **23.** A, B, C, D **24.** A, C **25.** A, C, D **26.** C **27.** A, B, D

Word Dissection

	Word root	Translation	Example
1.	kal	potassium	hyperkalemia
2.	natri	sodium	hyponatremia
3.	osmo	pushing	osmosis

Chapter 26 The Reproductive System

BUILDING THE FRAMEWORK

Anatomy of the Male Reproductive System

1. 1. When external temperature is low, they contract, drawing the testes closer to the warmth of the body wall. When the external temperature is high, they are relaxed, and the testes hang away from the body wall.

   2. The myoid cells surround the seminiferous tubules. They are presumed to contract rhythmically to propel sperm and testicular fluids through the tubules.

   3. This venous plexus absorbs heat from the arterial blood, cooling the blood before it enters the testes.

2. The spongy tissue that becomes engorged with blood during erection is the corpus cavernosum and spongiosum of the penis. The portion of the duct system that also serves the urinary system is the urethra. The structure that provides ideal temperature conditions for sperm formation is the scrotum. The structure that is cut or cauterized during a vasectomy is the ductus deferens. The prepuce is removed by circumcision. The glands producing a secretion that contains sugar are the seminal vesicles.

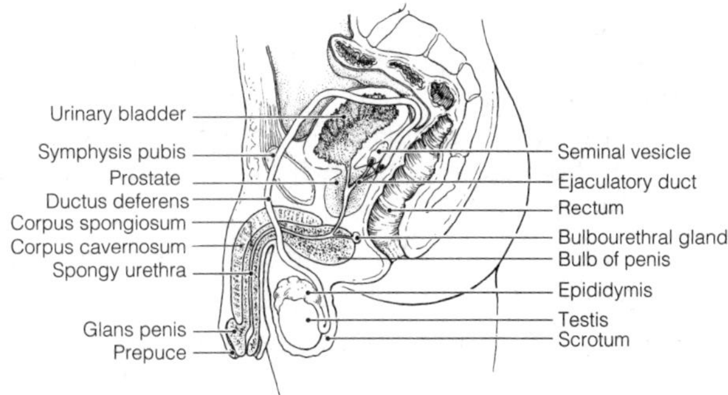

Figure 26.1

3. 1. E 2. K 3. C 4. L 5. A, G, H, K 6. I 7. B 8. F 9. G 10. H 11. A 12. J
4. seminiferous tubule → rete testis → epididymis → ductus deferens
5. The site of spermatogenesis is the seminiferous tubule. Sperm mature in the epididymis. The fibrous coat is the tunica albuginea.

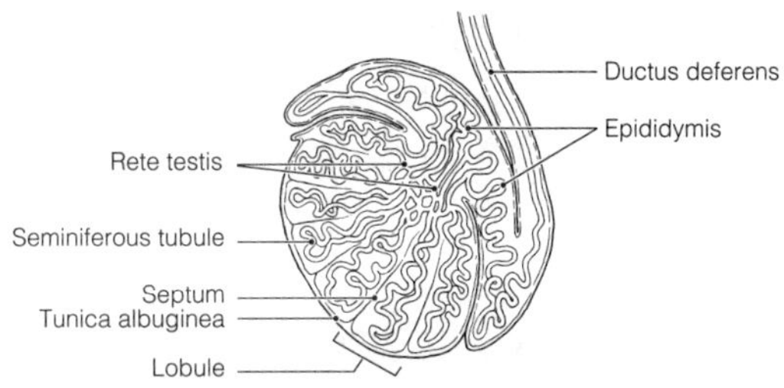

Figure 26.2

Physiology of the Male Reproductive System

1. 1. A 2. B 3. C 4. A 5. B 6. A 7. A 8. B 9. C 10. B 11. B 12. B
2.

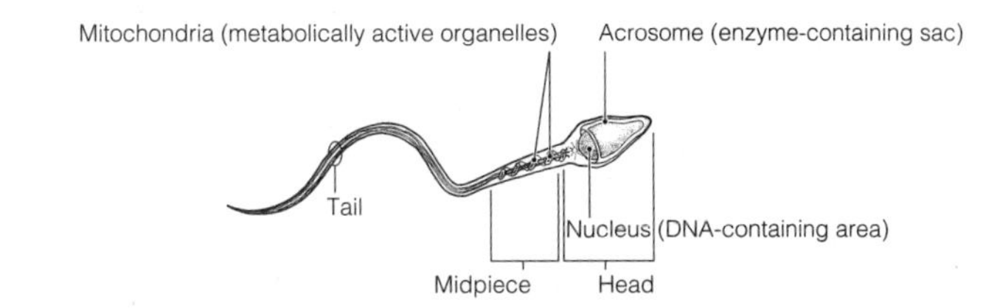

Figure 26.3

3. 1. D 2. C, E, F 3. C 4. F 5. E 6. A, G

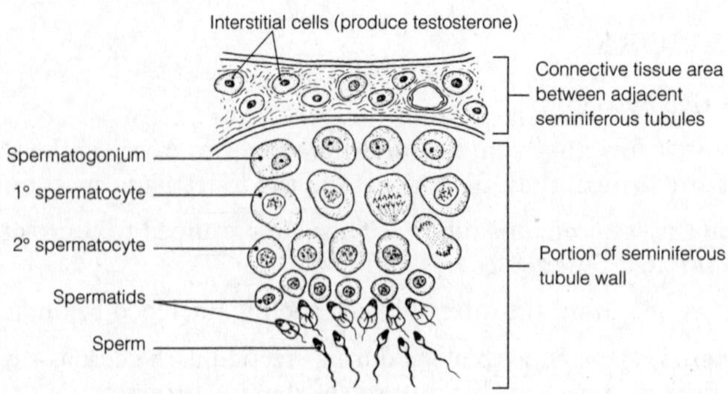

Figure 26.4

4. 1. A 2. E 3. C 4. D, F

5. 1. Spermiogenesis involves sloughing off of superfluous cytoplasm, fashioning a penetrating device (acrosome) over the nuclear "head," and forming a tail for propulsion. 2. sustentacular (Sertoli) cells 3. The blood-testis barrier is an unbroken barrier, formed by junctions between sustentacular cells, that divides the seminiferous tubule wall into two compartments. It is important because it prevents the antigens of sperm (not previously recognized as self by the immune system) from escaping through the basement membrane into blood, where they might promote an autoimmune response.

6. 1. deepening voice 2. enlargement of skeletal muscles 3. increased density of skeleton 4. increased hair growth, particularly in axillary and genital regions

7. 1. Acidic 2. Type B cell 3. Spermatogenic 4. Parasympathetic nervous system 5. Inhibin 6. Vasoconstriction

Anatomy of the Female Reproductive System

1. The clitoris should be colored blue, the hymen should be colored yellow, and the recess enclosed by the labia minora, red.

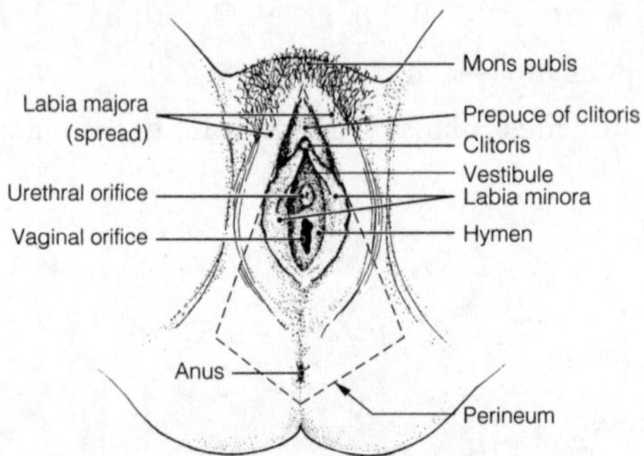

Figure 26.5

2. The egg travels along the uterine tube to the uterus after it is released from the ovary. Several ligaments suspend the reproductive organs; the one shown in Figure 26.6 is the round ligament. The ovary forms hormones and gametes, and the labia majora are the homologue of the male scrotum.

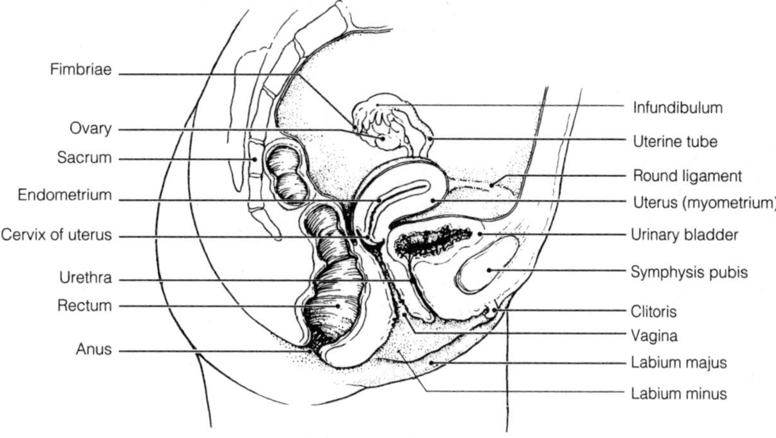

Figure 26.6

3. Because of this structural condition, many "eggs" (oocytes) are lost in the peritoneal cavity; therefore, they are unavailable for fertilization. The discontinuity also provides infectious microorganisms with access to the peritoneal cavity, possibly leading to PID.
4. 1. uterus 2. vagina 3. uterine tube 4. clitoris 5. uterine tube 6. hymen 7. ovary 8. fimbriae
5. The alveolar glands should be colored blue, and the rest of the internal breast, excluding the duct system, should be colored yellow.

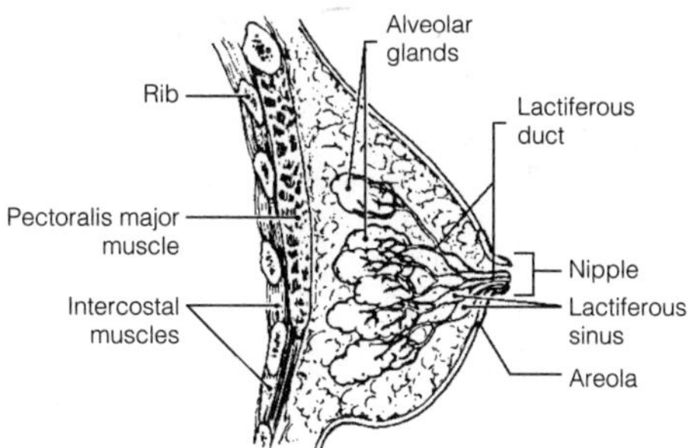

Figure 26.7

Physiology of the Female Reproductive System

1. 1. FSH 2. LH 3. estrogen, progesterone 4. estrogen 5. LH 6. LH
2. The granulosa cells produce estrogen. The corpus luteum produces progesterone. Oocytes are the central cells in all immature follicles and are the larger cells that are pushed to one side in the mature vesicular follicles.

   1. ovulation 2. peritoneal cavity 3. when sperm penetration and mixing of the gamete chromosomes occurs 4. ruptured (ovulated) follicle, which first becomes a corpus hemorrhagicum 5. one ovum, three polar bodies 6. They (the polar bodies) deteriorate because they lack nutrient-containing cytoplasm.

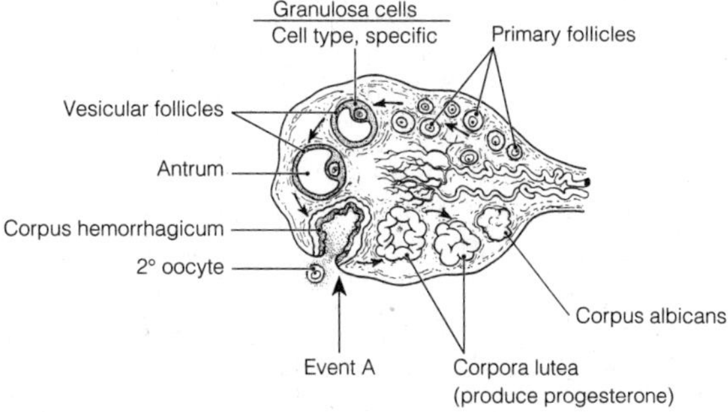

Figure 26.8

3. 1.C 2.D 3.D 4.B
4. From left to right on part C, the structures are: the primary follicle, the secondary follicle, the vesicular follicle, the ovulating follicle, and the corpus luteum. In part D, menses is from day 0 to day 5, the proliferative stage is from day 6 to day 14, and the secretory stage is from day 14 to day 28.

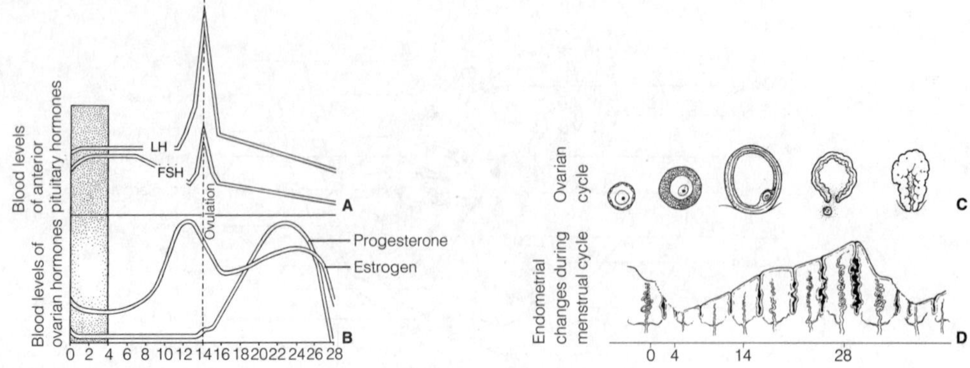

Figure 26.9

5. 1. A, B 2. B 3. A 4. B 5. A 6. A 7. B
6. 1. appearance of axillary and pubic hair 2. development of breasts 3. widening of pelvis 4. onset of menses
7. The erection phase is similar in both males and females. Blood pressure rises, the genitals (and breasts for females) become engorged with blood, and reproductive secretions are released. The male penis and the female clitoris and nipples become erect. The second phase, referred to as orgasm, involves pleasurable sensations and rhythmic muscle contractions. Ejaculation occurs in males but not in females.

Sexually Transmitted Infections

1. 1. B 2. E 3. D 4. A 5. B 6. A 7. B 8. A 9. E 10. C

The Incredible Journey

1. 1. uterus 2. ovary 3. fimbriae 4. ovulation 5. 2° oocyte 6. granulosa cells (corona radiata) 7. peristalsis 8. cilia 9. sperm 10. acrosomes 11. meiotic 12. ovum 13. second polar body 14. dead 15. fertilization 16. zygote 17. endometrium 18. vagina

CHALLENGING YOURSELF

At the Clinic

1. Testicular cancer.
2. He developed an inguinal hernia from the heavy lifting.
3. Enlarged prostate gland; prostatitis, prostate cancer.
4. Possibility of cervical cancer.
5. Pelvic inflammatory disease; the causative organism must be identified to determine the proper antibiotic.
6. Ovarian cancer, an extremely serious condition.
7. Uterine prolapse.
8. STIs (sexually transmitted infections); these infections can be asymptomatic in females; lack of treatment can lead to sterility and PID. In addition, these STIs often go hand in hand with infections caused by HIV, the AIDS virus.

Stop and Think

1. Testosterone and estrogens contribute to maintenance of bone density and muscle mass as well as metabolic rate and (in females) low cholesterol level.
2. Mumps would increase scrotal temperature transiently. This would have no effect before spermatogenesis begins, but would impair sperm production in an adult male. If inflammation led to a breakdown in the blood-testis barrier, antibodies to sperm could be produced (see next answer).
3. The blood-testis barrier is maintained by the tight junctions between adjacent sustentacular cells. This prevents exposure of the immune system to sperm antigens, which are not produced during the period of development of immune (self) tolerance and would be recognized as foreign by the immune system. A breakdown in this barrier, by trauma, poisoning, or inflammation, would be necessary to allow sperm antigens to escape into the interstitial spaces. Once exposed to sperm antigens, B lymphocytes would be activated and mount the immune attack.
4. No, because nearly all of the volume of semen is provided by the male accessory glands, which release their secretions beyond the vasectomy site.

5. The testes produce testosterone which, at puberty, stimulates the secondary sex characteristics of the male including enlargement of the larynx, which results in lowering of the voice. No testosterone, no lowering of the voice.

COVERING ALL YOUR BASES

Multiple Choice

1. D **2.** A, C **3.** A, D **4.** A, C **5.** A, B **6.** B, C **7.** C **8.** C **9.** D **10.** A, D **11.** B, C, D **12.** B **13.** B, C, D **14.** C, D **15.** B, C, D **16.** C **17.** B **18.** A, B **19.** A, C **20.** C **21.** A, B **22.** C **23.** C **24.** D **25.** D **26.** B **27.** B **28.** B **29.** D **30.** A **31.** A **32.** B **33.** A **34.** A

Word Dissection

	Word root	Translation	Example		Word root	Translation	Example
1.	acro	tip	acrosome	**20.**	mamma	teat	mammary gland
2.	alb	white	tunica albuginea	**21.**	mei	less	meiosis
3.	apsi	juncture	synapsis	**22.**	men	month	menopause
4.	arche	first	menarche	**23.**	menstru	monthly	menstrual cycle
5.	cremaster	suspenders	cremaster muscle	**24.**	metr	womb, mother	endometrium
6.	didym	testis	epididymis	**25.**	oo	egg	oogenesis
7.	estro	frenzy	estrogen	**26.**	orchi	testis	cryptorchidism
8.	forn	arch	vaginal fornix	**27.**	ov	egg	ovary
9.	gamet	spouse	gametes	**28.**	pampin	tendril	pampiniform plexus
10.	gen	hear, produce	genitalia	**29.**	penis	tail	glans penis
11.	gest	carried	progesterone	**30.**	pub	of the pubis	puberty
12.	gono	seed, offspring	gonads	**31.**	pudend	shameful	pudendal arteries
13.	gubern	a rudder; govern	gubernaculum	**32.**	salpin	trumpet	mesosalpinx
14.	gyneco	woman	gynecology	**33.**	scrot	pouch	scrotum
15.	herp	creeping	herpes	**34.**	semen	seed, sperm	seminal vesicle
16.	hymen	membrane	hymen of the vagina	**35.**	uter	womb	uterus
17.	hyster	uterus	hysterectomy	**36.**	vagin	sheath	tunica vaginalis
18.	inguin	the groin	inguinal canal	**37.**	vener	of Venus	venereal disease
19.	lut	yellow	corpus luteum	**38.**	vulv	covering, wrapper	vulva